# RESEARCH ON HUMAN SUBJECTS

## ETHICS, LAW AND SOCIAL POLICY

**Elsevier Science Internet Homepage** – http://www.elsevier.nl

Full catalogue information on all books, journals and electronic products.


**Related Journals**

*Free specimen copy gladly sent on request: Elsevier Science Ltd, The Boulevard, Langford Lane, Kidlington, Oxford OX5 1GB, UK*

International Journal of Law and Psychiatry
Journal of Criminal Justice
Child Abuse and Neglect – The International Journal
Aggression and Violent Behaviour – A Review Journal
Social Science and Medicine

# RESEARCH ON HUMAN SUBJECTS

## ETHICS, LAW AND SOCIAL POLICY

**Editor**

David N. Weisstub

*Philippe Pinel Professor of Legal Psychiatry and Biomedical Ethics*
*Faculté de médecine*
*Université de Montréal*

**Preface by**

David C. Thomasma          and          Edmund D. Pellegrino
*Loyola University of Chicago*                    *Georgetown University*

1998
PERGAMON
An imprint of Elsevier Science

ELSEVIER SCIENCE Ltd
The Boulevard, Langford Lane
Kidlington, Oxford OX5 1GB, UK

**British Library Cataloguing in Publication Data**
A catalogue record from the British Library has been applied for.

Library of Congress Cataloging-in-Publication Data

Research on human subjects : ethics, law, and social policy / edited
  by David N. Weisstub.
      p.   cm.
   Includes bibliographical references.
   ISBN 0-08-043434-7 (hc.)
   1. Human experimentation and medicine--Moral and ethical aspects.
2. Human experimentation and medicine--Law and legislation.
3. Human experimentation and medicine--Social aspects.  4. Medicine-
-Research--Moral and ethical aspects.  5. Medicine--Research--Law
and legislation.  6. Medicine--Research--Social aspects.  7. Medical
ethics.   I. Weisstub, David N., 1944-   .
R853.H8R46  1998
174'.28--dc21                                               98-38306
                                                              CIP

First edition 1998

ISBN: 0-08-043434-7

∞ The paper used in this publication meets the requirements of ANSI/NISO
Z39.48-1992 (Permanence of Paper).

Printed in The Netherlands.

# TABLE OF CONTENTS

v

# CONTRIBUTORS

## ABOUT THE EDITOR

**David N. Weisstub** — Philippe Pinel Professor of Legal Psychiatry and Biomedical Ethics, Faculté de médecine, Université de Montréal. Editor-in-Chief of the International Journal of Law and Psychiatry since its inception in 1977, Prof. Weisstub has also served on various editorial boards in the fields of health law, legal psychiatry and bioethics. For over two decades he has acted as a consultant to many governments on health policy and law reform. Recently he chaired two Enquiries for the Government of Ontario: Mental Competency (1990) and Research Ethics (1995). Prof. Weisstub is Honorary President of the International Academy of Law and Mental Health and Founding President of the Académie internationale de droit et de santé mentale pour la francophonie. In 1996 he was the recipient of an honorary doctorate from the Université de Liège.

## ABOUT THE AUTHORS

**Margaret Allars** — Associate Professor, Faculty of Law, The University of Sydney. She is the author of *Introduction to Australian Administrative Law* (Butterworths, 1990) and *Administrative Law: Cases and Commentary* (Butterworths, 1997). In 1993, she was appointed by the federal government in Australia to Chair the *Inquiry into the Use of Pituitary Derived Hormones in Australia and Creutzfeldt-Jakob Disease*, which reported in 1994. During 1995-96 she was Chair of the National Pituitary Hormones Advisory Council.

**Julio Arboleda-Florez** — Professor and Head, Department of Psychiatry, Queen's University. Dr. Arboleda-Florez is also Psychiatrist-in-Chief of the Hôtel Dieu and Kingston General Hospitals. A forensic psychiatrist and epidemiologist, Prof. Arboleda-Florez is an international lecturer and supervisor of fellows and students at universities in several countries. He is author of many books and chapters, and over 100 papers in peer-reviewed journals.

**M. Gregg Bloche** — Professor of Law, Georgetown University Law Center and Adjunct Professor, Johns Hopkins University. Dr. Bloche writes on health law and policy, biomedical ethics, and health and human rights, and he is the recipient of a 1996 Robert Wood Johnson Investigator Award in Health Policy Research (for the years 1997-2000). He has been a consultant to the World Health Organization, South Africa's Truth and Reconciliation Commission, and numerous government agencies and private organizations in the U.S. Dr. Bloche also serves on the Board of

Directors of Physicians for Human Rights and has participated in several international inquiries into complicity by health professionals in human rights abuse. He is a graduate of the Yale University Schools of Medicine and Law.

**Christian Byk** — Deputy Chief Justice of one of the Paris area circuit courts. During the past years he was a legal adviser for European and International Affairs, Ministry of Justice of France (1983-1998). Having served as special counsel to the Secretary-General of the Council of Europe on Bioethics, Dr. Byk is also Secretary-General of the International Association of Law, Ethics and Science and Vice-President of the Council for International Organizations of Medical Sciences. He has been responsible for many reports written for the European Union, the Council of Europe and the French Government. Widely published in the fields of medical ethics and human rights, he is the Editor of the *International Journal of Bioethics* and serves on the editorial committees of numerous French and other journals relating to medical, international, and comparative law. He is Associate Professor in the Faculty of Law at the Université de Pothiers, and member of the board of the French Centre of Comparative Law as well as the French Committee of International Private Law.

**Marie-Luce Delfosse** — Lecturer in Philosophy at the Facultés Universitaires Notre Dame de la Paix (Namur, Belgium). She serves as a member of the Inter-Faculty Centre on Law, Ethics and Health Science of her university. Her research interest has principally been in the area of bioethics. Her recent publications include: *L'expérimentation médicale sur l'être humain. Construire les normes, construire l'éthique* (De Boeck Université, 1993); and, as editor, *Les comités de la recherche biomédicale. Exigences éthiques et réalités institutionnelles.* (Belgique-France-Canada et Québec, Namur, Presses Universitaires, 1997).

**Severn S. Fluss** — Special Professor, Department of Nursing, Faculty of Medicine and Health Sciences, University of Nottingham. In 1965, Mr. Fluss joined WHO's Health Legislation Unit in Geneva, where he served for a period of thirty years, the latter ten as Chief of the Unit. In 1995, he joined the Office of the Executive Administrator for Health Policy and Development at the WHO, where he served as Programme Manager for Human Rights. He is currently Special Adviser to the Secretary-General of CIOMS. Mr. Fluss is known for his publications in health legislation and bioethics.

**Hanfried Helmchen** — Professor and Chairman, Department of Psychiatry, Free University of Berlin. Dr. Helmchen is known for his widely diversified interests in psychiatry and has published extensively. He is a member of the German Academy of Ethics and Medicine and numerous editorial boards and ethics committees, including the Central Ethics Committee of the Federal Republic of Germany and the Ethics Committee of the World Psychiatric Association.

**Heather Holley** — Associate Professor, Department of Community Health and Epidemiology, Queen's University. Previously, she was associated with the

University of Calgary, Department of Community Health Sciences and the Calgary Regional Health Authority. Dr. Holley is a socio-epidemiologist with the bulk of her publications being in the areas of psychiatric epidemiology and mental health services research.

**Charles R. McCarthy** — Senior Research Fellow at the Kennedy Institute of Ethics, Georgetown University. Dr. McCarthy has taught at The Catholic University of America and George Washington University. Dr. McCarthy served fourteen years as Director of the U.S. Office for Protection from Research Risks and was responsible for promulgation and implementation of policies for the protection of human research subjects. He recently served as a consultant to the President's Committee on Human Radiation Research.

**Paul M. McNeill** — Associate Professor of Law and Ethics, School of Community Medicine, Faculty of Medicine, University of New South Wales. He has made an international study of research ethics review committees (RECs — known as IRBs in the United States and REBs in Canada) and hospital ethics committees in Australia. His work on RECs is published in his book, *The Ethics and Politics of Human Experimentation* (Cambridge University Press, 1993).

**Anne Moorhouse** — Assistant Professor, Faculty of Nursing, University of Toronto. A member of the University *Joint Centre on Bioethics*, Dr. Moorhouse specializes in the fields of bioethics, mental health and community nursing. She is a member of various ethics committees.

**Norbert Nedopil** — Professor and Chairman, Department of Forensic Psychiatry, University of Munich. Dr. Nedopil is widely published in specialized areas in the field of forensic psychiatry, including both assessment and treatment. He is Chairman of the Committee on ethical questions relating to psychopharmacological research of the German Association of Psychopharmacology. He is also Chairman of the Task Force on the subspecialization on forensic psychiatry of the German Psychiatric Association.

**Tore Nilstun** — Assistant Professor, Lund University. His academic interests include moral philosophy, research ethics, philosophy of science and medical ethics. His articles frequently appear in Swedish medical journals. Among the topics he has addressed are access to health data in epidemiology, paternalism in health behaviour, preventive medicine and priority-setting in health care.

**Edmund D. Pellegrino** — John Carroll Professor of Medicine and Medical Ethics and Director of the Center for Clinical Bioethics at Georgetown University. Dr. Pellegrino is the former Director of the Kennedy Institute of Ethics and the Center for the Advanced Study of Ethics at Georgetown. He is a Master of the American College of Physicians, Fellow of the American Association for the Advancement of Science, member of the Institute of Medicine of the National Academy of Sciences, and recipient of forty honorary doctorates in addition to other honors and awards. Dr. Pellegrino is the author of over 500 published items in medical science, philosophy, and ethics and a member of numerous editorial boards. He is the author or co-author of seventeen books, and the Founding Editor of the *Journal of Medicine and Philosophy*. His research interests include the history and philosophy of medicine, professional ethics and the physician-patient relationship.

**Povl Riis** — Professor Emeritus, Faculty of Medicine, Copenhagen University. Chairman (1979-1998) of the National Research Ethics Committee for Denmark, he has served in many leadership capacities at various hospitals and national medical associations in Denmark. Author and editor of many works related to research, health care and ethics, he currently serves on the editorial boards of the *Journal of the Danish Medical Association and JAMA*.

**David J. Rothman** — Bernard Schoenberg Professor of Social Medicine and Director of the Center for the Study of Society and Medicine, Columbia University. Highly published as an historian, Dr. Rothman has achieved an international reputation for his publications in the area of human experimentation and medical ethics. Among his best known books are *The Discovery of the Asylum: Social Order and Disorder in the New Republic (1971), Strangers at the Bedside: A History of how Law and Bioethics Transformed Medical Decision-Making* (1991) and *Beginnings Count: The Technological Imperative in American Health Care* (1977). In 1987, he received an honorary doctorate from the John J. School of Criminal Justice.

**Michel Silberfeld** — Coordinator, Competency Clinic, Baycrest Centre for Geriatric Care, Toronto. He is a founding member of the *Centre for Bioethics* at the University of Toronto and Assistant Professor at the Department of Psychiatry. His publications have concentrated on the area of capacity and competency for personal and health care decision-making. He is the first-author of *When the Mind Fails*, published by the University of Toronto Press (1994).

**Jan Helge Solbakk** — Professor and Director, Centre for Medical Ethics, School of Medicine, University of Oslo. Having served as Director for the National Committee for Research Ethics of the Research Council of Norway from 1987-1996, Dr. Solbakk is known for his writings in the area of bioethics and philosophy of medicine. He is currently a member of the Executive Committee of the Council for the International Organization of Medical Sciences.

**Dominique Sprumont** — Deputy Director of the Institute of Health Law, University of Neuchâtel, Switzerland and Legal Advisor to the Intercantonal Office for the Control of Medicines in Berne. His main fields of interest are the ethics and regulation of biomedical research involving human subjects, the rights of patients and the relation of law and ethics in regulating biomedical progress. He has a number of publications in these areas, including *La protection des sujets de recherche, notamment dans le domaine biomédicale* (Berne, 1993).

**Bernard Starkman** — Senior Counsel in the Policy Sector, Federal Department of Justice, Canada. Formerly Associate Professor in the Faculty of Law at the

University of Windsor, he has for many years been advisor to the Department on medical law issues including matters relating to research with human subjects. He has chaired the Ontario Government's Committee on Guardianship and Advocacy and the Ontario Criminal Code Review Board, and co-authored in 1993 a book on *Mental Disorder in Criminal Proceedings*. Since 1987, Mr. Starkman has been the Canadian representative on the *ad hoc* Committee of experts on progress in the biomedical sciences and its successor, the Steering Committee on Bioethics, at the Council of Europe.

**David C. Thomasma** — Fr. Michael I. English, S.J. Professor of Medical Ethics and Director of the Medical Humanities Program at Loyola University Chicago Medical Center. He is Editor-in-Chief of *Theoretical Medicine and Bioethics* and Founding Co-Editor of *Cambridge Quarterly of Healthcare Ethics*. He has published 280 articles and 21 books, the most recent being *Health Care Ethics: Critical Issues for the 21st Century* (Aspen) with John Monangle and *The Influence of Edmund D. Pellegrino's Philosophy of Medicine* (Kluwer); *Helping and Healing: Religious Commitment in Health Care* (Georgetown University Press), *The Christian Virtues in Medicine* (Georgetown University Press) and *The Virtues in Medical Practice* (Oxford University Press) with Edmund D. Pellegrino; *From Birth to Death: Science and Bioethics* (Cambridge University Press) with Tomi Kushner; and *Clinical Medical Ethics: Cases and Readings* (University Press of America) with Patricia Marshall.

**George F. Tomossy** — Research Associate, International Journal of Law and Psychiatry. A graduate of the McGill University, Faculty of Law, he has written in the field of research ethics and been involved in projects associated with the Chaire de psychiatrie légale Philippe Pinel of the Faculté de médecine, Université de Montréal. He is presently attached to the Legal Branch of the Ministry of Health of the Government of Ontario.

**Janet Walker** — Assistant Professor, Osgoode Hall Law School of York University. Prof. Walker teaches and publishes in the areas of Conflict of Laws, International Business Law and Civil Procedure. She is currently a Special Advisor to the American Law Association for the *Transnational Rules of Civil Procedure*. In 1996, she assisted the Honourable Charles L. Dubin, QC in conducting *An Independent Review of the Canadian Protective Association*; and she was previously Research Coordinator for an *Enquiry on Research Ethics* conducted for the Government of Ontario.

**Claes-Göran Westrin** — Professor Emeritus, Department of Social Medicine, University of Uppsala. Dr. Westrin is noted for his publications in psychiatric health services, social medicine, medical research ethics and family medicine. He has served on many Swedish government committees and councils in a wide variety of public health areas. Editor of the *Swedish Journal of Social Medicine*, he has previously been the President of the Swedish Psychiatric Association and is also past President of the Swedish Society of Medicine.

# ACKNOWLEDGEMENTS

It is a pleasure to thank Prof. Eugenio Rasio of the Faculté de médecine, Université de Montréal, for his helpful and thoughtful collaboration. This volume evolved from a Symposium on Research Ethics generously sponsored by the Hôpital Notre Dame de Montréal, which Dr. Rasio co-chaired in 1996. Acknowledgement is made to the following journals for permission to reproduce material that has appeared in an earlier form: Acta Psychiatrica Belgica (Chapter 29), Canadian Journal of Psychiatry (Chapters 21 and 24), Health Law in Canada (Chapters 18 and 20), Health Law Journal (Chapter 4), International Journal of Law and Psychiatry (Chapters 8, 10 and 11), Revue de Droit de l'Université de Sherbrooke (Chapter 6), and the University of Ottawa Law Review (Chapter 17).

PREFACE

# MEDICINE, SCIENCE, SELF-INTEREST: VALUE SETS IN CONFLICT IN HUMAN EXPERIMENTATION

DAVID C. THOMASMA AND EDMUND D. PELLEGRINO

## INTRODUCTION

Fifty years ago the Nuremberg trials revealed to the world the infamy that results when biomedical science is pursued without ethical constraints, or perhaps more accurately, when there is pathological distortion of what those constraints should be.[1] If any good can come from such maleficence, it lies with the evolution since Nuremberg of the doctrine of informed consent,[2] and in the U.S. and other countries throughout the world of public surveillance and regulation of experiments involving human subjects. These public controls arose from the Nuremberg Code and the subsequent Helsinki Accords.

However, given the much wider scale on which human experimentation is now practiced, it is clear that a reassessment of the ethics of experiments with humans is again in order. This is why this volume, edited by David Weisstub, with its exhaustive study of the research enterprise, is so timely.

The complexity and extent of ethical issues in biomedical research today are amply evidenced in the scope of topics and problems covered by the authors of this volume. These chapters present us with the dilemma of writers of prefaces (and reviewers of anthologies we might add), namely, the dilemma of trying to comment cogently on each contribution. To do so and neglect some for others seemed unjust and prejudicial. To comment on all would be to take more space than we think

[1] E.D. Pellegrino & D.C. Thomasma, "Dubious Premises — Evil Conclusions: Moral Arguments at the Nürnberg Trials", Cambridge Quarterly of Healthcare Ethics 1998 [Forthcoming].
[2] R.R. Faden & T.L. Beauchamp, *A History and Theory of Informed Consent* (New York: Oxford University Press, 1986).

proper. Our preference, then, has been to select one aspect of biomedical research that is fundamental to all the others involving human subjects to which the different chapters in this collection bear relevance. That common ground we take to be the ethics of the relationships between the investigator and his or her subject, between being a physician and scientist simultaneously.

Our reflections prefacing this book therefore will concentrate not on new rules, regulations, public policies or international accords. These issues are extremely well covered by the essays in this book. These we take to be essential if the safety of human subjects is to be secured. But we think it equally important to focus attention on a fundamental issue that bears repeated analysis.[3] This issue is the inherent ethical conflict in all scientific research involving human subjects — namely, the conflict of obligations between three value sets: those deriving from science, from medicine, and from the self-interest of the investigators.[4] No matter how well-regulated our efforts to direct science conflicts in these three arenas will arise and continue to create ethical, legal, and political dilemmas.

We take this turn for several reasons. For one thing, despite the lessons we were presumed to learn from Nuremberg, egregious violations of research ethics have, since then, occurred in our country and elsewhere.[5] For another, investigators are increasingly under pressure to compete for support from Government grants and from a wider variety of sources. More and more of this support is coming from commercial sources so that the ethos of the marketplace is often commingled with that of science. Finally, the boundaries between experimental and therapeutic research are becoming less distinct and more difficult for patients and subjects to distinguish. Perhaps most tellingly, the research probings themselves touch on the most fundamental and intimate realms of human life, reproduction, the body, relationships, and the psyche itself.

As a result the investigators' obligations to the values intrinsic to good science, good medicine and self-interest are so commingled that the resulting moral nexus requires unraveling. This unraveling will require some priority ordering among three sets of values, each of which in isolation has legitimacy. In this reflection we will first examine these conflicting values sets, and then suggest what we consider an ethically mandatory order of priority among these sets when they are in unavoidable conflict.

### Values Sets and Obligations in Conflict

*The Values of Science*
If they are to be justified at all, experiments involving humans must at a minimum satisfy the canons of good science. Without this satisfaction, these experiments

---

[3] R.G. Spece, Jr., D.S. Shimm & A.E. Buchanan, *Conflicts of Interest in Clinical Practice and Research* (New York: Oxford University Press, 1996).

[4] Most often the conflict analyzed is the last, the self-interest of the researcher. See E.D. Pellegrino, "Character and the Conduct of Research", (1992) 2 Accountability in Research 1–12.

[5] E. Ernst, "50 Years Ago: The Nuremberg Doctors' Tribunal. Part 4: Nazi Medicine's Relevance Today", (1997) 147 Wien Med Wochenschr 30–33, 70–71. See also: *Final Report of the President's Advisory Committee, The Human Radiation Experiments* (New York: Oxford University Press, 1996).

cannot attain their major purpose, i.e., the production of valid, verifiable new knowledge about human illness or its treatment. This is not the place to detail the statistical design and interpretative criteria required for a good research design and subsequent outcome. Suffice it to say that without satisfying the canons of good science no justification can be made for exposing humans to even the most trivial danger or discomfort. To be sure, when a disease is universally fatal, one successfully treated case has epistemic validity. But the bulk of research involves evaluation of far more subtle and complex alterations in the outcome of the natural history of disease.

The clinical investigator, therefore, must be committed to the values of good science — to objectivity, accuracy and precision of observation and measurement, honesty in reporting, openness and sharing of information, receptivity to criticism of peers — in short, all those values that ensure that truth is attained to the degree the complexities of the question and the limitations of methodology allow.

The biomedical investigator then, whether a basic scientist or clinical investigator, must function within the sphere of science as much as the physicist or chemist. Viewed from this perspective, the human subject or patient must be viewed as an object of study, whether the focus is physiology, psychology or the socio-anthropology of illness or health.

The object of human experimentation cannot, however, be only an object. He or she is a human being and a patient as well, so that the investigator-patient relationship can never be guided solely by the canons of science. The investigator-physician, simply by virtue of being a physician, is compelled to be faithful simultaneously to another set of values, the values of medicine and of the physician-patient relationship.

This is where the Nazi physicians failed so utterly. They observed the canons of scientific objectivity. Indeed, they exalted their scientific commitment above their commitments as physicians. They had no compunction about treating their patients as experimental objects with no value beyond their utility in the service of scientific truth.[6]

*The Values and Obligations of Medicine*
Whether the investigator is simultaneously the caring physician, or someone else plays this role, some physician will be charged with safeguarding the welfare of the patient. We will comment later on the conflict between the physician investigator and the physician care roles. In either case, the value set of significance is the value set that characterizes the healing relationship. Here the central value is beneficence, not truth, the welfare of the patient and not the acquisition of knowledge.[7]

The oldest, most perdurable signature value for the physician-patient relationship

---

[6] Note, too, that the scientific truth was placed in service of nationalism and racism, as well as military triage. See G.J. Annas & M.A. Grodin, eds. *The Nazi Doctors and the Nurenberg Code* (New York: Oxford University Press, 1992).

[7] E.D. Pellegrino & D.C. Thomasma, *For the Patient's Good: The Restoration of Beneficence in Health Care* (New York: Oxford University Press, 1988).

is the primacy of the patient as a vulnerable human in distress and in need of help.[8] It is essential to examine this attribute more closely in the research setting because it is so acutely accentuated when a sick person becomes simultaneously a research subject as well as a patient.

There are at least three dimensions of vulnerability attending the fact of human illness: (1) the existential; (2) the relational; and, (3) the power balance.

The subject of illness or disease undergoes a change in his or her state of existence in the lived body and life world.[9] Sick persons are transformed by illness, their freedom limited, their life-world, family context, occupation, social standing, self-image, and future plans, all put into question until the nature of the illness is assessed and its full impact experienced.[10] The symptom, disability, pain and suffering take center stage in the patient's life. In fact, it is the constellation of these existential changes that make the sick person a "patient" — someone bearing a burden of suffering, with or without pain.

In addition to this existentially-experienced vulnerability, there is the vulnerability of a complex relationship of inequality with the health professional. The patient cannot help himself without help, the help of a professed healer. The doctor or nurse possesses the knowledge and skill the patient needs. The patient must place himself or herself in the power of the health professional. To be helped, patients are forced to trust in the competence and character of the doctor or nurse.[11] Whenever we trust, we become vulnerable to the possibilities of infidelity to that trust. We are eminently exploitable by virtue of our act of trusting. We lack the knowledge necessary to judge whether our trust has been violated.

Finally, there is what Foucault has called vulnerability to the power of the medical establishment.[12] This is the context of what he terms the "medical regime" which the patient enters over and above the inequality of her relationship with the individual practitioner. In this systemic context, the patient is redefined as a person, categorized, objectified, and made a case numbered item in a power system that controls the doctor as well as the patient.[13] This "medical regime" is increasingly ruled by economics, commerce, technology and utility in a competitive environment — none of which is automatically presumed to be responsive to the existential or relational vulnerability of the sick person.

These three vulnerabilities accompany the fact of illness and the existential reality of becoming and being a patient. When the patient is also a subject in a research

---

[8] E.D. Pellegrino & D.C. Thomasma, *A Philosophical Basis of Medical Practice* (New York: Oxford University Press, 1981).

[9] E.D. Pellegrino, "The Lived-World of Doctor and Patient: A Phenomenological Perspective on Medical Ethics" [forthcoming].

[10] J.V.M. Welie, "Authenticity as a Foundational Principle of Medical Ethics", (1994) 15 Theoretical Medicine 211–226.

[11] E.D. Pellegrino & D.C. Thomasma, *The Virtues in Medical Practice* (New York: Oxford University Press, 1993).

[12] M. Foucault, *L'Archeology du Savoir* (Paris: Gallimard, 1969) at 45–67.

[13] E. van Leeuwen & G.K. Kimsma, "Philosophy of Medical Practice: A Discursive Approach", in D.C. Thomasma, ed. *The Influence of Edmund D. Pellegrino's Philosophy of Medicine.* (Dordrecht/Boston: Kluwer Academic Publishers, 1997) at 99–112.

project, these three vulnerabilities are accentuated and their inter-relationships produce conflicts of obligations on the part of those who are physician-investigators or physician-carers cooperating with clinical investigators.

*Existential Vulnerability*

Most subjects, even the controls in a double-blind placebo-controlled study, are actually drawn from a patient population affected in some way by the research aims. The illness they suffer impels them to seek remedies that are not yet available, for example, in a Phase One trial of new therapy that holds promise for reversing peripheral neuropathy. Typically a Phase One Trial accepts from 20 to 100 volunteers to test safety and chemical action issues, perhaps dosage amounts. It would be unethical, and unproductive in this case, to test the drug on normal persons who do not suffer from this disorder. Volunteers would have to be sought from among a patient population already being clinically treated directly for neuropathy, or from those with diabetes and other diseases who suffer some substantial nerve death.

The disease state that has created what we call existential vulnerability predisposes the patient to volunteering. This is not *per se* a bad thing. All of us have some duty to participate in projects that may help future generations, since each of us participates in the ongoing finitude of human existence. If nothing else, we ourselves benefit from what medicine has learned about our disease from studies in patients in the generations preceding ours. Nonetheless, the fact of existential vulnerability from the disease impelling participation in research raises the ethical issue of protecting patients from harm.

First, and most obviously, particular care must be taken to emphasize that there is no intended benefit to the subject from a Phase One study. This is crucial because one can readily imagine volunteers with the disease secretly hoping some benefit will accrue even though their action is initially altruistic. The possibility of benefit cannot be excluded but it certainly cannot be promised, especially as an incentive for volunteering. Not only can it not be promised, but particular care must be produced to repeat the lack of intended benefit at every step of the decision-making process when a patient-subject signs on to the protocol.

Relational vulnerability is also accentuated if the patient becomes a research subject as, for example, in a Phase II Trial in which a sample of patients are studied for an extended period to evaluate a drug or device. These patients are usually already being treated by a physician who often recruits them for the project. Those who agree to participate automatically tend to think the researcher will look out for their best interests. He or she is someone the patient knows and trusts. The patient may even expect his or her disease to be ameliorated by participation despite disclosure about the aims of the study. Even if the patient's physician is not the investigator, the patient will transfer trust from one to the other, especially if the research recruitment occurs within the context of outpatient or hospital care under the trusted clinician.

*Power Vulnerability*
Both as a patient and then as a subject, persons in health care cannot help but be affected by the complex set of interacting interests that comprise today's medical system. In clinical investigation, the patient/subject too easily becomes a statistic in a matrix of other interests: corporate profits, gaining a competitive edge, entrepreneurship, advertising, and the contemporary commercialization of medicine. In the present climate of managed care, for-profit medicine and the "cooperative" arrangements between academic institutions and the pharmaceutical industry the patient seems only to have instrumental value. Phase II double-blind, randomized, placebo-controlled studies extended over several years' time make the completion of the protocol an end in itself. Unless the physician caring for the patient is vigilant and has the resources of personal character to resist, the welfare of the patient may well be compromised to interests far distant from the patient's or physician's own.

Thus far we have examined two competing sets of values: those of science and those of medicine, with special attention in the latter to the obligations of physicians to safeguard the several interlocking varieties of vulnerability a patient and a patient subject may experience. We turn now to a third value set: that comprising the physician's self-interest.

*The Values of Self-Interest*
A third value set overlaps the two sets already discussed, i.e., values of science and values of medicine. This third set comprises the values of self-interest. The investigator and the physician, like other human beings, are motivated by the values of self-preservation, survival and flourishing. These are legitimate values. They lie behind the need to make a living, to support a family, to advance in one's profession, to enjoy the satisfactions of discovery, recognition by one's peers, achievement and contribution to society, etc.

The structure of contemporary medical research puts some of these values of self-interest in opposition to the values both of science and of medicine. Scientific investigation in our day is in the public eye. It can bring prestige, financial rewards, emoluments like travel, consultancies, stock options, and interests in patents. More subtly, the intense competition for continued research support can unconsciously influence the selection of data, tempt to fraudulent results, or foster adherence to experimental protocols even when unwanted and unforeseen side effects begin to emerge. Scientific fraud is not simply the aberration of sociopathic scientists. Ordinary physicians under extraordinary pressures can rationalize risks to subjects or "fudging" data.

Scientific research and its reporting of results occurs in a cultural milieu which tempts to premature conclusion and exposure in the popular press. Physician-investigators are not immune to the intoxications of publicity and public acclaim. Today's cultural environment puts a premium on getting there first, establishing priority, and making the ever-popular public relations "breakthrough." None of this is conducive to the slow, deliberate, painstaking validation of data which solid research requires.

### Unraveling the Conflicts of Values

Given that there is legitimate substance to the conflict between the three value sets we have described, how do we resolve these conflicts ethically? They converge in every investigative effort involving human subjects. Each of them has corresponding obligations. The resolution of these conflicts of obligations is the foundation for an ethically defensible approach to human experimentation. Rules, regulations, policies, institutional review boards, the whole apparatus of surveillance of research with humans, ultimately must be grounded in establishing some priority among these three sets of obligations.

We propose the following elements of a research ethic which must be preserved no matter how exigent may be the utility of knowledge, nor how many lives might be saved were these elements set aside for the "good of the whole." We would argue that no utilitarian purpose can justify the violation of the humanity of individual research subjects. This is a crucial lesson from the Nazi experience, wherefrom the sometimes "enlightened" goals of public health and the good of society can lead to disastrous consequences for individuals who fall outside the defined social norms.[14]

The first priority, therefore, must be the welfare and safety of the research subject. This will be guaranteed only if the ethical values of the medical relationship take precedence over the values of science or self-interest. In human experimentation physicians, nurses and other health professionals will inevitably be involved. To be sure they must be committed to the values of good science or they expose humans to risk without reasons. This is second in the order of priority. Lastly is the investigator's personal or selfish self-interest.

To prioritize otherwise is to violate what it is to be a health professional. Like it or not, the subject will expect the physician to act in the subject's interests even while serving the scientific ends of the experiment. This will especially be the case when the investigator has been the subject's physician in the past or when the investigator is also acting as the attending or care-giver physician. For the latter, the conflict is an internal one and its outcome is dependent upon the firmness of character of the physician-investigator. This is why we have argued that a virtue ethic must be combined with one of principles, since the latter alone will not guarantee the proper prioritization of these conflicting value systems.

Where the investigator and the care-giving physician are different persons, the conflict is external to each as persons, but nonetheless exists. What is different is that the caring physician is better able to resist the powerful pull of scientific values and those of self-interest as well, since he or she has no formal responsibility for the

---

[14] H. Gallagher, "Managed Care: The German Experiment", (1994, July) Physicians for a National Health Plan Newsletter: 5–9.

scientific probity of the experiment. Obviously, this physician has a concurrent obligation to understand the science involved. His or her sole commitment, however, is as physician to the patient and not to the scientific rigor of the experiment or the personal rewards that go with a new discovery or marketable therapeutic agent.

For this reason, as well as the pressure to conform to the scientific-commercial model of so much of contemporary research, we think it best that the physician-investigator and the patient-caring physician be different persons. In this way responsibility for the welfare of the patient is located unequivocally in one identifiable member of the research team. When necessary that person must defend the patient's good against the good of science.

Second in the order of priority, are the values of science. Only if the canons of reputable science are observed can an experiment be justified in the first place. Inadequately designed, improperly observed, or pre-interpreted data collections are wasteful and misleading. They impose burdens on research subjects without proportionate justifying reasons. Values of science must yield to the welfare of the subject but not to the self-interests of the investigator This latter is an important source of conflict in today's research climate. Increasingly the paradigm is that of industrial rather than the academic model which had prevailed for so long.

In the industrial model, investigators are dependent on pharmaceutical companies whose major interest is product development, not knowledge *per se*. If research is to be supported, it must ultimately produce results that provide competitive or commercial advantage. Incentives for the researcher exist like retainer fees as consultants, foreign travel, stock options, patent rights and the opportunity to move out of academia into industry full-time or form one's own biotech company. To what degree these incentives have altered the choice of research topics, the interpretation of data, the directions of medical advancement, etc., is yet to be ascertained with accuracy. Suffice it to say that such fundamental values of science as openness, sharing of information, and peer review are constrained in a competitive commercial model of research.

The ethical center of research ethics has always been the character of the investigator. What the investigator does when no one is watching, or how she responds to the milieu of the industrial model is crucial and usually beyond the control of rules and regulations of monitoring committees. Contributing to the investigator's behavior is the moral climate of society itself. In an era in which self-seeking, moral relativism, and skepticism and the market ethos are dominant there is reason for worry.

### Reflections

It is time to revisit Bertrand Russell's opinion that, if a scientific civilization is to be a good civilization, it must be aimed at good human ends, something that science in itself cannot provide. Russell, no enemy of science, once noted:

> If a scientific civilization is to be a good civilization, it is necessary that increase in knowledge be accompanied by an increase in wisdom. I mean by wisdom, a right conception of the ends of life. This is something which science itself does not provide.[15]

Is it even possible to step outside of our scientific civilization, as neutral observers, and correct its course, its efforts to advance knowledge, when these become misguided to the extent that they begin to trample our commitments to the value of human life? Even if such a move were possible, from what pool of "scientifically-neutral" resources would we draw our proposals for the good human ends of our civilization? What checks on the advancement of scientific research would we impose based on this pool of resources, given our commitments to the intrinsic good of pushing back the boundaries of ignorance and promoting the progress of peoples everywhere throughout the world?[16]

Traditionally, and in Russell's view as well, the pool of resources best able to monitor and direct to good human ends our scientific civilization are the humanities, the disciplines particularly of theology and philosophy, literature and law. Disregard for the moment that these disciplines all benefit from and are influenced by the sciences, is it not naive to assume that they, the less powerful and less well-regarded fields of our day compared to technology and commerce, could check in any way the two-fold behemoth that couples science and technology with capitalism and national interest?

Our suggestion in this Preface has been to ground the checks and balances in the fiduciary relationship of physician with patient, and from that relationship anchor all duties to the subject, even when the investigator is not a physician, or perhaps is a physician, but not the care giver of the patient. The reasons we offered are that the patient today is lured somewhat into participation not only by altruism, but also by a naive acceptance or trust in the medical establishment that may no longer be warranted by the sources of support for the research. A caveat emptor principle has long governed the relationship of consumer to vendor. Something similar may now be required with respect to industry-sponsored research.

### *Conclusion*

All research guidelines are intended to protect the vulnerable population *vis-à-vis* the research project. In the past fifty years, there is widespread conviction that this process has moved forward. As this volume demonstrates, however, special vulnerabilities exist that demand exquisite and sophisticated reasoning, moral and legal, in order to establish safeguards and proper public policy.

This important volume details arguments for protecting the especially vulnerable children, elderly, mentally disabled, developmentally disabled, and prisoners. One of the exceptional notes of most of these populations is that they are institutionalized. As human beings under the care and supervision of others, their primary

---

[15] B. Russell, as quoted in D. Callahan, *The Tyranny of Survival*. New York: Macmillan, 1974.

[16] This commitment was shown in early 1998 by the US Congress in its refusal to ban human cloning, for fear that such a ban would impede scientific progress. See: W. Neikirk, "Senate Opts Not to Vote on Proposed Cloning Ban", The Chicago Tribune (Feb. 12, 1998), Section 1, 4.

vulnerability consists in incapacitation of some degree, either by age, disability, or enforcement. This incapacity, coupled with dependency on the institutions of family, hospital, nursing home, or jail, lends extra measure to the 3-fold vulnerability we have already sketched that belongs to all human beings facing consent to research.

As Professor Weisstub, Arboleda-Florez, and George Tomossy note in Chapter 18, such incapacities do not mean that persons so dependent are incapable of consenting to research, or that all members of a class of the specially vulnerable, say the demented, should be treated alike in restricting their access to participation. Our contemporary challenges lie in this arena then: how to properly protect especially vulnerable populations, while also permitting them to commit to and benefit from research? At the same time, we must keep the priority order among the conflicting value systems we have explored in this Preface.

CHAPTER 1

# THE ETHICAL PARAMETERS OF EXPERIMENTATION

DAVID N. WEISSTUB

## INTRODUCTION

There have been serious controversies in the latter part of the Twentieth Century about the roles and functions of scientific and medical research. In the decades following the Second World War, a liberal awakening emerged against the background of medical atrocities, prompting sharp responses to the possibility of abusive treatment of innocent or disadvantaged populations. Alongside a number of judicial decisions, these responses ripened into codes and ethical guidelines, all of which endeavoured to delineate the parameters of ethically justified interventions. Shortly into this evolution it became apparent that a global banning of research was an overstated reaction, and that a more considered approach was warranted. There began a cautious return to the notion that, in controlled circumstances, it could be shown to be in the best interests of certain disenfranchised populations, minorities, or specific groups within the population, such as children and the elderly, to pursue medical research.[1]

In undertaking to develop a conceptual framework from which to derive a precise logic of application, there is no simple philosophical set of norms around which we could rally such a pragmatic consensus apart from the evolutionary design of a standard of basic values enshrined in the global development of codification and

---

[1] It therefore becomes necessary, in certain instances, to conduct research even where it is not possible to obtain competent consent. The justifications for this are not always the same, as the populations to which we refer bring to our attention a diverse set of factors in the particularities of both their needs and capacities. Everything possible should be done to enhance the capacity of persons to achieve a level of autonomy and self-respect in the process of harnessing scientific endeavours to the needs of a specific population or illness while at the same time reserving the interest of society to protect these individuals or populations from maltreatment.

1

statutory enactment. Having said this, it is still a worthwhile investigation to reflect upon the classical modalities of value analysis upon which certain central claims are made about individual and social entitlements in the context of research ethics, such as the well-stated positions of utilitarianism and deontology.[2] Although there will be an attempt to demonstrate that there is no singular philosophical or ideological standpoint upon which to found our value base, we are prepared nonetheless to assert that between the Western Judeo-Christian ethic and a popular consciousness achieved in this century with respect to what can be understood as a liberal ethic, we can balance the individualistic-modernist morality of which we have become a part and the public interest and communitarian ethic upon which we justify some curtailments of absolute freedom in the name of a broader good.

The global realities of our industrialized economies have begun to force us to make difficult decisions about justifying a perceived social need to increase the efficiency of science in delivering care during a period when the allocation of resources presses upon us the need to prioritize care.[3] This is a weighty moral process, given that great medical costs, for example, are attached to the last years of life, to defects of birth and capacity, and to select populations who because of circumstance or genetics lay claim upon our limited social purse, given the state of our technological capacity. There is also a pronounced tendency in our contemporary societies to back away from some of the liberal rhetoric of the period of surplus economies, such as those experienced in the decades preceding the nineties, to a more conservative agenda in reaction to other vulnerable populations such as troubled youth or adult criminals, whose costs to society have escalated by threatening proportions.

Social attitudes towards the value of scientific and medical research have also gone through a process of change and reconstitution. The blind optimism about the capacity of science to redress social wrongs through rehabilitation and about the promise of medical science to regenerate physical or mental shortcomings accepted in the 1950s, has certainly waned with the passage of time. In the 1960s and 70s popular reactions against science and certain areas of medicine such as psychiatry, were connected to the consumerist movement, which began to question the capacity of elites to supervise and control delicate issues relating to health and personal welfare. Nonetheless, with the burgeoning of new diagnostic techniques and

---

[2] These positions are normally considered antithetical. As we will see later in this discussion, such claims are exaggerated, and often based on overly rigid statements of either position. But it still should be admitted that the theoretical premises upon which each position is based do have some significant differences, which have to be taken into account in understanding how certain values in the system are prioritized.

[3] We are concentrating here on the notion of applied science, medical science being one of the species. It is acknowledged that we may still make a claim as to the value and/or the existence of pure science, which in certain areas of intellectual pursuit achieves the Aristotelian concept of knowledge pursued as an end in itself. We do make the claim, however, that the mainstream of science as we know it in technological societies has been driven both in concept and by reason of social utility towards the pursuit of certain social ends, often heavily weighted in the direction of the expressed social needs of existing monetary and political powers. In the field of medical research, these pressures are highly in evidence at the current time.

technological revolutions within medicine, a renewed optimism, albeit an expensive one, has presented itself in our contemporary environment. Our dependency on expensive technology in shrinking economies has made decision-makers disquieted about making available these technologies where a perceived social payback is not in evidence. These pressures have put into question the charitable ethic associated with better economic times and the humanistic ethic associated with the preceding era. Balancing values and rhetorics has therefore become a real matter of conflict in such a charged emotional forum as the ethics of research involving vulnerable populations.

In the popular consciousness of our society, we seem to subscribe to the view that science as a pure endeavour may not be neutral, but with legal or social curtailment likely represents a "good" rather than a matter of questionable utility. Medicine is still regarded in its function as a science as a contributor to the welfare of our social order. Notwithstanding the abuses which occur, the pursuit of science *per se* is still widely perceived as a good in itself. This assumption should be subjected to further reflection because we should not be led to conclude that science is the highest value or truth for our society, nor that good science is necessarily good ethics.[4] Our choice of what we research and for whom is not a neutral inquiry. As Fagot-Largeault comments, "to the extent that researchers recognize that the acquisition of knowledge about human beings is not entirely neutral and innocent, they have to admit that there are circumstances in which acquiring knowledge may be untimely or pernicious in itself".[5] The choice of the subject-matter and method of research can have far-reaching consequences, justifying social interventions into the liberties of individuals, including populations who suffer from accidents of birth and are simply innocent parties in the scheme of nature.

We should not be too quick to conclude that the self-governance of our professional elites is sufficient to control the conflicting interests which present themselves in the hard choices we make about the nature and purpose of science. Whether we choose to research on animals, how we tolerate a level of risk, the manner in which we justify compromises of values like privacy, our perception of what level of threat we will tolerate as a society from members within the population

---

[4] We do not wish here to make any strong metaphysical assertions or ontological claims. Rather, it is to observe that science unrelated to a teleological construct involving some broad notion of commonality or a criterion of goodness is in danger of being used for untoward purposes. This does not mean that all forms of scientific research can be said to have no claim on neutrality, but rather that science understood as an endeavour with many dimensions and projects becomes a social undertaking which has to be related to a broader societal circumference. It is rare to locate scientific research projects which cannot be linked to sociological analysis. This is particularly true of medical research which is increasingly linked to analyses about health in the larger social sphere of public health, occupational health, and issues of political and social responsibility. However, it must also be recognized that many forms of teleological analysis lay us open to exploitation of minority interests once the larger constructs of commonality are linked to majoritarianism, justified through metaphysical claims. See D.N. Weisstub, "Law and Telos: Some Historical Reflections on the Nature of Authority" (1978) 12 University of British Columbia Law Review 225.

[5] A.M. Fagot-Largeault, "Epistemological Presuppositions Involved in the Programs of Human Research" in S.F. Spicker *et al.*, eds., *The Use of Human Beings in Research* (Boston: Kluwer Academic Publishers, 1988) 169.

before justifying radical responses, the accolades that we are prepared to give scientists as members of professional elites, and the determination of taboo areas of scientific inquiry, are all value-laden issues, which, regardless of the response we will make in legislation or law, will continue to haunt any democratic society of conscience. The choices we make about our vulnerable populations with respect to medical research test the fibre of the social values upon which we are prepared to base our society. If we are prepared to say that, in the public interest, we can make claims upon our vulnerable minorities, we must be prepared to relate our social policies to principles which reflect a consensus morality attached to certain core values, and to connect these principles to equitable procedures that facilitate a process which can be regarded as fair and just and not based simply on expediency or majoritarian interests.

We should not be naive about the capacity of codes or legislation to bring unanimity and predictability to the subject. It should be a clear lesson of history that the ethical guidelines proclaimed in Weimar Germany,[6] so well articulated in the liberal period preceding the Nazi regime, were no guarantee that vulnerable populations would be protected either against medicine or bad politics. As noted by Hans-Martin Sass:

> The Richtlinien remained binding law through the end of the German Reich in 1945. They were stricter and more detailed than the Nuremberg Code of 1947 and the Helsinki Declaration of 1964. The fact that such a government regulation actually existed, at a time when the Nazis were carrying out human experimentation in an irresponsible manner in concentration camps, underlines the irrelevance of legal regulation if they are not enforced by the authorities.[7]

In fact, we cannot, even in this day, improve on the moral sentiments expressed at that time. In more liberal periods of history, such as the one we have experienced over the past thirty years, and in the context of a society where we already began to order our research efforts according to carefully crafted guidelines, we can see a pattern of continued violations against vulnerable populations. Therefore, we are prepared to submit that any achievement we may accomplish with respect to statutory regulation is only a part of the richly needed raising of social and professional consciousness. This is so because the pressures have mounted in our decision-making about priorities done in the name of social justice.

---

[6] Reichsgesundheitsamt, Deutsches Reich: 1931, "Richtlinien für neuartige Heilbehandlung und für die Vornahme wissenschaftlicher Versuche am Menschen", *Reichsgesundheitsblatt* 6 (55th Yr.), Berlin.

[7] H.-M. Sass, "Comparative Models and Goals for the Regulation of Human Research" in S.F. Spicker *et al.*, eds., *The Use of Human Beings in Research* (Boston: Kluwer Academic Publishers, 1988) 47. See also M. Kater, *Doctors Under Hitler* (Chapel Hill: University of North Carolina Press, 1989); B. Muller-Hill, *Murderous Science: Elimination by Scientific Selection of Jews, Gypsies and Others, Germany 1933–1945* (Toronto: Oxford University Press, 1988); and R. Proctor, *Racial Hygiene: Medicine under the Nazis* (Cambridge, Mass.: Harvard University Press, 1988).

In periods of great social transformation, such as the current global trend of reactive discomfort against many of the radical liberal reforms of the 1970s and 1980s, we should be aware of what should be the minimal conditions that we as a society commit ourselves to preserving as representative of a long-term commitment to substantive and procedural standards of fairness and equity which are at the root of an ethically justified concept of health and research.

## THE NUREMBERG LEGACY

Although the Nuremberg trials and codification represent almost a mythic chapter in the history of our understanding of research ethics, the actual respect for the populations that Nuremberg addressed, despite the proclamation and the deference we have paid to it in our national cultures, presents in its details a rather more modest and puzzling past than we might have anticipated. Nuremberg is a distant collective memory, sometimes for perpetrators and victims alike, as well as for the generations which followed. For most of us, Nuremberg emerged both as a code and a symbol, both in its principles representing the foundations of civilized medical practice and research, and in its symbolization of the triumph of the democratic ideal over fascisim.

Despite this, regrettably, in the aftermath of Nuremberg, we observe the immediate reintegration of a sizeable number of German scientific technicians, including medical ones, into the American military or related industries by the Americans who were responsible for conducting the Doctor Trials under the authority of the U.S. military. For example, the American "Project Paperclip", which began immediately after the war and ended in 1955, employed approximately 765 Austrian and German scientists.[8] The U.S. military is recorded to have employed four of the accused who had been exonerated at the time of the Nuremberg trials. A number of the defendants were employed by the U.S. Air Force until their arrest. Among them was Hermann Becker-Frieyzing, who was sentenced for his experiments on freezing, high altitude, and sulphanilamide, among others. The case of Shaefer is particularly noteworthy. After he was acquitted, he was brought to Texas by the American Airforce under "Project Paperclip", but in 1951 was repatriated to Germany. There are other disturbing facts. Another defendant, Blome, who conducted plague experiments in the concentration camps, was given a job on August 21, 1951 with the U.S. Army Chemical Corps on Project 63. He could not take the job, however, because he was denied U.S. immigration clearance in Frankfurt. This did not prevent him from taking a camp doctor position at Oberusal at the European Command Intelligence Centre. Finally, it is no secret that a number of the accused were integrated into the burgeoning pharmaceutical industries of post-war Germany. The record of the American government in the Tokyo War Crime Trials was no improvement. Japanese doctors tested the effects of lethal biological weapons on American prisoners of war in China during World War II. These Japanese doctors were not prosecuted because they agreed to disclose the results of their experiments to U.S. military officials.[9]

The problem is deeper than a mere conspiracy of a small elite who have a protectionist interest in their professional colleagues. The moral compromises which unfolded are rather a tribute to the questionable tendency of humankind to identify with technical science and scientific elitism. This apparent willingness to forgive

---

[8] T. Bower, *The Paperclip Conspiracy: The Hunt for Nazi Scientists* (Boston: Little, Brown, 1987).
[9] A. Brackman, *The Other Nuremberg* (London: Collins, 1989); P. Williams & D. Wallace, *Unit 731: The Japanese Army's Secret of Secrets* (London: Hodder & Stoughton, 1989).

and forget in the name of scientific advance managed to coincide with the idea of social progress, to which the post-war worlds of western liberal democracies were fully committed. During this period, the United States led the world community in its vision of conjoined moral and economic progress. David Rothman, in reflecting on the social and economic forces that constructed a research *Weltanschauung* in America from the Second World War until the present, was led to comment:

> The Nuremberg trial of the Nazi Doctors, for example, received very little press coverage, and before the 1970s, the Code itself was infrequently cited or discussed in medical journals. American researchers and physicians apparently found Nuremberg irrelevant to their own work. They believed (incorrectly, as it turned out) that the bizarre and cruel experiments had been conducted not by scientists and doctors, but, by sadistic Nazi officers, and therefore, the dedicated investigators had nothing to learn from the experience.[10]

This apparent moral distancing of the research establishment in the U.S., coupled with the predisposition of government officials and the military to advance an "America first" policy during the Cold War years, did little to alleviate the mistreatment of research subjects, who were deemed relevant to the national interest.

With respect to the U.S. Supreme Court, it is a matter of record that the Nuremberg Code is referred to only once.[11] In the late 1950s, an army serviceman by the name of Stanley became a volunteer to test the effectiveness of protective clothing against chemical warfare. Without his knowledge, he was administered LSD and suffered considerable mental damage thereafter, leading to his discharge from the military in 1969. Stanley was not informed that he had been given LSD until he received a letter from the army in 1975, asking for his cooperation in the follow-up study on the volunteers who participated in the 1958 LSD research. Stanley was denied recovery in the U.S. Supreme Court in 1987. Justice Scalia, who wrote the opinion for the majority, concluded that allowing Stanley to sue the army would call into question military discipline and decision-making.

*Stanley* sits against a background of elaborate research conducted in the 1950s by the CIA under the code name MKULTRA. In 80 institutions, approximately 200 researchers were hired by the CIA to conduct studies on brainwashing techniques. It was not until 1982 that such experimentation was expressly prohibited by a presidential executive order, but the MKULTRA records were destroyed by the director of the CIA in 1973. When certain information became available by accident, prompting a number of plaintiffs to come forward to get additional information, the CIA continued to resist disclosure. In *Sims*, the Supreme Court protected the CIA, with all Justices in agreement, determining that the CIA activities were protected by dint of the fact that they were acting in the interest of national defence and security.[12]

---

[10] D.J. Rothman, "Ethics and Human Experimentation: Henry Beecher Revisited" (1987) 317 New England Journal of Medicine 1195.

[11] *U.S. v. Stanley*, 107 U.S. 3054 (1987) [hereinafter *Stanley*].

[12] *Central Intelligence Agency v. Sims*, 471 U.S. 159 (1984) at 162.

*Stanley* shows that the U.S. Army, along with the CIA, was very much involved with hallucinogenic drugs in the 1950s. George Annas has noted that, in the period between 1955–1967, a considerable number of research contracts were funded by the Army in this area. Although the Secretary of Defence, according to Annas, adopted the Nuremberg Code by 1953, the guidelines were classified as top-secret until 1974.[13]

One would have assumed, given the clandestine nature of the Army's history with respect to LSD experiments, that the Supreme Court would have taken the occasion of the *Stanley* case to differentiate Stanley's obligation, carried out in the course of military service, from the subjugation of his person. In a stinging dissent, Justice Sandra O'Connor proclaimed that,

> no judicially crafted rule should isolate from liability the involuntary and unknowing human experimentation alleged to have occurred in this case.[14]

In referring to the Nuremberg Code, Justice O'Connor asked that compensation be paid in the name of the Constitution's promise of due process. Justice Brennan, also in dissent, went on record saying that "the government of the United States treated thousands of its citizens as though they were laboratory animals".[15]

It is a remarkable fact that, as recently as 1987, after so much evolution in the law of informed consent throughout American judicial life, in a case as blatant as *Stanley,* with the strongest possible language used by distinguished members of the bench in dissent, the majority denied Stanley recovery. This underlines the extent to which the American judicial system was prepared to sacrifice what could be regarded as a clear violation under the veil of Nuremberg for the spurious larger umbrella of national security. The Supreme Court's position is indefensible because the persons who perpetrated the acts against Stanley were probably civilian, and in the American Army records themselves, which are noted in the case, the LSD experiments were covered up because the Army feared legal liability.

How restrictive is *Stanley*? We could rest more easily if we could convince ourselves that the Supreme Court's decision represents an aberration of judicial thinking since, in facing questions related to national security, it is not surprising to see establishment figures of the judiciary over-identifying with their establishment counterparts in the military. We could expect, therefore, that in cases relating to civilians, the record might be better. There seems to be no evidence to suggest this.

In one of the few cases on experimentation emanating from a court of distinction, the Ninth Circuit Court of Appeals rendered a decision in 1985, *Begay* v. *U.S.*,[16] where a group of Navajo Indian miners was denied recovery based on the failure to warn them about radiation dangers from uranium. In the course of conducting a research study, the U.S. Public Health Service did not warn the miners, even after

---

[13] G.J. Annas, L.H. Glantz & B.F. Katz, *Informed Consent to Human Experimentation: The Subject's Dilemma* (Cambridge, Mass.: Ballinger, 1977) at 305.
[14] See *Stanley, supra* note 11 at 3065.
[15] *Ibid.* at 3066.
[16] *Begay* v. *The United States*, 768 F.2d. 1059 (4th Cir. 1985) [hereinafter *Begay*].

the dangers were known. This was done in order to get the cooperation of the mine owners and to be sure to ensure the retention of the Navajo Indians for the study. The Court superseded the claim to a human rights violation by deciding in favour of the technical rule of the *Federal Tort Claims Act*, which protected the defendants as acting pursuant to a discretionary function.

The *Begay* case cannot be regarded as an isolated incident. Congressman Markey from Massachusetts revealed information in 1986 which exposed various experiments by the U.S. Government conducted from 1940–1971 on radiation exposure. These cases involved prisoners, patients with a limited life-span, and citizens at large. To our knowledge, there has not been a successful lawsuit brought by any plaintiff who suffered from these experiments. The radiation studies were sponsored by academic institutions and hospitals in some of the leading American centers.[17]

The histories indicate that even in a well-intentioned society with built-in mechanisms imposing checks and balances on professionals in positions of authority, it is part of the reality of the administration of authoritarian environments, such as the military and public institutions sustained by state contributions, that the subjects contained in them are rationalized as justified instruments of the public interest. It is imperative, therefore, to look carefully at the procedures that are most likely realistically to protect the populations perceived as vulnerable, given the natural tendencies of even the most democratic of societies. In the following section we explore a parallel reality which occurred in Canada, and which can arguably serve as a point of reference for our moral inquiry, because it raises some basic questions about the sociology of power as it unfolds in the public life of our hospitals and universities.

## A CANADIAN EXPERIENCE: THE CAMERON EXPERIMENTS

The primary Canadian case of reference, documented in the *Cooper Report*,[18] commissioned by the Federal Government of Canada, took place in Montreal at the Royal Victoria Hospital in the late 1950s under the auspices of Professor Ewen Cameron of the Allan Memorial Institute. His "depatterning" treatments and interventions have been the subject of severe criticism, not only with respect to the controversial techniques he used, but also with regard to his choice of patients. Among the notable critics was Sir Aubrey Lewis, head of the Maudsley Hospital in London, who in 1957 confided to Doctor Cleghorn, Cameron's successor, that he thought that Cameron's treatments were barbaric.

The Minister of Justice of Canada issued the following news release in 1992:

---

[17] For a discussion of the American radiation experiments, see Advisory Committee on Human Radiation Experiments, *Final Report* (Washington, D.C.: U.S. Government Printing Office, 1995). See also E.J. Markey, chair, *Report Prepared by the Subcommittee on Energy Conservation and Power of the Committee on Energy and Commerce, U.S. House of Representatives. American Nuclear Guinea Pigs: Three Decades of Radiation Experiments on U.S. Citizens* (Washington, D.C.: U.S. Government Printing Office, 1986).

[18] G. Cooper, *Opinion of George Cooper, Q.C., Regarding Canadian Government Funding of the Allan Memorial Institute in the 1950s and 1960s* (Ottawa: Supply and Services Canada, 1986) at 85.

Depatterning involved administering massive electroshock treatments to patients — something up to twenty or thirty times as intense as the "normal" course of electroconvulsive therapy treatments. In preparation for the treatment, the patient would be put into a prolonged sleep for a period of about ten days by using various drugs. At that point, the massive electrochock therapy would begin with the patient being maintained on continuous sleep throughout. At the end of up to thirty days of treatment, the patient's mind would be more or less in a childlike unconcerned state. Depatterning was then maintained for another week, with electroshocks being reduced to three per week and gradually to once per week. After this had occurred, the idea was to "repattern" the brain by trying to instil new and "correct" patterns of thinking in the patient's mind through various techniques.[19]

Judging by the standards of today, psychiatrists would conclude that these procedures were a failure, not only in terms of their efficacy as a medical treatment, but also because they represented an unjustifiable level of assault on the brain.

To date in the Cameron case, no plaintiff has been successful before a Canadian court. In the late 1980s, however, nine plaintiffs shared $750,000 (US) as a result of a settlement with the American government. There was no resolution, therefore, in an American court. The Canadian Minister of Justice offered an *ex gratia* payment of $100,000 to individuals who could identify themselves as having been subjected to Cameron's depatterning treatments; according to the Ministry, its contributions are charitable acts, and do not arise from any admission of legal liability. The Ministry received, to the author's knowledge, approximately three hundred communications.

The Cameron case raises a number of questions about the necessity of peer review: the vetting of funding; the requirements of informed consent; the rights of patients to withdraw in the course of research; the distinction between therapy, innovative therapy and therapeutic experimentation relevant to disclosure requirements; the role of family members in protecting their next of kin; the place of lay advocates and concerned parties to represent their case for protection before a neutral overseeing body; the burden of *Statutes of Limitations* on plaintiffs who have suffered in research contexts about which they gain familiarity many years after the fact; and the obligations that should be placed on disciplinary professional boards to police situations of abuse.

The fact that the experiments were conducted on non-institutionalized and non-military indvidiuals (i.e., ordinary voluntary psychiatric patients) does not alleviate the complexity of the issues. If anything, these subjects should represent the strongest case for recovery, given that their consent could not have been deemed to be vitiated through the circumstance of incarceration, which could arguably not be the case for military personnel.

The Cameron case raises hard and searching questions about the nature of the conflicts of interest that often arise in research, not only in universities, but also in the special settings of mental and forensic hospitals and prisons, where captive populations live under the surveillance of medical practitioners who have given

---

[19] Department of Justice, Government of Canada, News Release "Background Information — Depatterning at the Allan Memorial Institute" (17 November 1992).

oaths of allegiance to the state as much as to their patients. Doctors' roles are often unclear, and can become treacherous with respect to the avowed values of their profession, when the state finds reason to make specific demands in the name of the public interest.[20]

It is clear that Cameron's massive interventions in giving electroshocks and depatterning programs were, in his opinion, devices of therapy and not those of an experimenter. He held prestigious positions as a practitioner, was widely published, and was apparently not embarassed to receive a research grant for his treatments from the CIA. The mere fact that he received funding from the CIA does not, of course, mean that he did not believe in his therapies. It must be remembered that the late 1950s, when he conducted most of his research, was an optimistic period. Psychiatrists were vociferous in their praises of their abilities and "therapy" covered a wide variety of treatments.

The common law standard of informed consent in therapy was just at the onset of becoming popularly known in the legal literature of the mid-sixties. The issue then becomes, in the eyes of his professional peers, and also from the perspective of the legal profession itself, with repect to its own standards in civil liability, whether it is fair to judge Cameron by subsequent standards. It is submitted that, given the range of patients that Cameron treated, and the immensely powerful modalities of intervention utilized, the standard for disclosure in treatment should not have been differentiated from that of experimentation, even in the late 1950s. That is, complete disclosure of all risks, known and anticipated, should have been discussed with the patient/subject.

The Cameron case is useful to consider in detail because it contains the ingredients that, although not common, are representative of the subtle progression of research abuse that can be difficult to control. In each generation there are examples of innovative therapies developed under influential figures. Doctor Cameron himself was a dramatic, autocratic and charismatic pioneer in the world psychiatric community. In reviewing his record, all the experts called upon reflected on Cameron's stature and deferred to his reputation. The Dean of Medicine of the University of Saskatchewan, for example, wrote that:

> A review of his curriculum vitae reveals that he quickly climbed the academic ladder. His overall position in the profession is attested by the fact that he was elected to head three

---

[20] See D. Mechanic, "The Social Dimension" in S. Bloch & P. Chodoff, eds. *Psychiatric Ethics* (New York: Oxford Univerity Press, 1984) 46 at 49:

> The problem is in no way unique to psychiatry. Comparable problems occur in industrial medicine, in the determination of disability, and in many other activities involving general physicians. If psychiatry is different, it is primarily because the social control aspects of psychiatric practice are far-reaching, particularly in situations involving the patient's freedom: clinical assessment affecting involuntary hospitalization, parole of prisoners, release of soldiers from military service, criminal responsibility, and the like.

For an excellent discussion of moral conflicts for health care professionals, see E.D. Pellegrino, "Societal Duty and Moral Complicity: The Physician's Dilemma of Divided Loyalty" (1993) 16 International Journal of Law and Psychiatry 371. See also D.N. Weisstub, "Le droit et la psychiatrie dans leur problématique commune" (1985) 30 McGill Law Journal 221.

prestigious psychiatric organizations: the Canadian Psychiatric Association, the American Psychiatric Association, and the World Psychiatric Association. In short, whatever his shortcomings as a person, he obviously was a credible figure professionally.[21]

Cameron's professional status amongst his peers weighed heavily on their judgments. The testimony given to the Federal government was consistent in conflating peer status with professional ethical competence. It presumed that status would never be given in a professional organization to someone who would sacrifice his professional code of ethics.[22] Nevertheless, his professional colleagues were prepared to assert that:

(1) he was not a good researcher;
(2) he got carried away with a questionable theory;
(3) he applied the theory much too aggressively;
(4) in his commitment to the therapy he began broadening the net of research subjects, perhaps at the end even indiscriminately;
(5) he was irresponsible and indeed unprofessional in not measuring with any scientific rigour the effectiveness of his interventions;
(6) he showed a lack of normal feeling towards his patients, particularly those who he believed were not successful products of the treatment;
(7) he had a visionary sense of his own importance in the history of psychiatry, at the expense of recognizing even legendary figures in the field; and
(8) he had no reserve about lecturing and publishing his experiences about his misguided treatments.

However, from the point of view of his colleagues, all should have been forgiven, not in the name of science, but of therapy. With the benefit of the advances in psychiatry, Cameron's colleagues believed that no sanctions should have been held against Cameron because:

(1) the standards of the time were so low;
(2) experimentalism was rife in that period in psychiatry, with other damaging treatments then being in vogue;
(3) there were no ethics research committees of note reviewing research grants; and
(4) there were no legal decisions either in the common law provinces or in the civil law of Quebec that would have provided Cameron with a clear standard.

These arguments are not satisfactory. If other practices in the period violated patients as much as Cameron's, then they should be criticized as well. There were

---

[21] See Cooper, *supra* note 18.
[22] In defense of Cameron, we can note that he must have justified his conduct upon a more personal code of behaviour, which contemplated a more favourable result for his patients than was the case. That is, his paternalistic, even grandiose, idea of his capacity to reassemble the human psyche permitted him to take enormous risks. The question must be posed whether such a rationale is frequently in evidence when scientific malpractice and/or scientific frauds are committed.

clear principles in both the common and civil law traditions as to what constituted an unwarranted assault to the person (touching without consent) in addition to the requirements of the Nuremberg Code. It should also be recalled that Cameron was an American citizen who had frequent contacts with top professionals in universities and governments in the United States. Discussions on informed consent were already appearing in the 1950s. Furthermore, his treatments were most abusive at the end of his direction of the Allan Memorial Institute, in the 1960s, when commentators such as Beecher were pioneering their ethical critiques of unacceptable research practices.

To defend Cameron in the name of therapy demonstrates how language can be effective as a legitimating function. The doctors who have commented as experts on Cameron's case turned to legal language to excuse his behaviour. His excesses were described as errors or misadventures, straight out of the classic cases of negligence law, devices often used to protect and rationalize irresponsible conduct. Furthermore, even though there were readily available examples of people who were not "repatterned" after being reduced to a state of infantilism, the psychiatrists who testified adhered to yet another legalism in dismissing the allegations against Cameron, submitting with painstaking clarity that it is not clear how the patients would have fared otherwise. This is the classic protective point used in malpractice cases where psychiatric disorder is an issue. This is why so few psychiatric patients have been successful plaintiffs, even where there seems to be a *prima facie* case.[23] The psychiatrists called upon were all familiar with the language of the law used to defend cases against members of their profession. If the issue is one of malpractice, the defence is misadventure, understood nonetheless as practice in the interest of the patients. If the issue is ethics, there is said to be an absence of law.

It was arguable that in medical therapy, ethical standards as such had not been articulated in jurisprudence. Even with an absence of law, Cameron should have been bound by the ethics of his own profession. As far as the law is concerned, despite the submissions that there was no law on experimentation until *Halushka* v. *University of Saskatchewan*,[24] there is reason to believe that Cameron would have been held to the standard of assault and battery, both long-standing remedies within common and civil law systems, for his therapies, if not for his experiments. Given the gravity of the interventions there should not be a different standard for therapy than for experimentation.

In the mid-sixties, in the common law tradition in Canada, which not unincidentally was Cameron's tradition, the law of battery was used to deal with dignitary violations of the person, and not the law of negligence. What is even more damning to Cameron's position is the fact that there was clear law stated for therapeutic experimentation long before *Halushka*. The judgment in *Marshall* v.

---

[23] See L.R. Tancredi & D.N. Weisstub, "Malpractice in American Psychiatry: Toward a Restructuring of the Psychiatrist-Patient Relationship" in D.N. Weisstub, ed., *Law and Mental Health: International Perspectives, Volume 2* (Oxford: Pergamon Press, 1986) 83.

[24] (1965), 52 W.W.R. 608 (Sask, C.A.) [hereinafter *Halushka*].

*Curry*[25] is very precise about the requirements for consent in information exchanges between doctors and patients. The defendant's professional skill in this case was never questioned. The doctor was not challenged in the trial for his negligence, but for his lack of respect for the patient demonstrated by his failure to gain permission to treat his body in a certain way. Justice Chisholm wrote:

> The phrase "good surgery" has appeared in some of the cases. Its use is not helpful; it is general and vague and, I think, ambiguous. It may mean good execution by the surgeon, and in that meaning it does not touch the question of the surgeon's right to operate. In these emergency cases, it is not useful to strain the law by establishing consent by fictions — by basing consent on things that do not exist. It is not better to decide boldly that apart from any consent the condition discovered makes it imperative on the part of the surgeon to operate, and if he performs the duty skilfully and with due prudence, that no action will lie against him for doing so; as I have stated, that is the jurisprudence established in the Province of Quebec, and I think it can well be adopted in other jurisdictions.[26]

This Canadian case has been cited in American courts. It is a classic statement of the law of informed consent and it preceded Nuremberg. It establishes the fact that the matter at hand for therapeutic interventions with respect to information shared between doctor and patient about what will occur, refers to a basic ethical value given respect in law. Although a common law decision from Nova Scotia, it relied on the notion of doctors owing a higher duty to save the life of a patient found to be in a position of jeopardy to establish this point. In so doing it makes the idea clear that the value at stake is so significant that it crosses jurisdictional lines. It makes evident the fact that professionals, however talented and respected, must abide by the higher ethical rules of conduct in professional life. Cameron was subject to the laws of Canada and Quebec. It should also be presumed, that as a leader in his profession, he should have known of a mainline legal judgment concerning standards of medical practice.

What is disturbing about the *Cooper Report*,[27] then, was that in its response it treated Cameron's practices as a matter of experimentation to show that there was no established law, and as therapeutic interventions to demonstrate that there was no violation of professional standards. In this, Cooper received the support of professional experts, although we should bear in mind that it is well established that the consensus of a group of peers will not be an adequate defence when there is a serious allegation of unwarranted risk taking.

If what we observe to have occurred here is yet another example of the inherent corruption of power that occurs when professional elites have a fragile or passive population under their control, should we be led to a hasty conclusion that no matter which conceptual framework we evolve, the corrupt application of procedures is probable, and that therefore any philosophical analysis is an irrelevant digression? A response would be that it is essential for us to continue the dialogue of discovery

---

[25] *Marshall* v *Curry*, [1993] 3 D.L.R. 260 (N.S.S.C.) [hereinafter *Marshall*]. On the remedy of battery in Canadian common law in the area of medical treatment, we may also refer to the Alberta case of *Mulloy* v. *Hop Sang*, [1935] 1 W.W.R. 714 (Alta. C.A.).

[26] See *Marshall, ibid.* at 275.

[27] See *supra* note 18.

and clarification as part of the process of morally educating ourselves as a society. If we do not pressure ourselves to reflect upon the moral basis of the relations that occur in medicine and research, there will be less of a likelihood that new generations of researchers will be sensitized to a level of moral recognition of what their appropriate duties are to resist denigrations of human dignity. With this in mind, we turn our attention to building a case for a morality compatible with the overarching paradigm of a convenantal-moral bond between researcher and subject, concretized through the law of fiduciary obligations.

## CONCEPTUAL MODELS

### *Introduction: The Problem of Knowledge and Morality*

Our objective here is to consider the ethical discourses and legal modes which might be drawn upon to reconcile the problems posed by human freedom and human obligation. Our chief point of inquiry is twofold. Firstly, are there methods of moral discourse and accompanying legal modes that would provide an acceptable rationale and justification to allow non-therapeutic procedures to be performed upon vulnerable populations in the absence of an individual subject's free and informed consent? Secondly, what are the practical implications that follow from establishing an appropriate moral discourse with its accompanying legal modes? In framing these questions, we do so from the explicit premise that the law's chief purpose in this area is not to facilitate research *per se*, but rather to protect the individual and make the researcher and research community accountable, not only to the individual subjects who participate, but also to society as a whole.

To inquire into the nature of research is to ask yet a deeper question concerning the nature of knowledge itself. It is to realize, *inter alia*, that knowledge *qua* knowledge is both theoretical and practical and that theory leads to *praxis*. It is this relationship between knowledge and action that makes ethics a necessary condition for the right "ordering of actions and for regulating the power to act".[28] Without such ordering there is no normative restraint on human power, and both society and the individual are left with a moral nihilism based upon the simple axiom that knowledge is power and power is knowledge. In such a universe, value is predicated upon the will to power and its exercise by those who do so at the expense of those who do not. The objective reality of a moral order, the *nomos*, informs us of the difference between good and evil and indeed, forms the basis of law itself. However, this objective is not possible in the moral nihilism of knowledge as power.[29] If nihilism and its accompanying abuses are to be avoided, the question of knowledge must be addressed from the broader perspectives of its purposes and ends.[30]

---

[28] H. Jonas, "Technology and Responsibility: Reflections on the New Tasks of Ethics" in H. Jonas, ed., *Philosophical Essays: From Ancient Creed to Technological Man* (Englewood, N.J.: Prentice-Hall, 1974) 3 at 20.
[29] For a discussion of the moral difficulties posed by nihilism, see H. Jonas, *The Phenomenon of Life: Towards a Philosophical Biology* (New York: Harper & Row, 1966) at 211ff.
[30] See J.C. Smith & D.N. Weisstub, *The Western Idea of Law* (Toronto: Butterworths, 1981).

In the context of science, knowlege does not exist merely for its own sake. Human beings use knowledge for a variety of purposes; it is the purposes to which the knowledge is put that attract a moral judgment. Moreover, the manner in which knowledge is gained is also important, not only because it may tell us something about the nature of knowledge *qua* knowledge, but more importantly because it reflects back onto the actions of the human agent who is acting for a particular purpose. Depending on the nature of that purpose, moral condemnation or even legal sanctions may be used to regulate such conduct.

Regardless of how knowledge is obtained, it is connected to a moral situation, relating to personal, familial, social, and even metaphysical structures of values. The producers of such knowledge are all actors in these contexts, which give rise to inevitable conflicts between or among these groupings of values. In the process of clarifying these tensions, limits are set to the moral parameters of knowledge accumulation. In fact, it is through the ambiguity or ambivalence we experience in the roles that we play in acquiring and using the knowledge we obtain, that we begin to experience a moral identity as part of a scientific endeavour. Knowledge divorced from this process of self-inquiry leads to the kind of moral abstraction which, if accentuated, can lead to highly dissonant or repressive conduct against others.

In other words, there are limits which society is willing to accept in the quest for knowledge. Those limits reflect certain core values which society deems to be of greater importance than the mere acquisition of knowledge for either its own sake, or for the sake of some other goal, no matter how noble or socially beneficial. As Karl Deutsch has observed:

> Science itself depends for its life on the prior acceptance of certain fundamental values, such as the value of curiosity and learning, the value of truth, the value of sharing knowledge with others, the value of respect for facts, and the value of remembering the vastness of the universe in comparison with the finite knowledge of men at any particular moment.[31]

The limits and underlying values which have defined the doctor/patient relationship and the pursuit of medical knowledge have been constantly evolving and manifesting themselves in various moral and legal matrices through the millennia. In each case, the moral discourses which have shaped human conduct have had an impact on the legal modes which regulate that conduct. The next section of this chapter will discuss those discourses and modes in an attempt to understand the nature of the moral and legal problems associated with experimentation on human beings.

### *The Hippocratic Tradition*

The ethical/legal model which has traditionally regulated the doctor/patient relationship has been based upon the Hippocratic Ethic, best known for its credo of

---

[31] K. Deutsch, "Scientific and Humanistic Knowledge in the Growth of Civilization" in H. Brown, ed., *Science and the Creative Spirit* (Toronto: University of Toronto Press, 1958) 1 at 18.

*primum non nocere*. The ethical approach here is highly paternalistic, grounded in beneficence, and is aimed at regulating the physician's conduct in terms of the patient's best interest. In other words, the principle of *primum non nocere* places a duty upon the physician to act in the best interests of the patient and not to consider an object other than the medical/therapeutic needs of the patient. It presumes that the physician has the best knowledge available, and, in turn, is the most appropriate person to decide upon the best course of treatment. According to this tradition, while consultation with the patient may be regarded as clinically advisable, it should not be mandated either by ethical norms or legal sanctions. It should of course be noted that this particular aspect of the Hippocratic Tradition has been altered by the law as reflected in the evolution of the doctrine of informed consent.[32]

It is arguable that the Hippocratic Tradition represents a form of paternalism, which, while centered on the patient's best interests, does not allow the patient much (if any) autonomy in making treatment decisions. In addition, given the emphasis on a doctor/patient relationship that seeks to maximize the patient's best interests, the Hippocratic Tradition strongly suggests that it is best suited for the therapeutic relationship and not the field of non-therapeutic research, where the autonomy of the individual is paramount. Indeed, the importance of the autonomy principle as a cure for physicians who breach the *primum non nocere* rule is best illustrated by the *Nuremberg Code*[33] which enshrines that principle in its first precept.

It is in conjunction with this professional ethic that much of the law which defines, regulates, and facilitates the medical profession has developed. That relationship has come to be understood in law through a number of different paradigms, depending upon the nature of the legal question being examined.

At its most basic, the doctor/patient relationship itself has come to be understood in terms of a contractual paradigm of "contract for service". However, the actual means of defining the rights and duties of that relationship are not found exclusively in contract, but rather in the law of torts.[34] Nevertheless, the contractual paradigm as a means of initially defining the doctor/patient relationship has proved particularly attractive no matter what the underlying ethical discourse. The reasons for this are partly practical and partly doctrinal. The legal paradigm of contract fits well with our socio-political presumptions regarding individualism, self-interest and distrust of authority, which manifest themselves in a desire to move away from a purely paternalistic model, of which the Hippocratic Tradition represents the most concrete example. As such, contract offers a way to give some balance to a relationship that

---

[32] See, for example, the Supreme Court of Canada's decision in *Reibl* v. *Hughes*, [1980] 2 S.C.R. 880 [hereinafter *Reibl*], which recognizes a legal duty on the physician's part to disclose information. Such disclosure, however, is not absolute as therapeutic privilege continues to be recognized in certain circumstances. See *Videto* v. *Kennedy* (1981), 33 O.R. (2d) 497 (C.A.).

[33] The *Nuremberg Code* constituted part of the judgment in *U.S.* v. *Karl Brandt et al., Trials of War Criminals Before the Nuremberg Military Tribunals Under Control Council Law No. 10.* (October 1946– April 1949).

[34] C. Fried, *Medical Experimentation: Personal Integrity and Social Policy* (New York: American Elsevier Publishing, 1974) at 13–18.

may be characterized as somewhat unbalanced due to the informational advantage of the physician and the condition of necessity of treatment which the sickness of the patient has occasioned. In addition, contract also facilitates the very notion of informed consent by requiring the participation of two parties to the agreement. In this sense, contract is completely in agreement with the idea of individual autonomy and respect for the dignity of the individual. Finally, from a practical perspective, the contract paradigm offers the possibility of relatively easy enforcement in terms of the protection and accountability it affords to both doctor and patient.

What are the core values which underlie the Hippocratic Tradition? One of the primary values is reflected in the principle of *primum non nocere*, the protection of innocent and vulnerable human life. In addition, the value of health and its promotion are central to the duty imposed upon the medical profession as are the values inherent in the concept of dedicated service to one's fellows. The Tradition attempts to protect the individual human being through reliance upon paternalistic concern for their well-being. Other values implicit in the Hippocratic Tradition relate to the importance of the autonomy of the medical profession; the fostering of a personal relationship between the patient and doctor; and the promotion of the individual's autonomy through, ironically, a paternalistic approach that emphasizes the patient's best interests *vis-à-vis* treatment.

The Hippocratic Tradition's ethics, although conceived of in terms of the patient's best interests, nevertheless present a number of problems. The major difficulty with the Tradition is its paternalism. By relying upon a paternalism in which the physician is always to act in the best interests of the patient, the Tradition ignores the autonomy of the patient. This produces a tension between paternalism, on the one hand, and libertarianism, on the other.[35] In other words, autonomy is sacrificed in favour of the patient's best interests as determined by the physician. Such paternalism prevents the patient from being used as a means for some other purpose, but places too much reliance upon the physician to decide what the patient's prior wishes were concerning the course of their medical treatment. As such, the ethic is open to abuse unless the physician is aware of his cognitive and moral limitations in making such decisions.

Moreover, as alluded to above, the Hippocratic tradition, precisely because its focus is on the best interests of the individual patient, is best suited to situations of a therapeutic context. Non-therapeutic experimentation, which by definition is not for the immediate benefit of the subject, cannot be easily sanctioned, if at all, according to the canons of the Hippocratic Tradition. Accordingly, the tradition alone is not suited to provide a moral foundation in a research context in which the autonomy of the subject is of paramount importance. Having said that, it ought to be recognized, however, and emphasized that *primum non nocere* is a principle which transcends the therapeutic context of the Hippocratic Tradition and ought to continue to be placed as a fundamental requirement of any research protocol.

---

[35] L. Tancredi & D.N Weisstub, "Law, Psychiatry, and Morality: Unpacking the Muddled Prolegomenon" (1986) 9 International Journal of Law and Psychiatry 1.

## *Utilitarianism, Deontology and the Classic Liberal Tradition*

Given the obvious social concerns associated with issues of "public health" and the potential for enormous social benefit to be derived by society both in terms of present and future generations, it is not surprising that utilitarian analysis offers the best moral justification for experimentation with human subjects without their consent outside the therapeutic context.[36] Indeed, under a utilitarian ethic, statute, contract, tort, and criminal models might be used to regulate the doctor/patient relationship. However, unlike the Hippocratic Tradition, the Utilitarian ethic focuses not on the best interests of the patient but rather on the best interests of society as understood in terms of a cost/benefit analysis based upon a calculation of utility. The doctor's primary duty is no longer to the patient, but to society. The individual becomes valued not as an individual, but as a member of society. Under such circumstances, there is nothing preventing the law from being used as a means to accomplish socially useful goals and programs in which the individual is sacrificed for the benefit of many.[37] Accordingly, the legal paradigm that best facilitates a Utilitarian ethic is one of command, regardless of the particular legal mode or order that is used.

The values that underlie the Utilitarian ethic are closely associated with the maximization of social utility as a means of alleviating social misery and thereby creating the greatest happiness for the greatest number. The Utilitarian ethic values social institutions and the social consequences of individual actions which might impact upon those institutions. In its most sophisticated forms, one may conclude that Utilitarianism values equality over individual autonomy as a means of achieving that which is socially useful. However, the practical and social implications which flow from the failure to respect individual autonomy facilitate the justification of sacrificing the individual for the supposed benefits of the majority. This failing results in a propensity to view the individual not in subjective terms, but rather as pure object. This objectification necessarily regards the individual as being merely a means to an end. If the Hippocratic Tradition is problematic because of its paternalistic assumptions regarding the best interests of the patient, Utilitarianism is no less problematic for its assumptions regarding the best interests of society and the nature of the common interests to be promoted at the expense of the individual.

In its basic orientation, the deontological ethic stresses individualism and the unique individuality of each human being, and is thus most closely associated with

---

[36] The classic account of utilitarian thinking can be found in J.S. Mill, *Utilitarianism*, 4th ed. (London: Longmans, Green, Reader & Dyer, 1871). For one of the most salient critiques of Mill's position, see Sir W.D. Ross, *The Right and the Good* (Oxford: Oxford University Press, 1930). See also J.J.C. Smart & B.A.O. Williams, *Utilitarianism — For and Against* (Cambridge: University Press, 1973); R.B. Brandt, *Ethical Theories: The Problems of Normative and Critical Ethics* (Englewood, N.J.: Prentice-Hall, 1959) at 380–391; and A.K. Sen & B.A.O. Williams (eds.), *Utilitarianism and Beyond* (Cambridge: Cambridge University Press, 1982).

[37] For a discussion of sacrifice as a recurring theme within the history of utilitarian thought, see H. Jonas, *Philosophical Reflections on Experimenting with Human Subjects* (New York: George Braziller, 1970).

the autonomy principle.[38] One might argue, however, that Kantian metaphysics is not the only basis for individual autonomy, especially as understood in terms of individual rights. The classical liberal tradition that continues to animate much of our political and legal discourses also offers a justification for, and defence of, individual autonomy and its accompanying rights. The discourse of the classical liberal tradition, as evidenced in the writings of John Locke, Adam Smith, James Madison, and even John Stuart Mill, recognizes the importance of autonomy understood in terms of liberty. In other words, in the classical liberal tradition, human freedom is defined in terms of liberty, whereas the Kantian tradition understands the problems posed by human freedom in terms of autonomy. In discourses concerning medical ethics, where the emphasis is placed upon the *duty* of the physician as measured against the autonomy of the individual to regulate that duty, there has been the inclination to favour the Kantian understanding rather than the classical liberal tradition, if for no other reason than the latter, by concentrating upon liberty, underestimates the role that duty ought to play in moral, social and political conduct. Moreover, to the extent that the classical liberal tradition is based upon a radical subjectivism which makes universal moral postulates somewhat problematic, it does not provide the basis upon which to give full effect to the implications of individual autonomy at a universal level *vis-à-vis* a normative principle of duty.

Despite, or perhaps because of, the differences between the Kantian and classical liberal approaches to liberty and duty, there has been a blending of the two traditions in our contemporary moral and political debates. Thus the principles of individualism and autonomy have given rise to a rights/duties dichotomy in both political and moral discourses. From the perspective of the physician, a duty is placed upon him to obtain the patient's voluntary and informed consent before attempting any procedure, either therapeutic or non-therapeutic. From the perspective of the patient, the deontological ethic stresses the right of the individual to give consent to treatment or participation in research and the corollary right to be left alone.

The values that underlie the deontological ethic are centered upon respect for the individual *qua* individual. This respect, in turn, is evidence of the value placed in moral behaviour *per se*.[39] In addition, the value of individualism which recognizes the importance of individual responsibility by respecting the autonomy of the individual as an end in itself is an important postulate, preventing the objectification of the individual which would result in the individual being treated as an end, not as means to an end.

The major difficulty with a deontological approach to the question of research on non-competent individuals is that it overlooks vulnerable populations' legitimate

---

[38] See generally H. Jonas (1966), *supra* note 29; P. Ramsey, *The Patient as Person* (New Haven: Yale University Press, 1970); Sir W.D. Ross *supra* note 36; I. Kant, *Foundations of the Metaphysics of Morals*, trans L. W. Beck (Indianapolis: Bob-Merrill Educational, 1959).

[39] Characterized sometimes by the maxim: *Fait iusticiaruat caelum* — let justice be done though the heavens fall.

moral and medical needs. By its disposition towards autonomy and self-choice, the ethic can lead to a negative paternalism no less abusive than the one encountered under the Hippocratic Tradition. In other words, by relying upon individualism, the deontological position tends to be unable to account for those who cannot speak for themselves. This can lead to an over-protective paternalism which might be counter-productive in terms of recognizing legitimate social as well as individual needs. Such a situation presents the paradox of being unable to recognize and promote certain individual needs in the name of promoting individual autonomy.

## TOWARDS AN INTEGRATED MODEL

In the abstract specifications given by philosophers, we are sometimes overwhelmed by the differences or polarities represented by the Kantian and utilitarian points of view. In theory, we are led to believe that gross violations of human dignity flow directly from a utilitarian equation and that a Kantian preoccupation with the autonomous will cannot properly take into account the possibilities of action for persons suffering from a compromised rationality. Even in theory, refined versions of either theory indicate that, in their application, there has to be an accommodation to social and individual reality. There are elements of both theories to which most of us adhere that do not lead us into hedonism or a narrow duty-bound concept of ethics. In the Kantian format, some realistic consideration of means is necessitated so long as the notion of treating persons as ends in themselves prevails. In the case of Utilitarians, the measurement of the good, both in theory and application, leads us to calculating a utility where we take into account: the effects of an experiment, for example, on the society; a qualitative assessment of emotional and moral-like responses on the part of participants; and long and short term impacts on a range of groups involved in the process.

In defense of the theoretical distinctions between the classical models of utility and autonomy, we can say that there may be places where experiments would be warranted by Utilitarians while rejected by Kantians and vice-versa. But nonetheless in the application, that is, the calculation of what constitutes utility and what are, even for utilitarians, the parameters of unconscionable behaviour, albeit calculated on pragmatic grounds, credible positions held by either tend to enter the same circumference. As Macklin and Sherwin have observed: "In any case, it follows from either theory that we have a *prima facie* obligation to treat other persons honestly, to seek their consent in matters affecting them, and to be sensitive to their interest".[40]

It is unclear whether, in the decision-making models created for dealing with hard cases on the limits of ethically permissible experimentation, that pure form utilitarians or Kantians can be readily located. We might say that these theories both represent aspects of the value schemes that impact on conscionable behaviour in

---

[40] R. Macklin & S. Sherwin, "Experimenting on Human Subjects: Philosophical Perspectives" (1975) 25 Case Western Law Review 434 at 457–458.

civilized societies. If we are to guard against anything, then, it should be the perversions of the theories that have been put before societies under stress. An exaggerated Kantian view could lead to a rationalized justification of either excessive paternalism or the outright denigration or devaluation of humans of impaired capacity. If the highest regard of society is paid to rationality, and the life of emotions, suffering, or empathy are put in a secondary position, a Spartan society is a predictable outcome. The repression of emotions can, in such contexts, erupt with a savage denial of substandard human life. In theory, it is not inconceivable that a Kantian could will as a universal law that people of a certain genetic or racial characteristic should be denied access to certain social services or to the rights of ordinary citizenship. In the case of an extreme utilitarian postulate, we can be led equally into circumstances which would see the utility of using persons of limited capacity for the greater social benefit. If their utility can be registered as less than that of an ordinary citizen, the utilitarian might argue that our commitment should be to enhance the collective contribution of individuals to our social well-being, thereby protecting the most intelligent and, perhaps in certain circumstances, the healthiest and wealthiest members over and against those individuals who are most draining to society's resources.

In the context of an applied ethics we must be drawn to a position which can conjoin fundamental moral values with empathy. Without such an integration the protection of persons without full capacity will not be finally guaranteed.[41] Such a fundamental value can be seen to be a part of both the utilitarian and Kantian positions. It is, however, in the development of a humanistic ethic based upon empathy that the meaningfulness of such a basic commitment takes hold. Because it is rare in our actual systems to locate philosophical applications of values presented, defined and logically justified, it may be that, apart from a statement of general commitment of a tolerant humanism, our greatest protection for our civility and for the persons meant to be protected would be through a set of procedures, implemented by law, which demand our attention and obedience.

## ORDERING OUR NOTIONS OF INDIVIDUAL AND SOCIAL CONTRACTS

### *The Covenantal Bond*

In optimal conditions, the relationship between doctors, researchers, and their patients or subjects should be modelled according to the law of contract. Here we

---

[41] Rousseau writes "It is very certain . . . that pity is a natural sentiment which, moderating in each individual the activity of love of oneself, contributes to the mutual preservation of the entire species. It carries us without reflection to the aid of those whom we see suffer . . . it will dissuade every robust savage from robbing a weak child or an infirm old man of his hard-won subsistence . . .". The point here is simply that the protection of vulnerable populations can arguably be grounded in the visceral human emotion of pity or empathy even though we may not be able to locate a justification for such action if we rely solely on "reason". See J.-J. Rousseau, *The First and Second Discourses* (R.D. Masters, ed.) (New York: St. Martin's Press, 1964) at 132–133.

can find a supposed equilibrium of power and capacity, and each party, through an exchange of services, can make a free decision about whether or not to be involved. If certain contacts are viewed with moral suspicion by society they can be banned through social intervention. Otherwise the parties are free to trade in a marketplace of exchanges, for monetary or other types of compensation, including secondary gains, honours, satisfaction of sacrifice, or specific entitlements, such as better medical treatment, or an early release from incarceration.

Even if we admit that this contractualist model is the mainstay of the informed consent movement, its relevancy is limited in situations where the capacity of the recipient of treatment or research benefits is incapacitated, or suffers from a discernible limitation. It is useful therefore to explore whether we can locate a paradigm for understanding what is most suited to vulnerable populations while at the same time examining how the obligations between the parties, researchers and recipients should flow one to the other.

The idea of a covenantal relationship, although representing an agreement between two parties, can be distinguished from the ordinary contract. To begin with, we readily admit that in a covenantal relationship the parties need not be equal. Either implicitly or explicitly, there is the realization that the parties differ in what they bring to the relationship. This difference is reflected not only in terms of their relative power, but also with regard to their moral possibilities and responsibilities conceived under the penumbra of the covenant.

The idea of covenant is not a stranger to either the medical profession or to the Western legal or moral tradition. In law, originally, a covenent was simply a formal agreement, convention or promise of two or more parties, witnessed by deed in writing, signed, and delivered, by which either of the parties pledges itself to the other that something either be done or not be done. In addition, covenants could also be used to stipulate the truth of certain facts. At common law, the importance and solemnity of the covenant was attested to by the fact that such agreements were required to be under seal. The covenant has proven to be a remarkably flexible kind of legal agreement, even though its contemporary usage is usually confined to promises in conveyances or other instruments relating to real estate.

The medical profession itself has grown up with the idea of covenant as part of the Hippocratic Oath *vis-à-vis* the physicians' obligations and debt owed to his teacher and his progeny for the service of the knowledge given.[42] At a more general level, in the Western Tradition, the covenant is recognized in both the Code of Hammurabi and the Mosaic Law.[43] In the latter example, the nature of the covenant

---

[42] For a discussion of the covenantal nature of the Hippocratic Oath, see W.F. May, "Code and Covenant or Philanthropy and Contract?" in S.J. Reiser, A.J. Dyck & W.J. Curran, eds., *Ethics in Medicine: Historical Perspectives and Contemporary Concerns* (Cambridge, Mass.: The MIT Press, 1977) 65.

[43] The notion of convenant has its origins in the Old Testament. Abraham entered into a covenantal relationship with God. In this agreement, Abraham willingly gave up a certain element of freedom in order to have his prescribed destiny and role in history. In return, God gave up the freedom of having indeterminate choice. That is to say, at that historical point, God had committed himself to the promise of Abraham's actualization as a pivotal, key role in history. When this covenant crystallized into a social fabric, with a bond established between God and the people of Israel, a set of laws emerged.

differs insofar as "it places the moral duties of the people within the all-important context of a divine act of deliverance".[44]

The classical form of the covenant has a tripartite structure composed of the following elements: the exchange, the agreement and the response. The exchange is the original experience of gift, labour, or services, by one partner to the other before they enter into the covenant. It is an act that places the soon to be made covenant in a temporal context; a covenant is at heart a promise based upon the original, or anticipated exchange of gifts, labour, or services. The agreement is therefore the contractual/moral element of the convenantal relationship. The response is the ontological dimension to covenant which sets it apart from a mere contract. The response is indicative of a change in being that transcends the particular moment of the agreement (contract). As William F. May has observed:

> A contract has a limited duration in time, but a covenant imposes a change on all moments . . . when the professional is initiated, he is covenanted, and the physician is a healer when he is healing, and when he is sleeping, when he is practicing, and when he is malpracticing. A covenant changes the shape of the whole life of the convenanted.[45]

This discussion is applicable to physicians/researchers because they are members of a profession that has been covenanted. At one level, there has been an *exchange* between the physician and society in which the latter provides the former with the education and ability to practice medicine. At another level, there has been an *exchange* between the physician and the patient for the former to provide services to the latter; an *agreement* between the patient and physician that accepts the services offered; and, finally, a *reciprocity* or *response* in which the physician accepts the responsibility for the agreement. The notion that May articulates in the above-quoted statement is that a physician *qua* physician stands in a different relationship to society and his patient precisely because he is a physician. The subsequent duties and obligations imposed upon physicians as members of a learned profession, because of their specific knowledge, requires that that knowledge be used for the good of the covenantal relationship, for it is because of that relationship that the physician has been accepted to practice in the community.

In articulating a covenant paradigm, it is important to remember the differences between a contract and a covenant. Perhaps one of the most obvious distinguishing features is the nomenclature used to describe the participants. In contract, the normal description of "parties" to the contract is used and conveys a sense of their relative autonomy and individuality. In a covenantal relationship, however, the term "partners" is a more appropriate expression of the interdependence of the individuals involved and of their mutual continuing relationship.[46] In addition,

---

[44] See May, *supra* note 42 at 68.

[45] *Ibid.* at 69–70.

[46] See generally R.M. Veatch, *The Patient as Partner: A Theory of Human Experimentation Ethics* (Bloomington: Indiana University Press, 1989). The nature of covenant is something that also extends over and above negligence principles. It is an agreement between the parties, and therefore should be respected. If one respects the terms of the agreements of the parties, then the general principles of negligence law, except for those circumstances of intentional tort, should not be applicable. One does not apply negligence law because it requires the application of an externally defined set of criteria.

contract is most often associated with a commercial understanding of a particular legal relationship. It is driven by self-interest and self-gain, and it suppresses the moral element of gift and beneficence. Contract thus understood as mere commercial agreement overlooks the possibility that one of the essential elements to a relationship may be a moral component in terms of service and self-sacrifice.[47]

Contracts also differ from a covenantal paradigm insofar as a contract is based upon a categorization of particular services to be performed. In contrast, a covenantal relationship envisages a certain open-mindedness, a greater dynamic exchange between the partners that cannot be easily categorized. The idea of "professional services" is something that supersedes the "tit for tat" of contractual experimentation and requires a commitment that extends beyond the written word of a document to the spirit of the relationship that resists conventional categorization.

The ontological aspect of covenant also suggests an important distinction from mere contract, the difference being that contracts are external to the parties who make them. A contract does not normally require the parties to undergo an ontological shift in their life activity. It contains categories that are predetermined which place the duties and responsibilities of the parties at one moment in history, with no opportunity for flexible restructuring of the original terms. In a covenantal relationship, however, the agreement is internalized. The covenant affects the very being of the partners by prescribing new modes and orders of their social intercourse with each other; the covenantal relationship consists of an open-ended texture. That is, it allows for much more flexibility in the re-negotiation of its terms over the course of time. It provides a significant amount of room for discretionary decision-making on the part of both parties involved. It is here that there are significant differences between the nature of a covenant and that of a contract. The open-ended texture of the covenant, with its recognition of the disparities in the power prerogatives of the parties, allows for some restraining of the decision-making options of the stronger party. A fabric of values surrounds the decision-making that influences both the stronger and the weaker parties in the covenant. There is an assumption that trust rests in the weaker parties because they believe that the stronger will protect them through the values present in the social fabric.

### *Physician/Researcher as Fiduciary*

Insofar as the covenantal paradigm represents a moral construct depicting the relationship between researcher and subject as a duty-creating bond, it is necessary to pursue the logic of this to its legal conclusion. The fiduciary relationship can be seen as the concretization of this interaction. A fiduciary, the archetype of which is the trustee, is required in law to act in a completely selfless manner and not profit

---

[47] An effective critique of the modern commercial understanding of contract may be found in C. Fried, *Contract as Promise: A Theory of Contractual Obligation* (Cambridge, Mass.: Harvard University Press, 1981).

or gain as a result of holding such a position.[48] The fiduciary must therefore act in the interest of the beneficiary and in strict accordance with the terms of the trust traditionally ascribed to the position of trustee.[49]

It is important to understand that, in law, the fiduciary categories are not closed and subject to a continual process of review, as postulated in the Supreme Court of Canada's decision, *R. v. Guerin*.[50] While care must be exercised in attempting to expand the categories of fiduciary relations (sometimes beyond the point of recognition), the fiduciary model is pertinent because of the inherent trust that should exist between the researcher and subject. If realized more formally in law, this would make available a remedy for wrongful conduct which could represent a proper response to a dignitary violation. As Charles Fried has stated, "there is no reason why the doctor should not be held to be in a fiduciary relationship to his patient, and therefore, why the same fiduciary obligations that obtain for a lawyer, a money manager, a corporation executive or director should not obtain for a doctor".[51] Such a remedy would take its place as a complement to the already existing remedies in tort law and in the regulatory schemes pertaining to standards in professional practice. By accepting a covenantal paradigm enforceable and regulated through equitable constructs of fiduciary standards and liability, we would thereby reinforce our commitment to the fundamental value of respect for persons who, by virtue of incapacity, have an inherent limitation in their power to contract their will in relation to their own interests. The fiduciary model addresses the limitations that are part of the classical models of utilitarianism and Kantianism. Finally, it is unrealistic to foster an image of incapacitated individuals as failed human beings due to an incomplete rationality, or a fragile utility, in terms of an actual or potential contribution to society, once it is clear that a humanistic ethic requires our social concretization of a fiduciary link between researcher and subject. The fiduciary model is a legal response to the possibility of the perversion of our classical models that dominate our social thinking about the rights of mentally incapacitated individuals to protection in the area of research. Of course, it remains a condition of our commitment to individual self-determination that we do everything as members of a society of which we are capable to enhance the capacity of individuals to act on their own accord in making their wishes known about their participation in the research process. Where this is not possible, legal guardians should be held to a reciprocal fiduciary obligation of respect for the subject alongside those duties placed on researchers. There should be a monitoring of the subject's wishes in a perceived partnership of responsibility, at once a partnership between the researcher and subject.

---

[48] *Boardman* v. *Phipps*, [1986] 3 All E.R. 721 (H.L.). See also *LAC Minerals Ltd.* v. *International Corona Resources Ltd.*, [1989] 2 S.C.R. 574. For a recent American case pertaining to the application of a physician's fiduciary duty, see *Moore* v. *Regents of University of California*, 793 P.2d 479 (Cal. 1990), cert. denied 499 U.S. 936 (1991).

[49] See M.J. Mehlman, "Fiduciary Contracting: Limitations on Bargaining Between Patients and Health Care Providers" (1990) 51 University of Pittsburgh Law Review 365.

[50] [1984] 2 S.C.R. 335.

[51] See Fried (1974), *supra* note 34 at 34.

### The Limits of Social Contract

In societies where a social welfare system has delivered, with limited reference to financial ability, but largely based upon need, the necessary tools for physical and mental health, health has been held to be a fundamental right to which all citizens are entitled. In such a social order we can muster support for the notion that it is an acceptable requirement of citizens that they make themselves available to the health system for purposes of general betterment, in instances where only minimal risk is at stake, and the potential value of medical research is significant for general well-being. This line of reasoning can be presented forcefully to citizens more generally because it is not difficult to prove empirically that there is a disproportionate burden in medical experimentation placed upon minorities, the poor, and the institution-alized. Some philosophers have asserted that we should in the name of a cooperative ethic require citizens to take upon themselves the role of participant in research enterprises, based upon a presumed existence of a social contract.[52] Caplan argues that this notion of a moral duty based upon a reciprocal obligation incurred by us to past generations for sacrifices on our behalf is not a fair exchange because we did not enter into the exchange; research subjects have at times made a great personal sacrifice as a gift to humanity or indeed in other cicumstances have derived all sorts of benefits from their undertakings. In Caplan's view, no argument can be sustained on behalf of a contractual reciprocity.[53] Rather, he insists that fair play can serve as a generator of a moral basis for encouraging greater public interest and participation in medical research. The justification, according to Caplan, is that most citizens are activists in seeking out the benefits of health care and, because they are enterprising, they thereby become obligated to the society which serves them. This cooperativist ethic can be connected to a trade-off morality which can define the exchange value as money, or the contribution of a body useful to research. The choice is frequently made in health care, where we agree to certain inconveniences based on our calculation of the advantages that accrue to us in teaching hospitals and research centers.

Notwithstanding the fact that certain groups of society are disproportionately represented as research subjects and indeed Caplan argues the point that we therefore should be more sensitive as ordinary citizens to the disproportionate advantages which accrue to the general population. His notion of a cooperative ethic should still in principle apply to vulnerable populations. We could arguably prove that, despite their overrepresentation as research subjects, they are still taking a disproportionately large share of the economic pie in our protection of them. Does this mean therefore that we are justified in constructing a morality of obligation on their behalf, asking them, even if they are disabled and incapable of giving an informed consent, to be offered for minimal inconvenience to serve the interest of medical science? If we make it a necessary condition of our intervention that the

---

[52] See H. Jonas (1970), *supra* note 37.
[53] A.L. Caplan, "Is There an Obligation to Participate in Biomedical Research?" in S.F. Spicker *et al.*, eds., *The Use of Human Beings in Research* (Boston: Kluwer Academic Publishers, 1988) 236.

research be attached to the vulnerable population in question would that, then, be a sufficient condition to recruit their participation? It is often argued that we should first turn to competent and healthy members of the population, to see if their availability is present to test them on research which could be used in effect for a vulnerable population. Would a cooperativist morality necessitate such caution or priority? If that would be claimed, presumably that would be so because in our calculation of benefits we have decided that the general population has taken more as a group than the vulnerable population. Or alternatively, that the vulnerable population has suffered so much already, that to ask it to participate in research is to add insult to injury. In any event, such calculations are extremely difficult to make in reality, not to mention the problem of who is to make them. Furthermore, many research projects can be directly linked to the needs of the vulnerable populations in question. If we are to keep fidelity with the cooperativist ethic we may also be obligating ourselves to calculate in each and every case that the person upon whom we may wish to do research has been a personal beneficiary of the special attention that we as a society give to this group. Is it sufficient that the benefits can be potential, for special members of the group, or for future generations? Without answers to these hard moral questions the cooperativist morality can be a veneer to justify a harsh economic exchange system where the rich escape making contributions, even when the handicapped, and the poor may be required, in the name of a higher ethic, to sacrifice themselves to a common interest which has been defined in such a way as to work to their disadvantage.

Talk about the social contract has never been able to satisfy its critics who preoccupy themselves with the inequalities of birth and social destiny. Communitarian ethics which argue either the entitlements or obligations of special groups, and still argue a version of the social contract theory, run inevitably into the same murky waters. The difficulty remains for us as a society to determine the conditions under which we feel justified to demand of vulnerable populations that they participate even when there is no direct or indirect benefit to their own intermediate predicament. Research for future generations is difficult of course to calculate. No less is the difficulty in calculating benefits when the results of science cannot be known in advance. Without certainty, can we obligate vulnerable populations to unrequested medical interventions? The arguments for justification are not precisely the same for the different populations in question. But apart from their particularities there are still general questions of moral entitlement which persist.

One thing is certain. The tolerance for liberal contributions of public resources to vulnerable populations in need has diminished in the economics of scarcity that are part of the health delivery systems in all industrialized societies. Our careful claims, based upon sound moral judgements, for the selective participation of our vulnerable populations, protected by well entrenched procedures for their defense against possible abuses and manipulations, may be our best defense against the abandonment of these populations, which would leave them without an adequate level of health care, or the tools with which they can reconstitute themselves to a maximum capacity.

## CHARITABLE ACTS ON BEHALF OF OTHERS

Requiring that people, as citizens, make a contribution to society in exchange for the normal benefits received from living as part of the group is not generally perceived as an excessive demand. As in the family context, or any small group, the inclination to do so is natural and even instinctive. That is so because members of such social units usually have a defined role within a system of benefits and obligations, where the ongoing support of the group offers feedback or recognition for the contributions made. Even when this is not so, the overriding ideology of the group sustains the distribution of power and responsibilities such that members are embarrassed or commanded into habits of obedience. In the family, for example, children can be disciplined by parents in good conscience, or called upon to do chores, insofar as it is assumed that the inherent love for the child is beyond question. The corollary of this is that the child is expected to be protected fully by the family unit. Even in larger and more impersonal social structures such as the army, or large corporations, the family morality concept can be extended. People are asked to serve in the exchange system of benefits that are part of the organizational universe in which presumably they, as free agents, have decided to participate.

The apparent difference between families and armies is that in the former we are born or adopted into them for life, whereas in the military and the workplace an act of will (except for instances of conscription) is required in order to participate. Once the decision is made however, certain rights and duties ensue between the parties. Because the bond, however, is instinctive and therefore stronger within the family we expect that more can be expected to be given or sacrificed just because the obligations or reciprocities stemming from love or family relationships lead to greater sacrifices on the part of all the parties in question. In short, we expect less to be offered to us by the work environment or the state. Of course the distinction can be made between emotional and monetary forms of giving because we can readily point out in many instances that families may lack financial capacity to give aid, whereas the state may effectively wield that power. Nonetheless, if all things were equal we would expect to be entitled to ask more from family members than from states, except in a hypothetical universe where the state or corporations had emotively replaced the family. Arguably, this has been the case in closed environments, under special circumstances, such as warfare, or in totalitarian societies where families may be superseded, given an elitist ideology such as found in Plato's conception of society, or in fascist regimes where the state is depicted to embody the highest virtues or values to which citizens should give ultimate and complete obedience. In democratic societies we still preserve the image of the family as the unit of affection and protection giving rise to a special set of entitlements. This is especially relevant in instances where hard choices have to be made about the limits that we wish to place on justifying requests to enlist incapacitated citizens in making contributions to the large world of medical science.

In principle, there is arguably something objectionable to asking individuals who have underdeveloped moral systems of analysis, such as children and adults who suffer from some form of limitation of mental ability or illness, to participate in a moral fiction, where they are presumed to understand the exchange system which is part of the bonding or contracting in the familial and social orders of which they are a part. It seems that our objections reside in some sense of aesthetic or moral repulsion in taking the dependent or infirm among us and pushing them into acts of contribution when our love for them should be proven through our asking nothing in return for the affection and protective acts which we offer them. It would be as if, having given enormous attention to a dependent, we suddenly change the rules of the relationship and ask for a payment, or more than that, exact a penalty for the expenditures made.

With respect to our protection of animals which have come close to the family (i.e., a pet), most people would object if such animals were taken into harness to serve, unless the circumstances were dire, almost amounting to a defense of necessity, where the survival of the social unit in question was of issue. Our resistance to such a change of events is heightened emotionally when we might have to contemplate in a society the eating of pets during a period of famine. If our reactions can be shown to be charged emotionally with respect to animal members who live among us, then, it is not surprising that our resistance is strong to the notion that incapacitated, suffering human beings should not be given serious protection. It is felt, given the misfortunes of living that occur when persons are born with handicaps relating to mental capacity, that their inability to offer consent to charitable acts in research should not give rise to our making claims on them for even larger sacrifices for the common weal.

The sense that handicapped individuals have offered enough already lies at the crux of the moral angst that we feel in bringing them into the world of medical research. The sense of repulsion already alluded to is surely accentuated by the images of such individuals tormented and exploited by nefarious forces in history. Handicapped individuals have been harnessed into service in many societies, relegated to the most menial domestic tasks, treated as sub-human and utilized if they were able to serve the base instincts of oppressive overseers, and in circumstances beyond that, their annihilation has been justified by large social projects such as sterilization and euthanasia.

### On the Use of Moral Fictions

In many philosophical discussions we are called upon to engage in acts of imagination or game-playing. For example, we are asked to indulge the fiction that to be human is to be rational and that any notion of ethical autonomy should therefore be based on such a postulate. If such a world of abstraction leads to the discovery that there could be prejudicial attitudes taken on by those people who sense their own natural superiority and ask therefore that others serve their interest, we can resist such philosophical elitism by inviting citizens to indulge in the fantasy that we can all begin a hypothetical society under a veil of ignorance where no differentiations are allowed.[54] When metaphysical systems are complete, they usually afford us the opportunity to indulge our fantasies such that we locate ourselves on a trajectory of identifiers either with a perfect set of moral values, or with the ideal of perfection, such as we find in the classical works of Plato and Aristotle, or the evolutionary morality located in the Hegelian dialectic, or the teleological vision in the Thomistic tradition. When such systems are laid out we can participate in the fantasy of being in contact with ultimate truth or the divine. Remarkably however, none of these philosophical visions have led societies or even significant movements of decision-making to any clear pattern of protection for the handicapped.

Acts of charity without doubt preceded the development of philosophical systems.[55] Benevolence and caring treatment towards dependants do not and never have required careful philosophical rationalization. In fact, if instinct is not the basis for individual and social benevolence we might, even given the Western historical record, have cause for concern. In the Nazi society, for example, the Kantian and other metaphysical systems alluded to were widely known amongst the professional elites who administered deadly treatments against handicapped citizens. It is the submission here, as we have observed in our discussion about utilitarianism, that it would be facile to say that such a profound moral failing in a society came about because of the perversion of Kantianism in the face of a crass calculation of the relative social utility of its citizens. Society at large, and its professional elites, as leaders and representatives of a belief system about the relative values of groups within the social structure, prepared the groundwork for unspeakable atrocities. After the handicapped came the homosexuals, the Jews, and the Gypsies.[56]

To date, the failure of formalistic metaphysical philosophy to impact, in any discernible way, on socially protective actions of societies should be taken as a warning for our future practices. Kantians readily assert that it is inconceivable that any Kantian could justify treating the handicapped as means and not ends and that any notion to the contrary is a distortion of the Kantian moral vision. But given the formalistic requirements of the Kantian system we cannot say with assurance that a Kantian would be prevented from designating certain persons as subhuman and thereafter universalize their destruction. The only protection that we have against

---

[54] For the original statement of such a proposed veil of ignorance, see J. Rawls, *A Theory of Justice* (Cambridge, Mass.: Harvard University Press, 1971) at 136–142. We might suggest a defect in Rawls' theory in that it assumes that individuals to whom it applies are ". . . normal, active, and fully cooperating members of society over the course of a complete life". See J. Rawls, "Social Utility and the Primary Goods" in *Utilitarianism and Beyond, supra* note 36 at 168. Handicapped persons are not included among those who are envisioned to be found in Rawls' "original position" under the veil of ignorance. See also J. Rawls, "Justice as Fairness", in *Communitarianism and Individualism* (Oxford: Oxford University Press, 1992).

[55] Hobbes writes: "And whereas many men, by accident inevitable, become unable to maintain themselves by their labor, they ought not to be left to the charity of private persons but to be provided for, as far forth as the necessities of nature require, by the laws of the commonwealth". In other words, the state has an obligation to care for vulnerable populations. In this instance, their "vulnerability" seems to be restricted to disadvantage that is no fault of their own, i.e., "accident inevitable". See T. Hobbes, *The Leviathan — Parts 1 & 2* (New York: Liberal Arts Press, 1958) at 271.

[56] A. Mitscherlich & F. Mielke, *Doctors of Infamy* (New York: Henry Schuman, 1949).

such an application is to ask people to indulge in an imaginative experiment where they put themselves in the other person's shoes, or imagine that they, or a member of their family, themselves could be exposed by will, nature or accident to have a despised attribute. Usually persons when asked to participate in such an imaginative experiment are inclined to let go of some element of their prejudice. But there is no exactitude in this process. Other individuals might simply say that inferiority, as defined by the relevant system in question, has to be treated in every case in the same manner, which could lead to any number of radical social interventions.

In the hard cases of ethical decision-making found in circumstances where we are called upon to make choices about the participation of vulnerable populations in research activities, we test the moral mettle of our society at large. How we treat this minority of dependent citizens may expose much else about our society's thinking about questions of superiority and inferiority in the distribution of benefits and burdens. Consequently, we are inclined to argue that we should choose the strongest members of our society and researchers themselves as the first line of support for experiments involving high risk. It is because the fear is deeply felt that when we cross the line of utilization of dependent persons for the interests of others, there is a sharp tendency to pursue the logic into darker and deeper corridors. In the world where biological engineering has taken us beyond the doorstep of our imaginations we can contemplate a world where we could produce persons of limited intelligence from whom, for example, we could harvest organs for the use of stronger and more intelligent members of a new social order. Thus, it is critical that we take protective procedural steps to guarantee that social intervention are put under the surveillance of citizens whose duty it is to ensure that individuals not be sacrificed for higher ends or for the benefits of stronger persons.

### The Fiction of Making Whole the Fragile

If we are committed to the principle of respect for persons, and we have taken care to include even dramatically handicapped persons in our definition of personhood, a number of options present themselves for arguing a case on behalf of including vulnerable populations in research endeavours. There is the idea that, for purposes of moral action, we can give persons of limited capacity the advantage of being treated in our decision-making process as if they had a normal capacity for understanding the instinctive and learned exchanges of benefits experienced by other citizens. We attribute nobility of character to a sister who offers her kidney to a sibling; we have respect for citizens who risk their lives to enlist in a perceived just war; and we regard as heroic those experimenters who sacrificed their own bodies in order to discover a medical cure. Indulging in the fiction that we can take all human beings, indeed on their behalf, into the realm of being made whole as moral beings appears to attract some strong moral sentiment. In a relatively recent report of the U.K. Medical Research Council the following statement was made:

> ... it is not in the public interest for persons suffering from mental incapacity to be excluded from socially responsible behaviour purely through lack of consent competence. Where the risk attending participation in non-therapeutic research into mental disorder is

minimal and a reasonable person with that disorder but able to consent is likely to accept that risk, when told that such research might lead to advances in treatment, it would be strange if a person unable to consent because of that disorder should be imputed with a wholly different attitude to the welfare of the class of persons of which he is a member.[57]

It does not seem that this point of view is necessarily connected to a trade-off morality, namely, that if the society agrees to care for a dependent, that the least that can be done in return is to enlist such a person in seeking a solution for the illness or handicap with which the person is inflicted. It is a statement which rather appeals to a higher moral sentiment, that a reasonable citizen would want, if the risk involved is only minimal, to serve others. We assume this because it is part of what it means to be a human being living in society. Even if our bond with others is abstract in a society where differences do not drive citizens in any strong sense against each other, some version of a cooperativist ethic gets built into the system of moral life. People who do not participate can be seen to be anti-social and morally irresponsible. By saying that we will treat a person who, normally speaking, cannot attain this level of sensibility, we indulge the fiction that we will lift this person to a level of moral capacity for the purposes of living in the society, which if anything, enhances our tendency to treat this person as a real citizen among us, with a certain moral value, if not capacity. That is to say, the proof that we treat in practice limited persons with full respect is that we act on their behalf where there is no serious risk to them such that they are given full moral credentials in the actions that we initiate on their behalf. Our instincts, thus understood, are not base and do not fail to meet the criteria offered by the classical models of either utilitarianism or deontology. Respect for persons and social benefits are all fulfilled for a net gain morally and materially for all parties. In fact, what is interesting here is that the material benefits are likely increased for the charitable donor because of our thankfullness for the contribution made.

There is no advantage in actual treatment that accrues to persons of limited capacity when the consequence of our respect for them is our abandonment of their interests. This has occurred because, in our fidelity to the legal requirement of informed consent, we have under-researched the impact of medications on vulnerable populations. Finally, we have been reluctant as a society to thrust the goodwill of third parties either as substitute decision-makers or as persons obligated to defer to the best interests of their wards. For such reasons, we have produced a "therapeutic orphaning" of these populations.[58] This abandonment has been rationalized by society, *ex post facto*, because it feels that its contributions have already been excessive, given the level of dependency on social support systems by these populations.

The problem lies in where the steps of our fictions lead us. We know that in times of war, for example, the populace is called upon to make great sacrifices. Are we entitled in such periods of crisis to ask vulnerable populations to join us in the effort,

---

[57] Working Party on Research on the Mentally Incapacitated, *The Ethical Conduct of Research on the Mentally Incapacitated* (London: Medical Research Council, 1991) at 20.
[58] H.C. Shirkey, "Therapeutic Orphans" (1968) 72 Journal of Pediatrics 119 at 119–120.

and are we justified in indulging ourselves in a patronizing or paternalistic ethic which allows us to articulate the quantity of sacrifice to be exacted? Do we move easily in such a context from minimal to high risk experimentation? We can imagine a situation where a whole village would be under siege, all able bodies enlisted in the defense of the group, with the likelihood of death. To add to the difficulties, the group could in this scenario be threatened with a deadly virus, the cure of which could be hypothetically attached to the interventions of a research scientist, who called out for volunteeers. Would we be justified in enlisting vulnerable members, who could not otherwise engage in the defense of the group, in the experiment? It would be an unusual society living under such conditions where the members would have such a loyalty one to the other where they might not join the vulnerable members to such an act of sacrifice. In a tightly knit group in the most extreme cases such as the collective suicide in the historical fortress of Masada in the defense of the Judeans against the Romans, it would be odd for example to think that the group would have sacrificed itself in the entirety and left out mentally handicapped individuals. We take it that such an abandonment of them in such an extreme context would have been an act of profound disrespect, lowering their value to less than human. Asking, therefore, in extreme circumstances of group protection that individuals be volunteered into service should not come into conflict with the noble moral sentiments expressed by the history of deontological philosophy.

Apart from extreme cases where the public taking of property or life has been rationalized as the profound defense of the group, such as in the classical defense of necessity found both in criminal and torts law, our expectations of ourselves as citizens and the vulnerable parties called upon alter considerably. We do not feel justified in asking for more than mild forms of sacrifice in our modern heterogeneous democracies because the role of the state and public institutions have an uneasy and tenuous relationship with the citizenry. Aside from serious social crises, the only surviving unit which we might argue can lay claims upon its members for more than minimal contributions is the family. Insofar as we believe that the family is a truly functioning unit deserving of respect because it gives respect to its members, and thereby involves relationships of trust and loyalty, we are prepared to agree to the family acting on behalf of vulnerable members such that the family may be allowed to legitimate certain acts of charity. However, beyond minimal type levels of risk, even the family should not be warranted in initiating sacrifices on behalf of its members. When levels of risk are serious or the issue is a major physical intervention such as the donation of an organ, it should be for the judiciary as the most respected organ of social decision-making to take inputs from the family and thereafter to make the difficult determination.

It may be that a society may wish to deny, as a moral absolute, the possibility of major sacrifices being made on behalf of vulnerable populations. Such a reaction, although controversial, is understandable because it is akin to the reasoning attached to the prohibition, even in dire circumstances, of the eating of other humans. For in both cases, we believe that the threshold of civility has been crossed. It could be shown through situations, short of life-giving, such as the donation of organs, that although an idealized fiction, the moral elevation that occurs includes the vulnerable person in a life of caring and reciprocity. Needless to say, the morality of such a sacrificial ethic will be placed in jeopardy if not carefully controlled, put under continued social surveillance at the highest level, namely, through the judiciary, and made part of a rigorous ongoing public dialogue emphasizing our commitment to the protection in more general terms of our vulnerable populations, through necessary support systems pertaining to their health and welfare. Without such a dialogue and commitment our moral fiction should be regarded as a manifestation of inauthenticity or false consciousness.

CHAPTER 2

# BRINGING ETHICS TO HUMAN EXPERIMENTATION: THE AMERICAN EXPERIENCE

DAVID J. ROTHMAN

Although the principles that should govern human experimentation were enunciated well before World War II and were for the most part consistent with our current understanding, the actual practice of investigators paid them little heed. Even the publication of the Nuremberg Code in 1947 did not affect the conduct of clinical research in the United States. Nuremberg, it must be remembered, owed little to organized medical bodies. The prosecutors called physicians as witnesses and used them as consultants, but the document stood as the work of judges and its stipulations were realized through a court, not through professional medical bodies. In essence, the Code was external to medicine — which helps to account for both its weaknesses and strengths in the post-World War II period.

## THE IMPACT OF THE NUREMBERG CODE IN THE UNITED STATES, 1947–1966

The externality of the Code to the medical profession properly opens an analysis of why Nuremberg exerted so little impact on the conduct of human experimentation in the United States in the immediate aftermath of World War II. Although the Code was certainly known to investigators and government officials, references to it in the medical literature were sporadic. Even more important, investigators themselves paid little heed to the Code's principles. They continued the practices that had marked research during the period 1941–45, carrying out harmful non-therapeutic protocols on subjects incapable of giving consent. Accordingly, researchers transplanted cancer cells into demented old men in order to study the body's immune reactions; they fed hepatitis virus to children in institutions for the retarded

35

to analyze the etiology of the disease and try to create a vaccine. They injected patients with radioactive substances to evaluate the dangers of radioactive fallout to the population in the event of an enemy attack. Each of these experiments violated the opening provision of the Nuremberg Code: "The voluntary consent of the human subject is absolutely essential. This means the person involved should have the legal capacity to give consent . . . and should have sufficient knowledge and comprehension of the elements of the subject matter involved so as to enable him to make an understanding and enlightened decision". But that maxim did not serve as a barrier to the research.

Why should this have been so? Why did Nuremberg, even if it was produced by bodies external to medicine, have such limited influence? The answer is not that few people had heard of Nuremberg or read its provisions. Rather, the problem goes deeper. Part of the answer is that the urgency created by World War II gave way to an urgency created by The War Against Disease and the Cold War. The search for new knowledge seemed (as it always does?) so important as to justify overriding ethical precepts. In part, too, Americans presumed that the Code had no relevance to them, that is, to real scientists trying to advance the well-being of mankind. Nuremberg was written for Nazis, not for physicians — indeed, many Americans believed, altogether mistakenly, that the perpetrators of the gross misdeeds had not been doctors or authentic investigators, but madmen, political hirelings, or "pseudo-scientists", as Andrew Ivy called them.[1] No less important, however was the fact that the Code stemmed from a judicial process. Composed by lawyers, Nuremberg had no medical imprimatur or professional standing. Since it came from outside medicine, it carried no relevance to medicine.

Inseparable from the question of why researchers for two decades ignored Nuremberg is the issue of defining the ethical standards by which these investigators are rightly judged. The claim is often advanced, usually by physicians but occasionally by others as well, that to fault these investigators for ignoring the principle of consent is to impose the standards of the 1980s on actors and events in the 1950s and '60s, and to be guilty of *ex post facto* reasoning. But the glaring weakness of this contention should be apparent from the recognition, that the relevant principles went back at least to Bernard and Osler, and truly to Hippocrates. Moreover, by the mid–1960s, Nuremberg was not the only collective judgment on ethical research conduct. In 1946, the Judicial Council of the American Medical Association promulgated a code of research ethics which required "the voluntary consent of the person".[2] In 1964, the World Medical Association published its *Declaration of Helsinki*, insisting that subjects be informed of the aims, methods, benefits, and hazards of the research before they participated. A defense of "they knew not what they did" surely seems flimsy.

---

[1] A. Ivy, "Nazi War Crimes of a Medical Nature" (1947) 33 Federation Bulletin 133. Ivy himself recognized that only some of the investigators fit this category and that the medical profession itself in Germany had been undermined.

[2] (1946) 132 Journal of the American Medical Association 1090.

Nevertheless, the debate on the standards for evaluating the ethics of the 1950s and '60s research continues, most recently and prominently in the public attention devoted to the radiation experiments of this period. In 1994, a journalist in New Mexico, Eileen Welsome, identified by name several persons who had been purposefully injected with radioactive substances and her articles stimulated a flurry of media attention and, in short order, a government investigation spearheaded by a presidential task force. With the cooperation of government agencies, most notably the Department of Energy, the task force uncovered an extraordinary number of protocols involving radiation research over the period 1940–1974, including research involving plutonium on ostensibly terminal patients, radioisotope research on children, often mentally disabled, total body irradiation on cancer patients with advanced disease, testes radiation on prisoners, and environmental radiation in various communities.

Courtesy of this investigation, what do we know now that we did not know before? Perhaps the single most important and unexpected finding is that investigators and government officials were aware of the Nuremberg Code in particular and of the principles of research ethics, including informed consent, in general. Despite the minimal coverage given to the Code in the public and medical press, the leadership in the Department of Defense (DOD), the Army and the Navy, and the Atomic Energy Commission (AEC) were fully cognizant of its specific provisions and broader implications. In effect, the ethical standards by which to evaluate the conduct of the researchers were the standards that led up to and were encapsulated in the Nuremberg Code.

The AEC as early as 1947 appreciated that to pass ethical review, human experimentation had, in its own words, to carry an expectation of "therapeutic effect", and had to be "susceptible of proof that, prior to treatment, each individual patient, being in an understanding state of mind, was clearly informed of the nature of the treatment, and its possible effects, and expressed his willingness to receive the treatment". Thus the AEC documents insisted on the competence and "informed consent" (the AEC actually used the term) of the subjects, and ruled out research with no therapeutic benefit.[3]

When the AEC and the DOD were concerned about the risk to air crews posed by nuclear powered engines and considered using prisoners to test effects of exposure levels, one AEC official, Shields Warren, remarked, "[i]t's not very long since we got through trying Germans for doing exactly the same thing"; another investigator Joseph Hamilton, warned that the proposal "would have a little of the Buchenwald touch". Indeed, many armed services administrators (especially in the Navy) insisted that human experimentation proceed only with consent from subjects and some investigators did respect the principles.[4]

Nevertheless, and once again, the gap between established principle and deed was substantial. The historical record of radiation research is replete with dispiriting

---

[3] Advisory Committee on Human Radiation Experiments, *Final Report* (Washington, D.C.: U.S. Government Printing Office, 1995) at 87.
[4] *Ibid.* at 99.

examples of administrative failing to implement the ethics they espoused. Thus, the DOD incorporated the Nuremberg Code into its regulations and then, amazingly, classified the document Top Secret and would not even distribute it within its own network. The AEC banned non-therapeutic research, but did nothing to enforce it within its own ranks, including its contract research organizations. Whenever agency lawyers issued recommendations about the desirability of legislation to govern the conduct of human experimentation, other agency officials resisted. Some of them, reminiscent of physicians in the early Twentieth Century, were fearful of "unfavorable publicity"; others worried that "[t]o commit to writing a policy on human experimentation would focus unnecessary attention on the legal aspects of the subject", and narrow the discretion that researchers enjoyed.[5]

Perhaps most discouraging, clinical research often sacrificed the well-being of subjects even as researchers and government officials went to great lengths to keep the protocols secret. Thus, Ebb Cade, a 53 year old "colored male", who was hospitalized following an auto accident but otherwise enjoyed good health, was injected, without his knowledge or consent, with 5 micrograms of plutonium at the Oak Ridge Army Hospital as part of a project to study the effects of plutonium on workers. In all, as the presidential task force discovered, "at least twenty-two patients were administered long-lived isotopes in experiments". Nor were these subjects invariably terminally ill, a criterion that in itself is suspect but did carry the rationale of obviating a fear of long-term effects.[6] When the AEC considered the release of some of these research reports, an AEC declassification officer concluded that such a step was unthinkable:

> The document appears to be most dangerous since it describes experiments performed on human subjects, including the actual injection of plutonium into the body . . . It is unlikely that these tests were made without the consent of the subjects, but no statement is made to that effect and the coldly scientific manner in which the results are tabulated and discussed would have a very poor effect on the public.

There was no evidence for the assumption that consent had been obtained and all of the plutonium protocols violated the AEC ban on non-therapeutic research. As the president's task force rightly concluded: "[c]oncerns about adverse public relations and legal liability do not justify deceiving subjects, their families, and the public".[7]

Thus, in the research conducted as part of the Cold War, as with the research conducted in the war against disease, well-established ethical principles did not restrain or modify investigators' behavior. It was not that they were ignorant of the standards. But in their quest for knowledge — and for grants, prizes and fame as well — they transgressed them. Again, the subjects who were kept in ignorance and put at risk were almost always members of vulnerable groups. In the 1950s and '60s, as in the 1940s, those who suffered the most harm were mentally disabled or

---

[5] *Ibid.* at 101, 104.
[6] *Ibid.* at 243.
[7] *Ibid.* at 253, 255, 269.

prisoners or minorities. And again, those who broke the codes suffered no adverse professional consequences. To the contrary, professional organizations often praised and rewarded them for their work — as though the importance of their findings overrode violations of research ethics.

## IMPOSING ETHICS ON HUMAN EXPERIMENTATION: 1966 TO THE PRESENT

The critical changes that finally brought ethical standards into the practice of clinical research owe more to an aroused public than a troubled medical profession. Forces external to medicine helped to bridge the persistent and considerable gap that had for so long separated principle from practice. The catalyst was a series of exposés in the 1960s and the early 1970s that made the ethics of human experimentation into headline stories. The behavior of researchers appalled citizens and forced the federal government, specifically, the National Institutes of Health (NIH), to impose new regulations upon clinical research. Put most succinctly, it was scandal that finally brought ethics into the laboratory.

What constitutes a scandal? First, the behavior must be perceived as offensive to one's moral feelings. Second, it must be committed by someone who is trusted and looked to as an example. Third, it must be making public what had heretofore been a closely guarded secret, known to a small circle of participants but hidden from broader view. In this sense, an ordinary robbery by a thief is a crime, not a scandal. To enter the realm of scandal requires that the perpetrator be someone who has been trusted and the act itself must be a gasp of surprise and a shudder of revulsion.

It was the 1966 publication by Harvard Medical School professor Henry Beecher, "Ethics and Clinical Research" in the *New England Journal of Medicine*, that cast the record of human experimentation into the mold of scandal. Beecher described in capsule form 22 protocols of "dubious ethicality". He named no names and provided no footnotes; the *NEJM* had the citations, checked them, and agreed to publish the piece without references. To make certain that the story was not lost — remember that in the 1960s there were few medical journalists and they did not scrutinize the weekly medical journals as closely as they do now — Beecher alerted the popular press to the article, and his disclosures received enormous attention.

The 22 protocols were shocking to the conscience. They included the hepatitis experiments on the retarded and the cancer cell experiments on the demented old men. They also described military research that withheld known agents of efficacy against rheumatic fever and research at NIH itself that involved new methods of cardiac catheterization. None of the published articles suggested that the subjects had been informed about the experiments; indeed, many of subjects were incapable of giving consent because of their diminished capacity. When Beecher was asked whether the researchers might have obtained consent but neglected to say so in their publications, he aptly responded: "I have worked on the ward of a large hospital for 35 years [and] I know perfectly well that ward patients will not . . . volunteer for any such use of themselves for experimental purposes when the hazard may be permanent injury or death".[8]

The 22 protocols had appeared in prominent medical journals, including five in the *Journal of Clinical Investigation* and two in the *Journal of the American Medical Association*. They were funded by NIH, by drug companies, and by the armed services, and carried out at major universities, including Harvard and Case Western Reserve. The protocols, in other words, represented mainstream science at mainstream institutions by mainstream investigators — and no one, until then, had criticized the work.[9]

Why was the public so disturbed by the research? After all, the response might have been more calculating and self-serving. The subjects, after all, were marginal to society (retarded or senile), the research was very important (curing cancer or creating a vaccine against hepatitis), and by strictly utilitarian criteria, the good that would come might outweigh the injuries imposed. But that was not the position adopted. The popular identification was with the subject, not with the investigator.

One reason for this special angle of vision, a looking out onto the world from the vantage point of the underdog, was Nuremberg itself, the significance of the crimes committed by the Nazi doctors. The relative silence that surrounded Nuremberg in the 1950s and early 1960s had finally lifted. Why it took so long for Nuremberg to enter the public consciousness has never been clear — perhaps the events were so traumatic that a kind of psychological repression took hold. But first with the Eichmann trial and then with increasing scholarly and media attention thereafter, the events became a more common reference point. It was no longer possible to justify risks and injuries to human subjects on the grounds that the state required answers to pressing medical questions. Even German medicine, once so prestigious, had been corrupted by succumbing to an ideology that state interests trump other considerations.

If Nuremberg was one foundation for new perspectives towards human experimentation, the second was the social awareness that medical advances affected not only the individual patient but society more generally, and given the dimensions of the potential transformations, research had to reviewed and authorized by someone other than the single investigator. Transplant procedures were one case in point: should this society promote a medical technology that makes the body into a collection of spare and reusable parts? Moreover, physicians themselves are often eager to share responsibilities in decision-making about outcomes and resource allocation. The most noteworthy example was physicians in Seattle establishing a lay kidney dialysis committee (more popularly known as the "Who Shall Live Committee"), to decide who received the life-saving benefits when the machines were in very short supply. So too, the development of nuclear weapons also encouraged biologists to convene the Asilomar Conference and to delay

---

[8] Cited in D.J. Rothman, *Strangers at the Bedside: A History of how Law and Bioethics Transformed Medical Decision Making* (New York: Basic Books, 1991) at 75.
[9] *Ibid.* at ch. 4.

recombinant DNA research until a broader consensus about its safety was achieved.[10]

Finally, the public reaction was consistent with the sensibilities of the 1960s. This was the decade when the civil rights movement flourished, when most Americans came to view racial discrimination and segregation from the perspective of the disadvantaged minority. When brutish redneck sheriffs threatened peaceful protesters with snarling dogs, Americans identified with the victims, not with the authority figures. And by extension, when investigators took advantage of senility or retardation, Americans identified with the victims, not with the physicians. In all, the 1960s was a moment when the discretionary authority of parents, husbands, principals, wardens, and mental hospital superintendents lost legitimacy, when the rights of children, women, students, prisoners, and mental patients were advanced. In terms of principles, paternalism declined in appeal and autonomy rose. Inevitably, this reorientation entered medicine, promoting change in the conduct of human experimentation.

The impact of these several considerations was substantial because they easily and quickly led to structural changes. In the United States, more so than in Europe, the federal government was the primary funder of clinical research. The funds were distributed through the grants made by the NIH, and NIH, in turn, was dependent upon Congressional appropriations. Thus, public disapproval of the conduct of human experimentation came to the attention of Congressmen, who then expressed their displeasure to officials at NIH. These officials were acutely sensitive to Congressional reactions, recognizing that disfavor might reduce or cut off appropriations. At the same time, they had ample authority over grantees — whatever requirements NIH incorporated into its grant awards would be readily accepted. In brief, the fact that NIH was at once subordinate to Congress and superordinate to the investigators meant that exposés would affect policy. Put into the framework that we have been analyzing here, for the first time, the governance of human experimentation moved directly into medicine. In this way and in the wake of scandal, ethics and practice finally came together.

The essential regulatory mechanism was the Institutional Review Board. By federal regulation, all recipients of federal grants must secure IRB approval before conducting research on human subjects.[11] Every grant-receiving institution must establish an IRB, with a membership of no less than five persons, at least one of whom is not to be affiliated with the institution. The IRB's mandate is, first, to review clinical research protocols to determine whether the benefits of the proposed research outweigh the risks; second, it must make certain that the investigators have explained all the relevant issues to the subjects and received their informed consent. Although the regulations apply only to federally funded research, many states and

---

[10] This argument is presented more fully in H. Edgar & D.J. Rothman, "The IRB and Beyond: Future Challenges to the Ethics of Human Experimentation" (1995) 73 The Milbank Quarterly 489. See also D.J. Rothman & H. Edgar, "New Rules for New Drugs: The Challenge of AIDS to the Regulatory Process" (1990) 68 Milbank Quarterly.

[11] 45 C.F.R. (1991) §46.101ff.

the great majority of academic institutions require IRB review for any clinical research performed within their jurisdiction, no matter what the source of funding.

Although IRBs were imposed upon the research community, it is the research community that controls them. To look at the IRB only in terms of its formal structure and organizing principles, it would seem to be a paper tiger. The regulations do not on the whole protect against sloppiness or venality. The power to approve or disapprove research on ethical grounds is granted to a local institutional committee, controlled by members of the same institution that is seeking the funding. It is these insiders that dominate decision-making and they are themselves researchers who know that the standards they set for others will come back to affect them as well. Moreover, federal regulations are silent on how the institution makes appointments to the committee, how long members serve, and on what grounds a member may be dismissed or not reappointed. Hence, it is eminently possible that a member who takes a very hard and uncompromising position on the ethical issues will not be reappointed, and more permissive members will be.

These design features, however, also contribute to the strengths of the IRB. The committee is integral to the institution, its ethical standards incorporated into the ongoing research activities. Since it is colleagues, not outsiders, that make the decisions, a would-be investigator does not want to risk their derision or their rejection. For reasons of self-interest as well as a sensitivity to ethical concerns, they would not dare to propose purposefully infecting retarded inmates of an institution with a virus or radioactive substance, or transplanting cancer cells to senile patients. In the end, the quality of an IRB's work depends inordinately on the conscience and commitment of its members — which is why it appears at once so effective a mechanism to those within a medical institution and so frail a structure to those analyzing it from outside.

Thus, the history of human experimentation makes clear that left to itself, medicine in its collective capacity does not necessarily generate change, even when a sizeable gap separates professed ideals from actual practice. The impetus must come from outside, whether from a court, as with Nuremberg, or from federal regulation, as with the IRB. To be sure, individual physicians may play a critical role, Andrew Ivy at Nuremberg, Henry Beecher in the mid-1960s. And eventually external forces can generate internal conformity to the ethical standards, not as thoroughly as some might wish, but as the IRB itself still demonstrates, more successfully than before.

Still, the reluctance of professional medical bodies to move rapidly and effectively to confront challenges to medical ethics is made more worrisome in light of the challenges that clinical medicine faces today. The rise of managed care and other forms of corporate provider organizations are putting the ethics of clinical practice to new and hard tests; it is uncertain, for example, whether such fundamental principles as the physician's first and final responsibility to the well-being of the patient can survive in a health care system dedicated to generating profits for shareholders and company officers. Clearly the threat requires energetic and principled responses from professional medical bodies. In the terms we have been analyzing here, individual declarations of guiding principles will not suffice. What is needed is an assertion of collective responsibility, not singular expressions of dismay. Whether medicine will prove capable of mounting this effort is the question that will dominate medical ethics and medical care in the coming decades.

CHAPTER 3

# BEYOND CONSENT

M. GREGG BLOCHE

To revisit the ethics of medical research with human subjects is to return to the origins of contemporary American bioethics. Like so many reform efforts in the United States, the American bioethics movement drew its animating spark from shocking disclosures and public outrage. As David Rothman and others have noted, the emphasis on consent that has anchored clinical research ethics in the U.S. for the past quarter century began in response to revelations that American researchers had injected cancer cells into unknowing patients, withheld antibiotics from people with syphilis, and otherwise exposed uninformed subjects to the risk of serious harm.[1] American researchers who had dismissed the Nuremberg Code's insistence on consent as irrelevant outside the context of the Nazi medical horrors confronted their work's inherent potential for abuse. Pressed by the bioethics community and intent on restoring public confidence, the U.S. research community made informed consent part of its regular practice by the mid–1970s. During the ensuing decade, the rest of American medicine followed suit, incorporating the bioethics movement's emphasis on autonomy and consent into the ethical norms governing routine clinical work.[2]

Thus reports in late 1993 that the U.S. government had supported a large number of previously unpublicized experiments exposing unknowing subjects to dangerous doses of radiation brought the American bioethics movement back to its roots. Confronted by these reports, and by the ensuing outrage of victims and others, President Clinton did what American presidents often do when scandal breaks and media accounts stir passions. He appointed a commission. Over the next two years, the National Advisory Committee on Human Radiation Experiments and its dozens

---

[1] D.J. Rothman, *Strangers at the Bedside: A History of How Law and Bioethics Transformed Medical Decision-Making* (New York: Basic Books, 1991).
[2] P.H. Schuck, "Rethinking Informed Consent" (1994) 103 Yale Law Journal 899.

of staffers pored over thousands of pages of documents, conducted hundreds of interviews, and reconstructed the federal regulatory schemes and professional standards that governed medical research in the U.S. from World War II until the mid–1970s.

In so doing, the Committee and its staff replowed much old ground. For much had been disclosed and much had been condemned about pre-1970s American medical research practice. Yet the Committee deserves enormous credit for doing the unglamourous work of telling a story, about which much was known, in unprecedented detail. For the victims and their families, and for their descendants, the Committee's voluminous report ant its appendices represent a human dignity-affirming act of government truth-telling and public remembrance. The Committee acted in the spirit of such exercises in national healing as Argentina's 1983 National Commission on the Disappeared and similar bodies in South Africa, Chile, El Salvador, and Honduras.[3] To be sure, such enterprises suffer from the common failing that the telling of truth cannot ensure the doing of justice.[4] Yet the telling of truth, within a moral framework, can yield potent lessons for posterity, so long as the framework is sufficiently robust to bear upon present and future moral challenges.

In this regard, the National Advisory Committee on Human Radiation Experiments may have foregone an opportunity. As ethical analysis, the Committee's report plows little new ground. Rather, it is in no small part a celebration — a celebration of the bioethics revolution of the past quarter century and of the paradigm of consent that constitutes this revolution's centerpiece. To some degree, this is an oversimplification: the report offers many suggestions and criticisms regarding current research practice, and all those responsible for administering research that involves human subjects ought to review and reflect upon them. Yet the larger story told by the report's authors is that the bioethics revolution got it right, and that the paradigm of consent, which matured and spread for the most part *after* the radiation experiments reviewed by the Committee took place, remains our best answer to the threat of *future* abuses in medical research and practice. If this is the case, then we need do no more than tinker in the interstices of the received paradigm of consent, by refining our mechanisms of oversight and our understandings of the prerequisites for informed consent in diverse situations. There is much to be said for

---

[3] J.M. Pasqualucci, "The Whole Truth and Nothing But the Truth: Truth Commissions, Impunity, and the Inter-American Human Rights System" (1994) 12 Boston University International Law Journal 321.

[4] Indeed, the Radiation Committee may have interposed an obstacle to the pursuit of justice, by concluding that federal sponsorship of hazardous research on unknowing subjects did not violate then-accepted international human rights norms. The Committee argued that because the *International Covenant on Civil and Political Rights* (which proscribes medical research on human subjects without their consent) was not opened for signing until 1966, international law does not bear on pre-1966 failures to obtain research subjects' consent. The Committee thus declined to recognize the Nuremberg Code as a source of international law, or as a basis for interpreting American law, despite significant U.S. legal precedent to the contrary. See *In re Cincinnati Radiation Litigation*, 874 F. Supp. 796, 821 (characterizing the Code as "part of the law of humanity" and thereby applicable in U.S. civil and criminal cases); *U.S. v. Stanley*, 482 U.S. 669, 710 (O'Connor, J., dissenting) (citing the Code as a basis for construing the U.S. Constitution's "due process" clause to require that research subjects be asked for their voluntary and informed consent).

the power of the paradigm of consent. Some cautions, however, are in order about the paradigm's limits as a basis for securing personal dignity and liberty in the future.

To begin to get at these limits, it is helpful to recall briefly the historical circumstances of the paradigm's success. First, it is worth recalling that the Nazi medical atrocities that inspired the Nuremberg Code were not purely the product of the unrestrained sadism of their perpetrators. Rather, as commentators have observed, these atrocities were at least as much the product of their perpetrators' ideological and ethical commitments. The biological racism at the core of the Nazi belief system combined malignantly with the public hygiene-oriented tradition of pre-Nazi German medical ethics. The latter encouraged physicians to devote themselves to the well-being of communities and the nation, even when these might conflict with the interests of individuals. The former pushed its adherents toward a view of hated groups as subhuman, akin to infectious agents posing an epidemiological threat. The product of the two was the willingness of some physicians to treat despised people no differently than the medical profession does vectors of dangerous disease.

Some of the American abuses that inspired the bioethics revolution also reflected, to a degree, a failure to regard others as fully human. The racism that sustained the Tuskegee syphilis study is perhaps the outstanding example. But a more central explanatory theme, in accounts by the Radiation Committee and independent scholars, is the commitment of physician-perpetrators to purposes that fit awkwardly with the classic ideal of undivided professional loyalty to patients. These purposes included, alone and in combination, advancement of science, national security, and population-wide conceptions of health. The problem of conflicting clinical purposes is hardly unique to medical research. Such conflict pervades routine clinical practice, forcing myriad compromises with the ideal of complete fidelity to patients. Examples include reporting of contagious diseases to public health authorities, withholding of costly treatments disfavored by health care payers, and the exercise of gatekeeping authority with respect to myriad rights, responsibilities, and opportunities tied to health status.[5] The idea of consent has much appeal as a way to finesse the many moral questions presented by conflict between the purposes of medicine, both within and outside the context of clinical research. The Nuremberg Code's reliance upon consent averted the need to overtly judge Nazi doctors' claims about the legitimacy of their purposes, claims uncomfortably similar to the national security and other justifications later offered by American physician-researchers. This process-oriented finesse is no less enticing today — and no more illuminating as to the substantive legitimacy of the circumstances under which consent is sought.

An alternative response to the revelations that led to the bioethics revolution might have focused on the substantive moral questions presented by conflict

---

[5] Examples include eligibility for social insurance benefits, abortion, employment opportunities, military service obligations, criminal responsibility, and competence to sign contracts or stand trial.

between the purposes of medicine. Indeed, as the Radiation Committee's report notes in passing, a few early postwar commentators on the ethics of research with human subjects identified professional role conflict as the primary problem and turned their attention to the design of institutional mechanisms for reducing or otherwise coping with such conflict. In 1951, Otto Guttentag urged that the roles of researcher and primary clinical caretaker be separated. He proposed that each experimental subject be attended by both a member of the research team and a "physician-friend". The latter would have ultimate patient-care and counseling responsibility — and no role or interest in the research. Clinical interventions desired by the researchers would have to be approved by the "physician-friend". Guttentag's strategy sought to revitalize the Hippocratic ideal of professional loyalty and benevolence as protection against research abuses. While this approach did not preclude a role for consent and patient autonomy, it formulated the ethical dangers that inhere in clinical research as problems of divided clinical loyalty.

In the 1960s and 1970s, this approach was discarded and largely forgotten. As bioethics commentators, government regulators, and clinical researchers embraced the paradigm of informed consent, the idea of Hippocratic benevolence was increasingly cast as part of the problem, not the solution. At best, in the eyes of many, reliance upon physician benevolence as the primary safeguard against abuse slighted the discovery that patients have rights — and that they, not their doctors, should decide whether proposed clinical interventions represent acceptable or unwarranted intrusions. Failure to recognize that this ultimate line-drawing authority rests with patients, not their physicians, constituted, in the words of the Radiation Committee, "cultural moral ignorance", perhaps excusable, but not justifiable, as a consequence of Hippocratic paternalism. At worst, claims of Hippocratic benevolence were thin cover for the exploitation of patients by physicians with their own agendas.

The shift from dependence on Hippocratic benevolence to reliance upon the paradigm of consent fit well with the evolving culture of American medicine. In this regard, the makers of the bioethics revolution deserve credit for a coup of rhetorical brilliance. For in building a new moral foundation for doctor-patient relations upon the proposition that persons have rights discernible by reason, the creators of bioethics constructed a model with enormous appeal for the new postwar generation of scientifically-oriented physicians. Like the scientific method, the philosophical reasoning and rhetoric that gave rise to this model were dispassionate in tone and ahistorical in context.[6] The method of analytic philosophy, embraced by the bioethics movement's pioneers as an antidote to the tradition of faith in professional authority, presented its moral conclusions as discovered truths, akin in their verity

---

[6] There is today broad consensus among historians and philosophers of science that scientific method is intensely historical that the questions asked, experiments performed, and interpretations offered are shaped by a variety of cultural, economic, and other contingencies peculiar to time and place. My limited point here is that neither medical researchers nor bioethicists who work (as most do) within the tradition of analytic philosophy present their arguments and conclusions in a manner that acknowledges historical contingency.

to findings about human physiology. The ethics thereby derived rested on abstract principles, not practitioners' presumed goodwill; indeed, it distrusted physicians' benevolent feelings as potentially paternalistic. The Radiation Committee's report is of a piece with this approach. It derives the requirement of informed consent from a set of general principles, rooted in Kantian and utilitarian philosophy.[7]

For a generation of physicians accustomed to encountering patients as strangers, whether as subspecialist consultants or as primary care providers in hospitals and other large-scale, institutional settings, the abstraction and dispassion of this approach rendered it congenial. Not only did it mesh nicely with these physicians' self-understanding as scientist-clinicians: it did not put nearly the emphasis on affective engagement between doctor and patient as did the Hippocratic tradition of clinical benevolence. The bioethics movement did not counsel doctors that caring at the bedside was no longer necessary, but its adoption of the language and methods of analytic philosophy pushed the passionate aspects of care toward the periphery. By putting the ideas of autonomy and consent at center stage, the makers of the bioethics revolution further signalled the diminished ethical import of affective engagement in clinical work. For the physician whose specialized, bureaucratic work setting precluded anything resembling the personal engagement of the doctor of a century ago who rode on horseback to the homes of families he knew intimately, this new approach to clinical obligation reduced the awkward disjunction between the achievable and the ideal.

The bioethics movement's conception of autonomous consent also made for pragmatic reformist politics. It empowered its adherents to campaign for a measure of patient self-determination within parameters imposed by economic status, personal circumstances, and illness itself. It did so by manoeuvring artfully around an enormous problem for proponents of clinical self-determination. Put simply, the life circumstances of sick patients – their symptoms, clinical prognoses, finances, and family and social situations — exert enormous influence on their medical preferences. The patient presented with a choice between disfiguring surgery and death within months from an aggressive tumor, or the person unable to afford a potentially lifesaving treatment offered at no cost to participants in a clinical trial, is not likely to experience her consent as an affirmation of autonomy. The frequency of such pressure in clinical practice challenges the proposition that patients can choose freely.[8]

The pioneers of bioethics responded to this challenge by crafting conceptions of autonomous action that made a patient's medical and financial circumstances

---

[7] The report sets forth six general principles: (1) do not treat people as means for achieving other people's ends, (2) do not deceive people, (3) do not inflict harm or the risk of harm, (4) promote welfare, (5) treat people with equal respect, and (6) promote self-determination. See Advisory Committee on Human Radiation Experiments, *Final Report* (Washington, D.C.: U.S. Government Printing Office, 1995). These principles differ in form, albeit not in general thrust, from the classic "four principles" of bioethics – autonomy, beneficence, nonmaleficence, and justice. See T.L. Beauchamp and J.F. Childress, *Principles of Biomedical Ethics*, 4th ed. (New York: Oxford University Press, 1994).

[8] F.J. Inglefinger, "Informed (but uneducated) Consent" (1972) 288 New England Journal of Medicine 465.

irrelevant to the validity of her consent. There is a risk of oversimplification here. No single conception of autonomy dominates in bioethics commentary; indeed the question of what is required for choices to be autonomous in clinical settings has been the subject of lively and sustained debate.[9] Yet the bioethics thinking that has made its way into court opinions, government regulations, and the reports of presidential commissions over the past 25 years has almost uniformly appraised the voluntariness of consent without reference to the pressure of medical or economic circumstances. The scholarly writing of the Radiation Committee's chair, Ruth Faden, constitutes a particularly clear and richly developed example. Faden begins with the proposition that informed consent given by mentally competent persons is autonomous and thus valid, absent coercion or manipulation by a human "agent of influence".[10] Coercion, which always renders consent non-autonomous, occurs when a human agent issues a threat that is intended to influence a decision-maker.[11] Manipulation likewise entails a human agent with such intent. The resulting proposition that influence cannot preclude autonomous choice unless it is *intended* by a *human agent* renders decisions autonomous regardless of the medical or social circumstances from which they arise, so long as these circumstances are not themselves the product of human intent.[12]

This tolerant approach to unintended circumstances focuses attention, and reformist energy, on the actions of those in position to exert influence intentionally at the bedside — physicians and other health care providers. It channels attention away from economic and social influences that powerfully affect personal choice but may well be too deeply entrenched to be susceptible to serious challenge. One might imagine alternative conceptions of autonomy-negating influence — conceptions broad enough to incorporate the pressure of difficult circumstances, unreflective compliance with professional advice,[13] or even the operation of unconsciously experienced influence.[14] But the approach to autonomous choice developed by the pioneers of bioethics and exemplified in Ruth Faden's work has proven remarkably successful as a vehicle for criticizing and controlling the behavior of research clinicians and others with the power to exploit purposefully the fears and vulnerabilities of sick people. Contemporary disclosure and consent procedures,

---

[9] See G. Dworkin, *The Theory and Practice of Autonomy* (New York: Cambridge University Press, 1988); N. Daniels, *Just Health Care* (New York: Cambridge University Press, 1985); Beauchamp & Childress, *supra* note 6.

[10] R.R. Faden & T.L. Beauchamp, *A History and Theory of Informed Consent* (Oxford: Oxford University Press, 1986). Within this scheme, *all* coercion but only some manipulation (that which is sufficiently difficult to resist) renders consent non-autonomous.

[11] This formulation also requires that the threat be both "credible" and "irresistible".

[12] Although Faden and Beauchamp do not precisely define what they mean by "intent", they make it plain that they have in mind something akin to purpose, and that mere knowledge of a set of circumstances (e.g. voters' or political leaders' awareness of poverty and its constraints on choice) does not constitute intent, even if action (e.g. antipoverty programs) by those with such knowledge could change the circumstances.

[13] See e.g. J. Katz, *The Silent World of Doctor and Patient* (New York: Free Press, 1984).

[14] W. Gaylin, "On the Borders of Persuasion: A Psychoanalytic Look at Coercion" (1974) Psychiatry 37.

IRB review, and federal regulatory oversight are ethically grounded on this approach.

There is, nonetheless, a large conceptual problem inherent in the application of this approach. This problem has remained largely invisible when the voluntariness of consent has been subjected to evaluation in traditional research and clinical settings. But the problem threatens the power of bioethics theory (and the regulatory mechanisms it has spawned) to protect vulnerable persons in a variety of new and emerging contexts. Such contexts include the conduct of clinical outcomes research by managed health plans; the offering of free high-technology care in poor, Third World countries to participants in research protocols; the conduct (as opposed to mere financing) of clinical research by for-profit firms; participation by mentally incompetent persons in research protocols; and the management and disposition of sensitive medical information (e.g. genetic susceptibilities to serious illnesses) acquired in the course of clinical research.

The problem of which I speak is that prevailing bioethics theory casts the question of consent as a purely empirical matter — a factual inquiry into whether the conditions under which a person chooses meet abstract, formal requirements for autonomous decision-making. Ruth Faden's work nicely illustrates this. To preclude autonomous consent, Faden holds, an external influence must not only constitute a *threat, intended* by a *human agent* to affect the chooser's decision; it must also be both *credible* and either *irresistible* or *difficult to resist*.[15] Faden presents these tests as purely factual inquiries. Yet the answer to each depends critically on normative judgments about the influence of the agent's behavior and the chooser's situation. Whether a choice situation intentionally created by a human agent is a threat or an offer, consistent with autonomous action, cannot be answered except by implicit reference to some moral or other normative notion of what constitutes a legitimate or acceptable choice situation.[16]

Often, tacit agreement on such underlying normative matters averts conflict whether a choice situation constitutes a threat. A classic example is the armed robber's proposition, "your money or your life". Although the robber presents a choice, we feel strongly that the person who empties her wallet in response has been threatened, and thereby coerced. Underlying this belief is our powerful sense that this choice situation is morally intolerable. Similarly, the physician who says to his patient, "participate in my new clinical trial or I'll stop caring for your leukemia, and I'll see to it that my colleagues know you're an uncooperative patient", is, we are led to think, issuing a threat. Underlying this characterization is our sense that the patient's dilemma is morally unacceptable. But what about the worker whose employer provides a single medical coverage option – company-subsidized enrollment in a managed health plan that uses its subscribers as subjects for a comprehensive program of clinical outcomes research? If the plan discloses this fact

---

[15] Faden & Beauchamp, *supra* note 9.
[16] See C. Fried, *Contract as Promise: A Theory of Contractual Obligations* (Cambridge, MA: Harvard University Press, 1981); A. Wertheimer, *Coercion* (Princeton, NJ: Princeton University Press, 1987).

to all prospective enrollees before they sign up, does the worker's decision to enroll constitute voluntary consent to participation in the plan's outcomes research? The employee has made a choice – she could have declined her employer's subsidized plan and either gone uninsured or purchased unsubsidized, individual coverage. Whether we see this choice as voluntary consent to being a research subject depends on our normative judgments about the acceptability of her medical coverage options and the legitimacy of "bundling" consent to research into a single package that includes the entire gamut of mutual commitments made by the plan and its subscribers. Absent implicit agreement on these underlying normative issues, agreement on whether the worker has given valid consent to participation in the plan's research is impossible.

Lack of implicit agreement on the moral acceptability of the circumstances under which people choose underlies controversy over the voluntariness of consent in myriad situations, including the conduct of clinical research on impoverished Third World populations, clinical studies involving the mentally incompetent, and sharing and disclosure of health information acquired in the course of research. By failing to acknowledge that the question of voluntary consent is a function of underlying normative judgments about the circumstances of choice, bioethics commentators (and the authorities that invoke them) render themselves unable to do more than assert their conclusions as to the voluntariness of an actor's consent. This analytic disability is of little practical import when agreement prevails, in the law and at the bedside, about the voluntariness and validity of a purported act of consent. But this disability incapacitates bioethics discourse from contributing meaningfully to an understanding of disagreement about the voluntariness of purported consent in novel and controversial choice situations. Absent a language for discussing underlying differences about the legitimacy of the circumstances under which the actor gives purported consent, bioethics discourse on the question of consent is reduced to an exchange of conflicting, analytically disconnected claims.

A similar analytic problem besets the inquiry into resistibility. If a researcher states a proposition — e.g. "Be a subject in my clinical trial and I will give you free care" – and her patient agrees to the proposition, what does it mean to say that the patient *could have resisted* (and declined) this proposition, either easily or with difficulty? The patient has *in fact* accepted the proposition – that is, the patient did *not* resist it, at least successfully. As Douglas Hofstadter has pointed out, there aren't "degrees of didn't-happen-ness".[17] If a particular person, under particular circumstances, agreed to a proposition, then the proposition was irresistible, as an *operational* matter, to that person under those circumstances. This is not to say that the inquiry into resistibility urged by Faden and Beauchamp is pointless. On the contrary, the question of resistibility, posed after the fact, serves a *moral* function: it channels critical attention to the actions of the chooser. In particular, it calls upon us to consider whether the person who agrees to a proposition should be deemed responsible, as a moral matter, for making a choice. If we say that a proposition was

---

[17] D.R. Hofstadter, Gödel, *Escher, Bach: An Eternal Golden Braid* (New York: Basic Books, 1979).

resistible, we represent the chooser as a responsible moral agent. We do so by creatively contemplating a counterfactual prospect — that the chooser *might* have said no and *perhaps* should have done so. If we say that the proposition was irresistible, we decline to contemplate this counterfactual possibility. Instead, we represent the chooser as not morally responsible, and we portray his acceptance of the proposition in deterministic, non-autonomous terms. We thereby suggest that moral responsibility for his action lies elsewhere, perhaps with the individuals who posed the proposition.[18] Faden and Beauchamp give us no analytic basis for choosing between these characterizations. Nor could they do so if they tried, since this choice turns on prior normative beliefs about the proper allocation of responsibility. An ethics discourse that fails to make a place for these prior beliefs can do no more than reiterate conflicting claims about a proposition's resistibility.

To contribute meaningfully to the resolution of controversies over the adequacy of consent to participation in research, or to anything else, bioethics discourse must expand to encompass discussion of conflicting views about the propriety of the circumstances under which patients make purported choices. By itself, the ideal of patient autonomy tells us little, perhaps nothing at all, about the legitimacy of a claim that someone has consented to serve as a research subject. Rather, the ideal of autonomy and the paradigm of informed consent draw their content from prior normative beliefs about the myriad circumstances under which people act. This is hardly to say that the idea of autonomy lacks moral import. On the contrary, the respect for persons engendered by deference to choices as autonomously-made both expresses the dignity of individuals and safeguards a sphere of human action against the intrusive scrutiny that deterministic explanations invite.[19] My limited point here is that the outer bounds of this sphere cannot be defined except by reference to prior judgments about the circumstances of choice.

The practical implication of this point is that conflicting claims about the voluntariness of consent in novel and controversial settings are likely to get us nowhere, and that it would be highly profitable for bioethics discourse to look beyond autonomy, toward the moral acceptability of circumstances under which consent is sought. When a pharmaceutical firm provides free, high technology care to patients in Third World countries who participate in clinical trials, debate over whether they have consented voluntarily should be broadened into discussion of such matters as the moral acceptability of the care they would have received absent the clinical trial, the firm's duty to provide some minimum level of care to patients who decline to participate, and the appropriate level of communication to patients regarding risks, benefits, and alternatives. When managed health plans in the U.S. perform clinical outcomes research, the question of whether subscription to a plan constitutes consent to participate in such research should be explored via

---

[18] If we say that the proposition was resistible only with difficulty, we hedge to some degree on the question of moral responsibility, suggesting that a measure of responsibility lies both with the chooser and the framers of the proposition.

[19] M.G. Bloche, "Clinical Counseling and the Problem of Autonomy-Negating Influence" in R.R. Faden and N.E. Kass, eds., *HIV, AIDS & Childbearing* (New York: Oxford University Press, 1996).

consideration of such matters as the moral sufficiency of subscribers' health coverage alternatives, the appropriate level and timing of disclosure to plan members, and the acceptability of "bundling" consent to research into a single package of precommitments between the plan and its subscribers. When researchers cite advance directives or assent by relatives as grounds for employing mentally incompetent persons as subjects, argument over whether these steps constitute adequate consent should be expanded into exploration of such questions as the degree (and acceptability) of risk, suffering, and infringement upon human dignity[20] entailed by the research and the moral import of family members' preferences and patients' *ex ante* preferences.

A turn toward substantive moral inquiry of this sort entails acknowledgment that the paradigm of consent cannot finesse deep substantive differences over the ethics of research on human subjects. Because the Radiation Committee failed to acknowledge this, it failed to provide a blueprint for the management of conflict over the acceptability of new and emerging experimental designs and institutional and social settings for research on human subjects. The successes of the paradigm of consent rest upon tacit agreement on such underlying moral matters as the right of members of all racial, ethnic, and other groups to equal respect as persons and the presumptive priority of physicians' patient-centered duties over such ends as national security, advancement of science, and fostering the health of populations. Unless anchored to such agreement, the paradigm of consent cannot do the moral work expected of it by the Radiation Committee and the bioethics movement more generally.

Open acknowledgement of this limitation is both unsettling and empowering. Consideration of questions of substantive justice and fairness may to many seem more painful than continuing the quest for a process-oriented finesse. Yet this substantive focus holds out the potential for bioethics discourse that clarifies core beliefs, identifies disturbing trade-offs, and thereby deepens our common understanding of the basis for our moral differences. Out of this deepened understanding may come an extra measure of mutual respect. Moreover, reorientation toward substantive justice and fairness could inspire creative efforts to achieve mutual advantage and to mitigate the harshness of the inevitable moral trade-offs.

The pervasive problem of divided clinical loyalty presents a range of opportunities in this regard. Otto Guttentag's proposed division of the roles of researcher, on the one hand, and of clinical caretaker and counsellor (Guttentag's "physician-friend"), on the other, ought to be revisited and incorporated into current thinking about the governance of clinical investigation, especially in settings where physicians deeply committed to academic careers encounter prospective research subjects as strangers, made vulnerable by debilitating illness. When the dangers posed by a disease, its standard treatment, and/or the experimental protocol at issue are high, the moral and psychological gains from bedside trustworthiness that the

---

[20] Human-dignity concerns are especially important when research on mentally incompetent persons involves clinical interventions designed to influence mental functioning.

Guttentag approach promises surely outweigh the small cost of the additional counseling role. Separation of caretaking and counseling functions from forensic and other tasks that physicians perform for third parties should also be pursued, to the extent feasible.[21]

The knotty dilemmas posed by the conduct of clinical investigation in Third World countries present additional opportunities for creative problem-solving. The offering of free health care, food, or shelter to participants in clinical trials might, for example, be made less problematic by a commitment to make the drug or vaccine being tested available at no charge to impoverished residents of the country or region, should the trial prove successful. Such a commitment would surely be relevant to moral assessment of the offer — and thus to the question of whether the offer constitutes coercion. More generally, health promotion programs, diagnostic screening efforts, and multi-tiered pricing structures for new medical interventions might be developed to benefit populations put disproportionately at risk by the research that gave rise to these interventions. Approaches along these lines could also be applied to disadvantaged ethnic and other groups in industrialized nations that participate disproportionately in risky clinical investigation that yields society-wide benefits. The degree to which the reciprocity implicit in suggestions of this kind ought to weigh in favor of the ethical acceptability of particular clinical trials is a matter beyond my scope here. My limited point is that heightened attention to substantive justice and fairness can engender creative, conflict-reducing responses to the ethical problems presented by controversial uses of human subjects in clinical investigation. Preoccupation with the question of consent distracts attention from the possibilities for moral inventiveness of this sort.

## CONCLUSION

The paradigm of consent has, on the whole, served medical ethics well over the past several decades. By linking intervention to the expressed preferences of patients, the requirement of consent, however plastic, has affirmed the dignity and integrity of individuals. Where rough agreement prevails concerning the propriety of the circumstances constituting a choice situation, the plasticity of the consent requirement does not present a problem. Such agreement makes it possible to discern and apply prerequisites for autonomous consent in particular situations. Moreover, it may even be the case that focus on the question of consent in such situations reduces acrimony by diverting attention from lingering disagreements about the morality of the circumstances of choice.

The Radiation Committee's celebration of the paradigm of consent is in large measure justified by these achievements. Informed consent doctrine has extended unprecedented protection to vulnerable people in a broad array of research and clinical settings. Nevertheless, absent underlying agreement about the morality of

---

[21] A.A. Stone, "Paradigms, Pre-Emptions, and Stages: Understanding the Transformation of American Psychiatry by Managed Care" (1995) 18 International Journal of Law and Psychiatry 353.

the chooser's circumstances, the plasticity of the paradigm of consent disables it from speaking decisively to the propriety of a clinical practice. Current controversies over the adequacy of consent in research settings reflect this rarely-acknowledged difficulty. By failing to address the need for undergirding moral judgments about the circumstances under which consent is sought or inferred, the Radiation Committee sacrificed a high-profile opportunity to push bioethics discourse toward more robust and explicit engagement with substantive justice and fairness concerns.

Engagement of this sort will be essential if the bioethics movement is to contribute meaningfully to the resolution, or at least the management, of current and emerging controversies over the ethics of research on human subjects in non-traditional clinical settings. Such engagement carries some risk. Open discussion of passionately-felt substantive moral differences could generate rancor and thereby feed public incivility. The bitter public debates over abortion and medically assisted suicide in the United States and Europe illustrate this risk. Indeed, the classic appeal of procedural finesses lies in their ability to obscure acrimonious substantive differences. The potential gains, on the other hand, are substantial. Beyond the intrinsic virtues of open and honest moral discourse, substantive engagement of the kind I urge here carries with it possibilities for enhanced mutual understanding (and even empathy) and moral and institutional creativity in response to conflict.

The proliferation of clinical research in such non-traditional settings as managed care organizations, medical technology companies, Third World locales, and facilities for the mentally disabled demands this kind of creativity. Arguments about whether various groups of vulnerable people — from HMO subscribers to residents of third world countries to persons with intermittent psychotic illness — can give valid consent to present or future participation in research will remain a sterile endeavor without the substantive engagement and moral inventiveness I call for here. The need for this inventiveness is heightened by the internationalization of clinical research and the resulting ethical challenges posed by social, cultural, and moral diversity. Among nations with contrasting moral traditions as regards individual rights, fairness, and social justice, understandings of consent are bound to differ widely.

For at least the past 50 years, since the issuance of the Nuremberg Code and its accompanying judgments, international law has barred scientific experimentation on human beings without their consent.[22] Ironically, this proscription has become a matter of increasingly disparate interpretation even as it has won unprecedented global acceptance. If these disparate interpretations are not to dissolve the power of the idea of consent as an affirmation of human dignity and liberty, then the substantive moral questions that undergird them will need to be acknowledged and engaged.

---

[22] Any early, post-Nuremberg doubts about the international legal status of the Code's consent requirement were erased by the promulgation and widespread ratification of the International Covenant on Civil and Political Rights, which specifies that "no one shall be subjected without his free consent to medical or scientific experimentation".

CHAPTER 4

# ROLES IN CLINICAL AND RESEARCH ETHICS

DAVID N. WEISSTUB

## INTRODUCTION

In the earliest period of writing and reflection associated with bioethics, which
gathered force in the late 1950s, thoughtful generalists alongside theologians and
philosophers, began to pose serious questions about the limits of technical medicine,
given the humanistic burdens inherited from the monstrous period of dehumanization during the Second World War that showed the capacity of medicine in general,
and the research enterprise in particular, for gross violations against humanity. These
pioneer efforts were not unconnected to attacks against medical elitism and became
associated in the 1960s with a pervasive negativism in the popular culture about
"hard" medicine and the self-serving patterns of professional conduct connected to
corporativism in the health system.[1]

In the second generation of bioethics formation, more generalized philosophical
efforts were sustained, contributing to the foundation of special centres of reflection
such as Hastings and Kennedy. Since that time in the early 70s, the movement has
come a far way. Our point here is to query whether this rapid expansion of power
and territory is altogether clear in its production of philosophical mantras and power
elites. Our purpose here is not to debunk the seriousness of the overall endeavour,
given, as we have already observed, the great importance of its target area of
inquiry; nor do we wish to question the integrity of an association of intellectuals
and practitioners who are bent on cooperating as multi-disciplinarians in some
resolution of moral dilemmas presented in medical practice and research. Rather, it
is to ask whether the achievements of the movement lead us in any particular way
to an improved notion of how to make meaningful ethical decisions, how to fashion

---

[1] See S. Toulmin, "Medical Ethics in its American Context: An Historical Survey" (1988) 530 Annals of
the New York Academy of Sciences 7 at 7–15.

56

laws with an appropriate medico-ethical content and direction, and how to train advocates of bioethics, who are so fulsomely present in our hospitals and committees.

Bioethics entered by the late 1970s into the precincts of medical research, and represented the major influence on studies undertaken by the US government to work out the territory of the general principles of the discipline. In 1967, the National Institutes of Health produced the system of surveillance for institutional review boards, and by the mid-1970s, a full-scale presentation was made available to the public of how to begin to think as experts about the rights and wrongs of medical ethics in research.

The picture we are confronted with since the late 1980s, and fully matured in our *fin de siécle*, is vastly different from the eager and expansive contributions of earlier decades. Although bioethics experts can now be found everywhere, and are present as the key experts to advise research ethics committees, we must take time to reflect on the criteria and the role models guiding these experts. Our notions of the common good, responsibility, attitudes towards vulnerable persons, perceptions about communications styles, transformations of moral sensibilities on end-of-life matters, restructuring of familial models, and an overall tendency towards more egalitarian exchanges of information between professional elites and consumers, have all to be factored into our recalculation of benefits and risks in both therapy and the research enterprise.

The world of research ethics cannot be dissociated from its foundations in the bioethics movement as seen from a broader environment. And it is through the importation of core bioethics principles that research ethics has been dominated and fashioned as a way of perceiving moral thresholds and decisions about priorities. Our manner of thinking, informed by the Kantian and utilitarian polarities afforded by philosophy,[2] has led us to assume that failure to achieve the expressed value formulations leads directly to moral turpitude and/or pragmatic failure. This results from the position of our philosophical differences in rigid and uncompromising frameworks. In reality, our ethical thinking proceeds along the lines of our attempting to defer, in balance, to treating persons as ends and not as means, and to measure the good, both in theory and practice, when we take into account outcomes of an array of health interventions.[3] We should conclude that representative theories of pure form are appreciably present for our consideration, but that our actual decision-making models, for example in dealing with difficult cases of ethically permissible experimentation, our analyses and conclusions are foreshadowed by aspects and conflations of the value schemes that are associated with conscionable behaviour in liberal democracies.

---

[2] L.W. Sumner, "Utilitarian Goals and Kantian Constraints" in B.A. Brody, ed., *Moral Theory and Moral Judgments in Medical Ethics* (Boston: Kluwer Academic, 1988) 15.
[3] E.H. Loewy, "Kant, Health Care and Justification" (1995) 16:2 Theoretical Medicine 215; D.C. Thomasma, "Assessing Bioethics Today" (1993) 2:4 Cambridge Quarterly Healthcare Ethics 519; S.H. Furness, "Medical Ethics, Kant and Mortality" in R. Gillon, ed., Principles of Health Care Ethics (New York: Wiley, 1994) 159.

When we are forced to live and decide about cases in the domain of applied ethics, we must challenge ourselves to conjoin empathy with fundamental values, for without such, the social integration of persons without full capacity will not be guaranteed. Rather than emanating from any existing philosophical system, with preferred value schemes attached, our ethical humanism would be better located within the experience of empathy. In fact, when we are in the midst of actual ethical decision-making, it is rare even in our professional and cultured social environment to see value systems presented, defined, and logically justified apart from general statements of commitment to tolerance and respect for certain key values such as autonomy and personhood. Having committed in principle to such values, we must advance our protections through affording thorough and far-reaching guidance-rules upon which we can base a legal framework for regulation.

In my final report, as Chairman of the *Enquiry on Research Ethics* for the Government of Ontario, I asserted the following:

> Neither the established principles which dominate bioethics discussions on research, such as beneficence, respect for persons, and justice, nor legal remedies such as informed consent, have been found to be of any substantial practical use in resolving the conflicts of interests that present themselves in the field of research ethics. In ideal terms, these principles and legal remedies, to which we wish to subscribe as a matter of moral commitment and preferred values, do not translate easily into effective remedies. The most generalized moral principles have little predictive value. In the field of the protection of vulnerable populations, beneficence, which motivates us to interventionist policies, does not sit well with the complementary principle of respect for persons which should instantiate policies to enhance autonomy and self-determination, even at the expense of the best interests of subjects. The world of moral principles comes into contact with the universe of legal principles insofar as the law has developed certain instruments, or doctrines, to protect the autonomous life of subjects. Insofar as the application of legal principles has resulted in such autonomous actions causing harm to individuals and their extended social bodies such as the family, the conflict of principles of law has become the subject of scrutiny and disquiet in parallel with the conflicts experienced at the more abstract levels of discussion by philosophers, theologians, and ethicists. With regard to legal remedies it may be seen that the doctrine of informed consent has not been followed with any credibility and its cost in the quality of human relations or economy has not been empirically documented in any significant way to date. In fact, legal remedies have been highly selective and irregular, are after the fact, and have in any event not proven to protect the special populations to which such doctrines were initially directed.[4]

It is against this background perception that we pursue our further inquiry here. What we will attempt to show is that the dilemmas that we are facing in dealing with making priority decisions about the elderly and other high-cost groups of citizens, including specific groups of vulnerable persons, has led us quite naturally to create, arguably in good faith, a set of both moral and legal fictions in order that our social needs have a respectable "justificatory" ethics. We are not suggesting here that such fictions are mere rationalizations. On the contrary, much of what happens in the life of our moral and legal fictions falls into the domain of what we can term "white lies" and "soft fictions". These are meant to balance our need to fulfill different goals,

---

[4] *Enquiry on Research Ethics: Final Report* (David N. Weisstub, Chairman. Submitted to the Honourable Jim Wilson, Minister of Health, Ontario, 28 August 1995) at 2.

some of which are in implicit or explicit conflict with one another. The resolution of these conflicts can be directly linked to how we view the doctor/patient or researcher/subject relationship, and to our perspectives about the extent to which we can hold citizens morally liable to act on behalf of others, or for that matter, expect a society to accept the burden of incapacitated citizens. Insofar as philosophers and legal doctrine have afforded us some of the tools that assist us in our public discussion, we must remain both grateful and accommodating. However, unquestioning reliance on such expertise or legal regimes can not only be misleading, but can also be a contributing factor to the kind of dependency where moral failures can be most pronounced. Disfavouring a real public debate on controversial matters related to medical ethics, for example in research, is a sacrifice that no civilized democracy should allow. We turn then to consider the role models that are part of the active world of bioethics, as we have constituted it, and thereafter to the field of research ethics as the applied terrain upon which bioethics has already left indelible marks and patterns.

## ROLES IN CLINICAL ETHICS

In ethics consultations, the deliberations of committees, and informal exchanges which occur among professionals and their patients, there is a high degree of incertitude about two matters: firstly, whether there is an integrity to the body of knowledge called "bioethics"; and secondly, whether from this existing material, there emerges a clear directive about the best method of training people to benefit a potential group of clients in making, or having made for them, the most "beneficial" ethical decisions. Often, the jump is made quickly into the second category, where a polemicism has a tendency to replace scientific inquiry, because, given where people are situated, often out of self-interest in this professional arena, there is the assumption that choosing the wrong model, either in professional role-playing or in a concept of education, will lead to adverse ethics.

Without entering into a full inquiry into the integrity of the body of knowledge underlying this movement, we might have a more modest beginning to an improvement of our understanding by turning to the distinctive roles that emerge in the ordinary course of discussion found among clinicians. The three dominant models could be described as: (i) the ethicist as tranquilizer; (ii) the ethicist as advocate; and (iii) the ethicist as a conscience surrogate. In all these modalities, there are a set of assumptions and implications for what is perceived to be correct professional-ethical behaviour. It may be seen that to choose one course of action is not necessarily mutually exclusive with correlative roles, but certainly in many circumstances, the roles stand in opposition and lead to different ways of conducting relationships between professionals and their clients, and to profoundly different views about how to educate a generation of informed professionals and a specific group of experts.

At the root of these modalities is the presumption that we have in mind an ideal type or product. If, in the case of searching for an integrated paradigm of expertise, we think inevitably of specialized knowledge or experience, we must probe in the case of each of our modalities the link between taking on a role and a certain species of professionalism, either by way of background or according to the building-block model of achieving a justified hybrid. Put simply, questions must be posed about whether the "ethicist" is the goal or the problem; whether clinical experience is the necessary pre-requisite or condition for acting responsibly when ethical conflicts are presented; whether clinical ethicists are best placed or found in the context of committees, multidisciplinary consultation groups, one-to-one hands-on consultations with families or patients, in an informal structure within a hospital or clinical environment; or finally, whether they find their best ultimate role in an educational process meant to improve the dialogic exchange between health professionals and their subjects.

One thing is certain: the need for clinical ethicists is pronounced in hospitals and in the health environment at large, due either to the extended life span realized in most industrialized societies (influenced by the utilization of new technologies), or as a necessary input to stem the tide of social inquietude because of the evaporation of a consensus morality previously found in more cohesive social orders, driven by collective appreciation of religious or political values. The pressures towards reliance upon clinical ethicists have arguably been recent and never stronger than now. It is unclear, however, whether our preoccupation about having a need for clinical ethicists will be long-term (and that their evolution will be concretized and better defined), or whether clinical ethicists, after a generational trend has been exposed, will be rendered obsolete and regarded as ethical transition objects in a period of major social upheaval. Arguably, clinical ethicists may eventually be viewed in a similar way to the advocacy group which dominated the thinking of minorities and consumers, until replaced by more autonomous, independent, and politically assertive self-advocates who rejected the notion of dependency on experts.

One of the dynamics which lies behind all the options of role-playing is the presence or absence of the ethicist in terms of the directness, either visual or emotional, which occurs between professional and subject. This raises the question of whom the ethicist serves. Is the role of the ethicist to facilitate the value-formation of a group of professionals who, in a forum of a committee structure, decide ethical outcomes among themselves? How important is the act of communicating and assisting in the clarification of a value which is subject-oriented? Many professionals may even fear opening up imponderables relating to pain, life-threatening decisions, and the management of one's loved ones and personal affairs with a subject, or a family group in distress. Ethicists may often be poorly trained in the emotional dimension of the process of communicating values. On the other hand, mental health professionals, or physicians for that matter, may be modest in their understanding of the kinds of values which are at stake, or should be articulated in order to arrive at meaningful decisions. People faced with moral dilemmas want to be assisted with respect to gathering relevant knowledge, enhanced in their capacity for self-reflection, placed in a position to articulate shared values, and prepared to assess the impact of these values (and decisions flowing from them) on parties relevant to their own constituted sense of personal or social well-being. It seems as if one of the greatest threats to all of us is our sense of inadequacy in, on the one hand, being able to reflect on intra-psychic conflicts that are part and parcel of our sense of physical or spiritual health, and on the other, equipped with such self-reflection, being able to move out again into the world, with a sense of freedom and lack of shame, to share thoughts, and eventually decisions, with others who have been, are, or will be, part of a reconstituted universe which has a life beyond any mortal unit.

In the face of the pressures and requirements placed on the shoulders of ethicists who live the daily life of decisions, there is a need to explore how roles represent a set of reference points around which professional practices occur and are justified.

### The Ethicist as Tranquilizer

It may well be that the overwhelming purpose and function of clinical ethicists is to calm an environment which is already inflated emotionally due to the need for making a decision, where every alternative is connected to, or will result in, serious pain (physical or emotional) to a person, family, or group. The call for an ethicist is catalyzed by the absence of a pastoral or professional authority figure, who could in the circumstances present a moral resolution to the issue at hand. In some instances, there will be a great sense of satisfaction to see the appearance of an expert, perhaps in the actual physical guise of a doctor, robed in white, and under the command of a beeper unit, indicating the dependency of patients on the immediate availability to respond as a professional to a call of urgency. The content or ethical rationalization in such a picture is surely secondary to the pronounced emotional need from the dependent party. Depending upon the communication orientation of the clients, professionals often learn to fulfil the mandate with considerable social skill. In some cultural contexts, certainty is the measure of professionalism, whereas in others, an affected modesty and soft approach to merely facilitate an exchange through the role of arbitrator or mediator represents an appropriate comfort zone to the consumers. In any event, the ethicist is there to create the conditions for scaling down conflict, and to facilitate leading the parties to resolution.[5]

Some professionals see their roles as "ethics clinicians" as the donning of a mask, or worse, as participation in a fraudulent undertaking. A want or need does not, in and of itself, produce integrity for professionals who have serious intellectual doubts about their mandate and about the body of knowledge upon which advice or guidance is based. From the intellectual point of view, some ethicists who do applied work, still persist in the view that without a set of principles, which are the mainstay

---

[5] See also J.J. Glover, D.T. Ozar & D.C. Thomasma, "Teaching Ethics on Rounds: the Ethicist as Teacher, Consultant and Decision-Maker" (1986) 7:1 Theoretical Medicine 13; D.C. Thomasma, "The Role of the Clinical Medical Ethicist: the Problem of Applied Ethics and Medicine" in Braide *et al.*, eds., *The Applied Turn in Contemporary Philosophy* (Bowling Green: Bowling Green State University Press, 1983) 137.

of any operational activity, there will always be the gnawing disbelief in being able to establish moral connectors between thought and practice. The black box problem in clinical ethics in its embryonic stage may not be dissimilar to the conditions of medicine at an earlier time. Medicine, in the pre-industrial era, although limited in science, still had a calming-placebo effect, even where the fear level of the party in need was not at issue. But whereas medicine has always demonstrated some basis in science for offering effective remedies where something had to be proved causatively, ethicists, who are called upon to make linkages between knowledge and practice, are arguably on weaker ground. This is not to say that we need to achieve a positivistic ethics, which is measurable or provable, in order to be able to function in a useful and credible manner as ethicists, nor do we wish to suggest in our expressed difficulties here that we are prepared to declare moral bankruptcy because, as a society, we should admit that we suffer from the extremities of moral relativism. Such radical conclusions are neither what we should necessarily admit, nor accept, as a given.

Rather, in our search for general principles, we may clarify the limitations of the state of our science and what it leads to thereafter practically. Otherwise, we can only turn in circles on some form of professional intuition, which is dangerously close to an appeal to authority or paternalism.[6] Such elitism is out of synchrony with the increasing level of popular education about health matters and with the heightened awareness about life-and-death matters in medicine which are widely discussed in our popular culture. Therefore, we should conclude that regardless of how much the population, even when educated and aware, wants to be tranquilized by a new elite corps of professionals, such professionals and their critics are well-advised to exercise caution before putting practice ahead of reflection simply because the process has become self-aggrandizing and self-justifying, based upon emotional need.

### The Ethicist as Advocate

There is an ongoing ambiguity about which set of skills best serves the enhancement of empowerment, which has become the mainstay of the deontological movement that views the ideal subject as a fully rational and autonomous party who should always strive to maximize principled decision-making. With such a model in mind, the challenge to the ethicist is to bend to the legal will of the rational citizen, an orientation which also affects our view of how to treat incapacitated or dependent individuals. Thus, some ethicists find themselves functioning as "in-hospital" lawyers, and armed with an effective legal apparatus, prepare themselves to act as intermediaries or ombudspersons, between or among complainants in the health system. With exposure, it is rare that legally trained persons can avoid a natural inclination towards collusion where there is an identification, if not over-

---

[6] B.A. Brody, "Autonomy and Paternalism — Some Value Problems; a Utilitarian Perspective; a Deontological Perspective" in B.A. Brody, *Ethics and Its Applications* (New York: HBJ, 1983) 159.

identification,[7] with a caring or paternalistic ethic that is in conflict with the traditional role of a legal professional. It is not unusual to see ethicists with a legal background functioning through and for a mindset normally associated with the medical model, where intervention is directed to protect the patient from self-destructive decision-making. Ironically, the legally trained professional, who identified with a rational deontological substructure, is transformed into a paternalistically-oriented intermediary within a goal-oriented health system, and for which the advocate has had no significant or even relevant preparation. Interestingly, however, as we have observed in the case of ethicist as tranquilizer, the need of patients to be secure in a foreign and threatening environment may be equally strong *vis-à-vis* the lawyer as a symbol of certitude and rectitude.

Legalities have everywhere intruded into the pristine domain of both "ethics" and "professional practice". There is such a vast jurisprudence of health law, which carries with it elaborate judicial reflections on social values pertaining to health, that it is impossible for anyone presuming to function as an applied ethicist to forget, even for a moment, the momentous impact that legal decisións have on our ethical thinking about health matters and on the way in which we conduct our professional practices. These legal decisions are part of a larger picture of codification and legislation as well as an extensive literature where lawyers and legal philosophers associate their thinking with the major trends extant in philosophy and political theory. To think of law as isolated in the health area from ethics and politics is to associate either law or practical philosophy with abstractions, or other social eras, inapplicable to our current time and place.

The advocacy movement, apart from matters of theory, is a mirror of political attitudes which burgeoned in the 1960s and led to dramatic reinvestigations into our belief in professional elites. The association therefore of a newly created elite group of ethicists with the advocacy movement represents an odd alliance. For at once we may see that the advocacy movement, which was meant to subvert elitism, evolved in its various guises into a specific group of experts, many of them para-professionals, who then led their legally oriented expertise in the direction of taking on tasks which emerged from the health system, from a different, and even contradictory, set of premises from those of the advocacy movement. Now we are faced with advocates and ethicists who share two sets of problems: firstly, how to define their roles professionally; and secondly, how to orient themselves from a knowledge-base in such a way that a sense of professional integrity is preserved from an intellectual point of view in the definition and purposiveness of the ongoing roles encountered.

The dialectical play between advocacy, ethics clarification, and health system paternalism presents a dramatic challenge to all participants who wish to locate a professional role. As teamwork is so strongly part of the consultation process in which health advocates participate, any restrictive view of a legalistic role ceases to

---

[7] D. Barnard, "Reflections of a Reluctant Clinical Ethicist: Ethics Consultation and the Collapse of Critical Distance" (1992) 13:1 Theoretical Medicine 15.

make practical sense. Once the legally-oriented advocate begins to apply the craft of informing parties about the law, putting steps into motion to assess mental capacity or dealing with claims on the system for rights and benefits, advocates cannot separate their tasks from social judgment, accountability to health providers, and responsive behaviour towards family members or public institutions. Finding one's way back to a professional role, or learning how to justify morally a hybrid of activities, can become conflictual, challenging, and deeply confusing. Bluntly stated, it is often difficult for advocates who have found their way into the universe of clinical ethics to know whether it is in the undoing of their professional training or in the efficacy of a mature application of this background that good is brought to patients who depend on their counsel.

### The Ethicist as a Conscience-Surrogate

Even stronger than the need for tranquillity or advocacy is the desire on the part of persons to find a reference towards moral superiority. It is not true that we are able to operate in a moral vacuum. On the contrary, there is great moral anxiety in the modern world where acts of violence and disrespect are present and threaten our social order. Coupled with the prolongation of life and the vulnerability of an aging population is the celebration, in popular culture, of physical health and youthfulness. With fiscal crises and the privatization trend in health care services, something near a panic-level of concern can be seen, where there is a social recognition that in the process of prioritizing health entitlements, we will need to find a ground for ethics that will be collectively tolerated while remaining respectful to traditionally stated deontological systems which underline the need to respect persons as ends in themselves. Such universal principles are being tested and attacked insofar as clinical ethics comes into contact with scarcity of resources. As the tensions rise in this sector, the requirement for the ethicist to take on the role of conscience-surrogate accelerates.

In this way, the ethicist is the replacement and embodiment of the traditional role of the Church. In search of conscience, hospitals and medical institutions, no less than governments, seek experts to set the limits of individual sacrifice required by a novel communitarian ethic. Clinical ethics in this light is pulled between crass pragmatism and the impulse to preserve a humanistically-centered liberalism. As we realize that liberalism as a political philosophy is on the wane, and that in advanced industrialized societies we have a stark mainstream commitment to laissez-faire social arrangements and the privatization of publicly-centred institutions, clinical ethicists find themselves at the heart of social transformation. They are often consulted as a new species of experts of last resort. Although it is burdensome in the extreme to be the moral conscience of a confused society, it is also enticing, even heady. Therefore, it should be the mandate of clinical ethicists to guard against moral narcissism because, time and time again, even the most distinguished professionals in medicine will join patients in moral desperation to locate an expert to relieve either or both of the parties from exercising moral judgment when there is a limited understanding of the values involved, and even less, of the values implicated in the outcome of the decision arrived at. To be the moral conscience in such situations is not as difficult as might appear at first glance because there is little investment in requesting parties for the details involved in justifying acts of conscience. Rather, the commitment is to find relief from the guilt engendered by even a self-reflected move which is not grounded in consensus. Reliance on ethicists often springs from the depths of human-moral misery and inadequacy. Occupying the field ethically is no different than mental health professionals being overly directive in circumstances where their patients are truly fragmented and dependent. It is necessary, then, for clinical ethicists to define their relationship with the recipients of their delivery of moral goods, and to make sure that such pronouncements are virtuous and empowering of patient autonomy, rather than manipulative and dependency-creating.

Clinical ethicists should be trained in moral modesty. The creation of the clinical ethics movement can be directly attributed to the fact that contemporary society has nowhere else to turn. Because our society has allowed physical health to transcend spirituality, and hedonism to remain the dominant ideal of our social fabric for a number of generations, clinical ethicists will not easily be able to convince the public that they, as ethicists, are the embodiment of desirable codes of principles and behaviour practices. Clinical ethicists are part of the conditioning of society at large. How then, in such a universe, can clinical ethicists assist patients and health professionals to enter into a dialogue of mutuality and concern? A cautious answer is in order. To create the conditions among medical trainees and other professionals to approach patients so that there can be an honest exchange of risks and benefits, of a moral as well as of a medical nature, is what conscience should dictate. The role of clinical ethicists should be, through training and facilitating a process of dialogue between health professionals and patients, to raise the level of consciousness in our health-related decision-making to the point where conscience is heightened among all parties. The ethicist should not remain disembodied or disengaged. Clinical ethics, if it is to have meaning, should be the living process of locating conscience through dialogue, namely through caring communication. In this way, humanism or liberalism can be re-enlightened, even in economies of scarcity, and where political and social values have been found wanting.

## COMMITTEES, MULTI-ETHICS AND INFORMED CONSENT

We turn now to consider the role and function of clinical ethics in the specific domain of research, where we can observe the testing out and crystallization of a mature formation of a generalized movement of ethics training and concept building. After a number of decades during which philosophers and clinicians attempted to formulate a core of organizing concepts, the research industry became the recipient of a framework according to which, presumably, "applied ethics" could be named with a requisite level of respectability. The remaining analysis will concentrate on two focal points: firstly, research and multi-ethics, a consideration of issues raised by our preferred model of group decision-making; and secondly, a reflection on informed consent as an organizing concept and its attendant myths and misunderstandings.

### Research and Multi-Ethics

The idea of committee decision-making sits well with our democratic notion that a paternalistic "one man show" is marked with the arrogance of a bygone era. The "Doctor knows best" concept, which permeated ethical decision-making until only a few decades ago, has been bypassed, however, by a dependency of a similar nature on technical expertise. Resort to experts is subject to the qualification that, in matters which cast up moral differences, society should guarantee that the dialogue be a fair one, ruled by principles of administrative justice that obligate researchers to adhere to professional standards of disclosure and care.

Putting together a committee whose mandate is to occupy itself not only with whether the research under review has achieved a minimal level of scientific integrity, but as well, whether or not "ethics are fulfilled", we are warranted in questioning how the team effort contributes to the process of acquiring and disseminating a consensus or informed morality based upon some appreciation of expertise. In the discourse of research ethics committees, there is an almost startling repetition of certain catch phrases: beneficence, respect for persons, and justice. Normally, members of such committees are not particularly well-informed about the philosophical content surrounding these principles and it is not clear whether referring to them in any significant way advances the goal of attaining shared moral precepts in making hard decisions. In fact, the reverse may be true, such that these terms have become slogans behind which experts provide justifications where the tie-ins between principles and actions are not even questioned by committee members. We should not underestimate the investment that committee members have in avoiding being held responsible for ethical dilemmas which are confusing and guilt-producing.

Committee members acting on research matters bring with them some mode of professional training and carry a normal citizen's array of ethical impulses, arising from familial, environmental, cultural, professional, and socio-political inputs. Given that we are not a homogeneous society with respect to the format of these influences, it is not surprising that our impulse on these committees is to avoid opening up Pandora's box, lest we be prompted to enter debates which would expose disturbing matters for exchange. These would include such issues as the extent to which the committee in question is already weighted in the direction of a particular outcome, whether the research industry or a selected interest within the research endeavor is overly represented in the group, the style of decision-making dominant in the committee (for example, whether it is adversarial or mediational), and the key expert's orientation in theology or philosophy, which could make a serious difference in influencing the social attitudes of the other participants and the eventual outcome of the committee decision. Once the prism of such questions is expanded to allow for identification of how the ethics formation of each of the committee members really presents itself, we would have achieved a rare dialogue about "ethics controversies and priorities", which is more frequently avoided than broached.

In many institutions we can still find committees ruled by the researchers themselves, where the ethics could be best described as those of self-interest. In others, there is an excessive reliance on the oracular presence of an ethicist. In some instances, we can still find provocative examples of members of extended research teams sitting in judgment on their own research. In such extreme cases, we have a clear lack of ethical integrity and should disallow such practices. But in the great majority of cases, persons of goodwill are meant to grapple as a team in search of a viable consensus morality. Reasonable persons should not come to unreasonable disagreements, except in the minority of circumstances, which present true moral quandaries. The question is how we prepare committees to locate the right brand of ethics in passing judgment when confronted with such inevitable divisions of emotion and opinion.

The multi-ethical dimension of committees is both good and bad. In a more positive light, we have a better representation of the spread of public opinion, presumably revealed in the divergence of personalities appointed for the task. This is only a presumption because we certainly are not surprised to find an in-house quality in many of the existing committees. In most cases, we will not find members of the subject groups of research. Nor do we have any significant feedback from subjects about how they have been handled by research ethics committees, even though we might assume, given the ubiquitous reliance on the mantras of autonomy and respect for persons, that the researcher/subject relationship would have been the first priority. From this point of view, the ethical ideal of the multi-ethics committee structure seems to have been a failure.

Quite apart from the lack of consistency in analysis of ethical principles, and how these principles can be and are applied in practice, there is the endemic problem of uniformity among research committees in hard cases. The issue is whether we should aspire to such consistency given the difficult nature of the problems. From the point of view of planning and research, of consumer expectations about entitlements, and of access to research where there may be a correlative therapeutic benefit (short or long term), consistency becomes a meaningful requirement. Also, if it is the case that some research ethics committees are considered lax with respect to ethical standards which affect certain institutional practices, we can expect research funds to flow in an overly determined fashion to certain institutions, provinces or countries.

In order to realize a desirable level of consistency, guidelines should be formatted by overseeing bodies, and regulated by regional political authorities. We are not prepared to submit that rigid national standards should be put into place because a new set of issues arises once research ethics decision-making is taken over by any aloof and overly regimented body that might fail to take into account justified local differences. Nevertheless, given the uncertainty factor caused by leaving research committees to their own devices, due to the lack of clarity in how we make connections in research ethics between theory and practice, and to the variances of decision-making styles among professional elites, we should avoid *ad hoc* reliance on the committee structure. Perhaps the greatest value of the best-constituted committees is that they afford a period of reflection for interested and affected parties to meet face to face in order to make ethics a result of a collective exchange rather than collective interest. The strength of the committee is not, and should not be, in the unanimity of ethics, but rather that in the hard cases, which are the only truly relevant ones for reflection, there be a meaningful and truthful division of opinion expressed, attracting controversy and thereafter human-centred compromise.

### Informed Consent as an Organizing Concept: Myths and Misunderstandings

The "core issue" according to which research ethics committees make decisions can be tied directly to the primary input from the bioethics tradition as it was conceived in the 1960s, namely the doctrine of informed consent. Through this instrument, the core values of the bioethics movement have been deemed to be realized, so much so that informed consent should rightfully be regarded as the high watermark of the great period of bioethics expansionism in the 1970s and '80s. Despite the fact that informed consent represents a major improvement over the preceding era of overt medical paternalism, we should be cautious in overcelebrating this achievement in relationship to its originally stated goals and with regard to the most important groups that it was aimed to protect, namely vulnerable persons. To some extent, we might say that informed consent is, at least partially, a myth produced by the bioethics movement, and which has even regrettably distracted research ethics committees from confronting moral realities.

As an example of practical and legal ethics, consent mushroomed as an all-pervasive line of defense against the prospects of paternalistic abuse. In a brief period of twenty years, from courtroom to ethics committees, it became a basic assumption that if the conditions of a legal consent were realized, both legal and moral dimensions had satisfied an acceptable threshold of conduct. To accomplish this, an elaborate system of legal decisions drew the lines for outright bans on certain forms of medical practice, while drafting for other cases the permissible rationales for defining mental capacity, best-interests, guardian and substitution rules, maturation levels for children, and more recently, criteria for providing advance consents. All this work was surely not in vain, much of it motivated by a real desire on the part of courts and legal commentators to ensure that, to the best of our institutional abilities, we were prepared to enshrine the consent doctrine as a protection against the recurrence of past evils.

Without wishing to return to a pre-consent universe, we might nevertheless expose the mythical nature of the doctrine of informed consent. Such debunking is necessary since the doctrine is useful, though only within limits, much like the abstract philosophical systems that in their application seem to fail in delivering adequate protections. Over-reliance on the law of consent in fact carries with it the risk that we may wilfully blind ourselves to the real exchanges of power and benefits between researchers and subjects. To begin with, consent is a model based on a contractual relationship, which in its pristine form is rare both in medical practice and research.

Recently, Schuck observed:

> *A priori*, there are strong reasons to suspect that informed consent, at least the law in books, is often honored in the breach and almost impossible to enforce as a practical matter. Most of the existing empirical studies on informed consent support this intuition. These studies reveal three related impediments to implementation of informed consent doctrine: (1) most physician-patient discussions appear to be rather perfunctory and reinforce physician control; (2) the treatment context discourages patients from exploiting the information that physicians do provide; and (3) the nature of the tort system makes it difficult for patients to establish an effective legal claim.[8]

The practical limitations of the informed consent doctrine have now been fully documented by empirical research which assesses the link to subject understanding.[9] By and large, we have established the fact that subjects are often unaware that they are even participating in experimentation despite well-worked legal requirements.[10] This limitation persists even where institutional review boards have taken their supervisory role seriously.[11]

Apart from the problem that our procedures systematically disfigure the cognitive capacity that would make informed consent meaningful, there is also the fact that, more often than not, the informed consent process is managed by the subordinates of researchers. This is not to say that, as communicators, subordinates are inferior to their supervisors; the contrary may sometimes be true, and indeed revealing.[12] As an ideal we should commit ourselves to the prospects of moral relationships occurring between researchers and subjects in as personal a manner as possible in order to avoid the depersonalization of the research enterprise.

Model informed consent presumes that a rational decision-maker has been mobilized after being exposed to the relevant information according to which risks and benefits can be assessed in the light of a specific set of personal values. This idealization can have the impact of misdescribing not only reality, but also the

---

[8] P.H. Schuck, "Rethinking Informed Consent" (1994) 103 Yale Law Journal 899 at 933.

[9] There has been considerable criticism of these studies based on the conceptual and methodological deficiencies they contain. While this may tend to limit the practical usefulness of these studies in terms of constructing an adequate practical model of informed consent, the studies nonetheless point to serious problems with the way in which informed consent currently operates. A comprehensive critique of the informed consent research may be found in A. Meisel & L.H. Roth, "Toward an Informed Discussion of Informed Consent: A Review and Critique of the Empirical Studies" (1983) 25 Arizona Law Review 265. See also P.S. Appelbaum, L.H. Roth & C.W. Lidz, "The Therapeutic Misconception: Informed Consent in Psychiatric Research" (1982) 5 International Journal of Law & Psychiatry 319; P.S. Appelbaum & L.H. Roth, "The Structure of Informed Consent in Psychiatric Research" (1983) 1 Behavioral Sciences and the Law 9; and P.R. Benson, L.H. Roth & W.J. Winslade, "Informed Consent in Psychiatric Research: Preliminary Findings from an Ongoing Investigation" (1985) 20 Social Science & Medicine 133.

[10] P.R. Benson & L.H. Roth, "Trends in the Social Control of Medical and Psychiatric Research" in D.N. Weisstub, ed., *Law and Mental Health: International Perspectives, Volume 4* (New York: Pergamon Press, 1988) 1 at 24.

[11] B.H. Gray, *Human Subjects in Medical Experimentation* (New York: Wiley-Interscience, 1975). A comprehensive survey of the consent research can be found in Benson & Roth, *ibid.* at 24–31.

[12] See Appelbaum & Roth, *supra* note 9.

requirements of effective decision-making in the system.[13] After having made critical decisions about "trustability", research subjects are frequently called upon to commit to a process at large, and often without the requisite details at various stages in order to be part of an ongoing commitment to a particular course of action.[14]

It is often difficult to establish how subjects use information to make actual participation decisions given that the overwhelming variable is a matter of predisposition and the primacy of emotional factors rather than any logical or rational processing. Even when rationality is available for our assessment and deconstruction, further reflection reveals a high level of distortion owing to psycho-pathological factors.[15] The aforementioned difficulties amalgamate to make informed consent, in many diverse contexts, meaningless, abstracted, or misguided.

> All of these factors, it can be argued, interfere sufficiently with the rational deliberation that is at the core of at least some autonomy-related conceptions of informed consent that the purpose of the doctrine is vitiated. Informed consent then comes to be considered a "myth" [according to Fellner and Marshall's 1970 study], which can only act as an impediment to the delivery of health care without producing any benefits of its own.[16]

If we squarely face the power imbalance between researchers and subjects, it is difficult to consider informed consent as anything more than an open-ended legal mechanism for checking and controlling unsuspected or untrammelled power.[17] This redressing of power imbalances should be both the aim and outcome of the informed consent doctrine, rather than the provision of a detailed contract of disclosure, which is the normal ideal supported by diversified legal doctrines that have come under its broad umbrella. If we see our ideal as the creation of power equilibria and regard the matter as one of process rather than one of conclusion, it is through a commitment to the idea that parties should be involved in an ongoing exchange that protections can be realized. This inevitably will raise questions of competing interests for different groups found within the system.

The largest problem which looms above and beyond informed consent doctrinalism is the fact that health services, now under severe constraints and revamping, leave increasingly less time for health care providers to develop meaningful dialogues with the consumers of health treatments or research

---

[13] C.H. Fellner & J.R. Marshall, "Kidney Donors - The Myth of Informed Consent" (1970) 9 American Journal of Psychiatry 79.

[14] See R.R. Faden & T.L. Beauchamp, "Decision-making and Informed Consent: A Study of the Impact of Disclosed Information" (1980) 7 Social Indicators Research 314; Fellner & Marshall, *ibid.*

[15] W.C. Thompson, "Psychological Issues in Informed Consent" in President's Commission for the Study of Ethical Problems in Medicine and Biomedical and Behavioral Research, *Making Health Care Decisions: The Ethical and Legal Implications of Informed Consent in the Patient-Practitioner Relationship*, vol. 3 (Washington, D.C.: U. S. Government Printing Office, 1982) 83.

[16] P. Appelbaum, "Informed Consent" in D. N. Weisstub, ed., *Law and Mental Health: International Perspectives, Volume 1* (New York: Pergamon Press, 1984) 45 at 78. The author cites C.H. Fellner & J.R. Marshall, *supra* note 13.

[17] R.A. Burt, *Taking Care of Strangers: The Rule of Law in Doctor-Patient Relations* (New York: The Free Press, 1979). See also J. Katz, *The Silent World of Doctor and Patient* (New York: The Free Press, 1984).

protocols.[18] We are forced to question whether the doctrine is an actual disincentive to the effective use of resources that can objectively be linked to positive outcomes for recipients.[19] The problem is seen therefore as structural, showing an "informed consent gap . . . [which] reflects the constraints imposed by human psychology, the physician-patient relationship, the tort law system, and an increasingly cost-conscious health care delivery system — and that these constraints are largely intractable".[20]

Certain arguments have been represented by Veatch to show that the law of informed consent in the treatment area is shamefully inadequate to cope with anything more than a trivial exchange of facts in relation to risk. As Veatch rightly indicates, there is much beyond the shadow of facts in consent decision-making. Of greater importance is the host of values that make up a person's constellation of beliefs concerning family life, religious upbringing and belief systems, and attitudes about fundamental values that pertain to the fabric of one's social and political environment. Until physicians understand their own inadequacy to articulate what is relevant to disclose for a meaningful response, in the largest sense of the word, choice will remain an empty abstraction. In fact, Veatch suggests that informed consent is based upon liberalism, namely, the respect for a high degree of autonomy, and is a myth in a vacuum without the support structure of liberal values. Veatch argues that over time, a more mature idea of decision-making about medical matters will emerge as patients choose institutions and medical practitioners based on a rooted understanding of how they as partners share a value-scheme reaching to the heart of critical, human conflicts and their resolution.[21]

An unsympathetic accounting of such a partnership or covenantal bond[22] would note that the inherent imbalance in power would place the stronger party in a paternalistic role over the subject. Such an accounting would fail to appreciate the significant attribute of the fiduciary and evolutionary nature of the covenantal concept. It is the obligation of the stronger party to make every effort at all points in the relationship to maximize the autonomous reach of the assertive power of the subject; failure to do so is a violation of the covenant. The subject is also required to push back the paternalistic arm to the maximum extent possible, while showing

---

[18] T.P. Duffy, "Agamemnon's Fate and the Medical Profession" (1987) 9 West New England Law Review 21.

[19] See Schuck, *supra* note 8.

[20] *Ibid.* at 905.

[21] R.M. Veatch, "Abandoning Informed Consent" (1995) 25:5 Hastings Center Report 5.

[22] The idea of covenant is not a stranger to either the medical profession or to the Western legal or moral tradition. The medical profession itself has grown up with the idea of covenant as part of the Hippocratic Oath *vis-à-vis* the physicians' obligations and indebtedness owed to his teacher and his progeny for the service of the knowledge given. See W.F. May, "Code and Covenant or Philanthropy and Contract?" in S.J. Reiser, A.J. Dyck & W.J. Curran, eds., *Ethics in Medicine: Historical Perspectives and Contemporary Concerns* (Cambridge, Mass.: The MIT Press, 1977) 65. See also L.R. Tancredi & D.N. Weisstub "Malpractice in American Psychiatry: Toward a Restructuring of Psychiatrist-Patient Relationship" in D.N. Weisstub, ed., *Law and Mental Health: International Perspectives, Volume 2* (Oxford: Pergamon Press, 1986) 83 at 90–94. See also the discussion on the concept of the covenantal bond in research in Chapter 1.

a realistic appreciation of the other person's mandate. Rather than seeing the relationship as one between parent and child, the covenantal archetype is closer to that of a parent preparing and dealing with the emancipatory movements of an adolescent approaching adulthood whose space increases until free choice is respected and acknowledged. In this way, the second party of the covenant can be held morally responsible for making meaningful choice possible.

It is interesting to explore the parameters of Veatch's observations about the limitations of informed consent as we know it within the reality of the research context. Because his analysis is even more compelling when surrogate decision-making is put into place, when one turns to research, often dealing with incapacitated or profoundly vulnerable parties, one has to ask how realistic it is to speak about choices and pairing of value systems between researchers and subjects. As an ideal type or model, Veatch's point of view is compelling. Classical defenders of the doctrine of informed consent will nonetheless continue to advocate the viability and relevance of its protection. But even they are forced to admit that short of thoroughly integrating informed consent into a meaningful process of exchange, reflected in the ethos of medicine, the law of informed consent is destined to remain nothing more than a fairy tale.[23] In actuality, as a safeguard against the tyranny of elitist decision-makers, the doctrine of informed consent, based on a complete disclosure of facts relating to risk in the area of research, should continue to be regarded as a protective device. Indeed, for many who take part in research protocols, to speak of an "articulated value-system" is a luxury which is not supported by fact.

In research, safeguards are critical, and the rationalizations made on behalf of others, however much motivated by the ideal of values in concert, are a luxury we are ill-advised to entertain before achieving in our society a threshold level of listing, exchanging, and clarifying the nature of the risks that subjects may endure. Of course, values come into play in research and are at times even more important than in normal cases of therapy. But realism about the history of violations of rights and abuses with regard to subjects should lead us to extreme caution in experimenting with the concept of individual, professional, and societal values working in a synchronized relationship. Rather, we should continue to emphasize the foreseeable, the need for protection through laying out the level of risk measured against the perceived benefit to be obtained from a positive research outcome in as forthright a manner as possible.

---

[23] See J. Katz, "Informed Consent — Must it Remain a Fairy Tale?" (1994) 10 Journal of Contemporary Health Law & Policy 69 at 90–91. However, he argues that we should allocate the requisite resources and generate the necessary commitment for the doctrine to fulfill its mandate effectively.

CHAPTER 5

# THE CONCEPT OF GOODNESS IN MEDICAL RESEARCH: AN ACTION-THEORETIC APPROACH

JAN HELGE SOLBAKK

## INTRODUCTION

In this paper an account of "medical goodness" will be set out; that is, an attempt
will be made at providing a study of the concept of goodness in its different varieties
within medical research. In designing such an account, i.e. an account where the
multiplicity of the uses of the word "good" constitutes the focus of attention, I shall
apply an action-theoretic approach. In this I am indebted to the Norwegian
philosopher and bioethicist Knut Erik Tranøy, who may be viewed as the modern
architect and instigator of an action-theoretic approach to science and ethics. This
view implies a change of focus from the *product* of scientific inquiry to the *process*
that produces scientific knowledge. Furthermore, it implies a view of science as a
"normative system", i.e. as a finite and ordered set of norms and values for groups
of people doing science. According to Tranøy's definition,[1]

scientific research is the systematic and socially organized
(1) search for,
(2) acquisition, and
(3) use or application of knowledge and insight brought forth by acts and activities
involved in (1) and (2).

Giving priority to such an approach implies a focus on the fundamental relations
that exist between science and ethics in medicine, between medical epistemology

---

[1] K.E. Tranøy, "Science and Ethics. Some of the Main Principles and Problems" in A.J.I. Jones, ed., K.E.
Tranøy, *The Moral Import of Science. Essays on Normative Theory, Scientific Activity and Wittengenstein*
(Bergen: Sigma Forlag, 1988) 111–1 14.

73

and medical ethics. Consequently, the answer to the question of what *goodness* in medical research is should primarily be sought in the field of *medical epistemology*, from which *ethical* considerations then may be derived. "*Medical epistemology*" is an account or theory about:

(1) *how* we gain knowledge in medicine,
(2) *how we know* that various judgements are true or false, and
(3) *what kind* of knowledge we have in medicine (i.e. the question about the *nature* of medical knowledge).

Although the quest for *theory* has been a shared interest in medical ethics and epistemology for more than 20 years, there have been few attempts at taking into consideration the question of *whether*, and eventually *how* and to *what extent*, medical epistemology and ethics relate to each other. That is, in spite of the admission that medical knowledge is a *necessary* condition for moral deliberations in medicine, there has been a tendency towards considering the question about the need for some larger theoretical framework in medical ethics as a question belonging to the pure realm of ethical theory; as well, many consider dealing with this question to be the exclusive responsibility of ethicists.

In medical epistemology the quest for theory has seldom reflected any awareness of — or *theoretical* interest in — ethical issues[2]. In his excellent "Introduction" to the book, *Moral Theory and Moral Judgements in Medical Ethics*,[3] Baruch Brody says that we must go beyond bioethical principles if we are to find the appropriate *epistemological foundations* for bioethics. I agree with Brody about this and about the need for some larger theoretical framework in medical ethics. However, Brody leaves the reader in bewilderment, because he does not give any indications about how *far beyond* these principles he would be willing to go — and in *which direction*.

In this paper I shall argue that what we need today is an action-theoretic framework capable of rendering an account of the *interrelation* between medical epistemology and medical ethics, and between methodological norms and ethical norms. Such a framework, I claim, would be more fruitful than searching for a more refined moral theory, both in order to solve the "specification problem" and "the theory-to-concrete-judgment-problem" which Brody defines as common problems for all existing moral theories in medical ethics.[4] Besides, I hold that such a theory would be able to handle a third problem: the problem of *identifying* or *assessing* whether an issue dealt with in medical ethics today actually is an ethical problem. My claim is that there are problems treated as "ethical" problems, which are not ethical problems in the ordinary and pure sense of the word "ethics", but problems related to medical epistemology — that is, to a conflict between different

---

[2] One important exception is S. Gorovitz & A. MacIntyre, "Toward a Theory of Medical Fallibility" (1976) 1 Journal of Medicine and Philosophy 55.
[3] B. Brody, "Introduction" in B. Brody, ed., *Moral Theory and Moral Judgments in Medical Ethics* (Boston: Kluwer Academic Publishers, 1988) 2.
[4] *Ibid.*

conceptions of medical science or between different interpretations of medical knowledge and empirical data. Consequently, these problems *cannot* be adequately solved within existing theoretical frameworks. Or to put it more bluntly: the solutions offered are at best pseudo-solutions.

Before we proceed to investigating the concept of goodness within the particular field of medical research, I think it may be useful — in fact a "good" idea — first to take a look at an interesting philosophical attempt at coming to terms with a *general* account of goodness.

## GOODNESS IN GENERAL — A VIEW FROM OXBRIDGE

In *The Varieties of Goodness*,[5] the Finnish analytical philosopher Georg Henrik von Wright advocates that "a useful preliminary to the study of the multiplicity of the uses of the word "good" is to compile a list of familiar uses and try to group them under some main headings". For reasons of simplicity I shall rephrase his proposal in a somewhat condensed way.

We are all, says von Wright, familiar with everyday expressions such as "a good knife", "a good horse", "a good hammer", "a good razor", "a good meal", "a good wine", "a good cigar", "a good road", "a good carpenter", "a good chess player", "a good politician", "a good scientist", "a good lover", "a good lung", "a good heart", "good eyes", "good sight", "good memory", "a good head", "a good man", "a good act", "a good friend", "good knowledge", "good research", "Good-day". We also tend to believe that we understand what is meant by these different expressions. And in fact, most of us do. The list above illustrates the *multiplicity* and *variety* of uses of the concept of good in our everyday language. That is, the fact that wines, lungs and friends may all be termed good, demonstrates how *different* good things can be.

If we try to group the different expressions under some main headings, we may let the first group contain all the uses of "good" that are of an *instrumental* character. That is, we speak of a good knife, a good hammer, a good razor, and other artifacts, which are used as means (i.e. instruments) for various purposes.[6]

A common characteristic of the members of the second group of "goods", according to von Wright, is that they are all *good at* something. That which the members of the second group are good at is some activity or art, for which a man or a woman may possess a natural talent, but in which the person in possession will also have to undergo some special training before he can excel in it.[7] For example, in spite of a possibly inborn talent, a medical scientist has to undergo some basic training before she can succeed as a *good* scientist. For the kind of excellence characteristic of this group, the name of *technical* goodness seems to be suitable.[8]

A third group of the uses of "good", according to von Wright, is that kind of goodness that is "an attribute of *organs* of the body and *faculties* of the mind: for

---

[5] G.H. von Wright, *The Varieties of Goodness* (Bristol: Routledge, 1993) at 8.
[6] *Ibid.* at 9.
[7] *Ibid.*
[8] *Ibid.*

example, when we speak of a good heart, of good eyes, good sight, good memory". To these uses of "good", von Wright attributes the name of *medical* goodness.

A fourth group or form of goodness is also worth mentioning. It is described by von Wright:[9]

> Medicine is *good for* the sick, exercise for the health, manure for the soil, lubrication for the engine, to have good institutions is good for a country, good habits for everybody. Generally speaking, something is said to be *good for* a being, when the doing or having or happening of this thing affects the *good of* that being favourably.

For this form of goodness he reserves the name of the *beneficial*, constituting a subcategory of the *useful* or of that which may also be called *utilitarian* goodness.[10]

To these four groups of goodness von Wright also adds a separate form labelled *hedonic* goodness. That is, we speak of a good smell or taste, a good meal, a good wine, a good cigar, a good holiday or time, good weather.[11] The pleasure of having a paper accepted for publication would probably also fit very well into this group.

A last form of uses of "good" proposed by von Wright refers to matters of conduct and character, i.e. uses of good related to the so-called *moral* life of men. Some of the examples mentioned above do not at all fit in among the growing uses of "good", or seem to fall somewhere between the groups having been identified, i.e. "a good friend", "a good act", "good knowledge", "good research", "Good day"!

In spite of the lack of completeness, this kind of mapping of different uses of "good" has a value beyond conveying "the semantic multiplicity" and "the logical wealth" of the phenomenon called the "varieties of goodness"; von Wright asserts that it demonstrates the inadequacy and artificiality of schematisms such as the traditional classification of all good into *two* main types, i.e. "good as a means and good as an end, instrumental and terminal, extrinsic and intrinsic good".[12]

In the attempt at coming to terms with the concept of medical goodness in its different varieties I think it may be useful not only to keep in mind von Wright's mapping of goodness in general but also his problematisation of the relation between a *moral* sense of "good" and non-moral uses of the word, as well as his claim that the so-called moral sense of "good" is a derivative, which must be explained in the terms of non-moral uses of the word.[13]

For the field of medical research this seems to represent a return to the Socratic-Platonic dictum that goodness is knowledge and knowledge is goodness. Before turning to an analysis of the concept of goodness in contemporary medical research, we shall therefore in a short anamnestic sketch try to clarify where Plato stands as regards the relation between knowledge (*episteme*) and goodness (*arete*), and then

---

[9] *Ibid.* at 9–10.
[10] *Ibid.* at 10.
[11] *Ibid.* at 11.
[12] *Ibid.* at 11–12.
[13] Von Wright's own proposal is an account of moral goodness (as an attribute of acts and intentions) which defines it in terms of the beneficial. For this, see von Wright, *ibid.* at 114–135

see how he applies this view within his own account of medical knowledge and virtue. From this, it will also become clear that Tranøy's action-theoretic attempt at theorizing the fundamental relations between science and ethics has some family resemblance to Plato's organic vision of knowledge and goodness.

## GOODNESS AND KNOWLEDGE — A PLATONIC VIEW

According to Gosling, Plato seems to have four puzzling things to say about the relation between goodness and knowledge.[14] First, he says that a person is in the possession of knowledge only if he can distinguish between good and bad; second, if one can talk of a good (or bad) X, then X is something of which we can have knowledge;[15] third, real knowledge is of *the* good; fourth, *the* good is responsible for knowledge.

Within Plato's account of medicine the relation between knowledge and goodness is spelled out in the following way:

(1) Medicine, he says, is a *technical* science — a *techne* to use the Greek notion — which has investigated the nature of the subject (subject matter) it treats, i.e. the body of the healthy and the diseased, and is capable of rendering account of its specific activity.

(2) The physician's technical *know-how* is a guarantee of *virtuous* action in medicine. That is, to possess medical knowledge means knowing how to distinguish between good and evil in medicine. In other words, medical knowledge is sufficient for virtue as well as for virtuous action in medicine.

(3) To possess medical knowledge, however, is not only a prerequisite of *moral virtue* in medicine, but also of vice: Because he knows *how* to heal the sick, he is also capable of working harm in medicine. That is, a doctor who works harm does not do so because he is overcome by pain or pleasure, since it is always *knowledge*, not something else, that governs him. Therefore, it remains with the one who possesses medical knowledge or virtue also to become a bad doctor.

From this short "anamnesis" of Plato's medical account of the relation between goodness and knowledge, we shall now move to Tranøy's action-theoretic definition of science and ethics.

## GOODNESS IN MEDICAL RESEARCH — AN ACTION-THEORETIC VIEW

### *Search for Knowledge and Insight.*

As to the first step of this definition, it is important to distinguish between at least "two kinds of systematic and socially organized *search* for truth": on the one hand,

---

[14] J.C.B. Gosling, *Plato* (London: Routledge & Kegan Paul,1973) 55ff.

[15] *Ibid.* at 57: "In other words, it is taken as read that if one can talk of good and bad states of affairs or conditions of things, then there is a techne devisable in accordance with which one could discriminate between the good and the bad".

we have the search in the science policy sense; on the other hand, the search in the sense of *research*:[16]

> For the latter kind of search — as it occurs for instance in the double blind and randomized, controlled clinical trial in medical research — can only be organized and implemented by qualified researchers. The policy kind of search presupposes a (far from "value-free") political decision-making capacity and responsibility in the first place, in addition to participation from professional researchers.

I agree with Tranøy regarding the importance of making such a distinction. I also agree with him that the two kinds of search for knowledge reflect a not-always very clear distinction between *external* and *internal* norms of science.[17] Still, I believe he is exaggerating the difference between the two kinds of search when he says that search of the design-of-projects kind can be very nearly value-neutral in relation to social and political values.

While the *internal* norms of science guiding the activities of qualified scientists and scholars comprise the *methodological* norms and values of systematic and socially organized inquiry[18] as well as of *freedom of inquiry*,[19] the *external* norms relate to personal and cultural values of many kinds. According to Tranøy external norms of inquiry are necessary for "science policy purposes, as well as for the application of knowledge to practical human concerns, and for educational purposes:"[20]

> Activities for which responsibility must be shared between professional inquirers ("insiders") and "outsiders" are above all the following two. First, search for knowledge in the science policy sense of search; and secondly the dissemination of knowledge and the application of it to a wide variety of human concerns and, again, educational concerns not least.

Besides internal and external norms of inquiry a normative system of science also contains a different set of norms, including requirements of *fruitfulness* and *relevance*. They represent normative points of contact — "linkage-norms" so to speak — between the research community and the community at large, between internal and external norms.[21]

### *Acquisition of Knowledge and Insight.*

In this second step of "systematic and socially organized *acquisition* of knowledge and insight", attention is given to the products of scientific behaviour, i.e. "to the

---

[16] K.E. Tranøy, "Ethical Problems of Scientific Research: An Action-Theoretic Approach" (1996) 79:2 The Monist 186.

[17] See Tranøy (1988) *supra* note 1 at 118–1 19.

[18] Among such norms we count for example "testability", "intersubjective", "controllability", "honesty", "sincerity", "exactitude", "completeness", "simplicity", "order", "coherence", "consistency" and "objectivity". For a systematic view of this collection of methodological norms and values, see K.E. Tranøy, "The Foundations of Cognitive Activity: An Historical and Systematic Sketch" in A.J.I. Jones, ed., K.E. Tranøy, *The Moral Import of Science. Essays on Normative Theory, Scientific Activity and Wittengenstein* (Bergen: Sigma Forlag, 1988) 121–1 36.

[19] See Tranøy (1988), *supra* note 1 at 118.

[20] *Ibid.* at 119.

[21] *Ibid.* at 119–1 20.

'findings' or discoveries resulting from the search". Thus, *acquisition* of knowledge Tranøy refers to the three basic cognitive acts of accepting, rejecting, and withholding judgment:[22]

> It is precisely when the search for knowledge is brought to an end — temporarily or definitively — that the acts of accepting, rejecting, or withholding judgment follow, and must follow even if what we have found is not truth in the strictest and literal sense.

From this it also follows that a search which did not ever terminate in a certain kind of "finding" would not only be of doubtful value; it would hardly deserve the label "scientific search". In other words, although findings may turn out to be *negative* in relation to what is expected at the start of the search, the research process should at least end up with something that can be labelled a "finding".

### Use or Application of Knowledge and Insight.

By *use* or *application* of knowledge and insight brought forth by acts and activities involved in (1) and (2), what is meant is primarily "the acts of stating or *asserting*" the "finding", including the verbal acts of lecturing, reporting, conversing, etc. and the written acts of publishing. The acts of asserting, denying and keeping silent constitute the "assert-family" of cognitive acts, and these acts, according to Tranøy, are structurally isomorphic to the "accept-family" of (2), i.e. "the cognitive acts of accepting, rejecting and withholding judgment".[23] Thus we have an action-theoretic model comprising two "cognitive families" and six different cognitive acts which can be forbidden, permitted or obligatory.

## GOODNESS IN MEDICAL RESEARCH — ACTION-THEORY APPLIED

The great advantage of Tranøy's action-theoretic model of science is that it makes it possible to distinguish between three different levels of "goodness" in scientific research. For the field of medicine we can therefore now ask whether there is knowledge which it is

(1) good to seek or good to forbid,
(2) good to accept, reject, or neither accept nor reject, or
(3) good to use and apply and to publish and communicate to others.

### Goodness in Search for Medical Knowledge and Insight

"The first and most evident reason for doing medical research is to generate possibilities of treatment of diseases for which there is no cure today or for which

---

[22] See Tranøy (1996), *supra* note 16 at 186.
[23] *Ibid.* at 187. For this, see also Tranøy, "Norms of Inquiry: Methodologies as Normative Systems" in G. Ryle, ed., *Contemporary Aspects of Philosophy* (London: Oriel Press, 1977) 1.

there are only unsatisfactory cures available".[24] Most people would probably agree that medical *research* of the kind that aims at generating *useful* medical interventions may legitimately be labelled "good" medical research. According to Lie we have few possibilities of identifying the kind of research that would satisfy this kind of *utility*-requirement. We do not possess the knowledge *how* to do it, nor do we know *who* would be best suited for such a task. That is, we do not know whether it would be better to leave such questions to the researchers themselves to solve or to consider them questions of the science policy kind of search. Consequently, it seems to be difficult to operate with *utility* as a criterion or principle of prioritizing within research. Moreover, there are good reasons to believe that utility represents a by-product of the attempt at reaching true scientific knowledge. In fact, says Jon Elster, utility is *essentially* a by-product of research, in the sense that research gives rise to less utility if it takes utility to be its aim.[25] But scepticism in relation to a general possibility of identifying useful research does not necessary rule out such a possibility in every case. As a rule, however, it will not be possible to identify useful research. Furthermore, one should also be sceptical in the face of claims about potential values of utility of research projects. According to Lie, this raises the difficult question about what criteria or principles could then serve as measures of goodness in medical research. It then seems that we have to rely on members of the "core family" of internal norms of science such as *originality*, *dissimilarity* and *"interestingness"* and of linkage-norms such as *relevance* and *fruitfulness*. From this may be drawn two implications for establishing a science policy definition of "good" medical search. The first is that priority should be given to researchers and research groups or communities that have the courage to propose something new and different. That is, medical research programs and projects distinguished by their originality and dissimilarity deserve the label of "good" medical *research*. A second and somewhat paradoxical implication seems to follow from hearkening to the linkage norms of relevance and fruitfulness: priority should be given to *basic* medical research. This implication can probably best be explained by recalling the important distinction made at the outset of this chapter between the *product* of scientific inquiry and the *process* of producing scientific knowledge. That is, the medical relevance and fruitfulness of basic research is not primarily linked to the product part of it, but to the *process* of scientific "individuation" that Ph.D students are subjected to while doing their laboratory work. Why is this experience of such relevance and importance? The answer is partly empirical, partly of a pragmatic nature. That is, we know that a very high percentage of Ph.D students in medicine work in the field of *basic* medical research. We also know that basic research on the average is less time-consuming than clinical research. Consequently, there is a higher *production* of doctoral dissertations in the field of basic research than in other areas of medical research. In addition, we know that less than 30% of

---

[24] R.K. Lie, *Generelle prinsipper som bør styre prioritering og ressursgordeling I biomedisinsk forskning (General principles of prioritization of biomedical research)* [unpublished, 1991].

[25] J. Elster, "Grunnforskning i humanistiske fag" (Basic Research in the Humanities) in *Yearbook of The Norwegian Academy of Science and Letters* (Oslo: Oslo University Press, 1981).

PhD students continue their postdoctoral careers as scientists.[26] Most of them end up working as clinical doctors treating patients. In this way clinical units get populated by "refugees" from research, i.e. doctors who are trained in reading scientific journals critically. For patients, this represents a kind of "brain supply" which may lead to better, safer, as well as more updated treatment possibilities, simply because they are taken care of by clinicians who know how to handle and implement new scientific knowledge that may improve medical practice.

Before leaving the level of search, it is also necessary to pay attention to the question of whether there is knowledge which it would be good to *forbid*. That is, are there examples today of morally acceptable prohibitions and restrictions on search for medical knowledge? One such restriction of crucial importance is that "drug trials which carry a modest to high risk of serious harm to research subjects should be regarded as forbidden forms of search for medical knowledge".[27] This restriction is formulated in the following way in the *Declaration of Helsinki*:[28]

> Biomedical research involving human subjects cannot legitimately be carried out unless the importance of the objective is in proportion to the inherent risk to the subject.

In the *Declaration* it is also stressed that concern for the interests of the research subject must always prevail over the interests of science and society.[29] Today this represents a prohibition with world-wide acceptance.

It is also worthwhile mentioning here that, in order to give the Universal Declaration of Human Rights of 1948 legal as well moral force, the United Nations General Assembly adopted in 1966 the International Covenant on Civil and Political Rights, of which Article 7 states that:

> No one shall be subjected to torture or to cruel, inhuman or degrading treatment or punishment. In particular, no one shall be subjected without his free consent to medical or scientific experimentation.

As to other forms of search which are banned in most countries, experimental therapy on human gametes represents an outstanding example. The principal argument for banning this kind of *research* makes reference to the transgenerational risks involved. In fact, very few non-consequentialist arguments are present in the literature about this issue. In the research field of xeno-transplantations and human cloning, in contrast, non-consequentialist arguments about human dignity, human integrity, etc. seem to dominate.

---

[26] The National Committee for Medical Research Ethics, *Etiske sider ved prioritering og ressursfordeling i medisinsk forskning* (The Ethics of Resource Priorities in Medical Research) (Oslo: Falch Hurtigtrykk, 1995).

[27] See Tranøy (1996), *supra* note 16 at 189.

[28] See Basic Principle 4 of the World Medical Association, *Declaration of Helsinki*. Adopted at the 18th World Medical Assembly in Helsinki in June 1964. Amended at the 29th World Medical Assembly in Tokyo in October 1975; at the 35th World Medical Assembly in Venice in October 1983; and at the 41st World Medical Assembly in Hong Kong in September 1989 [hereinafter *Declaration of Helsinki*].

[29] See Basic Principle 5 of the *Declaration of Helsinki*, *ibid*.

From the above-mentioned examples we shall now move to three "grey zone" areas of *research*-bans, i.e. to restrictions on the search for knowledge involving vulnerable groups. Medical research involving *prisoners* represents one such ban, and is held to be mandatory in most countries today. The underlying argument for barring prisoners from serving as subjects of medical research is that the consent of prisoners cannot be valid in that it may be "influenced by the hope of rewards and other expectations, such as earlier parole". Advocates of *allowing* prisoners to participate in research, however, argue that the incarcerated are "particularly suitable" research subjects "in that they are living in a standard physical and psychological environment; that unlike fully-employed or mobile populations they have time to participate in long-term experiments; and that they regard such participation as a relief from the tedium of prison life, evidence of their social worth, and a chance to earn a small income".[30]

Although none of the international declarations contain an *explicit* ban on involving prisoners as subjects of research, the counter arguments, however, seem to preclude an internationally agreed-upon recommendation on this issue.

It is worth mentioning, though, that the Commentary on Guideline 7 of CIOMS' *International Ethical Guidelines for Biomedical Research Involving Human Subjects*, contains a warning against *arbitrarily denying* prisoners with serious illness, or at risk of serious illness, access to investigational drugs, vaccines or other agents that show promise of therapeutic or preventive benefit".[31]

In the Commentary it is emphasized that Guideline 7 is not intended as an endorsement of involving prisoners as research subjects.

Another form of ban on search for knowledge relates to *pregnant or nursing (breastfeeding) women*. According to paragraph 11 of the CIOMS Guidelines: "Pregnant or nursing women should in no circumstances be subjects of non-clinical research unless the research carries no more than minimal risk to the fetus or nursing infant and the object of the research is to obtain new knowledge about pregnancy or lactation. As a general rule, pregnant or nursing women should not be subjects of any clinical trials except such trials as are designed to protect or advance the health of pregnant or nursing women or fetuses or nursing infants, and for which women who are not pregnant or nursing would not be suitable subjects".[32]

A last form of search-for-knowledge ban that involves vulnerable groups relates to fertilized eggs and human embryos. This is a ban which is implemented in some countries, for instance in Norway and in Germany, while other countries accept research on fertilized eggs and spare embryos up to the age of ten to fourteen days.

Among advocates of this kind of *research*, as well as among its adversaries, arguments of a non-consequentialist nature seem to dominate. Proponents of the

---

[30] For these arguments, see Council for International Organizations of Medical Sciences, in collaboration with the World Health Organization, *International Ethical Guidelines for Biomedical Research Involving Human Subjects* (Geneva: CIOMS, 1993) 24.

[31] *Ibid.* at 24.

[32] *Ibid.* at 33.

restrictive position could be named "partisans of a biological criterion"[33] because the person is conceived as coextensive with the organism — from conception until the last breath.[34] The Roman Catholic Church is the most outstanding advocate of this position: "From the moment of conception all human life should be absolutely respected".[35] Advocates of the permissive position, on the other hand, take a "postconceptionalist" stand. Here, human life, in the sense of personhood, does not occur at conception, but is conceived afterwards. ". . . [H]uman ontogeny begins with the body as a 'what' or, at most a 'who' in the sense that an animal is a who", says H. Tristram Engelhardt, Jr., "which then, because of the development or new properties, becomes a who in a personal sense".[36] According to some advocates of this position, a person's coming into being is identified as the moment of capacity for reflection or self-reflection.[37] From their different answers to the metaphysical question, "When does human life begin?", the advocates and adversaries of research on human embryos draw opposite conclusions while using the same conceptual structure and logic. That is, the conceptionalists consider the human embryo *persona* non grata in medical research so as to protect it from involvement in experimentation, while the postconceptionalists try to legitimize medical research on human embryos by arguing that they are not persons, which actually means considering their *personhood* non grata in medical research.

### Goodness in Acquisition of Medical Knowledge and Insight

To investigate the concept of goodness in *acquiring* medical knowledge and insight is to enter the very "heartland of scientific research",[38] i.e. the "laboratories" of *scientific* acceptance decisions, decisions of rejection and decisions of abstention (or suspension) from making scientific judgments. These are decisions and judgments guided by the internal or *methodological* norms and values of science and represent some of the core activities of researchers.

In our goodness investigation on this level we shall move forward by recalling the internal norms and values that guide scientific acquisition of knowledge and try to evaluate their relevance in acquiring medical knowledge and insight. It then seems natural to start with the norms of *truth* and *probability*, since they represent norms of constitutive function in science. Tranøy says the following about the constitutive function of truth acceptance in science:

> It is difficult to conceive of any kind of systematic cognitive inquiry in which the values of accepting true (well-grounded) propositions and of rejecting false (unsupported) ones

---

[33] J.H. Solbakk, "Why is the Human Embryo Considered Persona Non Grata in Medical Research Today?" (1990) 41–42 BioLaw 483 at 486.

[34] A.M. Fagot-Largeault & G.D. de Parseval, "Les Droits de l'Embryon (Foetus) Humaine, et la Notion de Personne Humaine Potentielle" (1987) 93:3 Revue de Metaphysique et de Morale 364.

[35] The Sacred Congregation for the Doctrine of the Faith, 1987.

[36] H. Tristram Engelhardt, Jr., "The Ontogeny of Abortion" in S. Gorovitz, ed., *Moral Problems in Medicine* (New Jersey: Prentice Hall, 1976) 324.

[37] For references to this literature, see Solbakk (1990), *supra* note 18 at 488.

[38] See Tranøy (1996), *supra* note 16 at 190.

(i.e. the "acceptability values") are not indispensable basic values; they are among the values which are constitutive of cognitive inquiry.[39]

Although new knowledge, new insight and new truths seem to be mandatory for ascribing the notion of *goodness* to research, it is, nonetheless, equally important to acknowledge that the "truth-commitment" is not, and never can be sufficient in science to achieve the justifications required.[40] Neither in *medical research* can truth be said to represent a self-sufficient value; it always has to be supplemented by internal norms and values such as *originality, dissimilarity* and *"interestingness"* as well as by linkage-norms such as *relevance* and *fruitfulness*.

Before we proceed to the next subset of internal norms of relevance for acquiring medical knowledge and insight, we must look into Tranøy's unveilment of an ambiguity related to the notion of truth, which as truth, "in the strict sense of the word" is seldom available in medical research. In the great majority of cases decisions of acceptance and rejection "rest on degrees or estimates of *probability*". Thus it only makes "sense to say that such decisions are more or less justified by reference to reasons — to evidence for and against $p$ — evidence that, to be sure, is in the possession only of qualified researchers in the field".[41] By this it also becomes clear that in acquisition of medical knowledge and insight, *acceptability* represents a fairly vast and broad area of decision-making, situated between the extreme poles of decisions of logical necessity and decisions of rejection.

Another subset of norms and values involved in the acquisition of medical knowledge and insight is made up of *simplicity, exactitude* and *completeness*, as well as norms such as *testability* and *intersubjective controllability*. That is, one should strive for acquiring scientific statements and propositions that comply with these norms.

To act in accordance with these norms and values, says Tranøy, also requires an attitude of *open-mindedness* towards critique and legitimate counterarguments, and willingness to change one's mind in the light of new and relevant information. How difficult it may be to live up to such requirements of intellectual openness and critique in a community of researchers, has been vividly stated by J. Ziman:[42]

> "Damn fool!", "Idiot!", "Nincompoop!". The writing of a review article so as to do justice to the truth whilst preserving the friendship of one's colleagues requires a good deal of tact and art, and most of us balk at the stiffer jumps. It is best simply to ignore folly completely than to attempt to expose it; and are we not all vulnerable in this respect?

A second form of openness of relevance in the social process of acquiring knowledge and insight is openness to colleagues — about ongoing ideas and projects of research. This represents a vulnerable ideal of goodness, says Tranøy, since the social structure of science favours "priority". To be the first one to come up with a good idea is praised as of almost higher value than the idea itself! This is

---

[39] See Tranøy (1988), *supra* note 18 at 134.
[40] *Ibid.* at 123 and 134.
[41] See Tranøy (1996), *supra* note 17 at 190–1 91.
[42] J. Ziman, *Public knowledge. The Social dimension of Science* (London: Cambridge University Press, 1968) at 122–1 23.

also the reason why stealing colleagues' ideas is considered such a fatal, though commonly practised sin in science.

A last subset of internal norms to be considered represents a kind of linkage between the level of acquisition and the level of *use* or *application* of knowledge and insight: *honesty* and the norm of making scientific statements *public*. We shall therefore now proceed to the *assert*-family of cognitive acts, and the question of goodness in use or application of medical knowledge and insight.

### Goodness in Use or Application of Medical Knowledge and Insight.

"There is neither a legal nor a moral right to state, publish or communicate whatever we know simply for the reason that we know it".[43] In science, the situation seems to be quite different. That is, says Tranøy, regardless of what field of research we are dealing with, scientific statements should not be concealed: they should be made *public*, so that their scientific validity can be tested and checked by *other scientists*. That is also the reason why a sponsor's wish to keep secret the results of a research project does not comply with the rules of good scientific conduct on the application-level of medical research. When pharmaceutical companies act as sponsors of research this kind of moral challenge often seems to be involved.

During recent years we have also learnt that lack of *"scientific honesty"*, often conceptualised as *"scientific misconduct"* or *"scientific fraud"*, is not as rare as we thought.[44] The following list is not exhaustive but illustrates the wide scope of dishonest ways of *communicating* scientific knowledge. In the list I have not included the forms of scientific misconduct that belong to the level of *acquisition* of scientific knowledge and insight;[45] only those forms of dishonesty that, strictly speaking, belong to the level of *use* and *application* of scientific knowledge and insight are included:[46]

— plagiarism of the results of entire articles of other researchers,
— distorted representation of the results of other researchers,
— wrongful or inappropriate attribution of authorship,
— covert duplicate publication and other exaggeration of the personal publication list,
— presentation of results to the public, thus bypassing a critical professional forum in the form of journals or scientific associations,
— omission of recognition of original observation made by other scientists,
— exclusion of persons from the group of authors despite their contribution to the paper in question.

---

[43] See Tranøy (1996), *supra* note 16 at 191.
[44] For these terms, see D. Andersen *et al.*, *Scientific Dishonesty and Good Scientific Practice* (Copenhagen: The Danish Medical Research Council, 1992) at 19.
[45] Scientific dishonesty on this level includes "deliberate fabrication of data", "selective and undisclosed rejection of undesired results", "substitutions with fictious data", "erroneous use of statistical methods with the aim of drawing other conclusions than those warranted by the available data" and "distorted interpretation of results or distortion of conclusions".
[46] For a fairly complete list, see Andersen *et al.*, *supra* note 44 at 20–21.

Although a detailed analysis of the entire list would be desirable, I shall limit my scope to the so-called "*sins of omission*".

"Sins of omission" was originally employed in relation to scientific dishonesty by I. Chalmers, in his article, "Underreporting Research Is Scientific Misconduct".[47] By using this phrase, Chalmers managed to draw attention to forms of dishonest communication of medical knowledge of a more *structural* nature:[48]

> Scientific misconduct is commonly conceptualized as deliberate falsification of data — a sin of commission — but sins of *omission* may be even more important. . . . [A] tendency exists among investigators, peer reviewers, and journal editors to allow the direction and statistical significance of research findings to influence their decisions regarding submission and publication, and that about one in two trials initially reported in summary form is never reported in sufficient detail to permit an informed judgement about the validity of its results. Both of these phenomena should be regarded as forms of scientific misconduct.

In a public lecture held in Oslo in November 1993, "Ways in Which Research Ethics Committees May Behave Unethically", Chalmers went further, accusing the biomedical research community of applying a kind of "double moral standard". The biomedical research community, he said, currently applies an illogical and dangerous double standard: it usually *demands* that scientific principles be *observed* when people conduct *primary research*, but it usually *acquiesces* when scientific principles are *ignored* by people conducting *secondary research (reviews)*.[49] According to Chalmers the research ethics committees could play an important role in preventing "sins of omission" of this kind; by requiring from those who propose new research to present *systematic, scientifically-defensible reviews of relevant previous research*, showing that the proposed new research is both necessary and appropriately designed. This highlights the responsibility of research ethics committees in promoting good use and application of medical knowledge and insight.

## CONCLUDING REMARKS

In this chapter I have argued that we are in need of a theoretical framework capable of tracing and identifying the fundamental relations that exist between epistemology and ethics in medicine, between medical knowledge and goodness, and between methodologic norms and ethical norms. An action-theoretic approach to science (knowledge) and ethics (goodness) has been deployed as capable of offering such a

---

[47] I. Chalmers, "Underreporting Research Is Scientific Misconduct" (1990) 263:10 Journal of the American Medical Association 1405–1 408.

[48] *Ibid.* at 1405.

[49] I. Chalmers, "Ways in which Research Ethics Committees may Behave Unethically" (1993) (Unpublished Manuscript). For this, see also J. Savulescu, I. Chalmers & J. Blunt, "Are Research Ethics Committees Behaving Unethically? Some Suggestions for Improving Performance and Accountability" (1996) 313 British Medical Journal 1390–1 393.

framework. By this, I think, it also became clear that what goodness consists of in medical research depends on what we mean by medical knowledge and medical science. However, what medical science turns out to be, is in the last resort a question of ethics.

CHAPTER 6

# DRAWING THE DISTINCTION BETWEEN THERAPEUTIC RESEARCH AND NON-THERAPEUTIC EXPERIMENTATION: CLEARING A WAY THROUGH THE DEFINITIONAL THICKET

SIMON N. VERDUN-JONES AND DAVID N. WEISSTUB

**INTRODUCTION: WHY IS THE DISTINCTION IMPORTANT?**

The literature dealing with the ethical and legal dimensions of biomedical experimentation is replete with references to the fundamental distinction between *therapeutic* and *non-therapeutic* procedures. However, the *practical* task of drawing this distinction in the context of particular research protocols involving human subjects has traditionally proved to be profoundly challenging. The thesis of this chapter is that much of the difficulty associated with this vital task may be alleviated by abandoning the situation in which a biomedical experiment is considered to be primarily therapeutic or non-therapeutic on the basis of the *subjective* intent of the researcher(s) involved. Instead, it is our contention that a *functional* definition of therapeutic and non-therapeutic experiments should be developed. This definition should be based on an *objective* assessment by an *independent* research ethics committee of the risks and benefits (if any) to the patients or subjects concerned. Such an approach should prove to be of great utility in addressing a broad range of legal and ethical issues that arise in the context of biomedical experimentation.

Why is it important to construct a working definition of therapeutic and non-therapeutic experiments? Undoubtedly, the task of definition must always be undertaken in the light of the practical consequences that may flow from engaging in this critical exercise. As Hoffmaster has suggested:

> How an issue is treated is a function of how that issue is posed. What considerations are deemed relevant, how much weight is attached to various considerations and how

88

resolutions are extracted from a welter of conflicting considerations depend crucially on how issues are conceived.[1]

The practical consequences of drawing a distinction between therapeutic and non-therapeutic experiments become very clear in the interlocking spheres of law and ethics.

### Legal Consequences

One of the most critical *legal consequences* of identifying a research protocol as being primarily non-therapeutic is that there is a significant legal barrier to the involvement of those research subjects whose capacity to give an informed consent to participation is in doubt. Such subjects generally belong to one or more of the following groups, whose members have historically been regarded as being particularly vulnerable in the context of non-therapeutic experimentation: children, the elderly, the mentally disordered, the developmentally disabled, and prisoners. Individual members of these groups are potentially in a precarious position because there may be serious questions about their capacity to give an informed consent to participation in experiments and there may also be grave doubts, particularly in the case of those individuals who are institutionalized, as to whether any decision to become involved in biomedical experimentation is truly *voluntary* on their part.

The legal barrier to the participation of incapable members of these vulnerable populations in non-therapeutic experimentation exists because, in Canada — apart from the Province of Quebec[2] — there is considerable doubt as to whether a valid consent can be given on their behalf by a third party. Whereas it is widely accepted that third-party consent may be given to participation in research that may be considered as *beneficial* to an incompetent person's health, the converse is true when the research is considered to be non-therapeutic in nature.[3] In the absence of a potential therapeutic benefit to the subject, "it cannot be said to be in the ward's best interests [to participate] and thus the guardian probably has no authority to consent to the procedure even if it involves no risk of harm".[4]

Another important *legal consequence* of making the determination that research is non-therapeutic in nature is the stringency of the requirements for obtaining an informed consent to participation. Indeed, in *Halushka v. University of Saskatch-*

---

[1] B. Hoffmaster, "The Medical Research Council's *New Guidelines on Research Involving Human Subjects*: too Much Law, too Little Ethics" (1990) 10 Health Law in Canada 146 at 149.
[2] See generally Chapter 10.
[3] See Chapter 8.
[4] See G. B. Robertson, *Mental Disability and the Law in Canada*, 2d. ed. (Toronto: Carswell, 1994) at 163–164. Robertson states that it is possible that a "guardian may be able to authorize non-therapeutic research, if one applies a substitute judgment test rather than a best interests one". However, this has not been confirmed in the case law. The suggestion of applying a substitute judgment test originated with Dickens. See B.M. Dickens, "Substitute Consent to Participation of Persons with Alzheimer's Disease in Medical Research: Legal Issues" in J. M. Berg, H. Karlinsky & F. H. Lowy, eds., *Alzheimer's Disease Research: Ethical and Legal Issues* (Toronto: Thomson Professional Publishing, 1991) 60.

*ewan*,[5] the Saskatchewan Court of Appeal held that "the subject of medical experimentation is entitled to a full and frank disclosure of all the facts, probabilities and opinions which a reasonable man might be expected to consider before giving his consent".[6] This is considered to be a more rigorous standard than that applied in the context of consent to medical treatment.[7] Indeed, in the more recent Quebec case, *Weiss v. Solomon*,[8] the Court suggested that the "highest possible" standard of disclosure will be enforced and that a research subject must be informed of even the most remote risks. In this connection, it should be noted that, while from a legal point of view there may be some place for the so-called "therapeutic privilege" (which permits a physician to withhold information that may be detrimental to a patient's health) in the context of therapeutic research, there is manifestly no justification for asserting such a privilege in the context of non-therapeutic research.

### Ethical Consequences

From an ethical point of view, it is vitally necessary to draw a distinction between therapeutic and non-therapeutic procedures in those situations in which it is sought to recruit members of a vulnerable population as subjects in biomedical experiments. If there is a question as to the competence of these individuals to give an informed consent, then, with a few exceptions, it is remarkably difficult to find a valid justification for involving them in biomedical procedures that will not benefit their health. When research is perceived as being possibly, or probably, beneficial to the subjects concerned (i.e. therapeutic research), then it may be justified on paternalistic grounds; "interference is warranted because it will promote the welfare of the subjects themselves".[9] However, if it is not possible to identify a potential benefit for research subjects who cannot give informed consent, then some other justification must be found for their participation in biomedical experiments. For example, it may be possible to justify the involvement of incapable persons drawn from vulnerable populations where the non-therapeutic experimentation concerned

---

[5] (1965), 52 W. W. R. 608 (Sask. C.A. ). See also E.I. Picard, *Legal Liability of Doctors and Hospitals in Canada*, 2nd ed. (Toronto: Carswell, 1984) at 118–119; L.E Rozovsky & F. A. Rozovsky, *The Canadian Law of Consent to Treatment* (Toronto: Butterworths, 1990) at 80; G. Sharpe, *The Law & Medicine in Canada*, 2nd ed. (Toronto: Butterworths, 1987) at 79–85. See also *Cryderman* v. *Ringrose*, [1977] 3 W. W. R. 109 (Alta. S. C. ), aff'd [1978] 3 W. W. R. 608 (Alta. C. A. ); *Morrow* v. *Royal Victoria Hospital* [1990] R. R. A. 41 (Que. C. A. ).

[6] See *Halushka, ibid.* at 615–616.

[7] See *Rozovsky & Rozovsky, supra* note 7 at 80; Sharpe, *supra* note 7 at 80.

[8] [1989] R. J. Q. 731 (S. C. ) [hereinafter Weiss]. See generally B. Freedman & K. C. Glass, "*Weiss v. Solomon*: A Case Study in Institutional Responsibility for Clinical Research" (1990) 18 Law, Medicine & Health Care 395.

[9] R. Macklin & S. Sherwin, "Experimenting on Human Subjects: Philosophical Perspectives" (1975) 25 Case Western Law Review 434 at 451. See also Hoffmaster, *supra* note 1 at 148. Furthermore, in such circumstances, it is generally accepted that a third party may give consent on behalf of the incapable person. Such substitute consent will primarily be based on what is perceived to be in the subject's "best-interests" and may clearly be given in relation to participation in therapeutic research since it is designed to improve the subject's state of health.

can only be conducted using such subjects; this situation would arise where the research project is concerned with the very condition that affects these subjects. In these circumstances, the benefits of the research will accrue to others who are in the same category as they are (and there is even a possibility that the subjects themselves may ultimately benefit from any advances in medical knowledge that may result).[10] Another justification would be in the case of a prior directive consenting to participate in a non-therapeutic experiment, which was prepared at a time when the author of the directive was mentally (and legally) competent to do so.[11]

It is axiomatic that we need to conduct continuous and well-designed biomedical experimentation with human subjects;[12] if we do not, we clearly risk losing the potential benefits of new remedies, or "poisoning ourselves" either with "insufficiently tested new remedies" or with "accepted but unsound old remedies".[13] However, any society needs to pay very close attention to the process by means of which human subjects are recruited as volunteers for biomedical experiments. As the oft-cited *Belmont Report* points out, the fundamental ethical principle of justice requires that there be fairness in the way that we select human subjects for participation in biomedical research procedures. In particular, "social justice" requires that a distinction be drawn between groups of potential subjects who ought, or ought not, to be involved in such types of research.[14]

## DEFINING BASIC CONCEPTS

There is no doubt that the process of designating a biomedical experiment as non-therapeutic has significant legal and ethical consequences. However, a formidable problem faces those decision-makers whose task it is to draw the distinction between therapeutic and non-therapeutic procedures in the context of the application of legal and/or ethical principles. While the distinction between therapy and non-

---

[10] See Chapter 18.

[11] See Chapter 11.

[12] See Law Reform Commission of Canada, *Working Paper No. 61: Biomedical Experimentation Involving Human Subjects* (Ottawa: Law Reform Commission of Canada, 1989) at 1 [hereinafter *LRC Working Paper No. 61*]. See also H. Helmchen, "Ethical and Practical Problems in Therapeutic Research in Psychiatry" (1982) 23 Comprehensive Psychiatry 505 at 505, where it is asserted that the need for human subjects is particularly necessary in psychiatry because "adequate disease models of most human specific psychiatric diseases do not exist".

[13] C. Fried, *Medical Experimentation: Personal Integrity and Social Policy* (Amsterdam: North-Holland Publishing, 1974) at 4.

[14] National Commission for the Protection of Human Subjects of Biomedical and Behavioral Research, *The Belmont Report: Ethical Principles for the Protection of Human Subjects in Research* (Washington, D. C.: U.S. Government Printing Office, 1978) at 7 [hereinafter The Belmont Report].

therapy is very clear at the conceptual level, it is very difficult to draw in practice, because many research projects contain elements of both therapy and non-therapy. This makes it necessary to institute a process to ensure that any research proposal designated as "therapeutic" does, *from an objective point of view*, offer the research subject a reasonable degree of likelihood that the procedures employed will prove beneficial to his or her health. Furthermore, such a process must safeguard the research subject from the possibility that the proposal will require that he or she be submitted to an unacceptable number of incidental procedures designed to serve the interests of the researcher rather than the health needs of the subjects. Achieving this goal involves the most difficult of balancing acts and certainly helps to explain why the distinction between therapy and non-therapy has proved to be so intractable in both literature and practice. The starting point for the development of such a process must be the clarification of basic concepts with a view to establishing *objective* measures for the designation of a particular biomedical procedure as being therapeutic or non-therapeutic: clearly, classification of a protocol as being either therapeutic or non-therapeutic cannot rest on the *subjective* intent of the researcher.

### Research and Experimentation

*Research* is frequently used interchangeably with *experimentation*. However, the precise meaning and application of these two terms in the literature has been neither clear nor consistent. In some cases, *research* is used as a much more general concept in which experimentation, and human experimentation in particular, is considered to be but one method of carrying out research.[15] *Research* may be modified by the term *biomedical* to indicate that the research is related to health.[16] In turn, biomedical research has been defined as including not only "studies designed primarily to increase the scientific base of information about normal or abnormal physiology and development" but also "studies primarily intended to evaluate the safety, effectiveness or usefulness of a medical product, procedure, or intervention".[17] For the purposes of the analysis in this chapter, both terms, incorporating the notions discussed below, will continue to be used.

On a broad level, research and experimentation have been distinguished from other medical procedures and health care on the basis of the element of *uncertainty*

---

[15] A. Brett & M. A. Grodin, "Ethical Aspects of Human Experimentation in Health Services Research" (1991) 265 Journal of the American Medical Association 1854 at 1855.

[16] Council for International Organizations of Medical Sciences, in collaboration with the World Health Organization, *International Ethical Guidelines for Biomedical Research Involving Human Subjects* (Geneva: CIOMS, 1993) at 11 [hereinafter CIOMS *Guidelines*].

[17] Office for Protection from Research Risks, National Institutes of Health, *Protecting Human Subjects: Institutional Review Board Guidebook* (Washington, D.C.: U.S. Government Printing Office, 1993) at 5–1.

they imply.[18] Undoubtedly, both research and experimentation involve a departure from *standard medical practice*, which, according to the *Belmont Report*, refers to "interventions that are designed solely to enhance the well-being of an individual patient or client and that have a reasonable chance of success".[19] Whereas the purpose of medical practice, is to "provide diagnosis, preventive treatment or therapy to particular individuals",[20] research has been broadly defined by the Law Reform Commission of Canada as "a scientific activity directed to the advancement and systemization of knowledge, which seeks to benefit society as a whole".[21]

In essence, therefore, *research* involves the generation of data that goes beyond what is necessary for the patient's immediate well-being[22] and consists of "systematic investigation, including research development, testing and evaluation, designed to develop or contribute to generalizable knowledge".[23] *Experimentation* has been specifically defined by the Law Reform Commission of Canada as "the attempt to increase human knowledge through the systematic use of experiments". It is a technique or a process that "makes it possible to verify certain facts by creating favourable conditions for their realization".[24]

Cowan has usefully combined the various elements that characterize the nature of research activity into the following model:

> (1) a departure from standard medical practice, (2) having untested or unproved efficacy or no therapeutic intent, (3) designed to test a hypothesis, and/or (4) an intent to develop new knowledge. An additional characteristic of research and one that distinguishes it from practice is the use of the human as a subject rather than as a patient.[25]

Cowan's model clearly emphasizes the importance of identifying the *purpose* of an experiment that is, whether the objective is to treat a patient, or to study the effects

---

[18] E. Kluge, *Biomedical Ethics in a Canadian Context* (Scarborough: Prentice-Hall Canada, 1992) at 163. In a certain sense, even routine medical treatment can be considered a potential step into the unknown and, therefore, *experimental* since the response of any given patient can never be predicted with total certainty. See B.M. Dickens, "What Is a Medical Experiment?" (1975) 113 Canadian Medical Association Journal 635 at 635; J.-L. Baudouin, "L'expérimentation sur les humains: un conflit de valeurs" (1981) 26 McGill Law Journal 809 at 811; Fried (1974), *supra* note 13 at 25. Indeed, Katz has stated that ". . . since vast uncertainties and ignorance about effectiveness and risk-benefits are ubiquitous in the practice of medicine, every medical intervention, therapeutic or investigative in intent, constitutes an experiment". See J. Katz, "Human Experimentation and Human Rights" (1993) 38 Saint Louis University Law Journal 7 at 12. Helmchen has noted that psychiatric treatment, in particular, may be considered experimental in the sense that there is frequently uncertainty as to the best way to proceed in any given case. See Helmchen, *supra* note 12 at 507.

[19] See *The Belmont Report*, *supra* note 14 at 3.

[20] *Ibid.* at 3.

[21] See LRC *Working Paper No. 61*, *supra* note 12 at 4.

[22] See Medical Research Council of Canada, *Guidelines on Research Involving Human Subjects* (Ottawa: Supply & Services Canada, 1987) at 7 [hereinafter *MRC Guidelines*].

[23] 45 C. F. R. 46 (1991) §46. 102 (d).

[24] See LRC *Working Paper No. 61*, *supra* note 12 at 3.

[25] D. H. Cowan, "Innovative Therapy Versus Experimentation" (1985) 21 Tort & Insurance Law Journal 619 at 622–623.

of a procedure on a subject. Where a physician administers a medical procedure with an unknown outcome, but does so with the specific intent that it have an impact on the illness afflicting the particular patient, the experiment might be described as *therapeutic*. On the other hand, if a particular procedure is carried out only with a view to promoting and advancing scientific knowledge, the experiment might be classified as *non-therapeutic*.[26] The determination of whether an experiment is *therapeutic* or *non-therapeutic* has often been based on the stated intention of the researcher. Regrettably, such a system can lead to unfortunate results in cases where the integrity or judgment of the researcher is deficient. Furthermore, experiments often consist of both therapeutic and non-therapeutic aspects.[27] For these reasons, it is our contention that an *objective* determination of the purpose of an experiment should be made by a research ethics committee and that this process should be centered around a *functional*, rather than absolute, distinction between therapeutic and non-therapeutic experimentation.

Our approach is, of course, based on the premise that it is feasible to make an objective classification of the primary purpose of an experiment as being either therapeutic or non-therapeutic and that this task may be accomplished without placing reliance on the stated intent of the researcher(s) involved. It is important, however, to acknowledge the presence of *uncertainty* in the process of scientific inquiry and the need to take it into account whenever an attempt is made to apply a functional definition of experimentation or research. Viewed in this light, if an intervention presents a useful therapeutic benefit to a person and falls within the *generally accepted boundaries of medical certainty*, that intervention constitutes a *treatment*. However, where there is a *departure from standard medical practice*, raising an element of uncertainty which exceeds the level normally accepted in

---

[26] See *LRC Working Paper No. 61, supra* note 12 at 4; Fried (1974), *supra* note 13 at 25. Some commentators have taken experimentation to apply only in the non-therapeutic sense, limiting its application to those experiments which are of no personal benefit to the subject and are undertaken for the sole purpose of gaining scientific knowledge. See A. Schafer, "The Ethics of the Randomized Clinical Trial" (1982) 307 New England Journal of Medicine 719. See also *LRC Working Paper No. 61, supra* note 12 at 5. The Commission states that the term *experimentation* could be considered as being both a part of research and a part of therapy; however, it contends that the term should be used only in the *non-therapeutic* sense. Some analysts have limited the use of the term even more narrowly by defining it in such a way as to include only those procedures in which the subject is not a patient but, rather, a healthy volunteer who cannot expect to derive any personal benefit from it. See J.-L. Baudouin, "Biomedical Experimentation on the Mentally Handicapped: Ethical and Legal Dilemmas" (1990) 9 Medicine & the Law 1052. Other commentators, however, take a totally different tack, arguing that it is essential to recognize that experimentation in the non-therapeutic sense may be carried out on patients as well as on healthy volunteers. Patients constitute a pool of potential subjects and it is important that there be no suggestion that experimentation on them may be disguised because of the existence of the ongoing therapeutic relationship with their health-care providers. See M.A. Somerville, "Therapeutic and Non-Therapeutic Medical Procedures — What are the Distinctions?" (1981) 2 Health Law in Canada 85 at 87–88. For the sake of consistency, the term will be used to denote non-therapeutic experimentation that may be undertaken in relation not only to healthy volunteers, but also subjects who may have a separate, and entirely distinct, therapeutic relationship with the researcher.

[27] See Fried (1974), *ibid.* at 25. See also Article 21 of the *Civil Code of Quebec* which refers to "experimentation" that may be "carried out on one person alone" if "a benefit to the health of that person may be expected". Article 21, therefore, provides that experimentation can be therapeutic.

clinical decisions, the intervention becomes an *experiment* or *innovative therapy*.[28] Whether that experiment is subsequently characterized as primarily therapeutic or non-therapeutic does not alter the fact that the intervention constitutes an uncertain procedure, the goal and/or result of which is to further the process of scientific inquiry.

### Innovative Therapy

*Innovative therapy* is a term frequently used in the literature and is generally applied to *those medical procedures administered for the benefit of a specific patient but that have uncertain outcomes*. Of course, an innovative therapy may eventually become an accepted treatment and a physician may then be obliged to bring it to the attention of a patient who may benefit from it.[29]

One of the more comprehensive discussions of innovative therapy focuses on the fact that, unlike research, *it is not designed to test a hypothesis or to develop new knowledge*:

> Innovative therapies generally represent uncontrolled, often single, interventions intended to manage or solve particular clinical problems. They are not ordinarily designed to test hypotheses. Additionally, they are not undertaken in order to gain new knowledge beyond the needs of the patient. Although the use of innovative therapies may lead to the development of new knowledge, this consequence is secondary to their primary purpose of benefiting patients.[30]

Other definitions have tended to underscore the critical element of uncertainty of outcome. For example, the Law Reform Commission of Canada defines *innovative therapy* as "a treatment in the true sense of the word, an act performed for the direct and immediate benefit of the recipient, but not fully proved in scientific terms".[31]

It is not immediately clear, however, whether *innovative therapy* refers only to specific types of medical procedure that are already in use in other therapeutic contexts, or whether it is synonymous with new therapeutic procedures in general. For example, Kluge states that the term innovative therapy applies specifically to "procedures or modalities that are already known from another area but which are

---

[28] Where special populations are concerned, the determination of whether an intervention constitutes an *experiment* or *innovative therapy* must be made through an external ethical review. Kluge notes that, "to our mind, what distinguishes research and experimentation from all other forms of medical treatment and health care is the element of *uncertainty* that surrounds the nature of the activity itself and the intent to remove that uncertainty". It is true that a degree of uncertainty is also present in everyday clinical practice; however, it is "when this element of uncertainty exceeds the norms of usual practice", that "the situation becomes one of research and experimentation". See Kluge, *supra* note 18 at 162–163.

[29] See H. L. Prillaman, "A Physician's Duty to Inform of Newly Developed Therapy" (1990) 6 Journal of Contemporary Health Law & Policy 43.

[30] See Cowan (1985), *supra* note 25 at 623.

[31] See *LRC Working Paper No. 61*, *supra* note 12 at 4–5. See also P. H. Osborne, "Informed Consent" in B. Sneiderman, J. Irvin & P. H. Osborne, eds., *Canadian Medical Law: An Introduction for Physicians and Other Health Care Professionals* (Toronto: Carswell, 1989) 49 at 62. For case-law definitions of innovative therapy in the common law jurisdictions of Canada, see *Zimmer v. Ringrose* (1981), 124 D. L. R. (3d) 215 (Alta. C. A.); *Coughlin v. Kuntz* (1987), 17 B. C. L. R. (2d) 365 (S. C.).

adapted to a new context without there being as much certainty as there is in their standard use".[32] Baudouin, on the other hand, asserts that the term should apply more generally to any procedure that is "given to the patient both in the hope that it might have a curative effect on the disease and with the intent to determine whether or not it is really effective", provided that "the main intent is a therapeutic one".[33]

However, it has been pointed out that the discretion of physicians to employ innovative therapies is subject to certain restrictions. Dickens stated that such techniques or procedures may only be employed when "no orthodox treatment exists for the patient's condition". If an orthodox therapy is available, and the physician administers a new treatment to see if it will prove to be more successful, then this is *experimentation* rather than *innovative therapy*.[34] This is a vital distinction because, while proposed experiments must obtain prior approval from a research ethics committee, this is not normally the case where innovative therapies are concerned. Indeed, as Helmchen notes, innovative therapies are often beyond the control of ethics committees and may be "introduced or accepted before being tested because of the suggestive power of personal experience, ideology or a persuasive, warmhearted and engaged initiator".[35]

Dickens has, therefore, emphasized that the process of distinguishing between innovative therapy and experimentation should not be undertaken by trying to ascertain the physician's "primary motive" in administering the new treatment. Rather, the criterion should be objective and the appropriate approach is to search for "any signs of an investigation motive". Even minor experimentation should be exposed to prior ethical review and should not be permitted to escape such scrutiny by being "concealed within the interstices of orthodox therapy".[36] Indeed, as McNeill has demonstrated, there are certain cases where the courts may actually go so far as to ignore the physician's declaration of therapeutic intent. These cases arise where a proposed medical procedure is very risky and the prospect of the subject benefiting from it is very slight; in such circumstances, a court is very likely to find that the procedure is really an experiment — even though the physician may intend to benefit the patient.[37]

The *Civil Code of Quebec* recognizes the need to ensure that the line between experimentation and innovative therapy is not drawn by the physician. Article 21, which applies to minors and incompetent adults, states that "[c]are considered by the ethics committee of the hospital concerned to be innovative care required by the

---

[32] See Kluge, *supra* note 18 at 164.

[33] See Baudouin (1990), *supra* note 26.

[34] See Dickens (1975), *supra* note 18 at 636.

[35] See Helmchen, *supra* note 12 at 509.

[36] See Dickens (1975), *supra* note 18 at 636. *The Belmont Report* also took the view that "radically new procedures" should be made the object of a formal research proposal so that they can be scrutinized with a view to determining whether the procedures are "safe and effective". See *The Belmont Report, supra* note 14 at 3.

[37] P. M. McNeill, *The Ethics and Politics of Human Experimentation* (Cambridge: Cambridge University Press, 1993) at 127

state of health of the person submitted to it is not an experiment".[38] Experimentation for the benefit of the health of an individual minor or incompetent person must receive the approval of a court. On the other hand, if the proposed procedure is more properly defined as "innovative care", then it is sufficient that the physician obtain the consent of the authorized substitute decision-maker.[39] However, it is significant that the *Civil Code* stipulates that an *ethics committee* must make the determination that the proposed care is indeed "innovative care" as opposed to an experiment.

It is our view that an *innovative therapy* should be defined as either a novel clinical application of existing medical knowledge or as an unusual combination of normal forms of treatment, which, in either case, is appropriate to the particular needs of a patient and motivated by the specific intention to treat the patient. It is important that physicians exhaust all orthodox treatments before attempting an innovative therapy. Normally, proper disclosure of the nature of the novel procedure, as well as the availability of other viable alternatives would be sufficient for a competent adult. However, members of special populations may not understand the significance of an innovative therapy, nor may they be able to voluntarily or competently consent to submit to such a procedure. Therefore, with respect to the members of special populations, further restrictions are needed to ensure that such persons are protected. It is our contention that all proposals to administer *innovative therapy* to children or to adults whose capacity to consent may be in question should, except in cases of genuine emergency, be submitted for scrutiny to a research ethics committee. The procedure should be treated as an *experiment* if the research ethics committee concludes that any of the following circumstances exist:

— the proposed intervention does not have a realistic chance of benefiting the patient's health;
— an orthodox treatment for the patient's condition has not yet been attempted;
— the treating physician's motivation is investigational (this may be inferred from the nature of the procedures that are to be carried out over and above the proposed "innovative therapy" and that may, therefore, be more suggestive of an *experiment* than a *treatment*); or
— the risk is more than negligible.

## THERAPEUTIC AND NON-THERAPEUTIC RESEARCH

The manner in which the terms *therapeutic* and *non-therapeutic* are defined will have a profound effect upon the nature and effectiveness of regulations governing biomedical experimentation with special populations. However, the formulation of practical definitions in this context constitutes a task that is by no means as simple as it appears to be on the surface.

---

[38] Art. 21 C. C. Q.
[39] Arts. 14–15 C. C. Q.

At first blush, the distinction between *therapeutic* and *non-therapeutic* research or experimentation seems to be fairly obvious. The conventional view is that a *therapeutic* procedure can be broadly defined as a procedure that has as its object, both from the perspective of the patient and the person applying the procedure, the well-being of the particular patient, and it is only with this object that the procedure is undertaken.[40] By way of contrast, a *non-therapeutic* procedure is one where the primary objective of the investigator is the furtherance of scientific knowledge in general. Putting it simply, "experimentation is clearly non-therapeutic when it is carried out on a person solely to obtain information of use to others, and in no way to treat some illness that the experimental subject may have".[41]

Unfortunately, the conventional theoretical approach of identifying an experiment as either *therapeutic* or *non-therapeutic* is seriously deficient insofar as it might suggest that the two categories are mutually exclusive, when this is manifestly not the case. Indeed, many research protocols that involve procedures that may be deemed *therapeutic* also contain elements that may be categorized as non-*therapeutic* when one applies a strict view of the specific health needs of the individual subjects concerned. While certain aspects of a research project may be designed to meet the health needs of the individual patient-subjects involved, other aspects are designed solely to advance scientific knowledge. To this end, certain tests will be administered that extend beyond the immediate health needs of the subject-patients and may even expose them to additional risk, inconvenience, or discomfort that are not justified by any corresponding benefit.[42]

The observation that most research falls on a spectrum between "pure therapy", at one end, and "pure research", at the other,[43] emphasizes the artificiality of an antonymous distinction. Research will perforce contain both therapeutic and non-therapeutic elements. In light of such considerations, a number of bodies have recommended that the term *therapeutic research* be dropped from the medical

---

[40] Dickens points out, for example, that patients submit to treatment, in the *therapeutic* context, when they intend the medical intervention concerned to return them to the condition that is considered "healthy" for a person of their age or, alternatively, to alleviate the effects of a pathological condition so that they may live with it as best as they can. See Dickens (1975), *supra* note 18 at 635.

[41] Fried, *supra* note 13 at 25. Howard-Jones underscores the importance of protecting the members of vulnerable populations from inappropriate involvement in non-therapeutic experimentation by casting the essence of the therapeutic-versus-non-therapeutic distinction in somewhat dramatic terms:

> The primary purpose of the therapeutic experiment is to test whether a new treatment favourably influences the course of a disease. The only purpose of the non-therapeutic experiment is to test whether a new treatment produces disease in a previously healthy subject. See N. Howard-Jones, "Human Experimentation in Historical and Ethical Perspectives" (1982) 16 Social Science & Medicine 1429 at 1438.

[42] See Schafer, *supra* note 26; Fried, *ibid* at 26.

> It has been suggested that non-therapeutic experimentation may contain elements that are therapeutic for the subjects who participate. For example, such subjects may become the focus of medical attention to which they may not normally have access: see, for example, Fried, *ibid*. However, this "side effect" of participation would not justify the recruitment of members of vulnerable populations and, indeed, may raise serious questions about the level and quality of medical care that these individuals are receiving in the normal course of events.

[43] See Somerville, *supra* note 26 at 88–89.

lexicon.[44] Indeed, the Law Reform Commission of Canada has suggested that the use of the term may be "dangerous" insofar as it may result in a protocol not being subjected to the more stringent requirements concerning consent and disclosure of risk that apply to research as opposed to standard medical treatment:

In this perspective, to describe an act as therapeutic research is inaccurate and dangerous. It is inaccurate because the word "therapeutic" is then understood in a broad, derivative sense, i.e., the research may lead to treatment, but its *primary goal* is the advancement of knowledge, not the cure or relief of the experimental subject. Dangerous because the investigator's primary role is as a researcher and not as a therapist. Using the two words together may give a false impression of the real situation and lead the subject to consent to an experiment in the mistaken belief that it will be of immediate personal benefit.[45]

In more recent years, there has been a tendency in the literature to use the term *clinical research* instead of *therapeutic research*. Use of this term clearly reflects the reality that such research involves procedures that are both therapeutic and non-therapeutic for the individual patient/subject. As the Council for International Organizations of Medical Sciences points out, in all clinical research, there will be elements of the protocol that are not designed to be diagnostic, prophylactic, or therapeutic for the individual patient-subject (e.g. the administration of placebos and the performance of tests that go beyond the strict requirements of medical care).[46]

Unfortunately, merely changing terminology does not solve any of the practical problems that are faced by those involved in the task of regulating biomedical experimentation. Rather, it is the manner in which such terms are applied, and the rationale underlying their application, that deserve our attention. Upon declaring a procedure to be an experiment rather than an innovative therapy, a research ethics committee should declare the purpose of the protocol to be either *primarily therapeutic* or *primarily non-therapeutic*. These terms should be viewed as labels which acknowledge the artificiality of the puristic therapy-versus-non-therapy distinction and take into consideration the notion that experimental protocols may

---

[44] See *The Belmont Report, supra* note 14 at 2–4; *MRC Guidelines, supra* note 22 at 9.

[45] See *LRC Working Paper No. 61, supra* note 12 at 5.
"George Annas has referred to the juxtaposition of the terms, "therapy" and "research" as a form of "doublespeak" that is chillingly reminiscent of George Orwell's book, 1984:

> . . . even a cursory history of modern human experimentation demonstrates the pervasiveness of three doublespeak concepts: experimentation is treatment, researchers are physicians, and subjects are patients. Indeed, we have encapsulated all three into a "newspeak" word, "therapeuticresearch" (although we retain a space between the c and the r). This doublespeak allows us to use double standards as they suit our purposes. It permits us to treat truth as negotiable and then allows us to act irrationally. We act in the best interests of patients. The experiment is justified as therapy or potential therapy. But if the experiment produces harm, it was after all, only an experiment and thus nonetheless a "success" because we learned something from it that could benefit others . . .

See G. J. Annas, "Questing for Grails: Duplicity, Betrayal and Self-Deception in Postmodern Medical Research" (1996) Journal of Contemporary Health Law and Policy 297 at 300–301.

[46] See *CIOMS Guidelines, supra* note 16 at 11.

include both therapeutic and non-therapeutic procedures.[47] Any classification scheme must indicate which end of the spectrum between "pure treatment" and "pure research" best approximates a given experiment, and thus indicate which regulatory regime (with corresponding legal and ethical standards) applies.

A research ethics committee can assess a protocol submitted for classification as either therapeutic or non-therapeutic from the perspective of the researcher, the patient/subject, or an objective third party. The viewpoint of the researcher, used to determine the primary purpose of an experiment, can only be measured by that individual's stated intent: to provide a treatment or to further scientific knowledge. As will be discussed below, the perspective of the researcher may be compromised owing to the potential conflict of interest in certain situations. The goal of the researcher should become apparent subsequent to an external review of the experiment, which must include an objective assessment of its therapeutic value — independent of the researcher's opinion.

From the perspective of the patient/subject, a non-therapeutic procedure serves some interest other than the preservation of health.[48] As Dickens suggests, where the situation is one of pure experimentation, then the subject's participation may actually "appear to serve some interest of the investigator's rather than of his own, but his informed consent to this will be based upon his philanthropic or commercial motivation.[49] The fact that such participation serves the interests of others is precisely the reason why non-therapeutic experimentation is so problematic when there are doubts about the capacity of the subjects concerned to give an informed consent on their own behalf. For such subjects, it makes little sense to talk about "philanthropic or commercial motivation". Such a consideration may, however, be important to a substitute decision-maker acting on behalf of such an individual, and therefore should not be ignored entirely. Similarly, the motivation for a patient/ subject to participate in a therapeutic experiment is obvious and would significantly influence the decision of a substitute decision-maker.[50] In either situation, by utilizing the perspective of the patient/subject, the research ethics committee would be called on to make a value judgment, something which it may be unable to do

---

[47] A research ethics committee, faced with the task of defining what constitutes therapeutic research might approach the issue in three ways: (1) the research is only therapeutic in the general sense; (2) the research is only therapeutic in the experimental subject-patient centered sense; or (3) the research is *primarily* therapeutic in the experimental subject-patient centered sense — but may also have some objectives that are non-therapeutic. Similar distinctions may be applied to the determination of what constitutes non-therapeutic research. Indeed, most definitions of these terms found in the literature can be grouped into this analytical scheme. As indicated in the text, the third approach is preferred.

[48] See Kluge, *supra* note 18 at 166; Dickens (1975), *supra* note 18.

[49] See Dickens, *ibid.* at 635. The potential influence of such a motivation is at its strongest where there is no possibility of any therapeutic benefit. As one moves closer to what we would classify as therapeutic (either primarily or secondarily), then the philanthropic or commercial motivation becomes of diminishing importance.

[50] *Ibid.*

owing to the usual lack of personal knowledge of the subject.[51] As such, it is highly doubtful that a truly representative patient/subject perspective could be applied.

Any system for classifying an experiment as either therapeutic or non-therapeutic must attempt to reconcile the fundamental desire to protect the research subject with the ever-present need to be sensitive to the ongoing burdens facing those who are actually engaged in conducting research; indeed, for the latter, administrative simplicity and efficiency will always be important considerations. Since the perspectives of both the researcher and the patient/subject are primarily subjective, however, they provide little practical assistance to research ethics committees. Instead, the determination that the primary purpose of an experiment is therapeutic or non-therapeutic should be based on the merits of the procedures themselves and should be readily apparent from an objective point of view. In this respect, Kluge suggests that the distinction should be drawn in the following manner:

> There should be an objective correlation between an expectation of therapeutic effect and the experimentation in the first place. Therefore, if a particular protocol is appropriate for a particular condition; if, furthermore, it has as its intended purpose a therapeutic effect on the patient; and if, finally, that purpose is substantiated by acceptable data detailing an expected level or type of therapeutic effect, then the experiment may be considered therapeutic. Otherwise, it should be considered non-therapeutic.[52]

The first requirement of this model concerns the "*appropriateness*" of an experiment, taking into consideration the particular condition of a research subject. It should be a categorical requirement for ethically permissible experimentation on members of vulnerable populations that an "appropriate" relationship exist between the nature of the scientific knowledge that is sought to be gained by an experiment and the particular conditions or circumstances that directly affect potential subjects or the class of vulnerable individuals to which they belong. At this stage of the inquiry, research ethics committees will focus on the notion of "appropriateness" solely as a means of identifying the primary purpose of a particular experiment; whether the experiment will ultimately be considered to be ethically permissible is a question that can only be answered by the analysis of a number of additional factors.

In conducting a general assessment of the purpose of a particular research protocol, the research ethics committee must identify the type of knowledge sought by the researcher and the potential therapeutic benefit(s) that may benefit the proposed subjects. Various considerations may be relevant in the conduct of this process. These may include the following:

— Is the objective of the experiment to understand a disease in general terms, to measure the effectiveness of various dosages of a medication or to identify potential side-effects of a particular therapy?

---

[51] The presence of a research directive, however, would greatly assist a research ethics committee in applying a specific subject-patient perspective. However, the subjective evaluation of experimental protocols from the perspective of each prospective subject is an impossible task.

[52] See Kluge, *supra* note 18 at 166.

— Is the procedure, as a whole, appropriate in *clinical* terms and, if so, to what degree?

— Does a therapeutic benefit present itself *directly* or *indirectly*?

— Does such a benefit present itself in a major or minor way and is it an essential, or merely adventitious, outcome of the experiment?

— To what extent are the various elements of the protocol in accord with the scientific objective of the experiment?

Any assessment of whether or not an experiment is primarily therapeutic must focus on its predicted effects on the research subjects themselves.[53] In order to justify designating an experiment as primarily *therapeutic* in nature, it is important that the subjects can expect to receive a reasonably foreseeable benefit that can be measured objectively. This approach would tend to restrict the classification of an experiment as *therapeutic*, in most situations, to those protocols involving *therapies* (and, in certain cases, to diagnostic measures), where a therapeutic gain can be reasonably expected.[54] However, it is also critical that a research ethics committee focus on the question of the *immediacy of application* of the therapeutic benefit to individual research subjects: in other words, the committee should ask to what extent the anticipated benefit is likely to be immediate in its impact on the subjects and whether it is *directly* linked to the proposed therapeutic intervention. Adoption of this approach by research ethics committees should prevent researchers from pointing to a remote linkage between their protocols and future clinical applications as a justification for classifying an experiment as being primarily *therapeutic*.

The research ethics committee's assessment of the "immediacy of application" must be guided by acceptable scientific data. In so doing, the committee can also ensure the validity of any claim of a potential therapeutic benefit. Since the classification of an experiment as either therapeutic or non-therapeutic will profoundly affect the legal and ethical restrictions that apply, a high standard must be met before an experiment should be classified as *therapeutic*. As discussed above, an element of uncertainty is, by definition, inherent in all experiments. However, "possible", "hypothetical", or "speculative" therapeutic benefits should not be sufficient in the present context. Rather, a therapeutic benefit must be "likely", "probable" or "reasonably foreseeable". If this standard cannot be achieved, then the experiment must be classified as *non-therapeutic*. It is important to emphasize that it is not the desired purpose or objective of an experiment, but rather its reasonably expected effect on individual research subjects, which should influence research ethics committees in making this crucial assessment.

Finally, assuming that it is decided that potential subjects may indeed expect an immediate and reasonably foreseeable therapeutic benefit, then the research ethics committee must still balance the conflicting objectives of a biomedical experiment:

---

[53] Indeed, the benefit to the subject forms the core of the distinction between *therapeutic* and *non-therapeutic* provided by the Law Commission (UK). See The Law Commission, *Mental Incapacity* (London: HMSO, 1995) at 96–97.

[54] This approach would, therefore, exclude those procedures designed to provide no more than a basic understanding of the disease process.

namely, to further scientific knowledge or to treat a particular patient. The strength of the expected therapeutic benefit must be measured against the sum of the other procedures specified in the experimental protocol. This step is not to be confused with the process of balancing benefits and risks. Certainly, a substantial risk might make a proposed procedure less desirable; however, the presence of an unusual degree of risk in no way diminishes the therapeutic potential of that procedure. At this stage in the process of assessing exactly where an experiment falls along the spectrum between "pure research" and "pure therapy", the research ethics committee should focus its attention on the nature of the inter-relationship between, and the relative importance of, the incidental, non-therapeutic or diagnostic procedures, on the one hand, and the (purportedly) therapeutic procedure(s), on the other. It is important, furthermore, that this task be carried out in the broader context of how all of these various procedures relate to the therapeutic and research objectives of the experiment *as a whole*.

In general, research ethics committees should require that the number of incidental and diagnostic procedures be kept to a minimum. The presence of an excessive number of incidental procedures, that are not really mandated by the therapeutic objective, will tend to indicate that an experiment is more concerned with the process of scientific inquiry than with the treatment of patients and that it is, therefore, to be considered primarily *non-therapeutic* in nature. Similarly, an intervention whose therapeutic effect is entirely unrelated to the scientific goal of the experiment would tend to lead to the same conclusion, unless the non-therapeutic procedures are of such an incidental nature as not to interfere with the treatment objective of the protocol.[55]

The final issue which must be addressed by the research ethics committee in this process is the balancing of the risks, generated by the non-therapeutic procedures required by the experiment, with the strength of the expected therapeutic benefit. If

---

[55] The following hypothetical example is provided to illustrate this process. An experiment designed to study the effects of a recognized, anti-inflammatory drug on the alleviation of symptoms associated with Alzheimer's Dementia is submitted for review to a REC. The committee finds that the drug, although an acknowledged therapeutic agent on its own, does not have a reasonably foreseeable therapeutic effect in alleviating symptoms suffered by Alzheimer's patients. The drug is administered on a regular basis, with a normal dosage. The only incidental procedures involve the occasional taking of blood and behavioural observations. The study is to be conducted on elderly persons who also suffer from arthritis. In such a case, a subject would receive an immediate and valid therapeutic benefit from participating in such an experiment: relief from arthritic pains. However, this benefit is not related to the objective of the research. How should the REC proceed? Both the therapeutic and scientific objectives are clear. The requirement of a valid and immediate therapeutic benefit for the subject is also satisfied. In this case, the non-therapeutic procedures are truly incidental (with some concern being attached to the frequency and amount of blood taken) and do not overshadow or interfere in any manner with the quality or effectiveness of the treatment. Consequently, the experiment would be classified as *therapeutic*. If, on the other hand, a lower-than-normal dosage is administered, with the result that the amount is insufficient to have any real effect on the patients' arthritic pains, then the therapeutic benefit of the experiment to the subject would disappear, despite the therapeutic value of the drug itself. The primary objective would no longer be therapeutic and the experiment would be classified as *non-therapeutic*. Similarly, if a higher-than-normal dosage is administered, so that the subject is exposed to the risk of side-effects associated with the medication, the therapeutic objective is no longer foremost, demonstrating that the primary objective of the experiment is really *non-therapeutic*.

on the basis of the above-mentioned criteria the experiment is deemed to be primarily *non-therapeutic*, risk assessment then becomes part of the process of determining the ethical permissibility of the experiment. If, however, an experiment purportedly has "treatment" as its primary objective, any risks that go beyond those normally associated with the therapeutic intervention in question are difficult to justify and, indeed, may compromise the therapeutic worth of the experiment *as a whole*. Research ethics committees should pay particular attention to those aspects of a therapeutic research protocol that would subject participants to tests and procedures that go beyond the strict needs of their health. Where members of a vulnerable population are involved, the non-therapeutic elements should be kept to the bare minimum that is necessary for the scientific integrity of the research and in no event should these non-therapeutic elements pose a risk to the subjects' health or result in significant inconvenience or discomfort. In other words, any risks associated with the non-therapeutic procedures of a therapeutic experiment must be *negligible*. Otherwise, the procedure must be classified as non-therapeutic.

## THE RANDOMIZED CLINICAL TRIAL (RCT) AS AN EXAMPLE OF MIXED THERAPY AND NON-THERAPY IN CLINICAL RESEARCH

The *randomized clinical trial* (RCT) serves as an excellent example of the extent to which research and clinical practice may become intertwined, with the consequence that both therapeutic and non-therapeutic elements may be present in the procedures administered to patient-subjects. The RCT, widely regarded as "the gold standard for the evaluation of therapeutic agents",[56] involves a research design in which the patient-subject is randomly allocated to a particular treatment category, which may be one of a number of alternative therapies that are available, or, in some cases, to a no-treatment category in which a placebo is administered.[57]

The RCT illustrates the extent to which the therapeutic/non-therapeutic distinction may be blurred by the numerous gradations that exist between the purely experimental and the purely therapeutic ends of the spectrum. The problem is that clinical researchers necessarily have mixed objectives. While an essential aim of therapeutic research is to benefit the patient-subject, there are other objectives as well, such as the desire to contribute to scientific knowledge or the financial gain accruing to the investigator or the research unit. The scientific goal may result in procedures being undertaken that are not strictly necessary for the individual patient-subjects' health and that may expose them to additional risk, discomfort or inconvenience.[58]

---

[56] R. J. Levine, "Uncertainty in Clinical Research" (1988) 16 Law, Medicine & Health Care 174 at 174. See also R. K. Whyte, "Clinical Trials, Consent and the Doctor-Patient Contract" (1994) 15 Health Law in Canada 49. For a rather controversial finding that brings into question the assumption of "clinical equipoise", or neutrality of the physician-researcher in randomized controlled trials, and hence it also indicates the lack of supposed objectivity and bias-free nature of these trials, refer to M.C. Klein *et al.*, "Physicians' Beliefs and Behaviour During a Randomized Controlled Trial of Episiotomy: Consequences for Women in Their Care" (1995) 153 Canadian Medical Association Journal 769.

[57] See Fried, *supra* note 13 at 7.

[58] See Schafer, *supra* note 26 at 44.

On the one hand, it may well be contended that participation in an RCT provides the patient-subject with the treatment necessary to be cured of his or her medical condition. On the other hand, however, the fact that the treatment cannot be tailored to the specific needs of the patient, and the fact that the patient's well-being is not of exclusive concern to the investigator, both suggest that the interests of the researcher, rather than the patient, may play a dominant role in shaping the nature of the procedures that are administered in the course of the research.[59] Katz, for example, suggests that inherent conflicts of interest permeate the conduct of all RCTs:

> In the conduct of such clinical trials, conflicts between the interests of patients and science are ever-present and are all too readily swept aside by viewing patient-subjects less as subjects and more as patients who can only benefit from participation in clinical trials.[60]

Following up on this notion, Schafer has drawn a vivid picture of the inevitable ethical dilemma facing physicians who are involved in clinical trials. On the one hand, they must, before enrolling a patient as a subject, undertake an assessment of what is the best treatment strategy for that particular individual and, if this strategy is not followed, they are, in essence, sacrificing the interests of that individual patient on the altar of science. On the other hand, if physicians heed this requirement, then they may well increase the possibility of selection bias or the elimination of too many patients from the pool of potential research subjects.[61]

Various commentators have addressed the problems raised by RCTs. In particular, there has been a vigorous attempt to ensure that the impact of the non-therapeutic elements of such trials is reduced to a minimum. For example, it has been emphasized that a control group for an RCT should never consist of those patients from whom a treatment of known benefit is being withheld in order to advance the purposes of the study.[62] Furthermore, it has been contended that, "ideally", randomization should not be resorted to unless there is a situation where there is a real division of opinion concerning the relative merits of the therapies that are to be evaluated: this ensures that "no one is being deprived ex ante of the 'better therapy.'"[63] Alternatively, it has been suggested that the RCT may only be considered truly therapeutic in nature where the treatments that are involved are all acknowledged to be orthodox treatments for the particular medical condition

---

[59] *Ibid.*
[60] See Katz, *supra* note 18 at 17.
[61] See Schafer, *supra* note 26 at 44.
[62] See Howard-Jones (1982), *supra* note 50 at 1438.
[63] See Fried, *supra* note 13 at 8. The practical problem with this approach, however, is that physicians will almost certainly have a treatment preference of one kind or another, frequently based on anecdotal rather than scientific grounds. Presumably, most patient-subjects would not agree to participate in an RCT if they knew that they may receive a form of treatment that is different from that which is favoured by their physician. In this respect, Schafer (1982) has noted that "it is generally unethical to solicit consent from prospective subjects for a randomized trial without telling them how their treatment will be selected" and goes on to suggest that guidelines need to be developed to indicate when it is ethically permissible for a physician to withhold information about his or her treatment preferences from the patient-subject. See Schafer, *supra* note 26 at 48.

affecting the patient-subjects.[64] Other commentators have stressed that participation in an RCT should not be considered therapeutic unless it is preceded by a detailed inquiry by the physician-researcher as to whether this would constitute the best treatment plan available for this particular individual.[65]

The ethical dilemma that arises as a result of the juxtaposition of both therapeutic and non-therapeutic elements in the same protocol may be seen in a particularly stark form when the RCT involves the administration of a placebo as one of the alternative "therapies". Insofar as a placebo, by definition, does not have a biological effect, and as such, should not have any impact on the physiological process of the illness, the suggestion has been made that its use would necessarily remove the RCT from the realm of therapeutic research.[66] In fact, Rothman & Michels assert that the use of placebos in RCTs is not only scientifically unnecessary, but may yield deceiving results by providing an inflated measure of the strength of the association which is subject to considerable statistical error. Furthermore, these two authors indicate that the use of placebos in RCTs runs contrary to the stipulations of the *Declaration of Helsinki* by depriving patient-subjects of "the best proven diagnostic and therapeutic method".[67] On the other hand, it has been argued that a placebo should never be administered when alternative standard therapies are available to treat the condition of the subjects.[68] Howard-Jones, for example, asserts that a control group of subjects should never be deprived of a treatment that is known to have a certain therapeutic benefit unless it is reasonable to believe that a new treatment will be better or, at least, of equal efficacy. The administration of a placebo to a control group would, in his view, be ethical only where there is no known treatment for the disease concerned[69] or, as an alternative view, when available treatments have failed to improve specific troublesome symptoms of a particular illness.

Therefore, research ethics committees should make the determination whether a specific RCT is primarily therapeutic or non-therapeutic on the basis of the objective approach recommended above. The analysis should be applied to each treatment option available under the RCT, including the administration of a placebo. The Law Commission (UK) takes the following view:

---

[64] See Dickens (1975), *supra* note 18 at 636.

[65] See Schafer, *supra* note 26 at 44.

[66] See Dickens (1975), *supra* note 18 at 637. However, Helmchen has pointed out that up to 40% of subjects taking a placebo show a positive response of some kind and, therefore, suggests that this can be considered a therapeutic effect. See Helmchen, *supra* note 12 at 510. This point, however, ignores the ethical issue of permitting 60% of subjects taking a placebo to run the risk of their condition becoming worse.

[67] See K. J. Rothman & K. B. Michels, "The Continuing Unethical Use of Placebo Controls" (1994) 331 The New England Journal of Medicine 394.

[68] See Somerville, *supra* note 26 at 87. This view has recently been endorsed by the Tri-Council Working Group in its *Code of Ethical Conduct for Research Involving Humans* (July, 1997), Part 2 at 35: "The use of placebos in clinical trials is ethically unacceptable where clearly effective therapies or interventions are available".

[69] See Howard-Jones, *supra* note 50 at 1438.

> The label [of "therapeutic"] can also be applied to "randomised controlled trials", where neither researcher nor participant knows whether a particular person is receiving the established treatment, a placebo or the experimental treatment. The ethical case for such trials is that the researcher genuinely cannot say whether the old treatment, no treatment or the new treatment is preferable, and is therefore asserting that all options are equally liable to be in the best interests of the patient.[70]

However, as stated above, the mere fact that the protocol has been designated as therapeutic by the researchers themselves is not sufficient, particularly when the proposed subjects are members of vulnerable populations. Therefore, in cases where a placebo *is* administered, the RCT should not automatically be classified as non-therapeutic although this may often be the case.

## ETHICAL ISSUES ARISING FROM THE DISTINCTION BETWEEN THERAPEUTIC AND NON-THERAPEUTIC RESEARCH

The conduct of so-called *therapeutic* or *clinical* research raises fundamental ethical questions about the role of the physician who combines health care with biomedical research. Indeed, the dramatic increase in medical research during the past forty years has made the distinction between physician and scientist increasingly difficult to draw.[71] This situation may lead to potential conflict when the conduct of research and the provision of health care occur simultaneously, particularly when both are undertaken by the same individual.[72]

In the capacity of physician, the individual providing the treatment makes a commitment exclusively to the patient, and undertakes to hold the patient's interests as paramount. In this respect, it is significant that the *MRC Guidelines* emphasize the point that, in the context of therapeutic research, "the patient is always entitled to the best clinical judgment of the physician, and research considerations must never displace this".[73] By way of contrast, as scientific investigator the very same individual has a commitment to promote the acquisition of scientific knowledge, and

---

[70] The Law Commission, *supra* note 53 at 96. However, the Commission, in a footnote to the above statement, goes further by stating:

> If the researcher believes one of the treatments being given is in fact better than the others then this analysis will not apply. In our view, the trial would then fall into the category of "non-therapeutic" research. In most cases, the administration of a placebo, however, will likely result in the classification of the protocol as non-therapeutic.

[71] P. R. Benson & L. H. Roth, "Trends in the Social Control Of Medical and Psychiatric Research" in D.N. Weisstub, ed., Law and Mental Health: International Perspectives, Volume 4 (New York: Pergamon Press, 1988) 1 at 3.

[72] See Brett & Grodin, *supra* note 15 at 1857. See also Dickens (1975), *supra* note 18 at 635–636; Benson & Roth, *ibid.* at 29–30.

[73] See *MRC Guidelines*, *supra* note 22 at 9. The Tri-Council's Code of Ethical Conduct for Research Involving Humans also requires that researchers "separate their role as researcher from their roles as therapists, caregivers", etc. and emphasizes the need for a researcher, who is performing "dual roles", to disclose this fact to potential participants in the research enterprise. Significantly, the Tri-Council's draft Code prescribes that "researchers must articulate, during the recruitment process and throughout the project (to the satisfaction of the participant and the REB), how they have dissociated their role as researcher from that of caregiver". See Tri-Council Working Group, *Code of Ethical Conduct for Research Involving Humans*, (July 1997), Part 2 at 9.

consequently owes loyalty to the scientific integrity of the research project.[74] A potential conflict thus arises between the goals of therapy (the well-being of the patient) and the goals of research (the advancement of scientific knowledge). As articulated by Katz:

> In therapeutic encounters, unlike research encounters, physicians are expected to attend solely to the welfare of the individual patient before them. Throughout medical history this expectation has given physicians considerable discretion and authority to make decisions on behalf of patients . . . . .

In clinical research, on the other hand, patient-subjects are also being used for the ends of science. One cannot dismiss with impunity the implications of this difference. In these situations investigators are committed both to real, present patients and abstract, future patients. Individual patient-centered therapy gives way to a collective patient-centered endeavor in which the abstraction of the research question tends to objectify the person-patient.[75]

Traditionally, the basic ethical problem arising from the conflicting nature of the dual roles of physician and researcher, revolved around the fact that in practice, the decision as to whether a particular procedure should be classified as therapeutic or non-therapeutic was made largely on the basis of an after-the-fact appraisal of the intent of the doctor who administered the procedure. However, it is difficult to assess intent except by asking the person,[76] which implies that determining whether the procedure should be classified as therapy or research has rested heavily on the integrity and capacity for self-insight of the individual physician/investigator.

The MRC has unequivocally stated that "no one should interfere with the therapeutic freedom of the treating physician" and that the "resolution of the therapy/research question" should rest on "the integrity of the physician-researcher".[77] Without in any way raising questions about the integrity of those physicians who conduct research with patients, we do query whether the distinction between therapeutic and non-therapeutic research can best be drawn by individuals who are serving conflicting roles and who may, therefore, have mixed motives. The line could perhaps be more appropriately drawn by moving it from the motivation of the physician, or from the intended product of the intervention, to the training and role of the actor. A distinction should be made between physicians and researchers. Physicians are clinicians who treat patients; researchers are scientists who undertake research. Perhaps such a differentiation would enjoin clinicians who do not have research qualifications not to venture into experiments-*cum*-research. It could also resolve several, already identified, conflicts of interest for the clinician/researcher. Professional careers are sometimes dependent on the ability to conduct experiments which yield "interesting" results, which can lead to publications or other secondary

---

[74] See Schafer, *supra* note 35 at 43.
[75] See Katz, *supra* note 18 at 14–16.
[76] See Kluge, *supra* note 18 at 166.
[77] See *MRC Guidelines*, *supra* note 22 at 9.

professional gains.[78] This career pressure may make it very difficult, if not impossible, for the researcher to stand back and implement only those parts of a research project that are truly therapeutic for each individual patient-subject.[79] Equally, such a differentiation could resolve the financial conflict of interest faced by many clinical researchers involved in RCTs when they are perceived as funneling patients into clinical trials because of the financial rewards obtained per patient enrolled. In this respect, it is significant that Annas has suggested that physicians should actually be prohibited from "performing more than minimal risk research on their patients" and that clinicians who are also researchers should be required to recruit the patients of *other* clinicians for participation in their research protocols. The advantage of this policy, which may admittedly be difficult to implement in practice, would be that it would eliminate the blurring of roles that can confuse patients and lead them to make the mistaken assumption that their own physicians are concerned exclusively with their patients' best interests. Even if it is not feasible to implement this policy in certain situations, Annas nevertheless points to the possibility of enlisting the assistance of retired or semi-retired physicians to act as independent patient advocates whose only role would be to protect the welfare of patients in the research environment.[80]

Another vital ethical issue concerns the need for a physician-researcher to make crystal-clear to potential human subjects that experimentation necessarily involves a degree of "uncertainty and risk".[81] This basic ethical principle assumes even greater significance when it is clear that most patient-subjects do not understand the difference between treatment and research and imagine that a physician will always act in the interests of their health.[82]

## CONCLUSIONS

It is our view that that all proposals to administer innovative therapy to children or to adults whose capacity to consent may be in question should, except in cases of genuine emergency, be submitted for scrutiny to a research ethics committee. The procedure should be treated as an experiment if the committee concludes any of the following:

— the proposed intervention does not have a realistic chance of benefiting the patient's health;

— an orthodox treatment for the patient's condition has not yet been attempted;

— the treating physician's motivation may be considered *investigational*, insofar as this is indicated by the procedures that are to be carried out in addition to

---

[78] See Kluge, *supra* note 18 at 45.
[79] *Ibid.* at 166.
[80] See Annas, *supra* note 45.
[81] See Kluge, *supra* note 18 at 165. See also Helmchen, *supra* note 12 at 508.
[82] C. W. Lidz *et al.*, *Informed Consent: A Study of Decisionmaking in Psychiatry* (New York: Guilford Press, 1984) at 235.

the proposed "innovative therapy" and that may be more suggestive of an *experiment* than a *treatment*; or
— the risk is more than *negligible*.

The determination as to whether an experiment is primarily therapeutic or non-therapeutic should be made on the basis of an objective assessment of the entire research protocol, rather than on the basis of the stated intent of the researcher(s) involved. Such determinations should be made by research ethics committees.

A protocol should be considered therapeutic only if it meets the following conditions:

— the patient-subject can expect to receive an immediate, reasonably foreseeable therapeutic benefit that is directly linked to the proposed therapeutic intervention and can be substantiated by the available scientific data;
— the therapeutic objective is not compromised by the non-therapeutic elements of the experiment; and
— the risks associated with the non-therapeutic procedures required by the experiment must be deemed *negligible*.

If a protocol fails to meet all of these conditions, it must be treated as a *non-therapeutic experiment*.

Research ethics committees should determine whether a specific randomized clinical trial (RCT) is primarily therapeutic or non-therapeutic on the basis of the objective approach recommended above. The analysis should be applied to each treatment option available under the RCT, including the administration of a placebo. In cases where a placebo is administered, the RCT should not automatically be classified as non-therapeutic.

CHAPTER 7

# THE DISTINCTION BETWEEN "CLINICAL PRACTICE" AND "RESEARCH": THE CASE OF PITUITARY DERIVED HORMONES AND CREUTZFELDT-JAKOB DISEASE

MARGARET ALLARS

**PART I: INTRODUCTION**

Codes of ethics applying to experimentation on human beings require that such research be approved by an independent ethics committee and that informed consent of the subjects of research be obtained. Codes also require that research proposals be scientifically viable, conducted by scientifically qualified persons under the supervision of a clinically competent medical person, and that they be conducted with respect for the integrity of the subjects and without exposing them to unacceptable risk. The *Declaration of Helsinki* has provided a model for many other codes. A feature of this code is that the level of protection of subjects is adjusted according to the nature of the research. Within the sphere of biomedical research on human subjects,[1] distinctions are drawn in the *Declaration of Helsinki* between clinical research and non-clinical research, and between therapeutic and non-therapeutic research. I shall argue in this chapter that these distinctions are incoherent and in any event unnecessary.

There is a need for distinctions to be made, but not between different types of research. The critical distinction which must be made is between research and clinical practice which is not research. I shall argue that a code of ethics must attempt to define research, otherwise it is uncertain in the very ambit of its application. Defining research presents a formidable challenge. When does

---

[1] Scientific research which is purely laboratory based, without involving the collection and testing of personal information or human tissue, falls outside this sphere.

111

innovation in clinical practice amount to research? Do the uncertainties of diagnosis and treatment in clinical practice inevitably involve an element of experiment? If a line cannot be drawn between research and the clinical practice of medicine, then all clinical practice, innovative or non-innovative, is research.

This would mean that the codes are over-inclusive. If the codes cannot be confined by a definition of research, should they be abandoned? These are questions whose resolution is assisted by testing practical examples against possible definitions of research. This chapter explores the distinction between research and clinical practice, and other associated distinctions, by examining the code of ethics adopted in Australia. The Australian code is the National Health and Medical Research Council *Statement on Human Experimentation (NHMRC Statement)*. It is based on the *Declaration of Helsinki*. The operation of the distinctions drawn in the *NHMRC Statement* is analysed by reference to examples drawn from findings of the *Report of the Inquiry into the Use of Pituitary Derived Hormones in Australia and Creutzfeldt-Jakob Disease.*[2] Part II of the chapter describes the roles played by the distinctions in the *Declaration of Helsinki* and the *NHMRC Statement*. Part III of the chapter provides the background to the Inquiry into the Use of Pituitary Derived Hormones in Australia and Creutzfeldt-Jakob Disease. This enables examples to be drawn from the twenty years of clinical practice and research associated with pituitary derived hormone therapy in Australia in order to test the distinctions between clinical and non-clinical research and between therapeutic and non-therapeutic research. The examples assist in the analysis in Part IV of these distinctions. As a result of this analysis it is possible to make some initial suggestions regarding a fresh approach to the distinction between research and clinical practice.

Part V suggests that research is to be defined by reference to purpose, and acknowledges the difficulty of drawing distinctions where the purposes are multiple. The difficulty may be tempered by examination of principles of administrative law, which focus upon the purpose for which a power is exercised or for which information is collected. It is argued that like statutory power-holders, doctors should exercise power conferred in a clinical setting for proper purposes and without a personal interest in the outcome of the decision.

## PART II: ETHICAL PRINCIPLES

### *The Declaration of Helsinki*

The *Declaration of Helsinki*, adopted by the World Medical Assembly in 1964 and revised in 1975, states principles to govern biomedical research involving human subjects. The purpose of such biomedical research is:

> to improve diagnostic, therapeutic and prophylactic procedures and the understanding of the aetiology and pathogenesis of disease.

---

[2] *Report of the Inquiry into the Use of Pituitary Derived Hormones in Australia and Creutzfeldt-Jakob Disease* (Canberra: AGPS, 1994) [hereinafter *Allars Report*].

The *Declaration* recognises in Basic Principle 5 that these purposes of biomedical research are purposes of a public interest type, that the public interest and the subject's interest may conflict and that in the event of such conflict the interests of the subject should prevail:

> Concern for the interests of the subject must always prevail over the interest of science and society.

In its Introduction, the *Declaration of Helsinki* distinguishes between two categories of biomedical research which differ according to their purposes:

> medical research in which the aim is essentially diagnostic or therapeutic for the patient, and medical research the essential object of which is purely scientific and without direct diagnostic or therapeutic value to the person subjected to the research.

This difference in purpose provides the basis for a distinction between therapeutic and non-therapeutic research which is reflected in the four-part structure of the *Declaration of Helsinki*. The first part is the "Introduction", which is followed by "I. Basic Principles", then "II. Medical Research Combined with Professional Care (Clinical research)" and finally "III. Non-therapeutic Biomedical Research Involving Human Subjects (Non-clinical biomedical research)". The Basic Principles are intended to apply to all biomedical research, with particular elaborations or qualifications of those Basic Principles being made with respect to the different categories of biomedical research described in Parts II and III. Whether the Basic Principles are qualified in Parts II and III is considered later.

Part II of the *Declaration of Helsinki* contemplates that research may occur in the course of a patient's therapy. This idea is captured in the title of Part II, which equates "clinical research" with "medical research combined with professional care". Part II commences with the assertion in Principle 1 that a doctor must be "free" to use new diagnostic and therapeutic measures in treating a sick person "if in his or her judgment it offers hope of saving life, re-establishing health or alleviating suffering". Principle 6 in Part II states:

> The doctor can combine medical research with professional care, the objective being the acquisition of new medical knowledge, only to the extent that medical research is justified by its potential diagnostic or therapeutic value for the patient.

Basic Principles 9, 10 and 11 relating to informed consent of subjects to research are assumed to apply to clinical research. Basic Principle 10 has particular force in such research, as it requires that the doctor be "particularly cautious" if a subject is "in a dependent relationship to him or her". In such circumstances, Basic Principle 10 requires that the informed consent be obtained by a doctor who is not engaged in the research project and who is "completely independent" of the existing doctor-patient relationship. Basic Principle 10 is also the foundation for a more particular principle in Part II of the *Declaration of Helsinki*. This is Principle 4, which states that where the doctor seeks consent of the patient and the patient refuses the new diagnostic or therapeutic measure, this refusal should not interfere in the doctor-patient relationship.

While Part II appears to require more stringent regard for the well-being of the patient than does Part III, it also allows for exceptions from the requirement of informed consent in the case of clinical research. Principle 5 in Part II provides that if the doctor "considers it essential" not to obtain informed consent, the specific reasons for this should be stated in the experimental protocol forwarded to the independent committee. Although not spelt out, this exception appears to cater to circumstances where use of an innovative therapy in an emergency situation may save the life of a patient already in the care of the doctor.

### The NHMRC Statement on Human Experimentation in Australia

The *NHMRC Statement* endorsed by the NHMRC in Australia in 1966 and subsequently revised, owes its origin to the *Nuremberg Code and the Declaration of Helsinki*.[3]

As the predominant organisation funding medical research in Australia, the NHMRC in practice from 1968 made compliance with the *NHMRC Statement* a precondition to consideration for NHMRC funding.[4] As a result of revision of the *NHMRC Statement* in 1976, applications for funding were formally required to be submitted to an institutional ethics committee for approval. In 1985 the NHMRC accepted a recommendation of its working party of 1982 that compliance with the *NHMRC Statement* be a precondition to a successful application for a research grant involving human experimentation.[5] From 1984 the NHMRC required compliance with *Supplementary Note 1 (1992) — Institutional Ethics Committees to the Statement. Supplementary Note 1* sets out details of the constitution and procedure of institutional ethics committees. Compliance with these provisions is a precondition to any institution's receipt of NHMRC funding for research. Review by an institutional ethics committee is also required by other major funding organisations[6] and where a clinical trial of a drug is conducted under the Clinical Trial Notification Scheme.[7]

The *NHMRC Statement* is briefer than the *Declaration of Helsinki*.[8] It includes a list of thirteen principles which require consideration of protocols by institutional ethics committees, respect for persons, that research be conducted by suitably qualified persons, observance of ethical considerations and informed consent. Many

---

[3] P.M. McNeill, *The Ethics and Politics of Human Experimentation* (Cambridge: Cambridge University Press, 1993) at 47, 70.

[4] See generally *Allars Report, supra* note 2 at 447–454.

[5] See *Supplementary Note 1* of the National Health & Medical Research Council, *Statement on Human Experimentation and Supplementary Notes* (Canberra: NHMRC, 1992), note 1(a) [hereinafter *NHMRC Statement*]; McNeill, *supra* note 3 at 73, 75.

[6] For example, the Australian Research Council, which also requires social science research projects where personal information is collected by survey to be reviewed by university human ethics committees.

[7] See *Supplementary Note 3 — Clinical Trials of the NHMRC Statement, supra* note 5; Allars Report, *supra* note 2 at 454, 717–9.

[8] However, *Supplementary Notes* dealing with particular aspects of research have been added to the *NHMRC Statement*, making it a much lengthier document.

of these principles correspond with those in the *Declaration of Helsinki*.[9] The *NHMRC Statement* makes no attempt to define experimentation or research, but makes passing reference to different categories of research. The opening paragraph states:

> The collection of data from planned experimentation on human beings is necessary for the improvement of human health. Experiments range from those undertaken as a part of patient care to those undertaken either on patients or on healthy subjects for the purpose of contributing to knowledge and include investigations on human behaviour. Investigators have ethical and legal responsibilities toward their subjects and should therefore observe the [thirteen principles] . . .

One further reference to the concepts of therapeutic and experimental procedures is found in Principle 7 of the *NHMRC Statement*, which provides:

> New therapeutic or experimental procedures which are at the stage of early evaluation and which may have long-term effects should not be undertaken unless appropriate provision has been made for long-term care, observation and maintenance of records.

The opening words of the *NHMRC Statement* appear to be influenced by the *Declaration of Helsinki*, but rest content with envisaging a spectrum of types of research, some of which may occur "as part of patient care". The disjunction of "therapeutic" and "experimental" in Principle 7 does little to explain what is meant by these terms. Like the *Declaration of Helsinki*, the NHMRC Statement assumes that the reader, and therefore the researcher, knows what research is.

The *Supplementary Notes* to the NHMRC *Statement* must also be taken into account. *Supplementary Note 1 (1992) — Institutional Ethics Committees* expects that "all research projects involving human subjects and relating to health" must be approved by an institutional ethics committee. This is a broad conception of biomedical research on human beings, provided "research projects" is given a wide ambit. *Supplementary Note 1* provides no definition of the expression "research projects". In circumstances where a doctor commences an innovative clinical practice without submitting a proposal to an institutional ethics committee seeking approval, it is unclear whether the doctor's practice falls outside the expression "research project". In the absence of a definition of the expression "research projects", the scope of the jurisdiction of the institutional ethics committees, and also the scope of the application of the *NHMRC Statement*, remain undefined.

A distinction similar to that between therapeutic and non-therapeutic research found in the *Declaration of Helsinki* is explicitly adopted in *Supplementary Note 2 - Research on children, the mentally ill, those in dependent relationships or comparable situations (including unconscious patients)*. Principle 5 in *Supplemen-*

---

[9] The principles in the *NHMRC Statement* correspond roughly to the Basic Principles of the *Declaration of Helsinki*. Adopted at the 18th World Medical Assembly in Helsinki in June 1964. Amended at the 29th World Medical Assembly in Tokyo in October 1975; the 35th World Medical Assembly in Venice in October 1983; and the 41st World Medical Assembly in Hong Kong in September 1989) [hereinafter *Declaration of Helsinki*] as follows, with the *Declaration of Helsinki* Basic Principles in brackets: 1(2), 2(12), 3(7), 4(1), 5(6), 6(3), 8(9), 9(9), 10(11). NHMRC Principle 11 corresponds roughly with principle 3 in Part II of the *Declaration*.

*tary Note 2* requires that the risks of research on children be considered in terms of firstly "therapeutic research (where the procedure may be of some benefit to the child)" and secondly, "non-therapeutic research (where the procedure is of no direct benefit to the child)". With regard to research on unconscious and critically ill patients, a similar dichotomy is described, between "experimental intervention" which "is intended or expected to benefit the person", and experimental intervention which "is intended or expected to yield important scientific information but is not intended or expected to benefit the person". *Supplementary Note 2* gives as an example of the latter type of experimental intervention, the taking of a blood sample for studies not directly relevant to the diagnosis or treatment of the patient. More stringent requirements are placed on the latter type of research, in particular that "the requirements of the research do not influence the procedures that are clinically indicated".[10]

Despite their attention to categorisation of different types of research, neither the *Declaration of Helsinki* nor the *NHMRC Statement* defines "research". The importance of distinguishing between biomedical research and clinical practice was, however, recognised in the *Belmont Report* in 1978.[11] The *Belmont Report* acknowledged that the distinction was blurred, but sought to maintain it by defining clinical practice by reference to interventions designed to enhance the well-being of the patient. This test is examined in Parts 4 and 5.

The distinctions in the *Declaration of Helsinki* and the NHMRC *Statement on Human Experimentation* having been outlined, a particular area of medicine — pituitary hormone therapy — is described in the next Part, in order to provide the background for exploration of the distinction between research and clinical practice.

## PART III: MEDICAL PRACTICE AND RESEARCH IN THE PITUITARY HORMONE PROGRAM IN AUSTRALIA

### *Innovative Therapies*

The earliest research on human growth hormone (hGH), conducted by Raben in the United States, was a report in 1958 of the first case of growth in a pituitary dwarf using hGH manufactured from pituitaries collected at postmortems.[12] Endocrinologists in Australia began to extract hGH in their hospital laboratories, the first clinical use occuring possibly as early as 1963 and certainly by 1965.[13] Children with growth

---

[10] The idea that less stringent requirements be placed on non-therapeutic research on vulnerable populations is also discussed in Chapter 6.

[11] National Commission for the Protection of Human Subjects of Biomedical and Behavioral Research *The Belmont Report: Ethical Principles for the Protection of Human Subjects in Research* (Washington, D.C.: U.S. Government Printing Service, 1978) at 3 [hereinafter *Belmont Report*].

[12] M. Raben, "Treatment of a Pituitary Dwarf with Human Growth Hormone" (1958) 18 Journal of Clinical Endocrinology 901; *Allars Report, supra* note 2 at 131–2.

[13] See *Allars Report, supra* note 2 at 134–5. The early work was done by the Victorian Pituitary Group. See H. Burger *et al.*, "The Investigation and Treatment of Pituitary Dwarfism: Experience with Human Growth Hormone" (1967) 1 Medical Journal of Australia 739.

deficiency, who would otherwise have been pituitary dwarves, grew. Meanwhile, in Europe in 1958 Gemzell reported the results of stimulation of the ovaries of anovulatory women with follicle stimulating hormone (FSH) manufactured from pituitaries.[14] From late 1962 onward, researchers in Melbourne began to extract from pituitaries not only hGH but also human pituitary gonadotrophin, or hPG (comprised of FSH with some luteinising hormone, or LH). From the mid-1960s researchers in Sydney, Melbourne, Adelaide and Brisbane were publishing the results of ovulation induction using hPG manufactured in hospital laboratories.[15] Anovulatory women were as a result able to conceive and many had children. Australia was unique in manufacturing fertility hormone from pituitaries. Overseas researchers extracted human menopausal gonadotrophin from the urine of postmenopausal women. However the biochemist in Melbourne who pioneered the extraction of hPG recognised that it was easier to collect pituitaries than the urine in Australia.

These new therapies for pituitary dwarfism and infertility were not scrutinised by institutional ethics committees. Although the *Declaration of Helsinki* was adopted in 1964, and the *NHMRC Statement* in 1966, the system of institutional ethics committees which now exists in Australia was not in place. Some of the larger hospitals had research advisory committees, but this was not common until the second half of the 1970s.

On the initiative of royal colleges, treating doctors and the federal government, a national program for the production of pituitary derived hormones was established. From the late 1960s until 1985 Commonwealth Serum Laboratories (CSL) manufactured hPG and hGH from pituitaries collected from cadavers at post-mortems conducted throughout Australia. Over a period of twenty years about 2,100 people were treated with the hormones. At some stage during this period of twenty years an innovative clinical practice introduced by a handful of doctors at the forefront of research in endocrinology became a routine and orthodox therapy administered by many doctors.

Because of the innovative nature of the therapies, in the late 1960s the federal government gave an expert committee of doctors the responsibility for developing and monitoring the national program. The statutory provisions by which Departmental decision-makers normally tested and evaluated new drugs and approved pharmaceutical benefits were bypassed. The obstetrician-gynaecologists and endocrinologists who were members of this expert committee, the Human Pituitary Advisory Committee (HPAC) and its Subcommittees, were eminently qualified to provide expert advice to government on clinical or research issues. The HPAC and its Subcommittees monitored the safety and efficacy of hPG and hGH, made guidelines for assessment of suitable patients, approved individual patients for treatment pursuant to the guidelines, and also approved the distribution of hormone for research purposes.

---

[14] C.A. Gemzell, E. Diczfalusy & G. Tillinger, "Clinical Effect of Human Pituitary Follicle-stimulating Hormone (FSH)" (1958) 18 Journal of Clinical Endocrinology 1333.

[15] See generally *Allars Report*, *supra* note 2 at 166–200.

### Recognition of Risk

In 1985 the Australian pituitary program was stopped, as a result of a decision of the HPAC. A young man in the United States had died of a slow virus, the neurodegenerative disease Creutzfeldt-Jakob Disease (CJD). His doctor recognised a link between the disease and childhood therapy with hGH.

CJD is a very rare disease belonging to a group of unconventional slow viruses, or spongiform encephalopathies, and occurring at a rate of one in a million in its sporadic form. The average age of death at which it appears is 60 years. CJD can also be contracted iatrogenically by corneal transplant, transplant of dura mater tissue, or use of contaminated surgical instruments.

The pituitary is close to the brain and it was possible that in the processing of the hormone with which the patient in the United States was treated, infectious material from a diseased pituitary had not been eliminated. A new means of iatrogenic transmission of CJD was hence identified: transmission by hormone injections.

The tragedy was that due to the long incubation period of CJD, of 12 to 35 years, many people had been exposed to risk in a number of countries, including Australia, the United States, France and the United Kingdom. The knowledge of transmission of CJD by transplant was a highly specialised knowledge within the specialty of neuropathology, with which obstetrician-gynaecologists and paediatrician-endocrinologists could not be expected to be familiar. During the twenty years of operation of the Australian program, neuropathologists and pathologists had failed to make the link between pituitary derived hormone treatment and CJD. They believed that the processing rendered the hormone safe.

Between 1988 and 1991 four Australian women in their late thirties and early forties died of CJD.[16] They were anovulatory women who had received pituitary hormone therapy between 1973 and 1978, in hospital clinics in Adelaide, Melbourne and Perth. These deaths called for explanations.

### Findings of the Inquiry

In May 1993 the Federal Minister for Human Services and Health established the Inquiry into the Use of Pituitary Derived Hormones in Australia and Creutzfeldt-Jakob Disease. The terms of reference of the Inquiry required it to report on the decisions to commence and continue the program in the light of the scientific knowledge of the risks and benefits, the collection of pituitaries, the manufacturing process, treatment and informed consent, and the response made in 1985 when the Australian program ceased.

The Inquiry tabled its *Report* in federal Parliament in June 1994.[17] The findings relevant to the issues raised in this paper were as follows.

On the basis of a confluence of factors in the late 1970s, including the commencement of a study in the United Kingdom to address concerns about the

---

[16] A fifth death, of an hGH recipient in 1991, was identified in 1995.
[17] *Allars Report, supra* note 2.

safety of the hGH manufactured there, and new research regarding the nature of the infective agent in other spongiform encephalopathies, the Inquiry concluded that at a meeting of the HPAC in 1980 when the issue of a risk of slow virus contamination arose, HPAC failed to act appropriately in the light of scientific knowledge. The HPAC should have stopped or reassessed the program and certainly hormone recipients should have been advised of the risk of CJD.

The Inquiry found that treatment regimes for hPG differed slightly in each ovulation induction centre. Most hPG recipients were not told of the source of the hormone. While most were aware of the risk of multiple births associated with ovulation induction, most were not told of the risk of ovarian hyperstimulation or of its seriousness.

As well as approving patients for treatment under the guidelines for allocation of the hormone as a pharmaceutical benefit, the HPAC approved the issue of hormone for a range of research projects. Hormone was issued for use in Australia's first IVF research in the early 1970s, in breach of the HPAC's own guidelines.[18] These guidelines required that the patient be anovulatory. The IVF patients were ovulatory. They were hyperstimulated so as to allow recovery of eggs for implantation. This was an innovative use of hPG. No research protocol was required by the HPAC or by the research committee of the hospital where they were treated. The HPAC did not concern itself with whether fully informed consent was obtained from patients. Some of the hormone the IVF patients received was not manufactured at CSL but in a laboratory at Royal Women's Hospital, Melbourne, by the biochemist who pioneered the extraction of hPG. He had been the most successful producer of hPG in Australia, supplying it to doctors in Victoria and elsewhere, including overseas, from late 1962. When the official national program started in Australia he continued to manufacture hormone in his laboratory, using a process different from CSL's, under pressure from local doctors, particularly those entering IVF research. He continued in this way despite his being a member of the HPAC, whose function was to maintain a national program.

The Inquiry did not have the benefit of evidence from patients who entered the IVF program. It is doubtful whether these patients were fully informed as to the source of the hormone.

### Recommendations and Implementation

The Inquiry made recommendations for funding of research into CJD, counselling of hormone recipients, coordination of support groups, tracing of recipients, and a national committee to advise the Minister on the needs of recipients. All the Inquiry's recommendations have been or are being implemented by the federal government. A National Pituitary Hormones Advisory Council was established in 1995, comprised of hormone recipients, expert legal, medical and counselling members and ex officio Departmental representatives, with the function of advising

---

[18] *Ibid.* at 190–97.

the Minister on the needs of recipients. Funds were allocated for research into CJD in Australia. Legal actions brought against the Commonwealth and CSL by the widowers of the women who died were settled, but those brought by recipients claiming damages for nervous shock are still pending.

Amongst the Inquiry's recommendations relating to the future regulation of medical programs, was a recommendation for revision of the NHMRC *Statement on Human Experimentation* to provide guidance with regard to decisions as to whether treatment in a therapeutic setting constitutes an experiment, and to develop a procedure by which such decisions as to what research is are scrutinised, and not left entirely to the treating doctor. A separate *Review of the Role and Functioning of Institutional Ethics Committees* was conducted by Professor Don Chalmers during 1995, partly in response to the Inquiry's recommendation and partly in response to concerns about consent forms used in trials of RU486 in Australia. In addition, a recommendation has been made by an NHMRC *Working Party on the Long-term Effects on Women from Assisted Conception* that the NHMRC review the defintion of "experimental" and "non-experimental" treatment to clarify for patients the status of drugs, devices and procedures, particularly those used in IVF programs.[19]

### Core Ethical Problem

The core ethical problem in the operation of the Australian pituitary program arose from the government's choice of a special mode of regulation of a particular area of medical practice. The HPAC was formed in the midst of innovative clinical practice which was freely acknowledged in medical circles at the time to be "experimental". The doctors who were innovating became the regulators, as members of the HPAC. In a sense the initial *raison d'être* of the HPAC was to increase scientific knowledge about the efficacy of pituitary hormone therapy. This remained the rationale of the mode of regulation for twenty years. The HPAC exercised its regulatory power to pursue research objectives through the clinical practice of its own members and other obstetricans and paediatricians in Australia. The HPAC also released hormone for use in research projects on human beings on an *ad hoc* basis as proposals were made to it.

As clinicians the doctors who comprised the HPAC had a strong interest in maintaining the program and adequate supplies of hormone to fulfil patient needs. As researchers, holding academic posts and with international reputations, they had an interest in maintaining a program which offered them valuable research opportunities. As regulators they had power to ensure achieving what was in their own professional and personal interests. Clinical practice, research and regulatory power were intertwined. The HPAC failed to respond to warnings of risk because of the momentum of the program, fuelled by these professional and personal interests. The core ethical problem was one of conflict of interest.

---

[19] See *Report of the Review of the Role and Functioning of Institutional Ethics Committees*. Report to the Ministers of Health and Family Services 1996 (Commonwealth, 1996); NHMRC, *Long-term Effects on Women from Assisted Conception* (Canberra: NHMRC, 1995), c.6.

## PART IV: ANALYSIS OF THE DISTINCTIONS

### *Distinction between Clinical and Non-Clinical Biomedical Research*

The titles of Parts II and III of the *Declaration of Helsinki* are based upon a distinction between clinical and non-clinical research. These categories of research may be distinguished by reference to the subject of research. Part II contemplates a situation where a doctor wishes to use a new diagnostic and therapeutic measure on a sick person who is in his or her professional care for the purpose of treatment for the condition which is being researched. This is understood to be "clinical research". Part III contemplates that the subjects of "non-clinical research" are healthy volunteers or patients with an illness unrelated to the "experimental design".[20] The division between Parts II and III thus appears to turn upon whether the subject of research is being treated by the researcher in respect of the very illness to which the research relates.

Judged on the basis of the principles in each of Parts II and III, the distinction between clinical research and non-clinical research is blurred. There are two reasons. Firstly, both categories of research may occur within a clinical setting. Secondly, such research always occurs within a dependent relationship.

If the subject of "non-clinical research" is a hospital patient with an unrelated illness, selected in an arbitrary fashion and invited to participate as a subject, the physical context is still a clinical one. This is also true of a patient who attends an individual doctor's clinic and is invited to be the subject of research in relation to a condition from which he or she does not suffer. The context is a clinical one both with regard to the physical location and the reason for the subject being there. It should not make a difference that the research occurs in a location different from the physical location of the hospital or clinic where the subject has sought treatment, been identified as a potential research subject, and entered into the research project. Research on healthy volunteers also occurs in a clinical setting. A "clinical setting" connotes a context where a human being undergoes medical testing or treatment under the professional care of a doctor or other health professional. If it occurs within a clinical setting, then the research is clinical research.

More importantly, whether research is into the patient's illness or an unrelated illness, there is a dependent relationship between the doctor and the patient. Basic Principle 10 applies.[21] Principles 1, 3 and 4 in Part III of the *Declaration of Helsinki* stress that the well-being of the subject is to take precedence over the continuation of the research. Principle 1 states that it is "the duty of the doctor to remain the protector of the life and health" of the subject. A healthy volunteer also enters into a doctor-patient relationship with the "clinically competent medical person" who must, according to Basic Principle 3, be ultimately responsible for the project. It makes no difference that the taking of samples, testing and other involvement with the subject is carried out by other health professionals. The doctor has the ultimate

---

[20] See *Declaration of Helsinki, supra* note 9, part III, principle 2.
[21] See *supra* note 9 at 5.

responsibility for the research and for the manner in which it is conducted. The subject expects that the doctor will stop the research when it endangers his or her well-being. Consent forms reassure healthy volunteers that they are entitled to hold such an expectation.

The argument so far suggests that all biomedical research occurs in a clinical setting where there is a relationship of dependence between doctor and subject of research. All biomedical research is clinical research. Does it follow that all clinical treatment is biomedical research? No. Such a conclusion would involve logical error. However, the difficulty with the *Declaration of Helsinki* and with the *NHMRC Statement* is that each assumes we already know what research is, and proceeds to lay down ethical principles governing such activity. I will return below to consider what sort of activities in the clinical setting amount to research.

At this stage, the attempt to distinguish between "clinical research" and "non-clinical research" in Parts II and III of the *Declaration of Helsinki* appears to be misconceived. The expressions "non-clinical research" and "clinical research" in the titles to Parts III and II should be jettisoned. The expression "non-clinical research" would be better retained to describe the sort of research with which the Declaration is not concerned, namely scientific research which is purely laboratory based, without involving the collection and testing of personal information or human tissue.

### Therapeutic and Non-Therapeutic Research

The expression "non-therapeutic research " also appears in the title to Part III of the *Declaration of Helsinki*, where it is equated with "non-clinical research". This equation now also appears misconceived. What needs to be explored is whether biomedical research done in a clinical setting may be classified as either therapeutic or non-therapeutic, and whether such a distinction is useful.

Examination of a possible distinction between therapeutic and non-therapeutic research may be assisted by returning to the distinction drawn in the "Introduction" to the *Declaration of Helsinki*.[22] The Introduction identifies the "essential" object or purpose of research as being either "diagnostic or therapeutic" or "purely scientific and without direct diagnostic or therapeutic value" to the subject of the research. Commentators have made a similar distinction, with slightly different glosses. Therapeutic research is defined as research aimed at improvement in the diagnosis or treatment of a patient or group of patients in the care of the doctor.[23] Non-therapeutic research aims to increase scientific knowledge, irrespective of improving the diagnosis or treatment of any particular patient or patients. The knowledge has a potentially wider application than just immediate patient care.[24]

The distinction is problematic. Other commentators have criticised it for failure to cater to many types of research which cannot be clearly classified as either

---

[22] See *supra* note 9 at 4.
[23] J.K. Mason & R.A. McCall Smith, *Law and Medical Ethics* (London: Butterworths, 1991) at 348–349.
[24] *Ibid.*

therapeutic or non-therapeutic.[25] There are three areas of difficulty. The first is concerned with multiple purposes, the second with defining purpose, and the third with identifying the benefits of research.

Firstly, "essential" purposes may not be easily identified. An essential purpose is certainly a necessary condition in the set of factors which result in the research being undertaken. It can be said that the research would not have been undertaken "but for" the presence of this purpose. But "essential" could also be understood to mean the set of necessary and sufficient conditions for the research to be undertaken.

Secondly, the notion of purpose is in any event slippery. Therapeutic research has a purpose of direct improvement in the treatment of one patient or a group of patients, while non-therapeutic research has a purpose of obtaining scientific knowledge with wide implications. Both purposes concern increase in scientific knowledge about diagnosis and treatment of illness. One is modest and the other ambitious. However, in relation to the diagnosis or treatment of a particular medical condition, both purposes are located upon the same sliding scale of increasing knowledge in the area. Improved knowledge in relation to one patient or a group of existing patients may indicate the scientific viability of a controlled experiment involving larger groups of subjects, or even provide an immediate increase in scientific knowledge with wide implications, say with regard to transmission of disease. An ethical approach to scientific endeavour suggests that a purpose of increasing scientific knowledge with wide implications is not only accompanied by the researcher's purpose of achieving personal fame and fortune, but also by a purpose of application of that knowledge to improve patient care.[26] This is the problem of multiple purposes in a new form. The purposes can be characterised as distinct ones, pursued simultaneously. They may also be characterised as broad and narrow expressions of one purpose, somewhat like concentric circles. Distinguishing between therapeutic and non-therapeutic research by reference to purpose may dissolve into a mere matter of semantics.

A third theme emerging from the distinction between therapeutic and non-therapeutic research concerns the contrast between a direct benefit of the research to a patient or group of patients in the care of the doctor, and a benefit with wider application to patients in the future. If therapeutic research is understood to provide an immediate and direct benefit to its subjects, then it may be justified with respect to vulnerable populations, such as children and the mentally ill, whereas non-therapeutic research on such subjects cannot.[27] Professor R.M. Hare makes the following comments about the distinction:

---

[25] R. Levine, *Ethics and Regulation of Clinical Research* (Baltimore: Urban and Schwartzenberg, 1981) at 6; B. Gaze & K. Dawson, "Distinguishing Medical Practice and Research: The Special Case of IVF" (1989) 3 Bioethics 301; J.-L. Baudouin "L'expérimentation sur les humains: un conflit de valeurs" (1981) 26 McGill Law Journal 809 at 810–2.

[26] For an interesting catalogue of motives of researchers, ranging from winning the Nobel Prize to enjoyment, but excluding improvement in patient care, see C. Freeman & P. Tyrer, *Research Methods in Psychiatry: A Beginners Guide*, 2d ed. (London: Royal College of Psychiatrists, 1992) at 2.

[27] See Chapter 6.

> Some people have thought the distinction between [therapeutic research and non-therapeutic research] so important they have made it the basis of proposed restrictions, saying that non-therapeutic research on children should be banned absolutely. The distinction is not entirely clear, however. Therapy as such — I mean the cure of the individual patient — is never the purpose of research. The two aims are distinct though often combined. If, therefore, we ask for reasons for engaging in research of any kind, the cure of the individual patient cannot be given as one of them. We may hope to learn more about the disease and thus become able to cure other patients; but the results of the research, except in extremely rare cases, will not be known until it is too late to use them to help *this* patient.[28]

In this passage the criterion of direct and indirect benefits of research collapses into the criterion of the purpose of research. The benefits of an innovative therapy and of any research project cannot be measured in advance. If a code of ethics claims to distinguish between different categories of research and formulates different principles within each category, it is only workable if the categorisation can be done before the research commences. Since benefits can only be estimated as probabilities of what may be realised in the future, the code will have to make do with identifying the benefits the research aims to achieve. This is a test of the purpose of the proposed activity. The argument has returned to the second area of difficulty discussed above.

The distinction between therapeutic and non-therapeutic research may be acknowledged to be problematic and a matter of degree, but worthy of attempted salvage because of the important role it may play in limiting the scope of research on vulnerable populations.[29] A more sophisticated test of probable benefit may be developed. Therapeutic research may be defined as research where a reasonably foreseeable benefit capable of objective measure is immediately and directly linked to the therapeutic intervention. However, this basis for distinction encounters difficulty.[30] Research whose benefits to a patient are mere byproducts may fall within the test of research which has reasonably foreseeable benefits capable of objective measure. Yet it would not be prudent to include such activities by doctors within the ambit of therapeutic research, to which consent may be given by those responsible for the care of vulnerable persons. The beneficent purpose of the intervention in relation to the particular patient is what excuses such research from the more general presumption against the capacity of carers to consent to research on vulnerable populations.

The argument about the distinction between therapeutic and non-therapeutic research has not only collapsed into one about purpose; it has also begun to slide into an argument about what is research and what is clinical practice. Hare has the advantage of beginning with a definition of research before he proceeds to consider categories of research. Research is activity which excludes any purpose of cure of an individual patient. Hare's approach leaves no room in a code of ethics with

---

[28] R.M. Hare, *Essays on Bioethics* (Oxford: Clarendon Press, 1993) at 134.
[29] See Chapter 6.
[30] *Ibid.*

respect to experimentation on human beings for a category of therapeutic research. For Hare therapeutic research is a misnomer.

It is time to abandon the distinctions made in the *Declaration of Helsinki*, echoed in a muted form in the *NHMRC Statement*, and attend to the real issue. This is the distinction between what is, and what is not, research.

### Research and Experiments

What is research? When does a doctor step over the line between clinical practice and research? Must there be an element of the experimental? Is innovation the key to an experiment? Can any line be drawn?[31]

The dividing line between research and clinical practice may be explored in two areas:

- Collection of patient information, whether by taking the history, performing diagnostic tests or accessing medical records.
- Innovation in therapy.

### Collection of Patient Information

In clinical practice patient information is normally collected for the composite purpose of diagnosis and determination of the appropriate therapy for that patient. This activity would not in common parlance be described as "research". An obstetrican gynaecologist may arrange for a series of tests on a new patient presenting with infertility. On the basis of tests such as radiological examination of the pituitary fossa, gonadotrophin and prolactin assays to measure pituitary function, hormone assays to measure adrenal and ovarian function, basal temperature charts, endometrial biopsy and oestrogen and pregnandiol assay to establish anovulation, tests to establish tubal patency and viability of partner's semen, the doctor forms a diagnosis and proceeds to treat, usually initially with clomiphene.

However, patient information may be obtained and recorded for purposes other than diagnosis and determination of appropriate therapy.

By characterising the range of purposes for which a doctor may collect patient information, the distinction between clinical practice and research may be clarified. Three general situations can be identified where information concerning a patient is collected:

i.   obtaining information solely for the composite purpose of diagnosis and determination of appropriate therapy for the patient;

ii.  obtaining information for the composite purpose of diagnosis and determination of the appropriate therapy for the patient and for a purpose or purposes not related to diagnosis or treatment of the patient; and

iii. obtaining information solely for a purpose or purposes not related to diagnosis or treatment of the patient.

---

[31] For some responses by treating doctors to this question see *Allars Report, supra* note 2 at 719–722.

The circumstances described in (i) amount to clinical practice. The circumstances described in (iii) amount to research.

The circumstances described in (ii) include elements of clinical practice. However, the presence of additional purposes makes it a controversial question whether this is clinical practice or research. Category (ii) is highly problematic, presenting a danger of circularity. How are the purposes behind the collection of information about a procedure which is innovative or unorthodox to be ascertained and characterised? Information relating to innovative or unorthodox diagnosis or treatment may be categorised as information obtained for the purpose of diagnosis and determination of appropriate therapy for the patient, or as information obtained for additional purposes. If there are multiple purposes, when is the situation justifiably described as clinical practice rather than as research?

This distinction between clinical practice and research turns upon purposes, but does not replicate the distinction considered earlier between therapeutic and non-therapeutic research. Here the attempt is to distinguish between a doctor's purposes or motives in engaging in certain activities. It is not assumed that these activities are research. The concept of purpose is used to identify research, not to identify categories of research.

No distinction is drawn between research and experimentation. These expressions have the same denotation. Each is a trial or procedure for testing a hypothesis or a known fact, usually under conditions determined by the tester.[32] Of course in the biomedical context the expression "experiment" has a more pejorative connotation than does the expression "research".

Where a person under observation has given informed consent to be the subject of research, situations (ii) and (iii) do not raise concern, unless other ethical principles are violated. The danger is that some information may be collected in situations (ii) or (iii) from a person who has consented as a patient to enter treatment, but who has not given informed consent to be the subject of research. Codes of ethics applying to experimentation on human beings need to make it clear that they apply to (iii) and in appropriate circumstances to (ii).

The following examples drawn from the Australian pituitary program assist in exploring the controversial area of multiple purposes reflected in situation (ii).

*Dosage of hPG.*

> In the early years of hPG therapy the responses of patients were very variable. There was real uncertainty about dosage and therefore a very high risk of hyperstimulation. In the course of the early therapy in Europe, one of Lunenfeld's patients died.[33] When therapy commenced in Victoria, the "Brains Trust", a self-named group of doctors in Victoria, met regularly to discuss individual patient responses. In this period the assay technique for measuring oestrogen in 24-hour urine collections was refined to the point where results could be obtained within one day.[34] As the assay techniques improved, the risk to the patients decreased. However, throughout the Australian pituitary program, treating doctors

---

[32] See *infra* at note 38
[33] *Allars Report, supra* note 2 at 157.
[34] *Ibid.* at 186.

tested the responses of individual hPG patients to dosages of hPG as a routine element of the treatment regime. There was uncertainty as to how each individual patient would react to the drug. Patients were required to provide 24-hour urine collections so that assays of oestrogen and progesterone levels could be carried out to determine the appropriate dosage of hPG the next day and avoid hyperstimulation.

Did this collection of patient information by testing amount to clinical practice within (i), or research within (iii), or a multiple purposes situation as in (ii)? The dosage of many drugs must be adjusted in accordance with the differing responses produced by the physiology of different patients. However, in the case of the pituitary program the answer is not uniformly that this was clinical practice. In the early days of therapy the testing appears to be appropriately categorised as research. Only with the passage of time did the role of the assays in the treatment regime become routine, so that the testing fell within situation (i), clinical practice. Identifying the point of historical time at which research became clinical practice is a matter of judgment.

*Dosage of hGH.*

The acceptable minimum dosage of hGH was a controversial issue for many years. From 1968 to 1972 the Human Growth Hormone Subcommittee of the HPAC carried out a study of growth responses using different dosages of hGH.[35] The Subcommittee decided that patients residing in Victoria and Tasmania would receive 2.5 mg of hGH three times a week and patients in other states would receive 2.5 mg of hGH twice a week. At that time the combined number of patients undergoing therapy in Victoria and Tasmania roughly equalled the number of patients in the remaining states. At subsequent meetings, the Subcommittee varied the dosages for patients in specified states as the balance of the number of patients changed. The aim was to place patients residing in one group of states on different dosage regimes from those residing in the other group of states, for the purpose of determining an acceptable balance between minimising the dosage and maximising the growth response. During the peak response period of the first year of therapy, the study showed no difference in growth response to the different dosages.[36]

The Subcommittee's decision has to be understood in its context. The production of hGH at CSL depended upon the regular supply of pituitaries. Since the supply was during some periods inadequate to meet the needs of patients already in therapy, it was vitally important not to waste the hormone. The Subcommittee had to make difficult decisions not to approve patients who were marginal in meeting the guidelines for approval and who could have probably benefited from treatment. The Subcommittee was in a position to ensure an unequal supply, to require the results of treatment to be provided to it, and hence to engage in a comparative study. If the study showed that a lower dosage was equally effective as a high one, the Subcommittee could approve more patients and distribute the hormone more widely. The information which the study would provide was likely to benefit existing and potential new patients.

The presence of multiple purposes in obtaining the patient information places this case within (ii). Treating doctors would in any event have taken patient information

---

[35] *Ibid.* at 137–9.
[36] *Ibid.* at 138.

on growth response, for the purpose of determining future treatment. The Subcommittee forced on them a particular dosage regime which was determined by additional purposes. Dosage was determined on the basis of factors extraneous to individual patient responses, namely place of residence, and hence the information obtained as to patient response to the dosage served additional purposes. This factor of systematic decision-making regarding therapy on the basis of the patient's membership of a group determined on a basis unrelated to diagnosis or appropriate treatment, suggests categorisation of this situation (ii) as research.

*Publication of Case Studies.*

> When a patient presents with a rare condition, the treating doctor may write up the case, and submit it to a medical journal for publication. Such case studies are normally published without seeking the consent of the patient. Early published studies of pituitary hormone treatment invaded privacy by referring to patients by their initials. Even in the absence of such explicit personal information, publication of a case study allows for ready personal identification when the patient consults another super-specialist within the field in Australia.
>
> In one case which came to the notice of the Inquiry, an hPG recipient who had suffered severe ovarian hyperstimulation sought treatment many years later at the same hospital for another gynaecological condition. To her surprise, she was identified by name by a nurse who had studied a photograph labelled "Mrs X", showing the "grapelike" shape of the hyperstimulated follicles. Without the knowledge of the patient, this photograph was regularly used for teaching purposes in the major hospital where she had attended an ovulation induction clinic.

Was the taking of the photograph, situation (i), (ii) or (iii)? Suppose the photograph was initially taken for the purpose of enhancing and clarifying the determination of therapy, but that a fresh purpose later emerged of using the photograph for research and teaching purposes. There were multiple purposes for obtaining information. The diagnostic and treatment purposes for taking the photograph were exhausted when the patient recovered. Thereafter the only purposes were additional. Alternatively, the additional purposes may have been present from the start. In either event, the presence of the additional purposes rendered this situation (ii). The controversial question is whether the presence of those additional purposes justifies categorisation of the taking of the photograph, or its retention, or its use, as research.

*Collation of personal information.*

> In another class of case, information obtained and recorded in the course of treatment of one patient may be added to information obtained and recorded in the course of treatment of one or more other patients to provide a picture of a pattern of responses. Throughout the pituitary hormone program in Australia, doctors published articles based upon such collations of information.

As in the case of the last example, the fact that the additional purpose was present at the time of treatment may make us more inclined to categorise this situation (ii) as research. As in the case of the last example, the fact that the additional purpose was present at the time of treatment may make us more inclined to categorise this situation (ii) as research. Suppose the doctor does not collect excess information, but utilises what is obtained for multiple purposes, within category (ii). In such a case, situation (ii) appears to be properly called research, particularly where the additional purposes involve systematic comparison of responses of groups of patients.

If the additional purpose exists at the time of treatment it is likely that the doctor may collect information in excess of what is necessary for diagnosis and determination of treatment, in order to improve the quality and range of the comparative data. A possible approach is to treat the excess information as severable from that which is strictly necessary for the purpose of diagnosis and treatment. The excess information then is collected purely for additional purposes, and falls into situation (iii), a research activity.

The removal and use of human tissue can be understood in terms of the three situations. For example, blood samples are commonly taken for the purpose of obtaining information to permit diagnosis and determination of appropriate treatment of the patient at that time. The results of such tests may be used for diagnostic purposes at a later point in time. Results may also be used for additional purposes, such as comparative studies of patients with the same condition. But where excess blood is taken from a patient for the additional purposes of obtaining information from laboratory tests relating to the biochemistry of the patient's medical condition, there are strong arguments that the taking of excess blood is research.

*Accessing medical records*

> When medical records are accessed in epidemiological studies, new purposes are found for information which was originally collected purely for the purpose of diagnosis and determination of treatment, within situation (i). At the later point in time, when patient information is used for additional purposes, situation (iii) arises. Supplementary Note 6 to the NHMRC Statement recognises that such activity is research.[37]

An interim conclusion with regard to the collection of patient information is that (i) is clinical practice, (iii) is research and (ii) may be research depending upon the resolution of the question in what circumstances the presence of purposes additional to diagnosis and determination of treatment render the doctor's activity research. The analysis of multiple purposes is developed further in Part V.

### Innovation in Therapy

In contrast to the collection of information, innovation in the use of a therapy more vividly suggests the idea of research. The expression "innovation" has a similar connotation to the expression "experimentation" and therefore almost assumes the conclusion that the relevant activity is research.

An experiment can be defined as a test, trial or procedure for testing a hypothesis or a known fact, usually under conditions determined by the tester.[38] Professor

---

[37] See *Supplementary Note 6 — Epidemiological Research* of the *NHMRC Statement, supra* note 5.
[38] Definition adapted from *Concise Oxford Dictionary* (Oxford: Sykes, 1990) and *Butterworths Medical Dictionary* (London: Critchley, 1978).

Dickens has argued that when orthodox therapy is available, and a new treatment is administered to see if it will prove more successful, use of the new treatment is experimental.[39] This argument has to be accepted. It is irrelevant that the new procedure, on proving successful, later becomes the new orthodox treatment. Professor Dickens regards the evolutionary process by which a novel treatment becomes orthodox as a separate question, perhaps one of convention within the practice of a particular medical specialty. However, this does not exhaust the categories of situations where innovation in therapy occurs.

Innovation occurs in circumstances where:

(a) an established procedure or tested drug used in different circumstances is used in place of the orthodox treatment.
(b) an established procedure or tested drug is used in circumstances where there is no orthodox treatment.
(c) a procedure or drug not previously tested in any circumstances is used to treat a condition for which there is no orthodox treatment.

An example of innovative use of a procedure, occurring in the Australian pituitary program, was the use of a combined hPG and clomiphene regime in ovulation induction.[40] The orthodox procedure was to treat with hPG when clomiphene had proven unsuccessful, rather than to use the drugs in combination. However, the HPAC approved a research allocation of hormone to enable the combined procedure to be tested. (This was one of many hypotheses relating to hPG and hGH therapy tested at different centres in the course of twenty years. During that time at a general level the procedure for hGH therapy quickly became standardised, while hPG therapy differed in some respects from centre to centre for a significant period.) It was surmised that the different ovulation induction regime would be as successful or more successful than what was increasingly becoming the orthodox regime, and the procedure for testing this hypothesis fell within situation (a). Judged according to Professor Dickens' analysis, it arguably constituted research.

In situation (c) a procedure or drug not previously tested in any circumstances is used to treat a condition for which there is no orthodox treatment. The issue of what is research seems easier with regard to innovative treatment where there is no orthodox treatment. The paradigm case of the "experiment" in this area is the clinical trial of a new drug.

The term "clinical trial" is generally understood as the testing of a drug in humans. The Therapeutic Goods Administration of the federal Department of Human Services and Health in Australia defines a clinical trial as "an experiment conducted in humans in order to assess the effects, efficacy and/or safety of a

---

[39] B.M. Dickens, "What is a Medical Experiment?" (1975) 113 Canadian Medical Association Journal 635 at 636. Dickens also classifies as experimental the repetition of an orthodox therapy where this is unnecessary for therapy but is conducted to analyse the methodology of the procedure or to validate or refute its findings.

[40] *Allars Report, supra* note 2 at 475–6.

substance, product or procedure".[41] In *Supplementary Note 3 — Clinical Trials*, the NHMRC defines a clinical trial as "a study done in humans to find out if a treatment or diagnostic procedure, which it is believed may benefit a patient, actually does so".[42]

Unexpectedly, with regard to situation (c), Professor Dickens argues that innovative therapy is not experimental when orthodox medicine provides no adequate treatment for a given condition:

> *If no orthodox treatment exists for the patient's condition* (either because of the condition's novelty or because the orthodox treatment has become discredited by advances in medical knowledge) *the physician's innovation will be nonexperimental.* Indeed, in the absence of medically accepted treatment, and perhaps in the presence of knowledge of attempted treatments that have failed, the doctor must have recourse to novel procedures.[43]

If this view were to prevail, where no orthodox procedure exists, the testing by an innovative procedure of any hypothesis, even an extremely radical or risky one, would not be counted as research. Such activity falls outside the ambit of a code of ethics applying to human experimentation. There are many medical conditions for which orthodox medicine can provide no remedy. On this approach if there is no treatment available to remedy a medical condition, a new treatment may be given without seeking informed consent and approval of an ethics committee.

If Dickens' view prevailed, the early use in Australia of hPG and hGH to treat conditions for which there was no remedy would not have amounted to experimentation or research. However, the later use of hPG in combination with clomiphene as an innovative use of a now orthodox treatment (situation [a]), would qualify as research.[44] The result is paradoxical. The pituitary hormone treatment involved use of a new drug, which today would of course be assumed to be research and subject to requirements applying to clinical trials of new drugs, including approval of the protocol by an institutional ethics committee. Is it plausible to adopt a definition of research which accepts that use of a new drug is not an experiment? If use of a new drug is not an experiment, then neither is innovative treatment with a new procedure or untested drug where no orthodox procedure or drug exists.

The Inquiry interviewed 77 hormone recipients. It was important to them to know whether they had been part of an experiment. The Inquiry was not requested in its terms of reference to answer this question. No express finding was made as to whether the hormone program as a whole involved an experiment.[45]

However, the Inquiry documented a number of statements made by government decision-makers and doctors when the treatment was first introduced, which

---

[41] Therapeutic Goods Administration, Department of Community Services and Health, *Clinical Trials of Drugs in Australia* (May 1991) (as amended by Addendum of August 1991).
[42] See *Supplementary Note 3 — Clinical Trials* of the *NHMRC Statement, supra* note 5. *Supplementary Note 3* was adopted in November 1987.
[43] B.M. Dickens, *supra* note 39 at 636.
[44] See *supra*, text accompanying note 40.
[45] See *Allars Report, supra* note 2 at 519–520.

acknowledged that the treatment was experimental.[46] These included a statement in Parliament by the Minister for Health, and admissions made by doctors in published research papers that their use of the hormones was experimental. Pituitary hormone treatment continued in Australia until 1985. This became a routine therapy for growth deficiency and anovulation. Did the therapy nevertheless remain experimental for twenty years? Was there a point in time when it was no longer true that there was no orthodox treatment for anovulation and that ovulation induction with hPG had become orthodox? In the light of the practical reality of twenty years of therapeutic intervention of this nature, it is difficult to deny that such a point of time was reached, where what had once been innovative became orthodox, removing the therapy from situation (c). But it is impossible to pinpoint a precise point in time.

The category of innovative therapy described as situation (b) is illustrated by the use of hPG in the first IVF work done in Australia.[47] The treatment of anovulatory women with hPG had probably become orthodox by the mid 1970s. Its use on ovulatory women in an IVF program was an innovative use of the hormone. The early IVF program illustrates situation (b), where an established drug is utilised as part of a set of procedures to treat a different condition for which there is no orthodox treatment. Situation (b) is not specifically identified by Professor Dickens. This example, I would argue, was research. It involved the testing of a hypothesis that eggs could be recovered by stimulating normal ovaries with hPG and that these eggs could later be implanted.

While it is easy to say on the basis of the examples that (a), (b) and (c) represent research rather than clinical practice, these conclusions really rest uneasily upon some judgments of degree which may be more difficult to make in other cases. To what extent is a procedure orthodox and to what extent is a new use innovative in departing from previous practice?

As in the area of collection of patient information, the problem of multiple purposes has to be recognised. The presence of a purpose of testing a hypothesis points to the activity being research rather than clinical practice, just as the purpose of obtaining information for the additional purpose of testing a hypothesis points to research. When such a purpose is completely absent there is no doubt that the activity is clinical practice. The *Belmont Report* recognised that even a significant departure from a standard or accepted practice need not automatically be characterised as innovation amounting to research.[48] It depends upon how radical the departure is.

Is the degree of departure from accepted practice to be measured by the number of occasions on which the innovation occurs? It might be argued that testing of a hypothesis which amounts to innovation requires the use of a group of patients, or a group plus a control group. Yet some new treatments are tried out initially on one human being — for example, the use of baboon marrow in a patient suffering from

---

[46] *Ibid.*
[47] *Ibid.* at 190–197.
[48] See *Belmont Report, supra* note 11 at 3.

HIV. Such an intervention is commonly regarded as experimental yet it does not involve comparison of responses in groups of patients. Of course where a patient consents to the experimental drug or procedure, the concern about the distinction is removed. However, it shows that the presence of systematic group comparison is not a necessary feature of innovation, and hence not a necessary feature of research.

It can also be argued that since uncertainty as to diagnosis and appropriate treatment is typical rather than unusual in clinical practice, testing of a series of hypotheses is almost a necessary feature of clinical practice. The testing of a more general hypothesis, particularly where innovation is involved, is however, very different from testing a hypothesis as to which of several diagnoses is correct with a view of intervening with an orthodox treatment.[49]

An interim conclusion is that situations (a), (b) and (c) are research rather than clinical practice where the innovation includes a purpose of testing a general hypothesis under conditions determined by the doctor.

## PART V: CONCLUDING REMARKS AND ADMINISTRATIVE LAW PRINCIPLES

### *The Distinction Which Matters*

The distinctions drawn in the *Declaration of Helsinki* between clinical and non-clinical research and between therapeutic and non-therapeutic research were described in Part II and analysed in Part IV. The analysis was assisted by examples drawn from the *Report of the Inquiry into Pituitary Derived Hormones in Australia and Creutzfeldt-Jakob Disease* whose relevant findings were outlined in Part III. The analysis suggests that the distinctions are incoherent and unnecessary. The distinction which needs to be made is between what is research and what is clinical practice rather than research. The *NHMRC Statement*, which is modelled upon the *Declaration of Helsinki*, fails to define research. The task of drawing the distinction was briefly addressed in the *Belmont Report* and in more detail in 1995 in the *Enquiry into Research Ethics* in Canada,[50] but remains an urgent one. The need for a distinction is illustrated not only by the story of the pituitary hormones program in Australia but also by other recent examples of innovation which caused irreparable harm to individual patients, such as the use of deep sleep therapy in Australia, contrary to orthodox practice,[51] and the unorthodox treatment of cervical cancer in New Zealand.[52]

The concept of purpose proved to be unhelpful in forging a distinction between therapeutic and non-therapeutic research.[53] Yet purpose unavoidably appears to be

---

[49] See Chapter 6.

[50] *Enquiry on Research Ethics: Final Report* (David N. Weisstub, Chairman. Submitted to the Honourable Jim Wilson, Minister of Health, Ontario, 28 August 1995).

[51] *Report of the Royal Commission into Deep Sleep Therapy (Chelmsford Report)* (Sydney: 1990).

[52] *The Report of the Cervical Cancer Inquiry. Report of the Committee of Inquiry into Allegations Concerning the Treatment of Cervical Cancer at National Women's Hospital and into Other Related Matters* (Auckland: 1988) (*Cartwright Report*).

[53] See *supra* at 20–25

the key to distinguishing between what is research and what is not. In Part IV the focus shifted from purposes of different types of research to the purposes of different activities of a doctor. The strategy is to allow purpose to operate as an independent criterion for distinguishing research from clinical practice. Professor Hare writes:

> The key to understanding the distinction between therapeutic and non-therapeutic research is to notice that both research and therapy are marked out as such by their aims, which are different from each other. Research aims at the advancement of knowledge, therapy at the cure of the patient. Therapeutic research cannot be research which is therapy or therapy which is research (that is impossible); it is, rather, an activity which has both aims. In the purest case, the very same intervention on a patient may be intended both to cure and to discover something. In less pure cases interventions may take place which have a therapeutic intention but which are modified in some way as an aid to research (for example, a few extra millilitres of blood are taken when doing a diagnostic sample, so that the additional blood can be used for research). Therapeutic research is thus an activity which has only a research and not a therapeutic aim.[54]

Hare's recognition of the multiple purposes of doctors in engaging in activities is valuable. It reinforces the analysis in Part IV of the three situations in which patient information is collected and recorded. Situation (i) is clinical practice. Situation (iii) is research. Hare's example of the excessive blood sampling truly belongs in situation (iii), rather than in situation (ii) of multiple purposes as Hare's aside in the passage above would suggest. Situation (ii) is difficult to categorise as research or clinical practice. In Hare's view it is research. From Hare's discussion it can be concluded that the expression "therapeutic research" should be jettisoned, since the term "therapeutic" in it is misleading. The simple expression "research" should be used.

There is a need for analysis beyond Hare's conclusion to determine whether situations of multiple purposes, whether in the collection and use of patient information or in innovative therapy, inevitably amount to research rather than clinical practice. Administrative law principles can assist in this endeavour.

### *Administrative Law Principles*

The problem of multiple purposes is a familiar one in Australian administrative law. Administrative law principles assist in the task of unravelling multiple purposes and discerning which purposes matter with regard to the legal validity of a decision. Principles of abuse of power, applied in judicial review of administrative action, require that statutory powers not be exercised by public officials for improper purposes. Exercise of a power for an improper purpose, namely a purpose other than the purpose of the statute, renders the administrative decision invalid.[55]

The courts have grappled with the question whether an administrative decision is invalid on account of the presence of one improper purpose amongst other purposes

---

[54] Hare, *supra* note 28 at 134.
[55] *R. v. Toohey (Aboriginal Land Commissioner); Ex parte Northern Land Council* (1981) 151 CLR 170.

which are proper. One judicial approach is to require that the improper purpose be a dominant one if the decision is to be struck down.[56] Another approach requires that the improper purpose was necessary, in that the decision would not have been made but for the improper purpose.[57] The courts have declined to strike down a decision purely on the basis of the presence of one improper purpose, no matter how minor.

Administrative law approaches to multiple purposes may assist in determining when decisions of doctors with regard to diagnosis or therapy are so affected by additional purposes that they amount to research and are subject to codes of ethics. Dickens, using the term "motive" rather than "purpose", argues for a strict test for determining whether a doctor's activities involve research.[58] He rejects a test of dominant purpose in favour of a test of whether there is any improper research purpose at all. He justifies the strict test as follows:

> This properly attempts to protect the patient from even minor experimentation being concealed within the interstices of orthodox therapy, exposing it to the light of peer review and ethical (including legal) assessment. Cautious and restrained development of an orthodox procedure the investigator suspects to be inadequate will rank as experimental unless so minor as to come within the flexible limits of the orthodox.[59]

Is an activity research rather than clinical practice simply because of the presence of one, possibly, very minor, additional purpose, either in collecting patient information or in innovative therapy, of testing a general hypothesis? Must the additional purpose dominate over the therapeutic purposes in order that the activity be described as research? Or must the additional purpose be necessary in that certain aspects of the clinical work would not have been undertaken but for that purpose?

Other principles of administrative law concerned with purpose are the Information Privacy Principles (IPPs), which apply to federal agencies in Australia.[60] The IPPs require that individuals be advised of the purposes for which their personal information is collected, and that the information should not be collected or used for other purposes, in particular disclosure to third parties. The rationale of the IPPs can assist in understanding why collecting and using patient information for purposes other than diagnosis or determination of treatment, or using innovative therapy for the purposes of testing a general hypothesis, is a research activity rather than clinical practice.

Although not formulated in terms of purposes, the bias rule of procedural fairness may also assist in dealing with the conflicts of interest which arise from multiple purposes of doctors. According to the bias rule a government decision is void if a reasonable apprehension may reasonably be engendered in the minds of those who come before the decision-maker or in the mind of the public, that the decision-maker did not bring to the decision-making process a fair and unprejudiced mind.[61] A

---

[56] For example, in *Westminster Corp v. London and Northwestern Railway*, [1905] AC 426.

[57] For example, in *Thompson v. Randwick Municipal Council* (1950) 81 CLR 87.

[58] B.M. Dickens, *supra* note 39 at 636.

[59] *Ibid.*

[60] *Privacy Act 1988* (Cth).

[61] *Stollery v. Greyhound Racing Control Board* (1972) 128 CLR 509.

doctor who acts in the twin roles of clinician and researcher, as when there are multiple purposes, must not prejudice clinical judgment and duties to the patient as a therapist on account of the presence of a research interest.[62] The situation of multiple purposes in obtaining patient information is inherently one of conflict of interest if the doctor conceals from the patient the additional purposes for which the information is obtained and used. The same is true of innovative therapy. In the case of the pituitary hormone program the potential for conflict of interest, and hence prejudice in exercising judgment, was further complicated by the role played by the doctors as regulators.[63]

A patient who consents to medical treatment confers power on the treating doctor. That power must be exercised for proper purposes. To determine treatment by reference to extraneous factors such as the successful conduct of a research project is to travel beyond the scope of the power conferred by the patient upon the doctor to engage in clinical practice. The doctor steps over the line between clinical practice and research. It is an exercise of clinical power for a research purpose and hence an exercise of power for an improper purpose. Collecting and using personal patient information for the purpose of more accurate diagnosis or improving the therapy for that patient lies within the scope of the power conferred by the consent. Collecting and using such information purely for additional purposes falls outside the scope of that power if consent of the patient to the use of the information for that purpose is not obtained.

Whether every situation of multiple purposes, whether in the collection and use of patient information or in the use of innovative therapy, constitutes research is the challenging question requiring further examination. A purpose of testing a general hypothesis is almost inextricably part of innovative therapy. It will be rare that this purpose will be absent from innovative therapy so as to preclude categorisation of the activity as research. Administrative law principles do not lightly excuse the presence of an improper purpose and provide an instructive guide to development of a concept of abuse of power in the biomedical context. Unless codes of ethics such as the *Declaration of Helsinki* and the *NHMRC Statement* address the distinction between research and clinical practice, their ambit, and therefore their authority, will remain uncertain.

---

[62] See B. M. Dickens, "Conflicts of Interest in Canadian Health Care Law" (1995) 21 American Journal of Law & Medicine 259.
[63] See *supra* at 28–34.

CHAPTER 8

# "CONSENSUAL" RESEARCH WITH COGNITIVELY IMPAIRED ADULTS: RESOLVING LEGAL SHORTCOMINGS IN ADULT GUARDIANSHIP

GEORGE F. TOMOSSY AND DAVID N. WEISSTUB

## INTRODUCTION

It is a well-established ethical and legal principle that experiments should not be conducted without first obtaining a subject's free and informed consent. In the case of cognitively impaired adults, however, meeting this standard is difficult, if not impossible, as such persons often lack the requisite mental capacity to make autonomous and contemporaneous decisions with respect to their personal welfare. The need for concern is heightened when involving research that is non-therapeutic, that is, provides no direct medical benefit to the subject. A simple solution would be to ban such experiments entirely; however, scientific progress in the understanding and treatment of conditions specific to vulnerable populations is of great social utility, and an absolute prohibition would be an excessively paternalistic response. Understandably, attempts to accommodate this social incentive raise difficult ethical and legal dilemmas.

The doctrine of informed consent, although serving well the interests of fully competent persons, falls short when applied in the context of individuals with compromised decision-making ability. Viable alternative methods for obtaining a subject's consent, such as the use of advance directives and substituted decisions, must therefore be considered. Neither of these options is currently available (in the context of research) in Canada's English-speaking provinces, where guardianship statutes, and also the common law, fail to provide a satisfactory solution.

It is our intention here to address these deficiencies in the present legal order and to suggest much-needed reforms in adult guardianship laws.

137

## DEFINITIONS

*Research* and *experiment*, often employed interchangeably in the literature, are distinguished from *treatment* by the respective absence or presence of a benefit accruing to the subject/patient. This benefit is calculated by applying predictions made in accordance with established standards of medical practice. The descriptors *therapeutic* and *non-therapeutic* can be rationalized in a similar fashion. As antonymous labels, reflecting the often puristic nature of definitions, these terms are somewhat artificial, given that a research protocol is described most appropriately as falling along a spectrum between two poles, therapy and non-therapy.[1] Nevertheless, despite the seeming arbitrariness, being able to categorize an experiment is vital, because the duties and responsibilities owed by the researcher to the subject will be determined largely by the nature of their relationship, that is, whether it is therapeutic or not. Our present discussion will be restricted to research that is categorized as *non-therapeutic*,[2] meaning that the protocol is *primarily* non-therapeutic, based on an objective appraisal of the experiment as a whole rather than on the stated intent of the researcher.

*Cognitively impaired adults* refers to a heterogeneous group of individuals, including those with mental disorders and developmental disabilities, such as dementing or psychotic disorders and mental retardation, and persons suffering from sudden physical or mental traumas.[3] Cognitive impairment may vary in degree and in kind, potentially arising suddenly, developing over a period of time, fluctuating from day to day, or having existed prior to reaching adulthood. The common feature shared by all members of this diverse group is a diminished capacity to make decisions, or in the present context, the inability to properly understand, make choices about, or communicate decisions regarding participation in research.[4] This definition takes into account the modern trend in competency assessment, which requires that mental incapacity be determined on a functional basis, that is, with respect to the specific matter in question.[5] Incapacity in one area of decision-making does not automatically imply incapacity in another; for although belonging to the class of cognitively impaired adults, a person cannot be presumed to be incapable of making the decision of whether or not to become a subject in an experiment. Consequently, and because of the heterogeneity of the population, any protective legal régime must be sufficiently flexible to accommodate the different needs of individuals that may vary not only from group to group, such as persons with mental disorders as opposed to the developmentally disabled, but also within a group, where needs may vary on a case by case basis.

---

[1] See Chapter 6.

[2] The discussion will also generally be confined to clinical research, although some aspects will certainly bear relevance to research in the social sciences.

[3] The definition is derived from technical definitions of mental disorder and mental retardation. See American Psychiatric Association, *Diagnostic and Statistical Manual of Mental Disorders*, 4th ed. (Washington, D.C.: A.P.A., 1994) at xxi-xxii, 39-42. See also the discussion in Chapter 21.

[4] American College of Physicians, "Position Paper on Cognitively Impaired Subjects" (1989) 111 Annals of Internal Medicine 843 at 843.

[5] D.N. Weisstub, *Enquiry on Mental Competency: Final Report* (Toronto: Queen's Printer, 1990).

## JUSTIFICATIONS AND INCENTIVES

If we decide, as a society, that we are justified in exposing vulnerable persons to risks inherent in non-therapeutic research, however minute, for benefits to be reaped by others, then we must avoid the pitfalls of legal and moral fictions that may result from our efforts to appease our collective moral conscience. An appeal to a sacrificial ethic that presumes a person's desire (and perhaps consent) to participate in the research endeavour, if given the opportunity, would be a valiant model for citizenship. However, such a model would in fact be immoral if applied to persons who, because of their inability to act as autonomous agents, can neither affirm nor deny such a presumption. To speak in terms of a "duty" to accept the role of experimental subject would likewise be inappropriate.[6]

There is also the issue of a right to participate in biomedical experiments. Such a "right" is supported by arguments which hold that a person should not be paternalistically restrained from risk-taking.[7] Individuals have always enjoyed the freedom to participate in activities that do not provide any direct benefits other than moral or personal satisfaction, and which may even expose them to substantial risks of harm.[8] Although control over our bodies is a fundamental value insofar as it reflects our interest in preserving individual autonomy, this right cannot be without limits. Indeed, the principle of personal inviolability must at times supersede that of autonomy, including in the context of research.[9] This limitation becomes all the more relevant in the case of a person whose ability to make decisions is compromised. Someone with a serious developmental disability, for example, may lack the life experiences necessary to formulate certain moral choices. The decision of whether or not to expose oneself to risks for the benefit of others is very much a matter of morality. To appeal to a "right" in such cases as a basis for justifying participation would lead us to embark upon a "pathway of fictionalizing the moral enhancements of vulnerable populations",[10] which could in fact lend credence to

---

[6] To couch a presumption of consent in terms of an alleged social duty would be a good example of a moral fiction that should not be accepted in the research context. See Chapter 4.

[7] A second argument in support of a right to participate in research stems from the principle of beneficence, whereby persons should not be prevented from access to benefits of research, including certain drugs or therapies. See M.L. Elks, "The Right to Participate in Research Studies" (1993) 122 Journal of Laboratory & Clinical Medicine 130 at 131. Although arguments favouring a right to participate in research may be justified when concerning persons who are fully competent, such as in cases involving innovative treatments where no other options are available, they are difficult to apply in the context of non-therapeutic research involving cognitively impaired adults whose ability to make autonomous decisions is diminished, or perhaps even nonexistent. The notion of a "right" becomes relevant, however, in those cases where it is possible to express choices prior to the point at which autonomous behaviour is no longer possible.

[8] In Canada, discrimination on the basis of mental disability runs contrary to s. 15(1) of the *Canadian Charter of Rights and Freedoms*, Being Part I of the *Constitution Act, 1982*, being Schedule B to the Canada Act, 1982, c. 11. However, it is doubtful that the infringement of a "right to participate in non-therapeutic research", if in fact it could be demonstrated that such a right exists, would actually attract the censure of the courts.

[9] For a discussion of the principle of inviolability as it applies to human experimentation, see Chapter 10.

[10] See Chapter 4

slippery slope arguments favouring a ban on all research involving vulnerable persons.

We must therefore acknowledge that any decision to condone experiments involving cognitively impaired adults has at its root a societal need, rather than imagined altruistic motivations of mentally incompetent subjects, which we could not, in all fairness, attribute to ourselves in all cases.

The decision to permit non-therapeutic research is not merely a question of morality or social policy, but one of necessity. Researchers studying cognitive disorders, including Alzheimer's Dementia, are faced with a lack of alternative animal models, and consequently must employ human subjects.[11] Also, as a result of the changing demographics of our population, the importance of such research is increasing; and in fact, the growing number of elderly persons has created an important therapeutic market.[12] A decision *not* to conduct relevant research would be neglectful of the special needs of this population.[13] However, these necessities have not been recognized consistently. Both the *Nuremberg Code*[14] and the *International Covenant on Civil and Political Rights*[15] effectively prohibit non-therapeutic experimentation with persons who are unable to provide consent. This stance is maintained by the United Nations Human Rights Committee,[16] and is an understandable reaction to historic incidents of scientific misconduct which include, in addition to the horrendous abuses of World War II, the infamous Tuskegee Syphilis Study,[17] the Willowbrook Hepatitis Experiments,[18] the American Radiation Studies,[19] and the Cameron Affair.[20]

A further justification for restricting research is founded upon slippery slope arguments, which hold that procedural safeguards can never suffice to forestall the potential harms that can occur as a result of making what may seem at the outset to

---

[11] E.W. Keyserlingk *et al.*, "Proposed Guidelines for the Participation of Persons with Dementia as Research Subjects" (1995) 38 Perspectives in Biology & Medicine 319 at 319.

[12] See e.g. Health and Welfare Canada, *Fact Book on Aging in Canada* (Ottawa: Minister of Supply and Services, 1983) at 14, 25. The population aging phenomenon is one of Canada's major social issues, having as its implication an increasing demand for health care and other social services. See R.M. Gordon & S.N. Verdun-Jones, *Adult Guardianship Law in Canada*, Rel. 2 (Scarborough: Carswell, 1995) at 1-12 — 1-13.

[13] See e.g. C.G. Swift, "Ethical Aspects of Clinical Research with the Elderly" (1988) 40 British Medical Journal of Hospital Medicine 370.

[14] The *Nuremberg Code* constituted part of the judgment resulting from *U.S. v. Karl Brandt et al., Trials of War Criminals Before the Nuremberg Military Tribunal Under Control Council Law No. 10.* (October 1946-April 1949).

[15] G.A. Res. 2200 (XXI), 999 U.N.T.S. 171 (1966), art. 7.

[16] Human Rights Committee, 53rd Sess., 1413rd Mtg., CCPR/C/SR. 1413, (6 April 1995), paras. 21, 35.

[17] J.H. Jones, *Bad Blood* (New York: Free Press, 1981).

[18] See H. Beecher, "Ethics and Clinical Research" (1966) 274 New England Journal of Medicine 1354; M.A. Grodin & J.J. Alpert, "Children as Participants in Medical Research" (1988) 35 Pediatric Clinics of North America 1389.

[19] Advisory Committee on Human Radiation Experiments, *Final Report* (Washington, D.C.: U.S. Government Printing Office, 1995).

[20] See Government of Canada, *New Release: Background Information — Depatterning at the Allan Memorial Institute* (Ottawa: Department of Justice, 17 November 1992); G. Cooper, *Opinion of George Cooper, Q.C., Regarding Canadian Government Funding of the Allan Memorial Institute in the 1950s and 1960s* (Ottawa: Supply and Services Canada, 1986).

be a minor moral concession. This position, however, can be countered on two grounds. Firstly, the populations, if not the individuals, to be called upon to assume risks would also be the beneficiaries.[21] Secondly, our tools for surveillance and protection, if carefully harnessed, can indeed provide the requisite monitors and safeguards, assuming that the actors, including political ones, are not acting in bad faith.[22] An absolute ban, therefore, should not be the chosen course of action. Rather, there should be a balancing of society's interest in conducting important and promising research with the interests of the potential subject.[23] The social incentive to expand scientific knowledge, even at the cost of exposing individuals with impaired decision-making capacity to risks of harm, cannot be without bounds.[24] Over the years, the extended dialogue on this topic has yielded an international consensus which condones experimentation involving vulnerable populations, provided that appropriate safeguards are introduced, and which also recognizes the social utility in increasing our understanding of illness and disease. This view is upheld by the *Declaration of Helsinki*[25] and by the recent *International Ethical Guidelines for Biomedical Research Involving Human Subjects* promulgated by the Council for International Organizations of Medical Science in collaboration with the World Health Organization.[26]

## REGULATING RESEARCH

The historical record supports the unfortunate conclusion that members of certain populations, including cognitively impaired adults, may always be prone to exploitation and abuse in the research setting. Society must remain forever vigilant in order to protect its vulnerable members. This can only be accomplished by enforcing well-defined standards of ethical conduct. The question which lies before us is whether this goal can best be realized through internal or external regulatory mechanisms. Although a moot point in France, given its extensive legislation

---

[21] The need for action is heightened further by pressures to develop more efficient tools for low-cost health protection for these vulnerable populations.

[22] This, in any event, is never ultimately guaranteed in any civil society, including democratic ones. In attaining these standards of surveillance, the position taken here is that self-regulation or highly discretionary decision-making is unacceptable.

[23] See also V.L. Melnick *et al.*, "Clinical Research in Senile Dementia of the Alzheimer Type: Suggested Guidelines Addressing the Ethical and Legal Issues" (1984) 32 Journal of the American Geriatric Society 531 at 535.

[24] The *Declaration of Helsinki* clearly states that "[i]n research on man, the interest of science and society should never take precedence over considerations related to the well-being of the subject". See Medical Association, *Declaration of Helsinki*, Adopted at the 18th World Medical Assembly in Helsinki in June 1964. Amended at the 19th World Medical Assembly in Tokyo in October 1975; the 35th World Medical Assembly in Venice in October 1983; and the 41st World Medical Assembly in Hong Kong in September 1989.

[25] *Ibid.*

[26] Council for International Organizations of Medical Science, in collaboration with the World Health Organization, *International Ethical Guidelines for Biomedical Research Involving Human Subjects* (Geneva: CIOMS, 1993) [hereinafter *CIOMS Guidelines*].

governing the area,[27] the question remains a source of contention in Canada where, with the exception of Quebec, experimentation has been "regulated" only through the use of ethical guidelines, such as those promulgated by the Medical Research Council of Canada.[28] Although professional self-regulation and medical education both play an important role in instilling ethical behaviour, the mere availability of ethical guidelines, from a legal perspective, is of limited worth as a regulatory authority.[29] They are insufficient in establishing a uniform set of rules that are legally binding, "unless buttressed by overarching political and social institutions".[30] This point is illustrated by a recent Canadian case in which both the researcher and the university-affiliated hospital (for its research ethics committee) were held to be liable for the death of a subject in a non-therapeutic experiment.[31] The case turned on the failure of the researcher to disclose certain risks to the subject, and did nothing to clarify the legal status of professional guidelines as sources for defining standards of conduct, which if adhered to would absolve researchers and their affiliated institutions from liability. Although the case refers to the *Declaration of Helsinki*,[32] there is no mention of the 1978 Medical Research Council of Canada Guidelines which were being followed at the time the events took place. The judgment was therefore criticized on the basis that, in establishing a legal standard of professional conduct, a court should begin by examining the practice of the profession, regardless of whether this proves inconclusive or deficient.[33] However, even if professional guidelines are followed, they will not automatically provide legal protection in negligence actions. As in the United Kingdom, decisions by research ethics committees do not make researchers' actions lawful.[34] The Medical Research Council of Canada is a corporation created by a federal statute,[35] and committees adhering to its guidelines lack the power to authorize conduct which is not otherwise sanctioned by law. Therefore, it is in the interests of subjects, researchers, and their institutions, that a statutory basis prescribing ethical and legal conduct in biomedical research be established.

---

[27] C. Huriet, "La loi française relative à la protection des personnes qui se prêtent à des recherches biomédicales: origine et histoire" (1992) 43 Recueil International de Législation Sanitaire 414. See also the discussion in Chapter 9.

[28] Medical Research Council of Canada, *Guidelines on Research Involving Human Subjects* (Ottawa: Supply & Services Canada, 1987). These guidelines have been superceded by the Tri-Council Working Group, *Code of Ethical Conduct for Research Involving Humans* (July, 1997).

[29] P.R. Benson, "The Social Control of Human Biomedical Research: An Overview and Review of the Literature" (1989) 29 Social Sciences & Medicine 1.

[30] P.R. Benson & L.H. Roth, "Trends in the Social Control of Medical and Psychiatric Research" in D.N. Weisstub, ed., *Law and Mental Health: International Perspectives, Volume 4* (New York: Pergamon Press, 1988) 1 at 5.

[31] *Weiss v. Solomon* [1989] R.J.Q 731 (S.C.).

[32] *Ibid.* at 741.

[33] B. Freedman & K.C. Glass, "*Weiss v. Solomon: A Case Study in Institutional Responsibility for Clinical Research*" (1990) 18 Law, Medicine, & Health Care 395 at 401-402.

[34] The Law Commission, Mental Incapacity (London: HMSO, 1995) at paras. 6.29-6.33. This view is shared by the Queensland Law Reform Commission. See Queensland Law Reform Commission, *Assisted and Substituted Decisions: Decision-making by and for People with a Decision-Making Disability, Vol. 1* (Brisbane: Queensland Law Reform Commission, 1996) at 64.

[35] *Medical Research Council Act*, R.S.C. 1985, c. M-9.

The tendency in many countries has been to enact legislation or to promulgate official regulations, with prominent examples including the Department of Health and Human Services Regulations in the United States[36] and amendments to the French *Code de la santé publique.*[37] This trend is further reflected in recent law reform initiatives in Canada,[38] the United Kingdom[39] and Australia.[40]

## LEGAL NECESSITY

As mentioned earlier, Quebec is the only Canadian jurisdiction that specifically regulates human experimentation.[41] In the rest of Canada, neither existing adult guardianship statutes nor the common law provide a satisfactory solution. Before discussing much-needed reforms, it is important to identify the deficiencies of the present legal order, beginning with the failure of the doctrine of informed consent as a protective régime for vulnerable adults.

### *The Doctrine of Informed Consent*

Informed consent is one of the great canons of medical law, which includes the law on human experimentation. Indeed, the duty of disclosure of potential risks, irrespective of how remote they may appear, is greater in the context of research than of treatment;[42] this principle is clearly enunciated in Canadian case law.[43] Because individuals who submit to non-therapeutic experiments are exposed to risks without the promise of receiving any complementary benefits, other than perhaps a sense of satisfaction, participation must never result from consent that is not genuine, or worse yet, coerced. What remains to be seen is whether the doctrine of

---

[36] Title 45 Code of Federal Regulations Part 46 (18 June 1991).

[37] *Loi No.88-1138 du 20 décembre 1988* (J.O. December 22 1988); *Loi 90-86 du 23 janvier 1990,* (J.O. 25 January 1990); *Loi 90-549 du 2 juillet 1990,* (J.O. 5 July 1990); *Loi 91-73 du 18 janvier 1991,* (J.O. 20 January 1991); *Loi 94-630 du 25 juillet 1994,* (J.O. 27 July 1994).

[38] Law Reform Commission of Canada, *Working Paper No. 61: Biomedical Experimentation Involving Human Subjects* (Ottawa: Law Reform Commission, 1989); Law Reform Commission of Canada, *Toward a Canadian Advisory Board on Biomedical Ethics* (Ottawa: Law Reform Commission of Canada, 1990); *Enquiry on Research Ethics: Final Report* (David N. Weisstub, Chairman. Submitted to the Hon. Jim Wilson, Minister of Health of Ontario, Aug. 28, 1995).

[39] See The Law Commission, *supra* note34.

[40] See Queensland Law Reform Commission, *supra* note 34.

[41] Arts. 20-26 C.C.Q. It is worthwhile to note that, despite the existence of the *Civil Code* articles governing research, current regulatory models in Quebec were criticised as being inconsistent and inadequate in protecting research subjects while at the same time maintaining scientific integrity. It was recommended, therefore, that a more elaborate regime be developed, one which properly takes into consideration the ethical, scientific, and financial aspects of research, and that a permanent regulatory authority be created. See Comité d'experts sur l'évaluation des mécanismes de contrôle en matière de recherche clinique, *Rapport sur l'évaluation des mécanismes de contrôle en matière de recherche clinique au Québec,* (Chairman: Pierre Deschamps, Submitted to the Hon. Jean Rochon, Minister of Health and Social Services, Province of Quebec, June 9, 1995).

[42] See e.g. R. Delgado & H. Leskovac, "Informed Consent in Human Experimentation: Bridging the Gap Between Ethical Thought and Current Practice" (1986) 34 UCLA Law Review 67; K.C. Glass, "Informed Decision-Making and Vulnerable Persons: Meeting the Needs of the Competent Elderly Patient or Research Subject" (1993) 18 Queen's Law Journal 191.

[43] See *Halushka* v. *University of Saskatchewan* (1965), 52 W.W.R. 608 (Sask. C.A.); and *Weiss* v. *Solomon, supra* note 31.

informed consent can satisfy this directive fully when applied to adults with diminished cognitive capacities.

The exercise of empowering rights, in particular those which promote an individual's right to self-determination, relies upon the ability to communicate choices effectively, which involves the capacity to understand properly all relevant factors that may influence a given decision. A cognitively impaired adult may be unable to satisfy this requirement.[44] The doctrine of informed consent, which appeals to the principle of autonomy, is only useful in protecting *competent* persons from exploitation and abuse, and hence can only serve the interests of those who are able to fend for themselves without assistance or intervention by others.[45] Total reliance upon autonomy-based principles would therefore be unwise in the context of research involving persons with impaired decision-making abilities. Perhaps a more appropriate basis for the development of legal protections would be the principle of personal inviolability, which can be equated to an aspect of the ethical maxim of respect for persons.[46] This course was adopted in the recently revised *Civil Code of Quebec*, which restricted the participation of even fully competent adults in non-therapeutic research.[47] Legal reform in common law jurisdictions should reflect, at least in part, this protectionist approach.[48] However, the ideal foundation for developing legal safeguards should be a middle ground between these two positions, that is, one that seeks to maximize decision-making capacity while proscribing behaviour that could compromise an individual's mental and bodily integrity.

The doctrine of informed consent has been criticized for its inability to reconcile personal values and beliefs, which leads to its failure to elicit truly meaningful responses.[49] Also, from a practical perspective, the usefulness of the doctrine has been questioned in the light of the lack of a proper allocation of the resources necessary to fulfill its mandate effectively.[50] Nevertheless, informed consent should not be abandoned in its entirety. Indeed, the impetus behind the historical evolution of the doctrine, the reaction against paternalistic patterns of behaviour in the doctor/patient relationship which became unacceptable in an increasingly sophisticated and

---

[44] It is important to reiterate that a cognitively impaired adult's inability to make meaningful decisions in the experimentation context must not be presumed, but rather, as discussed in our definition above, determined on a functional basis.

[45] See M.A. Somerville, "Label versus Contents: Variations Between Philosophy, Psychiatry and Law in Concepts Governing Decision-Making" (1994) 39 McGill L.J. 179 at 193. An alternative view is to treat informed consent as a "gatekeeping" device, which can be used to distinguish those persons who are capable of making their own decisions from those who require additional protection. See R.R. Faden & T.L. Beauchamp, *A History and Theory of Informed Consent* (Oxford: Oxford University Press, 1986).

[46] *Respect for persons* calls not only for the promotion of self-determinative rights, but for the protection of persons with an impaired decision-making capacity. See *CIOMS Guidelines*, *supra* note 26 at 10.

[47] See Chapter 10.

[48] Ethical research involving vulnerable populations should fall within clearly stated boundaries of permissible behavior, whereby the actions of both prospective subjects and persons acting on their behalf (i.e. legal guardians) are carefully circumscribed. See Chapter 18.

[49] R.M. Veatch, "Abandoning Informed Consent" (1995) 25:5 Hastings Center Report 5.

[50] J. Katz, "Informed Consent — Must it Remain a Fairy Tale?" (1994) 10 Journal of Contemporary Health Law & Policy 69. See also P.H. Schuck, "Rethinking Informed Consent" (1994) 103 Yale Law Journal 899.

litigious culture of health care consumers, is equally relevant, if not more so, in the research context. We must forestall the abuse and exploitation of vulnerable persons by researchers who, whether well-meaning or for unscrupulous designs, may take advantage of their positions of trust and/or authority in order to expedite a preferred decision. This goal can be realized by obtaining a free and informed consent to participate, a choice which, as mandated by the doctrine of informed consent, must be reached through a dialogic exchange of information. Traditionally, the burden of this decision would fall upon the prospective subject, and therein lies the insufficiency of the doctrine as a protective régime when applied to non-autonomous persons. Indeed, blind reliance upon a notion of informed consent, without properly accommodating the special needs of incapable subjects, such as providing for the legal intervention of a guardian or advocate, can jeopardize the ethical and legal integrity of research. A solution to the problems raised by non-therapeutic experimentation with cognitively impaired adults cannot therefore be founded solely upon the doctrine of informed consent, regardless of its importance in the course of making decisions, without the support of complementary legal institutions.

### *Adult Guardianship Law*

The principal objective of adult guardianship law is to provide assistance to adults who cannot act for themselves. It governs many aspects of daily life, and can be broadly divided into two categories: the management of property and of the person, with the latter being concerned primarily with health matters.[51] The issue of participation in research is either specifically excluded, not mentioned at all, or if referred to, dealt with in an ambiguous manner.[52] Indeed, existing guardianship laws are generally poorly suited to resolving questions that cannot be answered easily through the application of a "best-interests" calculation.[53] Non-therapeutic experimentation, and indeed any other activity that does not lead to a concrete benefit for the subject, throws the proverbial wrench into the machinery of substitute decision-making. It is difficult enough for guardians, and also for the judiciary, to rationalize exposing an incompetent adult to risks, however minute, for a hypothetical treatment or cure, let alone in those cases where the benefits will never accrue to the subject, but rather to others with the same affliction or disability. This effort is frustrated further because it entails placing the interests of society ahead of those of the subject, which may constitute a breach of the guardian's cardinal duty to protect his ward.[54] Therefore, if we agree that we are justified in exposing incompetent

---

[51] See generally Gordon & Verdun-Jones, *supra* note 12.

[52] For example, in Ontario it is stated specifically that "nothing in this Act affects the law relating to giving or refusing consent on another person's behalf to a procedure whose primary purpose is research". See *Substitute Decisions Act, 1992*, S.O. 1992, c. 30, s. 66(13). British Columbia is the only province where guardians might be able to make decisions relating to research, insofar as the definition of "health care" includes "participation in a medical research program approved by an ethics committee designated by regulation". See *Health Care (Consent) and Care Facility (Admission) Act*, S.B.C. 1993, c. 48, s. 1. [not yet in force]

[53] As will be discussed below, a "substituted judgment" test is also inappropriate. Rather, a hybrid of both standards should be applied in consideration of each particular situation.

[54] See e.g. *Re Leeming*, [1985] 1 W.W.R. 368 (B.C.S.C.).

members of our population to carefully controlled risks for the benefit of others, then we must provide in our legal system for exceptional situations in which guardians may act for reasons that do not produce tangible benefits for their dependents. Granted, such an allowance raises a host of ethical and legal concerns; these will be discussed further below. For the present, we will restrict ourselves to identifying the deficiencies of current guardianship laws, as applied to the research context.

One substantial criticism relates to the "all or nothing" approach,[55] which is reflected in "the plenary nature of most guardianship orders and the requirement that a guardian be appointed only when there is evidence of total mental incompetency or incapacity".[56] This state of affairs runs contrary to the view that "incompetence is not to be understood in any global sense, but rather as reflecting incapacities with respect to specific decisions or areas of decision".[57] The modern model requires a functional assessment of competency, which would allow for a finding of partial incapacity and an award of corresponding powers to the guardian.[58] The rigidity of plenary guardianship appointments can be mitigated when appointing personal guardians; however, the courts appear to be unwilling to implement this option where not expressly authorized by statute.[59] Although the courts' *parens patriae* jurisdiction can be used to fill in gaps in legislation, and thus act to restrict the authority of guardians,[60] this common law power, at least in Canadian jurisprudence, is not likely to prove useful in the context of experimentation.[61]

The problem of inflexibility impacts directly upon the research endeavour. The abandonment of the plenary approach is only the first step. Each class of cognitive impairment requires a different level and kind of intervention, depending upon whether incapacity increases or fluctuates over a period of time, arises suddenly, or is the result of a developmental disability. With the latter two cases, there is often no alternative other than to rely upon a substitute decision-maker; however, guardians are not presently empowered by statute to make such decisions. In all other situations where a person can anticipate future incapacity, that individual may conceivably, while still competent, prepare an advance directive for research. Although legislation governing the use of enduring powers of attorney was enacted in all provinces, only some provinces provide for the creation of *health care* directives.[62] Even so, with the possible future exception of British Columbia, the specific application of advance directives for research is not presently supported in Canada's common law provinces.

---

[55] G.B. Robertson, *Mental Disability and the Law in Canada*, 2d. ed. (Toronto: Carswell, 1994) at 118.

[56] See Gordon & Verdun-Jones, *supra* note 12 at 1-16 — 1-19.

[57] See Weisstub (1990), *supra* note 5 at 35.

[58] See e.g. *Substitute Decisions Act, 1992*, S.O. 1992, c. 30, and also text at note 89.

[59] See Robertson, *supra* note 55 at 119.

[60] *Ibid.* at 170-171.

[61] A critique of the common law will follow.

[62] See Robertson, *supra* note 55 at 177-178; Gordon & Verdun-Jones, *supra* note 12 at 3-129. The provinces that provide for health care directives include Manitoba, Nova Scotia, Ontario, and Newfoundland, with legislation coming into force in British Columbia and in preparation in P.E.I.

If non-therapeutic research is to be facilitated, various elements of existing guardianship statutes will require modification, the extent and substance of which will depend upon their respective levels of sophistication. These areas include: the manner by which guardians are appointed, the nature and scope of their powers, the criteria upon which they are required to base their decisions, provisions for research directives, the degree to which participation of the subject in the consent process is facilitated, termination of guardianship, and various other factors, including mental competency assessment, the confidentiality and accessibility of written directives, and the review of decisions made by either the subject or the guardian. Current guardianship laws are generally lacking in one or more of these respects.

### The Common Law

In Canada, in the absence of guiding legislation, the *parens patriae* jurisdiction empowers the superior courts, when necessary, to protect those who cannot care for themselves. It is well established that such interventions must be based upon a "best interest" or "welfare" standard, for the benefit of the person concerned and not others.[63] The leading case, *Eve*, addressed the issue of whether the courts could approve the "non-therapeutic" sterilization of a mentally retarded woman who was incapable of giving an informed consent to the procedure. The request, made by Eve's mother, was denied on the grounds that non-therapeutic sterilization constituted a highly invasive procedure that would cause irreversible physical damage while providing questionable advantages, and led the court to conclude that "it can never safely be determined that such a procedure is for the benefit of that person".[64] In so doing, the court chose to apply a narrowly interpreted best-interests standard, and declined to adopt a substituted judgment test. This decision, although supported by some,[65] was criticized by others for various reasons. For example, the court prioritized the "privilege" of procreation over other factors, such as "the freedom to form satisfying human relations and experience sexuality without risking pregnancy or paternity",[66] and effectively restricted legally permissible medical interventions (at least within the context of sterilization) to those that provide a clear therapeutic benefit. The House of Lords, distinguishing both *Eve* and an earlier decision,[67] arrived at a different conclusion, choosing instead to adopt the substitute

---

[63] *E. (Mrs.) v. Eve* [1986] 2 S.C.R. 388 at 426,427 [hereinafter *Eve*].

[64] *Ibid.* at 431.

[65] See e.g. M.D.A. Freeman, "Sterilising the Mentally Handicapped" in M.D.A. Freeman, ed., *Medicine, Ethics and the Law: Current Legal Problems* (London: Stevens, 1988) 55.

[66] M.A. Shone, "Mental Health — Sterilization of Mentally Retarded Persons — *Parens Patriae* Power: *Re Eve*" (1987) 66 Canadian Bar Review 635 at 640. See also Manitoba Law Reform Commission, *Report on Sterilization and Legal Incompetence* (Winnipeg: Law Reform Commission, 1992).

[67] In an earlier case involving a minor, the House of Lords decided not to authorize a non-therapeutic sterilization under the *parens patriae* jurisdiction. See *In Re D (A Minor) (Wardship: Sterilisation)*, [1976] 1 All ER 326 (H.L.). Although acknowledged by subsequent cases to have been correctly decided on its facts, this decision was not upheld. See e.g. *In Re B (A Minor)*, [1987] 2 All E.R. 206 (H.L.) [hereinafter *In Re B*]. In Australia, however, it was recently decided that neither the *parens patriae* power nor existing legislation dealing with decision-making for persons with impaired capacity could authorize parents to consent to a non-therapeutic sterilization of their child. See *Secretary, Department of Health and Community Services v. J.W.B. and S.M.B. (Marion's Case)* (1992), 175 C.L.R. 218.

judgment test, and to reject both the non-therapeutic/therapeutic distinction and the conclusion that sterilizations should never be authorized for "non-therapeutic" purposes.[68] These cases, with their conflicting results, demonstrate the difficulty faced by the judiciary in arriving at consistent decisions when applying discretionary powers, including the *parens patriae* jurisdiction, in morally troubling cases.

The analysis in *Eve*, if applied to the context of non-therapeutic research, could lead to a similar conclusion, although the outcome would likely depend upon the level of risk involved. The Court stated that the *parens patriae* jurisdiction "must at all times be exercised with great caution, a caution that must be redoubled as the seriousness of the matter increases".[69] *Eve* focused on the irreversibility and certainty of physical harm. It is therefore possible that experiments with low or negligible risks would not attract the censure of the courts. After all, parents, when exercising their common law authority as the natural guardians of their children, are not prevented from involving them in sports activities, some of which may give rise to risks that are in fact quite substantial.

It is also conceivable that in the case of a person who was previously competent in life, such as someone suffering from Alzheimer's Dementia, a substitute judgment test could be adopted and legitimate a guardian's decision to enroll a dependent in non-therapeutic research. This would call for the guardian to believe that the now incompetent ward would have consented if competent.[70] Such an opinion would be based on knowledge of the values, views, and beliefs previously held by the incapable subject. Thus, mental incompetence itself would not preclude the use of the substitute judgment test.[71] *Eve*, however, involved a person with a serious mental disability who was never able to express views, values or beliefs. As such, a substitute judgment test would have been meaningless and purely speculative, and the Supreme Court was correct in rejecting it.

---

[68] This English case involved a minor, and is of interest as it involved the application of the *parens patriae* power to authorize a non-therapeutic sterilization. However, in his criticism of La Forest J. in *Eve*, Lord Hailsham of St. Marylebone was also censured for his statement that "[t]o talk of the 'basic right' to reproduce for an individual who is not capable of knowing the causal connection between intercourse and childbirth, the nature of pregnancy, what is involved in delivery, unable to form maternal instincts or to care for a child appears to me wholly to part company with reality". The unfortunate inference that can be drawn from such an assertion is that a person who is incapable of understanding or expressing a right may not be entitled to it. See *In Re B, ibid.* at 213. In another case, this time involving an adult, the House of Lords applied the common law concept of necessity, as the *parens patriae* jurisdiction could at that time no longer be applied to adults in England. See *F. v. West Berkshire Health Authority* (1989), 2 All E.R. 545 (H.L.). The application of this approach was criticized as tenuous, given that "necessity" should apply only in cases of genuine need, with non-therapeutic sterilization being a dubious example. For a detailed discussion of these cases and the points mentioned above, see D. Tomkin & P. Hanafin, *Irish Medical Law* (Dublin: Betaprint, 1995) at 192-200.

[69] See *Eve, supra* note 63 at 423.

[70] See B.M. Dickens, "Substitute Consent to Participation of Persons with Alzheimer's Disease in Medical Research: Legal Issues", in J.M. Berg, H. Karlinsky & F.H. Lowy, eds., *Alzheimer's Disease Research: Ethical and Legal Issues* (Toronto: Thomson Professional Publishing, 1991) 60. The additional requirement added by Dickens is that there would have to be no risks associated with the research.

[71] See *Eve, supra* note 63 at 425.

It is possible that subsequent cases could distinguish *Eve* either on the basis of the nature of cognitive impairment or on the relative level of risk or harm, thereby not forcing a superior court to invoke its *parens patriae* powers to invalidate a guardian's decision to enroll a dependent in non-therapeutic research. However, it is difficult to predict the judicial outcome of such fact-specific cases, which rely so strongly on the court's discretion. In fact, because of the emphasis on "necessity" and "benefit" in determining whether or not to invoke the *parens patriae* power, which should only be applied in favour of the individual concerned and not for the sake of others, it is more likely than not that Canadian courts would refuse to sanction a guardian's decision to submit a ward to an experiment.[72] Indeed, the Supreme Court forthrightly rejected the invitation to formulate social policy (at least in the context of sterilization), preferring to concern itself with the immediate interests of the incompetent person.[73] Given that our motivation to enroll incapable adults in non-therapeutic research is socially driven, it is likely that the courts would choose to prohibit the practice in order to encourage a legislative resolution.

We must also question whether the judiciary is the appropriate arbiter in such cases,[74] given that the review of research protocols requires multidisciplinary expertise, which would therefore require the participation of persons trained in the health sciences and legal medicine. There may be a place for the judiciary in the approval process; however, a specialized tribunal with quasi-judicial powers would likely be more effective.[75] The position taken here is that a centralized statutory body should possess the investigative powers necessary to review the decisions of local research ethics committees[76] and also the power to promulgate regulations that would carry the force of law, with the corresponding ability to impose sanctions. Conduct authorized under such a régime, such as the use of substituted consent, would thus become lawful. The central body could also serve a consultative role or act in a quasi-judicial capacity to resolve difficult cases or situations of conflict.[77] A supervisory/regulatory emphasis is preferred for two reasons: firstly, the desire to create a system that will ensure the observance of minimal requirements for ethical

---

[72] The view that the *parens patriae* jurisdiction is unlikely to permit the authorization of non-therapeutic experimentation is shared by the Queensland Law Reform Commission, *supra* note 34 at 64.

[73] See *Eve, supra* note 63 at 427.

[74] *Ibid.*

[75] The Queensland Law Reform Commission and The Law Commission (UK) concurred with this view, recommending that non-therapeutic research with mentally incapacitated adults be authorized only by a specialized statutory authority. The UK Commission, however, makes the additional requirement of either court approval, the consent of an attorney or manager, a certificate from a doctor not involved in the research that the participation of the person is appropriate, or designation of the research as not involving contact. See Queensland Law Reform Commission, *supra* note 34 at 393; The Law Commission, *supra* note 34 at paras. 6.33, 6.37.

[76] Local research ethics committees should be formally constituted, certified and subject to review by a central statutory authority. The composition of committees should include not only persons with scientific and legal training, but also, on an *ad hoc* basis, representatives of the subject or the class to which the subject belongs. The committee would be required to apply specific and official guidelines, and be free of institutional and personal bias. See Chapter 17.

[77] The central board would also reserve an exclusive right to make certain decisions, such as with regard to research involving substantial risks.

conduct while leaving research ethics committees considerable discretion for their application in the light of local conditions and current developments in science and medical technology; and secondly, for reasons of administrative efficiency. Therefore, the decision to enroll a cognitively impaired adult in research, and the experiment itself, would require the approval of a research ethics committee that would ultimately be responsible to the central board. Making the initial decision with regard to participation, however, preferably based on the prior wishes of the subject, should be entrusted to the legal guardian, who is presumably in the best situation to assess the desires and needs of the incapable subject. The duties of guardians would therefore have to be defined clearly, not through inconsistent rulings by the courts or according to standards of uncertain legal merit as described in professional guidelines, but through detailed legislation. Otherwise, researchers, their institutions, and guardians who enroll cognitively impaired adults in non-therapeutic biomedical experiments, may be open to liability.[78] Legal reform in common law Canada is thus necessary in order to clarify the existing situation and to establish a viable legal régime that is compatible with the research endeavour.

## THE ELEMENTS OF REFORM

We must first adopt an appropriate philosophical approach, one which requires us to abandon the exclusive reliance on benign paternalism that lies at the root of early (and many existing) adult guardianship laws.[79] Such laws are problematic in that they rely upon the assumption that, owing to their incapacity, cognitively impaired adults are to be treated like children and "are either denied or lose most of the powers and fundamental rights and freedoms enjoyed by others".[80] This perception is no longer compatible with our modern culture of rights and entitlements, which requires not only the preservation of personal autonomy, but of a right to the least restrictive and intrusive forms of assistance and protection.[81]

In order to further this ideal, we must entrench in legislation various fundamental presumptions. The first relates to mental competence. Upon attaining the age of adulthood, a person normally acquires full legal rights to manage personal affairs and to make decisions related to health and welfare. Adults should therefore be presumed to be competent unless proven otherwise. Secondly, it is well acknowledged that incompetence in one area of decision-making does not

---

[78] H. Sava, P.T. Matlow & M.J. Sole, "Legal Liability of Physicians in Medical Research" (1994) 17 Clinical & Investigative Medicine 148 at 165.

[79] This generalization, although currently applicable to many of Canada's common law provinces, no longer applies to current Australasian models, nor indeed to many jurisdictions elsewhere in North America.

[80] See Gordon & Verdun-Jones, *supra* note 12 at 1-28.

[81] *Ibid.* at 1-29; See also Robertson, *supra* note 55 at 118-123.

necessarily extend to another.[82] Thus, a person's inability to consent to participate in experiments should not be presumed, despite the presence of a mental disorder or disability. The presumption of competence, however, does not discharge the obligation to conduct a functional assessment of capacity when there is a reason to do so.[83] Finally, acceptance should not be assumed; and neither "assent" nor the lack of an objection, although important prerequisites to participation in research, should be sufficient to imply consent.[84]

It is also important that certain fundamental conditions defining the parameters of permissible experimentation be satisfied.[85] First and foremost is a requirement for the review and approval of all research protocols by a research ethics committee. Briefly, a committee must be satisfied that an experiment is scientifically valid, of significant value, and involves an acceptable level of risk that is proportionate to the potential benefits. The use of cognitively impaired adults must not only be justified by the experimental design, but also be essential to its stated objective. Finally, the committee must ensure that a valid consent to participate was given, either by the subject or by the legal guardian where the former is impossible to obtain.[86]

On this note, legislative reform must also be geared towards developing a régime that is flexible enough to accommodate the different classes and degrees of cognitive impairment. Alternative modes for obtaining a legal consent must be provided for. We refer, of course, to decisions made in advance or by a substitute. Although the latter can be applied to any situation of incapacity, use of the former is preferred.[87] There will always be cases where the only option is to elicit the consent of a surrogate, particularly in cases of serious unforeseen or developmental disabilities. Therefore, reforms must proceed along both lines. However, in providing mechanisms for maximizing personal autonomy, such as through the availability of research directives, it is important that bases for intervention, such as by a guardian, be available to protect individuals even if contrary to wishes made contemporaneously or in advance. Hence, safeguards founded on the principle of personal inviolability must complement those promoting individual autonomy.

---

[82] To take a classic example, a person may lack testamentary capacity, but may be sufficiently competent to marry.

[83] For an in-depth discussion of mental competency and its assessment, see Weisstub (1990), *supra* note 5.

[84] See e.g. Title 45 Code of Federal Regulations 46 (18 June 1991) §46.402 (b), 46.408 (a).

[85] See e.g. Queensland Law Reform Commission, *supra* note 34 at 391-393; The Law Commission, *supra* note 34 at para. 6.34.

[86] For further discussion of these points, including a synthesis of the various codes, guidelines, legislation and regulations promulgated to date, see Chapter 18.

[87] The Alberta Law Reform Institute and the Newfoundland Law Reform Commission each went one step further in proposing that guardians should not be permitted to enroll their dependents in non-therapeutic research unless authorized through a valid advance directive. It is our view that, at least for the present, and until the use and application of research directives becomes sufficiently widespread, this requirement would unduly inhibit research. See Alberta Law Reform Institute, *Advance Directives and Substitute Decision-Making in Personal Health Care* (Edmonton: Alberta Law Reform Institute, 1993) at 41; Newfoundland Law Reform Commission, *Discussion Paper on Advance Health Care Directives and Attorneys for Health Care* (St. John's: Newfoundland Law Reform Commission, 1988) at 51.

### The Legal Response

Advance directives provide a means for individuals, while mentally competent, to document and project their will into a future time where they anticipate a state of impaired decision-making ability. These devices yield an opportunity for persons to preserve their dignity by ensuring that the manner in which they will be treated is consistent with their wishes. Likewise, research directives would allow someone with decreasing or fluctuating cognitive capacity to provide an advance consent or refusal to participate in non-therapeutic experimentation. Guardianship statutes should be amended to specifically include research among the array of options available to the authors of health care directives. Those jurisdictions that do not yet provide for health care directives must enact corresponding reforms. It is not our intention here to review the ethical and legal considerations relating to the use of research directives, as this area is discussed in detail elsewhere in this Volume.[88]

Research directives provide an ethical solution to the dilemma of consensual research involving cognitively impaired adults only in those situations where the adult was previously competent. Otherwise, such as with adults who possess severe mental disabilities and are therefore unable to express wishes in advance, the only alternative is to obtain the consent of a substitute decision-maker. Guardianship laws should therefore be amended in order to provide for this alternative.

The powers of guardians should only be granted in accordance with a functional determination of incapacity.[89] Specifically, a finding of incompetence would be based on evidence of an inability to understand and make decisions in respect of participation in experimentation. It is important to reiterate that such choices belong in a special class of decisions that is distinct from those normally made in relation to personal health and welfare. This brings us back to the original therapy/non-therapy dilemma: treatment provides a benefit; non-therapeutic research does not. To condone the latter forces us to entertain an area of decision-making that is for the most part alien to traditional notions of guardianship.[90] It is therefore necessary that, along with appropriate safeguards and restrictions, special provisions allowing for the appointment of a "guardian for research" be incorporated into adult guardianship laws.

---

[88] See Chapter 11.

[89] This approach would be consistent with a "least restrictive and intrusive" means of protection, which was supported most recently by the Law Commission of the United Kingdom, *supra* note 34 at para. 3.14. See also Weisstub (1990), *supra* note 5. A practical example can be found in Ontario, where:

> A person is incapable of personal care if the person is not able to understand information that is relevant to making a decision concerning his or her own health care, nutrition, shelter, clothing, hygiene or safety, or is not able to appreciate the reasonably foreseeable consequences of a decision or lack of decision".

See *Substitute Decisions Act, 1992, supra* note 58, s. 45. Full guardianship is awarded only if there is an incapacity in all of the areas listed above. Otherwise, partial powers will be granted. (ss. 59(1), 60(1)).

[90] An exception to this rule includes the occasional need for a guardian to order the commitment of a dependent for the protection of others.

The guardian of choice should be the person identified by the cognitively impaired adult, ideally through a research directive containing additional instructions to assist the guardian. Failing such, a close family member or friend should be appointed. The prospective guardian should be familiar with the individual, and should understand and agree to respect the dependent's wishes, values, and beliefs. However, a close family member may not always be the best choice. For example, the inheritability of certain cognitive disorders, such as Alzheimer's Dementia, may lead to a conflict of interest.[91] Care must also be taken in situations where participation in research would require institutionalization throughout the course of an experiment. The promise of a reward for the guardian, or of special treatment for the dependent, in exchange for participation in research is unethical and should not be allowed. Indeed, before accepting the role, all guardians must declare that they are free from bias and conflict of interest. As an additional precaution, the decision by a guardian to enroll a dependent in an experiment should be subject to the scrutiny of a research ethics committee, which would address the individual concerns raised by family members, the treating physician, the dependent (even if incompetent), and independent third parties (such as advocate groups).[92] If the guardian breaches his fundamental obligations to his dependent, steps should be taken, which could be through an appeal to a judicial or quasi-judicial authority for the termination of guardianship.

Guardians for research should be empowered not only to enroll their dependents in research protocols, but to withdraw them as well. The power to withdraw should be exercised without fear of reprisal for either the guardian or the dependent, and a decision to exercise it should not be overruled. Irrespective of the existence of a research directive, withdrawal from an experiment is justified in the event of unforeseen risks, the availability of new treatments, or a change in the nature of the experiment. In so doing, the guardian fulfills his obligation to act in the dependent's best interests. This duty, which is the primary responsibility of all guardians, should not be interpreted restrictively. It includes the duty to honour and respect the subject's known wishes, values, and beliefs, even if incompatible with those of the guardian. This charge can be rationalized through a prioritized system of substitute decision-making.

A person's wishes would be given the first priority, including the instructions contained within a research directive and others expressed previously. When this is not possible, the guardian must act in the person's best interests, which involves taking into consideration current wishes (if they can be ascertained), and the values and beliefs known to have been held by the person when competent, which it is

---

[91] Children might have a personal stake in the outcome of research to be conducted on their elderly parents, and would therefore be unsuitable guardians in this respect.

[92] The guardian for research would not, therefore, be operating in a vacuum and without accountability. Moreover, as indicated earlier, research ethics committees would also be subject to review by a central statutory authority. See note 76 and accompanying text.

believed would still be acted upon if he or she were capable.[93] At all times, the guardian must strive to maximize the dependent's role, even if incompetent, in the decision-making process. This approach effectively incorporates advance instructions into a hybrid model of best-interests and substituted judgment decision-making. For each of these principles, although individually attractive in theory, apply poorly in real-life situations.[94] The discomfort faced by the Supreme Court in *Eve* illustrates this point well. There are obvious problems in applying either standard on its own, given that, in the application of each, one is forced to rely, at least in part, upon elements of the other. For how can we balance benefits that are instinctively calculated on a subjective plane, such as the right to procreate *versus* the freedom to experience sexuality without fear of pregnancy, and claim that we have arrived at an objective determination of another person's best interests? In deciding that one option outweighs another (and therefore lies in the person's best-interests), are we not in effect making a substituted judgment? Indeed, we can no

---

[93] Although specifically excluded from application to the context of research, this approach is embraced by the current legislation in Ontario. See *Substitute Decisions Act, 1992, supra* note 58, ss. 66(3-4):

(3) The guardian shall make decisions on the incapable person's behalf to which the *Health Care Consent Act, 1996* does not apply in accordance with the following principles:
　1. If the guardian knows of a wish or instruction applicable to the circumstances that the incapable person expressed while capable, the guardian shall make the decision in accordance with the wish or instruction.
　2. The guardian shall use reasonable diligence in ascertaining whether there are such wishes or instructions.
　3. A later wish or instruction expressed while capable prevails over an earlier wish or instruction.
　4. If the guardian does not know of a wish or instruction applicable to the circumstances that the incapable person expressed while capable, or if it is impossible to make the decision in accordance with the wish or instruction, the guardian shall make the decision in the incapable person's best interests.
(4) In deciding what the person's best interests are for the purpose of subsection (3), the guardian shall take into consideration,
　(a) the values and beliefs that the guardian knows the person held when capable and believes the person would still act on if capable;
　(b) the person's current wishes, if they can be ascertained; and
　(c) the following factors:
　　1. Whether the guardian's decision is likely to,
　　　i. improve the quality of the person's life,
　　　ii. prevent the quality of the person's life from deteriorating, or
　　　iii. reduce the extent to which, or the rate at which, the quality of the person's life is likely to deteriorate.
　　2. Whether the benefit the person is expected to obtain from the decision outweighs the risk of harm to the person from an alternative decision.

It should be noted that the last set of criteria, clearly relevant in the context of treatment-based decisions, would not apply in a decision with respect to research.

[94] We refer to the Advance Directive Principle, the Substituted Judgment Principle, and the Best Interests Principle. It is important to recognize that surrogates will use a combination of these standards in their reasoning. The degree to which advance directives will influence decision-making will depend upon the evidentiary weight of the person's expressed wishes: as the level increases, one approaches a substituted judgment; as the level decreases, one must rely on a calculation of best interests. See D.W. Brock, "Good Decision-making for Incompetent Patients" (1994) 24:6 Hastings Center Report Supplement S8 at S9.

more readily divorce ourselves from our personal values and moralities than guess those of another. What, then, is the difference between a "substituted judgment" and a decision made in a person's "best interests"?[95]

The Court in *Eve* was concerned about the irreversibility and non-necessity of the sterilization procedure, and thus chose to err on the side of caution by disallowing it. But was this decision truly in Eve's best interests? Could she not have been exposed to even greater risks of harm resulting from an unplanned pregnancy with the associated psychological trauma?[96] A best-interests decision, especially when superimposed upon a person's previously expressed wishes, is therefore nothing more than benign paternalism. Furthermore, both the best-interests and substituted judgment standards rely intrinsically on value judgments made by a person other than the incompetent adult for whom the decision is being made. The former involves an external assessment and balancing of interests, while the latter requires the decision-maker to "stand in the shoes" of the incompetent person, an act which is in fact rooted in fantasy.[97] The difference is purely semantic, and in reality, any decision made for a mentally incompetent person will inevitably rely on a combination of the two patterns of decision-making. In other words, a true substituted decision would logically be in the individual's best interests from that person's perspective. Conversely, any attempt to make a best-interests decision can only be accomplished by emulating the person's unique style of decision-making, which is in effect a substituted decision. We must therefore abandon any attempts to effect an idealized "best-interests" or "substituted judgment" model, preferring instead to adopt the system advocated above.[98]

In the course of performing their duties, guardians may find themselves in conflict with the various participants in the research process. For example, the treating physician may seek to intervene if participation in an experiment would interfere with the subject's treatment plan, researchers may wish to discourage a guardian from withdrawing the dependent once committed, or the now incompetent subject may have consented in advance to participate in an experiment which has taken an unexpected turn but that subject does not wish to withdraw. The overriding responsibility of a guardian for research in all cases must be to protect the interests of the dependent, which entails balancing various obligations:

---

[95] Indeed, this observation was made in response to a series of American cases involving the application of the substituted judgment test in treatment-related decisions for cognitively impaired persons. See T.G. Gutheil & P.S. Appelbaum, "Substituted Judgment: Best Interests in Disguise" (1983) 13(3) Hastings Center Report 8 at 11.

[96] See also Shone, *supra* note 66.

[97] As a "construct of imagination", the doctrine of substituted judgment in fact constitutes a dangerous legal fiction, whereby an assumption can easily lead one to forget the underlying reality of decisions made in such a manner: that one person is rational, and therefore in a position of control, while the other, because of mental incapacity, is neither. See L. Harmon, "Falling Off the Vine: Legal Fictions and the Doctrine of Substituted Judgment" (1990) 100 Yale Law Journal 1 at 70-71.

[98] Indeed, the Law Commission of the UK decided upon a "best interests" criterion which included an element of "substituted judgment". The exception, of course, is where it is impossible to ascertain a person's prior wishes, in which case, one must rely upon a best-interests determination. See The Law Commission, *supra* note 34 at para. 3.24-3.28.

- to act in the dependent's best interests, which includes carrying out previously expressed wishes, such as instructions found in research directives;
- to remain informed of the progress of an experiment and to be prepared to withdraw the subject;
- to assist and support the dependent in whatever manner required, including the maximization of the ward's participation in the consent process, both prior to and throughout the experiment;
- to interact with family members, public advocates, the researcher and the treating physician in order to ensure that the dependent's wishes are properly observed and that the person is not exposed to unreasonable risks; and
- to remain free of conflicts of interest, and to refrain from permitting personal values and beliefs to supersede those of the dependent.

Finally, it is important to bear in mind that experiments are by nature fraught with uncertainty, and always possess an element of risk. Guardians who act to fulfill their duties towards their dependents in good faith should be exempt from liability for unfortunate accidents or harms befalling their dependents despite their efforts. This protection, and all others discussed here, should be guaranteed by statute.

## CONCLUSION

The conduct of non-therapeutic biomedical experimentation involving cognitively impaired adults raises fundamental ethical, moral, and legal concerns. At the root of these issues lies the dilemma in resolving the conflict between the social incentive to advance scientific knowledge and the need to protect vulnerable persons from exploitation. Indeed, the historical record on abuses in research warrants caution in developing any policy that deviates from an outright prohibition. The problem is further compounded by the fact that non-therapeutic research, by definition, yields no benefit for the person who is placed at risk for the sake of others, a situation that is even more troubling when involving non-autonomous subjects. Nevertheless, the international consensus shows that biomedical research involving this population is both necessary and desirable. Extensively debated models defining the boundaries of ethical permissibility are available to assist us in developing a sound policy towards the regulation of research. Two goals must form the heart of any such initiative: to protect subjects, and to facilitate the research endeavour.

The law on experimentation requires that researchers meet a high standard of disclosure when recruiting subjects. Cognitively impaired adults, however, may be unable to properly understand, make choices or communicate decisions about participation in research, and therefore be unable to provide a direct and contemporaneous consent. The doctrine of informed consent, as a tool designed to preserve rights of self-determination, is insufficient in protecting non-autonomous persons, including adults who possess diminished, fluctuating, or non-existent decision-making capacities. Alternative means of obtaining a valid consent to participate in research must be made available to researchers. These include decisions made in advance or by a substitute.

Unfortunately, for example, in Canada, neither existing adult guardianship laws nor the common law provide a satisfactory solution. Not all provinces provide for health care directives, and no province explicitly includes research among the decisions that can be made by guardians or authors of directives. The common law on the subject is unclear, having adopted a restrictive interpretation of the best-interests principle in the context of non-therapeutic sterilization. Although it is possible that the court's *parens patriae* jurisdiction may be used to validate a decision to enroll an incompetent person in low-risk research or where the desire to serve as a subject had been expressed in advance, reliance upon the common law will not deliver a positive response in all cases, particularly where the prospective subject was never previously able to formulate wishes, values, or beliefs.

Consequently, adult guardianship laws require extensive reform in order to accommodate this special category of guardianship powers. Appropriate safeguards and regulatory mechanisms must be established in order to protect subjects, to govern the activities of researchers, including the review of all elements of the research process, and to provide regulations concerning the use of research directives and the behaviour of substitute decision-makers. Reforms must be flexible so as to provide for an array of cognitive impairments, endeavouring not to sustain an approach based on benign paternalism, but rather one that respects individual rights and entitlements.

In adopting provisions for the use of research directives, it is necessary to establish formal requirements for their creation, limitations on their scope, and rules governing their review, implementation, and revocation. These include the notions of a restrictive interpretation of ambiguous instructions, a prohibition of the use of directives for consenting to high risk procedures, especially in the context of Ulysses Directives, and a requisite threshold of competence for their preparation that is in general higher than that required for consent to treatment. Adult guardianship laws must abandon archaic models which fail to recognize the notions of functional incapacity, and thus allow for the appointment of a "guardian for research", whose authority would apply to the experimentation context. Guardians for research must be free of conflicts of interest and act in the best interests of their wards, applying a prioritized system of decision-making which includes in their calculation the previously expressed wishes, values, and beliefs held by their dependents. They must therefore respect the instructions contained in research directives, but always be concerned with the safety of their dependents and prepared to withdraw them from an experiment if circumstances so require. Finally, guardians who fulfill their obligations in good faith should be absolved of all liability in the event of unforeseen mishaps.

Only in keeping with these guiding principles, and through the application of strictly enforced ethical and legal requirements, can society, in good conscience, benefit from the use of cognitively impaired adults as subjects in non-therapeutic biomedical experimentation.

CHAPTER 9

# FRENCH LAW AND BIOMEDICAL RESEARCH: A PRACTICAL EXPERIMENT

CHRISTIAN BYK

Reflections on "medical research ethics" often seem to accept as a basic premise that research is comprised of an objective set of data while ethics is rather an ongoing interrogation. However, it is not clear that this has always been the case and, if one may posit as a necessity the reflection upon research ethics, it is appropriate, by way of preliminaries, to offer some of the justifications underlying this hypothesis. These justifications will shed light on the role that law is expected to play in this research, particularly in France.

## THE NEED FOR MEDICAL RESEARCH ETHICS

The Hippocratic oath and the international and national texts it has inspired serve as a reminder that in just over a century medicine has progressed from prehistory to modernity, with revolutionary changes in our understanding of diseases, the incidence of large-scale endemic plagues, medication, scanning technology, and surgical techniques.

This prodigious step forward has not occurred without a real, if more subtle, revolution in the nature of medical activity. The benefits that new technology brings are in fact the result of an increasing integration — one might even say symbiosis — between medicine and research. Thus, there are two sides to modern medicine: the traditional, personalized doctor-patient relationship, and the more recent phenomenon, the medical researcher, who is expected to take part in the systematic development of knowledge from which society hopes to benefit. Although the doctor-patient relationship may be permanently and universally entrenched in the principles doctors have vowed to respect ever since the time of Hippocrates, these principles may no longer be valid when a doctor takes on the role of researcher. It would even seem reasonable to suggest that, to some extent, research may distort the founding principles of medical ethics.

Indeed, how can the alliance between trust and conscience be respected when the patient is no longer treated for his or her own sake but rather becomes the means to obtaining new knowledge, often without even the hope of deriving any personal benefit?

Since the logic of research is propelled by the dynamics of an industrial-scientific society, its consequences present grave risks for the individual — and I am not thinking solely of the abuses committed by a few, or the crimes perpetrated during the Second World War — and therefore the issue of research ethics must be raised. This consists of an interrogation of the legitimacy of the privileged status that has increasingly been acquired by medicine — and therefore the doctors who have a monopoly on it — such that doctors may operate on the human body without incurring the risk of criminal sanctions provided that no fault is committed in the exercise of their medical art.

There is, then, an obvious relationship between research and law. We will turn to law, not quite as a source of the elements of research ethics, but more as a means of translating these elements into a legal régime. In this way, behaviour that conforms to ethical standards will also be found to conform to standards in positive law. Nevertheless, this process of legal harmonization is complex, since, as in any attempt to rectify inconsistencies, what is rectified is thereby brought forth, meaning that the modified rule is no longer, and indeed has never been, exactly correct. French law has tended to fall into this very trap.

Contrary to what some have argued, French law has never relegated biomedical research to a legal void; rather, the law required in a narrow fashion that research practices conform to medical ethics and the law. In other words, research was only possible if it had a direct therapeutic outcome for the patient, if the advantages to the patient outweighed the risks involved, and if the patient had given his consent.

In this way, the principles of French law neither quite "scorned" nor "tolerated" — as many have said — scientific and industrial reality, but turned a blind eye to it. The law did not deal with biomedical research on humans in drug manufacturing conditions, while at the same time paradoxically there existed specific regulations which established an obligation to undertake clinical trials.

Under these conditions, it was especially difficult to argue that the French pharmaceutical industry could be competitive in the international market, since any tests conducted in France were illegal if the research did not directly benefit individual patients.

It was in this context, which, incidentally, was certainly taken into account in European Community policy-making decisions, that a solution to the "research ethics" problem in French law was provided in the *Loi du 20 décembre 1988*.[1]

---

[1] *Loi n° 88–1138 du 20 décembre 1988 relative à la protection des personnes qui se prêtent à des recherches biomédicales*, J.O., 22 December 1988 (1988 Law relating to the protection of persons who become biomedical research subjects) [hereinafter 1988 law].

Necessity makes law, and it therefore fell upon legislators to make previously illicit biomedical research legal and to set the general conditions under which the research was to be carried out (I). In order to ensure the protection of research subjects, the new law on biomedical ethics created an exception to the principle of the inviolability of the human person, a principle which has since been reaffirmed by the legislature in 1994 (II).

Although it has been modified several times, the 1988 law no longer appears to be contested in principle, thereby demonstrating the existence of a certain interaction between this new "legal framework" and practices in the field of biomedical research.

## I — NECESSITY MAKES LAW: LEGALITY AND GENERAL CONDITIONS OF BIOMEDICAL RESEARCH INVOLVING HUMAN SUBJECTS

Jurisprudential recognition of the legitimacy of "direct benefit" biomedical research could have led the legislature to deal with research involving exclusively healthy volunteers.

This was in fact the intention of the French government. Following the release of Professor J. Dangoumau's report and of the National Ethics Council's opinion of 9 October 1984, the government prepared a draft bill in 1985, which was later abandoned due to fears of provoking hostile reactions in public opinion that could have jeopardized the upcoming election campaign. Eventually, a combination of various factors — including a new European drug policy initiative, the report by the *Conseil d'Etat* on the transition from ethics to law and the "Amiens scandal" in which experiments were done on comatose patients — led the government to accept for debate a parliamentary bill which led to the adoption of the *Loi du 20 décembre 1988*. The new law attempted to reassure people and to provide for their protection. In order to do so, the law (A) covers all biomedical research, thereby (B) submitting it to a system of general and specific rules.

## A — RESEARCH GOVERNED BY THE LAW

Despite the scope of the 1988 law's application, the law does not ignore the classical distinction between research that has a direct benefit and that which does not. However, before we analyze the legal entrenchment of this distinction (2), it is incumbent upon us to make some preliminary remarks regarding the type of research governed by the law. These observations will highlight the interaction between positive law and research practice (1).

### *1. Biomedical Research*

The law covers only biomedical research, defined as: "organized tests or experiments involving human beings for the purpose of developing biological or medical knowledge". Thus, the law excludes experiments and trials that do not involve human beings as well as those that do not aim to advance biological or medical knowledge.

With respect to the first point, researchers quickly initiated an amendment to the law that made the terms of the definition even more benign. The initial text included the term "studies", thereby bringing epidemiological studies, for example, under the conditions set out in the law. Where such studies do not affect the physical integrity of participants while making use of personal information gathered from them, it is not subject to the 1988 law on biomedical research. This does not mean, however, that research activities are not subject to other provisions, such as those regarding the protection of privacy found in the *Loi du 1 juillet 1994*.

It should also be noted that certain doctors will attempt to escape legal constraints by suggesting that the "research" they conduct on their patients is not "organized" in any strict sense, but that instead it consists simply of an innovative treatment. Such an argument becomes irrelevant as soon as the doctor publicly discloses any medical conclusions based on this type of "innovative experimentation".

With respect to our second point, two practical problems have emerged. On the one hand, behavioural research did not fall clearly under the *Loi du 20 décembre 1988*, despite the fact that this kind of research involves human subjects. This resulted most notably in the 14 October 1993 decision of the *Comité consultatif national d'éthique pour les sciences de la vie et de la santé* (National Consultatory Committee on Ethics for the Life and Health Sciences [hereinafter CCNE]), which suggested that Parliament should pronounce itself on the possibility that the law be extended to cover behavioural research. This was done in the *Loi du 25 juillet 1994*. On the other hand, practical experience has demonstrated that questions will almost always remain as to whether or not certain activities may be characterized as research having a biological purpose.

For instance, is research on cosmetics, food colouring or para-pharamaceutical substances biomedical, or is it simply consumer testing? The answer obviously depends on the purpose of the research (for example, is it to determine toxicity levels in a certain product, or simply to test the product's taste?). Businesses and institutions that are concerned about such questions may turn to an administrative body, such as the Departments of Health or Industry or the Drug Agency, in their attempt to find a solution in these borderline cases. It is possible to imagine that careless entrepreneurs might try to avoid such questions, while at the other extreme, some might submit their inert products to biomedical research regulations in order to obtain a label certifying quality.

## 2. The Distinction between Direct Benefit and Non-direct Benefit Research

Although both types of research are covered by the law, it is important to distinguish between them, given that the second category is subject to more thorough scrutiny.

*(a) Research of Direct Benefit to the Individual*

Even before the *Loi du 20 décembre 1988* was passed, French jurisprudence recognized the legitimacy of this type of research; more precisely, it accepted as experimental only those treatments from which a direct benefit to the patient was expected. The cost-benefit analysis had to weigh in the favour of the patient, who, furthermore, had to give his or her free and informed consent to the proposed treatment.

Although the 1988 law in some ways simply confirmed the existing positive law in this regard, it made two additional specifications.

Even though direct-benefit research takes place within the framework of a classic doctor-patient relationship, it is subject to general research conditions in order to ensure that the patient is clearly aware of the experimental nature of the treatment and its ramifications.

The immediate or short-term benefit need not be purely therapeutic but may also be diagnostic or preventative. This was the rationale underlying the legislative amendment made in 1990. Moreover, according to the Department of Health (see DPHM, circular 5, October 1990), trials involving placebos are still considered part of direct-benefit research, in that at least some participants will benefit from the trials.

*(b) Research without Direct Benefit to the Individual Patient*

This type of research must be classified according to its purpose, rather than according to the characteristics of the participants. Such research implies that the subjects will not necessarily derive any medical advantage from their participation. This category includes basic and applied research, with drug testing constituting most of the applied research (known as phase 1 and 2 testing).

Whatever its purpose, this type of research must now be carried on within the framework of a regulated research project.

## B — REGULATED RESEARCH PROJECTS: THEIR GENERAL STRUCTURE

Although biomedical research is now organized and even controlled by public authorities, it remains first and foremost a private activity of scientific and often industrial nature that is principally regulated by research contracts or agreements.

### 1. Research Contracts

As a legal activity, biomedical research gives rise to a great variety of agreements the content of which varies as a function of the roles played by the principal participants in the research process.

*(a) Participants in the Research Process*

The *Loi du 20 décembre 1988* recognized that the doctor-patient relationship was too narrow a paradigm for describing research relationships. As a result, the law outlines the complexity of these relationships by designating the three main participants involved:

(1) The research subject: this is self-evident and requires no particular comment;

(2) The person who initiates the research: the developer. It is clear that the legislature had the role of the pharmaceutical company in mind (the law provides that the developer may be a physical or legal person), although practice has shown that the developer may also be a public institution. In any case, given that promoting a research project generally requires making the necessary financial and material investments, the current system has not functioned well in the context of academic research and studies regarding products with a limited market ("orphan" drugs), thus leaving the researcher with little access to material and financial resources.

(3) Our third participant, the researcher, is usually of key significance, since he or she will often be both the sole link between the research subject and developer, and the person through whom the whole process advances. Researchers — who can only be physical persons — embody the two faces of modern medicine: they must constantly focus on the patient, while at the same time being fully integrated into the research project. The researcher therefore directs and supervises the project in two capacities: that of experimenter under *article L. 209–1 C. santé publ.*, and that of the expert charged with the clinical supervision of his/her subjects, as per *article L. 209–3 C. santé publ.*

It must be said that this double logic imposed on a single person — though it was the wish of doctors who did not want the patient to get the impression that the contract of care had been altered — is one of the law's weakest points.

It should be noted, in addition, that two legislative amendments have taken specific problems into consideration:

- Odontological research may only be carried out under the direction and supervision of a dental surgeon or a specialized doctor (*loi du 18 janv. 1991*).
- Behavioural research may be carried out under the joint direction of a doctor and a qualified behavioural scientist (*loi du 25 juill. 1994*).
  Clearly, there are also other participants — the "secondary characters" (insurers, hospital administrators, etc.) who are involved only to the extent that one or other of the types of formal legal agreements is entered into at the moment a research program is established.

*(b) Categories of Research Contracts*
In principle, there exist two types of contracts: between the subject and the researcher and between the researcher and the developer.

*The Researcher-Subject Contract*

Before the adoption of the *Loi du 20 décembre 1988*, direct benefit research undoubtedly would have fallen within the definition of a medical contract. Now that the legislature has explicitly authorized non-direct benefit research, a contract having as its object the placement of a human body at the disposal of a researcher is no longer invalid. The body has therefore become the object of a juridical act (although it is not supposed to be a commercial good), subject to the protection of provisions regarding the rights to liberty and integrity of the person.

Nevertheless, the researcher-subject contract must also have a moral and legal objective in order to be valid. Legality, as defined by the *Loi du 20 décembre 1988*, encompasses not only therapeutic ends, but also medical and cognitive purposes (*art. L. 209–2, 4è al.C. santé publ.*). In the domain of morality, the principle of a gratuitous undertaking by the participant is the key indicator.

The principles found in article L.209–8 of the *Code de la santé publique* and reinforced by the Civil Code in its incorporation by reference of 1994 bioethics laws, have been further clarified and limited. The clarification concerns the reimbursement of costs (transportation and, potentially, compensation for loss of remuneration). The limitation concerns specific cases of non-direct benefit research.

Given that the legislature did not wish to diminish the pool of so-called "healthy volunteers", it has allowed developers to pay "compensation for inconveniences suffered" (*art. L. 209–15 C. santé pub.*). (As of a 1994 law the developer is no longer obliged to make this payment.) In order to avoid abuses, that research which is conducted on minors and protected persons of full age, or on persons admitted to a sanitarium or to other institutions, is not remunerated. This stipulation was made in an attempt to avoid a situation in which legal representatives might find themselves in a conflict of interest. As a result, however, such persons have conveniently been placed free of charge at the disposal of developers. Finally, in an attempt to avoid the professionalization of volunteers, the legislature established a maximum annual indemnity (25 000 French Francs as of the 21 February 1994 Order) and a prohibition on participation in more than one non-direct benefit research project at a time. However, the enforcement of these provisions through a national trials registration list will not necessarily provide a foolproof means of ensuring that people comply with these stipulations.

In passing, it should be noted that the developer — the person with the most financial resources — also appears incidentally in the researcher-subject contract to the extent that s/he is held directly liable for indemnification of the subject where damages are incurred as a result of participation in the research project.

Without pronouncing on the legal nature of the contract between researchers and their subjects (contract of loan, contract of lease, or *sui generis* convention), which the subject may breach at any time, it may nevertheless be suggested that it implies that the human body is a commercial good even if, in some sense, it remains outside the normal flow of commerce.

*The Agreement between Developer and Researcher: The Research Contract*
This private-law contract defines the obligations of the researcher with respect to research projects undertaken for the developer's profit; in exchange, the contract specifies the consideration that the developer will pay to his or her co-contractant.

In principle, the protection of the research subject does not arise directly in this contract. This presumably explains why the law only gives explicit treatment to this contract in a very limited number of cases.

First, there is concern that the medical profession's ethical standards regarding remuneration should be respected and the question of liability should be clarified. In addition, physicians are obliged to give notice of the agreement to the *Ordre des médecins* [Order of Physicians] and to inform the hospital director (and possibly the head of the unit) about the organization of the research.

Second, where research employs hospital equipment, the developer must cover any additional expenses incurred by the institution, as well as fulfilling the obligation to provide free of charge any drugs or products used in experiments. This arrangement forms part of an agreement between the developer and the institution in which the researcher is carrying out his or her work.

It is unfortunate that the law does not go further in regulating the research contract, given that if the researcher is paid on a *pro rata* basis by the number of useable observations made by the end of the trial period, there is little incentive to be prudent or to exclude subjects who might be exposed to the greatest risks from the experiments.

The question therefore arises as to whether public authorities should also be part of the research process.

## 2. The Role of Administrative Bodies in the Organization of Biomedical Research

Although committees for the protection of persons will increasingly play an essential role in the regulation of research, they are not endowed with decision-making powers, which lie with other authorities.

(a) The *Comité consultatif de protection des personnes qui se prêtent à la recherche biomédicale* (Consultatory Committee on the Protection of Biomedical Research Subjects [hereinafter CCPPRB]).

As a first step, every research proposal must pass through one of fifty-six existing regional committees. The researcher must obtain the appropriate regional committee's evaluation before undertaking any research. The structure of these committees plays an important role in the proper execution of their mandate. Since they were put into place in 1991 (the law had been in force since 1988), the organizational structure of the committees has provoked two major criticisms, to which research practices and the legislature have aimed to respond.

The first criticism concerned the means by which the twelve committee members were designated. They were to be picked at random from category lists proposed by various groups. This meant that the committee members would tend to lack long-term experience, and that the members who had been co-opted would lose their seats.Under the *Loi du 25 juillet 1994*, however, the regional Prefect now selects the twelve members from lists proposing three names for each position.

The second criticism concerned the way the committees functioned. Over a period of many months — sometimes up to one and a half years — these committees did not receive the funds they required to function, as developers paid a fixed fee of 9500 Francs per file submitted. Moreover, administrative practices varied greatly among the committees (requirements ranged from the submission of twenty-four copies of the file for the twelve members and twelve replacement members, to only two files for the Chairperson and the Rapporteur). Most of these problems have now been rectified; in addition, in order to increase the experience acquired by each committee (and therefore the number of proposals submitted), committees have been established at the regional, indeed, inter-regional level.

Since the committees play a key role in the economic aspect of the research process, they must be efficient and give their recommendations within five weeks (as of 1994). This requires that the CCPPRBs be particularly clear about their mandate.

*The Mandate of the CCPPRBs*

The legislature has also cut short the battle — where power is at stake — between the research developer (who does not participate officially in this process) and committee members.

The *Loi du 25 juillet 1994* makes it clear that the committee only rules on the research's validity regarding the protection of the subjects involved. It is therefore not an ethics committee; its function is to apply a regulation rather than adding any supplementary ethical conditions. Nor is it a scientific committee; it must not make any pronouncement regarding the relevance or methodology of the research unless it concerns the protection of persons.

The committee does, however, have the right to monitor the research process: its opinion is not to be limited to a pronouncement on projects concerning information given to subjects, to the modes of acquiring their consent, to potential compensation amounts or to the qualifications of the researcher. Rather, the committee must also be able to give its judgment as to the way in which subjects are protected and informed throughout the research process. How is this to be achieved in practice? Could the committee revise and overturn its recommendation? Would that have any impact given that the committee has no decision-making power and cannot order the termination of a research project?

Indeed, the rule in the French system is that, once the committee's evaluation has been made, each party must assume its responsibilities. This is because, on the one hand, the committee's approval does not release any party from fulfilling its obligations, and, on the other hand, even in case of a negative evaluation, the research may proceed. Only two obligations exist: first, the Committee must communicate all negative evaluations to the appropriate administrative authority; second, the developer, who "reappears" at this point, must transmit a letter of intent together with the committee's evaluation to the same authority for each potential research project. Research proposals that have received a negative evaluation can only be undertaken after a two-month waiting period. This permits the authority to exercise its powers of supervision (this is also the reason behind the obligation to inform the competent authority in case of any major incident or new fact regarding the security of persons involved).

*(b) Supervisory Authorities*
In light of the creation in 1993 of the *Agence du médicament* (Drug Agency) and the powers delegated to its director,

- the administrative authority that receives the above-mentioned information will at times be the Drug Agency, and at other times it will be the Department of Health (Directorate of Hospitals regarding research on medical devices that require approval; the Directorate of Health for other research);
- supervisory powers, whether used to suspend or prohibit a research project, are exercised along the same jurisdictional lines by the Minister of Health and the Director of the Drug Agency.

A group of experts on biomedical research assists these authorities in their respective areas (established in an Order of 28 April 1994). In particular, the group coordinates the CCPPRB's operations, considers the negative effects of research, and gives its opinion on security conditions in research facilities. This expert group also includes representatives of the pharmaceutical industry.

Finally, it should be noted that certain research projects will also be subject to other administrative procedures. This is true particularly in the cases of genetic therapy and products resulting from genetic engineering. At the request of industry representatives, a parliamentary report has recommended that procedures for these two types of research be unified and simplified by harmonizing them with procedures for drug trials.

France was therefore suffering prior to 1988 not so much from a legal void as from a lack of practical ethical guidelines. This situation prompted the legislature's intervention in the field of biomedical research, bringing about the various legislative and regulatory amendments aimed at resolving certain problems concerning drugs. The 1988 law was intended to be economically efficient. Its apparent "assimilation" by the pharmaceutical industry seems to show that an equilibrium has been found on this point. Does the law, though, provide sufficient protection for people who participate as research subjects?

## II — THE PROTECTION OF RESEARCH SUBJECTS

Let us at once mention that, paradoxically, it was not until the moment at which the human body entered legal commerce that France's legislature, following Quebec's lead, conferred in the Civil Code legal status upon the human body — a status that has associations to the principles of the inviolability and the integrity of the person.

On this level, it is also noteworthy that the legislature declined to mention research in the Civil Code's list of legitimate exceptions to these principles. Article 16–3 of the new Civil Code says that "the integrity of the human body may only be violated in a case of therapeutic necessity". It is true — *generalia specialibus non derogant* — that the *Loi du 20 décembre 1988* had already established this exception. For this exception to be invoked, however, biomedical research must be preceded by the necessary preliminary clinical testing. As the National Consultatory Committee on Ethics has pointed out in several of its rulings, it is not clear that this requirement will be respected; indeed, it is often easier to undertake research on human beings than on animals! In order to provide adequate protection for people, the 1988 law thus includes provisions that are intended to prevent the violation of bodily integrity, as well as sanctions for any violations which actually occur.

## A — PREVENTION

Prevention relies on objective measures that aim at providing maximum protection of personal security and also on subjective conditions specific to each person participating in a research project.

### 1. *Objective Measures*

Administrative authorities attend in a preventative manner to the personal security of participants by requiring that non-direct benefit research be conducted in specially authorized facilities; these authorities are also to be kept informed of the developers' intentions and the CCPPRB's evaluations so that, where necessary, they may exercise their supervisory powers.

Having already explored the means by which information resulting from the CCPPRBs' initial evaluations is communicated, we will limit the following discussion to procedures of authorization for research facilities.

We should first specify that we are not concerned here with research having a direct benefit for the individual. This is because the legislature has decided that it is unnecessary to treat direct-benefit research — whether it takes place in a hospital or an out-patient clinic — any differently from other health-care activities where the patient's security is governed not by regulations but rather by general rules of medical liability. It may be that this system overlooks the fact that in certain cases — e.g., the use of placebos in experiments conducted on patients with severe depression — the research patient is in need of special supervision.

However, given that "healthy volunteers" would not be admitted as hospital patients, tests performed on them presupposed by their very methodology the use of specific facilities. Furthermore, *Art. L. 209–18 C. santé publ.* provided that such facilities should have equipment and techniques adapted to the research context. In addition, they must conform to security requirements, and in particular be able to provide for clinical supervision, emergency care, and even, where necessary, immediate transfer to hospital. In fact, a 26 May 1994 circular adds hospitals where research is performed on sick volunteers to those facilities receiving healthy volunteers, which are obliged to meet the requirements that there be a pharmacologist and a respirator on the premises, as well as being subject to detailed inquiries. In the case of hospital services performing research involving subjects who are ill, authorization will only be given for the area of specialization of the services. The authorization is given by the Minister of Health.

### 2. Subjective Measures

Two basic principles govern biomedical research and protect the subject: the principles of reasonable proportionality and of consent. Their application differs considerably depending on whether or not the subjects are "vulnerable persons".

*(a) Non-vulnerable Persons*

The principle of reasonable proportionality is borrowed from moral theology. It is expressed in a general manner in *art. L. 209–2 C. santé publ.*: "the foreseeable risk to persons ... must not be disproportionate to the benefit they will gain or the importance of the research".

Where research directly benefits the individual, the risks and chances for benefit concern one and the same person. Reasonable proportionality will therefore be calculated as it would for routine medical acts (*Cass. civ. 1ère 18 déc. 1979, Bull; civ. n° 323, p. 263*).

In contrast, in the case of non-direct benefit research, individual benefit is replaced by a collective benefit. The risks to the individual must therefore be weighed against the importance of the research. However, there is a risk threshold that cannot be exceeded by the weight given to the deemed importance of the research: this threshold is called "serious foreseeable risk" (*art. L 209–14 C. santé publ.*). Should this risk be calculated *in abstracto* according to the type of research being conducted, or *in concreto* for each person concerned?

On the basis of *Art. L. 209–14 C. santé publ.*, which provides that research is conditional on a prior medical exam for all people who will become research subjects, there is a tendency to favour *in concreto* analysis. The results of such an assessment must be communicated to the people involved before they can give their consent.

Consent, which is protected by the Civil Code, is not only a manifestation of autonomy to dispose of one's body as one wishes; it is also the means by which a person can protect him or herself by refusing to agree to take part in the research.

It is for this reason that great importance is placed on information given to potential research subjects apprising them of the relevant risks.

The information must be clear, honest and precise, and it should avoid being either too meager or too abundant. Furthermore, it must pertain to the purpose of the research, its duration, the stages involved, the foreseeable constraints and risks — including the risk of early termination of the research. The information must also include the CCPPRB's evaluation and a summary of basic human rights norms, including the right to withdraw one's consent at any moment without incurring any liability (*art. L 209–9 C. santé publ.*).

There are two limitations placed on this information: in accordance with the medical profession's Code of Ethics, a doctor has the choice to withhold information to a patient where a grave diagnosis or prognosis exists; and in the case of psychological research with no foreseeable risk of harm, the information given to patients may be quite concise (*art. L 209–9 C. santé publ.*) — the totality of the information need only be imparted at the end of the research period.

In terms of formal requirements, even if the information is given to the patient orally — although along with a written summary — consent must in principle still be given in writing. Where written consent is not obtainable, it must be given orally in the presence of an independent witness. The law does not distinguish between different types of research, which should lead doctors, in theory, to ask their patients for written consent, even when they are performing direct-benefit research. Certain practical difficulties may, however, arise in light of customary medical practices in France. In the end, it is up to researchers, or the physicians representing them, to obtain consent.

*(b) Vulnerable Persons*

First, a preliminary remark: the 1988 law applies only to persons or, at the very least, to those who once were persons. Research conducted on embryos therefore does not fall under this law. Where research is conducted on an embryo or a foetus *in utero*, the *Loi du 20 décembre 1988* will apply inasmuch as the research concerns a pregnant woman (cf. below). The *Loi no. 94–654 du 29 juillet 1994* prohibits *in vitro* research — only studies are permitted, and they do not come under the watch of the CCPPRBs; instead, they require positive evaluation from the National Commission on Medicine and Reproductive Biology (*art. L 152–8 C. santé publ.*).

Nevertheless, ever since the *Loi du 25 juillet 1994* was enacted, the statute now applies to research conducted on a person considered to be clinically brain dead. Such research had hitherto been rejected by the National Consultatory Committee on Ethics (CCNE), which had stated that "the person must have consented either directly or through his or her family's testimony" (*art. L 209–18–1 C. santé publ.*).

In principle, in the case of vulnerable persons, the requirement of reasonable proportionality is more stringent, and defective consent is guarded against.

*The Risk/Advantages Ratio*

Certain categories are excluded, either entirely or partially, from non-direct benefit research:

- People who have been deprived of their liberty, hospitalized without their consent and unprotected by the law, as well as patients in emergency situations, cannot be the subjects of non-direct benefit research (the last two categories were added in 1994).
- Minors, protected persons of full age, persons admitted to a mental or social institution for reasons other than research, pregnant women, women giving birth, and breast-feeding mothers may only be subjected to non-direct benefit research if three conditions obtain: there exists no serious foreseeable risk, the

research will benefit people presenting similar characteristics, and the research cannot be effected in any other manner.

Specific measures for obtaining consent only arise with regard to persons lacking competency and to persons in an emergency situation.

### Persons Lacking Competency: Minors, or Adult Persons Protected by the Law

Consent given by the legal guardian is sufficient to allow research presenting no serious foreseeable risks and having a direct benefit for the individual. Does this mean that research is similar to an ordinary treatment for which one of the parents may act alone? The issue is far from settled. In other cases, consent is given by the guardian, as authorized by the "conseil de famille" (a group of family members entrusted by law to look over the interests of the incapacitated person) or a judge.

Finally, the consent of minors or protected persons of full age must also be sought whenever they are capable of expressing their will. No research may be undertaken where such persons refuse or withdraw their consent.

In effect, these rules would seem to limit the types of research trials which may be conducted on minors or persons who lack competency.

### Persons in Emergency Situations

If procedures do not allow consent to be obtained prior to the experiment, for example in the case of people who are unconscious, it is possible to foresee soliciting the consent of family members who are present (until 1994, the law also included the person's partner under the term "close relations"). The patient is to be informed as soon as possible and his or her consent is to be obtained before any further steps are taken (*art. 209–9 dernier al.*).

The manner in which consent may be given are the same as those outlined above.

While the law specifies quite clearly who may participate in research and under what conditions, thereby offering a guarantee of protection for persons, this is reinforced by a strict régime of compensation for damages as well as by a system of penal sanctions.

## B — SANCTIONS AND COMPENSATION FOR DAMAGES

In contrast to the general sanctions that have been established by derivation from the diversity of the French *jus commune*, the régime of compensation established in the *Loi du 20 décembre 1988* is quite specific.

### 1. The Specific Régime for Compensation of Injuries Resulting from Research

Despite the possibility of subjecting the doctor-researcher to ordinary legal rules of contractual liability, the 1988 law recognizes a direct right of action for the research subject against the developer. In order to complete the system, an obligation to acquire insurance was also established.

*(a) The Basis of the Action Differs Depending on whether the Research is of Direct or Non-Direct Benefit*

*Non-Direct Benefit Research*
Given that public interest has led the legislature to allow a person to accept health risks without any expectation of receiving a direct benefit, the participant must be assured of indemnification in the event that injuries are visited upon him or her, even in the absence of fault.

Liability is therefore objective: neither the intervention of a third party, nor even the subject's withdrawal, may be used as a defence by the developer.

The damage must, however, have as its cause the research in question. Thus the debate — largely between experts — will no doubt be concerned primarily with issues of causality. This reaffirms the importance of the medical examination required for every subject of non-direct benefit research (*art. L 209–14 C. santé publ.*). The results of this testing are communicated to the subject through the intermediary of a doctor, and serve as proof of the subject's prior physical condition. One might also suggest that, although it is not required for direct-benefit research, the examination might be useful there as well.

*Direct Benefit Research*
The underlying principle here is that when a person stands to benefit from the research, then negligent behaviour must be able to be counted against him or her. Should the legislature have gone further, and allowed developers to exonerate themselves whenever the damages are imputable to a third party (the triangular subject-researcher-developer relationship)? The research subject currently benefits from a presumption of fault: the burden of proof is therefore reversed and falls on the developer, who must also be insured.

*(b) The Obligation to Insure*
Since the developer is the economic beneficiary of the research, the obligation to insure falls upon him or her (*art. L 209–21 C. santé publ.*), although coverage may be limited and does not exclude liability for compensation.

*The Limits on Insurance Coverage*
The conventional limitations are of three types:

*Time Limitation*
Insurance will only cover claims brought against the insured during the ten years following the end of the research; this may turn out to be shorter than the ten year prescription period (cf., *art. 2270–1* of the Civil Code, which refers to *art. L 209–22 C. santé publique*), as the period begins at the point that the injury manifests itself or becomes aggravated.

*Monetary Limitation*
Although the legislature set basic amounts in an apparently protective spirit (*Décret du 14 mai 1991*: 5 million Francs per victim; 30 million per research project; and 50 million per year), insurance company practice should, on the contrary, lead us to view these figures as ceilings.

*Limitation of Risks*
Injuries that result from research carried on in violation of the law are excluded from the régime. This includes research without medical supervision, without a CCPPRB evaluation, without a submission to the Minister, in a non-approved facility or where consent was obtained in violation of the law.

*Legal Recourse.*
The developer's insurer may, in accordance with the *jus commune*, exercise rights as against the researcher and/or his or her insurer. Damages will therefore eventually be shared between the insurers in proportion to their respective obligations.

Finally, it should be noted that *art. L. 209–22 C. santé publ.* gives civil courts (instead of administrative tribunals for cases involving public hospitals) exclusive competence to decide all actions for compensation.

## 2. Sanctions

Disciplinary and administrative sanctions exist, but the criminal sanctions are of particular interest.

*(a) Disciplinary and Administrative Sanctions*

*Disciplinary Sanctions*
These will be imposed primarily on doctor-researchers, since the law places every research project under the direction and supervision of a doctor. But does this therefore imply the exclusion of liability for those working in collaboration with the researcher, including the developer and, in all probability, the researcher's medical or pharmacological representative? Certainly not — but the question has to be decided on a case by case basis.

*Administrative Sanctions*
The medical and pharmacological health inspectors who make up the Drug Agency's body of inspectors are charged with ensuring that the law is observed and with the adoption of control measures. At any moment, within their respective fields, the Minister of Health and the Drug Agency's director may suspend or prohibit a research project or withdraw authorization for research facilities having no direct individual benefit.

Indeed, if they failed to do so, they could be susceptible themselves to criminal proceedings (see for example the tainted blood scandal).

*(b) Criminal Sanctions*
The law has instituted criminal provisions (though these do not exclude the application of the *jus commune*).

*Violations of the Biomedical Research Régime*
The most serious infractions include: performing prohibited research, failing to respect consent, and violating the protective provisions regarding certain categories of persons (three-years' imprisonment: *art. 223–8 C. pén.*).

Less serious infractions arise essentially from the failure to respect administrative obligations and are punishable by a one-year term of imprisonment.

*Infractions of the Jus Commune*
While the law permits an exception to the *noli me tangere* principle, the exception exists only to the extent that the strict conditions which have been imposed are respected. Where they are not respected, general criminal laws pertaining to assault will apply, and, where appropriate, they will apply in conjunction with one or several of the special penal sanctions outlined above.

The application of criminal law concerns the developer as well as the researcher, and any other party that may commit a criminal fault.

It should also be emphasized that under the new Criminal Code, in force since 1994, legal persons — pharmaceutical companies, hospitals or other authorized research facilities — may in some cases find themselves subject to the long arm of the criminal law.


## CONCLUSION

The Law for the protection of persons, the *Loi du 20 décembre 1988*, is not without its ambiguities. For instance, in order to provide better control over biomedical practices, the law proceeds by legitimizing those practices which infringe on individual rights without providing any direct benefit to any individual — a sign of the radical transformations that have taken place in medical practice. However, as far as the law is concerned, the redrawing of the boundaries between legal and illegal research makes a certain amount of sense in the light of a dynamic whereby the respective roles of civil and criminal law have changed, and where the regulation of practices has come to predominate over the "symbolism" of legal principles. One sometimes wonders if a political vision of the role of law can be satisfied by a situation in which the law appears to be nothing more than the formal attire of social practices, legitimate though these may sometimes be. As the law's aim is to resolve conflicts, it should not serve as their veil.

CHAPTER 10

# CONSENT TO HUMAN EXPERIMENTATION IN QUEBEC: THE APPLICATION OF THE CIVIL LAW PRINCIPLE OF PERSONAL INVIOLABILITY TO PROTECT SPECIAL POPULATIONS

SIMON N. VERDUN-JONES AND DAVID N. WEISSTUB

## INTRODUCTION: INFORMED CONSENT (LE CONSENTEMENT ÉCLAIRÉ)

In Canada, the protection of competent adults from unwanted medical treatment and biomedical experimentation is accomplished by the judicial application of the legal doctrine of informed consent. Medical practitioners who intentionally interfere with the body of a competent adult without that person's consent will, in the absence of legal authorization to the contrary, be liable to a criminal charge or a civil action in tort:[1] in Quebec civil law, furthermore, non-consensual medical interventions constitute an "unlawful interference with a person's physical integrity".[2]

In recent years, Canadian courts have not only required that consent to medical treatment be given by a competent adult on a *voluntary* basis (*le consentement libre*) but also have insisted that the adult's choice be *adequately* informed (*le consentement éclairé*).[3] However, the requirement of adequate information has been

---

[1] L.E. Rozovsky and F.A. Rozovsky, *The Canadian Law of Consent to Treatment* (Toronto: Butterworths, 1990) at 116; Law Reform Commission of Canada, *Working Paper 26. Medical Treatment and Criminal Law* (Ottawa: Ministry of Supply and Services Canada, 1980) at 17—37; E.I. Picard, *Legal Liability of Doctors and Hospitals in Canada,* 2nd ed. (Toronto: Carswell, 1984) at 25—40.

[2] B. Freedman and K.C. Glass, "*Weiss* v. *Solomon*: A Case Study in Institutional Responsibility for Clinical Research" (1990) 18 *Law, Medicine & Health Care* 395 at 397. See generally *Charter of Rights and Freedoms*, R.S.Q. c. C—12, a.1.; *Civil Code of Quebec*, L.Q. 1991, c. 64, Articles 10 and 11.

[3] *Hopp* v. *Lepp* (1980), 112 D.L.R. (3d) 67 (S.C.C.); *Reibl* v. *Hughes* (1980), 114 D.L.R. (3d) 1 (S.C.C.); Rozovsky and Rozovsky, *supra* note 1; G. Robertson, "Informed Consent in Canada" (1984) 22 Osgoode Hall Law Journal 139.

held to be pertinent to an action in the tort of negligence rather than the tort of battery.[4] In other words, a failure to perform the duty to exercise reasonable care in the provision of information does not affect the *validity* of the patient's consent and, therefore, does not give rise to an action in battery.[5]

The standard and scope of disclosure that is required for obtaining consent to medical treatment was delineated in 1980 by the Supreme Court of Canada in two landmark cases.[6] It was held that the basic standard of disclosure must be "geared to what the average prudent person, the reasonable person in the patient's particular position, would agree to, if all material and special risks of going ahead with the surgery or foregoing it were made known to him".[7] Although the Court based this standard on an objective test, it is clear that it must nevertheless be applied in the context of the patient's particular circumstances and situation.[8] At present, it is not entirely clear to what extent this modified objective test is compatible with Quebec Civil Law and the Quebec Court of Appeal has, in fact, articulated a test that can be described as "subjectivité rationnelle" or "raisonnabilité subjective".[9] This test is focussed on the reasonable response of the particular patient and not the response of the abstracted reasonable person referred to by the Supreme Court of Canada.[10]

Insofar as consent to experimentation (or non-therapeutic research) is concerned, the standard and scope for disclosure appears to be based on a judgment of Hall J.A. in *Halushka* v. *University of Saskatchewan*,[11] in which it was stipulated that "the subject of medical experimentation is entitled to a full and frank disclosure of all the facts, probabilities and opinions which a reasonable man might be expected to

---

[4] *Reibl* v. *Hughes, supra*; Rozovsky and Rozovsky, *supra* note 1 at 119—120.

[5] In *Rogers* v. *Whitaker* (1992), 109 A.L.R. 626, the High Court of Australia, citing *Reibl* v. *Hughes*, recently stated that the "amorphous phrase", "informed consent", should be treated cautiously and pointed out that it is misleading insofar as it "suggests a test of the validity of a patient's consent" (*per* Mason C.J. at 633). See, B. Milstein, "High Court Rules on Informed Consent: Rogers v. Whitaker (unreported, High Court 19 November 1992)" (1992/93) 1(4) Australian Health Law Bulletin 37.

[6] *Hopp* v. *Lepp, supra* note 3; *Reibl.* v. *Huges, supra* note 3.

[7] *Reibl* v. *Hughes supra* note 3 at 16 (*per* Laskin C.J.C.).

[8] Rozovsky and Rozovsky, *supra* note 1 at 8.

[9] *Pelletier* c. *Roberge*, [1991] R.R.A. 726 (C.A.).

[10] In the words of the Court (at 734):

> En effet, notre jurisprudence comporte une dualité d'opinions à ce sujet et ce n'est vraiment que dans nos trois derniers arrêts. dont deux d'ailleurs publiés postérieurement au jugement enterpris, que notre Cour fixe un test que l'on pourrait appeler de "subjectivité rationnelle" ou de "raisonnabilité subjective" qui consiste à déterminer et à apprécier, en fonction de la nature du risque et de la preuve, quelle aurait été la réponse raisonnablement probable *du patient en l'instance*, et non de l'homme raisonnable dans l'abstrait au sens de *Reibl c. Hughes*.

For further discussion of this issue, see J.-L. Baudouin, La résponsabilité civile, 4e éd. (Cowansville: Yvon Blais, 1994) at 277—278; and R.P. Kouri, "L'influence de la Cour Suprême sur l'obligation de renseigner en droit médical québécois" (1984) 44 Revue du Barreau 851.

[11] (1965), 52 W.W.R. 608 (Sask.C.A.); Picard, *supra* note 1 at 118—119; Rozovsky and Rozovsky, *supra* note 1 at 80; G. Sharpe, *The Law & Medicine in Canada, Second Edition* (Toronto: Butterworths, 1987) at 79—85. See, also, *Cryderman* v. *Ringrose*, [1977] 3 W.W.R. 109 (Alta. S.C.); affirmed, [1978] 3 W.W.R. 608 (Alta. C.A.); *Weiss* c. *Solomon*, [1989] R.J.Q. 731 (C.S.); *Morrow* c. *Royal Victoria Hospital*, [1990] R.R.A. 41 (C.A.).

consider before giving his consent".[12] It is considered to be a more rigorous standard than that which is applied in the context of consent to medical (therapeutic) treatment:[13] indeed, a recent case in Quebec, *Weiss* v. *Solomon*, suggests that the "highest possible" standard of disclosure will be enforced and that a research subject must be informed of even the most remote risks:[14]

> ... en matière de recherche purement expérimentale, le médecin doit révéler tous les risques connus même rares ou éloignés et, à plus forte raison, si ceux-ci sont d'une conséquence grave.[15]

It has been contended that *Weiss* v. *Solomon* established that "the obligation to inform would include not only the risks inherent in the actual experiment but also those pertaining to non-experimental examinations or tests which serve to monitor the outcome of the research itself".[16] Furthermore, it has been asserted that potential research subjects should also be informed of the scientific purposes of the experimentation concerned:

> It is evident that the reason for this prerequisite is to enable volunteers to avoid becoming involved in research which they may find morally repugnant.[17]

Naturally, the requirement of "full and frank disclosure" applies not only in relation to a potential research subject who is competent to make a decision whether or not to participate but also in relation to a substitute decision-maker who is empowered to give consent on behalf of an incompetent adult or a minor.

## LEGAL PRINCIPLES UNDERLYING THE DOCTRINE OF INFORMED CONSENT

It has been asserted that the development of the doctrine of informed consent (le consentement éclairé)[18] within the framework of the Civil Law system of Quebec has followed a somewhat different path from that taken in Canada's common law provinces.[19] Specifically, the contention is that, while the doctrine of informed consent in the common law jurisdictions has been focussed primarily on the central

---

[12] (1965), 52 W.W.R. 608 at 616—617. As to the distinction between therapeutic and non-therapeutic research, see M.A. Somerville, "Therapeutic and Non-Therapeutic Medical Procedures — What are the Distinctions?" (1981) 2 Health Law in Canada 85.

[13] Rozovsky and Rozovsky, *supra* note 1 at 80; Sharpe, *supra* note 11 at 80.

[14] See *Weiss* c. *Solomon*, *supra* note 11; see Freedman and Glass, *supra* note 2 at 398.

[15] [1989] R.J.Q. 731 at 743.

[16] R.P. Kouri, "The Law Governing Human Experimentation in Québec" (1991) 22 *Revue de droit, Université de Sherbrooke* 77 at 84.

[17] *Ibid* at 85. Kouri also notes (at 85—86) that, on this issue, there may be a divergence between the common law approach, reinforced by the Supreme Court of Canada in *Reibl* v. *Hughes* and *Hopp* v. *Lepp*, and the approach taken in Quebec Civil Law. The common law approach asks whether the reasonable person would have consented had s/he known of the scientific purpose of the research, whereas the Quebec Civil Law approach asks whether the particular subject would have consented had s/he known.

[18] W.F. Bowker, "Experimentation on Humans and Gifts of Tissue: Articles 2—23 of the Civil Code" (1973) 19 McGill Law Journal 161—194 at 167.

[19] M.A. Somerville, *Consent to Medical Care*, Protection of Life Study Paper, Law Reform Commission of Canada (Ottawa: Ministry of Supply and Services Canada, 1979) at 8—9.

principle of *individual autonomy*, its counterpart in the Civil Law system of Quebec has evolved with a greater degree of emphasis on the French principle of the *inviolability of the person.*[20]

The differences in nuance that may be uncovered in an examination of the evolution of the doctrine of informed consent within the common law jurisdictions, on the one hand, and the Quebec system of Civil Law, on the other, may acquire a significance that goes far beyond the subtleties of mere academic analysis. Indeed, the nature of the relationship that exists between the principles of autonomy and inviolability within the two different systems of law may well have a profound impact on the extent to which either system is able to shield those who are most vulnerable from the perils of exploitation and abuse that may be inflicted on them by the process of biomedical experimentation.

The principle of autonomy has traditionally been encapsulated in the famous aphorism of Cardozo J. in the American case, *Schloendorff* v. *N.Y. Hospital,*[21] to the effect that "(e)very human being of adult years and sound mind has a right to determine what be done with his (*sic*) own body". This view was later endorsed by Chief Justice Laskin, in a case before the Supreme Court of Canada, where he noted that every patient has the right to "decide what (if anything) should be done with his body".[22]

In recent years, a greater degree of emphasis has been placed in common law jurisdictions on the need to provide individuals with the opportunity to exercise their powers of choice within the medical context and, from the vantage point of so-called "therapeutic jurisprudence", it has even been asserted that there is a distinct therapeutic advantage to be gained from recognizing the need to foster this freedom of choice.[23] As Winick contends,[24] this is one of the major reasons why the "law strongly favors allowing individual choice rather than attempting to achieve public or private goals through compulsion".

In essence, autonomy refers to the individual's right of self determination or self governance[25] and, according to conventional wisdom, one of the most significant vehicles for the enforcement of this right is the requirement of informed consent as a precondition to the administration of any form of medical care or treatment.[26] However, by definition, the right to autonomy can only be employed as a means of

---

[20] A. Mayrand, *L'inviolabilité de la personne humaine* (Montréal: Wilson and Lafleur, 1975).

[21] (1914), 211 N.Y. 127 at 129.

[22] *Hopp* v. *Lepp supra* note 3 at 70.

[23] B.J. Winick, "The Right to Refuse Mental Health Treatment; A Therapeutic Jurisprudence Analysis" (1994) 17 International Journal of Law and Psychiatry 99.

[24] *Ibid* at 101. See, also, B.J. Winick, "New Directions in the Right to Refuse Mental Health Treatment: The Implications of *Riggins* v. *Nevada*" (1993) 2 William & Mary Bill of Rights Journal 205 and "Psychotropic Medication and the Criminal Trial Process: The Constitutional and Therapeutic Implications of *Riggins* v. *Nevada*" (1993) 10 New York Law School Journal of Human Rights 637.

[25] M. Hébert, "L'application des Chartes Canadienne et québécoise en droit médicale" (1989) 30 Cahiers de Droit 495 at 499. See, also, B.J. Winick, "On Autonomy: Legal and Psychological Perspectives" (1992) 37 Villanova Law Review 1705.

[26] Somerville (1979), *supra* note 19 at 7.

protecting a *competent* person from exploitation and abuse and therein lies the downside of an excessive emphasis on this right within the common law jurisdictions. Indeed, as Somerville suggests,[27] "there is danger in promoting the adoption of autonomy as a factor relevant to legal rights in relation to personal decision-making, because this could result in the invasion of the human rights of, a lack of respect for, and wrongful discrimination against, persons characterized as non-autonomous". The right of self determination, after all, may only be exercised by an individual who is considered to be legally competent to make decisions concerning treatment and/or experimentation. In this sense, "competence acts as a condition precedent to the exercise of the right of self-determination".[28]

On the other hand, the doctrine of inviolability is predicated on the "basic moral presupposition of respect for human life".[29] It involves the notion that the State has a legitimate interest in protecting its citizens not only from external threats to their physical integrity but also from their own choices where there is an unjustified risk of injury or death — hence the State has the right, for example, to prevent self-mutilation. This "paternalistic" approach is rooted in "moral precepts of the Middle Ages which espoused the principle of the totality of human physical integrity"[30] and clearly reflects the notion that, in certain circumstances, the sanctity of life may be a more important value to preserve than the autonomy of the individual person.[31]

Insofar as the value of inviolability refers to the right of the individual to refuse treatment as a means of exercising the right to self-determination, then it can be considered to be in perfect harmony with the value of autonomy. However, it has been contended that there may be a fundamental incompatibility between the values of autonomy and inviolability — at least, if the latter suggests a "principle that protects a person's physical and mental integrity against non-beneficial acts by the person himself, or others, when it is a preservation of life value".[32]

All jurisdictions recognize a principle of autonomy or self-determination;[33] however, as Somerville suggests that the limits placed on this principle depend on public policy (usually expressed through the medium of the criminal law) and that

---

[27] M.A. Somerville, "Label versus Contents: Variations between Philosophy, Psychiatry and Law in Concepts Governing Decision-Making" (1994) 39 McGill Law Journal 179.

[28] *Ibid* at 193.

[29] Somerville (1979), *supra* note 19 at 8—9.

[30] Law Reform Commission of Canada, *Working Paper 26. Medical Treatment and Criminal Law* (Ottawa: Ministry of Supply and Services Canada, 1980) at 62. The Commission notes at p. 126n. 329 that Thomas Aquinas, in his *Summa Theologica*, asserted that the mutilation of a limb by a private person was not permitted unless it was necessary to "improve the health of the entire body" and then only if there were no other means available to achieve this purpose.

[31] Hébert, *supra* note 25, states (at 501) that:

> Il apparaît très paternaliste de vouloir protéger l'adulte compétent contre ses propres choix en lui niant le droit de choisir la mort. Au nom de l'ordre public, certain jugements semblent mettre l'autonomie de l'individu au service de la protection et de l'intégrité de sa vie. *Plutôt que d'y voir des valeurs égales et complémentaires, certains y voient une opposition ou même une relation d'assujettissement.* (emphasis added)

[32] Somerville, *supra* note 19 at 5.

[33] *Ibid* at 6—7.

these limits tend to be wider in the common law jurisdictions than in their Civil Law counterparts:

> The Civil Law is less inclined to leave to chance the decision to act in a self-protective way and has a well-developed doctrine of inviolability of the human body, which arises from a basic moral presupposition of respect for human life.[34]

In one sense, the principle of inviolability may be seen as an adjunct to the principle of autonomy insofar as the former operates to prevent unwanted interference with one's body. However, the principle of inviolability may also be employed as a means of preserving life and health even when this goal may conflict with the individual's wish to the contrary.[35] Clearly, in this aspect, the principle of inviolability may have profound implications for the whole question of the extent to which a competent adult may risk his or her personal integrity in consenting to participate in research. Equally, the principle has undoubted value as a means of protecting those who are not competent to make such decisions but who may be submitted to the perils of experimentation by substitute decision-makers of one kind or another.

In fact, the *Civil Code of Quebec*[36] articulates a number of general principles that reflect not only the value of autonomy but also the value of inviolability. For example, Article 3 of the *Civil Code* provides that "(e)very person is the holder of personality rights, such as the right to life, the right to the inviolability and integrity of his person, and the right to the respect of his name, reputation and privacy" and that "(t)hese rights are inalienable". Significantly, Article 10 enshrines both the principle of personal inviolability and the principle of autonomy within one general statement:

> Every person is inviolable and is entitled to the integrity of his person.
>     Except in cases provided for by law, no one may interfere with his person without his free and enlightened consent.[37]

It has been suggested[38] that the potential conflict between the principles of autonomy and inviolability may be illustrated by making reference to two contrasting cases in which the common law courts and the Civil Law courts respectively arrived at fundamentally different decisions on the basis of granting primacy to the principle of autonomy in one case and to the principle of inviolability in the other.

The classic common law approach is illustrated by the *Astaforoff* case,[39] decided by the British Columbia Court of Appeal. In this case, the appellate court upheld the

---

[34] *Ibid* at 8—9.

[35] *Ibid* at 9.

[36] L.Q. 1991, c. 64, as amended by *An Act respecting the implementation of the reform of the Civil Code*, L.Q. 1992, c. 57.

[37] Another declaration of the principle of autonomy may be detected in Article 154, which provides that, "[i]n no case may the capacity of a person of full age be limited except by express provision of law or by a judgment ordering the institution of protective supervision".

[38] See, for example, Kouri, *supra* note at 852.

[39] *Attorney General of British Columbia* v. *Astaforoff*, [1984] 4 W.W.R. 385 (B.C.C.A.).

decision of the trial court to refuse to order the forcible feeding of an elderly hunger striker, who was being held in prison. The Court emphasized that the prisoner was competent to refuse nourishment and that, therefore, her wishes should be respected regardless of the fact that her life was threatened by the approach that she had adopted. The autonomy principle was clearly given precedence over the sanctity-of-life principle in these particular circumstances.

The Civil Law approach, on the other hand, is illustrated by the case of *Niemec*,[40] decided by the Cour Supérieure of Québec. In this case, the Court authorized the forcible medical treatment of a man held in non-criminal custody, pending deportation from Canada. This individual had deliberately inserted a steel wire into his oesophagus, thus creating a life-threatening situation. Despite the man's refusal to accept treatment (in fact, he indicated that he would rather die than be deported to his country of origin), the situation was deemed to constitute an *emergency*, in which there was a serious threat to the man's health and an urgent need to intervene medically. In these circumstances, therefore, the Court authorized treatment despite the fact that the man was deemed to be competent and his right of autonomy was apparently overriden by the principle of inviolability in its guise as bulwark of the sanctity of human life.[41]

While these cases do provide examples of the different emphasis placed on certain values within the common law courts as opposed to the Civil Law courts of Quebec, there is nevertheless a very real danger that these differences in approach may be exaggerated. In fact, there have been a number of recent cases in which the Supreme Court of Canada, although hearing appeals from common law jurisdictions, has given primacy to the inviolablity principle over the autonomy principle.

For example, in *Jobidon*,[42] the majority of the Court ruled that, as a matter of sound public policy, individuals should not be able to give a valid consent to the infliction of "serious hurt" or "non-trivial bodily harm" in the course of a fist fight or brawl. In the words of Gonthier J.,

> The policy preference that people not be able to consent to intentionally inflicted harm is heard not only in the register of our common law. The *Criminal Code* also contains many examples of this propensity. .... s. 14 of the *Code* vitiates the legal effectiveness of a person's consent to have death inflicted on him under any circumstances. The same policy appears to underlie ss. 150.1, 159 and 286 in respect of younger people, in the contexts of sexual offences, anal intercourse, and abduction, respectively. All this is to say that the notion of policy-based limits on the effectiveness of consent to some level of inflicted harms is not foreign. Parliament as well as the courts have been mindful of the need for

---

[40] *Procureur Général du Canada* c. *Hôpital Notre-Dame et un autre (défendeurs) et Jan Niemec (mis en cause)*, [1984] C.S. 426 (S.C.). See, generally, M.A. Somerville, "Refusal of Medical Treatment in 'Captive' Circumstances" (1985) 63 Canadian Bar Review 59; H. Savage and C. McKague *Mental Health Law in Canada* (Toronto: Butterworths, 1987) at 105—106.

[41] Hébert, *supra* note 25, states (at 501) that: "la Cour affirme que le respect de la vie, parce que conforme a l'intérêt même de la personne, primait sur le respect de la volonté du patient". Other examples of the extent to which the Quebec Civil Law accords primacy to the principle of inviolability over that of autonomy are *Institut Philippe Pinel de Montréal* v. *Dion* (1983), 2 D.L.R. (4th) (C.S.); and *Ville de Vanier* c. C.S.S.T., [1986] D.L.Q. 297 (C.S.).

[42] *R.* v. *Jobidon* (1991), 66 C.C.C. (3d) 454 (S.C.C.).

such limits. *Autonomy is not the only value which our law seeks to protect. . . .* All criminal law is "paternalistic" to some degree — top-down guidance is inherent in any prohibitive rule. That the common law has developed a strong resistance to intentional applications of force in fist fights and brawls is merely one instance of *the criminal law's concern that Canadian citizens treat each other humanely and with respect.*[43]

In similar vein, the Supreme Court of Canada recently rejected the assertion that, under the terms of the *Canadian Charter of Rights and Freedoms*, a terminally ill person has the right to an assisted suicide and upheld the validity of the existing *Criminal Code* provisions that render it an offence to provide such assistance. Indeed, in the *Rodriguez* case,[44] the Court recognized that the *Criminal Code* provisions effectively deprived the plaintiff of her autonomy and caused her both physical pain and psychological distress. However, a narrow majority (5–4) nevertheless ruled that the provisions were valid because they reflected the principle of the sanctity of life. According to Justice Sopinka,

> Section 241(b) has as its purpose the protection of the vulnerable who might be induced in moments of weakness to commit suicide  This purpose is grounded in the state interest in protecting life and reflects a policy of the state that human life should not be depreciated by allowing life to be taken.  This policy finds expression not only in the provisions of our *Criminal Code* which prohibit murder and other violent acts against others notwithstanding the consent of the victim, but also in the policy against capital punishment and, until its repeal, attempted suicide.  This is not only a policy of the state, however, but is part of our fundamental conception of the sanctity of human life . . . .[45]

According to Justice Sopinka, a blanket prohibition on suicide (such as that contained in section 241 of the *Criminal Code*) "is the norm among western democracies, and such a prohibition has never been adjudged to be unconstitutional or contrary to fundamental human rights".[46] Indeed, the purpose of such a prohibition is to uphold the respect for life and, in this sense, "it may discourage those who consider that life is unbearable at a particular moment, or who perceive themselves to be a burden upon others, from committing suicide". In his view, "to permit a physician to lawfully participate in taking life would send a signal that there are circumstances in which the state approves of suicide".[47] Clearly, the principle of inviolability of the person (the sanctity of life) was given precedence over the principle of autonomy (the right of citizens to do what they want with their bodies).

The Supreme Court of Canada has also indicated that the principle of inviolability outweighs the principle of autonomy where the subject is incompetent to make decisions. Indeed, in the *Eve* case,[48] the Supreme Court of Canada made it abundantly clear that the law must protect the rights of those who are deemed to be incapable and must ensure that their best interests and well-being are addressed at

---

[43] *Ibid* at 494—495 (emphasis added).
[44] *Rodriguez* v. *British Columbia (Attorney-General)* (1993), 85 C.C.C. (3d) 15 (S.C.C.).
[45] *Ibid* at 69.
[46] *Ibid* at 76.
[47] *Ibid* at 79.
[48] *E.* v. *Eve* [1986] 2 S.C.R. 388. See, R.P. Kouri, "L'arrêt *Eve* et le droit québécois" (1987) 18 Revue générale de droit 643.

all times; the autonomy of substitute decision-makers must be a secondary consideration in such circumstances. As Hébert notes, "ce qui est déterminant, c'est l'inviolabilité de la personne, même incapable, et non la liberté de ceux qui prennent la décision à sa place".[49]

However, it is clear that the inviolability principle has clear limits even within the context of Quebec jurisprudence. For example, in the *Nancy B.* case (1992),[50] Dufour J., of the Cour Supérieure granted an injunction to a patient who wished her physician to cease treatment with a respirator — even though this course of action would lead inevitably to her death. The patient had been afflicted with Guillan-Barré syndrome, an incurable neurological disorder which caused atrophy of her respiratory muscles; her life could only be prolonged by maintaining the support of the respirator. The trial Judge noted that the doctrine of informed consent had been deeply embedded in both statute and case law in Quebec and that "*the logical corollary of this doctrine of informed consent is that the patient generally has the right not to consent, that is the right to refuse treatment and to ask that it cease where it has already begun*".[51]

Dufour J. quoted the view of Jean-Louis Beaudoin (*sic*) to the effect that for "a competent person of the age of majority, the making of his (*sic*) own decisions with respect to his own body is the legal expression of the principle of personal autonomy and of the right to self determination".[52] However, the Judge noted that this author indicates that the ability to consent is not absolute. In fact, it is limited by two factors. First, an individual may not use his or her body in a way that might jeopardize the life or health of others. Second, public policy may dictate that there should be "limits on the right to do freely what one wishes with one's body". For example, the law prohibits individuals from disposing *inter vivos* of a vital organ or a part of the body that is not capable of regenerating itself. Nevertheless, "subject to these two limits . . . , one may consider that the right to autonomy and self-determination is absolute".[53]

Dufour J. reaffirmed the pivotal importance of the principle of autonomy, within the broader context of the criminal law, when he stated:

> What Nancy B. is seeking, relying on the principle of personal autonomy and her right of self-determination, is that the respiratory support treatment being given her cease so that nature may take its course; that she be freed from the slavery to a machine as her life depends on it. In order to do this, as she is unable to do it herself, she needs the help of a third person. Then, it is the disease which will take its natural course.[54]

---

[49] Hébert, *supra* note 25 at 502.

[50] *Nancy B.* v. *Hôtel-Dieu de Québec et al.* (1992), 69 C.C.C. (3d) 450 (Que.S.C.). See B.M. Dickens, "Medically Assisted Death: *Nancy B.* v. *Hôtel-Dieu de Québec*" (1993) 38 McGill Law Journal 1053.

[51] *Ibid* at 456 (emphasis in original).

[52] *Ibid* at 456. The reference to the work of Beaudoin (*sic*) is to a seminar, entitled *Le droit de refuser d'être traité*, given under the auspices of the Canadian Institute for the Administration of Justice (no date provided).

[53] *Ibid* at 456.

[54] *Ibid* at 457 (emphasis in original).

In sum, both the common law and Quebec Civil law jurisdictions place a varying degree of emphasis on both the principle of autonomy and the principle of inviolability. There is certainly some evidence to support the view that the principle of inviolability has traditionally been granted a greater degree of precedence in the Quebec courts than is the case in the common law courts. However, even in the common law courts, there has been a recent trend whereby the principle of autonomy has been decisively overriden where it was perceived that there was a serious threat to the sanctity of human life.

In terms of the relevance of this discussion to biomedical experimentation, it is clear that the protective aspects of the principle of inviolability should be given considerable emphasis in any legal régime that is designed to regulate the research enterprise. It is manifest that the principle of autonomy needs to be balanced with the principle of inviolability where it is a question of whether a *competent* adult should be permitted to consent to biomedical experimentation; however, it is undoubtedly the principle of inviolability that must be accorded primacy when the issue is whether an *incompetent adult or child* should be subjected to such experimentation. The extent to which recent legislative changes in Quebec reflect the influence of the principle of inviolability is the subject of the following analysis.

## CONSENT TO MEDICAL TREATMENT REQUIRED BY THE STATE OF A PATIENT'S HEALTH

In the specific context of medical (therapeutic) care, the *Civil Code of Quebec*[55] establishes the right of the competent adult to personal autonomy and establishes the framework for substitute consent in the event that competence is lacking:

> No person may be made to undergo care of any nature, whether for examination, specimen taking, removal of tissue, treatment or any other act, except with his consent.
> If the person concerned is incapable of giving or refusing his consent to care, a person authorized by law or by mandate given in anticipation of his incapacity may do so in his place.

As far as health care for *minors* is concerned, the *Civil Code* stipulates that consent may be given by the person having parental authority or his/her tutor. However, a minor of fourteen years or more may give a valid consent on his or her own.[56]

---

[55] L.Q. 1991, c. 64, as amended by *An Act respecting the implementation of the reform of the Civil Code*, L.Q. 1992, c. 57.

The framework for substitute consent is somewhat complex. Insofar as adults who are incapable of giving their consent to medical care "required by the state of (their) health" are concerned, then consent may be obtained from the mandatary (a decision-maker appointed by the patient), the tutor, or the curator (appointed by a court). If there is no such third party to give consent, then consent may be sought from the spouse or, failing that, from a "close relative or a person who shows a special interest" in the adult concerned. See Article 15.

[56] Article 14.

These provisions of the *Civil Code* must be read in tandem with important directives in the *Code of Ethics of Physicians*[57] which impose a duty to obtain informed consent from patients or their substitute decision-makers as well as an obligation to protect the "health and well-being" of those under their medical care:

> 2.02.01. The physician's paramount duty, in the performance of his medical functions is to protect the health and well-being of the persons he takes care of, both individually and collectively.

> 2.02.02 The physician must not, by any means, either directly or indirectly, interfere with the patient's freedom of choice of a physician.

> 2.03.28. Except in an emergency, a physician must, before undertaking an investigation, treatment or research, obtain informed consent from the patient or his representative or any persons whose consent may be required by the law.

> 2.03.28 A physician must ensure that the patient or his representative or the persons whose consent may be required by law receive suitable explanations on the nature, purpose and possible consequences of the investigation, treatment or research which the physician prepares to make.

Insofar as the principles guiding the choices of substitute decision-makers are concerned, the *Civil Code*[58] stipulates that the best interests of the patient are paramount, although the latter's wishes are to be taken into account wherever it is feasible to do so:

> A person who gives his consent to or refuses care for another person is bound to act in the sole interest of that person, taking into account, as far as possible, any wishes the latter may have expressed.
> If he gives his consent, he shall ensure that the care is beneficial notwithstanding the gravity and permanence of certain of its effects, that it is advisable in the circumstances and that the risks incurred are not disproportionate to the anticipated benefit.

These protective provisions plainly reflect the influence of the principle of personal inviolability and accentuate the notion that the principle of autonomy has no application to substitute decision-makers themselves — except insofar as they are able to take the *patients'* wishes into account.

Where the substitute decision-maker is "prevented" from giving consent or, without justification, "refuses to do so", then court authorization is required before care may be administered. Such authorization is also required where an adult who is incapable of giving consent "refuses to receive care, except in the case of hygienic care or emergency" or where a minor of fourteen years or more refuses care.[59]

---

[57] R.R.Q. 1981, c. M—9, r.4.

[58] Article 12.

[59] Article 16. However, in the case of an emergency, where such a minor's life is in danger or his or her integrity is threatened, then consent to care may be given by the person having parental authority or by the tutor.

## CONSENT TO MEDICAL CARE NOT REQUIRED BY THE PATIENT'S STATE OF HEALTH

Minors who are fourteen or over may give a valid consent to medical care *not required by the state of their health* (for example, cosmetic surgery) although the consent of the person having parental authority or the tutor is required "if the care entails a serious risk for the health of the minor" concerned and "may cause him grave and permanent effects".[60]

The *Civil Code* postulates the implementation of special, protective procedures in relation to minors under the age of fourteen or incompetent adults who are being considered for the administration of medical care that is not required by their state of health and that "entails a serious risk for health" or "might cause grave and permanent effects". In such circumstances, *court authorization* is required *in addition to the consent of the person who has parental authority or the consent of the duly appointed substitute decision-maker (mandatary, tutor or curator as the case may be).*[61] The prerequisite of court authorization in such cases represents a powerful affirmation of the principle of personal inviolability in the context of a vulnerable population.

The *Civil Code* also addresses the important issue of consent to the removal of human tissue (for the purpose of transplantation, etc.). An adult who is capable of giving consent may "alienate a part of his body *inter vivos*, provided the risk incurred is not disproportionate to the benefit that may reasonably be anticipated". However, a minor or an incapable adult may "alienate" a part of their body (e.g. donation of bone marrow) only if "that part is capable of regeneration and provided that no serious risk to (their) health results" and only if the appropriate substitute decision-maker gives their consent.[62] It is notable that the impact of the principle of personal inviolability is evident not only in relation to minors and incompetent adults but also in relation to fully competent individuals.

## CONSENT TO BIOMEDICAL EXPERIMENTATION INVOLVING HUMAN SUBJECTS: COMPETENT ADULTS

The *Civil Code* does not give even competent adults *carte blanche* to submit to (non-therapeutic) experimentation. Indeed, the protective aspects of the inviolability principle are clearly manifested in Article 20:

> A person of full age who is capable of giving his consent may submit to an experiment provided that the risk incurred is not disproportionate to the benefit that can reasonably be anticipated.[63]

According to Bowker, "experimentation" appears to refer to "scientific or non-therapeutic experiments" and, for this reason, he contends that the word, "benefit",

---

[60] Article 17.
[61] Article 18.
[62] Article 19.
[63] See *Weiss* c. *Solomon, supra* note; *Morrow* c. *Royal Victoria Hospital, supra* note 11.

means "future benefit to persons other than the one submitting to the experiment".[64] In this context, the "benefit" in question "takes the form of an increase in learning, in scientific knowledge" and the beneficiary is "society as a whole".[65]

Although Bowker also declares that the "weighing of risks and benefits" is an "understandable approach to take" and is one that is espoused by the *Helsinki Declaration*, the author nevertheless points out that the process is one that is fraught with inherent difficulty:

> The anticipated benefit may be great, or it may be slight; the likelihood of achieving the benefit may be high, or it may be remote. On the other hand, the risk may be great or it may be slight.[66]

Of course, it is difficult to conduct the risk-benefit analysis required by Article 20 when the potential detriment will be experienced by the individual subject, whereas any potential benefits will be enjoyed by others. In this respect, it has been suggested[67] that it would be preferable to adopt the approach manifested by legislation in France, which stipulates that:

> Art. L.209—2. Aucune recherche biomédicale ne peut être effectuée sur l'être humain: . . . si le risque prévisible encouru par les personnes que se prêtent à la recherche est hors de proportion avec le bénéfice escompté pour ces personnes *ou l'intérêt de cette recherche.*[68]

This approach would dictate that the risk-benefit analysis must be conducted in relation to the specific research project in question rather than the more abstract concept of scientific knowledge in general.

Furthermore, it has also been noted that, while a literal reading of this provision of the *Civil Code* might suggest that "almost any risk may be incurred" provided only that it is believed that the benefits outweigh the risks, the principle of inviolability enshrined in both the Quebec and Canadian *Charters* nevertheless limits the degree of risk to which subjects may actually be exposed: given these constitutional protections, it is said that, "the magnitude of any risk which may be assumed must remain quite limited".[69]

---

[64] W.F. Bowker, *supra* note 18 at 166—167. Bowker's analysis refers to the pre-1991 provisions of the *Civil Code*; however, many of the points made are equally applicable to the most recent version of the *Civil Code*. See, also, J.-L. Baudouin, "L'expérimentation sur les humains: un conflit de valeurs" (1981) 26 McGill Law Journal 809 at 819; Kouri (1991), *supra* note 16 at 81.
Bowker also notes that the "appraisal of risk and benefit must be determined at the time rather than subsequently" (at 167).

[65] Law Reform Commission of Canada, *Working Paper 61: Biomedical Experimentation Involving Human Subjects* (Ottawa: Law Reform Commission of Canada, 1989) at 33.

[66] *Ibid* at 167. It has been pointed out that section 12 of the *Canadian Charter of Rights and Freedoms* protects individuals from experiments that can be described as "cruel and unusual". M. Ouellette, "La Charte canadienne et certains problèmes de bioéthique" (1984) 18 Revue juridique Thémis 271—293 at 285. Section 7, of course, is also relevant insofar as it protects the "right to life, liberty and security of the person".

[67] Kouri (1991), *supra* note 16 at 107—108.

[68] *Loi no 88—1138 du 20 décembre 1988 relative à la protection des personnes qui se prêtent à des recherches biomédicales*, J.O. 22 déc. 1988, p. 1603, J.C.P. 1989. III. 62199 (emphasis added), as cited by Kouri (1991), *supra* note 16 at 107—108.

[69] Kouri, *ibid*. at 93.

## CONSENT TO BIOMEDICAL EXPERIMENTATION INVOLVING MINORS AND INCOMPETENT ADULTS

In 1991, major amendments were made to the *Civil Code of Quebec* in relation to the regulation of human experimentation involving minors and incompetent adults. As Kouri notes, prior to this date, the situation was that "adult mental incompetents" and "minors lacking discernment" could not "be submitted to experimentation, even with the concurrence of their legal representatives, for the simple reason that this type of activity is not in the immediate interest of the person unable to consent".[70] It is interesting that, from this commentator's point of view, this was considered to be a lamentable situation:

> It is disturbing to note that by restricting participation in purely scientific experimentation only to capable adults or minors having discernment, the present *Civil Code* seems to be somewhat out of touch with scientific imperatives. To begin with, certain highly useful experiments entail absolutely no risk to the research subject. Moreover, in the field of pediatric medicine, research on children remains essential . . .[71]

The new provisions of the *Civil Code* have now rendered it possible for an incompetent adult or a minor "lacking discernment" to participate in an experiment, *provided a substitute decision-maker gives his or her consent and provided various safeguards are respected.* This undoubtedly represents a departure from the formal approach taken in many other countries. As Baudouin notes, "le Québec devient un des rares pays au monde à légaliser l'expérimentation sur le malade mental, ce qui paraît en conflit avec certains textes nationaux et internationaux".[72] However, it must be conceded that the principle of personal inviolability has nevertheless exerted some degree of influence over the legislators and that the amendments do indeed establish a protective framework to ensure that the interests of this admittedly vulnerable group of subjects are respected.

According to Article 21 of the *Civil Code*, adults who are incapable of giving consent or minors may only be submitted to experimental procedures in very limited circumstances. In general, there must be an "absence of serious risk" to their health; their objections to participation must be respected; the appropriate substitute decision-maker (mandatary, tutor or curator, as the case may be) must give his or her consent; and there must be an *independent review* (by a court or an ethics committee) of the proposed research before any consent given by a substitute decision-maker will become effective. It is notable that the legislators did not stipulate that the risks associated with the research should be weighed against benefits that may accrue to others through the acquisition of scientific knowledge. Instead, the focus is on the risk to the potential research subjects themselves. Indeed, no experimentation may take place if there is any "serious risk" to their health. In

---

[70] Kouri (1991), *supra* note 16 at 87.
[71] *Ibid* at 90.
[72] J.-L. Baudouin, "Quelques aspects de la loi 20 et des droits de personnalité" (1987) 18 Revue de droit, Université de Sherbrooke 45 at 52.

principle, this is a more efficacious safeguard than exists in the case of a competent adult who is considering participation in an experiment.[73]

For the purpose of determining the precise form of external review that must be implemented before a substitute decision-maker's decision becomes effective, the *Civil Code* draws a fundamental distinction between research conducted on one person alone and research conducted on a group of subjects:

> An experiment may be carried out on one person alone only if a benefit to the health of that person may be expected, and the authorization of the court is necessary.
>
> An experiment on a group of minor persons or incapable persons of full age shall be carried out within the framework of a research project approved by the Minister of Health and Social Services, upon the advice of an ethics committee of the hospital designated by the Minister or of an ethics committee created by him for that purpose; in addition, such an experiment may be carried out only if a benefit to the health of persons of the same age group and having the same illness or handicap as the persons submitted to the experiment may be expected.
>
> Care considered by the ethics committee of the hospital concerned to be innovative care required by the state of health of the person submitted to it is not an experiment.[74]

In other words, the form that any external review takes depends on whether the experimentation is designed with a view to benefiting a specific individual, on the one hand, or the health of persons of the same age group and having the same illness or handicap as the proposed subjects, on the other. In the case of an experiment designed to benefit a specific individual, *court authorization* is necessary before the experiment may proceed, while, in the case of an experiment that involves a group of minors or incompetent adults, there must be prior approval of the research project by the Minister of Health and Social Services, who, in turn, is advised by an *ethics committee.*

It has been questioned whether it might not have been more desirable for the legislation to have required *judicial authorization* not only for research conducted on specific individuals but also for experimentation on groups of incapable individuals, since approval by the Minister and an ethics committee does not necessarily ensure that the rights of the vulnerable are protected in specific cases.[75] However, such a requirement may well impose considerable (and, perhaps, prohibitive) delays and expenditures on those who are engaged in the research enterprise and it may be more appropriate to protect vulnerable groups of potential research subjects by placing members of ethics committees under a clear legal duty to enforce all of the safeguards enacted by the legislators. In any event, it remains to be seen whether judicial authorization for research on a group of incompetent adults or minors will be considered a constitutional requirement under the Québec or Canadian *Charters of Rights and Freedoms* — regardless of any arguments based on cost and delay.

The wording of Article 21 is designed to prevent exploitation of incompetent adults and children by requiring that experimentation on individuals must be

---

[73] *Ibid* at 106.
[74] Article 21.
[75] See Kouri (1991), *supra* note 16 at 107.

conducted for the benefit of their health and that experimentation on a group of individuals must be justified by an expected benefit to the "health of persons of the same age group and having the same illness or handicap as the persons submitted to the experiment". As Kouri points out,[76] the corollary of this rule is that incapable persons cannot participate in an experiment "with regard to conditions or diseases which are not exclusive to their group". Consequently, a mentally deficient adult cannot be experimented upon in order to determine, for example, the causes and potential treatments for ulcerative colitis since colitis affects all segments of the population.

It is significant that, under the terms of Article 21, "care considered by the ethics committee of the hospital concerned to be 'innovative care' required by the state of health of the person submitted to it is not an experiment". Presumably, consent to "innovative care" must be obtained in accordance with the statutory requirements that apply to (therapeutic) medical care. The problem is that it is difficult, at a practical level, to decide whether a proposed biomedical procedure is "innovative care" or a "therapeutic experiment". It is nevertheless an important distinction because, insofar as minors under the age of fourteen and incompetent adults are concerned, therapeutic experimentation requires court approval, whereas, in general, innovative therapy, required by the subject's state of health, merely requires the consent of a substitute decision-maker.

The term, "innovative therapy" was defined by the Law Reform Commission of Canada as "a treatment in the true sense of the word, an act performed for the direct and immediate benefit of the recipient, but not yet fully proved in scientific terms".[77] This definition would appear to distinguish innovative therapy from therapeutic experimentation on the basis of the primary goal of the proposed procedure. If the main object of the person administering the procedure is to benefit the patient's state of health (with any contributions to scientific knowledge constituting a secondary consideration), then one is dealing with "innovative therapy". However, if the paramount objective is to advance scientific knowledge (with any benefits to the subjects' health constituting only an incidental byproduct), then the procedure may be considered to be "therapeutic experimentation". Unfortunately, attempting to identify the primary objective of a biomedical procedure may be considerably more difficult to accomplish in practice than it is in theory. The truth is that any test that is designed to draw a distinction between "innovative therapy" and "therapeutic experimentation" is unlikely to prove to be entirely satisfactory in practice and, for this reason, some authors have contended that innovative therapy and therapeutic

---

[76] *Ibid* at 107.

[77] Law Reform Commission of Canada, *supra* note 65 at 4–5 (emphasis added).

See, also, P.H. Osborne, "Informed Consent" in B. Sneiderman, J. Irvin and P.H. Osborne, eds., *Canadian Medical Law: An Introduction for Physicians and Other Health Care Professionals* (Toronto: Carswell, 1989) at 62: "Innovative procedures are medical advances directed toward therapeutic ends, whereas research is one aspect of scientific investigation that is designed to enhance the field of human knowledge."

In the context of the common law approach to innovative therapy, see, *Zimmer* v. *Ringrose* (1981), 124 D.L.R. (3d) 215 (Alta.C.A.) and *Coughlin* v. *Kuntz* (1987), 17 B.C.L.R. (2d) 365 (S.C.).

experimentation have so many similar characteristics that they should be regulated in exactly the same manner (although, ironically, there is no agreement as to whether they should be regulated as therapy or experimentation).[78] Unfortunately, the luxury of adopting this more rational approach is not an option that is presently available in Quebec given the wording of the amendments to the *Civil Code*. In the near future, the Québec courts will, no doubt, be required to develop a specific body of jurisprudence that articulates the boundaries between innovative care and therapeutic experimentation.

## MISCELLANEOUS SAFEGUARDS

The *Civil Code* has imposed a number of duties on courts that are requested to give their authorization for medical care or experimentation. For example, whenever a court is required to rule on an application for authorization to administer care, to alienate part of the body, or to conduct an experiment, it must obtain the advice of experts as well as that of the relevant substitute decision-makers. The court may also obtain the opinion of "any person who shows a special interest in the person concerned by the application". The requirement that expert testimony be considered is designed to ensure, *inter alia*, that the court hears independent evidence as to the nature of the risk that is posed by the proposed procedure.

Furthermore, the court is placed under a clear duty to obtain the opinion of the patient or research subject concerned, unless that is not feasible, and to "respect his refusal unless the care is required by his state of health".[79] The effect of this provision is to protect the autonomy of an incompetent adult or a minor by requiring that his or her opinion be sought and by enshrining an absolute right of refusal to engage in experimentation. However, the principle of personal inviolability has been accorded precedence whenever an incompetent adult or a minor refuses medical care that is required by the state of their health since, in these circumstances, their refusal will generally not be sustained. It has been noted that this principle has the consequence of effectively removing the right to refuse to participate in the context of therapeutic experimentation:

> By definition, therefore, therapeutic experimentation implies that in almost all situations, the incapable patient would never be able to refuse treatment.[80]

The *Civil Code* also establishes an additional safeguard by providing that any consent to care not required by the subject's state of health, to the alienation of a part

---

[78] See Kouri (1991), *supra* note 16; M. Somerville (1981), *supra* note 12; Law Reform Commission of Canada, *supra* note 65 at 405; Medical Research Council of Canada, *Guidelines on Research Involving Human Subjects* (Ottawa: Minister of Supply and Services Canada, 1987) at 9.

The Law Reform Commission, *supra* note 65 at 5, recommends that the term, "experimentation", be limited to "non-therapeutic biomedical experimentation". However, Article 21 of the *Civil Code* refers to "experimentation" that may only be "carried out on one person alone" if "*a benefit to the health of that person may be expected*". This clearly maintains a category of "therapeutic experimentation".

[79] Article 23.

[80] Kouri (1991), *supra* note 16 at 103–104.

of a person's body, or to an experiment must be given *in writing and may be withdrawn at any time* (even verbally).[81] The requirement of a formal signification of consent and the right to withdraw that consent in an informal manner offer a further degree of protection to subjects involved in experimentation.

Finally, the *Code* also prohibits the intrusion of a commercial motivation into either the research process or the practice of organ or tissue donation:

> The alienation by a person of a part or product of his body shall be gratuitous; it may not be repeated if it involves a risk to his health.
> An experiment may not give rise to any financial reward other than the payment of an indemnity as compensation for the loss and inconvenience suffered.[82]

This provision addresses the difficult issue of "altruism in experimentation", by requiring that consent to participation in research be "untainted by economic incentives".[83] As such, it is designed to reduce the threat of undue coercion or influence on the decision as to whether to give consent to participation in an experiment.

## CONCLUSIONS

There is no doubt that the principle of personal inviolability has played a critical role in the evolution of the Civil Law of Quebec insofar as the issue of consent to medical care and biomedical experimentation is concerned. Indeed, as we have seen, there have certainly been occasions upon which the principle of personal inviolability has been applied in Quebec Civil Law as a means of preserving bodily integrity even when this course of action has entailed a clear violation of the expressed wishes of a competent adult who elected to refuse life-saving treatment. On the other hand, in the common law jurisdictions, the courts have generally granted the principle of autonomy a somewhat greater degree of precedence in this area of the law and they have been considerably less amenable to overruling the wishes of competent adults who wish to spurn therapeutic medical intervention.

The influence of the principle of personal inviolability is also evident in the *Civil Code* provisions concerning consent to non-therapeutic experimentation. It is significant that the *Civil Code* actually limits the scope of consent that even a competent adult may give to participation in research by stipulating that the anticipated benefits to science must outweigh the risks to the subject before he or she may submit to such research. In the common law jurisdictions, in contrast, the focus has been primarily on the duty of the researcher to disclose all possible risks rather than on the validity of the subject's consent.

For the present, however, the central issue is the extent to which vulnerable populations, such as incompetent adults and children, should be submitted to non-

---

[81] Article 24.
[82] Article 25.
[83] Kouri (1991), *supra* note 16 at 82.

therapeutic biomedical experimentation. In the common law jurisdictions, it appears to be the case that a substitute decision-maker (such as a guardian or parent) does not have the clear legal authority to consent to such experimentation.[84] In this context, the principle of autonomy has only a limited degree of relevance since it presumes the existence of competence as a prerequisite for giving informed consent to participation in research and the common law jurisprudence dealing with substitute consent has primarily been constructed on the basis of the application of the "best interests" principle, which, by definition, precludes the giving of surrogate consent to non-therapeutic procedures.

However, in Quebec, recent amendments to the *Civil Code* now permit non-therapeutic experimentation on both minors and incompetent adults where substitute consent has been obtained. While permitting non-therapeutic research to be conducted in relation to these vulnerable populations, the Quebec legislature has nevertheless implemented the principle of inviolability by entrenching (what appear to be) strong safeguards against potential abuse and exploitation. The *Civil Code* provisions require, for example, not only substitute consent but also prior approval of an experiment by either a court or an ethics committee. However, the imprint of the principle of inviolability is most evident in the emphasis on the absence of "serious risk" as a pre-condition for participation in experimentation. As Moorhouse has contended, "vulnerable individuals can be subjects in non-therapeutic research without having their integrity threatened",[85] provided they are not exposed to any significant degree of risk. Surrogate consent is acceptable, in her view, provided "the research does not expose the subject to more risk than is associated with daily living".[86] Of course, it remains to be seen exactly how the term, "absence of serious risk", will be interpreted in the context of the new *Civil Code* provisions; the inviolability principle would apparently dictate that the subject should not be exposed to any procedure that poses more than a minimal degree of risk.

The enactment by the Province of Quebec of *Civil Code* provisions that permit the participation of minors and incompetent adults in non-therapeutic research represents a striking departure from the existing jurisprudence in Canada. However, in a humane and responsible society, it is clear that the involvement of vulnerable populations in non-therapeutic experimentation should only be permitted where there is a sound framework of protective mechanisms that will be effective in preventing abuse and exploitation of these populations. Unquestionably, the implementation of the new provisions in Quebec will attract a considerable degree of scrutiny elsewhere in Canada with a view to determining if they do, in fact, establish effective safeguards for the subjects of biomedical experimentation. The

---

[84] R.M. Gordon and S.N. Verdun-Jones, *Adult Guardianship Law in Canada* (Toronto: Carswell, 1995) at 4–5 — 4–16; Sharpe, *supra* note 11 at 75; Law Reform Commission of Canada, *supra* note 65 at 40–42.

[85] A. Moorhouse, "Ethical and Legal Issues Associated with Alzheimer's Disease Research and Patient Care: To Do Good Without Doing Harm" in S.N. Verdun-Jones and M. Layton eds., *Mental Health Law and Practice Through the Life Cycle: Proceedings from the XVIIIth International Congress on Law and Mental Health* (Burnaby, B.C.: Simon Fraser University, 1994) at 47.

[86] *Ibid* at 56.

long tradition of commitment to the principle of personal inviolability in Quebec jurisprudence, however, does suggest some cause for optimism that the new legislation will be implemented in a manner that will lend some support to the notion that the participation of vulnerable individuals in such experimentation is, in very specific (and limited) circumstances, sound social policy.

CHAPTER 11

# THE ADVANCE DIRECTIVE IN RESEARCH: PROSPECTS AND PITFALLS

DAVID N. WEISSTUB AND ANNE MOORHOUSE

## INTRODUCTION

Advance directives for health care, well-known in the popular press and among lay people as living wills, have become widely accepted as a necessary tool to guarantee the civilized management of one's life during increasingly protracted periods of aging and frail health. Although gaining wide public acceptance, these directives have received varying degrees of support from ethicists, lawyers and health care professionals.[1] Alongside these developments and with added complexities, research advance directives, hereinafter referred to as *research directives*, are offered as a concrete ethical solution to the need for conducting significant research with impaired members of special populations.

Simply put, research directives are devices that document individual preferences about the appointment of a substitute decision-maker (SDM) in the event of incapacity, while providing documentation about a person's wishes pertaining to research involvement. Seen in this light, research directives project an individual's autonomy into the future, crystallizing preferences in the light of an anticipated state of decreased or compromised mental capacity to provide consent to a research protocol. Of course, contemporaneous and direct consensual acceptance, both for matters of personal care and research, remains the preferred mode of providing

---

[1] See Advance Seminar Group, Centre for Bioethics, University of Toronto, "Advance Directives: Are they an Advance?" (1992) 146 Canadian Medical Association Journal 127 [hereinafter *Advance Directives*]; P. Singer, "The University of Toronto Centre for Bioethics Living Will" (1993) 60 Ontario Medical Review 35; D.W. Molloy & V. Mepham, *Let Me Decide* (Toronto: Penguin Books, 1992); President's Commission for the Study of Ethical Problems in Medicine and Biomedical and Behavioral Research, *Making Health Care Decisions: A Report on the Ethical and Legal Implications of Informed Consent in the Patient-Practitioner Relationship*, vol. 1 (Washington, D.C.: U.S. Government Printing Office, 1982) [hereinafter *President's Commission*].

195

consent. When such direct consent is unavailable, which is often the situation for difficult cases, it is submitted that substitute consent for research participation is sufficient when preceded by a specific advance directive stating the desire to become a research participant, or by the more specific advance authorization for a decision-maker to substitute the required consent. In either category, the advance decision must be valid and informed. As well, it should be noted that specific requests not to be included in certain species of research must override any other set of prevailing conditions.

It is the intention of this review to investigate the kind of ethical reasoning that can be used to support such recommendations. In addition, the merits, costs and appropriate use of research directives will be examined. Viewing advance directives as a hybrid of the professional domains of law, ethics and health sciences, as well as involving significant personal interests, it is necessary to analyze the philosophical arguments that can be mustered for and against research directives. This undertaking consists of three discrete tasks: (1) an analysis of why advance directives are desirable; (2) a consideration of arguments for and against the specific introduction of research directives; and (3) a critical appraisal of current solutions and future recommendations for their restricted use.

## THE SOCIAL UTILITY OF RESEARCH DIRECTIVES

### *The Clinical Need*

A typical profile is presented in clinical settings. Clinicians find themselves in often confusing dialogues with other health care professionals, researchers, family members, and (of course) prospective subjects. Even if we accept the researcher's ultimate and noble goal of pursuing knowledge to discover or enhance therapeutic interventions, it is understandable that researchers are dismayed when research approval is not given on the grounds that the research considered offers no promise of benefit to the specific subject who is incapable of consenting. To illustrate the problem, consider the situation of a research team studying senile dementia of the Alzheimer's type (SDAT). The team requires subjects with end-stage dementia to conduct its research. The use of these subjects would be clearly warranted because the research cannot be conducted without their participation. The purpose of the study would be to examine the effects on brain tissue, renal, and hepatic functioning of a medication that at a future date may reduce memory loss. A research protocol that would involve weekly injections would state clearly that the persons engaged in this research, while not benefiting directly, would contribute to developing effective interventions for persons suffering from this disease in the future. Although adverse effects are not expected normally in such research, subjects would likely receive monthly CT scans, weekly blood work, and liver biopsies before and after the experiment.

In a case such as the above, the research subject has no competency to give consent. Family members are often called upon by researchers, as well as by clinicians, to give substitute consent; they often find themselves in disagreement because of diverse ethical sensibilities, emotional factors, and self-interest (e.g., the fear of suffering from such an illness later in life). In such encounters, medical directors of institutions may have defensive reactions because of primary self-definitions as caregivers, and because of special sensitivity to the prospect of litigation and possible liabilities. In this context, we should also be aware that once an institutional discussion begins, there is not always uniformity among staff members, some of whom may hold the view that it is unfair in principle to test drugs on terminally and chronically ill patients. It should also not be forgotten that families are often put into highly pressured circumstances because they are highly stressed and dependent on the institutions which have been entrusted with the care of their immediate family members. Administrators are often troubled by the dramatic changes in personality which are experienced in an institutional setting with such force and abruptness that prior wishes are thrown into question. In addition, it is often unclear, due to the vagueness of legislation or the lack of legal clarity in institutional directives, what the limits of authority truly are for SDMs when there is some risk of harm to the prospective subject. When such conflicts arise, the disappointment and frustration of researchers, who have often passed through scientific review by a Research Ethics Board (REB) only to be refused ethics approval, should not be surprising.[2] The ethical justification for the refusal can be queried on the grounds that the opportunity costs are too high. The rising incidence and immense costs of caring for persons with chronic and terminal illnesses such as Alzheimer's Disease (not only for individuals and families, but also for communities and governments who are currently facing diminishing health care resources), weigh heavily when we contemplate a mandate to reduce the emotional, social, and economic costs of the disease.

There is an apparent value to the availability of a research directive when family members are confronted with the demands or needs for research participation. Without such directives, preferably containing specific instructions, family members are placed in the difficult role of defending presumed best interests, which are not obvious when no direct benefit is proffered. Presumably, with the tool of a research directive, family members would be mandated with a clear directive to overrule the preferences and objections of staff members, researchers, administrators, and indeed, other family members and concerned parties.

In noting the importance of clarifying research directives, it should be remembered that there is some advantage in informing persons about future prospects. Whereas even some health care professionals are reluctant to raise the sensitive topic of life-sustaining treatment with patients and families grappling with end-of-life decisions, health care professionals may find themselves more at ease in pursuing a dialogue about participation in research endeavours. Although living

---

[2] Ethical and scientific review of research proposals can be conducted by the same or separate committees. The trend is to combine committees because of the overlap between scientific review and ethical review of the research's design and methodology.

wills regarding treatment are predominant in the discussion of research directives, presently neither health care professionals nor public health institutions can profess extensive experience with their use.[3] This could be a positive feature affecting the prospects of developing research directives for wide use within institutions. However, it is a double-edged sword. It is equally plausible that institutionalized individuals could become highly susceptible to solicitation for inclusion in research projects through the extensive use of research directives. It is imperative, therefore, that uniform policies that address the interests and needs of researchers, subjects and the public be articulated and that research directives be defined in ways suitable for their implementation through legislation.

### Ethical Considerations

It is well-accepted in the vast treatment of the consent issue that consent must be informed and freely given by the prospective subject. It is a corollary that relevant information, including the nature of the research, expected risks and benefits must be understood and appreciated. Post-Nuremberg Western societies have enshrined the rule that persons have the right to withdraw from research at any time without penalty.[4] It is also almost universally accepted that if SDMs can be morally justified, because of their commitment to represent the best interests of subjects, their indirect consent can be sought when persons are found incapable.[5]

In the majority of Western industrialized states, REBs are mandated with evaluating both the scientific and ethical merits of research proposals. Ethical review of the design will include assessment of the process of subject selection, the risk-benefit ratio, and the monitoring of the research. When a study is not approved on account of its design, the principal reason is the protection of subjects from an unjustified exposure to risk. Inequitable distribution of the risks of being a research subject is to be avoided.

We must keep in mind that the overarching purpose of the research directive instrument is to project the autonomous wishes of persons into a future time in such a manner as to preserve and enhance our commitment to respect for persons. Such a careful attention to detail in terms of approaching the subject in the context of the research directive is based upon the attempt to assist the individuals in question (or the protectors of their rights) as much as is permitted in an imperfect society.

The above challenge lies at the root of the construction of research directives, leading to the question of what level of abstraction and specificity such devices

---

[3] See e.g. I. Rassooly *et al.*, "Hospital Policies on Life-Sustaining Treatments and Advance Directives in Canada" (1994) 150 Canadian Medical Association Journal 1265.

[4] The *Nuremberg Code* constituted part of the judgment in *U.S. v. Karl Brandt et al., Trials of War Criminals Before the Nuremberg Military Tribunals Under Control Law No. 10.* (October 1946–April 1949).

[5] World Medical Association, *Declaration of Helsinki*. Adopted at the 18th World Medical Assembly in Helsinki in June 1964. Amended at the 29th World Medical Assembly in Tokyo in October 1975; the 35th World Medical Assembly in Venice in October 1983; and the 41st World Medical Assembly in Hong Kong in September 1989. Reprinted in (1991) 19 Law, Medicine & Health Care 264.

should be put into place. A general statement of values reflecting the person's preferences is most certainly a relevant preamble, but in most cases will not provide sufficient guidance. The SDM needs more detailed instructions for substitute decision-making to be a meaningful representation of the choices related to another's life-history. Inevitably, there are qualitative questions which enter into any difficult choice about priorities in research, in addition to the need for balancing personal sensibilities about well-being against concern for others. Given that changes can occur rapidly in health care and that persons can undergo profound alterations of personalities and values as they grow older and perhaps have failing health, we can assume that the optimum conditions for research directives should be linked to anticipated specificities which will be unchallengeable regardless of these transformations. But as we will later observe, such changes indeed pose the most compelling question of all, namely, whether specificities decided upon in advance will and should hold when the immediate necessities and choices are effectuated by a changed and vulnerable personality.[6]

Another range of difficulties appears with respect to any responsibility vested in others about the interpretation of the recorded will of the prospective subject. A range of possibilities can be contemplated. For example, does the person wish to:

- participate in research investigating specific conditions that the person has developed but not for another illness?
- limit participation to certain levels of risk or discomfort?
- give *carte blanche* for involvement in all types of research?
- refuse involvement in any form of research?

It is not realistically possible to demonstrate fidelity to the integrity of the person in question without being in a position to reconstitute a world of values surrounding the decision-making options that could be entertained in light of the person's life-history. Therefore, any notion of substitute decision-making, which is *prima facie* incapacitated with respect to this knowledge-base, must be viewed with trepidation, if not with a view to rejection. However, to push a prospective subject into limbo, because a seasoned collaborator who can provide a facsimile of the personality of the prospective subject is unavailable, creates too high a demand on the system. The issue becomes what level of uncertainty is ethically permissible, given the limitations of knowledge or capacity for identity to be found among many SDMs. Over time, it is hoped that structures will be created for detailing the criteria

---

[6] It is an unattainable ideal that a general statement of values could be synchronized to a perfect level of co-identification between a statement of preferred life-values and attendant links to specific mandates. Nevertheless, it should be the stated goal of well-drafted research directives to re-assess them to the latest stage possible before the subject's apparent incapacity becomes undeniable. In the best of all possible situations, prospective subjects will be informed, or avail themselves, of up-to-date data on the status of research affecting their interests and the level of discomfort to be expected given the latest knowledge available with respect to the technology involved in the pertinent field of research. See J. Downie, "Where there is a Will, there may be a Way: Legislating Advance Directives" (1992) 12 Health Law Review in Canada 23; Singer (1993), *supra* note 1; Molloy & Mepham, *supra* note 1.

according to which prospective subjects wish their SDMs to decide; for the moment, these criteria are embryonic, if not unavailable.

When people have not been able to provide a direct consent either because of mental incapacity or structural deficiencies such as circumstances of duress, and moreover, when there is no specific and clear directive from the person to constitute a credible legal document, substitute decision-making is preferred. The SDM is obligated to act in the best interests of the person, commonly understood through considering the following criteria:

- will the person's condition be substantially improved?
- will the person's condition likely improve without any intervention?
- will the anticipated benefits outweigh the risk of harm to the person?
- is the treatment the least intrusive, while at the same time meeting the above criteria?

Central to this discussion is the issue of the "double-reference point". In dealing with research directives, we are caught between adhering to two potentially divergent objectives. On the one hand, there is the primary obligation to defer to the author's express wishes, which we can assume include a consideration of values by the prospective subject. On the other hand, if we are either administering the document or called upon to stand in for acts of interpretation, we are pressed to use the standard of protecting the person's best interests.[7] Because of altered circumstances, or a lack of guidance from the text of the research directive itself, a conflict can arise between fidelity to the two complementary principles. The most troubling problems arise when the subject is incapable and the research is non-therapeutic.[8] As will be seen, research directives can play a role in responding to this dilemma, but there must be restrictions on their use.

The main objective of completing advance directives for personal care and research participation is to promote autonomy. In the case of personal care directives, patients and health care professionals should not limit their discussions to exploration of the patient's wishes but should also consider the guidelines to be used by the health care professional when fulfilling them. Although the primary obligation of a health care practitioner is to respect a person's wishes, health care professionals must be given some latitude when following health care directives.[9] Regarding research directives, a SDM must not only safeguard the subject's welfare by evaluating the probability and magnitude of possible harms against the expected benefits to the subject and others, but must also endeavour to respect the subject's wishes. This task is not always straightforward. When there is exposure to risk and no promise of benefit to the subject, involvement in research is clearly not in the

---

[7] As is already the case in some jurisdictions, this would include taking into account the values and beliefs that the guardian knows the person held when capable and believes would still be acted upon if capable, and on the person's current wishes, if they can be ascertained. For further discussion of modern interpretations of the best-interests standard, see Chapter 8.

[8] See Chapter 6.

[9] *Advance Directives, supra* note 1.

person's best interests. The SDM made responsible by a research directive to consent to research must decide whether the obligation to respect prior wishes overrules the obligation to act in the person's best interests.

Current research guidelines for the ethical use of human subjects do not resolve these dilemmas. Although to date these guidelines have not seriously attempted to address the criteria or conditions for the use of research directives, as a precondition for their creation, they have dealt with the matter principally through the mechanism of proxy consent, but in some cases have also recognized prior directives to participate in research. Indeed, there is a logical progression from accepting proxy consent to approving advance directives for non-therapeutic research.[10] If a proxy can give consent to research not in the person's best interest but offering little risk of harm, then a proxy could follow instructions from the person, made when competent, that permit involvement in non-therapeutic research.

### Legal Considerations

In Canada, as in many common law jurisdictions, there is considerable incentive for developing legislation providing for the use of research directives given the present legal uncertainty associated with the validity of substitute consent for participation in experimentation. For example, with the exception of the Province of Quebec, there are considerable doubts about the legality of third-party consent to non-therapeutic research.[11] A strict application of the best-interests principle, as required by the common law, would preclude an SDM from allowing a dependent to be subjected to a procedure which would yield no direct benefit to that person. However, respecting an individual's values and beliefs is a benefit for that person. A person's research directive, being an "instruction applicable to the circumstances that the incapable person expressed while capable", should guide relevant decisions by the SDM. Nevertheless, projecting prior wishes onto a contemporary situation must be done with caution. The validity of the consent given in the research

---

[10] In Canada, the Medical Research Council identified research involving persons incapable of consenting as a serious ethical and legal problem. The MRC queried whether the right to integrity can be waived by a third party and acknowledged that the common law in Canada favours a ban on the use of mentally incompetent persons in non-therapeutic research. In contrast, the Law Reform Commission of Canada supported research with mentally incompetent subjects on the condition that a substitute consent was secured and strict guidelines to protect the safety of the subject were followed. In its revised *Code of Ethical Conduct for Research Involving Humans*, the Tri-Council Working Group continues to support the use of proxy consent for incompetent individuals and the use of prior directives to indicate a desire to participate in research. See Medical Research Council of Canada, *Guidelines on Research Involving Human Subjects* (Ottawa: Supply & Services Canada, 1987); Law Reform Commission of Canada, *Working Paper No. 61: Biomedical Experimentation Involving Human Subjects* (Ottawa: Law Reform Commission, 1989); Tri-Council Working Group, *Code of Ethical Conduct for Research Involving Humans* (Ottawa, July 1997). The Law Commission (U.K.) reached a similar conclusion. By approving and recommending the use of proxy consent when mentally incompetent persons are sought as research subjects, the LRC gave indirect support for the concept of proxy consent for mentally incompetent persons being considered eligible to be subjects in non-therapeutic research. See The Law Commission, Mental Incapacity (London: HMSO, 1995).

[11] See Chapter 8.

directive, the scope and effects of research directives, and the corresponding responsibilities of SDMs need to be carefully delineated.

## THE PROS AND CONS OF RESEARCH DIRECTIVES

### *Arguments in Favour of Research Directives*

The arguments in favour of using research directives flow from several fundamental principles. Some arguments are based on respect for the person. Others come from a desire to improve the efficiency of health care through health sciences research. Still others emanate from a belief that incapable persons can benefit from being involved in research classified as non-therapeutic.

*Self-Determination*
Research directives permit individuals to direct their care when they are unable to make decisions regarding their own welfare. Thus, an advance directive functions like a will, by which a capable person can arrange to dispose of their estate before death by writing a will that is witnessed and gives instructions. In addition to dispositions about property and assets, there may be directions about funeral arrangements and care of the body. By analogy, a capable person's advance directive can give instructions about his or her personal care and name an SDM to execute previously stated wishes. Similarly, it is argued that a competent person can give instructions about participation in research.

The cornerstone of research ethics is the *Nuremberg Code*, which in its first statement requires that research subjects give an informed and voluntary consent.[12] The *Code* prohibits research requiring the use of persons who cannot consent. However, researchers have found the *Code's* categorical position, that it is absolutely essential that subjects themselves provide informed, voluntary consent, to be too uncompromising. The need for indirect consent by an SDM was recognized in Art. 11 of the *Declaration of Helsinki*[13] which permits experimentation on mentally incompetent persons on the condition that the persons give consent in accordance with local research regulations. The Belmont Commission recommended supporting legalization of research involving mentally incompetent persons on the basis that a total prohibition would have a large opportunity cost for persons with the illnesses that led to their becoming mentally incompetent. The Commission recognized that mentally ill patients are at a high risk for exploitation and it wanted strict limits on the involvement of these patients. Therefore, the Commission recommended that only minimal risk is acceptable, and if at all possible, the person should give consent. If the prospective subject is incompetent to consent, the Commission decided that consent could be provided by a legally appointed guardian. In brief, in the post-Nuremberg years, consent for research participation is required from the subject; the international consensus has developed to allow proxy

---

[12] See *Nuremberg Code, supra* note 4.
[13] See *Declaration of Helsinki, supra* note 5.

consent under certain specific circumstances. Acceptance of research directives would extend the conditions for obtaining consent by which an incapable person could be involved in non-therapeutic research.

*Utilitarian Argument*

It is uncontroversial and self-evident that health sciences research is an instrumental good contributing to discovering or learning more about the prevention and treatment of diseases and debilitating conditions. In addition to lengthening lives, quality of lives can be enhanced and human suffering diminished. At the macro level, more effective use of scarce resources is socially and financially beneficial.[14] The use of incapable persons in non-therapeutic research is justified on the grounds that health sciences research cannot proceed without research involving persons suffering from specific conditions, if progress is to be made in reducing morbidity and mortality rates of the diseases and conditions being studied. Ethical research practices require that human subjects not be used unless necessary, which is the situation with some SDAT, schizophrenia, HIV, and cancer studies. In such cases and under strict conditions, persons unable to consent directly are asked to be subjects in studies that cannot be conducted without their participation. Unless vulnerable and affected subjects are used, it is reasoned, the research will never be achieved. When proxies do not permit their wards to be subjects in non-therapeutic research, unquestionably, there is the cost of reducing the pool of research subjects. The question is whether the brake is always justified and when, if ever, one can justify its release. Another benefit of research directives is that SDMs, health care professionals, and researchers know whether the person wanted to be a research subject. The burden of trying to decide if the right decision is being made is relieved. In short, permitting proxies through prior directives to consent for their vulnerable wards to be research subjects can have important benefits for persons with the disease being studied, at a future date, and can therefore benefit society. There can also be considerable psychological, social and economic benefits for the person, his or her family, health care providers, and the research enterprise, broadly understood.

### Problems with Research Directives

The arguments against research directives are numerous and overlapping: a change in the person with the passage of time that renders the documents void; a denial that all research can benefit subjects; the excessive costs of permitting incapable persons to be subjects in non-therapeutic research; and a questioning of the proposition that

---

[14] See J. Bentham, *Fragments on Government and Introduction to the Principles of Morals and Legislation*, ed. by W. Harrison (Oxford: Blackwell, 1948); J.S. Mill, "On Liberty" in M. Warnock, ed. *Utilitarianism, On Liberty, Essay on Bentham, together with Selected Writings of Jeremy Bentham and John Austin* (London: Collins, 1962). Act and Rule Utilitarianism are distinguished from each other and discussed in R.B. Brandt, Ethical Theory (Englewood Cliffs, N.J.: Prentice-Hall, 1959); R.M. Hare, *Freedom and Reason* (London: Oxford University Press, 1963) and J. Narveson, *Morality and Utility* (Baltimore: John Hopkins Press, 1967).

the SDM's priority will be the subject's welfare. If these arguments are accepted, the consequences can be far-ranging. For instance, people with SDAT or a mental illness which prevents them from giving consent could not be involved in a randomized clinical trial, because their proxies would be unaware of whether their wards would receive the experimental medication or placebo. Receiving a placebo will not benefit the incapable person and is not in the person's best interests. Yet the study could ultimately lead to advances in the prevention and management of diseases and conditions that affect and could affect a large population.

*The Problem of Personhood*
People who are mentally incapable of consenting to research are in fundamental ways not the same people who earlier gave directions about what should happen in the event that they became dependent on a proxy for decision-making. The person who made decisions about quality of life made them when rational and autonomous. When another identity emerges, vulnerable and incapable persons are subjected to wishes made by a capable person. The persons whom they were made the decisions, but the persons they are today, did not. Projecting the wishes of a previously competent person onto the presently incompetent person is to impose the wishes of one person onto a different person.

Predictions about what will be a valuable use of time, or what will be painful or degrading, can change as the person's circumstances change. Brushes with illness and death can prompt a reassessment of a person's values and goals. With SDAT, the worry is that the pace of the illness can reduce the possibility of changing or reversing the research directive. By the time the person begins to experience fluctuating and diminishing capacity, labeled unfortunately as a "drain on resources", it is usually too late to change prior wishes. Furthermore, there is the possibility of a person being unable to express wishes when able to make decisions, including changing prior directives.[15] Under these circumstances, to impose previous wishes on a person "locked" into their body, unbeknown to the researcher and SDM, is to show disrespect for the person's autonomy and could result in seriously harming the well-being of the person who becomes the relevant term of reference. Once the person is incapable, the ethical priority is to protect the vulnerable person from harm. Therefore, health care professionals and proxies have the responsibility to act in the best interests of the vulnerable person's welfare and are not obligated to respect prior wishes. It is not possible to confirm whether the person agrees at a later time frame with wishes made at an earlier time, by a different person.

With respect to advance directives for health care and property, the self-determination argument, in conjunction with minor arguments about the secondary gains to families, health professionals, and society, has outweighed serious concerns about the personhood argument. In health care decisions, the thorny problems of personhood and the "locked-in-syndrome" are tolerated because of the considerable

---

[15] This situation is called, in lay terms, the "locked-in-syndrome".

benefits of living wills to the person, their family, health professionals, and society. However, the important philosophical issues raised by the personhood argument have not been addressed. With respect to non-therapeutic research, there is insufficient justification for overruling the personhood argument. When the intervention is treatment, a person at a later time has a reasonable probability of benefiting from prior instructions. However, a vulnerable person at a later time will not benefit from prior wishes giving consent for him or her to be associated with non-therapeutic research, and is exposed to harm. Therefore, the personhood argument cannot be outweighed by the benefits offered to the subject by participating in non-therapeutic research.

With therapy there are expected benefits, so the personhood argument has been allowed to be displaced. When there are no expected benefits for the subject, no longer a patient, it is argued that the risks to the person must be considered. Also, incapable people cannot request that they be removed from the study; this responsibility rests with the SDM and researcher or research team monitors. Allowing research monitors to cancel the person's participation is based on the recognition that vulnerable persons should not be exposed to unnecessary harm. It is argued that the same moral responsibility should be extended to proxies. If proxies receive such an extended power, then before consent is given, the proxy should be able to overrule prior wishes if the proxy thinks that the instructions would expose the vulnerable person to excessive risk.

Decisions about not only when to preserve life and to reduce suffering but also when to respect wishes about treatment and research participation rest on the concept of a "person". The question of what is a person and the nature of a person has been studied by philosophers for centuries, and more recently by psychologists and neurologists.[16] There are two aspects of being a person: intrinsic or essential qualities and relational qualities. Tooley summarized the six versions of the essential definition and found them to share the following necessary qualities: persons can have experiences that happen at different times and are linked by memory, enjoying states of intentionality or consciousness. Relational qualities include ability or potential to be conscious of the environment and to maintain relationships with others, which assumes that the person has some way to communicate with others. Tooley's preferred definition is that "[s]omething is a person if and only if it is a continuing subject of experiences and other mental states that can envisage a future for itself and that can have desires about its own future states". In other words, to be a person, that is to be capable of seeing a future for oneself as a person (a subject of experiences and other mental states), one must "be capable of having the concept of a continuing subject of experiences, and of recognizing oneself as such a

---

[16] The concept of a person and personhood has been discussed in the following articles: M. Tooley, "A Defence of Abortion and Infanticide" in J. Feinberg, ed., *The Problem of Abortion* (Belmont: Wadsworth, 1973); D. Parfit, "Later Selves and Moral Principles" in A. Montefiore, ed., *Philosophy and Personal Relations* (Montreal: McGill-Queen's University Press, 1973); M.A. Warren, "The Moral and Legal Status of Abortion" (1975) 57 The Monist 143.

continuing subject".[17] Whatever conception of a person or personhood is accepted, the position taken does not affect the proposition that the person, however conceptualized, changes with time. The person with the capacity for reasoning, self-consciousness, maintaining relations, and communication is not static and independent of his or her environment. Experiences, intentions, and relations are perpetually in flux. This is certainly not a novel idea. Heraclitus is said to have believed that all things pass and nothing abides; you cannot step twice into the same stream, whether of water or of consciousness.[18]

These discussions lead inevitably to questions about the sanctity and quality of life. The principle of the sanctity of life entails absolute commitment to preserve life with no attention to its quality. Life itself, regardless of the person's condition, must be valued and preserved. Acceptance of the concept of quality of life does not mean accepting that some lives are more valuable than others. In other words, assessments of the quality of a person's life should not be based on social worth criteria, but rather in reference to the individual's situation and values, including a change or diminishment in personhood.[19] Personhood, like autonomy and capacity to consent to treatment and research, is on a continuum. With different capacities for decision-making, awareness of one's environment, and ability to maintain relationships, there is a different person facing different circumstances.

In summary, besides the possibility that a person's wishes may change or become inappropriate, the strongest argument against claims that research directives represent self-determination is that they fail to consider the major changes in personal identity that occur when a person becomes irreversibly mentally incompetent or suffers from dementia. Arguably, projecting the stated wishes of a previously competent person onto a presently incompetent person is equivalent to imposing the wishes of one person onto a different person.[20] While the risk of such an occurrence cannot be avoided, the potential for harm can be minimized, and the personhood problem can be circumvented. Once again, the solution lies in properly regulating the use of research directives. Necessary safeguards must encompass the creation of research directives, including the preparation of instruction directives and the appointment of SDMs by durable powers of attorney. Limits must be placed on the scope of research directives, and it must be ensured that an individual's

---

[17] M. Tooley, "Decisions to Terminate Life and the Concept of a Person" in J. Ladd, ed., *Ethical Issues Relating to Life and Death* (New York: Oxford University Press, 1979) 84.

[18] See Plato, *Fragments* (12 and 91) in J. Burnett, ed., *The Works of Plato*, 3d ed. trans. B. Jowitt (Oxford: Oxford Classical Texts, 1982).

[19] Callahan defined personhood as follows:

> We value human life, and the personhood that is the crowning glory of that life, because it possesses certain capacities that are the distinguishing marks of human beings, those which separate humans from animals and lend them their power to reflect upon their condition.

See D. Callahan, Setting Limits: Medical Goals in an Aging Society (New York: Simon Shuster, 1987) at 177–179.

[20] *President's Commission, supra* note 1.

directives are observed and not misrepresented. Also, an allowance must be made for the possibility of overriding an individual's research directive in specific circumstances. Finally, safeguards regulating the use of research directives must be designed to protect vulnerable persons and to facilitate the ethical conduct of experiments with such individuals.[21]

### Durability of Informed Consent

There are several problems with the argument that prior, informed wishes must be respected by the SDM. First, the analogies with advance directives for health care and with wills are not exact. When writing a will, the capable person fears no physical suffering from the instructions. Only parties external to the testamentary process will benefit or be harmed by the will. In the case of research directives, the author/subject could be harmed by prior instructions. For example, a *carte blanche* directive could lead to the person's being involved in a study that presents a significant exposure to harm.

There are concerns with meeting the requirements by which a research directive is considered to have been given with informed consent. In addition to changes in personhood, the condition of the person who prepared the directive is another point in issue, both because the person changes with the ravages of the disease and because diminished quality of life may influence decision-making, including in respect of the assessment of risks. In the case of consent to treatment, the consent can become obsolete on account of changes in health care practices and new knowledge about the management of the condition; so too with research directives. Decisions made about the type of research involvement approved can rapidly become obsolete or remain current for several years. Whereas health care directives are intended to promote the patient's welfare interests, that same level of protection is waived by a research directive, hence the need to have limits on their scope and to monitor the effect of research participation on incapable persons.

In the realm of treatment decisions, an advance directive will trump the healthcare professional's discretion to act on what he or she may perceive to be in a patient's best-interests. It should not be assumed that this same principle ought to be transferred automatically to the realm of research directives. It is a fundamental ethical principle that the protection of a subject's welfare requires that possible harms be justifiable and outweighed by the expected direct or indirect benefits from the research.[22] According to the Nuremberg Code, the degree of risk to be taken should never exceed that determined by the humanitarian importance of the problem solved by the experiment. By this reasoning, even the contemporaneous altruistic wishes of a fully competent individual can be curtailed in the research endeavour. Hence, if capable subjects are prevented from exposing themselves to unreasonable risks, then it is logical that even more stringent limits should apply to research

---

[21] In particular, health care professionals, researchers and SDMs must be vigilant in looking for signs of the locked-in-syndrome: a capable person is trying to communicate despite their apparent inability to use verbal and non-verbal communication to express their wishes.

[22] See Chapter 18.

directives, which may result in the exposure of vulnerable persons to equally unreasonable risks of harm. It should therefore be possible to override a research directive in those situations that could lead to unreasonable or unanticipated risks of harm to the now incompetent subject.

*The Substitute Decision-Maker's Dilemma*
There are two ways that an SDM can face an ethical dilemma. In the first instance, the research directive allows involvement in a study with an unethical design. Although the study should not pass ethical review in the first place, the research directive should not provide a mechanism to leap over the requirements for an ethical design. If the study receives approval, the SDM has an obligation to refuse a dependent's participation if the experiment exposes the subject to an unreasonable risk. The SDM becomes the last line of defence for the best interests of the incapable person. To consent would expose an incapable person to risks to which a capable person could not be exposed.

The second possibility is that the study has been favourably reviewed, but that the study is non-therapeutic and the risk-benefit ratio is unfavourable for the affected person. The SDM is in the unenviable position described above, whereby the obligation to respect the dependent's wishes comes into direct conflict with the duty to protect the subject's welfare. In such instances, the latter must take priority.

An SDM is not merely the person who gives or withholds consent. The SDM as a moral agent must reflect on the nature of the research and the consequences of the research participation, and must also monitor the research to know if the person should be withdrawn from the study.[23]

## RESEARCH DIRECTIVES: SOLUTIONS AND RECOMMENDATIONS

### *Preparation of a Research Directive*

There are essentially two types of research directives. The first is an *instruction directive*, which can be substantive and specific or a mere statement of values used to guide the SDM. An instruction directive may state how, when, and why a person would want to be a research subject, or may express the person's attitude towards involvement in research based on clearly stated values and beliefs. The second type of research directive is a *durable power of attorney*, through which an SDM is appointed by the dependent person while competent. When the person becomes incompetent, the SDM must make decisions on the patient's behalf; that is, the SDM would have the responsibility of granting or withholding consent for the person's

---

[23] Ideally, a research directive should provide directions concerning withdrawal of a subject from a study. Even in the absence of such instructions, the SDM should be considered to have the authority to withdraw a dependent from research in the event that the parameters of the protocol change or if risks to the subject turn out to be greater than anticipated.

involvement in a particular research project.[24] Instruction directives and durable powers of attorney could be used separately, but should be used together. In all situations, research directives must be prepared by a person having the capacity to understand and appreciate the significance of their informed and freely made decisions regarding delegation of decision-making and the instructions.

One difficulty in developing health care directives is that generic schemes must be resilient enough to accommodate the widest possible population; however, we are equally pressed to take into account the specificities which occur with illnesses. In fact, some authors have advanced the notion that advance directives should be disease-specific.[25] Such directives may be especially helpful to SDMs who may not have the requisite level of knowledge to be able to interpret directives which become compromised by the level of specificities which come with a specific illness. On the other hand, individuals often experience multiple illnesses, and great specificity might curtail or shadow the ability of a decision-maker to act with meaningful fidelity to the document which has been drawn. There are no easy solutions to these questions. Only through careful planning can we be prepared for the problematic exercises of discretion which are inevitably part of the interpretation of advance directives.

In the case of research directives, there should be a number of necessary conditions established for their use and application. Guidelines developed for health care directives should apply equally to research directives. We should require that they be in writing and witnessed, and that they possess a high degree of specificity. This does not mean that all possibilities should be covered through detail, but rather that the warranting of certain practices must be established so that the specific intentions of the subjects can be honoured to the highest degree possible. It should be clear, nevertheless, that where no research directives are available, SDMs have limits to their authority, such that the research involved has to be specifically connected to the illness or condition suffered by the individual.

When preparing research directives, individuals should specify the details about the types of procedures to which they are prepared to consent, the levels of risk expected, and the protocols or types of protocols contemplated. If the research directives documented are of a general nature, specifying only a consent in principle to participate in biomedical experimentation, such directives should be read restrictively. In an ideal world, individuals would canvass their research directives with a host of relevant parties to clarify values, concerns about risks, and perhaps fears relating to the undertakings. Patterns of consultation, with relatives, friends, family doctors, and legal advisors, will not be consistent. Whatever the pattern, thorough discussions with intimate partners in life should be conjoined to informative discussions with medical and/or research experts to arrive at a

---

[24] A. Moorhouse, "Advance Directives for Research Purposes: What is Bad about Trying to Do Good?" (Paper presented at XIXth International Congress of the International Academy of Law and Mental Health, 1993) at 2–3.

[25] See e.g. P.A. Singer, "Disease-Specific Advance Directives" (1994) 344:8922 Lancet 594 at 594–596.

thoughtful decision[26] In the context of research environments, researchers should set aside the necessary time to facilitate the drafting of documents via thorough exchanges of information, preferably beyond the normal legal requirements for an informed consent. Volunteers and their SDMs' questions and concerns should be responded to in an honest, non-judgmental manner. When the prospective subject lacks the capacity to consent, depending on the context, the person should be included as much as possible in discussion of the study. Ways to enhance the person's understanding of information should be explored. More time may be necessary for the discussion; a private and quiet area reserved for the meeting and communication aids (large print or hearing devices) may be imperative.

In deciding about and preparing research directives, individuals must make choices about the parties best suited to implement these durable powers. Families are often preferred because of their close identification with the subject's welfare and because they are usually the social unit which is best informed about the long-standing preferences and values attached to the subject's life plan. Families, even despite major redefinitions of social roles in this century, continue to be the first order social group in our society, and are the source and overseers of basic values.[27]

In the context of research directives, society tends to defer to the subject's selection of an SDM, making the assumption that the person chosen will have the requisite interest and knowledge to implement decisions done in the best interests of the subject. Therefore, where there are instances of only instruction directives, and no SDM has been specified, the patient's spouse, parents, or guardians (if the subject is a child) should be preferred unless the patient objects. Nevertheless, family members might not prove to be the best SDMs where there is, for example, a vested interest which could lead to a conflict of interest with regard to the research in question. For example, in the case of SDAT, due to the inheritability of the disorder, family members might be unduly motivated to submit their wards to related research. In such cases, a friend of the incompetent person might serve as a better selection.

Finally, a system of registration for research directives should be considered. However, registration of research directives should not be an onerous process. In fact, the Alberta Law Reform Institute advised against the requirement that all advance directives be registered, fearing that a complicated bureaucratic system would discourage individuals from preparing directives.[28] The Queensland Law

---

[26] It is imperative that the directive be discussed with the delegated SDM. If the SDM is not a next of kin, it should be at the prospective subject's discretion to whom the contents of the research directive will be disclosed. Ideally, SDMs should be protected by confidentiality, but in light of recent decisions by the Supreme Court of Canada such is not likely to be the case without a statute specifically protecting them. See *R. v. O'Connor*, [1995] S.C.J. No. 98 (QL). Informing the SDM is also imperative so that the person can have the opportunity to understand the individual's stated wishes, values and attitudes, and to agree to serve as the SDM.

[27] See also D.M. High, "Families' Roles in Advance Directives" (1994) 24:6 Hastings Center Report Supplement S16.

[28] See Alberta Law Reform Institute, *Advance Directives and Substitute Decision-Making in Personal Health Care* (Edmonton: Alberta Law Reform Institute, 1993) at 17.

Reform Commission concurred with this perception, rejecting many arguments in favour of registration and reaching the conclusion that enduring powers of attorney for decisions other than financial matters should not be registrable. With respect to standard forms, the Commission recommended against the imposition by legislation of a prescribed form, owing to problems of inflexibility and potential invalidation on technical grounds, but rather that forms be developed by professional organizations in collaboration with consumer groups for use as guides.[29] The UK Law Commission, on the other hand, recommended that, in order to be valid, a continuing power of attorney should be registered after its execution.[30] The position held here is that registration should be required upon the invocation of a research directive. The process should bear no cost to prospective subjects and should be streamlined to avoid administrative delays. Of course, there would be the corresponding onus of confidentiality placed on the registering authority, which could by delegation be the research ethics committee itself.

### Assessment of the Capacity To Complete a Research Directive

To be considered valid, research directives must meet the standard criteria for informed consent. At the time of preparing their research directives, individuals must have sufficient information regarding the type of procedures they would be involved in, the associated risks and benefits, and the availability of alternatives. They must be capable of consenting or of making decisions, and the decisions must be voluntary. In addition, their research directives must meet standards specifically related to advance decision-making, which include the requirement that the documentation of preferences be made in such a way that accurately represents the individual's reasonably stable preferences.

A properly-formulated informed consent form contains information detailing the overall purpose of the study, any potential risks and benefits, alternatives to participation, an assurance of confidentiality, and the right to withdraw from participation. Similarly, research directives must provide evidence of understanding and of an appreciation of the specific information contained therein.[31] And most importantly, where a person intends to participate in an experiment when no longer capable of providing contemporaneous consent, the research directive must include an expression of both the desire to participate in research and of the desire that this wish be given effect at a time when the person is no longer competent.

---

[29] See Queensland Law Reform Commission, *Assisted and Substituted Decisions: Decision-making by and for People with a Decision-Making Disability, Vol. 1* (Brisbane: Queensland Law Reform Commission, 1996) at 148–159, 355–356.

[30] See The Law Commission, *supra* note 7.28–7.31.

[31] The concern with ensuring "true comprehension" on the part of mentally disordered persons was emphasized by M. Irwin *et al.*, "Psychotic Patients' Understanding of Informed Consent" (1985) 142 American Journal of Psychiatry 1351. The authors stated that "[t]he patients, all of whom were acutely psychotic were able to read the informed consent information, and most reported that their understanding of the information about antipsychotic medication was good. Objective measures, however, did not confirm their self-reports. Many simply affirmed understanding to mask confusion while reading the information about antipsychotic medications".

Though competency to consent to a research directive should generally be assumed, when the competence of an individual to provide valid consent is in doubt, an accurate competency assessment should be undertaken to determine the individual's capacity to provide a valid consent. It is well accepted that an incapacity to make decisions may be either general or specific; and either situation may apply to members of special populations such as the elderly, children and young adolescents, and persons with mental disorders.[32] When the completion of an advance directive is accompanied by an assessment of the person's capacity to do so, SDMs and health care providers can be more confident that the directive truly reflects the person's expressed wishes. Since an SDM has a fundamental obligation to respect a dependent's wishes, knowledge of the validity of instructions in a research directive, which includes the absence of coercion, is of paramount importance.

All research directives should be reviewed by research ethics committees prior to their endorsement, ensuring that both legal and ethical requirements are satisfied, including an assurance that the person possessed the requisite level of capacity to prepare the directive.[33] The capacity to complete an advance directive is distinct from the capacity to consent to treatment and the capacity to make a testamentary will.[34] Guidelines developed to assess capacity to consent to treatment will not adequately measure capacity to consent to a health care directive;[35] the same will be true for the capacity to consent to treatment. The choices made in advance directives deal with situations that will arise in the future. They are hypothetical choices which take effect only after the person has become incompetent to make treatment decisions. Furthermore, if a person is unable to complete an advance directive, no one else can complete the research directive. In contrast, contemporaneous decisions regarding treatment and research represent actual choices, which are relevant immediately if the person is still competent. Also, the decision to refuse treatment, is a choice. There may be little time to reconsider a contemporaneous treatment choice, whereas the choices made in advance directives (and similarly research directives) may be updated many times before becoming effective.[36]

Owing to the nature of non-therapeutic experimentation, the threshold of capacity for making a research directive should be higher than that required by either a health

---

[32] See D.N. Weisstub (Chair), *Final Report: Enquiry on Mental Competency* (Toronto: Queen's Printer, 1990).

[33] Evidence of the author's consultation with family members, a legal advisor, the treating physician, or the proposed guardian would assist in this determination.

[34] The concept of a threshold for capacity reflects the need to consider situational parameters affecting decisions, including increased complexity of information and the level of significance. This has been referred to as a "contextual" sliding scale and in no way interferes with a functional assessment of capacity, which requires that a determination be made on the ability to understand and appreciate both the reasons for and consequences of a specific decision while taking into account the vulnerabilities and special characteristics of the individual in the light of cultural and social factors. See Weisstub (1990), *supra* note 32.

[35] M. Silberfeld, C. Nash & P.A. Singer, "Capacity to Complete An Advance Directive" (1993) 41 Journal of the American Geriatric Society 1141 at 1141.

[36] *Ibid.*

care directive or a testamentary will. This higher threshold is a result of the presence of risk without any promise of benefit to the subject; and this fact should be reflected by the criteria for competency as applied to the context of preparing a research directive. A person should understand and appreciate the following concepts in order to be considered capable of completing a research directive:

- that there are differences between being a patient and a research subject;
- that research directives contain choices to be acted on in the future;
- that the choices will be acted upon at a time when the person is no longer capable;
- that some of the choices may involve medical treatments or procedures;
- that choices regarding participation in research may require that someone else, either selected or appointed, make decisions on one's behalf;
- that an SDM has certain discretion, in particular circumstances, to override a person's instructions *to* participate in an experiment, but may never override an express wish *not to* particpate;
- that the choice to participate in research has associated risks for the subject;
- that in the case of non-therapeutic experimentation, participation will carry no promise of a benefit to the subject;
- that a person should change the directive to accurately reflect any change in a person's choices; and
- that a person can revoke a research directive at any time, even while incompetent.[37]

### *Durability of Research Directives*

Research directives, like wills, are valid when completed in accordance with the established protocol and can then be valid indefinitely. However, it is strongly recommended that, just as advance directives regarding health care should be updated whenever people change their minds about their choices, so should research directives. Regular updating of the directive would make the persons responsible for its execution feel more confident that they were following the patient's actual wishes. Any changes to advance directives should be made known to all relevant parties. The directive should be updated whenever there is a change in the patient's clinical status (improvement or deterioration), when the patient is admitted to a health care facility, when it is deemed appropriate by the physician or lawyer to do so, or when a major event such as death or divorce occurs.[38] In sum, research directives should be prepared within a period of time reasonably proximate to the onset of cognitive impairment.

As stated above, a person should be able to revoke an advance directive at any time, regardless of whether or not the person is mentally capable at the time of

---

[37] See *ibid.*; A. Moorhouse, "Ethical and Legal Issues Associated with Alzheimer's Disease Research and Patient Care: To do Good Without Doing Harm" in S.N. Verdun-Jones and M. Layton, eds., *Mental Health Law and Practice Through the Life Cycle: Proceedings of the XVIIth International Congress on Law and Mental Health* (Burnaby, B.C.: Simon Fraser University, 1994) 55.

[38] See *Advance Directives, supra* note 1 at 130–131.

revocation. Therefore, the threshold for the capacity to revoke or refuse to participate should be lower than the capacity required for consent, either in advance or contemporaneous contexts. A valid question is whether a person lacking capacity to consent can refuse to consent. In other words, is the revocation or lack of cooperation a demonstration of an informed decision? The threshold being accepted is the standard of evidence of choice. This lower standard is accepted because of the obligation of the SDM; researchers and society must protect vulnerable persons from harm. If the person indicates through their actions and emotional responses that they find their participation unwelcome and harmful, then evidence that the incapable person finds their involvement harmful should be respected.

### *Fundamental Restrictions on Research Directives*

Halting SDMs from volunteering their dependents in any research project without the prior indication of a desire to participate in experiments would be overly restrictive. In many cases we might be warranted in the assumption that the prospective subject would volunteer if given the opportunity, but this assumption has to be balanced by the limitation of the level of risk to which such an assumption would apply. Although altruistic acts should be supported and respected, people should not be expected to submit to, nor should they be indiscriminately exposed to, unreasonable risks for the benefit of others.

An individual's research directive should be read restrictively as to intentions to participate in therapeutic versus non-therapeutic research. Take, for example, an individual who specifies an intention to participate in a particular research protocol which, at the time of preparing the research directive, was characterized as therapeutic. Such protocols would still be subject to an objective determination of where they fall on the spectrum of treatment as opposed to research. If a research ethics committee subsequently re-characterizes the protocol as non-therapeutic, then the individual could no longer participate in that research. In order to avoid such problems, research directives concerning specific protocols should include a disclaimer whereby the individual agrees to participate in the protocol regardless of whether or not classification of the research protocol changes at some time in the future.

Another restriction relates to the level of harm to which an individual can consent by way of a research directive. When the research offers no promise of directly benefiting mentally incompetent subjects, the question to consider should be whether their participation in the research will harm them in a significant way.[39] If the research poses more than minimal risk, then consent should not generally be given. Therefore research directives should be presumed to apply only to experimentation with a negligible or less than substantial risk; a statement not to participate in non-therapeutic research must be respected in all cases.

Given that research directives are made in anticipation of future incompetence, the person formulating them must express their wishes in light of present knowledge

---

[39] See Moorhouse (1994), *supra* note 37 at 17–18.

to be applied at a later time. Problems associated with this include the fact that medical science may have evolved between the time of drafting of a directive and its application, either rendering the research redundant or indicating a risk of harm beyond what was originally predicted. In the alternative, the patient's medical condition may have improved or deteriorated. Therefore, in spite of the difficulty in enumerating all of the possible risks associated with a research protocol, the development of significant unforeseen risks should result in the termination of research on an individual despite prior consent to participation. Continuation of a protocol in light of new risks which are less than substantial should only be permitted with the consent of the incapable person's SDM.[40] If the person's medical condition has changed to the point where participation in the research protocol would cause additional risks for the individual, the research directive should become null, and likewise if a new treatment were to become available, the use of which would preclude participation in the research experiment.

Situations may arise where the most ethical treatment choice is not permitted by the patient's treatment directive. Consequently, health care professionals may not be required to comply with patients' requests for treatments that they feel are ill-advised, harmful or futile. However, it is accepted that there must be limits placed upon the liberties that health care professionals may take in interpreting health care directives.

In the case of research directives, there should be little room for flexibility in interpretation. As stated above, research directives should be specific rather than general. Any need for interpretation should require consultation with the person's SDM. In the case of a disagreement between a researcher and the SDM, the decision of the latter should take precedence. A treating health care professional, as distinct from the researcher, should also be able to state an objection to a patient's participation in research if the research were to interfere with the patient's treatment plan (assuming that the patient did not specifically forego a given course of treatment as stated within a research directive). In such a case, an SDM should withdraw the person, despite any research directive to the contrary.

In the event of the development of unforeseen risks, a change in the subject's condition, or an objection expressed by the incapable subject or a concerned third party, the SDM must have the authority to override an individual's research directive and to withdraw the individual from an experiment. Only the best interests of the incapable person (which include following the research directive) should guide the SDM.

Therefore, an SDM should be strongly guided, but not completely bound, by a research directive. In other words, an SDM's obligation to respect the person's prior wishes is limited by the obligation to protect the person. The function of the SDM is to promote what subjects think are their best interests, which necessarily excludes

---

[40] This assumes that the unforeseen risks do not, in conjunction with the other risks associated with the protocol, cause a research ethics committee to re-characterize the entire experiment at a level of risk which would not be compatible with the person's research directive.

consenting to being intentionally harmed, or to being unreasonably exposed to the risk of harm.

It is difficult, if not impossible, to predict how pain and discomfort will be experienced at a future time when the person's circumstances, if not their personhood, will have changed. Also, some incompetent persons may not be able to communicate their desire to withdraw or object, and competent persons locked in their bodies cannot communicate their consent or refusal. If the subject is incapacitated during the experiment, research directives for non-therapeutic research with substantial risk of harm, pain, and discomfort should require the approval of an independent body such as a court, which may approve such experiments only in exceptional cases.[41]

Therefore, an important restriction on the scope of research directives involves their use in the form of so-called Ulysses Contracts.[42] Applied to the context of research, such contracts would require that once an experiment is underway and the subject has become incompetent, researchers should ignore any objection the subject might make to continued participation. The central argument supporting the use of Ulysses Contracts is based on the notion of respect for a person's autonomous rights, including the right to expose oneself to risks. However, as has been discussed earlier, the principle of personal inviolability limits this right. While arguments favouring the use of Ulysses Contracts may be acceptable in the context of personal care decisions, a strict application of these justifications to the context of non-therapeutic research would be incorrect. With treatment decisions, a person knows in advance that they are agreeing to be treated with the intention of helping themselves, and that care will be given by qualified health care practitioners, skilled at assessing the course of a treatment, and who will make sound clinical decisions about how to operate within the parameters established by a person's directive. In

---

[41] It has been argued that consenting to a higher level of risk should only be allowed if a subject has specified a willingness to undergo a higher level of risk in his or her research directive and was able to show that he or she has previously experienced a similar level of physical or psychological pain or discomfort. See E.W. Keyserlingk *et al.*, "Proposed Guidelines for the Participation of Persons with Dementia as Research Subjects" (1995) 38 Perspectives in Biology & Medicine 319 at 351. However, proving such a claim may be very difficult, if not practically impossible, primarily because "pain" is experienced on a highly subjective level, making it impossible to quantify levels of pain and discomfort accurately. While it may be possible to qualitatively categorize different levels of pain, any such system would still be unable to take into account subjective factors which may cause a given "level of pain" to be mild for some people, and severe for others. As such, inquiries into whether a person has previously experienced "similar" levels of pain should serve to fulfill more of an informational requirement, aiding in the assessment of whether the person's advance consent to participate in an experiment is genuine — that is, valid and informed. Otherwise, by requiring individuals to have experienced high levels of pain in the past in order to submit to such levels in the future would imply that individuals fortunate enough to live "pain-free" lives are unable to make the decision to submit to higher levels of risk, pain or discomfort.

[42] In the *Odyssey*, Ulysses, wishing to hear the songs of the sirens without jeopardizing his life and those of his crew, instructed them to tie him to the mast of his ship and not to release him regardless of his orders or his suffering. The term "Ulysses Contract" is used to describe the situation where a patient, fearing incompetence, leaves binding decisions to be applied should this ensue. See A. Macklin, "Bound to Freedom: The Ulysses Contract and the Psychiatric Will" (1987) 45 University of Toronto Faculty Law Review 37.

the case of non-therapeutic experimentation, however, a person is agreeing in advance to participate with the intention of helping others, without personal benefit. Therefore, Ulysses Contracts consenting to experiments involving a substantial level of risk, harm or discomfort should not be permitted.

### Executing Research Directives

In a recent study, it was found that among those family physicians who had used health care directives, 56% said that they had not always followed the directions contained there. Family disagreement, inappropriate or unclear wording of the directive, an illness that was not terminal, the likelihood that the patient preference expressed in the directive would change if aware of the clinical situation, and the obsolescence of the directive, were given as reasons for noncompliance.[43] All of these reasons are equally applicable to research directives. However, the concerns regarding non-compliance with research directives can be alleviated if sufficient care is taken to draft the directive and if the above restrictions regarding scope and level of risk or harm, as well as registration are satisfied.

Another study of health care directives found that directives were often left behind in the institutional long-term care facility when patients were transferred to an acute care hospital and that, despite the fact that the directives specified treatment choices for a variety of circumstances, they were not always followed.[44] Ensuring that a research directive is followed requires that the appropriate persons become aware of the research directive. For this reason, it was recommended above that research directives be registered with a central authority.

When the document is available and valid, problems can persist. The SDM may be faced with deciding whether to consent to the person being in non-therapeutic research rated at higher levels of risk, and the directions can be vague and outdated. For reasons discussed earlier, when ranking obligations the duty to do no harm and respect the welfare of the vulnerable person should take priority. Ways to reduce incidence of inappropriate and unhelpful research directives have been discussed. Researchers ought to be bound by the requirements of an ethical design for selecting incapable subjects to be involved in non-therapeutic research. In the event that the research is approved, there ought to be limits to the obligation to respect prior wishes. SDMs are moral agents that should not passively follow directions. Today's heroic measures can become standard treatments in a few years. Similarly a benign wish can become an approval to be engaged in a high risk research project.

Although a research directive prepared by a person can be clear and of immense assistance to an SDM, a standard form enhances the possibility of receiving clear instructions. Without a standard form, the language could be vague and the instructions ambivalent. Also, the document could be silent on key issues including

---

[43] D.L. Hughes & P.A. Singer, "Family Physicians' Attitudes Toward Advance Directives" (1992) 146 Canadian Medical Association Journal 1937.
[44] M. Danis *et al.*, "A Prospective Study of Advance Directives for Life-Sustaining Care" (1991) 324 New England Journal of Medicine 882.

whether the person wants to be involved in all types of research regardless of the risk-benefit ratio or only in research investigating specific conditions (e.g., the disease or condition from which the person suffers). In respect of treatment advance directives, these forms require the authors to consider different states of health and the level of care they want, given their condition. Also, the authors are asked to amplify or clarify their values and goals in a written statement. The documents are witnessed and dated. These forms can be of great assistance to the SDM, even if the person's condition does not match identically the possibilities listed in the treatment directive. With this form in hand, the SDM is better able to decide what the person would have preferred in a given situation. Standard forms should be used in preparing research directives, and could include the following information: conditions under which a person wishes to be a research subject, the fields of research that the person wants to contribute to (e.g., only studies examining the very condition that the person has or all clinical research studies), and restrictions on involvement in research (e.g., no blood work). A place for additional information about preferences should be completed.

When designing or reviewing research studies involving vulnerable subjects, the following criteria should be met:

- There should be a specific, scientifically valid reason to conduct the proposed research.
- It must be demonstrated that there is no alternative to involving mentally incompetent persons as research subjects because there is no other means of obtaining relevant, similar or identical information. In other words, using animal subjects or other human subjects are not available options.
- At the review process, research proposals with an unacceptable level of risk of harm should not be approved. If approved, the decision of whether a research intervention has an acceptable level of risk should be passed to the SDM. The probability and magnitude of risk must not be disproportionate to the expected benefit for others. Vulnerable persons should not be exposed to a substantial probability and magnitude of harm, whatever the expected benefits to others.
- The involvement of mentally incapable persons should be monitored closely to ensure that there is equity in selecting vulnerable persons to be approached to be subjects. For example, elderly people in low socio-economic groups should not be targeted to sign research directives.
- It is preferable that recruiting researchers asking an SDM for consent not also be health care professionals caring for the prospective subject. When the research and health care provider wear the same hat, there is a conflict of interest that the SDM may not recognize, or may even result in coercion. Therefore, it is recommended that the roles be distinct. Frequently in clinical settings, the scarcity of resources and expertise in the field of investigation prevent this division of responsibilities. Thus, the investigators have a responsibility to assess how to minimize coercion given the particular circumstances. For example, a member of the research team who does not

provide direct care could seek consent from the person when competent, or from a guardian at a later date.

- When consent is requested, the guardian must understand the nature of the research, the promise of benefits and the risk of harm and be able to assess the benefits and harms of research participation. Then the SDM must decide whether granting consent would be contrary to a directive's purpose, which includes respecting both the person's wishes to help investigators acquire new knowledge and restricting the person's exposure to harm. The SDM must understand that the subject can be withdrawn at any time without penalty and that the SDM may exercise this option.
- The SDM should have the responsibility to keep informed of the person's condition and to determine the effects of being a research subject on the person's mental and physical status. If significant adverse effects are observed, the guardian should withdraw the person from the study. Neither the SDM nor the subject should be penalized for this decision.
- Health care professionals and investigators should keep the SDM informed of the subject's response to being a research subject. When adverse responses are observed, health care providers should respond to the subject's condition, and the SDM should be alerted as soon as possible to discuss whether the person should remain a subject, and if so, under what circumstances. If the subject experiences serious harm or is at risk of doing so, health care providers or the investigators must withdraw the person immediately from the study and inform the SDM promptly of the subject's status.

### Possible Objections

It is submitted that guidelines and a legislative response to the use of research directives are needed. Without strict guidelines, vulnerable persons remain at risk of being harmed through involvement in research for several interdependent reasons. The coupling of the moral obligation to conduct research with the difficulties in obtaining consent creates the ideal conditions for an erosion of informed consent for research involving mentally incapable persons. Approval of the research design by scientific and ethical committees should not be deemed sufficient. Research directives, even subject to regulations, can be vague and difficult to interpret and apply. Without regulation, research directives will not have to meet consistent standards regarding discussion of the type of research and acceptable risk-benefit ratio. To protect the prospective subject and to assist SDMs to respect intended wishes, regulation of research directives is recommended.

Presumably, another objection will be given that the proposals are too costly. Developing standard forms, establishing a protocol, and maintaining a registry will be an additional expense. Given the benefits accruing to health care professionals and society, these costs should be seen as justifiable. It is imperative that these costs not be borne by the person preparing the research directive, nor the SDM, nor the subject. On the other hand, transferring costs to research funding agencies could establish a conflict of interest for the research funding agencies and their researchers.

Further objections relate to the view that legislation and regulations regarding research directives will halt or hinder much-needed research. It must be admitted that the proposed recommendations will limit the scope of research involving incapable persons in non-therapeutic research. However, the lot of incapable persons in therapeutic research should be improved. The recommendations benefit researchers wanting to recruit incapable persons in non-therapeutic research, which presents ethical and legal problems. The proposed recommendations permit SDMs to consent to non-therapeutic research if directed to do so by a properly completed research directive. Substitute consent can be given on condition that the research design will not expose the subject to a substantial risk of harm. Hence, the recommendations would allow guardians to consent to non-therapeutic research, which for ethical and moral reasons was hitherto ethically problematic.

The cost of not conducting research exposing incapable persons to excessive risk for the benefit of others is not known with certainty. For the sake of argument, if it could be established that the gains would justify harming others, would the price be a fair and acceptable one? We have enough examples from recent history to demonstrate what happens when society adopts as a general rule that vulnerable individuals can be both used as research subjects and made to endure physical and psychological duress for the advantage of others. Accepting this application of the utilitarian calculus as a standard of conduct has a deleterious effect on society. More specifically, acceptance of this general equation has consequences in terms of the value we place on the lives of the mentally impaired, and on the types of relationships we are to have with dependent individuals.

A final objection could be that the recommendations permit SDMs to overrule previously stated wishes. It is allowed in this discussion that research directives can be overruled, but only under specific conditions designed to protect the subject's welfare.

## CONCLUSION

Preventing SDMs from volunteering their dependents in any research project without the prior indication of a desire to participate in experiments would be overly restrictive. Presumably, the prospective subject would volunteer if given the opportunity, but this assumption has to be balanced by the limitation of the level of risk to which such an assumption would apply. Although altruistic acts should be supported and respected, people should not be expected to submit to, nor should they be indiscriminately exposed to, unreasonable risks for the benefit of others. An open research directive does not resolve this ethical problem of whether the proxy should respect prior wishes to be a subject in research. Instead, the proxy now has an ethical dilemma: should he or she consent for the mentally incompetent person to be a subject in a study that probably will expose the subject to excessive risk of harm on the strength of a prior wish that the person cannot revoke? The use of research directives under specific conditions is recommended. The position advanced is that research directives should not generally permit the involvement of incapable persons in non-therapeutic research rated as involving substantial risk. In any event, SDMs ought not be required to respect wishes that will lead to exposing the vulnerable person to unacceptable risks.

To put the questions about research directives in context, they are extensions of the discussions in bioethics about the viability of the doctrine of informed consent in light of the fiscal constraints on health care. Granted that respect for persons, beneficence, and non-maleficence provide the ethical basis of health care, great strides in research and health care have raised pressing questions about the limits of respect for self-determination and how to equitably distribute scarce resources. Although there may be a consensus on the importance of the ethical principles that are to serve as the foundation of health care, their application is riddled with controversy. Similarly with research, applying the ethical principles to the designing of ethical research is complicated because of competing responses to what is fair, what is in the subject's best interests, what cost subjects should bear to benefit others, and how strong and durable is informed consent, whether given in advance or contemporaneously.

Advance directives for health care have been well established. General arguments in their favour, however, are not sufficient to justify their use for research purposes. Research is fundamentally different from health care, and the rights ascribed to research subjects are jeopardized when subjects are incompetent. To ask proxies to volunteer their dependents to be research subjects, exposed to harm without any promise of significant personal benefit, requires a redefinition of the moral responsibilities of proxies who are asked to respect advance instructions about research participation. It is hoped that with the restrictions proposed, the preparation and use of research directives provide a resolution of the conflicting values of respecting the self-determination of an autonomous person and respecting the best interests of the vulnerable person.

CHAPTER 12

# THE REGULATION OF HUMAN EXPERIMENTATION: HISTORICAL AND CONTEMPORARY PERSPECTIVES

SEV S. FLUSS*

## INTRODUCTION

In 1990, the distinguished Secretary-General of the Council for International Organization of Medical Sciences, Professor Zbigniew Bankowski, received an invitation from Professors Michael Grodin and George Annas to contribute a chapter on certain events leading to, or that occurred after, the promulgation of the *Nuremberg Code* on 20 August 1947, for inclusion in a then forthcoming book on the Doctors' Trial in Nuremberg. The author and two of his colleagues (one an intern, Sharon Perley, currently with the Division of Civil Rights in the US Department of Justice) joined Bankowski and a chapter entitled "The *Nuremberg Code*: an international overview" was duly included in a book[1] that will remain a classic on the subject, at least in the English language, for years to come. Numerous sources were used for the preparation of this chapter, including in particular, the

---

* I am grateful to the following, in particular, for information and documentation used in the preparation of this paper (they are listed here in alphabetical order); Dr. Zbigniew Bankowski (CIOMS, Geneva); Dr. Thomas Gerst (Federal Chamber of Physicians, Cologne); Professor Larry Gostin (Georgetown University, Washington, DC); Dr. Stuart Nightingale (Food and Drug Administration, Rockville, MD); Professor Qiu Renzong (Chinese Academy of Social Sciences, Beijing); and Mr. Dominique Sprumont (University of Neuchâtel, Switzerland). I am also indebted to Professors Michael Grodin and George Annas (Boston University) and Ms. Sharon Perley (US Department of Justice, Washington, DC) for some of the original impetus, ideas, and research without which this paper could not have been written. I am also deeply grateful to Ms. Isabel Monreal (Annecy, France) for her invaluable assistance in finalizing this paper.
[1] G.J. Annas & M.A. Grodin, eds., *The Nazi Doctors and the Nuremberg Code: Human Rights in Medical Experimentation* (New York and Oxford: Oxford University Press, 1992).

fruits of original research conducted by the late Norman Howard-Jones[2] (a British physician who served as the first Director of WHO's Division of Editorial and Reference Services, and as a mentor for the many WHO staff members, including the author, who were privileged to serve under him). Other archives, some hitherto unexplored, in the United Nations Library were exploited by Perley, including the 1947 *travaux préparatoires* of the Commission on Human Rights of the United Nations that led to the *International Covenant on Civil and Political Rights*, as well as commentaries on the drafting of the human experimentation-related provisions of the four Geneva Conventions of 12 August 1949 and of Article 11 of Protocol I of the Additional Protocols of 8 June 1977 to the *Geneva Conventions*, as published by the International Committee of the Red Cross and the International Federation of Red Cross and Red Crescent Societies.[3]

In his extensive research on the history of human experimentation, Howard-Jones was fortunate enough to have had access to the unequalled collections of medical literature at the National Library of Medicine (Bethesda) as well as the not inconsiderable collections available in the WHO Library in Geneva. However, some historical elements appear to have gone unnoticed (at least in the English-language literature) and I propose to briefly touch upon them here before moving on to more recent developments, particularly during the period 1990–1996 (i.e. after the completion of our contribution to the Annas/Grodin book).

## EVENTS IN GERMANY, 1930–1949

We are indebted to Thomas Gerst,[4] the historian of the Bundesärztekammer (Federal Chamber of Physicians) in Cologne, for a recent account of the origins of the successive editions of the classic work by Mitscherlich and Mielke on the Nazi medical experiments, *Medizin ohne Menschlichkeit* (published in New York in 1949 under the title *Doctors of Infamy: The Story of the Nazi Medical Crimes*).

To go back to the pre-Nazi period, the Directive of 29 December 1900 of the Prussian Minister of Religious, Educational, and Medical Affairs on the conduct of medical experimentation is now well known to those interested in the history of human experimentation. Grodin describes it as probably the "first document dealing with the ethics of human experimentation that specifically recognizes the need for the protection of uniquely vulnerable populations such as minors or incompetents". He considers it to be "critical in the history of the development of human experimentation guidelines in that it not only states the substantive standards for the ethical conduct of research, but also contains specific procedural mechanisms to ensure responsibility for the experimentation".

---

[2] N. Howard-Jones, "Human Experimentation in Historical and Ethical Perspectives" (1982) 16 Social Science & Medicine 1429.

[3] International Committee of the Red Cross and International Federation of Red Cross and Red Crescent Societies, *Handbook of the International Red Cross and Red Crescent Movement*, 13th ed. (Geneva: ICRC, 1994).

[4] T. Gerst, "Der Auftrag der Ärztekammern an Alexander Mitscherlich zur Beobachtung und Dokumentation des Prozessverlaufs" (1994) 91 Deutsches Ärzteblatt-Ärztliche Mitteilungen 22.

Grodin discusses the 1900 Directive and the events leading up to the promulgation by the Reich Minister of the Interior, on 28 February 1931, of *Guidelines on Innovative Therapy and Scientific Experimentation*, a text described by Fischer and Breuer in 1978[5] as "clearer, more concrete, and more far-reaching than both the *Nuremberg Code* and the [1975 *Declaration of Helsinki*] recommendations". One of the authors cited by Grodin is Alfons Stauder,[6] and his paper contains numerous interesting, and at least one intriguing, observations and perspectives. His lengthy (but, unfortunately, not fully referenced) contribution concludes with the words:

> Kein Gesetz und keine noch so strenge Kontrolle wird Versuche am Menschen hindern, sondern nur eines, die Besinnung der Aerzte auf ihre Sendung und das lebendige Beispiel.[7]

Stauder cites a statement — with which he expresses full agreement — by the President of the Reich Health Office, at a session of the Reichstag on 26 March 1928:

> Experiments on dying children, if they have indeed occurred, are inadmissible. Faced with the elevated nature [Hoheit] of death, experimental science must stop.

Subsequently in his paper, Stauder makes an intriguing reference to a paper by Moll (not otherwise identified) entitled "Earlier ordinances issued by the Prussian Ministry of Ecclesiastical Affairs". This suggests that the 1900 Directive may have been preceded or followed by other legal instruments, and at the time of writing the author is seeking to elucidate this point; Moll's paper has not yet come to hand.

Moving now to the post-War period, significant events occurred in the resort town of Bad Nauheim on 14–15 June 1947. The so-called Nauheimer Meeting (*Nauheimer Tagung*) brought together representatives of the Chambers of Physicians of the American, British, and French Zones of Occupation in Germany. One outcome, relevant to this paper, was unanimous agreement on an Oath to be taken by every German physician upon graduation. It is clear from the debate that preceded its adoption that this measure was perceived within the context of the denazification of the medical profession in Germany and indeed, to quote one speaker (Haedenkamp), as reported in a German journal, *Südwestdeutsches*

---

[5] F.W. Fischer & H. Breuer, "Influence of Ethical Guidance Committees on Medical Experimentation — A Critical Appraisal" in N. Howard-Jones & Z. Bankowski, eds., *Medical Experimentation and the Protection of Human Rights* (Proceedings of the XIIth CIOMS Round Table Conference, Cascais, Portugal, 30 November — 1 December 1978) (Geneva: Sandoz Institute for Health and Socio-Economic Studies, 1979) 65.

[6] A. Stauder, "The Permissibility of Medical Experiments on Healthy and Ill Humans" (Paper presented at a meeting of the Reich Health Council on 14 March 1930) reproduced in (1931) 3 Münchener Medizinische Wochenschrift 107.

[7] An approximate translation: "No law and no control however rigorous will prevent experiments on humans, but only one thing, namely a reflection by the physician on its outcome and living examples". For perspectives on Stauder's later role, see M.H. Kater, *Doctors under Hitler* (Chapel Hill: University of North Carolina Press, 1989).

$Ärzteblatt^8$, as a "reaction to the Nuremberg Doctors' Trial" (then nearing its conclusion). The sixth of the nine paragraphs that make up the Oath reads:

> Gegen seinen Willen und auch mit seinem Verständnis werde ich weder am gesunden noch am kranken Menschen Mittel oder Verfahren anwenden oder erproben die ihm an Leib, Seele oder Leben schaden oder Nachteil zufügen könnten.[9]

Further research will be needed to determine whether this Oath was in fact administered thereafter in German medical schools, at least in what was to become the Federal Republic of Germany.

## EVENTS IN MANCHURIA, 1932–1945

Much previously inaccessible material on the medical experiments conducted by the notorious Unit 731 of the Japanese Imperial Army is presented and reviewed in a recent book by a US historian, Harris.[10] However, the author has still not succeeded in tracing the so-called Khabarovsk Rules, stated in some sources to be a *Nuremberg Code*-like document promulgated by a Soviet court following the trial in Khabarovsk, on 25–31 December 1949, of Japanese military personnel charged utilising human subjects in biological warfare research. No such rules appear in a Russian-language publication on the trial.[11] Harris points out that many of the archives on this episode, both in Russia and the USA, remain closed.

## SOME EVENTS IN THE USA, 1944–1974

Annas and Grodin have drawn attention to an important Memorandum, dated 26 February 1953, of the US Secretary of Defense on the use of human volunteers in experimental medical research. Originally classified as "Top Secret", this document was declassified in August 1975.

Another, perhaps less well-known, development in 1953 was the approval, by the Director of the Clinical Center at the National Institutes of Health in Bethesda, of a document entitled "Group Consideration of Clinical Research Procedures Deviating from Accepted Medical Practice or Involving Unusual Hazards". This occurred on 17 December 1953, and provided for review, by a committee or other mechanism, of clinical research involving "unusual hazards". No less than 22 years were to elapse before such a review system was introduced internationally, in the

---

[8] (1947) 2:7–9 *Südwestdeutsches Ärzteblatt 50.*
[9] An approximate translation: "I shall not administer or test products or procedures on either healthy or sick persons, against their will or even with their consent, if those products or procedures could harm or be potentially injurious to limb, mind, or life". The Oath is not included in the relevant Appendix of T. Reich, ed., *Encyclopedia of Bioethics*, vol. 5. (New York: Simon and Schuster MacMillan, 1995).
[10] S.H. Harris, *Factories of Death: Japanese Biological Warfare, 1932–1945, and the American Cover-Up* (London and New York: Routledge, 1994).
[11] *Materials from the Trial of the High-Ranking Military Staff of the Japanese Army Involved in the Preparations For and Use of Biological Warfare* (in Russian) (Moscow: State Publishing House for Political Literature, 1950). I am indebted to Mr Ralf Hägele (Meersburg, Germany) for making available a copy of this book.

*Declaration of Helsinki II* (adopted in Tokyo in October 1975). A noteworthy feature of this document is language that is evocative of current approaches to informed consent and the physician-patient relationship:

> *Extract from Section III (Principles governing physician/patient relationship)*
> The patient or subject of clinical study shall be considered a member of the research team and shall be afforded an understanding suited to his comprehension of the investigation contemplated, including particularly any potential danger to him.
> Each prospective patient will be given an oral explanation in terms suited to his comprehension, supplemented by general written information or other appropriate means, of his role as a patient in the Clinical Center [of the NIH], the nature of the proposed investigation and particularly any potential danger to him. After admission, the patient shall receive information in keeping with the development of a sound physician-patient relationship.
> Voluntary agreement based on informed understanding shall be obtained from the patient and, when appropriate, from responsible next of kin when the approved investigation includes procedures which deviate from accepted medical practice. In all such cases, a notation shall be made on the patient's chart of the essential points of the explanation and of the agreement obtained, together with any comment or problems raised by the patient.
> [The physician] shall be responsible for incorporating in the medical record the information given to the patient and the nature of the informed consent or agreement accomplished with the patient, including any comments, objections or general reactions made by the patient.

Approximately one year ago, an Advisory Committee on Human Radiation Experiments, set up by the Human Radiation Interagency Working Group of the USA, issued its final report.[12] Chapter 18 of this report is devoted to "Recommendations"; these relate to "remedies pertaining to experiments and exposures during the period 1944–1974". The first group of Recommendations deal with "Biomedical experiments" and include a series intended to "protect the rights and interests of human subjects in the future". Among the most important is Recommendation 9, which reads as follows:

> The Advisory Committee recommends to the Human Radiation Interagency Working Group that efforts be undertaken on a national scale to ensure the centrality of ethics in the conduct of scientists whose research involves human subjects.

The commentary on this particular Recommendation is of crucial importance. A "national understanding of the ethical principles underlying research and agreement about their importance" are stated to be essential to the research enterprise and the "advancement of the health of the nation". The historical record, it is stated, "makes it clear that the rights and interests of research subjects cannot be protected if researchers fail to appreciate sufficiently the moral aspects of human subject research and the value of institutional oversight". The Advisory Committee affirmed that it was not clear to it that scientists whose research involves human subjects are "any more familiar with the *Belmont Report* today than their colleagues were with the *Nuremberg Code* forty years ago". The distinction between the ethics of research

---

[12] Advisory Committee on Human Radiation Experiments, *Final Report*. (Washington, D.C.: U.S. Government Printing Office, 1995).

and the ethics of clinical medicine "was, and is, unclear", to quote further the commentary on Recommendation 9. Among many other interesting, important, and no doubt timely (in the US context, at least) recommendations, some address the issue of compensation for research injuries of future subjects of research funded by the US Government; it is stated that a system of compensation for research injuries had been contemplated since at least the late 1940s, when the US Army is said to have debated, but ultimately rejected, suggestions to establish a "uniform" programme for compensating prisoner volunteers who were injured during experiments involving malaria and hepatitis. (Major extracts from the Recommendation appear in Annex I.)

Reference should also be made to a recent press release concerning a report (issued on 30 January 1996) by a panel set up by the U.S. National Academy of Sciences/National Research Council on the implications of radiation experiments conducted by the U.S. Air Force in Alaska in 1955–1957.[13] These involved 102 native Alaskans and 19 U.S. military personnel. It is clear from the information contained in the press release that, to quote the Chairman of the panel (Professor Kenneth Mossman), the "informed consent process was flawed" and that certain other shortcomings occurred.

## THE INTERNATIONAL REGULATION OF HUMAN EXPERIMENTATION: THE CURRENT GLOBAL CONFIGURATION

Table 1 provides a listing of those international codes and other instruments (some at various stages of drafting) that have been identified. The number is significant, and indeed it may be no exaggeration to affirm that there is no other area of biomedicine that has attracted so many regulatory or para-regulatory initiatives. This is perhaps not surprising, given the array of legal, ethical, procedural, and public policy issues raised by human experimentation and, no doubt, its interface with the development of new drugs, medical devices, therapeutic modalities, etc. Another noteworthy aspect is the fact that certain countries, for example, in Scandinavia, simultaneously belong to several of the entities that have produced binding legal instruments or guidelines on the ethical conduct of human experimentation. At some stage, it may be warranted to examine any possible substantive (as opposed to linguistic) divergences between particular texts.

Table 2 identifies some of the key international and national "milestones" that have led to the current global patterns in the regulation of human experimentation.

## THE CONTRIBUTIONS OF WHO AND CIOMS

The author played a minor role in the process that led to the issuance of the 1982 Proposed International Guidelines for Biomedical Research Involving Human Subjects, and the 1993 text that superseded these Guidelines, viz., the International

---

[13] Associated Press, 30 January 1996 (V0043).

**Table 1**

**SELECTED INTERNATIONAL INSTRUMENTS REGULATING RESEARCH ON HUMAN SUBJECTS (as of July 1996)**

| *ORGANIZATION/ENTITY* | *TITLE* |
|---|---|
| MILITARY TRIBUNAL (Doctors' Trial, Nuremberg, 1946–47) | NUREMBERG CODE (August 1947) |
| INTERNATIONAL COMMITTEE OF THE RED CROSS | GENEVA CONVENTIONS I–IV (1949) (I and II — Article 12; III — Article 13; IV — Article 32); Protocol I (1977) (Article 11) |
| UNITED NATIONS GENERAL ASSEMBLY | INTERNATIONAL COVENANT ON CIVIL AND POLITICAL RIGHTS (1966) (Article 7) |
| WORLD PSYCHIATRIC ASSOCIATION | DECLARATION OF HAWAII II (1983) |
| COUNCIL OF EUROPE | EUROPEAN PRISON RULES (February 1987) (Rule 27) |
| WORLD MEDICAL ASSOCIATION | DECLARATION OF HELSINKI IV (September 1989) |
| NORDIC COMMITTEE ON MEDICINES (Nordic Council) | GOOD CLINICAL TRIAL PRACTICE (December 1989) |
| COUNCIL OF EUROPE | RECOMMENDATION NO. R (90) 3 TO MEMBER STATES CONCERNING MEDICAL RESEARCH ON HUMAN BEINGS (February 1990) |
| EUROPEAN COMMISSION | NOTE FOR GUIDANCE ON GOOD CLINICAL PRACTICE FOR TRIALS ON MEDICINAL PRODUCTS IN THE EUROPEAN COMMUNITY (July 1990) |
| ORGANIZATION OF THE ISLAMIC CONFERENCE | THE CAIRO DECLARATION ON HUMAN RIGHTS IN ISLAM (Article 20) (August 1990) |
| CIOMS and WHO | INTERNATIONAL GUIDELINES FOR ETHICAL REVIEW OF EPIDEMIOLOGICAL STUDIES (1991) |
| UNITED NATIONS GENERAL ASSEMBLY | PRINCIPLES FOR THE PROTECTION OF PERSONS WITH MENTAL ILLNESS AND FOR THE IMPROVEMENT OF MENTAL HEALTH CARE (December 1991) (Principle 11 [paragraph 15]) |

**Table 1 (Continued)**

| ORGANIZATION/ENTITY | TITLE |
|---|---|
| CIOMS and WHO | INTERNATIONAL ETHICAL GUIDELINES FOR BIOMEDICAL RESEARCH INVOLVING HUMAN SUBJECTS (February 1993) |
| COMMONWEALTH MEDICAL ASSOCIATION | GUIDING PRINCIPLES ON MEDICAL ETHICS (October 1994) (Principle 16) |
| WHO | WHO GUIDELINES FOR GOOD CLINICAL PRACTICE (GCP) FOR TRIALS ON PHARMACEUTICAL PRODUCTS (1995) |
| COMMONWEALTH OF INDEPENDENT STATES | CONVENTION ON HUMAN RIGHTS AND FUNDAMENTAL FREEDOMS (May 1995) (Article 3) |
| EUROPEAN FORUM FOR GOOD CLINICAL PRACTICE | GUIDELINES AND RECOMMENDATIONS FOR EUROPEAN ETHICS COMMITTEES (1997) |
| COUNCIL OF EUROPE | PROTOCOL (to Convention on Human Rights and Biomedicine) ON BIOMEDICAL RESEARCH (in preparation) |
| COUNCIL OF EUROPE | DRAFT PROTOCOL TO THE EUROPEAN CONVENTION ON HUMAN RIGHTS SECURING CERTAIN ADDITIONAL RIGHTS TO PERSONS DEPRIVED OF THEIR LIBERTY (Article 6) (in preparation) |

Ethical Guidelines for Biomedical Research Involving Human Subjects. The latter together with the 1991 International Guidelines for Ethical Review of Epidemiological Studies are perhaps unique among existing international texts on the subject in that they formulate the basic ethical principles in accordance with which all research involving human subjects should be conducted. In the 1993 Guidelines, these principles are stated to be "respect for persons, beneficence and justice" — non-maleficence (often expressed as a separate principle) being subsumed under the second of these. These are, of course, the principles developed by Beauchamp and Childress at Georgetown University, and there is no need to dwell on them in this paper.

Both the 1991 and 1993 Guidelines are intended to amplify and give effect to the World Medical Association's *Declaration of Helsinki IV.* This is also the case for the Guidelines for Good Clinical Practice for Trials on Pharmaceutical Products, recently published by WHO.[14] Developed in consultation with national drug

---

[14] World Health Organization, *Sixth Report of the WHO Expert Committee on the Use of Essential Drugs* (Geneva: WHO, 1995) (WHO Technical Reports Series No. 850). Large extracts are reproduced in the (1995) 46:3 International Digest of Health Legislation 404.

## Table 2

### SOME INTERNATIONAL AND NATIONAL MILESTONES IN THE HISTORY OF THE REGULATION OF HUMAN EXPERIMENTATION

| Date | Event |
| --- | --- |
| 1900 | Directive of the Prussian Minister of Religious, Educational and Medical Affairs |
| 1931 | Guidelines of the Reich Minister of the Interior |
| 1947 | Promulgation of the Nuremberg Code by US Military Tribunal |
| 1952 | French National Academy of Medicine defines position on clinical trials |
| 1953 (Feb.) | Memorandum of US Defense Department on Use of Human Volunteers in Experimental Research (originally classified as "Top Secret") |
| 1953 (Nov.) | US National Institutes of Health issues document on clinical research |
| 1954 | World Medical Association (WMA) adopts resolution on human experimentation |
| 1955 | Netherlands Guidelines on experiments in human beings |
| 1962–1963 | First UK Medical Research Council document on ethical aspects of research |
| 1963 | Swedish Circular on clinical trials |
| 1964 | Declaration of Helsinki I adopted by WMA |
| 1966 | American Medical Association issues Ethical Guidelines for Clinical Investigation |
| 1966 | Adoption by United Nations General Assembly of International Covenant on Civil and Political Rights (Article 7) |
| 1968 | New Zealand Medical Research Council endorses Declaration of Helsinki I |
| 1969 | First Austrian Guidelines on clinical trials |
| 1970 | First Swiss Directives (Swiss Academy of Medical Sciences) on human experimentation |
| 1973 | First Mexican legislation on human experimentation |

**Table 2 (Continued)**

| Date | Event |
|------|-------|
| 1974 | First US Federal Regulations on protection of human subjects in research |
| 1976 | First Australian National Health and Medical Research Council Statement on human experimentation |
| 1976 | First German legislation on human experimentation (Federal Republic of Germany and former German Democratic Republic) |
| 1976 | Hungarian Directive on institutional ethics committees |
| 1978 | Belmont Report issued in USA (National Commission for the Protection of Human Subjects of Biomedical and Behavioral Research) |
| 1978 | First Canadian Medical Research Council document on ethical aspects of research |
| 1979 | Argentine Regulations on human experimentation |
| 1980 | Indian statement on ethical issues in medical research (Indian Council on Medical Research) |
| 1982–1983 | Proposed Guidelines on biomedical research of the Council for International Organizations of Medical Sciences (CIOMS) |
| 1982–1983 | Series of reports on ethical issues in research issued by President's Commission in USA |
| 1987 | Ottawa G–7 Bioethics Summit on human experimentation |
| 1988 | First French statute on human experimentation |
| 1989 | Declaration of Helsinki IV adopted by WMA (in Hong Kong) |
| 1990 | Council of Europe Ministerial Recommendation on medical research |
| 1990 | European Commission guidance document on clinical trials |
| 1991 | Guidelines on epidemiological studies (CIOMS) |
| 1991 | Issuance in USA of common Federal Policy for Protection of Human Subjects |

**Table 2 (Continued)**

| Date | Event |
|------|-------|
| 1993 | Guidelines on biomedical research (CIOMS) |
| 1995 | GCP on clinical trials (WHO) |
| 1997(?) | Council of Europe Protocol on Medical Research |
| 1997 | First Netherlands statute on human experimentation |

regulatory authorities in WHO's Member States, their purpose is to "set globally applicable standards for the conduct of such biomedical research on human subjects". As the Introduction points out:

> The Guidelines are addressed not only to investigators, but also to ethics review committees, pharmaceutical manufacturers and other sponsors of research, and drug regulatory authorities. By providing a basis both for the scientific and ethical integrity of research involving human subjects and for generating valid observations and sound documentation of the findings, these Guidelines not only serve the interests of the parties actively involved in the research process, but protect the rights and safety of subjects, including patients, and ensure that the investigations are directed to the advancement of public health objectives.
>
> The Guidelines are intended specifically to be applied during all stages of drug development both prior to and subsequent to product registration and marketing, but they are also applicable, in whole or in part, to biomedical research in general. They should also provide a resource for editors to determine the acceptability of reported research for publication and, specifically, of any study that could influence the use or the terms of registration of a pharmaceutical product. Not least, they provide an educational tool that should become familiar to everyone engaged in biomedical research and, in particular, to every newly trained doctor.

## THE CONTRIBUTIONS OF OTHER INTERNATIONAL ORGANIZATIONS

The current configuration of international regulations governing the ethical and human rights aspects of research on human subjects is illustrated in Table 1. This indicates that the principal instrument currently under development is a Protocol on Biomedical Research, now in the final stages of drafting within the framework of the Strasbourg-based Council of Europe. This Protocol is a subsidiary instrument to the Convention on Human Rights and Biomedicine drawn up within the framework of the Council of Europe. It is likely that this convention, and the two Protocols (the second dealing with organ transplantation) will be finalized in the course of 1998. It goes without saying that States ratifying the Convention and Protocols will be bound by their provisions and many will no doubt have to introduce appropriate implementing legislation at the national or sub-national levels.

## SOME HISTORICAL PERSPECTIVES ON THE DEVELOPMENT OF THE CURRENT PATTERN OF INTERNATIONAL REGULATION OF HUMAN EXPERIMENTATION

It is not possible within the confines of this paper to describe the historical development of the CIOMS guidelines to which reference has already been made. Nor is it possible to overemphasize the extraordinary amount of valuable materials contained in the proceedings of the successive conferences organized under the auspices of CIOMS, in close cooperation and indeed in partnership with WHO. Readers are referred to these proceedings, and, in particular, those edited by Bankowski *et al.*[15] as well as those edited by Howard-Jones and Bankowski.[16] The reader is likewise referred to a pioneering contribution of Howard-Jones,[17] the contents of which still include lucid and cogent points on the clinical and biomedical contexts in which ethical and legal issues in human experimentation have arisen and, indeed, continue to arise.

## SOME CURRENT ISSUES IN THE REGULATION OF HUMAN EXPERIMENTATION

### *Research on Children*

A number of interns attached to WHO's Health Legislation Unit in Geneva in recent years have made significant contributions to the literature in certain areas relevant to the topic of this paper. Sujit Choudhry (then a Rhodes Scholar at University College, Oxford, and currently working at the University of Toronto) undertook, on the basis of documentation available to WHO, a survey of existing codes and legislation on paediatric research.[18] His conclusions were summarized in the following terms.

(1) There seems to be a general consensus that pediatric research is permissible but should only be conducted when research with adults cannot yield the same information.

(2) Documents differ to a great extent in how they approach children's involvement in the decision to participate in research. Some adopt a highly protectionist stance, relying solely on parental consent, while others adopt the dual requirement of parental permission and children's consent.

(3) Therapeutic procedures are subject to a favorable risk-benefit analysis, although some documents additionally look directly at the magnitude of the risk or benefit. Non-

---

[15] Z. Bankowski, J.H. Bryant & J.M. Last, eds., *Ethics and Epidemiology: International Guidelines* (Proceedings of the XXVth CIOMS Conference, Geneva, 7–9 November 1990) (Geneva: Council for International Organizations of Medical Sciences, 1991); Z. Bankowski, N. Howard-Jones, eds., *Human Experimentation and Medical Ethics* (Proceedings of the XVth CIOMS Round Table Conference, Manila, 13–16 September 1981) (Geneva: Council for International Organizations of Medical Sciences, 1982); Z. Bankowski & R.J. Levine, eds., *Ethics and Research on Human Subjects: International Guidelines* (Proceedings of the XXIVth CIOMS Conference, Geneva, 5–7 February 1992) (Geneva: Council for International Organizations and Medical Sciences, 1993).

[16] See Howard-Jones & Bankowski, *supra* note 5.

[17] See Howard-Jones, *supra* note 2.

[18] S. Choudhry, "Review of Legal Instruments and Codes on Medical Experimentation with Children" (1994) 3 Cambridge Quarterly of Healthcare Ethics 560.

therapeutic research is permitted, subject to a limitation on the magnitude of the risk faced. Exceptions to this ceiling of "minimal" risks require a weighing of risks and benefits.

(4) Codes give scant attention to the composition of research ethics committees that review proposals for pediatric research. Given that research with children raises unique issues, this is an area where further attention may be warranted.

There have of course been many other publications on this topic.[19]

### Research Involving Women

A Canadian intern of Scottish origin, currently based in a New York law firm, has undertaken a somewhat parallel study on this topic, which is currently receiving increasing attention in several countries.[20] Pending publication, it would be inappropriate to attempt to summarize the findings of this study, which examines the status of the issue in a number of countries, as well as in international codes and other instruments. It is perhaps noteworthy that the Beijing Platform of Action, adopted at the conclusion of the Fourth World Conference on Women (Beijing, 4–15 September 1995), includes provisions calling upon governments, the United Nations system, research institutions, etc., to:

Provide financial and institutional support for research on safe, effective, affordable and acceptable methods and technologies for the reproductive and sexual health of women and men, including more safe, effective, affordable and acceptable methods for the regulation of fertility, including natural family planning for both sexes, methods to protect against HIV/AIDS and other sexually transmitted diseases and simple and inexpensive methods of diagnosing such diseases, among others; this research needs to be guided at all stages by users and from the perspective of gender, particularly the perspective of women, and should be carried out in strict conformity with internationally accepted legal, ethical, medical and scientific standards for biomedical research.[21]

Here too there is an abundant, and rapidly growing, literature.[22]

### Clinical Research Conducted in Emergency Circumstances

On 21 December 1994, the US Commissioner of Food and Drug published a Notice in the Federal Register announcing the holding of a public meeting (co-sponsored by the FDA and the National Institutes of Health) on the above topic. The meeting was duly held (on 9–10 January 1995) and the Executive Summary of the Report, issued in May 1995, is reproduced as Annex II to this paper, as the issues addressed are

---

[19] See e.g. M.A. Grodin & L.H. Glantz, eds., *Children as Research Subjects: Science, Ethics & Law* (New York and Oxford: Oxford University Press, 1994).

[20] N. Cummings, "Laws and Research Codes Governing the Participation of Women in Clinical Drug Trials" (Paper presented in Stockholm at a Symposium on Gender-Related Health Issues during the International Congress of Pharmacy, 28 August 1995) [unpublished]. See also A.C. Mastrioanni, R. Faden & D. Federman, eds., *Women and Health Research: Ethical and Legal Issues of Including Women in Clinical Studies*, vol. 1. (Washington, D.C.: National Academy Press, 1994) [Vol. 2 was not available to the author].

[21] United Nations, *Platform for Action and the Beijing Declaration: Fourth World Conference on Women, Beijing, 4–15 September 1995* (New York: Department of Public Information, United Nations, 1996) at para. 109(h).

[22] See e.g. Mastrioanni, Faden & Federman, *supra* note 18.

likely to be of wide interest. It is known that attention is being given to ethical considerations raised by the research under emergency circumstances in at least some other countries (for example, Switzerland).

### *Research on the Use of Investigational Drugs and Vaccines in Certain Military Situations*

A few months ago, the FDA issued an Interim Final Rule "permitting the use of investigational agents for prophylaxis/treatment with waiver of informed consent in battlefield or combat-related situations" (contained in Part 50 of Title 21 of the US Code of Federal Regulations).[23]

### *Research in the Field of Gene Therapy*

The development of ethical principles for research on gene therapy is receiving substantial attention both at the national level (notably in Australia, Brazil, Canada, France, Germany, Japan, Norway, South Africa, Switzerland [Canton of Geneva], the United Kingdom, and the USA) and at the international level. Internationally, there have been some developments within the framework of UNESCO and the European Community. In a recent review, Mauron and Thévoz[24] argued for "non-exceptionalism" for the regulation of experimental gene therapy. The abstract of their paper reads as follows:

> Current trials in the field of gene therapy represent an increasingly diversified reality both as regards the diseases concerned and the therapeutic paradigms utilized. Analysis of the ethical dimensions and the formulation of regulations must take account of diversity. This paper summarizes recent developments in the field, discusses the arguments presented in the ethical debate as to the legitimacy of these therapies, and formulates guidelines for their regulation. It is desirable that the regulation of the gene therapies is coherent with the standards regulating clinical experimentation in general, notably in such allied fields as, for example, experimentation on vaccines and immunobiological products. It is important to avoid gene therapies being treated entirely separately from biomedical research; rather, one should seek to identify the ethical and biosafety dimensions that are specific to these therapies by drawing in extensive interdisciplinary expertise.[25]

## CONCLUSIONS

The ethical dimensions in medical experiments on human subjects clearly warrant continued vigilance — witness the opening Chapter (entitled "A History of Unethical Experimentation on Human Subjects") in a recent and important contribution to the literature.[26] This volume itself hopefully casts light on some of

---

[23] The background to this text was set forth in a presentation by Stuart L. Nightingale (the FDA's Associate Commissioner for Health Affairs) before the Presidential Advisory Committee on Gulf War Veterans' Illnesses, on 12 January 1996. For a more general paper see S.L Nightingale, C.A. Kimborough & P.H. Rheinstein, "Access to Investigational Drugs for Treatment Purposes" (1994) American Family Physician 845.

[24] A. Mauron & J.-M. Thévoz, "Thérapie génique: faut-il une réglementation spécifique?" (1995) 39 Cahiers Médico-Sociaux 165.

[25] Informal translation from the French.

[26] P.M. McNeill, *The Ethics and Politics of Human Experimentation* (Cambridge: Cambridge University Press, 1993).

the subsisting and emerging problems in this area, as will, no doubt, the series of international conferences scheduled to mark the 50th Anniversary of the *Nuremberg Code* (Nuremberg, October 1996; Washington DC, December 1996; and Freiburg-im-Breisgau, October 1997). Nuremberg, 1947, was, and remains, a milestone in societal efforts to safeguard ethical principles in biomedical research.

# ANNEX I

## ADDITIONAL RECOMMENDATIONS OF THE U.S. ADVISORY COMMITTEE ON HUMAN RADIATION EXPERIMENTS (FINAL REPORT, OCTOBER 1995)
### (*Extracts*)[27]

### Recommendations for the Protection of the Rights and Interests of Human Subjects in the Future

While we were constituted to consider issues related to human radiation experiments, in critical (but not all) respects, the government regulations that apply to human radiation research do not differ from those that govern other kinds of research. In comparison with the practices and policies of the 1940s and 1950s, there have been significant advances in the protection of the rights and interests of human subjects. These advances, initiated primarily in the 1970s and 1980s, culminated in the adoption of the Common Rule throughout the federal government in 1991. Although the Common Rule now affords all human subjects of research funded or conducted by the federal government the same basic regulatory protections, the work of the Advisory Committee suggests that there are serious deficiencies in some parts of the current system. These deficiencies are of a magnitude warranting immediate attention.

The Committee was not able to address the extent to which these deficiencies are a function of inadequacies in the Common Rule, inadequacies in the implementation and oversight of the Common Rule, or inadequacies in the awareness of and commitment to the ethics of human subject research on the part of physician-investigators and other scientists. We urge that in formulating responses to the recommendations that follow, the Human Radiation Interagency Working Group consider each of these factors and subject them to careful review.

### Recommendation 9

**The Advisory Committee recommends to the Human Radiation Interagency Working Group that efforts be undertaken on a national scale to ensure the**

---

[27] Certain passages have been deleted here on account of space considerations.

**centrality of ethics in the conduct of scientists whose research involves human subjects.**

A national understanding of the ethical principles underlying research and agreement about their importance is essential to the research enterprise and the advancement of the health of the nation. The historical record makes clear that the rights and interests of research subjects cannot be protected if researchers fail to appreciate sufficiently the moral aspects of human subject research and the value of institutional oversight.

It is not clear to the Advisory Committee that scientists whose research involves human subjects are any more familiar with the *Belmont Report* today than their colleagues were with the Nuremberg Code forty years ago. The historical record and the results of our contemporary projects indicate that the distinction between the ethics of research and the ethics of clinical medicine was, and is, unclear. It is possible that many of the problems of the past and some of the issues identified in the present stem from this failure to distinguish between the two.

The necessary changes are unlikely to occur solely through the strengthening of federal rules and regulations or the development of harsher penalties. The experience of the Advisory Committee illustrates that rules and regulations are no guarantee of ethical conduct. The Advisory Committee has also learned, in responses to our query of institutional review board (IRB) chairs, that many of them perceive researchers and administrators as having an insufficient appreciation for the ethical dimensions of research involving human subjects and the importance of the work of IRBs. The federal government must work in concert with the biomedical research community to exert leadership that alters the way in which research with human subjects is conceived and conducted so that no one in the scientific community should be able to say "I didn't know" or "nobody told me" about the substance or importance of research ethics.

The Advisory Committee recommends that the Human Radiation Interagency Working Group institute, in conjunction with the biomedical community, a commitment to the centrality of ethics in the conduct of research involving human subjects. We urge that careful consideration be given to the development of effective strategies for achieving this change in the culture of human subjects research, including, specifically, how best to balance policies that mandate the teaching of research ethics with policies that encourage and support private sector initiatives. It may be useful to commission a study or convene an advisory panel charged with developing and perhaps implementing recommendations on how best to approach this challenge for the research community.

**Recommendation 10**

**The Advisory Committee recommends to the Human Radiation Interagency Working Group that the IRB component of the federal system for the**

**protection of human subjects be changed in at least the five critical areas described below.**

1. **Mechanisms for ensuring that IRBs appropriately allocate their time so they can adequately review studies that pose more than minimal risk to human subjects. This may include the creation of alternative mechanisms for review and approval of minimal-risk studies. . . .**
2. **Mechanisms for ensuring that the information provided to potential subjects (1) clearly distinguishes research from treatment, (2) realistically portrays the likelihood that subjects may benefit medically from their participation and the nature of the potential benefit, and (3) clearly explains the potential for discomfort and pain that may accompany participation in the research. . . .**
3. **Mechanisms for ensuring that the information provided to potential subjects clearly identifies the federal agency or agencies sponsoring or supporting the research project in whole or in part and all purposes for which the research is being conducted or supported. . . .**
4. **Mechanisms for ensuring that the information provided to potential subjects clearly identifies the financial implications of deciding to consent to or refuse participation in research. . . .**
5. **Recognition that if IRBs are to adequately protect the interests of human subjects, they must have the responsibility to determine that the science is of a quality to warrant the imposition of risk or inconvenience on human subjects and, in the case of research that purports to offer a prospect of medical benefit to subjects, to determine that participating in the research affords patient-subjects at least as good an opportunity of securing this medical benefit as would be available to them without participating in research.**

In research involving human subjects, good ethics begins with good science. In our Research Proposal Review Project, the Advisory Committee was unable to evaluate the scientific merit of a significant number of proposals based on the documents provided by institutions. We suspect that this occurred in part because there is ambiguity about the role that IRBs should play with respect to evaluation of scientific merit and, thus, that documents submitted to IRBs may be inadequate in this area. The Advisory Committee also heard dissatisfaction with this ambiguity in our interviews and oral histories of researchers and from chairs of IRBs. If the science is poor, it is unethical to impose even minimal risk or inconvenience on human subjects. Although the fine points of the relative merit of research proposals are best left to study sections and other review mechanisms specially constituted to make such judgments, IRBs must be situated to assure themselves that the science they approve to go forward with human subjects satisfies some minimal threshold of scientific merit. In some cases, the IRB may be the only opportunity for this kind of scientific reviews.

In our Subject Interview Study interviews with patient-subjects, we confirmed that patient-subjects often base their decisions to participate in research on the belief that physicians, and research institutions generally, would not ask them to enter research projects if becoming a research subject was not in their medical best interests. For these patients, even the most candid, clearly written consent form affords little protection, for both the consent form and the consent process are of little interest to them. For patient-subjects whose decisions to participate in research are based on trust, and not on an assessment of disclosed information, the IRB review is of special importance. It is the only source of protection in the *federal* system for regulating human research positioned to ensure that their participation in research does not compromise their medical interests. Such a determination, however, often requires more specialized clinical expertise than any one IRB can possess. Federal policy must make it clear that IRBs have the responsibility to make this determination, but it must also allow mechanisms to be devised at the local level that permit this responsibility to be satisfied in an efficient and effective manner.

## Recommendation 11

**The Advisory Committee recommends to the Human Radiation Interagency Working Group that a mechanism be established to provide for the continuing interpretation and application of ethics rules and principles for the conduct of human subject research in an open and public forum. This mechanism is not provided for in the Common Rule.**

Issues in research ethics are no more static than issues in science. Advances in biomedical research bring new twists to old questions in ethics and sometimes raise new questions altogether. No structure is currently in place for interpreting and elaborating the rules of research ethics, a process that is essential if research involving human subjects is to have an ethical framework responsive to changing times. Also, for this framework to be effective, any changes or refinements to it must be debated and adopted in public; otherwise, the framework will fail to have the respect and support of the scientific community and the American people, so necessary to its success.

Three examples of outstanding policy issues in need of public resolution that the Advisory Committee confronted in our work are presented below:

(1) Clarification of the meaning of minimal risk in research with healthy children, including, but not limited to, exposure to radiation.
(2) Regulations to cover the conduct of research with institutionalized children.
(3) Guidelines for research with adults of questionable competence. Of particular concern is more-than-minimal-risk research that offers adults of questionable competence no prospect of offsetting medical benefit.

Current regulations permit the involvement of children as subjects in research that offers no prospect of medical benefit to participants when the research poses no more than minimal risk. An important question that has come to the Advisory

Committee's attention, both in the literature and in our Research Proposal Review Project, is whether research proposing to expose healthy children to tracer doses of radiation constitutes minimal risk. The uncertainty surrounding this issue calls into question the adequacy of the federal regulations, as currently formulated, in providing guidance for this category of research. This is a policy question that ought to be discussed and resolved in a public forum at the national level, not left to the deliberations of individual IRBs.

Current regulations do not provide any special protections for children who are institutionalized unless they are also wards of the state. Thus, researchers and IRBs have no more guidance from the federal government on the ethics of conducting such research than was available at the time of the Fernald and Wrentham experiments, decades ago.

The Advisory Committee also confronted in its Research Proposal Review Project another issue of research policy deserving public debate and resolution in a public forum. This is the issue of whether and under what conditions adults of questionable capacity can be used as subjects in research that puts them at more than minimal risk of harm and from which they cannot realize direct medical benefit. It is important that the nation decide together whether or under what conditions it is ever permissible to use a person toward a valued social end in an activity that puts him or her at risk but from which the person cannot possibly benefit medically.

## Recommendation 13

**The Advisory Committee recommends that the Human Radiation Interagency Working Group take steps to improve three elements of the current federal system for the protection of the rights and interests of human subjects — oversight, sanctions, and scope.**

(1) **Oversight mechanisms to examine outcomes and performance. . . .**
(2) **Appropriateness of sanctions for violations of human subjects protections. . . .**
(3) **Extension of human subjects protections to nonfederally funded research. . . .**

## Recommendation 14

**The Advisory Committee recommends that the Human Radiation Interagency Working Group review the area of compensation for research injuries of future subjects of federally funded research, particularly reimbursement for medical costs incurred as a result of injuries attributable to a subject's participation in such research, and create a mechanism for the satisfactory resolution of this long-standing social issue. . . .**

## ANNEX II

## REPORT OF THE PUBLIC FORUM ON INFORMED CONSENT IN CLINICAL RESEARCH CONDUCTED IN EMERGENCY CIRCUMSTANCES

## EXECUTIVE SUMMARY

A Public Forum on Informed Consent in Clinical Research Conducted in Emergency Circumstances was held on 9–10 January 1995, in Bethesda, Maryland. Co-sponsored by the Food and Drug Administration and the National Institutes of Health, the purpose of the Public Forum was to explore the ethical, legal, and operational aspects of obtaining informed consent in research conducted in emergency circumstances. Examples of the conditions studied in emergency circumstances include traumatic brain injury, occlusive stroke, cardiac arrest and arrhythmia, myocardial infarction, hemorrhagic shock, pulmonary embolism, status epilepticus, and poisoning. Patients with these conditions are acutely and gravely ill and face death or severe disability. In almost all circumstances they are cognitively and physically unable to give legally effective informed consent to participate in research protocols. To be effective, many of the experimental therapies for these conditions must be initiated as soon as possible after the onset of the condition, generally within a few hours and often within minutes. Legal representatives of prospective subjects often cannot be identified or located within these stringent time limits. In most cases, obtaining consent in advance is not feasible. Not only is the onset of the condition sudden and unexpected, but potential subjects are not practicably or readily identifiable in advance of the onset of the condition.

Since the publication of the Federal regulations for the protection of human subjects of research in 1981, a steadily increasing number of clinical research protocols in emergency and acute care medicine have been presented to institutional review boards (IRBs) for review and approval. Some of these protocols also proposed waiving informed consent requirements owing to the nature of the research. It has been increasingly apparent in recent years to IRBs, clinical investigators, the pharmaceutical and device industries, Congressional committees, and Federal agencies that the regulations governing informed consent have been interpreted differently at different times by IRBs, investigators, and Federal agencies. Some participants contended that, as a result of varying interpretations of the regulations, the conduct of some research protocols has been seriously delayed while the applicability of the Federal regulations to the research has been analyzed. In some cases IRBs have delayed or disapproved protocols calling for waiver of informed consent on the basis of their interpretation of the Federal regulations. In other cases IRBs have approved protocols, only to be instructed by a Federal agency that the protocols do not comply with the requirements of the regulations. Some clinical investigators related experiences in which the requirements to obtain informed consent from subjects in emergency circumstances made subject accrual difficult and protracted, precluding some important research projects.

Participants at the Public Forum affirmed the need to protect research subjects while allowing this important class of clinical research — the benefits of which are directly applicable to the conditions of the subjects — to go forward. Some expressed the view that the current regulations value the principle of autonomy excessively and at the expense of the principles of beneficence and justice. They contended that when the expected outcome of standard therapy is dismal, the principle of beneficence, that is, seeking what is best for the class of subjects, should outweigh the principle of autonomy. Participants discussed the ethical, regulatory, and operational challenges faced by potential subjects, IRBs, and emergency and acute care researchers, as well as ideas for meeting these challenges. A specific proposal for change was put forward by a broad coalition of academic/professional medical groups.

Presentations and discussions at the January Public Forum were wide-ranging and lively. Although the Forum was not a consensus conference, there was general agreement among the participants on a number of issues and strong support for some proposed courses of action. Briefly, participants broadly agreed that:

- the relevant Federal regulations need to specifically address clinical research performed under emergency conditions where it is not feasible to obtain prospective consent;
- it is important to establish special safeguards for protecting vulnerable research subjects while at the same time facilitating this class of research;
- the Federal regulations pertaining to this area of research should be congruent or at least compatible;
- there is a need for careful local IRB review and more frequent monitoring of clinical research, especially research initiated without consent;
- a return to a previous broad interpretation of the definition of minimal risk on the part of the Office for Protection from Research Risks would provide a short-term solution that would facilitate this class of research while long-term solutions are sought; and,
- serious consideration should be given by Federal Agencies to the issues, concerns, and recommendations set forth in the consensus statement of the Coalition of Acute Resuscitation and Critical Care Researchers.

These issues, and other topics on which there was a wider range of opinions, are described in more detail in the meeting summary.

CHAPTER 13

# INTERNATIONAL TRENDS IN RESEARCH REGULATION: SCIENCE AS NEGOTIATION

PAUL M. McNEILL

Internationally there are changes in the way in which science and research are conceptualised and conducted. These changes reflect changing views about what counts as knowledge, acceptable ways to arrive at knowledge, and the status of that knowledge in relation to other views and other human activities. This paper explores the idea that these changing perspectives on epistemology and ontology are related to changes in the regulation of research. They have a bearing on two basic questions in the approval of research: "Who should decide on what is ethical in science and research?" and "How should scientific activity be allowed to proceed?"

These two questions are explored in relation to changes in various countries to the way in which research is regulated. The first of the questions is relevant to changes in the composition of review committees in various countries. The inclusion of more community members is seen as part of a movement toward greater representation of the interests of research participants: the people who volunteer as research "subjects". The question "Who should decide?" is also considered in relation to new models in the United Kingdom for the review of research conducted in multiple locations. The second question "How should scientific activity proceed?" is considered in relation to extending review processes beyond medical research to include fields such as psychology, sociology, and anthropology. This extension challenges the assumptions on which research is conducted in those fields. In turn however, the very different assumptions about the nature of science, the conduct of research, and knowledge itself, particularly resulting from changing perspectives within anthropology (and philosophy), bounce back to challenge basic assumptions within medicine and the so-called "natural sciences".

If there is one trend to which all these changes are related it is that no assumptions are sacrosanct any more. Scientific research, like all other human activities, must be negotiated and accepted as one of many human activities. Its privileged status is now open to review.

## BACKGROUND

Most countries rely on committees to regulate research involving human participants. The system of review by committee was developed in the United States in the 1960s and adopted by most other developed countries subsequently.[1] It has been endorsed by international convenants, most notably by the Council for International Organizations of Medical Sciences (CIOMS).[2]

The predominant model of research regulation is that any researcher, before embarking on any research involving human participants, must seek and gain approval for that research from a committee. The committee, specially constituted to consider the ethics of research, has the power to recommend modifications to the proposed research, reject the application outright, or approve of the research. Committees worldwide reject very few proposals although the application process results in modifications being made to a proportion of them.[3] The setting for the committee is different in different countries, but with few exceptions, committees are based in an institution, usually a research institute within which the research is conducted, or are established within a regional authority, such as an area health board.

From the late 18th Century onwards there were numerous examples of doctors of medicine and bio-scientists callously disregarding extremes of suffering which they inflicted on human beings in the name of science. The experiments conducted in Nazi concentration camps are the best known of these and resulted in the Nuremberg trials including the trial of doctors and bio-scientists for crimes against humanity.[4] The court enunciated ten principles (now known as the *Nuremberg Code*) which "must be observed in order to satisfy moral ethical and legal concepts".[5] None of these principles include review by committee. It was health bureaucrats and politicians in the United States, in reaction to revelations of abuses of human

---

[1] P.M. McNeill, *The Ethics and Politics of Human Experimentation* (Cambridge: Cambridge University Press, 1993) at 37–50.

[2] Council for International Organizations of Medical Sciences (CIOMS) in collaboration with the World Health Organization (WHO), *International Ethical Guidelines for Biomedical Research Involving Human Subjects* (Geneva: CIOMS, 1993) [hereinafter *CIOMS Guidelines*].

[3] In Australia, for example, 63% were approved on first consideration and only 2.7% were rejected. A further 14.7% were approved after clarification and 17.3% were approved after modification by the researcher. See McNeill, *supra* note 1 at 89.

[4] J. Katz, *Experimentation with Human Beings: The Authority of the Investigator, Subject, Professions, and State in the Human Experimentation Process* (New York: Russell Sage, 1972) at 292–305. See also G.J. Annas & M.A. Grodin, eds., *The Nazi doctors and the Nuremberg Code* (New York and Oxford: Oxford University Press, 1992).

[5] See Katz, *ibid.* at 305. The *Nuremberg Code* is also printed in R.J. Levine, *Ethics and Regulation of Clinical Research*, 2d ed. (Baltimore-Munich: Urban & Schwarzenberg, 1986) at 425–426.

subjects in that country, who insisted on rules for the review of research by committee. These rules, derived from a system of in-house review within the USA's National Institutes of Health, included prior review by committee of any research proposed with human participants before that research could be eligible for public funding. Essentially in-house peer review became the model.[6]

## SELF-REGULATION OR INDEPENDENT REVIEW

There are concerns about whether professional associates of a researcher can be relied on to review without bias. This is the issue of self-regulation or independent review. The major issue concerning review by one's peers is the same as for any process of self-regulation: "can you trust the advertising agents, television broadcasters, the police, or any group to regulate themselves and put the interests of others ahead of their own interests?" In this context, can researchers or staff of research institutes be trusted to put concerns about the welfare of the human participants ahead of the "need for research" and collegial loyalties? After all, a requirement for review was part of recognising that medical researchers, unlike medical practitioners, could not be trusted to put the interests of their patients ahead of their interests in research. Why then should we suppose that colleagues from the same institute, with the same biases toward research, can be relied on to give an impartial review? Is this putting Dracula in charge of the blood bank? In New Zealand an ethics committee, composed of only peers of the researcher, was found by an official inquiry to have approved unethical research which led to the deaths of women with cervical cancer. These were deaths that could have been prevented with adequate review.[7] The chairman of that committee was subsequently found guilty of professional misconduct.[8]

The position taken by international and national codes of ethics is that the interests of the research subject must be paramount. For example, the *Declaration of Helsinki* states that: "[c]oncern for the interests of the subject must always prevail over the interest of science and society".[9] If the interests of the research institute

---

[6] See McNeill, *supra* note 1 at 58.

[7] P.M. McNeill, "The Implications for Australia of the New Zealand Report of the Cervical Cancer Inquiry: No Cause for Complacency" (1989) 150 Medical Journal of Australia 264 at 266.

[8] "Editorial: Medicolegal" (1994) 103:901 New Zealand Medical Journal 547 at 547–549. The New Zealand Medical Council found Professor Bonham guilty of "professional misconduct" (in part) for his role in approving the research and "disgraceful conduct" (in part) for his failure to take adequate steps to monitor the trial and to protect the women affected once it became obvious that the trial was ethically unsound.

[9] World Medical Association, *Declaration of Helsinki: Recommendations Guiding Medical Doctors in Biomedical Research involving Human Subjects*. Adopted at the 18th World Medical Assembly in Helsinki in June 1964. Amended at the 29th World Medical Assembly in Tokyo in October 1975; at the 35th World Medical Assembly in Venice in October 1983; and at the 41st World Medical Assembly in Hong Kong in September 1989, amended principle 1.5. See also Australian National Health and Medical Research Council, *Statement on Human Experimentation and Supplementary Notes* (Canberra: NHMRC, 1992), supplementary note 1.6 as reproduced in McNeill (1993), *supra* note 1 at 261.

dominate on a review committee it is possible that "concern for the interests of the subject" may be over-shadowed by other needs, such as the need to meet research deadlines and to ensure a flow of research money to the institution. In these circumstances review by peers can be mere "window-dressing" which (like the ethical codes of some professions) serves the predominant role of deflecting public criticism without significantly altering the behaviour of individuals within the profession.[10] The United Kingdom Royal College of Physicians' guidelines state that committees should "provide reassurance to the public".[11] That is a reasonable objective if the public can be assured that the review process is effective in protecting review participants. It is a concern however if reassurance of the public, rather than sound review, is the *raison d'être* of a committee.

Given this concern it is not surprising that rules governing review committees in most countries have added various people to the in-house review team. In the United States members of institutional review boards (or IRBs as they are known) were originally described as "associates" of the researcher.[12] Subsequent regulations required the addition of a non-scientist, (such as a lawyer, ethicist, or member of the clergy) and a member not otherwise affiliated with the institution (who became known as the community member).[13] The UK College of Physicians' guidelines added a nurse and lay members to the basic peer review team.[14] The Australian National Health and Medical Research Council initially required "at least one member not associated with the institute" in addition to the "peer group" and subsequently added a minister of religion, a lawyer and two lay members.[15]

The question is whether the addition of community members (and other members from outside the research institution) ensures impartial review. In Australia early defences of the system of ethics review by committee emphasised the importance of community members contributing to decision-making within these committees and suggested that this ensured unbiased review.[16] Findings from research into Australian ethics committees, however, reveal that the additional members made little difference on committees, constituted by the research institute, with a staff member as the chair person and members of staff in the majority. Our research found that community members ("lay members" as they are called) were in the minority, had the least to say within their committees and were regarded (by themselves and

---

[10] Moore and Tarr use the term "window-dressing" in relation to voluntary "self-regulation". A.P. Moore, & A.A. Tarr, "Regulatory Mechanisms in Respect of Entrepreneurial Medicine" (1988) 16:1 Australian Business Law Review 4 at 9.

[11] See paragraph 2.1 of The Royal College of Physicians of London, *Guidelines on the Practice of Ethics Committees in Medical Research involving Human Subjects*, 2d ed. (London: College of Physicians, 1990) at 3 [hereinafter *UK College of Physicians' Guidelines*].

[12] W.J. Curran, "Governmental Regulation of the Use of Human Subjects in Medical Research: The Approach of Two Federal Agencies" (1969) 98:2 Daedalus 542 at 577. Reprinted in P.A. Freund, ed., *Experimentation with Human Subjects* (New York: George Braziller, 1970).

[13] See Levine, *supra* note 5 at 398.

[14] See *UK College of Physicians' Guidelines*, *supra* note 11 at 9–10.

[15] See McNeill (1993), *supra* note 1 at 70–73.

[16] R. Lovell, "Ethics at the Growing Edge of Medicine — The Regulatory Side of Medical Research" 3:9 Australian Health Review 239 at 240.

by other members of the committees) as the least influential.[17] These Australian findings supported earlier findings in the United States in which non-scientists "reported themselves to be less active and less influential than other IRB members".[18] It was clear that the power in most committees lay with the researchers and members from the institute.

We also found examples of community members on some Australian research ethics committees being ex-employees of the research institute, next-door neighbours of the committee chairman, and the hospital surgeon's wife. Most commonly community members were appointed by the chairperson from people known to committee members and thought by them to be a "suitable person". It is likely that appointments made on this basis will be from people who are similar, in terms of social standing and education, to other members of the committee. It could be expected that the primary allegiance of people appointed on this basis would be to the chairperson (or other members of the committee). A consequence is that these members are more likely to align themselves with institutional perspectives and goals rather than the protection of research participants. As Evans and Evans have written, "the 'tokenism' of the resultant situation is obvious".[19] There is evidence, even now, that committee review may not be effective. Institutional Review Boards in the United States are considered by Annas to have "betrayed the research subjects they are charged to protect".[20]

Evans and Evans consider that one of the prime requirements for good review is *independence* of the committee. Whilst allowing that committees need to include members with professional expertise, these authors emphasise the importance of the members with professional expertise keeping separate their professional insights and "any mutual interests of researchers and reviewers".[21] Even if we admit that this separation is possible there still remains the difficulty that committees are composed of members appointed according to very different criteria: those appointed for their professional expertise and those included simply as members of the community.

## STILL NEEDED: A RATIONALE FOR COMMITTEES

More than 20 years ago Veatch made a fundamental criticism of the review of research. He considered that there was no consistent rationale for committees comprised of professional people and community members. In his view such

---

[17] P.M. McNeill, C.A. Berglund & I.W. Webster, "How Much Influence do Various Members Have Within Research Ethics Committees?" 3 Cambridge Quarterly of Healthcare Ethics: The International Journal for Healthcare Ethics Committees 522.

[18] National Commission for the Protection of Human Subjects of Biomedical and Behavioral Research, *Report and Recommendations: Institutional Review Boards* (Washington, D.C.: U.S. Government Printing Office, 1978) at 58; National Commission for the Protection of Human Subjects of Biomedical and Behavioral Research, *Appendix to Report and Recommendations: Institutional Review Boards* (Washington D.C.: U.S. Government Printing Office, 1978) at 1–60 to 1–63.

[19] D. Evans & M. Evans, *A Decent Proposal: Ethical Review of Clinical Research* (Chichester: John Wiley & Sons, 1996) at 110.

[20] G.J. Annas, "Questing for Grails: Duplicity, Betrayal and Self-Deception in Postmodern Medical Research" (1996) 12 Journal of Contemporary Health Law and Policy 297 at 324.

[21] See Evans & Evans, *supra* note 19 at 122.

committees derived from two different and incompatible models of review: the "professional review panel" and the "jury model". The professional review panel is composed of people with appropriate expertise to assess a problem presenting complex or technical issues. A jury, on the other hand, is presented with evidence (including expert opinion where necessary) and asked to make a decision on the facts from a common sense perspective. Relevant expertise is not necessary and indeed may well disqualify someone from serving on a jury.[22] From this perspective, ethics review committees fall between those two models in that some members are appointed for their expertise and some are appointed to represent the community common-sense perspective. This analysis helps to explain why it is that community members of committees (and ministers of religion) often feel overawed, or not qualified to speak, in the company of experts[23]. The perspective is also consistent with our finding that lawyers on committees, unlike community members and ministers of religion, were seen as both important and influential. They have an expertise which was perceived as relevant to the review of research.[24]

Veatch had mentioned a "representative model" but had not developed the idea.[25] My suggestion was that committees be composed of representatives of the two major interests in review of research: representatives of researchers and representatives of the human participants of research.[26] The rationale for the suggestion was that the predominant role of a committee is to balance between the need for research and the need to protect research participants. It is not the community as such that needs representation in these discussions but members who can adequately represent and consider the two major interests in achieving a balance between legitimate community concerns. Furthermore, the inclusion of people from the community with no relevant expertise had been shown to be ineffective (as is described above).

Commentators however have rejected the suggestion that ethics review committees be composed of representatives, at least in the sense of representing a particular constituency or community group. The perception is that this could be divisive. They are concerned that the agenda of external groups would intrude into a discussion of the ethics of research. Evans and Evans for example consider it important to avoid the "representation of sectional interests".[27] A more recent report on Australian

---

[22] R.M. Veatch, "Human Experimentation Committees: Professional or Representative?" (1975) Hastings Center Report 31 at 31.

[23] See McNeill *et al.*, *supra* note 17 at 526–528. See also McNeill, *supra* note 1 at 91–94. Evans and Evans quote W.H. Auden as saying, "[w]hen I find myself in the company of scientists, I feel like a shabby curate who has strayed by mistake into a drawing-room full of dukes". Evans & Evans, *supra* note 19 at 136.

[24] See McNeill *et al.*, *ibid.* at 528. See also McNeill, *ibid.* at 91–92.

[25] Veatch described a representative as requiring "a generalized skill not found in the jury member. The representative ought to be skilled in perceiving and communicating the views of his constituents". Veatch, *supra* note 22 at 31.

[26] See McNeill, *supra* note 1 at 207–216.

[27] Evans & Evans, *supra* note 19 at 113. They also emphasise "the importance of individual members' freedom from having simply to champion the prior and already laid-down concerns of some constituency of professional opinion, which would of course reduce the member to being little more than a delegate with a mandate". *Ibid.* at 108. On the same page they refer to members speaking freely "without confining themselves to prepared positions".

review committees commissioned by the Minister for Health took the view that "members must be impartial and must not see themselves as representative advocates for a particular group".[28] This was a reiteration of the Australian NHMRC guidelines which state that "[m]embers shall be appointed as individuals for their expertise and not in a representative capacity".[29] The concern in the Australian Report on Committees seems to be that committees will become bogged down in "adversarial debate".[30] An explicit rejection of my proposals for representation was offered in the Australian Social and Behavioural Research Report.[31] Each of their objections boils down to a concern about representatives from particular nominated groups within the society: either in terms of finding a suitable representative from a community group, or the narrowing of interest that may occur with a "professional" representative.[32] As I have stated in my book, I believe this concern is based on "stereotypes of 'single-issue' protest groups taking extreme action to gain media attention".[33] It may also be based on observation of parties opposed in legal disputes. There is increasing evidence, through "conflict resolution" processes, that parties with vested interests are very capable of achieving equity between their claims.[34] Discussion between representatives of relevant interests need not be adversarial and, properly conducted, is a way of ensuring that all the relevant concerns are considered.

The prohibition against representation in itself introduces a bias in that it does not inhibit research members of the committee from representing the concerns of their research colleagues, the institution, and their professions. It does however effectively disempower the community members from a potential source of support in representing the interests of the human participants of research. Protection of the human beings who volunteer for research should be the main concern of a

---

[28] D. Chalmers *et al., Report of the Review of the Role and Functioning of Institutional Ethics Committees: A Report to the Minister for Health and Family Services* (Canberra: Department of Health and Family Services, 1996) at 44 [hereinafter *Australian Report on Committees*]. It is also stated that "there should be no need for a separate patient advocate or research representative on the committee". *Ibid.* at 47.

[29] National Health and Medical Research Council [NHMRC], *Statement on Human Experimentation and Supplementary Notes* (Canberra: NHMRC, 1992), note 2.4 (v). Similarly the *UK College of Physicians' Guidelines* state that "it is essential that members should serve on the Committee as individuals and not as delegates taking instruction from other bodies or reporting to them". *UK College of Physicians' Guidelines, supra* note 11 at para. 6.10. Similarly the United Kingdom Department of Health Local Research Ethics Committees' guidelines state that: "LREC members are not in any way the representatives of [any] groups. They are appointed in their own right, to participate in the work of the LREC as individuals of sound judgement and relevant experience". United Kingdom Department of Health, *Local Research Ethics Committees* (London: HMSO, 1991) at paragraph 2.6 [hereinafter *UK Department of Health Guidelines.*]

[30] *Australian Report on Committees, supra* note 28 at 44.

[31] S. Dodds, R. Albury & C. Thomson, *Ethical Research and Ethics Committee Review of Social and Behavioural Research Proposals: Report to the Department of Human Services and Health* (Canberra: Australian Government Publishing Service, 1994) at 46 [hereinafter *Australian Social and Behavioural Research Report*].

[32] *Ibid.*

[33] McNeill, *supra* note 1 at 202.

[34] H. Cornelius & S. Faire, *Everyone Can Win: How to Resolve Conflict.* (Sydney: Simon & Schuster, 1989).

committee but effective representation is undermined by assuming that all members (even those with a clear bias toward research) are capable of countering that bias and giving a greater weight to the concerns of the human participants.[35] There is an inconsistency in the logic of (1) acknowledging the need for inclusion of community members to counter any bias; and (2) suggesting that researchers and members from the research institute are just as capable of representing the interests of the subjects.

Given the almost universal reaction to the argument for representation of particular groups however, I concede (as I have previously) that representation of the interests of the human participants of research is difficult and that a practical solution may be to nominate appropriate "knowledgeable people" who can represent the interests of potential research participants without being directly accountable to a particular community group.[36] This is to accept a modified notion of representation which consists of appointing community members from nominations of suitable people who would inform themselves of concerns of potential subjects of research. Ideally, they would be nominated by some body from outside of the research institute or, at the very least, be appointed from responses to open advertisement by the other community members of that committee.

One remaining concern is with the way in which the role for the community member is described. It is often stated that it is the primary function of *all* committee members to give precedence to protection of research participants.[37] I accept the wisdom of this mandate. Indeed researchers, and other members with technical expertise on the committee, may well foresee potential harms that community members fail to anticipate. The "expert" members will at times be the best advocates for protection of the research participants. However this observation should not mislead us into assuming that members from the research institute are able to set aside any bias toward research. If this was the case there would be no need for community members on review committees. The brief history with human experimentation (given above), and observation of self-regulation in other industries, indicates that this is not a facility that can be relied on. For this reason I do not concede that the roles of the community members and professional experts on committees are the same. Representing the concern for protection of research participants (free from any bias toward research interests) is the main reason for inclusion of "community members". This needs to be explicit. Without that function they have no clear role, no recognised expertise, and are effectively disempowered in relation to professional and institutional members of the committee.

The arguments against representation of the interests of the participants are based on concerns about representatives from specific community groups. If these

---

[35] McNeill, *supra* note 1 at 195–203.

[36] *Ibid.* at 210.

[37] The *Australian Report on Committees*, *supra* note 28 at 44, states that "all members must look primarily to the welfare of the research subjects". Whilst this statement is unobjectionable as an ideal, any implication that suggests members from the research institute can overcome a bias toward research, and represent the interests of the research participants as well as community members, must be rejected.

concerns are accepted as valid, and a requirement for a specific constituency is dropped, there still remain very good reasons for membership of review committees based on representation (in the more general sense). I conclude this section by summarising and restating the rationale offered in answer to Veatch's critique. The rationale is based on recognising that the main function of a research ethics review is to balance between a society's interest in promoting research and society's interest in protecting human beings who may be harmed in the course of that research. Essentially then, committees need a balanced representation of members with research expertise and members with the necessary expertise to protect subjects of research. The community members should be free of any apparent bias toward research and be given the primary role of representing the interests of potential research participants. Given the international acceptance of the principle that protection of research participants should predominate over the research interests, there should be a majority of such members. As an additional safeguard of this important function, the chair person should be elected from one of the community members.

## TRENDS TOWARD STRONGER COMMUNITY REPRESENTATION

Research ethics review is an evolving field. In every country the basic peer review model adopted from the United States has been applied and developed in different ways. There are some commonalities in these developments, however. The question considered here is whether any of these developments are consistent with the rationale for review I have outlined above. This is not to suggest that developments within various countries are based on this rationale.[38] It is simply to look for any pattern which may be consistent. What I find in a number of countries is a trend toward stronger community representation on committees which is consistent with the rationale I have offered. This is evident in a number of ways:

(1) There is a trend toward recognising the need for a balance of community members (or at least members from outside of the institution) and members from the research institute. Denmark has long required an equal number of "lay members" and "medical/scientific" members.[39] New Zealand now requires that the "lay membership" should "approximate one half of the total membership".[40] In Australia a recent report to the Minister of Health notes that community members are often significantly out-numbered by institutional and medical committee members. It recommends that "not less than half the committee should consist of non-medical members from outside the institution".[41] Annas puts the case more strongly in saying that United States

---

[38] In fact relevant authorities in many countries have explicitly disavowed representation as the reason for the changes (as is stated above).

[39] McNeill, *supra* note 1 at 102.

[40] New Zealand Department of Health, *Standard for Ethics Committees Established to Review Research and Ethical Aspects of Health Care* (Wellington: New Zealand Department of Health, 1991) at para. 3.2 [hereinafter *NZ Standard*].

[41] *Australian Report on Committees, supra* note 28 at 46.

Institutional Review Boards [IRBs] "should be radically overhauled". After reviewing reports of radiation experiments on human beings and clear lapses in the functioning of IRBs, he argues that there is a need to "restructure IRBs so that their role is to protect the subjects of research (not the researcher)" and concludes that the minimum need is "democratising the IRBs by requiring a majority of members to be community members".[42]

(2) Recent descriptions of community members make it obvious that these members are not chosen as ordinary or typical members of the community. For example the UK College of Physicians' guidelines state explicitly that lay members should be "persons of responsibility and standing".[43] The recent review in Australia recommended that lay members "should be respected by the community" and they should have "the ability to represent the community (with current or recent community involvement) and to mirror community standards".[44] This is clearly not the same as choosing members for a jury. These people are chosen in a representational capacity. It is acknowledged that this is representation without the members having a specific constituency from which they are nominated and to which they report.

(3) There is a trend toward the appointment of stronger community representatives. These are clear indications that community members ought to be appointed from people who can withstand any pressure from other members to approve research with which they disagree. For example the UK College of Physicians' guidelines state explicitly that lay members should be "persons of responsibility and standing who will not be overawed by medical members". They also state that "individuals who are acquiescent and may be thought likely to give automatic approval are also not suitable members".[45] The Australian Report on Committees recommended that lay members should be "articulate, curious and able to advance an argument".[46]

(4) There is a recognition, at least from some bodies, that the special role of the community members is to be free of any bias and to be focussed on the research participants. For example, the recent review in Australia recommended that "laymembers should hold a non-institutional viewpoint and research subject focus".[47]

(5) There is an increasing recognition of the need for specific representation by particular community groups. The CIOMS guidelines, for example, offer the following recommendation:

> Committees that often review research directed at specific diseases or impairments, such as AIDS or paraplegia, should consider the advantages of including as members or

---

[42] Annas makes other suggestions also, including "opening all meetings to the public". Annas, *supra* note 20 at 324.

[43] *UK College of Physicians Guidelines*, *supra* note 11 at para. 6.5.

[44] *Australian Report on Committees*, *supra* note 28 at 46.

[45] *UK College of Physicians Guidelines*, *supra* note 11 at para. 6.3 and 6.5.

[46] *Australian Report on Committees*, *supra* note 28 at 46.

[47] *Ibid*

consultants' patients with such diseases or impairments. Similarly, committees that review research involving such vulnerable groups as children, students, aged persons or employees should consider the advantages of including representatives of, or advocates for, such groups.[48]

Similarly the Australian Report on Committees recommends that where a research ethics committee is in a locality which includes "ethnic" or Aboriginal populations, community members may be appointed from amongst those populations especially where the committee considers research "principally carried out amongst those groups".[49] In those circumstances the Australian Social and Behavioural Research Report also suggests the inclusion of Aboriginal people or people from "non-English speaking backgrounds".[50] The New Zealand Standard requires those forming review committees to "take account of the need for . . . knowledge and experience of . . . women's health, patient advocacy and Tikanga Maori".[51]

These guidelines (or recommendations) for review committees all require representation of the interests of particular (potential) research participants. The Australian Report on Committees gives a rationale in terms of appointment of "local" members. However the effect is toward greater representation of people who may participate in research.

(6) There is a trend for chairpersons (presiding members) of committees to be appointed from the community members or from outside the institution. New Zealand requires that the committee "shall be chaired by a lay member".[52] The recent suggestions in Australia do not restrict the chairperson's role to a lay or community person but suggest that the "[c]hairperson must have no involvement in the conduct or supervision of research considered by the committee". The report is a little more equivocal than the New Zealand Standard in suggesting that it is "desirable that the Chairperson not be employed or otherwise directly connected with the institution".[53] The UK Department of Health guidelines require that the chairman or the vice-chairman be a lay person.[54]

In summary therefore, there are international trends toward a stronger representation by community members. This consists in a growing trend for equal (or greater) numbers of community members on committees; chairpersons to be elected from

---

[48] *CIOMS Guidelines, supra* note 2 at 40–41.

[49] *Australian Report on Committees, supra* note 28 at 46. Further on the same page it is stated that "due regard should be paid to the ethnic backgrounds of the research subjects dealt with by the researchers". Read for "ethnic" (in common Australian parlance) "non-English speaking" immigrants to Australia or people who strongly identify with their "non-English speaking" origins.

[50] *Australian Social and Behavioural Research Report, supra* note 31 at 47.

[51] *NZ Standard, supra* note 40 at para. 3.7.

[52] *Ibid.*

[53] *Australian Report on Committees, supra* note 28 at 45.

[54] *UK Department of Health Guidelines, supra* note 29 at 8. The same requirement is included in the "Standard" for local research ethics committee issued on behalf of the Department of Health: C. Bendell, *Standard Operating Procedures for Local Research Ethics Committees: Comments and Examples* (London: McKenna, 1994) at 15.

the community members; the selection of people "of standing" as community members who are more capable of withstanding pressure from other committee members; and suggestions that community members represent research participants in general and particular groups of potential research participants (where research with those groups is likely to be considered by the committee). These trends are consistent with a rationale of representation of the interests of the research participants. The question asked at the outset was "Who should decide on what is ethical in science and research?" The answer being given internationally is: people with research expertise and an equal number of people who are independent and able to represent research participants.

## MULTI-LOCATION RESEARCH

The question "Who decides?" has been critical in relation to debates on the review of multi-centre research also. The term "multi-centre" research is a term used to describe drug trials conducted in a number of different centres. The term has also been extended to include population studies and epidemiological research conducted in different locations. These locations may, or may not, include centres. For this reason, Evans and Evans suggest "multi-location" research as a more appropriate term and I have adopted it here.[55]

The major problem for researchers wishing to conduct research in multiple locations is the work required in gaining approval from numerous research ethics committees. Each has its own process for application and approval, its own application forms, and typically, the decisions made across a number of committees are inconsistent with each other. Modifying a program to meet the requirements for one committee may not meet the requirements of another and on occasions it becomes impossible to reconcile the differences. Effective results from a national program may be invalidated by inconsistent modifications made to satisfy various local research ethics committees (or LRECs as they are known in the U.K.).

A further difficulty for researchers is the time, effort and cost of communicating with numerous committees and supplying essentially the same information in many different forms. Some committees also insist on having researchers present themselves for discussion and clarification. Whilst this may be reasonable, from the perspective of one committee, such requirements become onerous in mass. The time needed for approval, in addition to developing the proposal, applying for funding, conducting the study, and writing reports, can make it impossible to complete a research program within the time and resources available to many researchers. Researchers therefore, especially those testing new drugs and epidemiologists, have been keen to have the review process modified to make it less demanding.

On the other hand, from the committees' point (or points) of view, it is imperative that every proposal is considered thoroughly. Each committee has responsibility (including legal responsibilities) for approval of projects conducted within its institution or region. Committees usually take the view that they cannot legitimately

---

[55] Evans & Evans, *supra* note 19 at 169.

delegate this responsibility to any other body. From the perspective of an individual committee, the decision of another committee may well have overlooked something of ethical significance. There are also local differences (for example in the nature of the local population or in the nature of the resources available) which require modifications to programs before they can be approved as ethical within that institution or region.

In practice however, committees often find themselves faced with a "multi-centre" drug research proposal that has been developed by a major international pharmaceutical company. They are told that no variations will be tolerated and that if the institution is to be a part of the study, and receive benefits in academic and financial terms that will flow from the study, then the committee must approve it in the form in which it is presented. This puts enormous pressure on committees, which may approve a program notwithstanding misgivings.

To deal with these difficulties various suggestions have been made in different countries. Usually these propose either a national committee to review all multi-location research, or a "lead committee" whose consideration and decision is recommended as a lead to all other committees that might consider the research. Until recently neither of these proposals has been successfully implemented on a wide scale although individual committees have taken account of decisions of other committees on an *ad hoc* basis.

In the United Kingdom in 1992, the Department of Health commissioned a group from the University of Wales Swansea, to study these problems. This group made a number of proposals but suggested, in essence a "three tiered" approach. This included researchers of multi-location research applying to a central advisory committee that would handle all correspondence, and give "general scientific and ethical review of the protocol on an advisory basis". The next level was for proposals to be sent to a regional "lead committee". Once approved at this level, researchers were free to approach local committees, which could accept or reject the proposal for their region. Modifications could be suggested by local committees but only those modifications of a minor nature as appropriate for the research to be conducted within that locality. Local committees could not suggest major modifications to the study that would affect the overall design.[56]

What was subsequently agreed (late in 1995) between the Department of Health and representatives of LRECs was the creation of 12 regional councils with representatives of each of the local committees on each council. Once a program is approved by the regional committee, individual LRECs could opt out of the study or suggest minor amendments to suit local conditions, but no local committee could require changes to the design and conduct of the research program.[57] Whilst this may

---

[56] *Ibid.* at 187.
[57] Agreement reached on 24 October 1995. Information supplied by Mr. Clive Marritt, Department of Health in interview (22 September 1995). See also C. Marritt, "Review of Multi-location Research Proposals, A Training Conference on 'Ethical Review of Clinical Research'" (Robinson College, Cambridge, 26 September 1995). Agreement on the plan confirmed by Dr. Frank Wells, Association of British Pharmaceutical Industry, Sydney (October 1995).

look like a victory for centralised decision-making, there are two points to be observed about this scheme: (1) Agreement was achieved only by recognising the entrenched commitment to local decision making. The reason that regional committees were accepted is that their membership includes representatives from each of the LRECs in the region. In other words LRECs still retained some control, at least through representatives. (2) The conditions in which this compromise was achieved are peculiar to the United Kingdom. There is a close working relationship between representatives of the large pharmaceutical industry and the Department of Health.[58] LRECs are part of regional health authorities and are within the National Health Scheme, which is funded by the Department of Health. This gives the Department more influence than comparable bodies in other countries.

In Australia similar proposals have been rejected in the recent Australian Report on Committees.[59] The review panel had considered a number of proposals similar to those put forward in Britain. Whilst the Report suggests improved communication and cooperative arrangements between committees, and encourages administrative consistency, as well as the sharing of review of the scientific merit of studies, it did not favour the establishment of regional committees or a national committee and it did not favour delegating responsibility to another committee.[60] Essentially there is a commitment to individual institutions reviewing research within their own area of responsibility. Similarly in both Canada and the United States commentators have observed that the likelihood of individual committees giving up responsibility for the review of multi-location research to another committee or to a national or regional committee outside of the institution is slight.[61] Even though review committees within the National Health Scheme in the United Kingdom have accepted regional committees of review for multi-location research proposals, this scheme, which limits the responsibility of local committees, has to be seen as an exception in the general trend to maintain local responsibility for decision-making.

In practice there are no hard and fast rules by which one can determine that a research proposal is (or will be found to be) ethical. Each proposal has to be considered and approved by a committee within the relevant locality. The ethics of research and bioethics in general are often discussed in relation to principles or decision-making models. What occurs in practice, however, is that most decisions are made by people involved together with community representatives. These are judgments that result from a process of discussion and negotiation between different perspectives at a local level. Although regional committees, about to be established in the United Kingdom, appear to be contrary to that observation a closer analysis suggests that this exception underscores the general rule.

---

[58] For example the Association of British Pharmaceutical Industry funded workshops conducted by the Department of Health and the production of manuals for "Standard operating procedures for local research ethics committees" (Bendall, *supra* note 54). Information supplied by Mr. Clive Marritt, *ibid.*

[59] *Australian Report on Committees*, *supra* note 28 at 21–24.

[60] *Ibid.*

[61] Interviews with Professor John Last, University of Toronto (5 October 1995) and Professor Robert Levine, Yale University (11 October 1995).

## EXTENSION OF REVIEW TO SOCIAL SCIENCES AND OTHER RESEARCH

The original requirement for committee review of research took effect in the United States in 1966. It was made in general terms inclusive of all research but in practice it applied to medical research funded through the Public Health Service.[62] However, the Surgeon-General made it clear that requirements for consent, confidentiality of information and protection from misuse of information applied to all research whether behavioural, social or medical.[63] Publicity given to the Wichita Jury Study (a covert study of juries) in the 1950s and Stanley Milgram's studies of obedience (involving deception of research participants) in the 1960s demonstrated that social science research could also be unethical.[64] Consequently in the United States there was a greater awareness of the need to extend review of research to include all research with human participants.[65] The review process in the United States is more inclusive of other research than systems of review that developed in other countries and the rules, as they developed, were less obviously slanted toward medical research. The systems of review in the United Kingdom, Australia and New Zealand were developed by medical bodies, using medical terminology and the resulting review committees were often in hospitals and medical research institutes.[66] A study of IRBs in the United States in 1978 showed that the ratio of medical to non-medical studies was 2:1 whereas in Australia for example, a similar study in 1990 showed the ratio of medical to non-medical studies was 4:1.[67] Since that time one of the trends in Australia and New Zealand has been toward the development of systems of review which better accommodate non-medical research. The rationale for the need to review research, offered earlier in this paper, is that researchers' interests in the research process may bias them in ways that prevent them from adequately protecting from harm the participants in their research. This is true of non-medical research also, even though the harms may be of a different nature.[68]

Canada and Norway are the only countries to have developed separate systems of review for medical and social and behavioural research. The Norwegian system of review of research was established by three national committees with similar terms

---

[62] It required research proposals to undergo "impartial, prior peer review". McNeill, *supra* note 1 at 58.

[63] R.R. Faden, & T.L. Beauchamp, *A History and Theory of Informed Consent* (New York and Oxford: Oxford University Press, 1986) at 210.

[64] *Ibid.* at 172–176. For an account of other unethical social science studies see also R. Homan, *The Ethics of Social Research* (London and New York: Longman, 1991).

[65] The Surgeon-General was deliberate in choosing wording to include all research. Faden and Beauchamp, *ibid.* at 210.

[66] In Australia for example, the NHMRC simply added the words "investigations on human behaviour" to a Statement that was obviously written to regulate medical research. See also *Australian Social and Behavioural Research Report*, *supra* note 31 at 20–26, which makes the same point. In the United States the National Commission had given a great deal more thought to the need for regulations governing research to apply to medical and other research. This is reflected in language more appropriate to social science and other research as well as in the provisions for review committees (including membership). McNeill, *supra* note 1 at 72.

[67] McNeill, *ibid.* at 87.

[68] *Ibid.* at 83. See also *Australian Social and Behavioural Research Report*, *supra* note 31 at 24.

of reference and between them they are responsible for research in medicine; social sciences and the humanities; and law and theology.[69] In Canada the Medical Research Council of Canada set up a system of review for medical research and the Social Sciences and Humanities Research Council of Canada developed a separate system with quite different procedures and concerns.[70] In 1992 and 1993 the Medical Research Council held meetings to consider rewriting its guidelines to include a broader range of health research (including biomedical and population-based research).[71] In late 1993 it was decided to invite representatives of the Social Sciences and Humanities Research Council of Canada, and the Natural Sciences and Engineering Research Council to form a tripartite committee to develop a common set of guidelines for all research with human participants. Each of the latter two Councils added two members to the Medical Research Council group already in existence and this tripartite committee met for the first time in mid-1995. Its objective was to find a common approach to the regulation of research that would require researchers and their organisations to be more accountable for conducting research with human subjects ethically. This tripartite group is currently working towards a policy that will be appropriate for regulating all research with human participants within Canada. It is anticipated that a final document may be ready by the end of 1996.[72] Canada then, is moving in the direction of other countries, including the United States and Australia (as follows), in developing common rules.

An Australian Department of Federal Government (Human Services and Health) funded a study on the review of social and behavioural research proposals. The Report noted that "there was widespread concern" about the lack of expertise on research ethics committees and the lack of understanding of social research methodology and recommended that "[n]o one ethical, sociological, or ideological perspective should be exclusively represented" on a committee and that a "diversity of perspectives" be represented "so as to pitch one view against another".[73] More specifically it recommended including members with expertise relevant to studies being considered by a committee. This could require expertise in epidemiology, social research, and behavioural research.[74] Another report, the Australian Report on Committees, also noted that concerns had been expressed about the need to clarify Australian guidelines for non-medical research.[75] Given these influential reports, it can reasonably be expected that the relevant national committee in Australia will be

---

[69] McNeill, *ibid.* at 103–104.

[70] *Ibid.* at 81–82.

[71] The purpose was also to accommodate changes in fields such as genetics; and to require greater accountability in the review of research.

[72] Information supplied in interviews: Dr Francis Rolleston, Medical Research Council; Ms Nina Stipich and Mr Ron Clarke, Social Sciences and Humanities Research Council of Canada, and several members of the National Council on Bioethics in Human Research. (5 October 1995).

[73] *Australian Social and Behavioural Research Report, supra* note 31 at 45 and 46. McNeill, *supra* note 1 at 215.

[74] *Australian Social and Behavioural Research Report, ibid.* at 47.

[75] *Australian Report on Committees, supra* note 28 at 18 and 58.

prompted to press for changes to the national guidelines covering the review of social sciences and other research.[76] It could be expected, at the very least, that there will be changes to rules governing membership of ethics review committees to provide for relevant expertise (rather than specifically requiring medical research expertise) and changes to the guidelines so that the review process better accommodates social, behavioural and other research.

A question asked at the beginning of this chapter was: "How should scientific activity be allowed to proceed?" It is apparent that in the United States the answer, for research with human participants, is that *all* research should be subject to regulation and review. The approach has been to require that any research funded through any of 16 Federal Government agencies must be approved by an IRB.[77] The review processes in Canada and Norway are different but the answer is essentially the same. In Canada for example, research in the humanities and social sciences requires approval of a committee. The indications are that Canadian review will be covered by one policy statement in the near future. In Australia also, it is recognised that the review should be extended to include all psychological, sociological, anthropological and other research. The answer, then, to the question, is that *all* research that requires the participation of human beings should proceed only after a thorough consideration of the ethics of the study by a properly constituted committee. In some countries this represents a considerable widening of the net of research to be reviewed and an increasing workload for committees.

## CHALLENGING THE ASSUMPTIONS

At the beginning of this chapter it was said that the inclusion of a wider field of disciplines into the review process has resulted in challenges to the basic assumptions within medicine and the "natural sciences". A major challenge to the limited assumptions on which review has been based has come from anthropological research. Indigenous peoples have questioned the right of researchers to conduct research in accordance with academic traditions which are offensive to them. The taking of articles, body samples, or photographs, for example, is offensive to indigenous people in many parts of the world. The publication of details of secret rites and the discussion of the lore of particular tribes in public forums can transgress their rules. Furthermore any re-interpretation of behaviours, values, and beliefs within a reductionist framework (such as Social Darwinism) can be both demeaning and damaging to indigenous cultures. Even the publication of health statistics, disease patterns, educational standards or scores on intelligence tests can be damaging to the culture of the group identified in those reports. Much research work has proceeded on assumptions of superiority of the academic world-view, with little credence or respect given to the world-view of indigenous people. Researchers

---

[76] The relevant committee is the Australian Health Ethics Committee of the National Health and Medical Research Council. Its statutory role includes calling for submissions and revising guidelines for research.

[77] McNeill, *supra* note 1 at 64.

have been accused of giving little or no thought to benefiting the people studied from their point of view. From an indigenous perspective such research is exploitative and harmful.

In Australia a number of these concerns were published in a report on Aboriginal research.[78] As a result the National Health and Medical Research Council published guidelines for research review committees suggesting *inter alia* that committees ensure that researchers have properly consulted with Aboriginal communities before the committee approves any research, that there is evidence of community involvement (such as paid work for Aborigines), and that researchers discuss and negotiate issues concerning publication and ownership of data.[79] Whilst these principles of inclusion, consultation and negotiation have thus far been limited to a guideline for research on indigenous people in Australia, there is no defensible reason for that limitation. The cultural differences between "scientifically-educated" researchers and Aborigines (educated in their own culture) may be more obvious, but similar differences can be seen between researchers and many other groups in the Australian multi-cultural society. If researchers' assumptions are different from much of the rest of society (and I am suggesting that they are) then similar provisions are needed for all research to those adopted for research with Aborigines and Torres Strait Islanders. To accept the guideline for indigenous people, consultation is needed at all stages in the development and conduct of research. This includes the process of developing a hypothesis, the choice of research methods and tools, and assumptions about what counts as relevant observations, who is qualified to observe, what are appropriate tools of analysis and what is the status of any conclusions reached. The same point is made in the Australian Social and Behavioural Research Report. It drew attention to the whole culture within which research is developed, funded, conducted and interpreted. There are ethical considerations at each of these stages and between a number of players, including the research institution, the funders of the research, the researchers themselves, the participants in the research, and the people who make use of research findings and reports. The authors of this Report point out that a focus solely on the ethics review committee process overlooks important ethical issues in many areas of research and gives undue responsibility to committees to rectify problems.[80] These challenges, which derive in part from conceptual differences between fields of study, suggest changes to research methodology which are far reaching.

Some of the criticisms of the review process can be accommodated by changes to the guidelines. However they go much further. The critiques attack fundamental assumptions in research itself. For example, the Australian Social and Behavioural

---

[78] NHMRC, Medical Research Ethics Committee, *Some Advisory Notes on Ethical Matters in Aboriginal Research: Including Extracts from a Report of the National Workshop of Research in Aboriginal Health* (Canberra: NHMRC, 1988) See also McNeill, *ibid.* at 179.
[79] NHMRC, *Guidelines on Ethical Matters in Aboriginal and Torres Strait Islander Health Research* (Canberra: NHMRC, 1991). These guidelines were approved by the NHMRC in 1991 but not released until 1995 (after agreement from leading groups representing Aborigines and Torres Strait Islanders).
[80] Australian Social and Behavioural Research Report, *supra* note 31 at 1.

Research Report is critical of the assumptions inherent in the "natural science paradigm" which divides citizens into scientists who are active and observe, and "subjects" of research who are passive and are observed. The Report states that this model assumes an epistemology which is usually not shared by researchers working in fields other than medicine.[81]

At one level it is an issue of language and particular words have come under scrutiny. The term "subjects", for example, is commonly used by medical researchers to denote people on whom research is conducted. However the assumption of passivity is more than a question of words. It deprives researchers of an important source of understanding through a cooperative relationship with people assisting in the research program (as "subjects"). The assumptions are no longer appropriate in anthropological research for example. The obvious cultural bias introduced in such an approach to studying another culture would make any findings themselves suspect and of limited value. Even within medical research these assumptions have come into question. People with HIV and AIDS made it very clear that they were not prepared to comply with conditions of research designed along classical lines. They demanded participation in the process at all levels: including the planning of the research process, the manner in which the research was conducted, and the interpretation of the results.[82] It is my prediction that research in all fields will increasingly require negotiation between those who conduct the research and those who participate as volunteers. These views are not new. They hark back to views expressed more than 25 years ago. Ramsey, for example, described research participants as "joint venturers in a common cause" and Mead (whose views were derived from anthropological research) described "a research participant, who is also a collaborator of the research worker".[83]

At an even more obvious level, research is funded by many different groups for many different purposes. In Australia at least (and I assume that this is true in other countries), politicians intent on making a change (which accords with their ideology) commission research to justify the change. There is a tendency for researchers to find what it is that they are being paid to find. Furthermore the results of research which fail to support a politician's case are ignored or given a limited distribution. This observation also applies to other fields. Research is increasingly a commercial activity with the potential for generating enormous profits for companies and individuals. Given these commitments, it is naïve to support research on the assumption that research *per se* is a good thing. There is a need to ask more basic questions: "What research?" and "Who is the research being conducted by?" and "For what purpose is the research conducted?" We need to ask "Who stands to gain from this project?" and to be sceptical about claims for the potential benefits

---

[81] *Ibid.* at 27–28. In this model any effects of observation are assumed to be small.

[82] McNeill, *supra* note 1 at 182–183.

[83] P. Ramsey, *The Patient as Person: Explorations in Medical Ethics* (New Haven: Yale University Press, 1970) at 5. Extracts reprinted in Katz, *supra* note 4 at 589–91. M. Mead, "Research with Human Beings: A Model Derived from Anthropological Field Practice" (1969) 98:2 Daedalus 361 at 371. See also H. Jonas, "Philosophical Reflections on Experimenting with Human Subjects" (1969) 98:2 Daedalus 219.

arising from the research and claims about the minimal risks of harm to research participants.

An even more challenging critique is offered by "postmodernists" who have called into question both the assumption that there is a reality out there that is amenable to description and (assuming there is) that we are capable of describing it in terms free from cultural and perceptual biases. This has led to a rethinking of the social sciences as a negotiated perspective on human activity and cultures. However the postmodernist challenge is applicable to all scientific activity and research. Gone is the self-assurance that the world is as our scientists tell us it is. What science offers is at best a perspective which may be more useful and appropriate in particular circumstances. The recommendations for treatment, for example, arising from careful research might be preferred to those of a witch doctor on pragmatic grounds: they may be shown to work better. However ontological claims arising from that research have no privileged status.[84] It is likely that "truth" in all spheres involving human beings (including medicine) will be seen as multi-faceted, culturally determined, and the result of a process of negotiation.[85] Research can no longer be seen as scientific activity which is outside the play of power and politics that is a part of every other human activity. If this is the case then the manner in which research is conducted is a part of this negotiation.

One of the obvious implications of these challenges is that there are, and should be, limits to scientific research. This has long been recognised, but the reason for that limitation is not just to avoid harm to research participants. It is a limitation coming from a recognition that research is one (and only one) of many potential values in society. Whilst science and knowledge rest on important values, there is no privileged position for science. It is no longer appropriate to assume that "progress" is a good thing and that scientific research and the acquisition of knowledge is necessarily also a good thing. Hans Jonas issued this challenge in 1969 when he wrote of progress as a "melioristic goal that is in a sense gratuitous".[86] Science and progress need to be weighed in relation to other important values.

## CONCLUSIONS

Internationally there are many changes in the review of research: what counts as "research"; the extension of a requirement for review beyond medical research to include behavioural, social, anthropological and other fields; changes in arrangements for committees themselves, their composition, and whether one committee can act in the place of many. There are other trends also, which have not been explored in this chapter, like the tendency for governments to off-load responsibility

---

[84] Kuhn's work on paradigms, Gödel's theorem, and studies showing the limitation of our perceptual processes have undermined assumptions implicit in the modernist program.

[85] W.T. Anderson, ed., *The Truth about the Truth: De-confusing and Re-constructing the Postmodern World*, (New York: Jeremy P. Tarcher/Putnam, 1995). See also H. Bertens, *The Idea of the Postmodern: A History* (New York: Routledge, 1995).

[86] Jonas, *supra* note 83 at 230.

for evaluating drugs to committees. The idea explored in this chapter is that a deeper analysis suggests that these changes can be seen as a part of a more fundamental change happening in society, a change that manifests as a distrust of authority, and in particular a distrust of science and research as definers of reality and as instruments for progress. The assumptions of modernism, which put faith in reason and in science (untainted by human weakness), seeing them as above culture, have been called into question.

The predominant trend that I am identifying in this chapter is toward a view of science and research as negotiated perspectives between different possible views in society. Ethical review is ultimately a weighing of values and a part of negotiation for a better life. Necessarily, this must take into account the different views of what constitutes the good life in our increasingly multi-cultural societies. The most appropriate people to weigh those different values, and reach a decision about whether a particular research program should go ahead or not, should represent those most affected. This is ethical review in the postmodern world.

One of the consequences is committees less dominated by researchers and staff of research institutes. Another is the extension of the whole notion of research as a negotiated process from the initial conception to publication of results. Another is vigorous resistance to centralised decision-making for multi-location research. Whilst the predominance of science has been challenged before, postmodernism makes it easier to recognise that there is a trend away from accepting any sacrosanct view of reality including those constructed by scientists. Increasingly, the ethics of research and the ethics by which we live our lives will depend on negotiation. Stripped of the certainties of the past, we have to take responsibility for reconstructing the world and finding perspectives we can live with. This is a communal and political activity, in the broadest sense. This model of representative decision-making, along with effective communication, has the potential to accommodate the many ideals, values, beliefs and dreams we bring as human beings.

CHAPTER 14

# MODELS FOR REGULATING RESEARCH: THE COUNCIL OF EUROPE AND INTERNATIONAL TRENDS*

BERNARD STARKMAN**

## INTRODUCTION

Today there is general consensus that research involving human subjects ought to be regulated. The concern is with the form that regulation should take. Since the introduction of federal legislative regulation in the United States, apart from a handful of instances involving clinical drug research,[1] countries have been slow to adopt legislative measures. Criticism of ethical guidelines has usually resulted only in attempts to improve them. Yet at the same time a number of countries have rushed to bring in legal regulation of research on the embryo as part of wider legislation on new reproductive and genetic technologies (NRGTs).[2] In this area, perceived political necessity is overcoming bureaucratic preference for guidelines. But thus far

---

* In memory of Dr. Sydney Segal.
** The author was asked to review international models for the regulation of research. Any views expressed are those of the author, and are not intended to represent any official position of the Department of Justice of Canada.
[1] In Europe: Germany, 1976; Eire, 1987 (amended 1990); France, 1988 (amended in 1990 and 1991); Spain, 1990. I. Dodds-Smith, "Clinical Research" in C. Dyer, ed. *Doctors, Patients and the Law* (Oxford: Blackwell Scientific Publications, 1994) 140 at 164.
[2] In Europe: Spain, 1988; United Kingdom, Germany, 1990; Sweden, 1991; Austria, Denmark, 1992; France, Norway, 1994. In Australia, the States of: Victoria, 1984; South Australia, 1988; Western Australia, 1991.
  The U.S. Congress has annually since 1994 prohibited expenditure of federal funds for human embryo research. There is no federal regulation of *in vitro* fertilization or embryo research. A recent survey indicated that the laws (from 1989 to 1993) of ten States prohibited various forms of *ex utero* embryo research: L.B. Andrews, "State Regulation of Embryo Research" in National Institutes of Health, *Papers Commissioned for the NIH Embryo Research Panel*, vol. 2 (NIH Publication No. 95–3916, September 1994) 297.

there is no general political perception that the public is concerned to prevent future disasters arising out of unethical research with human subjects. This may help to explain why Britain has chosen to introduce extensive legal regulation of embryo research as well as the clinical practice of in vitro fertilization (IVF), while insisting that there should be no formal legal requirements for other research. The same pattern is being followed in Canada.

The reasons for favouring legal regulation of research on human subjects are not essentially different from the reasons for legal regulation of in vitro fertilization and embryo research. Moreover, the choice is no longer between ethical guidelines and legislation. It is between regulatory models that are effective and well adapted to their purpose, and those that are difficult to apply, interpret, or enforce.

## BACKGROUND

The post-Second World War regulation of research involving human subjects was in response to the medical horrors that had taken place under the Nazis. Many will be familiar with the basic principles of the Nuremberg Code of 1949, and with the restatement and extension of these principles in the Declaration of Helsinki, adopted at the 18th World Medical Assembly in June, 1964, and since revised in 1975, 1983, and 1989. While these principles, and the requirements and distinctions they introduced, played an important part in the development of ethical guidelines, they did not have the particularity necessary for the effective regulation of an activity where protection of the subject had become an increasing concern. This was recognized in the late 1980s by what is now the Steering Committee on Bioethics of the Council of Europe, which developed recommendations on medical research on human beings, including research on vulnerable groups.[3] The Council of Europe's work coincided with the growing interest of some European countries, as well as Canada, in legal regulation of biomedical research.[4]

At about the same time, the United States was following up on the change from policy guidelines to legislative regulation, which had taken place more than a decade earlier, by making the regulations of different federal departments and agencies conform to a common policy.

## REGULATION IN THE UNITED STATES

In 1966 the United States Food and Drug Administration (FDA) published a new regulation to clarify the policy on consent for the use of investigational new drugs.

---

[3] Council of Europe, Committee of Ministers, Rec. No. R(90)3 (1990).

[4] For example, Denmark, where Act No. 503 of 24 June 1992, entitled *Act on a scientific ethical committee system and the handling of biomedical research projects* was that country's first legislation on the ethical requirements for biomedical research on human subjects and the bodies for controlling this research. The Act came into force on 1 October 1992.

The Law Reform Commission of Canada recommended amending the *Criminal Code*, R.S.C. 1985, c. C–46 and enacting general federal legislation to recognize the legality of non-therapeutic biomedical experimentation. Law Reform Commission of Canada, *Working Paper 61: Biomedical Experimentation Involving Human Subjects* (Ottawa: Law Reform Commission, 1989).

It used the Nuremberg Code and the Declaration of Helsinki as general guidelines. Domestic disasters were the reason for the introduction of the regulation, the most notorious being the injection of live cancer cells into twenty-two patients at the Jewish Chronic Disease Hospital in Brooklyn, New York.[5] Earlier, in 1962, the U.S. Congress had passed drug amendment legislation which had also been the result of a disaster. The legislation, which gave rise to new FDA regulations the following year, was the direct result of investigations by Senator Estes Kefauver's Subcommittee on Antitrust and Monopoly.[6] There was concern with the FDA's testing and warning requirements, and also with monitoring and licensing; but the impetus for passage of the legislation was reports from Europe on the effects of the drug thalidomide on infants.[7] An important feature was compromise on the first consent provision in U.S. legislative history that required researchers to inform subjects of the experimental status of a drug and receive their consent. The compromise was for an exception if researchers deemed it not feasible or, in their professional judgement, contrary to the subjects' best interest to obtain their consent before starting an investigation. The exception was subsequently criticized on the ground of vagueness which left it open to abuse. As already indicated, it was clarified in 1966.[8]

Compromise also played an important role in changing what by 1970 had become the Department of Health, Education and Welfare's (DHEW) extensive grants administrative policies governing human research (which included committee review) into formal regulations for the entire department, including the National Institutes of Health (NIH). In 1974 Senator Edward M. Kennedy, chair of the Senate Subcommittee on Health of the Committee on Labor and Public Welfare, who had wanted to create a permanent, regulatory commission independent of NIH, agreed

---

[5] R.R. Faden & T.L. Beauchamp, *A History and Theory of Informed Consent* (New York: Oxford University Press, 1986) at 204. The Jewish Chronic Disease Hospital Case is discussed at 161–162. See also "The Jewish Chronic Disease Hospital Case" in J. Katz, *Experimentation with Human Beings* (New York: Russell Sage Foundation, 1972) 9.

[6] For an account of Senator Kefauver's efforts to get the bill through Congress, see R. Harris, *The Real Voice* (New York: The Macmillan Company, 1964). Kefauver told the Senate that he doubted ". . . whether under a parliamentary system the investigation [of the drug industry] would ever have been made [or] new and original remedies conceived. . . . The bill involved new thinking, new ideas. They came from a legislative committee. At the outset of the investigation, we were actually discouraged by top officials of the Food and Drug Administration. Not only had they no remedies for most of the problems with which we were beginning to be concerned; they did not even recognize them as problems . . .". *Ibid.* at 242–243.

[7] See Faden & Beauchamp, *supra* note 5 at 202–203. For the full story of thalidomide, which could not be published until much later, see The Insight Team of *The Sunday Times, Suffer the Children: The Story of Thalidomide* (London: Futura Publications Limited, 1980).

[8] Faden & Beauchamp, *ibid.* at 203–204. The 1966 regulation made it clear that consent must be obtained in all cases of research that is not done for the benefit of the subject. In treatment for diagnostic or therapeutic purposes, the exception is limited to (a) where immediate treatment is imperative and it is not possible to obtain consent because of inability to communicate with the patient or his representative, or (b) where communicating the information to patients would seriously affect their disease status. The regulation, entitled "Consent for Use of Investigational New Drugs on Humans: Statement of Policy", is reprinted in H.K. Beecher, *Research and the Individual: Human Studies* (Boston: Little Brown, 1970) at 299–300.

to accept an advisory commission, provided DHEW published satisfactory regulations under the authority of the *Public Services Act*. NIH did not want a regulatory commission, and the House supported NIH. The regulations were passed, and the 1974 *National Research Act* provided for an advisory, not a regulatory, commission.[9] Once again disasters (the Tuskegee Syphilis Study,[10] the Willowbrook State School Hepatitis Experiments[11]) were the motivating force for legislative regulation of research involving human subjects. But the Congressional response looked to Institutional Review Boards to provide protection. It did not extend to the use of statutory licensing bodies, which would be introduced in Australia a decade later to regulate research on the embryo.

The Senate Subcommittee on Health had held hearings on human experimentation[12] and also on what Senator Kennedy referred to as "serious projects that are being funded by HEW".[13] None of these matters were being regulated by legislation. In the case of a family planning program which provided voluntary sterilization operations, the Subcommittee heard testimony that 25,000 copies of guidelines intended to protect those who used the services of the program had been printed, but that none were ever distributed. The witness said he had been told that the reasons were political, and that the guidelines would not be issued until after the 1972 elections.[14] At the same time, in other testimony, DHEW's Deputy Assistant Secretary for Health and Scientific Affairs produced reasons for opposing the proposed legislation that have now become familiar:

> "Mr. Chairman, the administration is keenly aware of the current high level of public demand for action to provide for the protection of human rights in the field of human research.
>
> We are certainly in agreement with this high purpose. However, we feel that the passage of any legislation could tend to slow down the continuous evolution of public policy in this area.
>
> We feel strongly that administrative controls, regularly scrutinized and updated, provide the most effective and responsible means of protecting human rights and public interests".[15]

---

[9] Faden & Beauchamp, *ibid.* at 214. Senator Kennedy had introduced an unsuccessful bill to create a National Human Experimentation Board, a permanent body to regulate all federally sponsored research on human subjects. Senator Hubert Humphrey had introduced a bill to create "a separate federal agency with authority over research similar to the Securities and Exchange Commission's authority over securities transactions". *Final Report of the Advisory Committee on Human Radiation Experiments* (New York: Oxford University Press, 1996) at 103, and note 53 at 110. A commentator suggested that the review committee system under the 1966 NIH Guidelines which preceeded legislative regulation might be analogized to licensing hearings. W.J. Curran, "Governmental Regulation of the Use of Human Subjects in Medical Research: The Approach of Two Federal Agencies" in P.A. Freund, ed. *Experimentation with Human Subjects* (New York: George Braziller, 1970) 402 at 445.

[10] The Study is the subject of J.H. Jones, *Bad Blood* (New York: Free Press, 1993) The Study is also discussed in Faden & Beauchamp, *ibid.* at 165–167.

[11] Discussed in Faden & Beauchamp, *ibid.* at 163–164.

[12] United States Senate, Ninety-Third Congress, First Session. The hearings before the Subcommittee on Health were printed in four Parts under the title *Quality of Health Care — Human Experimentation, 1973* (Washington: U.S. Government Printing Office, 1973).

[13] *Ibid.* at 1466.

[14] *Ibid.* 1509–1510.

[15] *Ibid.* at 1461.

In the event, Congress did enact legislation, and DHEW produced regulations. Some related issues in areas (psychosurgery, sexual sterilization) examined by the Subcommittee were addressed in Canada by legislation and court decisions.[16]

The decision to use the law to regulate research resulted from pressure to resort to the most powerful means available for the purpose. A legislative régime would be more enforceable than administrative guidelines; and there was no reason why regulations should impede policy development. The form taken by the regulations made them suitable for the purpose and helped ensure their acceptance by the scientific and medical research communities. The institutional experience[17] was that the regulations were sufficiently flexible to accommodate the activities to which they were directed. An important reason for their acceptance was that the National Research Act required DHEW "to respond to requests for clarification and guidance with respect to ethical issues raised in connection with biomedical and behavioral research", a responsibility which was delegated to the newly formed Office for Protection from Research Risks (OPRR).[18] From the beginning, OPRR kept open the lines of communication with the institutions seeking to comply with the law, and was able to provide authoritative assurance regarding what constituted compliance.

The legal basis of the regulations provided an important rationale for insisting on responsible cooperation with the research review process. In the course of the author's observation of the proceedings of the Institutional Review Board (IRB) in one major medical research institution, there was an apparent instance of failure to cooperate with this process and to meet other requirements. A senior physician responsible for the administration of the IRB subsequently informed the author that the researcher would be leaving the institution. It was probably never far from the minds of medical administrators that compliance with the regulations helped provide protection from legal liability, and in a litigious society this was important. Before the regulations, various ethical guidelines had been used in institutions. They were often useful, but administrators were worried about liability.

## UNIFORMITY

One of the problems with informal regulation of research is the diversity of guidelines. They may vary with the various public or private funding agencies. The

---

[16] *Legislation: Sexual Sterilization Act* of Alberta repealed by S.A. 1972, c. 87; *Sexual Sterilization Act* of British Columbia repealed by S.B.C. 1973, c. 79; O.Reg. 986/78, s. 1 under *the Ontario Public Hospitals Act* (prohibiting sexual sterilization operations on patients or out-patients under the age of 16 years except where medically necessary for protection of their physical health); S.O. 1978, c. 50, s. 12 amending the *Ontario Mental Health Act* (providing that consent of an involuntary patient or his nearest relative to treatment does not include psychosurgery). *Court decisions: E. (Mrs.) v. Eve* [1986] 2 S.C.R. 388 (refusing to authorize non-therapeutic sterilization of a mentally disabled person under the court's *parens patriae* jurisdiction); and recently *Muir v. Alberta* [1996] 4 W.W.R. 177 (Alta. Q.B.) (awarding damages against the Crown for sexual sterilization in "circumstances contemptuous of the statutory authority [under the former legislation] to effect sterilization") (Summary at 183).

[17] As communicated to the author in visits to major U.S. biomedical research institutions some ten years ago. The comments in this and the following paragraph are based on the author's notes.

[18] See Chapter 16.

U.S. federal regulations on research involving human subjects, which were promulgated by different government departments, also varied. However, they all reflected the same legislative policy, and making them substantially similar would make it easier for researchers and institutions. An *ad hoc* Committee appointed in 1982, made up of representatives of the affected government departments and agencies, agreed with a 1981 recommendation of the President's Commission for the Study of Ethical Problems on Medicine and Biomedical and Behavioral Research[19] that uniformity was desirable "to eliminate unnecessary regulation and to promote increased understanding and ease of compliance by institutions that conduct federally supported or regulated research involving human subjects".[20] The Committee also agreed with the recommendation of the President's Commission that all federal departments should "adopt as a common core the regulations governing research with human subjects issued by the Department of Health and Human Services (HHS)".[21]

A common Federal Policy for the Protection of Human Subjects was published in 1991 as a final common rule and promulgated in regulations by sixteen federal departments and agencies that conduct or support research involving human subjects.[22] The Food and Drug Administration amended its regulations to make them conform to the Federal Policy to the extent permitted by the federal *Food, Drug, and Cosmetic Act* under which FDA operates. As the FDA pointed out, "The agency is committed to being as consistent with the final Federal Policy as it can be, given the unique requirements of the act and the fact that FDA is a regulatory agency that rarely supports or conducts research under its regulations".[23]

Legal regulation had long been well accepted. Flexibility of the regulations was never an issue. The achievement of the common Federal Policy in 1991 was to make them easier to use.[24]

---

[19] The successor to the National Commission for the Protection of Human Subjects of Biomedical and Behavioral Research established as a result of the efforts of Senator Kennedy's Senate Subcommittee on Health. The National Commission developed useful guidelines for research involving special populations.

[20] *Federal Policy for the Protection of Human Subjects*, 56 Federal Register 117 (1991) at 28004.

[21] *Ibid.* at 28004.

[22] *Ibid.* at 28002–28030.

[23] *Ibid.* at 28025.

[24] The language of the Preamble to the Council of Europe's Recommendation No. R(90)3, *supra* note 3, offers a striking parallel:
— Considering the aim of the Council of Europe is to achieve a greater unity between the members, in particular by the adoption of minimum common rules on matters of common interest;
— Considering that medical research on human beings should take into account ethical principles, and should also be subject to legal provisions;
— Realizing that in member states existing legal provisions are either divergent or insufficient in this field;
— Noting the wish and the need to harmonize legislation.

## REGULATION IN THE UNITED KINGDOM

The report of the *Enquiry on Research Ethics*, chaired by Professor David Weisstub, which was recently submitted to the government of Ontario,[25] noted that owing to the influence of the U.K., the Council of Europe's 1990 Recommendation on research stopped short of referring to the adoption of legislation as the only way of ensuring that its Principles would be carried out. It recommended "the governments of member states: (a) to adopt legislation in conformity with the principles appended to this recommendation, *or to take any other measures in order to ensure their implementation*".[26] Curiously, the Explanatory Memorandum,[27] which does not have the authority of the Recommendation, is not as clear. Paragraph 8 of the Explanatory Memorandum states that the legislation or other means will introduce "a series of binding rules"; and paragraph 9 explains that the Recommendation comprises "a series of Principles which member States are invited to introduce *into their national law* by legislation or by other appropriate means. The aim of these Principles is first of all to protect the human rights and health of persons undergoing research, as well as to establish clear *legal* rules on the duties of research workers and promoters of medical research". Finally, paragraph 14 states that "the Recommendation asks the member States . . . to adopt legislation in conformity with the Principles or to take any other measures *thus giving the Principles legal status at national level*". In fact, with the exception of embryo research,[28] the U.K. has no formal legal requirements. The Department of Health's 1991 Guidelines require District Health Authorities to establish Local Research Ethics Committees. The Guidelines were an attempt to ensure that research on National Health Service premises was impossible without Committee approval.

In his 1967 book *Human Guinea Pigs*,[29] M.H. Pappworth criticized ethical codes as ineffectual by themselves and called for legislation, pointing out that, ironically, animal research was strictly controlled by legislation, and not simply regarded as a matter of medical ethics.[30] But despite the shocking account of unethical clinical trials in this book, and subsequent deaths and injuries to patients and healthy volunteers in clinical trials of new drugs, it was not considered necessary to have statutory regulation of clinical research.[31]

Pappworth called for research committees responsible to the General Medical Council, which is the licensing body for physicians.[32] Shortly thereafter, in response

---

[25] *Enquiry on Research Ethics: Final Report* (Chairman: David N. Weisstub. Submitted to the Hon. Jim Wilson, Minister of Health of Ontario, Aug. 28, 1995). [hereinafter *Enquiry on Research Ethics*]

[26] *Ibid.* at 136. Preamble to Recommendation No. R (90) 3, *supra* note 3. [Emphasis added]

[27] Document AEXPLMEM903.

[28] Which is regulated under the *Human Fertilisation and Embryology Act 1990*, 1990, c. 37. (U.K.)

[29] M.H. Pappworth, *Human Guinea Pigs: Experimentation on Man* (Harmondsworth: Penguin Books, 1969).

[30] *Ibid.* at 247–249. "Before such experiments can be carried out licences must be obtained, and to obtain them the experimentors must state in detail the purpose and nature of what they are proposing to do" (at 247).

[31] See Dodds-Smith, *supra* note 1 at 164, note 1: "There is no legislation in the UK governing the actual performance of clinical research".

[32] See Pappworth, *supra* note 29 at 252.

to criticism and to developments in the United States, the Royal College of Physicians urged that research committees be set up, and issued some guidelines for their structure.[33] But as late as 1981 a survey by the British Medical Association of Local Research Ethics Committees discovered what has been described as "a totally lamentable state of affairs".[34] In 1985 a lawyer member of a London LREC suggested[35] that although the question of research ethics committees had traditionally been left to professionals by the government, the climate might be changing. There seemed to be an opinion that there was a more active role for the Department of Health to play.

At that time, the London LREC was using guidelines put out by the Royal College of Physicians (RCP) in 1984. The current version of these guidelines was recently published.[36] The principles in the *Declaration of Helsinki* remain influential, as does the Medical Research Council's 1992 guidance on Responsibility in Investigations on Human Participants and Material and on Personal Information, and its 1991 reports on The Ethical Conduct of Research on Children and The Ethical Conduct of Research on the Mentally Incapacitated.[37]

Significant problems remain. A 1992 study of research ethics committees in the U.K.[38] recommended the enactment of legislative regulation for a number of reasons, but principally to ensure that all research is submitted to the committees ("few RECs vet research other than that which is conducted on NHS patients"[39]), and to enable them to stop research they regard as unethical[40] — in other words, to provide an effective enforcement mechanism. It is significant that in its recent report on Mental Incapacity,[41] the Law Commission recommended the enactment of legislation to establish a Mental Incapacity Research Committee to approve all research involving mentally incapable persons, because "LRECs have no legal

---

[33] C. Faulder, *Whose Body Is It? The Troubling Issue of Informed Consent* (London: Virago Press Limited, 1985) at 96.

[34] *Ibid.* at 97.

[35] The opinion was expressed to the author in the course of an interview on the review of biomedical research.

[36] Royal College of Physicians of London, *Guidelines on the Practice of Ethics Committees in Medical Research Involving Human Subjects*, 3d. ed. (London: RCP, 1996). I am grateful to the Royal College for allowing me to see the final page proofs while the Guidelines were being printed.

[37] These reports were prepared by Working Parties established by the Medical Research Council in July, 1988. The report on The Ethical Conduct of Research on the Mentally Incapacitated contained an exposition of the legal position as well as a discussion of the ethical case for including mentally incapacitated persons in research. It concluded that a Law Commission Consultation Paper might lead to legislation "but not in the immediate future. In the meantime, in view of the legal uncertainty, the only sure legal protection for researchers concerned about their position would be for them to seek a declaration from the court that proposed research procedures are legal". *Ibid.* at 21.

[38] J. Neuberger, *Ethics and Health Care: The Role of Research Ethics Committees in the United Kingdom* (London: King's Fund Institute, 1992).

[39] *Ibid.* at 44.

[40] *Ibid.* at 8.

[41] The Law Commission, *Mental Incapacity* (London: HMSO, 1995).

standing, a decision by a LREC does not make a researcher's actions lawful, and statute cannot enable a non-statutory body to achieve such an end".[42]

Is legislation the answer? If it is, is it needed in Canada, where there is even less control over research than in the U.K.? Would a Canadian legislative model be similar to the U.S. federal model? To answer these questions, it is necessary briefly to examine events in Canada, and the pattern that is emerging there.

## REGULATION IN CANADA

The Medical Research Council of Canada (MRC) has shared with U.K. authorities a reluctance to consider that legislative regulation of research might be more efficient and effective than ethical guidelines revised from time to time. Until 1978, when MRC published the ethical requirements to which research involving human subjects should conform,[43] there were guidelines for research on animals, but not for research on humans. As described on the 1974 MRC application form, the responsibility for dealing with the ethical aspects of research applications was on the investigator, the institution, and the Council.

The Head of the Department in which the research was to be done had to convene a local institutional committee to review the application on ethical grounds. The report of the committee was to be signed by the Head and the Dean. If the committee found the proposed research unacceptable, the application should not be sent to the Council. In addition, the Council examined the ethical considerations as part of its review, and funds would not be provided unless the protocol was satisfactory. It was sometimes necessary for the President of the Council to advise Deans of Medicine that their colleagues' research procedures would have to be amended before their proposals could be funded. There was little or no monitoring. The sanction was to deny funding renewal if the research report offended against ethical concerns.

The 1978 guidelines continued to emphasize the three tiers of responsibility, but expanded on how investigators should carry out their responsibilities e.g., by specifying procedures for obtaining consent; and on the limits within which the research institutions should establish membership and procedural requirements for the review committees. The Council would continue to assess the ethical aspects of research proposals.

These guidelines were replaced in 1987.[44] After the earlier reports of the two United States commissions and the federal responses, it was considered that chapters on consent, research involving children and incompetent adults, and

---

[42] *Ibid.* at para 6.3.3. In a consultation paper, the Law Commission had suggested a judicial forum (court or tribunal) with a statutory jurisdiction to make declarations as to whether non-therapeutic research on incapacitated subjects was lawful. See Law Commission, *Mentally Incapacitated Adults and Decision-Making: Medical Treatment and Research (Consultation Paper No. 129)* (London: HMSO, 1993) at paras 4.4–4.9 and para. 6.29 .

[43] Medical Research Council, *Ethical Considerations in Research Involving Human Subjects* (Ottawa: Minister of Supply and Services Canada, 1978).

[44] Medical Research Council of Canada, *Guidelines on Research Involving Human Subjects* (Ottawa: Minister of Supply and Services Canada, 1987).

research involving fetuses and embryos were called for. The committees now became Research Ethics Boards (REBs), the question of risk was discussed, and there was a short discussion of confidentiality. The three levels of responsibility remained, but there was an important duty added — institutional monitoring by the REB. This was not limited to a researcher's undertaking to provide updates on the status of approved projects, and any new information that might affect the ethical basis of the work. It was to involve active monitoring "by periodic review of the research and of the factors involved in the ethics approval". The researchers and the institutions were to bear the costs of this day-to-day monitoring.[45] There appears to be no record of how this recommendation was received in 1987, but one might speculate on how it would be received now, when money is even scarcer.

In the 1978 guidelines it was recommended that the review committee should have access to professional legal advice to identify legal problems that may arise, but that it was not necessary that a lawyer be a permanent part of the review process. By 1987 the presence of a lawyer on the REB was seen as desirable "since common law and legislation often impinge directly on the decisions an REB will need to take".[46] But recognition of the need for legal expertise did not extend to the possibility of using legislation to regulate research on human subjects.

## THE QUESTION OF LEGISLATION

It was thought necessary for the final report from the Standing Committee to MRC containing the 1987 Guidelines to attempt to explain why legislation would not be preferable to guidelines. Unfortunately, the explanation revealed a lack of understanding of the nature, province, and function of law and legislation, and in effect no real attempt was made to examine the case for the latter. This was surprising, because after asking the question whether MRC should propose legislation rather than guidelines, the report said "in that way, the rules provided would carry the usual sanctions of law *and thereby might have more force*".[47] But nothing more was said about enforceability, one of the main concerns in this area, and one of the strongest arguments in favour of legislation. Despite recognition of the *Canadian Charter of Rights and Freedoms*, a fundamental constitutional document, as "a new and powerful legal force in Canada",[48] and the uncertainty of the law relating to research on special populations such as children and mentally disabled persons, the 1987 Guidelines expressed satisfaction "with the use of Guidelines within the current body of law".[49] That the law was not discussed in the Guidelines, and that much of the law in this area was unclear, was apparently not considered much of a disadvantage, because in the words of the Standing Committee

---

[45] *Ibid.* at 50.
[46] *Ibid.* at 46.
[47] *Ibid.* at 10. [Emphasis added]
[48] *Ibid.* at 12.
[49] *Ibid.* at 11.

> Guidelines, administered responsibly in an atmosphere of public openness and within a
> society that respects the judgments of its different parts, can be an effective instrument of
> social control. Indeed, the truly ethical quality of the assessments to be made may atrophy
> when judgments are directed by law.[50]

The idealism of this evaluation is reminiscent of the following passage from an improving book given to Bertie Wooster, the principal character in a P.G. Wodehouse tale:

> Of the two antithetic terms in the Greek philosophy one only was real and self subsisting;
> and that one was Ideal thought as opposed to that which it has to penetrate and mould. The
> other corresponding to our nature was in itself phenomenal, unreal, without any permanent
> footing, having no predicates that held true for two moments together; in short, redeemed
> from negation only by including indwelling realities.

Bertie's reaction is, understandably, one of puzzlement:

> "Well — I mean to say — what?"[51]

There are many reasons for regulating research through legislation rather than through guidelines. A number have already been mentioned, and others will be suggested below. Briefly, legislation would apply to all biomedical research. It is debated openly, its provisions are made public, and the process of legislation provides accountability.[52] It promotes uniformity and certainty. It is more readily enforceable. It can offer protection against liability. It is consistent with other fundamental law — for example, in Canada, the *Charter of Rights and Freedoms*. It makes it possible to clarify the conditions under which vulnerable persons can legally be the subjects of research. It provides the same powerful protection that countries are now choosing for regulating embryo research and in vitro fertilization. Legislation has all the advantages that have been claimed for guidelines, and none of the disadvantages.

## CRITICISM

For some reason, the 1987 final report to the Medical Research Council assumed that the federal *Criminal Code* would be used as the vehicle for any legislative regulation of research, and proceeded to criticize this unlikely possibility.[53] In fact, while prohibitions could be put in the *Code*, it is much more likely that a separate statute would be used for the actual regulation of research, as well as for prohibitions. A possible model, which includes a role for the provinces and territories, may be seen in the position paper which accompanied the recent introduction of Bill C-47, "An Act respecting human reproductive technologies and

---

[50] *Ibid.*.
[51] "Jeeves Takes Charge" in P.G. Wodehouse, *The World of Jeeves* (New York: Manor Books, 1973) 18.
[52] See Waller, *infra* note 78 at 32–33.
[53] *Guidelines on Research, supra* note 44 at 10–11.

commercial transactions relating to human reproduction".[54] The bill prohibits thirteen unacceptable uses of new reproductive and genetic technologies (NRGTs). The position paper[55] sets out the elements of a proposed legislative regulatory framework to deal with NRGTs.[56]

Another criticism of using the *Code* was heavy-handedness of control and the difficulty of proving wrongful intention beyond a reasonable doubt. While the Criminal Law power (which supports other legislation than the *Criminal Code*, for example the federal *Food and Drugs Act*) cannot be used for the primary purpose of regulating an area of activity, it can prohibit certain acts and omissions, and exceptions can be carved out from the application of the prohibitions. In this sense, no one would deny that the *Food and Drugs Act* has a regulatory effect. There would not necessarily have to be proof of wrongful intent; but if this was required, there could be inspection provisions for facilitating the collection of evidence, and there could also be flexibility of control through licensing.

Another criticism in the report was the putative difficulty of legislation in adjusting to evolution in the field of inquiry. This criticism, as related above, was raised unsuccesfully by a senior official of DHEW in opposition to proposed U.S. federal legislation on research involving human subjects. One can dismiss this objection by noting that, in fact, adjustment to changes does not seem to have been a problem with the U.S. legislation.[57] Nor has it been a problem with NRGT legislation anywhere.[58]

The report to MRC admitted that provincial legislation could accommodate the fine tuning of regulations, but it criticized the possible lack of national uniformity. Experience has shown that useful legislation in one province or territory has often led to its enactment in others where it was needed. Also, the work of the Uniform Law Conference of Canada has led to the production of model Uniform Acts in different areas of health, with subsequent adoption by the provinces and territories through legislation in some cases. There would therefore be a reasonable prospect of producing legislation on research with human subjects that would be substantially uniform across the country.

It can be argued that any ethical guidelines which are not designed for a specific area of practice are inherently unsatisfactory because they are too general and lend themselves to different interpretations. The National Council on Bioethics in Human Research (NCBHR), a private agency funded by the Medical Research Council and

---

[54] Bill C–47, *An Act respecting human reproductive technologies and commercial transactions relating to human reproduction.*

[55] *New Reproductive and Genetic Technologies: Setting Boundaries, Enhancing Health* (Ottawa: Minister of Supply and Services Canada, 1996). [hereinafter *New Reproductive and Genetic Technologies*]

[56] *Ibid.* at 26–33.

[57] "[T]he regulations work. Their adoption has not impeded research in the United States, and abuses of human subjects in research subject to the regulations occur infrequently". R.B. Dworkin, *Limits: The Role of the Law in Bioethical Decision Making* (Bloomington and Indianapolis: Indiana University Press, 1996), c. 7 at 155. To the same effect, see the discussion by McCarthy in Chapter 16.

[58] "In a modern democratic state, where the Parliament sits regularly, statutes may be altered, and altered swiftly — if the will is there". Waller, *infra* note 78 at 33.

Health Canada, and supported by the Royal College of Physicians and Surgeons of Canada, was established to help bring about some measure of consensus. A publication explaining the role of NCBHR[59] states that its advice on these issues rests on interpretation of the MRC Guidelines, "supplemented by other developments, approaches in other jurisdictions and societal concerns".[60] Workshops "promote exchanges on ethical issues in research with human subjects and build towards a consensus on standards".[61] But unlike legislation, the Guidelines are not regarded as prescriptive, and a great deal of interpretation continues to take place.[62]

There are other indications that the use of the National Council, in conjunction with guidelines, is not an effective substitute for legislation. The *Enquiry on Research Ethics* mentioned criticism of the effectiveness of NCBHR on a number of grounds: no public authority; advice limited to those who ask for it; accountable only to its sponsors; institutional and financial dependence; membership largely composed of representatives of the health care professions; no incentive to report the activities of REBs; and little public involvement.[63] It is difficult to determine how effective the system of ethical guidelines has been in preventing harm to the human subjects of biomedical research. However, it is useful to consider it in the light of the recommendations of the Council of Europe and other international trends.

## THE CONVENTION ON HUMAN RIGHTS AND BIOMEDICINE

For some time now the Steering Committee on Bioethics (CDBI) of the Council of Europe has been working on a Convention whose long title is "Draft Convention for the Protection of Human Rights and Dignity of the Human Being with Regard to the Application of Biology and Medicine". The CDBI completed and adopted the Convention at its meeting in June, 1996, replacing the short title "Bioethics Convention" with "Convention on Human Rights and Biomedicine". The Convention was adopted by the Committee of Ministers on November 19, 1996. It will be open for signature by non-member States as well as member States of the Council of Europe.

The scope of the Convention is of course wider than that of the Principles in the Council's 1990 Recommendation on research. While it shares a number of concerns with the earlier document, it is different in at least three respects. The first is that the Convention is expected to be followed by Protocols on certain specific areas of biomedical activity, for example, human organ transplantation, and research involving human subjects, and that it was necessary to include certain fundamental

---

[59] 2d ed. (Ottawa: NCBHR).

[60] *Ibid.* at 4.

[61] *Ibid.*

[62] Interview with a member of the MRC working group reviewing the 1987 MRC Guidelines, in a news report entitled "Research council revising ethics guidelines. Rapid Scientific Advances Keep Raising New Issues, Such as the Use of Vulnerable Elderly People in Medical Studies" *The [Toronto] Globe and Mail* (5 August 1995) A4.

[63] *Enquiry on Research Ethics, supra* note 25 at 132.

principles governing these areas in the Convention, as a legal condition precedent to further elaboration in the Protocols. The second difference is that the Convention addresses current topics which are seen as problematic and pressing, for example, interventions on the human genome, and research on embryos. The third is that the Parties to the Convention are to see to it that the fundamental questions raised by the developments of biology and medicine are to be made the subject of public discussion. It is considered essential to promote public debate on these issues and to give careful consideration to the responses.

These features have important implications for the choice of models for regulating research. The first is that with greater specificity, there will be expectations of effective enforcement, as well as protection from liability. The former is likely to remain problematic without legislation, and the latter is possible only with legislation. The second is that the common response of a number of States faced with the sensitive issue of embryo research has been to legislate. The third is that experience has shown legislation to be a good way to promote public debate on these issues.

## HUMAN RIGHTS

It has been pointed out that the Convention on Human Rights and Biomedicine also shares a number of concerns with the Council's 1990 Recommendations on research. The most prominent one is for the protection of human rights, which underlies all the work of the CDBI. Article 1 sets out the Purpose and Object of the Convention in a statement that the Parties "shall protect the dignity and identity of all human beings and guarantee everyone, without discrimination, respect for their integrity and other rights and fundamental freedoms with regard to the application of biology and medicine". And Article 23 provides that "the Parties shall provide appropriate judicial protection to prevent or to put a stop to an unlawful infringement of the rights and principles set forth in this Convention at short notice".

Another shared concern is for the protection of special populations. Although the Preamble to the Convention is not as specific as the Preamble to the 1990 Recommendation, which recited the consideration "that particular protection should be given to certain groups of persons", and despite the fact that specific reference is no longer made to the situation of pregnant or nursing women as subjects of research, nevertheless more focused protection is given to minors and mentally incapacitated adults in Article 6, and reference is made in the same Article to the importance of minors' opinions and the possible participation of incapable adults in the authorization procedure for a biomedical intervention. In other words, in keeping with current opinion, there is a preference for having vulnerable persons play whatever role they are capable of in the decision making process: there is a reluctance to ignore them and leave all power of decision making to their representatives.

These two matters — the protection of human rights and the protection of special populations — are clearly related. It has already been mentioned that in Canada human rights are constitutionally protected. In the case of a request for non-therapeutic sterilization of an incapacitated adult, the Supreme Court of Canada said that such an order was not within the Court's *parens patriae* jurisdiction. The courts have limited powers, under this jurisdiction, "to do what is necessary for the benefit of persons who are unable to care for themselves". To go beyond this would require legislation, which "will then, of course, be subject to the scrutiny of the courts under the *Canadian Charter of Rights and Freedoms* and otherwise".[64] Some research on mentally incapable persons that is not done for the benefit of the individual subject[65] may have a better claim to approval than the order requested in the above case; but, like non-therapeutic sterilization, it does not come within the *parens patriae* jurisdiction. Its legality therefore remains doubtful, and at best, uncertain. The qualification under Article 17 of *the Convention on Human Rights and Biomedicine* is that the research should entail only minimal risk and minimal burden to the individual concerned.

The attitude of the Supreme Court of Canada in the above case was in strong contrast with the view of the British House of Lords that in the absence of legislation, British courts can provide declaratory judgments approving serious interventions on a case-by-case basis. As mentioned earlier, the English Law Commission, after suggesting in a consultation paper that "a judicial body should have power to make a declaration that proposed research involving persons without capacity would be lawful", abandoned this proposal in its final report on Mental Incapacity in favour of establishing a new statutory committee.

The British courts' need to resort to declaratory judgments in the case of adults (the courts continue to have the *parens patriae* wardship jurisdiction over children) was the result of the revocation in 1960 of the Royal Warrant by which the Crown's *parens patriae* prerogative powers over mentally incapacitated persons were delegated to the Court of Chancery.[66] The use of declaratory judgments would not commend itself as a method of proceeding where the issue was research involving mentally incapacitated persons. It would scarcely be an improvement over the climate of uncertainty that currently prevails. Without legislative guidance, it may be difficult for even a court with *parens patriae* jurisdiction to deal with some requested interventions in a completely satisfactory manner. A good example may be found in the judgments of the members of the Irish Supreme Court in a recent

---

[64] *E. (Mrs.) v. Eve, supra* note 16 at 432.

[65] That is, research that is designed to contribute to generalizable knowledge. See R.J. Levine, "Commentary: Terminological Inexactitude" in D.M. Gallant & R. Force, eds. *Legal and Ethical Issues in Human Research and Treatment: Psychopharmacologic Considerations* (New York: Spectrum Publications, 1978) 85 at 92. See also "Unacceptable Terminology" in R.J. Levine, *Ethics and Regulation of Clinical Research*, 2d ed. (Baltimore-Munich: Urban & Schwarzenberg, 1986) at 10.

[66] See B. Hoggett, "The Royal Prerogative in Relation to the Mentally Disordered: Resurrection, Resuscitation, or Rejection?" in M.D.A. Freeman, ed. *Medicine, Ethics and the Law* (London: Stevens & Sons, 1988) 85 especially at 92 and 94.

case[67] where permission was sought to remove life support from a patient who was in a situation described as very close to a persistent vegetative state.

## THE NEW REPRODUCTIVE AND GENETIC TECHNOLOGIES

The area of medical research that is currently the focus of concern is the experimental aspects of the new reproductive and genetic technologies (NRGTs). Embryo research issues are the most controversial. Controversy over these and other issues helps explain why the 1987 Principles on human artificial procreation, prepared by the *ad hoc* Committee of experts on progress in the biomedical sciences (CAHBI), now the Steering Committee on Bioethics (CDBI), were not formally approved by the Committee of Ministers of the Council of Europe, but instead were published in a 1989 report described as an information document prepared by the Secretariat of the Council with a view to giving an overview of the work carried out during the period 1985 to 1987 by CAHBI.[68] The report pointed out that no comprehensive legislation on human artificial procreation existed at the time in any member state at the national level, and considered it desirable to harmonize regulation at the European level "because of the danger that exclusively national regulations might prove ineffective in practice, since it would always be possible to use techniques prohibited by one country in another having different regulations".[69] This reasoning also applies to research involving human subjects. The report seems to have contemplated legislative harmonization, and indeed it is difficult to see how informal regulation could be harmonized and made effective in either case, given its potential diversity even within countries, as well as its problems of enforcement.

The Principles prepared by the CAHBI were modelled to a large extent on the recommendations of the 1984 British Warnock Report,[70] which had been followed in the same year by a private member's bill introduced by the Rt. Hon. J. Enoch Powell. The Unborn Children (Protection) Bill had sought to prohibit all embryo research. Strong support (238 to 66) in the Commons in favour of a second reading of the bill in 1985 showed the strength of feeling against embryo research. The medical and scientific community had then attempted to convince public opinion "of the case for research carried out under controls". The controversy had forced the

---

[67] *Re a Ward of Court* [1995] 2 ILRM 401. See K.M. Doyle & A.J. Carrol, "The Slippery Slope" (1996) 146 New Law Journal 759. For a discussion of the use of the declaratory jurisdiction in an English case of a similar kind, see J. Stone, "Withholding Life-Sustaining Treatment: The Ultimate Decision" (1995) 145 New Law Journal 354.

[68] *Human Artificial Procreation* (Strasbourg: Council of Europe, 1989) at 3.

[69] *Ibid.* at 13.

[70] *Report of the Committee of Inquiry into Human Fertilisation and Embryology* (London: HMSO, 1984)

government to indicate an intention to bring forward legislation,[71] and it eventually introduced what became the *Human Fertilisation and Embryology Act 1990*.

The CAHBI Principles prohibited a range of acts similar for the most part to those which the Warnock Report had marked out for penal sanctions, with an even more restrictive approach to embryo research, and the European countries and Australian States that enacted legislation generally followed the Warnock example.

## EFFECT OF NRGT LEGISLATION

The way was carefully prepared for the U.K. legislation with the co-operation of the scientific and medical community. The year after the Warnock Report the Medical Research Council and the Royal College of Obstetricians and Gynaecologists set up the Voluntary Licensing Authority (VLA) to license and monitor all embryo research and IVF treatment. Five years later, in 1989, the name was changed to the Interim Licensing Authority (ILA), in the expectation that the government would eventually introduce legislation. Finally, the *Human Fertilisation and Embryology Act 1990* replaced the ILA with the Human Fertilisation and Embryology Authority (HFEA), a statutory licensing authority.

The Act contains certain absolute prohibitions, with criminal sanctions, and provides the Authority with comprehensive licensing powers and criminal sanctions for performing certain procedures except under license. There are provisions for inspection and enforcement. Directions may be given for any of the purposes under the Act, and these must be complied with. There is a requirement to maintain a code of practice "giving guidance about the proper conduct of activities carried on in pursuance of a license under this Act and the proper discharge of the functions of the person responsible and other persons to whom the license applies".[72] In addition, the Secretary of State may make regulations under the Act. The U.K. legislation, and other similar types of legislation, provide a sophisticated regulatory package which is far removed from the *Criminal Code* model which the 1987 Medical Research Council of Canada report assumed would be used for any regulation of research.

Are the issues raised by IVF and embryo research so different from the issues raised by other research that the reasons for legislative regulation of the former do not extend to the latter?

> Lurking in the background and brought into sharp focus by the advent of IVF was a deep-seated fear that irresponsible scientists might abuse the process, for example, to seek to predetermine human characteristics, to clone, to create hybrid creatures. Many speeches in the parliamentary debates on these subjects cited the activities of concentration camp scientists and doctors. Recently enacted German legislation expressly forbids manipula-

---

[71] The above is from the summary in R.L. Cunningham, "Legislating on Human Fertilization and Embryology in the United Kingdom" (1991) 12 Statute Law Review 214 at 221–222 (there were other bills on similar lines). The article is an authoritative account of prior developments as well as of the preparation and parliamentary stages of the *Human Fertilisation and Embryology Act 1990*, *supra* note 28. For a commentary on the Act and the reasons for it, see D. Morgan & R.G. Lee, *Blackstone's Guide to the Human Fertilisation and Embryology Act 1990* (London: Blackstone Press Limited, 1991).

[72] *Human Fertilisation and Embryology Act 1990*, 1990, *supra* note 28, s. 25(1).

tion of, or research on, embryos in any form: the ghost of Dr. Mengele still casts a long shadow.[73]

While "the willingness of the medical and scientific community to have statutory controls imposed on them" may have been "a particular response to popular concern about the . . . implications of IVF and embryo research",[74] the underlying public concern was about the potential for medical and scientific interventions that are ethically offensive.[75] Medical horrors under the Nazis extended to all kinds of so-called research, including genetic research, on vulnerable persons. There have been enough examples of both therapeutic and non-therapeutic research which was ethically or legally offensive to demonstrate the need for a legislative approach, which would enable us to prohibit or control research that is or has the potential to be ethically offensive. Prevention is preferable to legislating after the fact.[76] To-day we are committed to protecting the human rights of all individuals. There cannot therefore be any justification for limiting to IVF and embryo research what a number of countries by their recent actions have shown they consider to be the most powerful and effective form of control.

## IMPLICATIONS FOR CANADA

From the above, it seems clear that trends at the Council of Europe and internationally encourage adopting methods of regulating research involving human subjects which can guarantee fundamental human rights, and which offer effective protection to special populations of vulnerable persons. Concern over embryo research and other possibilities presented by NRGTs has led to the introduction of a considerable amount of legislation in a relatively short period of time. Some of this legislation provides sophisticated methods for ensuring oversight and control of the new technologies, as well as provisions for effective enforcement of prohibitions and sanctions for breach of licensing conditions. We have entered a new generation

---

[73] See Cunningham, *supra* note 71 at 216.

[74] *Ibid.* at 220. "On the other hand, increasing scientific knowledge about genetics could increase the need for more control over what society permits doctors and scientists to undertake. The scientific community has established a standing commission to consider bioethical issues across the whole field of science and medicine which suggests that we may not have heard the last of the Warnock approach".

[75] The recent Canadian federal position paper on a proposed legislative framework for NRGTs states that germ-line genetic alteration "has the potential to lead to the production of 'designer' children, whose genes are manipulated to produce certain desired characteristics (e.g., tallness, blue eyes)" (*New Reproductive and Genetic Technologies*, *supra* note 55 at 41). But it also states that the safety of this practice cannot be guaranteed. This seems to be the real reason for prohibiting it at the present time. Should it ever become safe, it might prevent the transmission of disease genes to future generations. Since this objective is not unethical, it would not likely be a matter of public concern, and legislation which provided for licensing on conditions could permit the practice for this purpose while otherwise prohibiting it.

[76] If at all. In Canada, the federal government established a limited compensation plan for the subjects of certain experiments at the Allen Memorial Institute in Montreal in the 1950s and early 1960s. Except for five articles in the Civil Code of Quebec which refer to experiments, there is still no legislation regulating research in Canada. In Ontario, the *Health Care Consent Act, 1996* "does not affect the law relating to giving or refusing consent on another person's behalf to . . . a procedure whose primary purpose is research". *Health Care Consent Act, 1996*, S.O. 1996, c. 2, s. 6.

of regulation, and the pace is being set by a number of legislative responses to the challenge of NGRTS. These responses are well adapted to their purpose, and effective. Moreover, governmental action through legislation is needed in order to engage the *Charter*, and thus guarantee for Canada the trends in the Council of Europe and internationally toward protection of human rights and special populations.

There is also the question of continuing dialogue with the public, something which, by themselves, public meetings to discuss proposed ethical guidelines do not provide. The membership of IRBs and REBs may include lay members, but they do not usually engage in dialogue with the public. However, the statutory licensing committees in the Australian States not only represent the public, but they often meet with interest groups. The latter inform these statutory bodies of their concerns in much the same way as they would their legislative representatives or a special committee of the legislature. From meetings with committees and interest groups in these States, it was the author's impression that the committees welcomed public input — an important factor in obtaining general acceptance of the regulatory process. In this connection, the former chair of the Victorian committee (now the chair of the new Victorian Infertility Treatment Authority[77]), recently said that the enactment of the original Victorian legislation was helpful in building public agreement. In his opinion, if there is public debate and the relevant legislation is heavily publicized, it must have an effect on public opinion.[78]

On the government side, it is of interest to note that the statutory licensing committees provide the responsible government minister with an important mechanism for careful consideration of difficult ethical issues, as well as with a helpful channel of information on the needs of the public and the scientific community, which is necessary for the development of policy. In addition, ministers may refer issues to the committees for study and recommendations. The interposition of statutory authorities between ministers and the public removes the need for the ministers to be involved directly in the granting of licenses and other day-to-day decision-making, although as demonstrated recently in the U.K., public alarm about the possibility of some NRGT practices may on occasion force a government to respond directly.[79]

## A NOTE ON LIABILITY

Liability is also seen as an important factor by researchers and institutions in the U.K. and in Canada. As in the United States, liability for injuries through research

---

[77] Established under the Victorian *Infertility Treatment Act 1995*.

[78] Professor Louis Waller, in discussion that followed his paper entitled "Australian Legislation on Infertility Treatments". (Paper presented at the International Symposium on Governing Medically Assisted Human Reproduction, 11 February 1996) [unpublished]. From the author's notes of the discussion. See also L. Waller, "Australia: The Law and Infertility — the Victorian Experience" in S.A.M. MacLean, ed. *Law Reform and Human Reproduction* (Aldershot: Dartmouth Publishing Company Limited, 1992) 17 at 31–33.

[79] See "Britain Outlaws Fetal Egg Transplants" (1994) 308 British Medical Journal 1062.

is left to be determined by the general law of torts, or civil wrongs. But researchers in the U.K. and Canada lack the protection that conformity with legal regulation may afford in the United States. A recent article claimed that contrary to the opinion of the Law Reform Commission of Canada, the 1987 MRC Guidelines do have the force of law.[80] The explanation was that breach of the Guidelines could constitute a breach of the funding agreement for government funding of a protocol, or breach of the researcher's contract of employment with a university or hospital, and that therefore legal remedies may be pursued under both public and private law. This may be so, in the same way that agreements between parties which refer to guidelines are enforceable as between the parties. However, it is a different matter to claim that guidelines have the force of law, a phrase which implies that the guidelines are authorized by or under legislation.

On a related topic, there are pharmaceutical association compensation schemes in the U.K., but they provide very selective protection. The new Royal College of Physicians' Guidelines referred to earlier explain the different compensation arrangements for healthy volunteers and for patients where the research is sponsored by pharmaceutical companies.[81] The Medical Research Council will give sympathetic consideration to making *ex gratia* payments to those suffering adverse consequences as a result of participation in research.[82] While apparently some universities have no-fault compensation insurance,[83] the general situation for compensation without legal proof of liability is said to be unsatisfactory.[84]

## CONCLUSION

There are currently two processes which have important implications for the form that regulation of research should take in Canada. One is the proposed legislative regulatory structure for NRGTs recently announced by the federal Minister of Health, which would permit any province or territory that wishes to do so to develop its own regulatory controls on a basis of equivalency. The controls would have to be "substantially the same as, but not necessarily identical to, the federal legislation in substance and enforcement".[85] There seems to be no reason why effective regulation

---

[80] B.M. Dickens, "Conflicts of Interest in Canadian Health Care Law" (1995) 21 American Journal of Law & Medicine 259 at 274.

[81] See Section 9 (Injuries During Clinical Research) of the *Guidelines, supra* note 36 at 42–46. See C. Hodges, "Harmonization of European Controls over Research: Ethics Committees, Consent, Compensation and Indemnity" in A. Goldberg & I. Dodds-Smith, eds. *Pharmaceutical Medicine and the Law* (London: RCP, 1991) 63. See also Dodds-Smith, *supra* note 1 at 162–164; and the commentary on the 1983 Association of the British Pharmaceutical Industry (ABPI) guidelines on compensation to patients for injuries incurred in clinical trials of drugs, in A.L. Diamond & D.R. Laurence, "Compensation and Drug Trials" (1983) 287 British Medical Journal 675 at 676–677.

[82] Guidelines, *ibid.* at para 9.11. Since making this statement the Medical Research Council has not had a single approach. Private communication to the author.

[83] *Ibid.* at para. 9.2.3.

[84] *Ibid.* at para 9.3.

[85] The absolute prohibitions would not be subject to equivalency agreements. *New Reproductive and Genetic Technologies, supra* note 55 at 33.

of research on human subjects should depend on whether or not research is caught by legislation on NRGTs. In the U.S., where there is no federal regulation of NGRTs, the reverse is the case: effective oversight depends on whether the activity is caught by the regulations on research. If a federal/provincial/territorial co-operative model is used in Canada to regulate NRGTs, a similar model could also be used to regulate research. There would be no question of imitating the U.S. federal regulations, which at any rate only apply to federally-funded research, since most research activity is within provincial or territorial jurisdiction. U.S. government funded research in Canada is currently subject to U.S. regulations,[86] and the procedural requirements in the common Federal Policy for the Protection of Human Subjects apply unless there is a determination that the procedures prescribed by a Canadian institution afford protections that are at least equivalent to that provided in the common Federal Policy.[87]

The other process referred to above is an attempt to regulate research by means of a Code of Conduct for Research Involving Humans.[88] A draft Code of Conduct was recently prepared by a Tri-Council Working Group which undertook the development of common guidelines for the ethical conduct of research involving human subjects. It would supersede the 1987 MRC Guidelines. The account of how the draft document was produced ends with the following statement: "The Councils require that institutions in which they fund research comply with the prescriptive elements and the spirit of the ethical principles expressed in this Code of Conduct".[89] Presumably the draft Code is an attempt to improve on the 1987 MRC guidelines by introducing sufficient precision to imply compulsory national standards for researchers.[90] But it is the treatment of issues almost entirely in terms of ethical principles which makes interpretation necessary.[91] The Code and the ethical commentary are more elaborate than the former guidelines, but the Code is no more successful in dealing with the fundamental fact that important legal issues must be addressed if national guidance and protection is to be provided. The potential conflict between the law and ethical principles is recognized, but presumably is left to the lawyer members of the REBs to deal with, along with the

---

[86] See *Enquiry on Research Ethics, supra* note 25 at 120.

[87] The determination must be made by a department or agency head, who may then approve the substitution of the Canadian procedures in lieu of the procedural requirements provided in the common Federal Policy. See 56 Federal Register 117 (1991) at 28013.

[88] Tri-Council Working Group, *Code of Conduct for Research Involving Humans* (Ottawa: Minister of Supply and Services Canada, 1996).

[89] *Ibid.* at ii. The Councils are the Medical Research Council, the Natural Sciences and Engineering Research Council, and the Social Sciences and Humanities Research Council.

[90] See Interview, *supra* note 62 at A4.

[91] At the same time, Canada is committed to adopt the useful, practical, goal-oriented ICH (International Conference on Harmonisation of Technical Requirements for the Registration of Pharmaceuticals for Human Use) Guideline for Good Clinical Practice once it is available in both official languages. The Guideline deals with clinical trials of drugs under development from the point of view of drug quality, safety and efficacy. It provides focused attention to the role of investigators, research ethics committees and sponsors. The guideline on Good Clinical Practice for trials on medicinal products in the European Community came into effect in 1991 and has legal status. The ICH Guideline was approved in June 1996, and is intended eventually to supersede the other national GCP guidelines.

many important *Charter* issues. There seems to be no indication that the Working Group was alert to the implications of the proposed statutory regulatory structure for NRGTs, announced by the Minister of Health, for the choice of an effective means of regulating research involving human subjects. *Plus ça change, plus c'est pareil.*

CHAPTER 15

# RESEARCH COMMITTEES AND THE PRINCIPLE OF JUSTICE: PUTTING ETHICS AND LAW TO THE TEST

MARIE-LUCE DELFOSSE

Since the 1970s, research committees have been created in most Western countries. These committees are a concrete embodiment of the changes in institutional dynamics brought about by the development of medical experimentation on human beings. Established to respond to the demands within the medical community and the public at large,[1] these committees share traits in common wherever they are found, whilst at the same time undergoing distinct evolutionary processes. Their evolution is closely linked to the manner in which they undergo institutionalization. Indeed, from the moment it is recognized that research on human subjects needs to be controlled, and that this mandate is given to research committees, one particular requirement — the submission of all research proposals involving human subjects to a committee — gives rise to forms of institutionalization that combine law and ethics in different ways, according to the specific characteristics of each country.

We will compare three typical cases of committee institutionalization: France, Belgium, Canada and Quebec.[2] The relationship between law and ethics existing in each of these countries will be described so that we may reflect on the advantages and disadvantages that arise in each case. In other words, and to be more precise, we will consider which type of norms, or which combination of different types of norms, allows the committees to fulfill their mandate most successfully. We will therefore focus on the legal, ethical and deontological nature of the norms, rather

---

[1] See especially: F.-A. Isambert, "Aux sources de la bioéthique" (1983) 25 Le Débat at 85.
[2] The situation in France and Belgium will be presented in greater detail given that the Canadian and Quebec contexts are described in various parts of this collection of essays.

than simply enumerate the requirements laid down in the countries under consideration.

Two observations underlie this investigation, allowing us to specify its content. First, history reveals that law, deontology[3] and medical ethics have all dealt with the issues raised by human experimentation. This has influenced research committees, which have, from the outset, been located at the crossroads between ethics and law in different forms in various countries. Furthermore, the combined intervention of these three normative systems has led to collaboration, and even exchanges, between them. This has in turn influenced the committees' work, inasmuch as it conditions the interpretation of the criteria laid down for the evaluation of research projects. Reflecting on the institutionalization of research committees therefore necessarily requires that we consider the status of these committees. More precisely, we must examine the type of normativity that provides the source of their authority, whether it be law, deontology or ethics. In addition, we must consider the types of norms to which the committees refer in the course of their evaluation of research projects.

The reflection undertaken here is placed under the aegis of the principle of justice, understood as the governing ideal societies strive toward in their search for better institutional organization, in other words, a better sharing through institutional means of the benefits and burdens inherent in social existence. With regard to this, our reflections stem from the analysis made twenty years ago in the *Belmont Report*.[4] Indeed, the *Report* demonstrated that research ethics, in particular as they pertain to human experiments, should be conceived of as a dialectic between respect, beneficence and justice — principles which stem from three distinct moral traditions. The principle of justice appears to be both the synthesis and complement of the two other principles. The necessary conditions for respect and beneficence are united in the principle of justice. Each of these principles is therefore necessary but none by itself is sufficient; one can only really do justice to medical ethical problems by taking all three into account and confronting their respective claims. By making the principle of justice one of the three basic axes of ethical reflections on human experimentation, the *Report* emphasized, moreover, that the researcher-patient or subject relationship poses social issues that will be ignored if we limit ourselves to an approach that considers only the interpersonal aspects of the relationship. The *Report* applied its inquiry concerning the principle of justice to the question of how the benefits and burdens of experimental testing are to be shared. This question was raised in particular in the context of the selection of research subjects, and specifically of people who are vulnerable due to their social or health condition. In order to define rules of conduct in this area, the *Report* showed that we must make use of a variety of precepts: "to each person an equal share"; "to each person according to individual need"; "to each person according to individual effort"; "to

---

[3] By deontology, we mean professional codes of conduct whose legal normativity is variable, depending on the value given to them by the legislative branch of government, see *infra* at 298 and note 21.

[4] The National Commission for the Protection of Human Subjects of Biomedical and Behavioral Research (U.S.A.), *The Belmont Report: Ethical Principles and Guidelines for the Protection of Human Subjects of Research*, Department of Health, Education and Welfare, no (OS)78-0012.

each person according to societal contribution"; and "to each person according to merit".[5] Each of these propositions expresses a specific aspect of the general principle of justice although, as the *Report* itself points out, they are not necessarily compatible.

The principle of justice therefore has its own internal dialectic, while at the same time it is linked dialectically to the principles of respect and beneficence. Beyond the formulae through which it is expressed, the principle of justice appears thus as a regulatory ideal of crucial importance, envisioned in conjunction with the two other fundamental principles of ethics regarding experimentation on humans. It is for these reasons that the principle of justice will provide the overarching norm for all of the comments that follow. We will be concerned with its application to possible modes for the institutionalization of committees, rather than to the selection of potential research subjects. This will enable us to determine the types of norms or combinations thereof that are most likely to allow the committees to fulfill their social mandate and to find, for each project, a balance between the special conditions required for research and the need to respect the physical and moral integrity of the people involved.

## DESCRIPTION OF THE SITUATION IN THREE COUNTRIES

### *France*

In France, from 1974 to 1983, ethics committees were established either independently or at the initiative of the *Institut national de la santé et de la recherche médicale* (National Institute for Health and Medical Research) [hereinafter INSERM] and the *Assistance publique des Hôpitaux de Paris* (Paris Public Hospital Service) [hereinafter AP-HP]. Beginning in 1983, they were created at the insistence of the *Comité consultatif national d'éthique pour les sciences de la vie et de la santé* (National Consultative Committee on Ethics in Life and Health Sciences) [hereinafter CCNE]. One of the tasks that committees assigned themselves was the review of research proposals involving human beings. They did so initially on their own initiative, then, as of 1984, in response to the obligations imposed on them by the CCNE.[6] The CCNE in fact required that trials involving human beings be submitted to five conditions that it outlined as follows: sufficient prerequisites; scientific value of the project; acceptable ratio of risks to benefits; free and informed consent of the persons involved; and a review of the experiment procedure by an ethics committee. These conditions were inspired by those found in international documents regarding medical ethics, such as the *Helsinki Declaration* and the *Proposed International Guidelines for Biomedical Research Involving Human Subjects* formulated in 1982 by the Council for International Organizations

---

[5] See the *Belmont Report* at 5, col. 1.

[6] Comité national d'éthique pour les sciences de la vie et de la santé (France), *Avis no 2: sur les essais de nouveaux traitements chez l'homme. Réflexions et propositions.* 9 October 1984 in *Xe anniversaire: Les avis de 1983 à 1993* (Paris, 1993) at 19–50 [hereinafter *Avis no. 2*].

of Medical Sciences (CIOMS) and the World Health Organization (WHO).[7] The conditions were intended to supplement the laconic *Code de déontologie* that was then in force and that devoted only one article to research on human subjects.[8] They also constitute a complete set of guidelines for experiments that have not yet been legally authorized and that are conducted on healthy persons.

Since December 1988, the situation has changed.[9] The Huriet-Sérusclat law established a legal framework for human experimental activities undertaken for research purposes and, within this framework, has established consultatory committees that concern themselves with the protection of persons in biomedical research (CCPPRB — "comités consultatifs de protection des personnes dans la recherche biomédicale"). These committees have been assigned the role of evaluating research proposals, a role that until 1988 had been assumed by the ethics committees. Each CCPPRB must give an advisory opinion regarding: the validity of research conditions relating to the protection of persons, and in particular that of the participants; the information that will be given out to these persons before and during the research project, as well as the means by which the subjects' consent will be obtained; the compensation owed in case of injuries; the general relevance of the project and the adequacy of the goals being pursued in relation to the means adopted to achieve them; and the qualifications of the researchers.[10] The law outlines several requirements relating to these conditions. The committee's advisory opinion must be transmitted in written form to the researchers within a five-week period. If it is favourable, the research may begin; if it is unfavourable, it must be transmitted to the competent administrative authority. This authority centralizes the documented declarations of research promoters; it may prohibit or suspend research projects; it also exercises a right of control by means of the *Inspecteurs généraux des affaires sociales* (Inspectors general of social affairs). Thus, the CCPPRBs have become one of the cogs in a state-controlled organization, with their mandate and the criteria by which to carry it out both fixed by law.

A significant number of these criteria correspond to requirements imposed by the two international medical ethics texts cited above, and by the CCNE. They include: in-depth knowledge of scientific literature; sufficient pre-trial testing; competence of those carrying out the trials; proportionality between the foreseeable risks to subjects and either the benefits accruing to them or the importance of the research; and free and informed consent from participants in the research. The law also

---

[7] Regarding these documents, see especially: *Idem.* at 39–49. We wish to emphasize that the CCNE adopts the dual idea proposed in the *Proposed International Guidelines for Biomedical Research Involving Human Subjects* of CIOMS and WHO regarding the review of research proposals: a set of grounds for review that follow a fixed order; insistence on the fact that wherever one condition is not fulfilled, there is no need to examine the subsequent conditions.

[8] *Idem.* at 49.

[9] *Loi n° 88–1138 du 20 décembre 1988 relative à la protection des personnes qui se prêtent à des recherches biomédicales, Journal Officiel de la République française*, 22 December 1988 (1988 Law relating to the protection of persons who become biomedical research subjects) [hereinafter *Loi n° 88–1138*]. The law has been modified several times, in particular most recently by the *Loi n° 94–630 du 25 juillet 1994*, J.O., 26 July 1994 [hereinafter *Loi n° 94—630*].

[10] Art. L. 209–12 (new).

incorporates certain requirements or suggestions specifically formulated by the CCNE, such as the rule that healthy volunteers be indemnified for inconveniences suffered, that volunteers not be remunerated, and that the committees be organized according to territorial boundaries. The law also draws on certain deontological principles, including, for instance, the requirement that the patient's medical state be taken into account, as well as current practices in the field of therapeutic medical activity, particularly in emergency situations.

Although at one time it was based on ethical rules, the French situation is now conditioned by the existence of a law that creates and organizes new institutions, the CCPPRBs, and which sets the conditions that must be satisfied by all research proposals. In order to do this, the law draws on medical ethics, professional codes of conduct and certain other criteria transposed from professional usage which thereupon acquire legal effect and hence, become legally enforceable.

### Belgium

In Belgium, the first ethics committees were created in 1974 at the behest of the *Académie royale de médecine* (Royal Academy of Medicine), and of its Flemish equivalent, the *Koninglijke academie voor geneeskunde*, and as of 1976, at the behest of the *Fonds scientifique de la recherche médicale* (Medical Research Science Fund) [hereinafter FSRM]. In 1984, the *Conseil national de l'Ordre des médecins* (National Council of the Order of Physicians) formulated rules that would complement articles 89 to 94 of the *Code de déontologie médicale*, and which are related to human experiments. The Council specified that proposals of research involving human subjects had to be submitted for approval to an ethics committee before they could be carried out. Certain rules regarding the composition and functions of the committees were also established. These rules, reviewed in 1992, led to the creation of an increasing number of committees.[11]

Since 1994, a royal decree requires that an ethics committee be established in every hospital or group of hospitals.[12] The decree establishes a minimal (and not exhaustive) set of rules regarding committee composition and functions. It also defines the committees' mandate: to advise and consult on ethical aspects of hospital care practices; to assist regarding ethical matters raised by decisions in individual cases; and to give opinions and make recommendations regarding all human trial research proposals. However, the decree gives no indication regarding the criteria the committees are to refer to in executing these mandates. With respect to the evaluation of proposals involving research on human subjects, thus, one must refer back to the rules set out by the *Conseil national de l'Ordre*. These rules, however,

---

[11] Conseil national de l'Ordre des médecins, "Expérimentation humaine. Règles déontologiques" (1983–1984) 32 Bull. off. de l'Ordre des médecins at 46–47 and (1992) 55 Bull. du Conseil national de l'Ordre des médecins at 32–34.

[12] *Arrêté royal du 12 août 1994 modifiant l'arrêté royal du 23 octobre 1964 fixant les normes auxquelles les hôpitaux doivent répondre*, Moniteur Belge, 27 September 1994.

are so general that it was judged necessary to interpret them in accordance with medical ethics texts.[13] They do specify, however, that: "Biomedical research proposals must describe the object, the subject and the risks posed by the experiments, the exact nature of the information to be given to the subject, the manner in which it will be given (oral or written) and the manner by which the subject's consent will be given (oral or written). The proposal must also indicate whether an insurance policy covers experimental risks and, if it exists, the nature of that policy".[14] The committee's advisory opinion must be transmitted to the subjects selected to participate in the research and be mentioned in all publications or communications. The committee must be informed of any subsequent modifications to the proposal, although there is no explicit requirement that these changes be reviewed. "An adequate summary of research progress and outcomes" must also be transmitted to the committee.[15]

In Belgium, the hospital ethics committees were formerly under the control of the body that had called for their creation, namely, the *Ordre des médecins*. Now, however, they are an institutionalized part of hospitals and their creation is a requirement of law. Nevertheless, to the extent that the relevant legislation leaves many questions unanswered, the hospital committees in fact find themselves under the obligation to respect the rules formulated by the *Conseil national de l'Ordre*, rules that themselves refer to medical ethics texts. The other committees existing outside of hospitals continue to be governed by the *Ordre*. There exists therefore a "mixed" situation in which medical ethics, deontology and law are combined in a fairly unique manner.

Nevertheless, with respect to the law, the current lack of clarity makes the situation similar in some ways to the one that prevailed in France prior to the enactment of the 1988 law. Indeed, the *Arrêté royal n° 78 du 10 novembre 1967 relatif à l'art de guérir* provides the framework for medical activity having a preventive or therapeutic aim. Since September 1992, a royal decree made pursuant to the 1964 law on drug testing incorporates EEC Directive 91/507 into Belgian law, thereby establishing the conditions for carrying out clinical drug testing.[16] The EEC directive requires that activities conform to the European Union's *Bonnes pratiques cliniques* (BPC) and respect the *Helsinki Declaration* principles which are thus incorporated into Belgian law and are hence enforceable. Research aimed at advancing knowledge is therefore authorized in the case of drug trials; it is not clearly authorized in other cases, however, and in certain forms (experiments without therapeutic purposes), it is even forbidden if no drug is being tested. In this

---

[13] "Avis du Conseil national : Essais cliniques de médicaments" (1983–1984) 32 Bull. officiel de l'Ordre des médecins at 45.

[14] "Expérimentation humaine. Règles déontologiques" (1992) 55 Bull. du Conseil national de l'Ordre des médecins Rule 6.

[15] *Idem*, Rule 1.

[16] *Arrêté royal du 22 septembre 1992 modifiant l'arrêté royal du 16 septembre 1985 concernant les normes et protocoles applicables en matière d'essais de médicaments à usage humain*, Moniteur belge, 5 December 1992.

context, deontological rules play a particularly important role, as they provide basic guidelines in a domain that is not governed as such and in its entirety by law.

The Belgian situation is therefore especially complex. Although at one time it was organized by rules found in deontology, themselves based on medical ethics in the matter of research involving human subjects, it is now characterized by selective legal interventions. These interventions have generated important changes, though their effect is sometimes hard to perceive, given that they sometimes do little more than transpose into law the conditions previously formulated by medical ethics and/ or deontology.

### Canada and Quebec

In Canada and Quebec, research ethics committees (REC) were created during the early 1970s. In 1978, their development was influenced by guidelines emanating from the Medical Research Council of Canada (MRC), the country's main medical research funding agency. These guidelines were intended to help committees with their evaluations of research proposals and were revised in 1987.[17] Currently, the Medical Research Council, the Social Sciences and Humanities Research Council of Canada (SSHRC) and the Natural Sciences and Engineering Research Council of Canada (NSERC) are working together to produce a *Code of Conduct for Research Involving Humans*, for the benefit of researchers, members of the RECs and the administrators of research facilities at which experimentation on humans takes place. A preliminary *Code of Conduct* has been distributed to members of the research community in order to obtain their comments prior to drafting the final version of the text. Given the current transitional situation, I believe it is preferable not to give a detailed outline of the *Code*'s contents and requirements, as they may yet be subject to further modification. Instead, I will highlight two general characteristics of RECs in Canada and Quebec.

First, the RECs' legitimacy is derived primarily from the reality of administrative practice: funding organizations will only provide money to researchers for proposals that have received a favourable committee advisory opinion. Secondly, the work of the RECs was formerly undertaken in the name of deontology, then of medical ethics. It now seems safe to assume, based on the current report drafted by the three national research Councils, that their work will emphasize medical ethics.

Canadian federalism has played a role in the choices that were made in this area. Nevertheless, they do not simply flow from the legal and jurisdictional differences that exist among the provinces. They also flow from the belief that the problems raised by human trials require value judgments and subtle considerations grounded more in ethics than in law. Ethics is perceived as being more flexible than the law, and is considered to have a broader scope inasmuch as it incorporates subjective criteria. Having said this, researchers may not violate legal provisions regarding the

---

[17] Medical Research Council of Canada, *Ethics in Human Experimentation*, Report No. 6 (1978) [*La déontologie de l'expérimentation chez l'humain*] and *Guidelines on Research Involving Human Subjects* (1987).

protection of persons. For instance, Article 21 of the new Quebec Civil Code mentions the ethics committees in the context of experiments conducted on minors and protected persons of full age.

The Canadian situation is therefore characterized by its recognition of the *de facto* existence of research ethics committees. The activities of these committees is structured by governmental organizations and are viewed as falling more within the realm of ethics than of law. Researchers must obtain a favourable appraisal in order to receive funding. In Quebec, the RECs benefit from incidental legal recognition under Article 21 of the new Civil Code.

* * *

We are therefore confronted with three very different situations. Each is influenced by the respective country's legal régime, power relations and cultural heritage. Nevertheless, there is no mechanistic relationship between each of these factors. Indeed, a society's institutional life is a never-ending quest for the best — or least displeasing — solutions to the problems encountered at any given time. Even if these solutions are significantly influenced by the context in which they emerge, they are not totally reducible to those contexts. Thus, comparing them produces a wealth of insight. By elaborating on and deepening the approach adopted so far, we will be able to highlight the issues raised by each of these situations and then to evaluate them in turn.

## COMPARISONS AND ISSUES

France's 1988 law represents a current of opinion that views law as the optimal extension of ethical norms, to the extent that it ensures an equality in the treatment of issues. According to this view, the law aims to cover the totality of issues relating to the protection of biomedical research subjects. The enactment of this law provides a new organizational structure having its own defined rules replacing pre-existing ethical rules and organization. The new law sets the conditions that must be fulfilled by research proposals and organizes the CCPPRBs to which it assigns the evaluation of these proposals.

Since all of the issues are brought within the legal order, French law unquestionably provides greater transparency and procedural uniformity. Despite these advantages, however, one might be tempted to suggest that this situation belies a reductionist conception of the work to be undertaken by the committees and therefore gives a restrictive interpretation of the conditions designed to protect persons involved in biomedical research. In sum, it seems this approach risks replacing an ethics-based inquiry with a system of control that simply assesses the extent to which each file conforms to certain legal requirements. In the procedures of many committees, even before the law was passed, there was already evidence of a formalist tendency, even though evaluations had to be made based on norms emanating from authorities in "ethics" and derived from medical ethics. This tendency seems to have been legitimated by a legal text that attempts to cover all aspects of human experiments and that, in consequence, gives credence to the idea that compliance with legal requirements is sufficient to ensure the protection of persons.

The assimilation of ethics and law is further reinforced by the fact that many of the evaluation criteria for research proposals that had been set out under medical ethics carry the same names as certain long established legal requirements that have been used to decide proximate, yet distinct issues: for example, consent and cost-benefit analysis were the bases upon which the legality of medical acts with a strictly therapeutic purpose were judged. Two types of assimilation therefore tend to take place: between acts of medical care and acts of research, and between medical ethics and the law.

The confusion between acts of medical care and acts of research tends to hide the differences between the two situations and therefore the different problems posed by each one. Research activity must be distinguished from care-giving activity. By its very existence, the law provides evidence of the recognition of this necessary distinction. However, it is unfortunate that the law maintains a global approach to different types of research and does not give any specific consideration to the situations in which care-giving and research are mixed, or where knowledge is pursued by means of acts that also have a therapeutic effect. Indeed, such situations require a special type of vigilance in order to ensure the protection of persons.[18]

The second type of assimilation — that of ethics and law — obscures the fact that norms are given a particular meaning and scope that depends on the normative system that posits them. The fact that the law encompasses the totality of issues within its own internal order presents the question of whether or not the law is capable of solving all of the issues raised by the evaluation of research proposals. France's legislature seems to have answered this question in the negative, since it has effectively required that the committees be composed so as to ensure diversity of expertise in the biomedical field and with respect to ethical, social, psychological and legal issues. This requirement follows the conditions set in its *Avis* by the CCNE, which has consistently linked ethics with the encounter of perspectives and disciplines and has therefore required that committees be pluralistic and multi-disciplinary.[19] Although one may envisage an effective exchange of viewpoints between biomedical and legal experts within the CCPPRBs, it seems difficult to

---

[18] Medical ethics texts regarding human trials give an account of the efforts that have been made to provide the clearest possible definition of research. The *Belmont Report* is exemplary on this issue, since it proposes a distinction between medical practice and research that considerably clarifies the situation. The 1993 *International Ethical Guidelines for Biomedical Research Involving Human Subjects* set out by the CIOMS and the WHO also provide evidence of the preoccupation with, and desire to provide, a distinction in the research domain between strictly cognitive activities and those with both a cognitive and therapeutic purpose. The CCNE also attempted to deal with these two types of situation and defined the particular conditions of each, specifically on the matter of consent (see *Avis sur les essais de nouveaux traitements chez l'homme, op. cit.*, at 21–22). It is unfortunate that the French law did not include this distinction.

[19] Comité consultatif national d'éthique pour les sciences de la vie et de la santé, *Avis n° 2* cited above; "Recommandations n° 13: sur les comités d'éthiques locaux" 7 November 1988, in *Xe anniversaire: Les avis de 1983 à 1993* (Paris, 1993) at 33, 39, 189, 195–196.

imagine the same taking place as between ethicists, psychologists and sociologists. Research proposals must be evaluated according to the conditions established by law which in turn requires the coordination of only two steps: scientific expertise and an appraisal of the consistency of the proposal with the conditions established by law.

The situation in Belgium raises three issues that acquire special significance when they are linked to those raised by reflections on the situations in France, Canada and Quebec. In France, one may be concerned about the risks of formalism that might be created by an exclusively legal treatment of the issues raised by research involving human subjects. In contrast, Belgium raises the possibility that by failing to establish any legal guidelines for the committees as they carry out their mandates, the royal decree requiring the creation of hospital committees may give rise to the demise of ethical reflection. Moreover, the French law may lead one to fear the disappearance of ethical, psychological and social approaches to the issues raised by human trials, despite some attempts to leave room for such approaches. In Belgium, on the other hand, due to the conditions of committee composition, one might be tempted to surmise that the royal decree requiring the creation of hospital committees will lead to the exclusion of these approaches altogether in favour of strictly medical and legal perspectives. Finally, the issue of the nature of ethical constraints is also raised in a somewhat different fashion when one compares the laconic way in which the royal decree describes the hospital committee mandates — broad as they are — to the detail with which both the French law and the *Canadian Code of Conduct for Research Involving Humans* describe the various tasks conferred on the CCPPRBs and research ethics committees respectively.

Let us consider the first issue. In Belgium, the diversity of philosophical opinions (characteristic of pluralistic societies) is very strongly organized, especially since certain common convictions underlie institutional networks. Belgium is indeed organized not only according to its different languages but also, in large part, as a function of its two social "pillars", one secular and the other Catholic. Almost all political parties, schools and hospitals are linked to one of these two pillars. Belonging to one or the other inevitably determines one's approach to medical ethics and codes of conduct. In many cases, differences may be relatively slight. However, they may become extremely important where issues are sensitive, such as those raised by embryos or terminally ill patients. By linking ethics committees to the hospitals, and thereby organically joining them to one or the other of the country's philosophical tendencies, there appears to be a definite risk of favouring the isolated development of two separate and competing ethics. This risk is even greater given that the hospital ethics committees must be composed of a majority of members from the hospital in which they were created, who are then given the three tasks outlined above. We may see in the not so distant future, however, that certain medical activities will require a legal framework that must be adhered to by all. In the absence of such adherence, the imposition of one side's views on the other may become necessary. There will therefore be a transition from ethics to law by way of a power struggle that will be far from ethical. This would surely be an unfortunate development. For several months now, however, Belgium has benefited from the creation, at the federal level, of a Consultative Committee on Bioethics after many years of deliberation. It seems probable that this Committee will provide the first forum for discussion concerning these issues. It is to be hoped that it will facilitate an effective dialogue, one going beyond the simple head-on clash of strongly held beliefs.

Turning now to the second issue, we will consider the risk that ethical reflection by hospital committees will be eliminated, particularly in their evaluation of research proposals. We mentioned above that the *Arrêté royal du 12 août 1994* establishes certain conditions regarding the composition of hospital ethics committees: each committee must be comprised of a majority of physicians from the hospital in question, at least one nurse from the same hospital, a general physician who has no link to the hospital, and a jurist. These conditions are not exhaustive; others may be added to this core group, although their presence is not necessary. To the extent that the required minimum composition will most likely be the one that exists in practice, it seems probable that hospital committee ethical practices will be based primarily on scientific expertise or professional wisdom. In other words, personal wisdom and professional competency are seen as interchangeable. This configuration does not warrant the label "ethical", since it does not give rise to the dimension unique to ethics that involves reflecting on norms and practices so as to clarify the interests at stake in relation to a vision of humanity and society. By requiring that a majority of the committee be medical professionals and deliberately making only vague provisions that might permit a more diversified approach to the issues raised by the evaluation of research proposals, the royal decree distorts the process in a manner conducive to the exclusion of ethical reflection.[20]

The laconic wording of the royal decree also leads us to consider the nature of ethics in another light, as mentioned above. Whether it is placed under the aegis of law or ethics, the apparently well-defined research proposal evaluation process in fact becomes quite complex. It may include the diverse aspects found in French law, or it may presume or imply other tasks (counseling, consultation, education) as found in the *Canadian Code of Conduct for Research Involving Humans*. Does such concise language really lead to creativity? Or will it simply promote a reductionist conception of the tasks damaging the scope and meaning of ethics that these tasks are guided by? This last question raises issues already discussed in connection with the conditions set by the royal decree regarding committee composition. The combination of these conditions with the lack of task definition effectively reinforces the two problems already highlighted. First, the process for research proposal evaluation is conceived of in a way that tends to reduce it to a control of the fulfillment of formal requirements set out by medical ethics, deontology or law. It does not leave room for contemplating the spirit of such requirements, even though this less formalistic mode of inquiry may be viewed as

---

[20] One may also note the paradox arising from the social act of creating ethics committees in Belgium, consisting in the absence of requirements relating to the participation of members of society who are not physicians, nurses or members of the legal community.

the hallmark of ethical reflection. Second, there is the risk of an undue assimilation of this type of reflection to the professional and personal attributes of physicians that definitely tends to obscure the need for diverse and multidisciplinary representation on committees.

The situation in Canada and Quebec also raises new matters for reflection, even if the law has had to respond to different demands than those encountered in France and Belgium. In Canada, the common law tradition is dominant, despite the fact that in Quebec it is combined with the civil law tradition. Without going into the details of the effects of this legal context, it should be noted that in Canada and Quebec the law intervenes to permit research ethics committees to fulfill their mandates by defining rules for the protection of persons, rather than to institute the committees themselves. These rules or conditions must also be combined with other scientific, administrative, professional or ethical criteria. The research proposal evaluation process has deliberately been placed within an ethical, rather than legal or deontological framework, even though it is of course morally obligatory to act consistently with legal requirements. In a context in which the existence of research ethics committees is legitimized, above all by administrative practice, the ethical dimension of research proposal evaluation is valued as an end in and of itself.

In contrast to the situation in France and Belgium, the Canadian experience leads us to consider whether the fact of placing research proposal evaluation within an ethical framework gives rise to the equitable treatment of all submissions. In France and Belgium, as we have seen, ethics is linked to a diversity of approaches and is therefore perceived as leading to differences among evaluations. In France, the law is viewed as guaranteeing greater uniformity in the processes it allows for. In Belgium, on the one hand, there are those who wish for legislative intervention that would reduce the current diversity in types of assessments, while, on the other hand, others have shown their preference for professional self-regulation that would allow this diversity to persist. In Canada and Quebec, the ethical approach is valued, over a purely legal one, to the extent that it allows for a greater cognizance of diversity. At the same time, it is expected that purely personal perspectives will be transcended through a specific training program for ethics committee members. The program will develop the members' abilities to analyze different types of situations and prepare them for the uncertainty they are likely to encounter, rather than relying on legal or professional norms that would decide matters by setting formal standards and conditions.

## REFLECTIONS

By comparing these three nations' paradigms of research committee institutionalization, we are in a position to indicate two risks threatening the quality of the work performed by the committees, even though we are not able to formulate an ideal model.

First, as we have observed, in each of the countries considered, the combination of various normative systems may, and often does, lead simply to the transposition of the requirements formulated within one system into another. These transpositions may take place from medical ethics, to deontology, to law, as well simply from deontology to law. The French *Loi n° 88-1138*, as amended, as well as EEC Directive 91/507 and the royal decree that transposes it into Belgian law, all illustrate the way in which conditions established under medical ethics may be imported into a legislative text that then makes them legally enforceable. The rules formulated by the *Conseil national de l'Ordre belge des médecins* also draw on requirements established by medical ethics and make them into professional obligations. Finally, the Belgian *Arrêté royal du 12 août 1994*, which legally entrenched committees that had hitherto come under the control of the *Ordre des médecins*, certainly incorporated some of the requirements formulated by the *Ordre*, converting them into obligations toward society in general and not just toward the *Ordre* itself.

This process creates the impression of a convergence in the way in which problems are perceived and treated. It also furthers a more general approach to the issues raised. However, the process masks the fact that the significance and scope of these issues are related to various conceptual structures, each being organized according to an internal logic specific to each particular normative system. Medical ethics and deontology aim to guarantee the quality of professional conduct through the definition of physicians' obligations. From this perspective, medical ethics makes moral recommendations to physicians — which appeal to the conscience but carry no concrete sanctions — whereas when deontology exists independently it is differentiated by its formulation of obligations that lead to disciplinary sanctions when violated.[21] Law, in contrast, pursues a different goal: the organization of relations between parties in question, taking into account their respective interests.

A brief analysis of the rules regarding consent is revealing in this regard: it allows us to become aware of the fact that this requirement, posited by each of the three normative systems, nevertheless carries a very different meaning and varies in scope among the three systems. I will demonstrate this phenomenon by recalling the results of research I undertook in Belgium regarding current human experimentation research norms.[22] I determined that medical ethics, inspired by a dual concern with ensuring that experiments are possible while at the same time protecting the subjects involved, attaches great importance to consent, although it also carves out several exceptions to the requirement that it be obtained. Despite the differences found in

---

[21] This general characterization would be incomplete without mentioning that the status of deontological rules differs among each of the three countries under discussion. In France, deontological rules are incorporated into the legal order by means of publication in the form of decrees. In Belgium, given the absence of royal signature, the *Code de déontologie médicale* has a *sui generis* position lying somewhere between ethics and the law: breaches may be sanctioned by disciplinary measures taken by professional bodies, but the rules are not legally binding. They may however be used in the process of judging a breach alleged under the *Code civil* or *Code pénal*. For their part, Belgian physicians view these rules as customary law: they constitute guidelines that are supplemented and corrected by opinions given in response to questions submitted to the *Conseil national de l'Ordre*. In Canada and Quebec, deontology does not constitute a normative class of their own, but rather are viewed as part of professional ethics.

[22] M.-L. Delfosse, *L'expérimentation médicale sur l'être humain. Construire les normes, construire l'éthique* (Bruxelles: De Boeck Université, 1993) at 131–247.

the many texts on the subject, it seems valid to maintain that consent is viewed as having two dimensions: protection of the subject's right to self-determination; and, given the obligation to inform implied by this protection, the creation of a relationship between the medical researcher and his or her subject. For its part, the *Code belge de déontologie médicale* gives only minimal importance to consent, which is required in only two types of experiments: those conducted on healthy volunteers and those that may give rise to problems for patients without their gaining any direct benefit. When limited to these conditions, the requirement of consent appears designed more for the protection of physicians than of patients. Fortunately, this approach has been supplemented by rules formulated in 1992 by the *Conseil national de l'Ordre* on ethics committees. These rules require that subjects be given information and that their consent be obtained. However, in their current state these rules are so general that they allow for highly diverse interpretations in certain cases where they may be difficult to satisfy. As for Belgian law, it "in principle" requires the free and informed consent of each subject in a clinical drug test, in addition to compliance with the conditions set out in the *Helsinki Declaration*. Other experimental activities are subject to the legal standards for medical acts with a therapeutic or preventive purpose. Within this framework, consent — which is demanded for any treatment that poses risks — takes on a particular significance. As the seal on the "medical contract", consent implies that the patient has accepted to bear the risks of any non-negligent act. It thereby gives expression to the patient's will, while also ensuring the protection of the physician.

Consent is therefore imbued with multiple meanings that reveal its many aspects. Of course, our analysis was only undertaken with reference to Belgium. We may nevertheless suppose that the conclusions drawn have broader implications. No matter what the characteristics of countries under consideration, the above analysis underlines the considerable impoverishment that would result if consent was understood solely as a function of just one of the normative systems. Furthermore, our conclusions reveal the problems, and even the risks, that arise where the requirements of one system are simply transposed into another. To ensure that the transposition is not reductionist in nature and that the requirement is properly received into the new system, there must also be a translation that takes into account the meaning and scope of any given requirement within its original system. Only in this way will the full scope of problems raised by research involving human subjects be understood and dealt with in a rigorous fashion. One might also ask, however, if fulfilling this condition will be sufficient or if it might not also be necessary to favour several diverse approaches, while at the same time seeking to establish and ensure that they are complementary. According to this point of view, the task is no longer just to determine which normative system must either take precedence over the others or assimilate their requirements. On the contrary, it now becomes necessary to recognize that each of these systems has a special role to play in developing a unique and specific approach. This also requires that we determine the nature of the relationships that exist between the different systems in order to ensure that they function in a complementary manner.

The above exposition and comparison of several national situations has led us to another problem related to the one just raised. We have observed that the work carried out by ethics committees is always subject to the risk that it will be given reductionist interpretations. On the one hand, their work is in danger of being too formalistic whenever it is undertaken in the name of evaluating research proposals in accordance with the conditions set out in the law. On the other hand, in light of the conditions imposed relating to committee composition, the work may become little more than the imparting of expert scientific opinion or the application of professional wisdom. Such interpretations do not follow only from the limitations imposed by various texts. They may also be the result of the minimalist language of the texts or, at the opposite extreme, a reaction to an overabundance of details. In any event, the committees have a crucial role to play. They allow research to proceed by ensuring respect for, and the protection of, the people involved. However, they also provide possible forums of reflection regarding the questions raised during the evaluation of research proposals. To this end, they may contribute to the refinement of norms already formulated in medical ethics, deontology or in the law, provided that bilateral exchanges between these normative systems and the committees are established. Once again, complementarity in the recognition of the diversity of competencies is shown to be necessary to ensure optimal treatment of the questions posed by experimentation on humans.

CHAPTER 16

# THE INSTITUTIONAL REVIEW BOARD: ITS ORIGINS, PURPOSE, FUNCTION, AND FUTURE

CHARLES R. McCARTHY

## BACKGROUND[1]

From its earliest days, the U.S. government has acknowledged, at least to a limited degree, a responsibility to provide health care for some segments of its society. Under President John Adams in 1798, the U.S. Congress authorized the establishment of the Marine Hospital Service for sick and disabled seamen as a division of the U.S. Treasury Department.

The decision to provide public support for research carried out to promote the public health of the nation did not develop for almost another century. It came with the establishment, in 1887, of the one-room bacteriological laboratory for investigation of cholera and other infectious diseases at the Marine Hospital, Staten Island, NY, under the direction of Dr. Joseph Kinyoun.

In the years that followed, the bacteriological laboratory conducted research into communicable diseases such as yellow fever, cholera, and smallpox. Congress gradually expanded the mission of the Marine Hospital Service and assigned new research tasks to its laboratory (designated the Hygienic Laboratory in 1891). The Marine Hospital Service was renamed the Public Health and Marine Hospital Service in 1902 and in 1912 it was reorganized into the Public Health Service (PHS).

In 1918 the Chamberlain-Kahn Act provided for the study of venereal diseases by the PHS. By that time the PHS had expanded beyond the confines of a small laboratory into a commissioned corps dedicated to promoting the public health

---

[1] C.R. McCarthy, "Research Policy" in W.T. Reich, ed., *Biomedical Encyclopedia of Bioethics* (New York: The Free Press, 1978) 1492. Most of the material dealing with the period prior to World War II is taken, with certain editorial liberties, from the cited article.

301

under the direction of a Surgeon General who was empowered by the Congress to initiate grants-in-aid of research. In 1918 the Surgeon General authorized awards for research to twenty five institutions.

The Second World War brought profound changes to health-research policy. President Franklin D. Roosevelt created the Committee for Medical Research to mount a research program aimed at reducing the effects of war-related disease and injury. In conjunction with the National Research Council of the National Academy of Sciences, the committee mobilized the community of biomedical scientists in an extraordinary outpouring of creative energy. Spectacular results were attributed to the work coordinated by the Committee including: testing and development of sulfanilimides; gamma globulin; adrenal steroids; cortisone; and many other drugs. A wide variety of treatment regimens and surgical techniques were initiated. Development of the capability to produce large amounts of penicillin led to the saving of the lives of thousands of American soldiers. The momentum generated by medical research efforts during the war years was the forerunner of unprecedented expansion of research efforts for more than two decades.

In 1944 the Congress enacted P.L. 78–410, landmark legislation known as the *Public Health Service Act*. The *PHS Act* consolidated all of the authorities under which the PHS operated and divided the PHS into the Office of the Surgeon General, the Bureau of Medical Services, the Bureau of State Services, and the National Institute of Health. The *Act* included authority for clinical research within PHS medical facilities. The *Act* gave the Surgeon General broad non-categorical powers to support research into the etiology, prevention, diagnosis, and treatment of human diseases and disabilities. It made the National Cancer Institute a division of the NIH. Virtually all future medical research authorizing legislation would come in the form of amendments to the *PHS Act*.

## THE LEGACY OF RESEARCH ETHICS FROM THE YEARS PRIOR TO THE END OF WORLD WAR II

David Rothman's history[2] suggests that concerns for the rights and welfare of research subjects involved in publicly funded research were, generally speaking, honored in research studies prior to World War II. He asserts that serious failures to obtain informed consent from human research subjects and other breaches of biomedical research ethics were exceptional prior to 1940, but they were commonplace during and after World War II. He may be correct. However, a careful

---

[2] D.J. Rothman, *Strangers at the Bedside* (New York: Basic Books, 1991) at 29. Rothman states that:

> In the end, neither Noguchi's research [reference to a diagnostic test for syphilis tested on subjects without their knowledge or consent] nor the other experiments on the retarded or mentally ill produced prosecutions, corrective legislation, or new professional review policies. Violations were too few; non-therapeutic research on captive populations was still the exception. And when the public learned about such incidents, objections quickly arose, reflecting a widely shared sense of what was fair and unfair in human experimentation. Had these norms held sway even as the methods of research changed, Beecher might not have been compelled to write his article.

reading of the Rothman text provides precious little evidence to support the view that, prior to World War II, serious breaches of research ethics were rare. It might be more accurate to suggest that the ethics of research prior to World War II were seldom mentioned, rarely documented, and have become, to a significant degree, historically invisible.

Reasons why the ethics of prewar medical research were difficult to discern are not difficult to understand. (1) Clear standards of research ethics were not systematically taught, or even well-defined, in the medical schools and the research community of that period. (2) Public funding of research involving humans constituted only a tiny part of the federal budget; and the number of publicly funded projects was relatively small. As a consequence, studies in which research subjects were involved did not attract much attention either from the Congress or the media. (3) A clear distinction between "research" and "care" was seldom made — either by the general public or by scientists who often played a dual role of physician/ scientist. (4) Public trust in the integrity of physicians appears to have been high. As a consequence, few subjects were inclined to criticize the ethics of research conducted by physicians.

Research investigators identified themselves primarily as physicians, rather than as research scientists, and because their subjects thought of themselves as patients rather than subjects, informed consent was rarely formally sought and subjects did not expect any formal process of consent. What changed in the course of World War II is, as Rothman notes, that research subjects were often unknown to investigators, and therefore had a much lower level of trust than the prewar subjects who usually considered the investigators to be their friends and private physicians.[3]

Responsibility for the ethical conduct of research was left almost entirely to the conscience of individual investigators who received little formal training in ethics. The Hippocratic traditions of respecting the privacy of patient/subjects and *"primum non nocere"* were not systematically taught, but were conveyed to investigators by mentors, role models, and by institutional traditions.

Despite Rothman's assertion that prewar research ethics were for the most part ethically acceptable, we are led to the uncomfortable conclusion that a documented account of biomedical ethics as an academic discipline or the quality of ethics practiced by those conducting research involving human subjects in the U.S. prior to World War II does not exist. Research and health care service were not distinguished by either researchers or their subjects, and therefore the investigators saw little reason to treat subjects any differently than patients, and subjects expected little more than administration of an acceptable level of care.

The absence of either federal or academic standards of conduct for research involving human subjects and the failure to recognize a need for such standards explains in part some of the moral lapses that occurred in World War II.

The federal support of research during World War II and the ethical aspects of that research, on the other hand, are comparatively well documented. President Franklin

---

[3] *Ibid.* at 51.

Roosevelt established the Office of Scientific Research and Development (OSRD) in 1941 to address a wide variety of scientific issues occasioned by World War II. OSRD in turn oversaw the Committee on Medical Research (CMR) that recommended funding of some 600 medical research projects (not all involving human subjects) to be conducted at 135 institutions throughout the country.[4] Although the NIH was already in existence, the CMR was the forerunner and model after which a more aggressive postwar NIH was fashioned. Rothman has rightly concluded that World War II was the event that forever changed both the volume and the methods of biomedical research involving human subjects. He says:

> The transforming event in the conduct of human experimentation in the United States, the moment when it lost its intimate and directly therapeutic character, was World War II. Between 1941 and 1945 practically every aspect of American research with human subjects changed. For one, a cottage industry turned into a national program. What were once occasional, *ad hoc* efforts by individuals now became well-coordinated, extensive, federally funded team ventures. For another, medical experiments that once had the aim of benefiting their subjects were now frequently superseded by experiments designed to benefit others — specifically, soldiers on the battlefront. For still another, researchers and subjects were more likely to be strangers to each other, with no necessary sense of shared purpose or objective. Finally and perhaps most important, the common understanding that experimentation required the agreement of the subjects — however casual or general the request or general the approval — was often superseded by a sense of urgency that overrode the issue of consent.[5]

Rothman appears to be on firm ground when he presents a number of examples and offers reasons why breaches of research ethics were commonplace during World War II. He presents the following statement as representative of the ethical reasoning of wartime research investigators:

> Researchers were no more obliged to obtain the permission of their subjects than the selective service was to obtain the permission of civilians to become soldiers. One part of the war machine conscripted a soldier, another part conscripted an experimental subject, and the same principles held for both.[6]

This argument, however appealing, does not fully explain the widespread failure to obtain and document informed consent from research subjects during the war. There is a long tradition that says that military personnel are not to be involved in research without their explicit informed consent. Those who likened research participation to the conscription of soldiers failed to recognize that health research did not involve, in most cases, issues of national security and safety. Furthermore, although military personnel surrendered many rights they enjoyed as civilians, the right to choose whether to be involved in research was neither explicitly nor implicitly waived by military personnel.

It seems more likely that failure to obtain informed consent from research subjects was already commonplace before World War II.

---

[4] Records of the Office of Scientific Research and Development, Committee on Medical Research, Contractor Records, National Archives, Washington, D.C. Record Group at 227.

[5] Rothman, *supra* note 2 at 30.

[6] *Ibid.*

## RESEARCH THROUGH THE 1950s

The unprecedented support and enthusiasm for medical research that characterized the postwar period represents a cultural shift in the way that American society viewed medicine. Although the public attitude toward medicine cannot be summarized in a few sentences, there seems to be considerable truth in Mark Frankel's statement that prior to World War II, medicine was viewed as the art of healing the sick and comforting the dying. After World War II the public view of medicine emphasized the conquering of disease and the promotion of health.[7] The meteoric rise of the NIH both reflected and facilitated that cultural shift.

The hoped-for benefits of research were frequently couched in military terms such as "conquest of disease", "war on cancer", "victory over suffering", "finding a silver bullet", "research strategies", and the like. These terms recalled popular military victories over the Fascist Axis forces that sought world domination. The "mobilization of efforts to combat disease" tended to create a somewhat naive hope for the early creation of a disease-free world.

As is often the case when a new technological capability is placed at public disposal, relatively little public attention was initially given to the subjects' inherent risks associated with biomedical research — risks to physical integrity, risks to social well-being, and risks to personal autonomy. Even less obvious was the risk that both the burdens and the benefits of research would be distributed inequitably across racial, ethnic, gender, and socio-economic segments of our society.

It was only near the end of the '50s that doubts concerning the benefits and the processes of new biomedical technology began to appear. Two events coalesced to make this happen. The first event was a series of hearings held by Senator Estes Kefauver, beginning in 1958 and continuing periodically until the passage of the Kefauver-Harris amendments to the *Food Drug and Cosmetic Act* in 1962.

As important as the amendments to the law were, the findings of the Kefauver Committee were perhaps more important in the long run. Kefauver publicly exposed the common practice of pharmaceutical houses at that time to provide samples of experimental drugs, whose safety and efficacy had not been established, to physicians who prescribed the experimental drugs without their patients' knowledge or consent. Physicians had agreed to serve as research investigators without informing their patients that they were research subjects.

The Kefauver-Harris amendments required, along with many other provisions, that informed consent be obtained in the testing of investigational drugs. For the remainder of the '60s, the FDA sought to implement the informed consent requirements required by law. Eventually, the creation of the IRB system enabled the FDA to implement the informed consent requirements of the law.

The other event that occurred was the documentation in 1961 of the thalidomide disaster in Europe, Canada and, to a lesser degree, the U.S. It was recognized that thalidomide was the cause of dreadful birth defects in newborn infants. By the time

---

[7] M. Frankel, *Public Policy Making for Biomedical Research, The Case of Human Experimentation* (Ph.D. dissertation, George Washington University, 1976) at 27. The attribution is slightly paraphrased.

of the thalidomide episode, television had spread to the majority of U.S. households. The visual impact of grossly handicapped "thalidomide" babies outraged and saddened viewers. More than any single stimulus in the early sixties, the thalidomide episode caused Americans to question whether the protections for human subjects involved in research were sufficient. In the U.S., thalidomide was not widely distributed because of questions raised by the FDA concerning the safety of the drug. Nevertheless, the thalidomide story added impetus to the Congressional effort to require oversight of safety and efficacy of drugs by the FDA.

## THE 1960s, A TIME OF UPHEAVAL

The tumult of the '60s sets that decade apart from all others in the 20th Century. Historians offer a variety of reasons to account for the turbulence of that period; whatever the reasons, it affected virtually every segment of American society. Pervasive lack of confidence and widespread mistrust of government were everywhere to be seen. A decline of public confidence in the military, the Congress, the Executive Branch, the Courts, organized religion, large business firms, labor unions, and institutions of higher learning characterized the period. Medicine and medical research did not escape the disillusion which extended to any enterprise that could be placed under the catch-all category of "the establishment".

Since the time of the Enlightenment and particularly in light of the writings of Francis Bacon, science had been viewed as a tool "for the relief of man's estate". The influence of the Enlightenment sharply declined in the '60s when the Cold War threat of imminent nuclear destruction of our planet was cynically seen to be the result of naive and uncritical reliance on science and technology to produce social progress, and the negative "payoff" for overconfidence in the structures of society to prevent disaster. Of course not all Americans were sympathetic to the new wave of social skepticism that was accompanied by unprecedented emphasis on individual freedom of expression and political activism, but a critical mass of Americans provided enough upheaval to change the course of American history.

## THE WINDS OF CHANGE IN RESEARCH INVOLVING HUMANS

Evidence began to accumulate that reliance solely on the integrity and the sound ethical judgments of research investigators did not provide adequate protection for research subjects. The transplantation of an animal kidney into a human subject at a Texas University without prior consultation, and with virtually no chance of providing therapeutic benefit or new scientific knowledge outraged NIH's Director, Dr. James Shannon.[8] One senior scientist who worked in the NIH intramural program at the time commented that "Shannon was not as deeply concerned about the rights and welfare of subjects as he was about the good name and public confidence placed in the NIH".[9] Senior officials recognized that what happened in

---

[8] *Ibid.* at 146.
[9] Private communication to author.

Texas could happen at most of the institutions that received NIH funding, and could easily happen in the NIH Clinical Center. "The absence of written guidelines on the use of investigative drugs or procedures on sick patients was no longer tolerable . . . a set of basic guidelines to govern this activity had to be developed".[10]

A three-year NIH grant to the Boston University Law-Medicine Research Institute revealed that in a study of 86 departments of medicine at American medical schools only 16 had written procedures for obtaining the informed consent of subjects, and only two had guidelines generally applicable to all clinical research.[11]

## THE FIRST FEDERAL EXTRAMURAL POLICY: THE BIRTH OF THE IRB

Discussions between NIH and the Office of the Surgeon General led to the creation of the Livingston Committee to recommend a suitable set of controls for the protection of human research subjects. A portion of the Livingston report was helpful to the development of policy, but the conclusion that "NIH is not in a position to shape the educational foundations of medical ethics . . .".[12] was entirely unsatisfactory to NIH's Director, Dr. Shannon. While these discussions were going on, two events occurred. First, the press revealed that research investigators at a New York Hospital had injected live cancer cells into an elderly and indigent population of subjects without their consent. Second, the World Medical Association issued the first version of its *Helsinki Declaration*. Although these incidents probably did little to shape the NIH policy, they added urgency to the process. A decision was made by Dr. Shannon and Surgeon General Luther Terry to bring the matter of a public policy for the protection of human subjects to the attention of the National Advisory Health Council (NAHC).

NAHC sent a recommendation to new Surgeon General William Stewart, who, on February 8 1966, issued Policy and Procedure Order 129 (PPO 129) requiring PHS grantee institutions to:

> . . . provide prior review of the judgment of the principal investigator or program director by a committee of his institutional associates. This review should assure an independent determination (1) of the rights and welfare of the individual or individuals involved, (2) of the appropriateness of the methods used to secure informed consent, and (3) of the risks and potential medical benefits of the investigation.

Although PPO 129 included few procedural details, nearly all of the ingredients of the mature policy for the protection of human subjects to be developed over the next fifteen years are found in an inchoate form in the document. PPO 129 required that institutions: (1) provide a written assurance of compliance with PPO 129 to the funding agency; and (2) conduct a review of the research prior to recruiting subjects

---

[10] Interview with Joseph S. Murtaugh cited in Frankel, *supra* note 7 at 146.

[11] *U.S. Congress Senate Committee on Government Operations*, U.S. Senate, 87th Congress, 2nd Session, 1962.

[12] R.B. Livingston, "Memorandum to the Director, NIH, Moral and Ethical Aspects of Clinical Investigation" (20 February 1964).

by an independent institutional committee. The institutional committee was to: (a) pass judgment on the appropriateness of the methods to be used to obtain informed consent from prospective subjects; and (b) determine what were the risks and potential benefits inherent in the proposed research.

The PPO document was brief and vague. It would be modified, elaborated, and expanded again and again in the years to come. Yet the basic structure of subsequent policies can be readily discerned in PPO 129.

PPO 129 lacked two major requirements that would not be supplied until 1981 after many further revisions had occurred. First, informed consent was not required in cases where the reviewing committee found no risk to subjects. The insight that subjects should not be "used" without their informed consent (irrespective of levels of risk) would not be reflected in the policy for many years. Second, PPO 129 and subsequent versions lacked a justice provision that would require the equitable distribution of the burdens and the benefits of biomedical research throughout society.

## ADMINISTRATION OF THE IRB POLICY

The NIH assigned only a small, part-time staff to exercise oversight over the new Policy for the Protection of Human Subjects. The Institutional Relations Branch (IRB) of NIH's Division of Research Grants (DRG) was assigned responsibility for implementing the policy. The number of research projects involving human subjects and the time involved in negotiating and approving detailed Assurance of Compliance documents with each awardee institution nearly overwhelmed the tiny DRG/IRB office. That office was charged to see that the Policy would be carefully applied by a properly constituted, well trained, institutional committee dedicated to the review of all research protocols involving human subjects (not merely those funded by components of the PHS). If events had not intervened, it seems probable that it would have taken many years, if not decades, before the Policy was seriously applied to human subjects research in a uniform way in all PHS awardee institutions that conducted research in the U.S. and abroad.

## CRITICISM OF THE ETHICS OF RESEARCH

The importance of the Policy was suddenly underscored by the publication of an article by Henry K. Beecher entitled "Ethics and Clinical Research" in the New England Journal of Medicine.[13] The Beecher article identified twenty-two published research studies involving human subjects that were, according to Dr. Beecher, ethically flawed. The broad sweep of the article indicted important figures in the biomedical research community, and criticized the medical journals that published the results of allegedly unethical biomedical research. Because Beecher was a

---

[13] H.K. Beecher, "Ethics and Clinical Research" (1966) 274:24 New England Journal of Medicine 1354.

highly respected Harvard Professor and a leading anesthesiology research investigator, his criticism of the ethical standards of his colleagues could not be easily dismissed. The Beecher article and its sequelae have been widely discussed in other contexts. It is cited here to illustrate the fact that the publicity surrounding the article lent credence to PPO 129 and alerted administrators within NIH and within research institutions across the country, that systemic protections for the rights and welfare of research subjects could no longer be ignored.

## POLICY DEVELOPMENT

The PHS Policy was expanded and revised in July of 1966, again in 1967 (to cover behavioral as well as medical research) and further revised in 1969. From the outset, the Policy assumed that the government would demand accountability, but the implementation of the Policy would be local. The basic thrust of the government's strategy was to issue procedural guidelines rather than a detailed substantive moral or ethical code. Dr. Shannon testified before Congress: "We do not feel that as a central body in Bethesda we can know all the circumstances which surround the local [research] procedure, and we think it would be unwise for us to make such a central judgment".[14] Eugene A. Confrey, Director of the Division of Research Grants was even more specific about the primary role of the institution in implementing the policy when he said, "[t]he government's role is to state general requirements as clearly and unambiguously as possible, but it should not try to tell an institution or investigator in meticulous detail how to meet these requirements ... the policy recognized both the scope and appropriate limits of federal action".[15]

Despite the fact that the Policy was clarified and reissued at regular intervals, implementation was very uneven. Forty site visits to randomly selected institutions revealed a wide spectrum of compliance ranging from full compliance to total disregard of the Policy. There was widespread confusion among committees about how to assess risks and how to weight possible benefits. Some research investigators refused to cooperate with their committees, and committees were unsure of their authority. In many cases the policy was met by indifference on the part of research administrators. Complaints of overworked review committees and requests for clarification and guidance came from all points of the compass. The tiny Institutional Relations Branch lacked legal authority and administrative backing to compel compliance. In the early years, the Policy was considered to require only a good faith pledge by awardee institutions that could be enforced by moral suasion. It took some time for the NIH to recognize that institutions could accept NIH awards

---

[14] J.J. Shannon, Testimony, *U.S. Congress, Senate Committee on Government Operations*, Subcommittee on Government Research, S.J. Res. 145, 90th Cong. 2nd Sess. April 2, 1968, 350.
[15] E.A. Confrey, "PHS Grant Supported Research with Human Subjects" (1968) 83 Public Health Reports 129.

only on condition that the Policy would be implemented, and any serious breach of the Policy could result in loss of funding.

> To many it was "an entirely new and strange concept",[16] and the PHS policy provided few guidelines ... institutions were permitted to review proposals at any time prior to their actual acceptance. Understandably, many institutions followed the practice of reviewing only after the actual awarding of a grant. While this was an administrative advantage for the institution as well as the investigator, it was a cause for concern among NIH officials.[17]

In 1970, Dr. Philip Lee, Assistant Secretary for Health and Scientific Affairs appointed a task force to review and revise the Policy. Dr. Eugene A. Confrey chaired the task force that confirmed the utility of the policy, but recommended changes. Institutional review committees were to be concerned primarily with (1) the rights and welfare of subjects; (2) the appropriateness and adequacy of efforts to obtain informed consent; and (3) the risks versus the potential benefits of the investigation or the importance of the knowledge to be gained. This statement indicating that research could be justified by the importance of the knowledge to be gained was new. It made clear that an informed subject could participate in research that did not hold out the prospect of direct benefit to the subject.[18]

Dr. Ernest Allen, Director of DHEW's Division of Grants Administration Policy, recognized that research was conducted and supported by agencies within the Department of Health Education and Welfare, but outside the health agencies of the Public Health Service. Allen urged that the applicability of the Policy be expanded to cover research conducted or supported by those agencies as well. Allen communicated his thoughts to John G. Veneman, Under Secretary for DHEW, who issued a memorandum supporting Allen's suggestion and calling for "a uniform policy, applicable to all its (HEW's) research support ...".[19] Dr. Allen, James H. Cavenaugh, Deputy Assistant Secretary for Health and Scientific Affairs, and Dr. John Sherman, Deputy Director, NIH, assigned responsibility for drafting a revised Policy for the Protection of Human subjects to Donald T. Chalkley, Ph.D., Special Assistant to the Director of DRG, NIH.[20]

The result of Dr. Chalkley's work was the production of the DEW Policy for the Protection of Human Subjects. (The original Policy had applied only to PHS agencies and the institutions that received awards to conduct research involving human subjects.) When the revised policy was published it was enclosed in a bright yellow cover, and it became widely known as the "Yellow Book". This 1970 version of the Policy contained both requirements and commentary on how the requirements

---

[16] D.T. Chalkley, "Developing Guidelines" (1973) 21 Clinical Research 777.

[17] Frankel, *supra* note 7 at 161.

[18] *Ibid.* at 163.

[19] J.G. Veneman, "Memorandum to the Commissioner of Education, the Administrator of Social and Rehabilitation Service, and the Commissioner of Social Security: Protection of the Individual as a Research Subject" (19 June 1969).

[20] Frankel, *supra* note 7 at 161.

were to be understood and implemented. Focus on the rights and welfare of individuals was emphasized, details concerning adequate informed consent were provided, risks were to be weighed against expected benefits to subjects and/or knowledge to be gained. Surrogate consent was described for children or others who were incompetent to consent. Review of grant applications and contract proposals by institutional committees was to take place *prior* to submission to NIH for peer review and funding. The Yellow Book provided the most enlightened U.S. policy statement that had been produced up to that time. Dr. Chalkley was appointed Director of the Institutional Relations Branch (IRB) and new staff was added.

The HEW Policy had matured to the point where it was workable and acceptable to most institutions and investigators within the research community. Assurances of Compliance were drafted and submitted to the Institutional Relations Branch of the Division of Research Grants at the NIH for approval. Admittedly, many of these assurances contained exculpatory language and enforcement loopholes that seemed to dilute the force of the policy. Many Institutional Review Committees (later designated Institutional Review Boards [IRBs] in amendments to the *Public Health Service Act*) began to function in a manner consistent with the Policy. Nevertheless, compliance was uneven. Some Assurances were written in such vague terms as to be essentially useless, many committees failed to provide rigorous or continuing review, and some seemed to think that if they required informed consent, their task was completed. Although the Policy began to work reasonably well in most institutions conducting research involving free, competent, adult subjects, it offered little guidance for research involving children, prisoners, cognitively impaired persons, and human fetuses. The Policy still allowed informed consent to be waived if review committees determined that the research involved "no risk" to the subjects. In order to avoid the necessity of obtaining written informed consent some committees found "no risk" in situations that were patently risky.

Nevertheless committees and research investigators were slowly learning a new way of conducting the business of research. The NIH was both leading the way and learning how to oversee a Policy designed to be enforced by a government agency, but administered at the institutional level.

The most serious shortcoming in the system was the lack of a commonly accepted set of standards or principles to guide the NIH, the institutions, and the committees in their judgment of the ethical aspects of research protocols.

But the situation was not hopeless. Biomedical ethics, as a discipline, was reaching a new level of maturity. Scientists gradually learned to invite ethicists to assist in developing research designs *before* the research was conducted. The NIH began a monthly series of bioethics seminars co-hosted with the newly established Kennedy Institute of Ethics at Georgetown University. Attendance grew from about four persons at the first seminar (including the speaker) to approximately 250 by the end of the decade. DHEW made it clear to awardee institutions that failure to discharge responsibilities for the protection of the rights and welfare of human subjects, whether or not DHEW funds were involved, could affect the eligibility of the institution to compete for awards.

## THE SENATE HEALTH SUBCOMMITTEE

In January, 1971, Senator Edward (Ted) Kennedy assumed chairmanship of the Subcommittee on Health of the Senate Labor and Public Welfare Committee. Kennedy teamed with Senator Jacob Javits (ranking minority member) and Senator Walter Mondale (who introduced Senate Joint Resolution 75 [S.J.RES 75] calling for a National Advisory Commission on Health, Science and Society), in focusing attention on biomedical research and its impact on society. Preparations for hearings on the need for an Advisory Commission were begun.

The November hearings were held. Only Assistant Secretary for Health, Merlin K. DuVal provided tepid opposition to the creation of the Advisory Commission on grounds that it was not needed because other mechanisms were able to provide adequate advice to the government. He also argued that a National Commission could threaten the autonomy of local institutional committees.

### *The Tuskegee Study*

It is difficult to exaggerate the political importance of the Tuskegee syphilis study that was publicly disclosed and exposed in 1971. The notorious study, begun in the 1930s, in which some 400 syphilitic black males were denied treatment in order to enable a study of the natural history of syphilis is well documented.[21] The ramifications of that study are still being felt today.

For our purposes it is sufficient to identify some of the political "fallout" that accompanied Tuskegee. The major political ramifications are listed below. Of course not all of these events were caused exclusively by the Tuskegee disclosures, but Tuskegee created a climate in which all of these events appeared to be natural sequelae.

> 1. Despite the fact that the PHS — later DHEW — Policy for the Protection of Human Subjects had been in place for six years, the Tuskegee study was exposed by a journalist rather than by a review committee. Although an institutional committee had, allegedly, reviewed the Tuskegee study it was not discontinued until a Commission was formed to evaluate it. Clearly the system of protections for human research subjects was not without serious flaws.
>
> 2. Dr. DuVal, Assistant Secretary for Health who had opposed the creation of a Bioethics Commission acted immediately to appoint a citizen's commission chaired by Dr. Jay Katz of Yale University to examine the study. Within a few weeks the study was discontinued.
>
> 3. No compensation scheme was available for subjects — or their survivors — injured in the Tuskegee study or any other research. [Note: Even today there is no program or system in place to provide benefits for subjects injured in the course of participation in research. In the few cases where compensation has been provided not as a matter of policy, but on a case by case basis.]
>
> 4. Senator Javits introduced a bill that would have made the DHEW Policy a regulation backed by Federal law. Senator Hubert Humphrey introduced a bill to create a National Human Experimentation Standards Board — a separate federal agency with authority over research similar to the Security and Exchange Commission's authority over fiscal

---

[21] J.H. Jones, *Bad Blood: The Tuskegee Syphilis Experiment* (New York: Free Press, 1981).

transactions. These and other Congressional proposals came late in the term. No action was taken.

5. Senator Kennedy held hearings, not only on Tuskegee but on a wide variety of other issues (some in the area of health care delivery), to illustrate the need for the National Commission.

6. Robert Q. Marston, Director NIH, delivered an address to the 1972 graduating class of the University of Virginia Medical School in which he called for a broad expansion of the DHEW Policy. He followed his talk in 1973 by creating a Study Group chaired by Ronald Lamont-Havers M.D., (soon to be) Associate Director for Extramural Affairs at NIH. Within a short time the committee was expanded to include representatives of all PHS agencies, and subsequently the entire DHEW. Its task was to examine the adequacy of protections for subjects involved in research. It was charged with reviewing the entire DHEW Policy and its administration. Dr. Lamont-Havers incorporated into the PHS Committee a small subcommittee formed by the National Institute for Child Health and Human Development (NICHD) under the direction of Dr. Charles U. Lowe. That committee had been formed to consider research involving pregnant women and human fetuses. Special attention was given to research involving prisoners, and the mentally infirm. Marston called for development of a compensation scheme for injured subjects and created a committee chaired by Seymour Perry, M.D., to address the matter. Although the Marston Address and the Study Group received little attention in the media, they were given high visibility within the PHS. A signal that the issues were to be taken seriously was the continuous presence of Richard Riseberg, new legal advisor to the NIH, and Joel Mangel, legal counsel for the PHS.

7. Dr. DuVal, recognizing that Congressional action was probable, issued a memorandum calling for a "posture of anticipation rather than defense . . .".

8. Controversy erupted over the ethical propriety of psychosurgery, described as "murder of the mind", in Congressional hearings on this topic in 1973.

9. NIH was accused in the public media of supporting research on live aborted fetuses. Rep. Roncallo (D.NY) circulated the charge throughout the Congress. Despite repeated denials and lack of evidence, the issue was linked with publicity concerning abortions, recently legalized by the *Roe* v. *Wade* decision of the Supreme Court.

10. Dr. Marston reorganized the Institutional Relations Branch of the Division of Research Grants (that had responsibility for oversight of human subjects protections). The Office was renamed The Office for Protection from Research Risks (OPRR) and was incorporated into the larger Office of the Director, NIH. Although OPRR remained in its former office space in the Westwood Building (several miles from the main campus of the NIH) and was given little new personnel or budget, it enjoyed considerably greater prestige and was perceived by the research community to be more powerful.

In a word, protections for human subjects had become a major and continuing issue for the DHEW, particularly the NIH, and for the public media. Television cameras recorded the procedures of the Senate Subcommittee hearings. Segments of the hearings were featured on all of the networks. Bioethics hearings in the Senate were among the first Congressional hearings to be televised. The public was fascinated.

Senators Javits, Mondale, Humphrey, Allen, and Sparkman introduced legislation pertaining to the rights of human research subjects at the outset of the 93rd Congress. Representatives Edward R. Roybal, and Paul Rogers introduced legislation in the House. Hearings were held by Senator Kennedy dealing with inappropriate use of *depo provera*, a contraceptive linked with cancer. The drug had been tested in poor Mexican women. Hearings were held by Senator Kennedy that suggested the pharmaceutical industry was preying on vulnerable populations in the Third World. A link was found between *diethylstylbesterol*, a drug used to quell morning sickness in pregnant women, and uterine cancer in female offspring of the women who had ingested the drug. The FDA had not approved the drug for use in pregnant women.

## LEGISLATIVE ACTION

The legislative maneuvering to produce a human research law in 1974 occupies close to one hundred pages of Frankel's excellent study. That process cannot be recorded here, or even adequately summarized. All that can be stated is that the final passage of compromise legislation occurred after DHEW had converted its Policy — first issued in 1971 — into regulatory form and published it in the Federal Register on May 30, 1974. Senator Kennedy had indicated to DHEW that he wanted to review the regulations before he took final action on the bill.

Title II of the *National Research Act* (P.L. 93–348), signed July 12, 1974, combined a number of features:

> 1. It required the Secretary, HEW, to issue regulations requiring each awardee institution to provide assurances satisfactory to the Secretary that it has established an Institutional Review Board to protect the rights of the human subjects involved in biomedical or behavioral research.
> 2. It required the Secretary to establish a program within the Department to respond to requests for clarification and guidance with respect to ethical issues raised in connection with biomedical and behavioral research. (This responsibility was delegated to the newly formed OPRR.)
> 3. It required establishment of a process for the prompt and appropriate response to information respecting incidences of violations of the rights of human subjects of research.
> 4. It established the National Commission for the Protection of Human Subjects of Biomedical and Behavioral Research, and charged the Commission to report on a variety of topics including: psychosurgery, research involving pregnant women and human fetuses, research involving prisoners, research involving the mentally infirm, research involving children, and a special report on ethical aspects of foreseeable future developments in technology.
> 5. It required the Secretary, HEW, to impose a moratorium on fetal research until such time as the Commission should make a report and recommendations concerning the conduct of such research.
> 6. It required the Commission to identify principles underlying the ethical conduct of biomedical and behavioral research.

## THE NATIONAL COMMISSION

The National Commission for the Protection of Human Subjects of Biomedical and Behavioral Research, created by the *National Research Act*, met for two days of each month for four years (1974–1978). It issued reports on fetal research, research involving prisoners, research involving children, informed consent and Institutional Review Boards. It also issued the *Belmont Report* that identified the principles of respect for persons, beneficence and justice as guiding norms under which all research involving human subjects should be conducted.

The Commission approved the fundamental thrust of the PHS regulations issued in 1974, but offered many recommendations for improvement. Through its method of public hearings and public debate involving the best scholars in the bioethics field at the time, the Commission provided credibility and respect to the DHEW system of protections for human subjects. For four years after the Commission completed its work, the DHEW, led by OPRR, struggled to incorporate the findings of the Commission into regulatory form. The notion that informed consent was not needed if there were no risks associated with the research disappeared and was replaced by a presumption that informed consent would be sought in all cases except where very narrow exceptions applied.

## COMPLETING THE REGULATORY FRAMEWORK

The major work of fulfilling the recommendations of the Commission was completed with the issuance of fundamental Regulations in 1981. Regulations governing research involving children were added in 1983. All but one of the Commissions reports was incorporated into regulatory form. The one glaring exception was the report of the Commission recommending special protections for "institutionalized mentally infirm" subjects. Although its recommendations were proposed in the *Federal Register*, opposition from many in the research community who claimed the proposed regulations were so stringent that they would stifle all research in the field, and from civil rights advocates who claimed that the proposed regulations were so lenient that they would allow institutionalized persons with cognitive impairments to be exploited. The Commission and the Department sought a middle ground in the debate, but were unable to produce a consensus. The upshot was that no federal regulations for the protection of research subjects who are institutionalized patients and who are cognitively impaired were ever promulgated. The federal regulations that provide protections for all subjects involved in research conducted or supported by the Department of Health and Human Services (HHS — formerly HEW)[22] of course pertain to cognitively impaired subjects, but, except for the fact that they empower IRBs to offer special protections to vulnerable subjects, they contain no special provisions for this category of research subjects.

From 1980 to 1983 The President's Commission for the Study of Ethical Problems in Medicine and Biomedical and Behavioral Research debated ethical issues in medicine. In 1981 that group recommended that all federal departments and agencies that conduct or support research involving human subjects follow a Common Rule modeled after the HHS regulations. The task of coordinating 16 federal departments and agencies and coordinating their efforts with the Department of State and the Office of Management and Budget (the White House) was once again led by the Office for Protection from Research Risks. This monumental task was finally completed on June 18, 1991 when the Common Rule was promulgated.

## THE PURPOSE OF THE IRB

Institutional Review Boards have always had many purposes. Nevertheless, there has never been any question that the paramount obligation and purpose of the IRB

---

[22] 45 C.F.R. 46 (1991).

is to protect the rights and welfare of human research subjects. The IRB accomplishes that goal directly through the review of informed consent documents and procedures, determining that risks are reasonable in light of expected benefits, and judging that the risks and benefits of research are equitably distributed across the population. Indirectly, the IRBs accomplish their work by requiring research investigators to address a whole range of ethical questions prior to submitting their protocols for review. Most investigators and most IRB members agree that, on average, the primary effect of the IRB is felt before they review a protocol. There is wide agreement, as well, that the IRB accomplishes its goal of protecting subjects primarily by educating the institutional community in which it does its work. The educational component of the mission of the IRB argues strongly for local rather than regional or national review bodies.

In June of 1991 the HHS Regulations for the Protection of Human Subjects were extended to research conducted or supported by all federal departments and agencies. IRBs are now required under the Common Rule by sixteen departments and agencies of the federal government. They are also required for research involving drugs, medical devices and biologics, regulated by the FDA. Their future seems assured. Since the regulations were revised in 1981 there have been many breaches of the regulations, but none has been anywhere as serious as the notorious Tuskegee Study. Judged simply in terms of research abuses, one must judge that the IRBs and the regulatory system that supports them have been a success.

## THE FUTURE OF THE IRB

The IRB system has evolved and matured in the U.S. for thirty years. Few biomedical research scientists conducting clinical investigations can remember a time when their work did not require review by an IRB. It has become a fixture in the firmament of U.S. medical research. Endorsed by the entire federal government and reinforced by the laws and policies of most state governments, it can now be said to be a part of the culture of biomedical research in the U.S. Perhaps the strongest indication of this fact is that policies on humane care and use of animals, regulation on recombinant DNA research, and proposed regulations for scientific integrity are all patterned after the regulations that govern IRBs.

Nevertheless, the system of IRBs in the U.S. faces many daunting challenges. First and foremost is the challenge of implementing a regulation, the oversight of which is now spread across sixteen departments and agencies of the federal government. The departments and agencies have different missions, different subcultures, and different experiences with regulations. The regulations apply to both civilian and military research. Can they be applied consistently and effectively in both settings?

The paradigm case for which the regulations were written was one in which a single investigator in a single institution with a small cohort of subjects carries out research under the watchful eye of a local IRB. Today, a single research protocol is often funded and conducted in twenty-five, fifty, or even a hundred different institutions in two, three, or more countries. The protocol itself may be designed by a consortium of government and industry scientists. In such a setting the role of a single IRB is uncertain and its ability to protect human subjects limited, since the local IRB seldom sees the data as it develops. Continuing review is often assigned to Data and Safety Monitoring Boards that serve in an advisory capacity to the sponsors (either government or industry) of the research. The system, as it was designed, still has much to recommend it, but it must be adjusted and made more adaptable if it is to continue to offer adequate protections for the rights and the welfare of human research subjects.

The AIDS epidemic has caused the FDA to produce treatment INDs and parallel track regulations that allow the distribution of drugs for which complete safety and efficacy data have not been collected and evaluated. The role of the principle of beneficence — the balancing of risks and benefits — has been diminished and the principle of autonomy is relied upon as the primary means of protection for desperate subjects who are often willing to try any drug that offers even a faint glimmer of hope. What is the role of an IRB in evaluating the release of drugs for treatment programs that are offered in the name of research — lacking the rigor of carefully designed trials?

Finding ways to provide minorities and women with equal access to biomedical research without enticing, entrapping, or recruiting under false pretenses is another problem facing research in which the IRBs must play both an educational and motivating role.

As research becomes ever more invasive, justice requires that reasonable and appropriate compensation for injured research subjects be provided. The federal government has wrestled unsuccessfully with this problem since 1975. The problem becomes ever more acute. The very credibility of IRBs may depend on finding a way to provide compensation for legitimate cases without exposing research to a flood of spurious cases.

The IRB has served the United States and many other countries well. It is here to stay. But unless it is considered to be an evolving and expanding mechanism, adapting to the problems of each period of history, it is in danger of becoming fossilized and ineffective. Administrators, research investigators, ethicists, regulators, the Congress and the general public bear the responsibility of creating mechanisms and methods for the IRB to continue to protect human subjects in a manner that is demanded by the highest principles of human ethics.[23]

---

[23] C.R. McCarthy, "Overview: Challenges for the 1990s" in R.H Blank & A.L. Bonnickson, eds., *Emerging Issues in Biomedical Policy: An Annual Review*, vol. 2 (New York: Columbia University Press, 1993) 156.

CHAPTER 17

# THE REGULATION OF BIOMEDICAL EXPERIMENTATION IN CANADA: DEVELOPING AN EFFECTIVE APPARATUS FOR THE IMPLEMENTATION OF ETHICAL PRINCIPLES IN A SCIENTIFIC MILIEU

SIMON N. VERDUN-JONES AND DAVID N. WEISSTUB

**INTRODUCTION**

The regulation of biomedical experimentation became an urgent issue after the horrors of the Nazi atrocities, committed upon prisoners during World War II, became widely known in the world community. The immediate outcome of these atrocities was the articulation of ethical and legal codes that were designed to regulate the process of biomedical experimentation with human beings.[1] The first, and undoubtedly most famous, of these codes was the so-called *Nuremberg Code* that emerged from the trial of the defendants involved in the Nazi experiments.[2] Subsequently, at an international level, the most significant codes of ethics that have been widely embraced are the World Medical Association's *Declaration of Helsinki* (first adopted in 1964 and revised on a number of occasions between 1975 and

---

[1] M.P. Dumont, "Book Review: *The Nazi Doctors and the Nuremberg Code*, by George J. Annas & Michael A. Grodin" (1993) 36 Social Science & Medicine 1519 at 1519–1520.
[2] *U.S. v. Karl Brandt et al., Trials of War Criminals Before the Nuremberg Military Tribunal Under Control Council Law No. 10.* (October 1946-April 1949).

318

1989)[3] and the Council of International Organizations of Medical Sciences' *International Guidelines for Biomedical Research Involving Human Subjects* (promulgated in 1993).[4]

In addition to the emergence of international legal and ethical codes, there have been parallel developments at the national level. In the United States, for example, federal regulations, embodied primarily in the *Federal Policy for the Protection of Human Subjects*, establish the basic requirements for biomedical experimentation, although these regulations are supplemented by state legislation and the requirements of local institutions.[5] In Canada, with the exception of the Province of Quebec, there is no comparable legislation; however, the Medical Research Council of Canada issued a set of ethical guidelines in 1987 that were applied to all research projects funded by the Council.[6] These guidelines were superceded in 1997 by a *Code of Conduct for Research Involving Humans* that was endorsed by the three national funding councils in Canada.[7]

While there appears to be an emerging international consensus as to the general nature of the basic ethical principles that should guide the conduct of biomedical experimentation with human subjects, there has nevertheless been a keen debate in relation to the critical question of exactly how these principles should be implemented in practice. This debate encompasses such issues as whether biomedical experimentation should be regulated, in whole or in part, by *legislation* or whether control of the research enterprise should be left primarily in the hands of the granting agencies, universities, the medical professions and researchers themselves through a process of *self-regulation*. At present in Canada, only the Province of Quebec has followed the legislative route; elsewhere, a pattern of self-regulation continues to hold sway.

This chapter addresses the fundamental question of how biomedical experimentation should be regulated in the Canadian context. The range of potential regulatory alternatives is discussed and the experience of the United States and the United

---

[3] World Medical Association, *Declaration of Helsinki*. Adopted at the 18th World Medical Assembly in Helsinki in June 1964. Amended at the 29th World Medical Assembly in Tokyo in October 1975; at the 35th World Medical Assembly in Venice in October 1983; and at the 41st World Medical Assembly in Hong Kong in September 1989. [Reprinted in (1991) 19 Law, Medicine & Health Care 264.] In 1974, the World Medical Association also adopted the *Declaration of Tokyo*, which prohibited physicians from involvement in any form of torture or cruel, inhuman or degrading treatment of prisoners or political detainees.

[4] Council for International Oganizations of Medical Sciences, in collaboration with the World Health Organization, *International Ethical Guidelines for Biomedical Research Involving Human Subjects* (Geneva: CIOMS, 1993). The Steering Committee on Bioethics of the Council of Europe has recently (June 1996) adopted a *Convention on Human Rights and Biomedicine*. If the Convention is approved by the Council of Ministers and is signed by member states, it is expected that it will be followed by, *inter alia,* a specific protocol concerning research with human subjects. See Chapter 14.

[5] See 45 C.F.R. 46 (1991).

[6] Medical Research Council of Canada, *Guidelines on Research Involving Human Subjects* (Ottawa: Minister of Supply and Services Canada, 1987). [hereinafter *MRC Guidelines*]

[7] Tri-Council Working Group, *Code of Conduct for Research Involving Humans* (July, 1997). The three funding councils are the Medical Research Council of Canada (MRC), The Natural Sciences and Engineering Research Council of Canada (NSERC) and the Social Sciences and Humanities Research Council of Canada (SSHRC).

Kingdom in this arena is examined with a view to informing the debate as to the nature of the appropriate regulatory policies that should be adopted in Canada. The central role played by research ethics boards (or committees) in the regulation of biomedical experimentation in Canada, the United Kingdom and the United States is examined and recommendations are made as to the nature and direction of reforms to the existing system. One of the most important goals of any regulatory process must be to protect human subjects (particularly those who are members of vulnerable groups) from the very real dangers of unethical research. However, such a process must also operate so as to ensure that ethically sound experimentation with human beings is facilitated and encouraged because, without such research, sorely needed advances in medicine may well not be accomplished.

## THE SOCIAL CONTROL OF BIOMEDICAL EXPERIMENTATION: THE RANGE OF ALTERNATIVE REGULATORY MECHANISMS

There is a marked diversity of mechanisms that are potentially available to assert regulatory control over the conduct of biomedical experimentation. These mechanisms include: (1) *intraprofessional controls* based on medical licensing and disciplinary proceedings administered by the appropriate professional bodies; (2) *judicial controls* based either on the law of torts and criminal law or on specific legislation dealing with experimentation; and (3) *independent boards of review* that administer ethical rules that have been officially sanctioned in one form or another. In the United States and an increasing number of other countries, there has been a dramatic departure, over the past several decades, from reliance on primarily intraprofessional methods of regulation that are based on broad ethical codes emerging from within the confines of the medical community itself. Instead, as Benson points out, there has been a determined shift towards adoption of a system of controls that is administered by bureaucratically organized boards of review that are, on the face of it, external to the medical professions themselves.[8] Indeed, most countries have adopted the system of research ethics boards (or committees)[9] and it has been given the seal of approval by such international agreements as the *International Ethical Guidelines for Biomedical Research Involving Human Subjects*, promulgated by the Council for International Organizations of Medical Sciences (CIOMS) in 1993.[10]

Why have ethical review boards assumed such a central role in the regulation of biomedical experimentation in so many countries? To some extent, their very dominance in this field reflects the failure of other regulatory mechanisms to establish themselves as viable alternatives. For example, the general view seems to

---

[8] P.R. Benson, "The Social Control of Human Biomedical Research: An Overview and Review of the Literature" (1989) 29 Social Science & Medicine 1 at 1. See also P.R. Benson & L.H. Roth, "Trends in the Social Control of Medical and Psychiatric Research" in D.N. Weisstub, ed., *Law and Mental Health: International Perspectives, Volume 4* (New York: Pergamon Press, 1988) 1 at 3–4.

[9] See Chapter 13.

[10] See *CIOMS Guidelines, supra* note 4.

be that intraprofessional controls are not very effective as a means of controlling abuses in biomedical experimentation. In particular, it has been suggested that such factors as medical education, peer influence, codes of ethics and disciplinary procedures are not *per se* adequate to prevent unethical research practices.[11] Similarly, the point has been made that it is scarcely surprising that there have been very few reported disciplinary cases in which questions have been raised concerning the propriety of physicians conducting biomedical experiments, because, in general, the available research suggests that disciplinary actions taken by licensing boards are of dubious efficacy and, in any event, disciplinary actions based on allegations of incompetence are rare.[12]

Social control of biomedical experimentation may also be achieved through litigation before the courts. In particular, a civil action may be brought by a research subject who alleges a cause of action recognized by the law of torts. In the context of biomedical experimentation, the most likely action is one based on alleged negligence.[13] In theory, the very existence of the tort system is supposed to act as a deterrent to negligent conduct or unethical conduct on the part of biomedical researchers.[14] However, there has been a marked paucity of court cases that have arisen in the specific context of biomedical experimentation and the few cases that have been brought to trial have focussed on the issue of informed consent.[15] They have not dealt with any of the issues that arise from biomedical experimentation with subjects whose capacity to consent is in doubt nor have they come to grips with the thorny problem of how one balances risk to the subject against the potential benefits to society that may flow from such experimentation.[16] While there is some evidence that this may be an area in which the courts may, in the future, become more involved as a regulatory mechanism,[17] it is clear that, to date, the role of the civil courts has been relatively minor.

The criminal law has generally not been used as a means of regulating biomedical experimentation. In Canada, for example, there is a number of provisions in the

---

[11] Benson, *supra* note 8 at 2–4.

[12] J.A. Goldner, "An Overview of Legal Controls on Human Experimentation and the Regulatory Implications of Taking Professor Katz Seriously" (1993) 38 Saint Louis University Law Journal 63 at 68–69.

[13] *Ibid.* at 70–88; Benson, *supra* note 8 at 6–7; Benson & Roth, *supra* note 8 at 12–15; P.M. McNeill, The Ethics and Politics of Human Experimentation (Cambridge: Cambridge University Press, 1993) at 121–138.

[14] Goldner, *supra* note 12 at 70.

[15] See e.g. McNeill, *supra* note 13 at 122. For cases in Canada, see *Halushkav. University of Saskatchewan* (1965), 53 D.L.R. (2d) 436 (Sask.C.A.); *Weiss v. Solomon* [1989] R.J.Q. 731 (S.C.). For cases in the United States, see, for example, *Kaimowitz v. Michigan Department of Mental Health*, Civil Action No. 73–19434-AW (Wayne County, Michigan Cir. Ct. 1973) in A.D. Brooks, *Law, Psychiatry and the Mental Health System* (Boston: Little Brown, 1974) at 902 ff.; *Blanton v. United States*, 428 F. Supp. 360 (D.C. 1977); *Burton v. Brooklyn Doctors Hospital*, 452 N.Y.S. 2d. 875 (N.Y. 1982); *Begay v. The United States*, 768 F. 2d 1059 (4th Cir. 1985), and *Moore v. Regents of the University of California*, 793 P. 2d (Cal. 1990), cert. denied, 499 U.S. 936 (1991).

[16] However, in *Muir v. Alberta* [1996] 4 W.W.R. 177 (Alta. Q.B.), damages were awarded against the Government of Alberta for the wrongful sterilization of a woman who had been in the care of the Province.

[17] See Benson, *supra* note 8 at 6.

*Criminal Code* that might be relevant to experimental abuses, particularly those provisions relating to assault.[18] However, in practice, the criminal law is not resorted to for this purpose. The Law Reform Commission of Canada made the following observations about the role of the criminal law in medical treatment and, in many respects, they are equally applicable to the field of biomedical experimentation:

> The impact of criminal law on the administration of treatment is largely overlooked in the Canadian context. Potential criminal liability is rarely considered by doctors or hospitals when seeking consent to or waiver of treatment. The handful of prosecutions in the last fifty years, as compared with the increasing frequency of civil litigation suggests to some that the Criminal Code is ineffective in this area because the type of harm contemplated in the administration of treatment is not of the degree to warrant the intervention of the criminal law. Indeed, the risk of criminal liability has been said to be more the "product of a fertile legal mind than a realistic possibility".[19]

What has been the role of legislation as a means of regulating biomedical experimentation? In the United States, the elaborate institutional review board structure has been established by regulations that have been issued by federal agencies such as the DHHS and FDA and have been codified in the Code of Federal Regulations.[20] Furthermore, a number of individual states have, since the 1970s, enacted legislation regulating various aspects of human experimentation,[21] with the most detailed and comprehensive statutory framework being established in the State of California, where the legislators included an "experimental subjects' bill of rights".[22]

However, to date, countries such as Canada, New Zealand, and the United Kingdom have made relatively little use of legislation as a means of regulating biomedical experimentation, although there have recently been a number of calls to give some form of legislative backing to the system of ethics review in those jurisdictions.[23] In Canada, while the Medical Research Council of Canada's *Guidelines on Research Involving Human Subjects*[24] are applied by most ethics review boards that scrutinize the conduct of biomedical experimentation, they do not have any independent legal force. Significantly, the Law Reform Commission of Canada has recommended the introduction of legislative controls, at the federal level, through amendments to the *Criminal Code* and a general federal statute on

---

[18] See, for example, *Criminal Code*, R.S.C. 1985, c. C–46, sections 266 (assault); 267 (assault causing bodily harm); 268 (aggravated assault); 269 (unlawfully causing bodily harm). For discussion of the issue of consent in the context of assault, see *R.* v. *Jobidon* (1991), 66 C.C.C. (3d) 454 (S.C.C.) and *R.* v. *Welch* (1995), 101 C.C.C. (3d) 216 (Ont.C.A.).

[19] Law Reform Commission of Canada, *Working Paper 26: Medical Treatment and the Criminal Law* (Ottawa: Minister of Supply and Services Canada, 1980) at 1.

[20] See 45 C.F.R. 46 (1991). Note also *Federal Policy for the Protection of Human Subjects: Notices and Rules*, 56 Federal Register 117 (18 June 1991).

[21] Benson, *supra* note 8 at 7.

[22] California Laws 1987, ch.. 1.3, codified as California Health and Safety Code, ss. 24170–24179.5.

[23] See McNeill, *supra* note 13 at 136–137. Legislative intervention in the area of medical ethics did occur in Australia: see *National Health and Medical Research Council Act, 1992*, 1992 Austl. Acts 225, ss. 35–36.

[24] *MRC Guidelines*, *supra* note 6. For a critical review of these *Guidelines*, see B. Hoffmaster, "The Medical Research Council's New *Guidelines on Research Involving Human Subjects*: Too Much Law, Too Little Ethics" (1990) 10 Health Law in Canada 146.

experimentation.[25] To date, no action has been taken on these recommendations. However, there is now one provincial legislature that has enacted statutory provisions that deal explicitly with biomedical experimentation, namely the National Assembly of Quebec. These provisions, the content of which will be addressed later, were enacted as amendments to the *Civil Code of Quebec*.[26] As is the case in the United States, the Quebec legislation articulates basic principles concerning the conduct of biomedical experimentation and establishes a legislative framework to underpin a system of regulation by research ethics boards.

Although a number of alternative regulatory mechanisms are available, there is little doubt that the most important component in the contemporary apparatus for the control of biomedical experimentation in many countries around the world is the research ethics board. In some jurisdictions, there is a legislative framework that supports the work of these boards while, elsewhere, they operate in the context of a non-statutory system of self-regulation administered by the granting agencies, universities, the medical profession and biomedical researchers themselves. It is to the nature and functions of the research ethics boards that the focus of inquiry will now turn.

## THE EXISTING STRUCTURE OF RESEARCH REVIEW: THE CENTRAL ROLE OF THE RESEARCH ETHICS BOARD

Institutional research ethics boards (or "committees") have become well-entrenched in a number of countries, with 140 in Australia, 240 in England and Wales, over 100 in Canada, and over 5000 in the United States.[27] The evolution of such quasi-legal decision-making mechanisms may be traced back to the 1960s but it acquired its most significant degree of momentum in the late 1970s with the promulgation of the revised *Declaration of Helsinki* in Tokyo in 1975.[28]

Specifically, research ethics boards (known, in the United States, as institutional review boards or IRBs) have evolved in a context in which biomedical research is carried out through "public or quasi-public medical research funding agencies, generally termed National Research Councils (NRCs)".[29] NRC research funds are generally disbursed through a system of competition-based grants and most NRCs stipulate that all governmentally funded research must undergo an ethical review before it may proceed. Unlike many other countries, in the United States, this process of ethical review is unequivocally required by legislation. In those countries where there is no such legislation underpinning the system of ethics review, the

[25] Law Reform Commission of Canada, *Working Paper 61: Biomedical Experimentation with Human Beings* (Ottawa: Law Reform Commission of Canada, 1989) at 61–63. [hereinafter LRC *Working Paper No. 61*]
[26] L.Q. 1991, c. 64, as amended by *An Act respecting the implementation of the reform of the Civil Code*, L.Q. 1992, c. 57. See articles 20 to 25. See Chapter 10.
[27] National Council on Bioethics in Human Research, "Protecting the Human Research Subject: A Review of the Function of Research Ethics Boards in Canadian Faculties of Medicine" (1995) 6 NCBHR Communiqué 3 at 5.
[28] See Benson & Roth, *supra* note 8 at 34.
[29] *Ibid.* at 34.

exact legal status of the ethics review boards is somewhat unclear and it may well turn out that the only effective sanction that may be imposed on a researcher who does not comply with the requirement of ethical review is the denial of government funding.[30]

The current Canadian system for regulating biomedical experimentation reflects a model of ethics review that is undoubtedly predicated on the pivotal role of review committees. The model consists of three general components: (1) a requirement or duty that medical studies involving human research pass before a local institutional research ethics committee; (2) national research ethics guidelines, principles or criteria, which are to apply to the evaluation of individual medical research proposals; and (3) broadly articulated goals or purposes to guide the functions of local research ethics committees.[31]

Before turning to a detailed examination of the operation of the nature and functions of research ethics committees in Canada, we shall establish a more general context in which to anchor this discussion by considering the development of the ethics review board in the United States, the trailblazer for this particular mechanism for the regulation of human experimentation, and in the United Kingdom, where the system of regulating biomedical research closely resembles that which exists in Canada.

### *Evolution of the Review Board System in the United States*

The need for regulation of biomedical experimentation should have become apparent during World War II, when the exigencies of war were used as a justification for subjecting human subjects to experiments that, it was argued, would assist the U.S. war effort even though today there would be little doubt that they would be considered unethical. During this era, prison inmates were routinely used in the testing of new drugs that might assist members of the armed forces in combat and some were deliberately infected with malaria in order to test the efficacy of available treatments. However, there was little, if any, contemporary criticism of the unethical nature of these practices. Indeed, the participation of prisoners was clearly regarded by many as a "patriotic" contribution to the American war effort.[32] At the time, emphasis was placed on the utilitarian justification for such experiments and their potential contribution to the struggle against the enemy rather than on such profoundly ethical considerations as whether these prisoners gave a free and informed consent to such experiments.[33]

---

[30] *Ibid.* at 34–35.

[31] See NCBHR, *supra* note 27 at 6.

[32] See K. Schroeder, "A Recommendation to the F.D.A. Concerning Drug Research on Prisoners" (1983) 56 Southern California Law Review 969 at 971 and C.M. McCarthy, "Experimentation on Prisoners" (1989) 15 New England Journal on Criminal and Civil Confinement 55.

[33] See D.J. Rothman, "Ethics and Human Experimentation: Henry Beecher Revisited" (1987) 317 New England Law Journal of Medicine 1195. The fact that these issues are still relevant in the contemporary world is illustrated by the contention that, in 1990, the F.D.A. and the Department of Defence of the United States Government permitted the use of experimental drugs and vaccines on U.S. troops involved in the Gulf War without the soldiers' consent; see G.J. Annas & M.A. Grodin, "Treating the Troops: Commentary" (1991) 21 Hastings Center Report 24.

However, it was not these wartime experiments with prisoners that ultimately led to the intervention of the federal government. In the 1960s, a number of scandals, involving the abuse of human research subjects came to the attention of the public. For example, in 1966, Henry Beecher published an article in which he provided information about some 22 cases of unethical research involving human subjects.[34] Such reports contributed towards an emerging consensus that there should be legislative intervention to control the enterprise of biomedical experimentation in the United States. Nevertheless, it took the discovery of one particular example of unethical experimentation for the U.S. Congress to build on this consensus and to implement a régime of legislated regulations.

The immediate spur to Congressional action was intense criticism of the so-called Tuskegee Study, which commenced in the early 1930s but did not come to the attention of the American public until 1972. This study involved a study of syphilis that was inaugurated by the United States Public Health Service (Venereal Disease Division). The study was designed to investigate the effects of untreated syphilis on the human body (assessed on the basis of autopsies). Approximately 600 black males, from Tuskegee, Alabama, were observed for an extended period in order to determine the natural progression of the disease: 400 of these men had the disease and 200 (the control group) did not. The major ethical issue that arose centred on the fact that, by the mid-1940s, penicillin had become readily available and was considered to be an unequivocally effective cure for syphilis. Nevertheless, subjects continued to be observed by the researchers without the administration of penicillin. The Department of Health, Education and Welfare finally called a halt to the study in 1972 and the whole sorry episode became the centre of Congressional hearings in 1973.[35]

The outcome of these hearings and the public furor that surrounded them was the enactment, in 1974, of the *National Research Act*, which established the National Commission for the Protection of Human Subjects of Biomedical and Behavioral Research. The Commission was given the mandate to review the "the problems and practices associated with protection of the rights and welfare of human subjects involved in various forms of biomedical and behavioral research".[36] Over a period of four years, the Commission produced nine separate reports that addressed a variety of topics including research involving prisoners, the "institutionalized mentally infirm" and children; the nature of informed consent; the selection of human research subjects; psychosurgery; and sterilization.[37]

---

[34] H. Beecher, "Ethics and Clinical Research" (1966) 274 New England Journal of Medicine 1354.

[35] See J.H. Jones, *Bad Blood* (New York: Free Press, 1981).

[36] See J.V. Brady & A.R. Jonsen, "The Evolution of Regulatory Influences on Research with Human Subjects" in R.A. Greenwald, M.K. Ryan & J.E. Mulvihill, eds., *Human Subjects Research: A Handbook for Institutional Review Boards* (New York: Plenum Press, 1982) 3.

[37] The most influential of the reports was the so-called Belmont Report. See National Commission for the Protection of Human Subjects of Biomedical and Behavioral Research, *The Belmont Report: Ethical Principles for the Protection of Human Subjects of Research* (Washington, D.C.: U.S. Government Printing Office, 1978).

On the basis of the recommendations made by the National Commission, the U.S. Department of Health, Education and Welfare ultimately prepared regulations for the conduct of biomedical experimentation with human subjects that were enacted in 1981. These regulations have since been amended, most significantly in 1991. The June 1991 revision encompassed the adoption of the *Federal Policy for the Protection of Human Subjects*, promulgated by the sixteen federal agencies that conduct, support, or are otherwise involved in the regulation of research with human subjects.[38] Adoption of this policy resulted in a uniform system for protecting human subjects in all relevant federal agencies and departments.[39] Furthermore, as Starkman points out, the common Federal Policy in 1991 rendered the legal regulations more user-friendly".[40] The federal regulations serve as the bedrock upon which the system for regulating biomedical experimentation with human beings is constructed; however, they have been modified and extended by legislative action in certain states and by the specific requirements of individual institutions that conduct research.[41]

At the heart of the regulatory system in the United States is the Institutional Review Board. The IRB in any given institution has the "authority to approve, require modifications in, or disapprove all research activities that fall within its jurisdiction as specified both by the federal regulations and local institutional policy".[42] As far as jurisdiction is concerned, it is clear that the regulations apply to "all research involving human subjects conducted, supported, or otherwise subject to regulation" by any federal department or agency that has adopted the human subjects regulations.[43]

Each institution involved in research covered by the regulations must give a written "assurance" to the federal department or agency concerned that it will comply with the requirements of the *Federal Policy for the Protection of Human Subjects*. Without this assurance and the prior approval of individual protocols by an IRB, the relevant department or agency will not conduct or support any research in the institution concerned.[44] The assurance must include: (1) a statement of the general ethical principles that will govern the institution in discharging its duty to protect the rights and welfare of human subjects in research; (2) designation of at least one IRB which will apply these principles and which will meet the federal requirements for a diverse membership base; (3) written procedures which the IRB will follow in reviewing research protocols and which require that any significant changes in a research activity will be promptly reported to the IRB; and (4) written procedures for reporting any unanticipated problems involving risk to subjects or

---

[38] See 45 C.F.R. 46 (1991).

[39] See Office for Protection from Research Risks, National Institutes of Health, *Protecting Human Subjects: Institutional Review Board Guidebook* (Washington, D.C.: U.S. Government Printing Office, 1993) at xix. [hereinafter *OPRR Guidebook*]

[40] See Chapter 14.

[41] See Benson, *supra* note 8 at 7.

[42] See *OPRR Guidebook*, *supra* note 39 at 1–1.

[43] See 45 C.F.R. 46 (1991) § 46.101 (a).

[44] *Ibid.* § 46.122.

any incidents of serious or continuing non-compliance to the Federal Government.[45]

It has been suggested that, apart from this process of negotiating an assurance with the Federal Government and the included requirements for reporting incidents of noncompliance, "there is no other formal mechanism whereby the activities of IRBs are in any way monitored by the federal government".[46] Effectively, enforcement of the regulations is left to the institutions themselves, because they stand to lose federal funding if the regulatory requirements are not met. Goldner notes that,

> ... most institutions generally make compliance with its policies on human research a condition of employment, or part of the faculty manual or contract so that the institution would be able to pursue disciplinary actions against an employee who failed to follow their policies.[47]

One major problem with the localized nature of the IRB system is that the boards themselves do not engage in general reflection about the nature of the broad ethical principles that should guide research in the United States. Katz, among others, has suggested that a need exists for a national approach to develop such broad principles. In particular, he has recommended the establishment of a National Human Investigation Board to formulate broad research policies and promulgate procedures that IRBs should use in implementing them:

> The policy questions underlying the tensions between the inviolability of subjects of research and advancing the frontiers of knowledge require more careful articulation and resolution than can be gleaned from the federal regulations so far enacted. The concerns I have raised and the recommendations I have made need to be examined, debated and decided by a national regulatory body to which the IRBs can also turn for advice and guidance on difficult problems that require resolution.[48]

The National Board would not only publicize decisions made by both itself and local IRBs but also function as a source of advice and guidance for IRBs. Local decisions tend to have low visibility, which prevents valuable experience from being made known and shared among those charged with implementing ethical standards. Furthermore, the current situation denies members of the public the opportunity to express their views on the practices followed in the course of biomedical experimentation. Indeed, Katz suggests that there is some truth to the contention that IRBs might currently be constituted to protect the institution and its researchers rather than the human subjects, because the majority of the members are on the faculty of the institutions to which the researchers belong.[49]

Each IRB is required to have at least five members, with "varying backgrounds to promote complete and adequate review of research activities commonly conducted

---

[45] *Ibid.* § 46.103.
[46] See Goldner, *supra* note 12 at 99–100.
[47] *Ibid.* at 103.
[48] See J. Katz, "Human Experimentation and Human Rights" (1993) 38 Saint Louis University Law Journal 7 at 39.
[49] *Ibid.* at 40–41.

by the institution". It must also be "sufficiently qualified through the experience and expertise of its members, and the diversity of the members, including consideration of race, gender, and cultural backgrounds and sensitivity to such issues as community attitudes, to promote respect for its advice and counsel in safeguarding the rights and welfare of human subjects". Where the IRB regularly reviews protocols involving "a vulnerable category of subjects, such as children, prisoners, pregnant women, or handicapped or mentally disabled persons, consideration shall be given to the inclusion of one or more individuals who are knowledgeable about and experienced in working with these subjects". Every effort must be made to ensure that no IRB consists only of men or women, or of members of only one profession. Each board must include at least one member whose primary expertise is in the field of science and one member whose primary concerns are non-scientific areas. In addition, each IRB must include at least one member who is not affiliated with the institution. The board may seek specialized assistance where necessary, but an expert who is invited to assist in this way may not participate in any vote.[50]

In spite of the apparent intent of the regulations governing membership of IRBs, most commentators have emphasized that, in practice, they are dominated by biomedical and behavioral researchers. As Goldner notes, "this dominance occurs both numerically, in terms of the composition of the typical board, and in terms of how the boards themselves usually function"[51] and the consequence of this dominance is a systematic bias in favor of conducting research. Indeed, Annas has contended that the IRBs have effectively "betrayed the research subjects that they are charged to protect".[52]

Goldner suggests that greater participation by non-scientists and persons not affiliated with the institutions actually conducting research would best be achieved by establishing a dual-committee system. A committee of professionals would first determine whether a research protocol is scientifically valid, and a second committee, consisting primarily of community-based members, would subsequently address the issue of ethical permissibility in light of community values, particularly those held in relation to the risk-to-benefit ratio presented in any protocol.[53] It has been contended that increasing the lay membership of the IRBs would be beneficial, because it would help to render the process of obtaining informed consent relevant to the needs of the particular subject group that is being approached: lay members are more likely to understand the average subject and to know what information is really important to him or her and whether the information will be presented by the researcher in a form that would be readily understood by such a subject.[54]

In October 1995, in the wake of a report on human radiation experiments carried out prior to 1975, President Clinton ordered a review of the regulation by federal

---

[50] See 45 C.F.R.46 (1991) § 46.107.
[51] See Goldner, *supra* note 12 at 106.
[52] G.J. Annas, "Questing for Grails: Duplicity, Betrayal and Self-Deception in Postmodern Medical Research", (1996) 12 Journal of Contemporary Health Law and Policy 297 at 324.
[53] See Goldner, *supra* note 12 at 107–108.
[54] *Ibid.* at 108.

agencies of research involving human subjects and established a National Bioethics Advisory Commission to oversee this review.[55] It is, no doubt, possible that this review will result in significant changes to the current regulatory system in the United States.

## The Role of Research Ethics Committees in the United Kingdom

Research ethics committees in the United Kingdom (RECs) date back to the mid-1960s when they were established partly in response to the establishment of the system of IRBs in the United States and partly in response to disturbing revelations of scandalous experiments. In 1967, a committee of the Royal College of Physicians, referring to U.S. experience, recommended that every institution in which biomedical research was conducted should form a group of doctors which would "satisfy itself of the ethics of a proposed investigation".[56] In the very same year, Pappworth published an influential study that exposed serious incidents of unethical clinical research in the U.K. Pappworth recommended that RECs should be created in every region, and that at least one lay member should serve on each committee.[57] Pappworth also recommended that the RECs should be made legally responsible to the General Medical Council (the body that is responsible for licensing physicians); significantly, this counsel has never been accepted in the United Kingdom.[58]

In recent years, RECs have been guided in their work by a series of guidelines issued by both the Royal College of Physicians (RCP) and the Department of Health (DoH). The RCP guidelines are contained in two publications that were most recently revised in 1990: *Research Involving Patients and Guidelines on the Practice of Ethics Committees in Medical Research Involving Human Subjects*. The most recent revision of the DoH guidelines are contained in a 1991 volume, entitled *Local Research Ethics Committees*. In a more recent development, the Medical Research Council published and endorsed the reports of two of its working parties: *The Ethical Conduct of Research on Children and The Ethical Conduct of Research on the Mentally Incapacitated*. The guidelines contained in the MRC reports, published in December 1991, also have a role in governing the decision-making of the RECs. The responsibility for establishing RECs rests on the District Health Authorities. Although some attempts have been made to place the DoH guidelines on a statutory basis, they remain purely advisory in nature.[59]

---

[55] See D.L. Wheeler, "Making Amends to Radiation Victims: Report on Controversial Cold War Experiments Calls for Tighter Ethical Standards for Research" (October 13, 1995) The Chronicle of Higher Education A10.

[56] J. Neuberger, *Ethics and Health Care: The Role of Research Ethics Committees in the United Kingdom* (London: King's Fund Institute, 1992) at 9.

[57] M.H. Pappworth, *Human Guinea Pigs: Experimentation on Man* (London: Routledge & Kegan Paul, 1967).

[58] Neuberger, *supra* note 56 at 9.

[59] *Ibid.* at 12.

With the sole exception of research concerning human embryos,[60] regulation of biomedical research in England and Wales operates without any formal legislative sanction or legal requirements.[61] This approach persists in spite of the attempt by the Committee of Ministers of the Council of Europe to encourage member states to give formal legal status to the principles that guide the conduct of such research.[62] However, in the absence of a basic legislated framework to direct the activities of RECs, there has been a perceived lack of uniformity in their nature and functioning and in the application of ethical standards across the country. For example, Nicholson's 1982–83 findings highlighted the marked discrepancies in the practices and composition of RECs and the lack of agreement as to their role.[63] Similar findings were published in 1989 by Gilbert, Fulford and Parker.[64]

Apart from providing valuable experience that illuminates the policy question of whether RECs should be placed within a legislated framework of some kind, it is clear that many of the basic issues surrounding RECs' functioning within the United Kingdom are of equal concern to those concerned with the operation of their counterparts in Canada. For this reason, it would be particularly appropriate to refer to a major study of the nature and role of RECs in the United Kingdom which was published by Neuberger in 1992.[65]

This study was based on a postal survey of REC members in England and Wales and on-site visits to twenty-five RECs. As in many countries, a critical issue is the optimal size of such committees. The study found that RECs varied considerably in size and were often larger than the suggested maximum of twelve members.[66]

---

[60] See *Human Fertilisation and Embryology Act 1990*, 1990, c. 37 (U.K.)

[61] See Chapter 13. Significantly, the Law Commission recommended, in 1995, that Parliament should enact legislation to establish a Research Committee to approve all research protocols involving participants who are mentally incapacitated. Among other considerations, an important reason for this recommendation was the perceived need to create a régime that would ensure that the approval of a protocol would render a researcher's actions lawful, if they were carried out in accordance with its stipulations. See Law Commission, *Mental Incapacity* (London: HMSO, 1995) at 99.

[62] Council of Europe, Committee of Ministers, Rec. No. R(90)3 (1990). This recommendation included a set of principles designed to "protect the human rights and health of persons undergoing research, as well as to establish clear legal rules on the duties of research workers and promoters of medical research". The Recommendation itself falls short of requiring legislation but calls on member states to "adopt legislation in conformity with the principles appended to this recommendation, or to take any other measures in order to ensure their implementation". The "explanatory memorandum" attached to the Recommendation indicates that the Recommendation asks the member states "to adopt legislation in conformity with the Principles or to take any other measures thus giving the Principles legal status at national level" (Paragraph 14). The memorandum does not share in the authority of the Recommendation but it does suggest an intention to ensure that the Principles are accorded legal status in the member states. The Steering Committee on Bioethics of the Council of Europe has recently (June 1996) adopted a *Convention on Human Rights and Biomedicine*. If the Convention is approved by the Council of Ministers and is signed by member states, it is expected that it will be followed by, *inter alia*, a specific protocol concerning research with human subjects. See Chapter 14.

[63] See R.H. Nicholson, ed., *Medical Research with Children: Ethics, Law and Practice* (Oxford: Oxford University Press, 1986).

[64] C. Gilbert, K.W.M. Fulford & C. Parker, "Diversity in the Practice of District Ethics Committees" (1989) 299 British Medical Journal 1437.

[65] See Neuberger, *supra* note 56.

[66] The recommended size is from 8 to 12 members. See Department of Health, *Local Research Ethics Committees* (London: HMSO, 1991).

Neuberger also found that the committees were "medically dominated"[67] in that more than half of the total membership of the RECs studied consisted of hospital doctors.[68] This finding reflects the situation elsewhere (e.g. in the United States).

Significantly, the Department of Health Guidelines for RECs require that membership should consist of both men and women; should be drawn from a wide range of age groups; and should include hospital medical staff, nurses, general practitioners, and two or more laypersons. The Royal College of Physicians' Guidelines make similar suggestions for membership, although they place more emphasis on the role of scientists and on the need for the physicians and nurses to work directly with patients.[69] In theory, these guidelines should ensure that all RECs have a reasonably well-balanced and representative membership. However, Neuberger found that, while a majority of the RECs did meet the requirements of these Guidelines, a substantial minority did not — "by having too many or too few members, or by having no GP member, no nurse, or insufficient lay members".[70] Furthermore, in 28% of the committees women constituted less than 20% of the total membership, and in only 7% were they either equally represented or in a majority.[71] Equally surprising was the finding that 44% of the RECs did not include either a pharmacist or a clinical pharmacologist in their membership, even though many of the research protocols involved testing drugs.[72]

Those who wish to make ethics review committees more reflective of the values of the community will be particularly concerned to ensure sufficient involvement of lay members. In this respect, Neuberger discovered that one third of the RECs had fewer than the minimum two lay members required by the guidelines, and only 18% had more than two. In half of the RECs studied, the lay members made up less than 20% of the total.[73]

A perennial difficulty for research ethics committees appears to be establishing effective procedures for the appointment of lay members. The Community Health Councils (and their equivalents in Scotland and Northern Ireland), which have been created to represent the interests of patients within the National Health Service, provide one source for the recruitment of lay members. In fact, the Department of Health Guidelines stipulate that lay members should be selected in consultation with the local Council.[74] However, only half of the RECs had lay members who were also members of a Community Health Council. In any event, the tactic of approaching such bodies as representative patient groups as a means of recruiting credible lay members could well serve as an invaluable precedent for other countries, such as Canada.

---

[67] See Neuberger, *supra* note 56 at 7.

[68] *Ibid*. at 17.

[69] *Ibid*. at 16. Citing Royal College of Physicians of London, *Guidelines on the Practice of Ethics Committees in Medical Research Involving Human Subjects*, 2d. ed. (London: RCP, 1990).

[70] See Neuberger, *ibid*. at 16–17.

[71] *Ibid*. at 17

[72] *Ibid*. at 18.

[73] *Ibid*. at 19–20.

[74] *Ibid*. at 20.

Insofar as lay representation is concerned, the Guidelines now require that either the chairperson or the vice-chairperson should be a layperson and, by 1992, nine of the twenty-eight RECs that Neuberger visited had lay chairpersons.[75] Again, this could establish a serviceable model to be embraced by ethics review committees both in Canada and other countries.

Neuberger also addressed the problem of multi-centre trials and pointed out increasing support for a national committee to deal with the special problems they raise. She suggested that a national committee could give conditional approval to a multi-centre proposal but that each local REC would have the opportunity to approve or reject it (but not to modify it).[76] This would avoid the situation in which a multi-centre study would potentially be hampered because each local ethics committee might request different, and even conflicting, modifications. While a national committee may be the appropriate body to undertake this function in the unitary state of England and Wales, it may well be the case that, in the Canadian (federalist) context, it should be carried out by a provincial Board.

Finally, English and Welsh experience with sanctions may well provide valuable lessons for Canada. Significantly, Neuberger states that the RECs in England and Wales do not have the power to impose sanctions against researchers who ignore their advice and that they generally do not see themselves as fulfilling a monitoring role. The Guidelines merely require that researchers notify the RECs of any significant changes to their protocols or any unusual difficulties in recruiting subjects.

As far as active monitoring of research is concerned, the study found that only one of the RECs had ever conducted a "spot-check" in relation to an ongoing project and only one had a computerized system for checking on the status of projects that were required to submit reports on a six-month or annual basis.[77] As presently constituted, the RECs in England and Wales have only an advisory role in relation to their respective District Health Authorities and other appointing authorities, and they are apparently very loath to assume a policing or monitoring role. Commenting on this critical aspect of the role of the RECs in the regulation of biomedical experimentation, Neuberger states that,

> However hard they work, however thorough their examination of research protocols on a case-by-case basis, however much better constituted and trained, and however well supported they may be administratively, unless they have the power to ensure that all research is submitted to them and to stop research that they regard as unethical, they will not be taken sufficiently seriously. For these reasons and others, this report . . . recommends that there should be proper legislation.[78]

This call for the enactment of regulatory legislation and the establishment of an effective mechanism for enforcement of the RECs' decisions runs counter to the prevailing approach in England and Wales. However, the position advanced does

---

[75] *Ibid.* at 21.
[76] *Ibid.* at 29.
[77] *Ibid.* at 34–35.
[78] *Ibid.* at 8.

reflect the views of an increasing number of commentators in this field and may be of particular relevance to other countries, such as Canada, which face similar problems of enforcement in the area of regulating biomedical research.

### The Structure of Regulation in Canada

In Canada, as in the United States and the United Kingdom, ethics review committees, known as Research Ethics Boards (REBs), have been assigned the task of assuring the public that only ethically sound biomedical research is being conducted and that the process of obtaining informed consent for participation in such research is acceptable. The REBs are affiliated with the institutions in which the relevant research is to be conducted. One of the more significant aspects of the policy developed by the Medical Research Council of Canada (MRC), between 1978 and 1987, is the emergence of the REB as the primary locus of responsibility for the maintenance of ethical standards in medical research. In the words of the *MRC Guidelines*, "[e]valuation by an REB (Research Ethics Boards) of a research protocol is the major step through which the community can be certain that its values are respected".[79]

At present, biomedical research is guided by the application of the Tri-Council's *Code of Ethical Research Involving Humans*.[80]

In common law Canada, no legislative framework underpins the system of REBs. In making their decisions, REBs have applied the *MRC Guidelines*, U.S. regulations (if funding is sought from the National Institutes of Health), and other ethical codes (such as the *Nuremberg Code* and the *Declaration of Helsinki*). The REBs have no independent power to require that research protocols involving human subjects be submitted to them for prior approval. However, the MRC requires that all biomedical research that it funds receive prior approval by a REB and, in practice, most institutions stipulate that such approval be obtained for all research protocols involving human subjects, regardless of the source of funding.[81]

In Quebec, however, the system of REBs does have a clear basis in legislation, and formal legal rules have been articulated in relation to the general circumstances in which biomedical experimentation may be conducted in the province. For example, article 21 of the *Civil Code of Quebec* sets out the limited circumstances in which children and incompetent adults may participate in biomedical experimentation. In general, they may participate if the appropriate substitute decision-maker gives their consent (the mandatory, tutor or curator, as the case may be) and there is an "absence of serious risk" to their health. However, Article 21 of the Civil Code also stipulates that:

---

[79] See *MRC Guidelines*, *supra* note 6 at 46.
[80] See Tri-Council Working Group, *supra* note 7.
[81] See NCBHR (1995), *supra* note 27 at 9. See e.g. the definition of the jurisdiction of REBs associated with the Faculty of Medicine at McGill University: McGill University, *McGill University Ethical and Legal Aspects of Research Involving Human Subjects Conducted in the Faculty of Medicine and Affiliated Hospitals: Policies and Procedures* (Montréal: Faculty of Medicine, McGill University, 1994) at 8.

> An experiment on a group of minor persons or incapable persons of full age shall be carried out within the framework of a research project approved by the Minister of Health and Social Services, upon the advice of an ethics committee created by him for that purpose: in addition, such an experiment may be carried out only if a benefit to the health of persons of the same age group and having the same illness or handicap as the persons submitted to the experiment may be expected.

Clearly, Article 21 not only mandates the involvement of an ethics committee but also requires the further approval of the Minister before any such biomedical experimentation may proceed. No such legislation exists in the other Canadian provinces and territories.

A precedent for legislated regulation of biomedical experimentation in Canada as a whole may well arise in the area of "New Reproductive and Genetic Technologies". In June, 1996, for example, the Minister of Health introduced a bill in the Canadian Parliament that would prohibit certain uses of these technologies[82] and also indicated that there was an intent to establish a system of regulation at some stage in the future.[83] Nevertheless, for the present, the universal mechanism of regulating experimentation in Canada is the local research ethics board that (except in Quebec) operates without legislative sanction.

The Tri-Council Working Group's proposed *Code of Conduct* devotes considerable attention to the nature and function of the REB and its analysis is useful as a means of identifying the mission of the REB in contemporary Canadian society. According to the Tri-Council Working Group, the primary reason for the establishment of the REB is "to help ensure that ethical principles are applied to research involving humans". The Group goes on to state that the REB has:

> . . . both educative and administrative roles: in its educative role, the REB serves the research community as a consultative body and thus contributes to education in research ethics; and in its administrative role, the REB has the responsibility for Independent review of the ethics of research to determine whether it should be permitted to start or continue.[84]

The Working Group also emphasizes the need to establish the authority of the REB on an unequivocal basis. Indeed, it suggests that:

> The REB must be vested by its institution with the authority to approve, reject, propose modifications to or terminate all proposed or ongoing research involving humans within the institution's jurisdictions on grounds of the ethical considerations set forth in this Code.[85]

As far as the terms of reference of the REB are concerned, the Working Group states that such terms should be adopted for each REB and that they must include:

(1) protecting participants from research harms;

---

[82] Bill C–47, *An Act Respecting Human Reproductive Technologies and Commercial Transactions Relating to Human Reproduction*. See also The Royal Commission on New Reproductive Technologies, *Proceed with Care: Final Report of the Royal Commission on New Reproductive Technologies* (Ottawa: Government Services, 1993).

[83] See Chapter 14.

[84] Tri-Council Working Group, *supra* note 7 at 16.

[85] *Ibid.*

(2) respecting the duties and rights of researchers; and

(3) reviewing proposed and ongoing research to ensure that it complies with this Code.[86]

Having identified the functions that the REBs are supposed to perform in terms of their general mandate, it is now necessary to examine the available empirical data illustrating how they operate in practice. Fortunately, comprehensive information concerning the nature, structure and functioning of REBs in Canada has become available through a report issued, in 1995, by The National Council on Bioethics in Human Research (NCBHR).[87] The report follows an ambitious three-year study of the functioning of REBs in the Faculties of Medicine at 16 Canadian Universities from 1990 to 1993. In addition to the data generated by questionnaires and site-visits, the report contains a number of recommendations by the Council to improve the current system.

One important finding of the study was the extent to which there may be multiple REBs associated with one institution in Canada. There were some 100 REBs affiliated with the 16 medical schools surveyed. In most cases, they operated within a hospital and were ultimately responsible to its president. There were varying degrees of communication between the different REBs affiliated with the same university; however, it is disconcerting that, in some cases, it was reported that such communication was "often random or absent".[88]

In submitting this potentially confusing situation to further analysis, the NCBHR suggested that there were generally two lines of authority and reporting responsibilities in Canadian universities. In the first, a university committee receives its mandate from the University President or Senate, usually through a senior member of the administration. This committee may fulfill a number of different functions, including serving as a review committee for a broad range of research activities or functioning as a review committee for non-medical human research only. In the second, a hospital committee reports to the president or board of trustees of the hospital, "usually through a medical advisory committee, although sometimes through a board bioethics committee or a quality assurance committee". Most of these hospital committees conduct ethical reviews, although a minority of them consider their role to be restricted to the review of resource-related issues.[89] To ensure coordination among multiple REBs, the NCBHR made the following recommendation:

> University REBs should consider taking responsibility for co-ordinating REB activities, especially when multiple hospital REBs exist. Such co-ordination should focus on providing policy direction, on considering difficult or generic problems, and on training and educational activities for REB members.[90]

---

[86] *Ibid.* at 16–17.
[87] See NCBHR (1995), *supra* note 27.
[88] *Ibid.* at 9.
[89] *Ibid.* at 17.
[90] *Ibid.* at 18.

The Tri-Council Working Group, in its *Code of Conduct for Research Involving Humans*, also stresses the desirability of all REBs within a single institution striving to "apply the same ethical standards" and the need to establish a mechanism to "coordinate the practices of all REBs within the institution".[91]

A critical issue concerns the extent to which REBs exercise regulatory control over the whole range of biomedical experimentation in Canada. Significantly, the NCBHR report states that more than 90% of the REBs indicated that no research could be undertaken at their institution without prior board approval.[92] This finding suggests that almost all biomedical experimentation is receiving prior ethical approval regardless of the source of funding. Similarly, the Tri-Council Working Group emphasized the need for each institution, that qualifies for research funding from one or more of the three granting Councils, to vest its REB with the "authority to approve, reject, propose modifications to or terminate all proposed or ongoing research involving humans within the institution's jurisdiction on grounds of the ethical considerations set forth in (the) Code".[93] The Working Group also recognized the fact that other organizations (such as "pharmaceutical companies, non-profit agencies, community groups or government research agencies") or "for-profit review bodies" (such as "private companies in the business of reviewing research projects involving humans") may establish their own REBs in order to determine the "ethical acceptability of research involving humans". However, in this respect, the Working Group made a plea for these REBS to espouse its (draft) Code of Ethical Conduct for Research Involving Humans:

> It is recognized that, in the absence of relevant legislation in Canada, other ethics review bodies are not required to adopt this Code. Because there are clear advantages in harmonizing the ethical standards for REBs evaluating research involving humans, it is hoped that other ethics review bodies will regard this Code as an appropriate model to follow. Adopting this Code will ensure that the minimal standards for research involving humans have been met. This, ultimately, will preserve and enhance the public trust and confidence in Canadian researchers.[94]

What ethical principles have been applied by REBs in Canada during the recent past? According to the NCBHR report, the vast majority (93%) of REBs indicated that they used the *MRC Guidelines* as the basis for their review of research protocols, although they also relied on other sets of guidelines, where appropriate (e.g. U.S. Federal regulations when funding was provided by the National Institutes of Health).[95] However, a number of REBs criticized the *MRC Guidelines* for being ambiguous; the NCBHR report suggests that "the perception that *MRC Guidelines* sometimes prove ambiguous and that they are flexible and not necessarily "binding" may also explain their less than universal use".[96] Of course, the problems raised by

---

[91] See Tri-Council Working Group, *supra* note 7 at 18.
[92] NCBHR (1995), *supra* note 27 at 9. 10% of the REBs indicated that research involving no patient contact (i.e. chart review) could be undertaken without a prior review.
[93] See Tri-Council Working Group, *supra* note 7 at 16.
[94] *Ibid.* at 24–25.
[95] *Ibid.* at 10. See also *OPRR Guidebook*, *supra* note 39 at I–1.
[96] NCBHR (1995), *ibid.* at 19.

multiple sets of ethical guidelines would be considerably ameliorated if the proposed *Code of Conduct*, developed by the Tri-Council Working Group were to be adopted as the basis for the ethical review of all biomedical experimentation in Canada.

An important procedural question about the operation of REBs in Canada concerns the extent to which there is any attempt to separate the function of determining the scientific validity of a research protocol from the function of reviewing its ethical permissibility. There have been determined calls, both in the United Kingdom and the United States, to introduce a bifurcated process of review in this respect. However, more than 80% of the Canadian REBs indicated that in the NCBHR study they carry out their own scientific evaluation of research proposals, while the other boards stated that they referred this task to external experts.[97]

The Tri-Council Working Group addressed this issue in its *Code of Conduct for Research Involving Humans*. The Code suggests that there are many different models that might be employed in the assessment of the scientific validity of a research protocol. For example, it suggests four potential models that might be adopted by an REB:

- an independent external peer review;
- a permanent internal peer review committee reporting directly to the REB;
- where a permanent internal peer review committee is not available or feasible, the REB may arrange for independent review on an *ad hoc* basis; and
- the REB may assume complete responsibility for the scholarly and/or scientific merit, which would require that it have the necessary expertise to carry out peer review of the research in question.[98]

Significantly, the Working Group does not indicate a preference for any particular model, leaving it to individual REBs to decide which one to adopt in light of its own peculiar circumstances. Howevever, the Working Group does emphasize the critical importance of ensuring that research protocols reflect minimum scholarly standards for the discipline in question. It comments that poor research may waste valuable resources, damage the credibility of the enterprise of research in general and place research subjects in the face of risk. As far as the process of peer review is concerned, the Group states that

> . . . Insofar as REBs evaluate research projects for scholarship, they should generally rely on the judgement of others by ensuring that an adequate peer review has been conducted. Insofar as REBs undertake the assessment of the scholarly aspects of the proposed research, it is absolutely essential that there be appropriate experts on, or added to, the REB for the purposes of reviewing the particular proposal.[99]

---

[97] *Ibid.* at 19. However, Meslin points out that the precise nature of the connection between scientific validity and ethical permissibility in REB review has not been clearly articulated. See E.M. Meslin, "Ethical Issues in the Substantive and Procedural Aspects of Research Ethics Review" (1993) 13 Health Law in Canada 179 at 181.

[98] Tri-Council Working Group, *supra* note 7 at 19 (Part 2).

[99] *Ibid.*, at 16 (Part I).

As far as the nature and outcome of the review process are concerned, the NCBHR study found that more than 60% of the REBs met on a monthly basis. Most REBs indicated that they reached decisions by consensus and that the average time for the processing of an application was one month, a factor of considerable importance to researchers.[100] It appears that most protocols are approved with only minor modifications being requested: more specifically, responses indicated that only 3% of proposals were rejected outright; 75% were approved with minor modifications; and 22% were approved as they were. More than half of the REBs had an appeal process available to researchers. Interestingly, the Tri-Council Working Group's (draft) Code refers to the right of researchers to request a reconsideration by an REB of decisions affecting their research.[101] In addition, it is significant that the Working Group refers to the role played by "appeal boards" in Canada. An appeal board is defined as "an REB mandated by a research institution to review decisions of other REBS in the same institution".[102] The Working Group stipulates that "the appeal boards should be required to meet the *minimal* membership requirements for REBs in order to ensure consistency in the application of scientific and ethical standards. Furthermore, the Working Group clearly articulates the rule that no institution should have the power to overrule REB decisions made on ethical grounds".[103]

As we have seen, perhaps the most controversial aspect of ethics review boards, in whichever country they operate, is the extent to which their membership and operation reflect a genuine balance between the interests of researchers, on the one hand, and the interests of members of the public and potential research subjects, on the other. As far as the overall size of the membership of the REBs was concerned, the NCBHR report established that there was a considerable range, with numbers varying between 3 and 21 on any given board. Almost half of the institutions that were surveyed indicated that their REBs had from 10 to 15 members and one-third indicated a size of 4 to 9 members. However, the critical question is which constituencies these members represented and whether they reflected a balance of interests and expertise.

In this respect, only 10% of the REBs met all of the compositional requirements set by the *MRC Guidelines*, which intend to achieve some kind of balance in Board membership. The Guidelines stipulate, for example, that REBs should contain "members who can reflect community values", preferably not affiliated with the institution concerned; at least "one specialist in the relevant discipline of the research"; scientists with a broad knowledge of research methodology; and nurses (where clinical research is involved). The *Guidelines* also indicate that a clinical psychologist or mental-health expert may "aid in assessing the subject's capacity to understand the protocol and exercise a free choice". In addition, the *Guidelines*

---

[100] NCBHR (1995), *supra* note 27 at 10.
[101] Tri-Council Working Group, *supra* note 7 at 24 (Part 2) (Article 2.11).
[102] *Ibid.* at 1 (Appendix E).
[103] *Ibid.* at 24 (Part 2).

suggest that "bioethicists, philosophers or theologians will also contribute greatly to the work of an REB".[104]

According to the NCBHR report, very few REBs (only 6%) consisted exclusively of researchers. The great majority of boards, therefore, consisted of a mixed membership of some kind and, although there was some degree of variation, the report concluded that the REBs are generally strong in the areas of scientific and medical expertise but there was a much greater degree of variation in the depth of their ethical expertise. A few boards included ethics consultants, specialists "in pastoral care", patient representatives, and lawyers who were competent in the field of bioethics. However, the Council noted that others had reported difficulty in recruiting members with such expertise.[105]

Clearly, the lack of expertise manifested in the specific area of ethics raises many questions about the adequacy of the balance of interests struck among the various members of the REBs. As in many other jurisdictions, they appear to be dominated by scientists and researchers, even though the fundamental ethical issues facing the REBs cannot be resolved by applying an exclusively scientific or technological expertise.

On the critical issue of lay membership, the Council took note of the fact that, while a number of REBs had active lay members, some boards had found it difficult to find such representation, while others had not made this a priority. Overall, about half of the REBs had some type of lay representative (usually no more than one) who had no direct affiliation with the institution.[106] Significantly, there was no specific procedure established for selecting this member; usually he or she was selected by the chair of the committee.[107] Indeed, the lack of any standard process for the nomination and appointment of any of the members to REBs is clearly perceived as being an ongoing problem of major proportions.[108] In this respect, the NCBHR states very candidly that,

> ... REBs are having difficulty meeting some requirements of MRC and US OPRR guidelines, since only a minority of REBs described a composition that would meet the standards set by these bodies.[109]

Generally, institutions require faculty, external and departmental representation on the REB. Others indicated for representation include individuals with research

---

[104] See *MRC Guidelines, supra* note 6 at 45-46.

[105] See NCBHR (1995), *supra* note 27 at 20.

[106] *Ibid.* at 20–21. In a National Workshop on Ethics Review, co-sponsored by the NCBHR and the MRC in April 1989, "discussants agreed that the lay member is a key part of the process". An "ideal lay member was defined as a mature person not connected with the institution, who was not doctrinaire, and who was willing to invest time and work hard". See J.N. Miller, "Ethics Review in Canada: Highlights from a National Workshop: Part 2" (1990) 23 Annals RCPSC 29 at 30.

[107] See NCBHR (1995), *supra* note 27 at 11.

[108] *Ibid.* at 21.

[109] Since the U.S. Department of Health and Human Services funds a considerable number of the research projects carried out in Canadian settings, many REBs attempt to meet the requirements for membership specified by the U.S. Office for Protection from Research Risks (OPRR). See e.g. McGill University, *supra* note 81 at 10. The general requirements for IRB membership are set out at 45 C.F.R. 46 (1991) § 46.107.

expertise, members of the legal profession, and a member of the Board of Trustees. *REBs, with few exceptions, seem unable to ensure lay representation.*[110]

The record of Canadian REBs in adding lay people to their membership compares quite unfavourably with that of the RECs in England and Wales. One potential explanation for this deficiency may lie in the fact that considerably more attention appears to have been paid in England and Wales to the actual process of identifying appropriate representative organizations to nominate community representatives to the RECs. This might well be addressed within the Canadian context at a provincial level.

The NCBHR report recommended that the Council should work with REBs to "encourage" compliance with *MRC Guidelines* on composition and to assist them to find individuals with the necessary expertise. In particular, the study recommended that the NCBHR should provide assistance in the area of recruiting lay representatives, who are seen as having "a primary role in assuring patient/subject protection". It was also recommended that the National Council should "assist REBs by defining the minimal ethics expertise that members should have for the adequate functioning of an REB".[111] This would presumably address the current paucity of ethicists on the REBs.

Whether the assistance of the NCBHR can resolve the long-standing membership problems of all the REBs that review biomedical research protocols in Canada remains to be seen. It may well be the case that one of the only efficacious methods to achieve a more balanced membership on REBs is to establish province-wide committees charged with assisting local REBs to find suitable members both from representative community groups and from the ranks of those professionals who have a specific expertise in the field of ethics. This is a theme to which we shall return in the concluding section dealing with potential reforms to the system of REBs in Canada.

For the present, however, it is clear that many REBs are making a serious effort to meet the requirements of a diverse membership. For example, the McGill University Guidelines for Ethical Review of Research Involving Human Subjects address the issue of committee membership by stating that:

> The committee should include a community volunteer(s); a person from other non-medical disciplines, such as a bioethicist, a lawyer (preferably not the Institution's counsel), or a theologian; and a patient advocate. The latter is particularly important when an REB regularly reviews research that involves a vulnerable category of subjects (children, elderly people, mentally ill persons, among others). DHSS requires and the Faculty of Medicine recommends, that one member of the REB (including his/her immediate family) not be affiliated with the institution served by the REB. In the McGill setting a non-affiliated member might be a physician or scientist who is not in the employ of the University or Hospital to which the REB reports.[112]

This statement is particularly significant insofar as it contains an explicit recognition of the need for REBs to include in their membership a representative of

---

[110] See NCBHR (1995), *supra* note 27 at 20.
[111] *Ibid.* at 21
[112] See McGill University, *supra* note 81 at 10

potential research subjects, particularly where they may belong to one of the populations perceived as being particularly vulnerable (e.g., the mentally disordered, the developmentally disabled, prisoners, children, and the elderly).

Perhaps one of the most vital issues of all is whether there is an adequate process for the *monitoring* of ongoing research projects that have been approved by an REB. Prior approval of a research protocol is only one step in a process intended to ensure the protection of human subjects through all phases of experimentation. There must also be measures to ensure that the protocol is followed in the actual conduct of the research. However, most REBs do not consider ongoing monitoring of research to be part of their mandate and they lack the time and resources for it,[113] even though the *MRC Guidelines* clearly recommend that REBs should establish a process for the ongoing review and monitoring of research protocols.[114]

The NCBHR report assesses the situation pertaining to ongoing monitoring of research projects by REBs in the following manner:

> Statistically, our survey found that it is obligatory in 88% of institutions to report any change in protocol design, yet because of time and resource constraints 25% have no monitoring mechanisms. Some institutions rely on individual departments to review ongoing research, and a few institutions are reluctant to monitor because they believe it suggests a lack of trust in the researcher, who is expected to monitor voluntarily and review ongoing projects. Progress reports for all human research are required by 50% of institutions; others require reports in specific instances. End-of-protocol reports are required by 30% of institutions; 66% require adverse incidents reports. In short, when the monitoring processes used by REBs in Canada are examined, the nature of the continuing review and monitoring varies. This appears to relate, in part, to a lack of resources for carrying out a monitoring function by the REBs and, in part, to a lack of specifics in the *MRC Guidelines*.[115]

Of particular interest is a statement made in the NCBHR report that there was very infrequent consultation between members of the REBs and the researchers once the protocols had been approved, and that "untoward incidents" were rarely brought to the attention of the REBs. This meant that the "REBs did not always have the most current information about the risks and benefits to human subjects".[116] Such information might conceivably be brought to their attention by feedback from the subjects themselves. In this respect, it interesting that, while more than 90% of the REBs stated that an aggrieved research subject would have access to their members, only about 18% indicated that this had ever happened.[117]

To ensure adequate monitoring, the NCBHR report recommended that both institutions and the REBs should "further develop and implement mechanisms for the thorough review and monitoring of human research". Among the measures recommended are (at least annual) reports dealing with such matters as significant changes to research protocol and unexpected incidents, as well as termination reports. All these reports should be reviewed by the REB and, for "sensitive

---

[113] *Ibid.* at 12.
[114] See *MRC Guidelines*, *supra* note 6 at 49.
[115] See NCBHR (1995), *supra* note 27 at 23.
[116] *Ibid.* at 15.
[117] *Ibid.* at 13.

protocols", the boards should require a more "frequent and rigorous" process of review, potentially including external monitors.[118]

The solution to the monitoring problems proposed by the NCBHR involves close collaboration between the Council and the REBs. However, the important issue at stake is whether the public interest in protecting the rights of human research subjects can be left to the REBs and the NCBHR or whether there should be an independent body that carries out this vital function. Coburn, for one, argues that routine monitoring of research protocols must be undertaken by an independent body if the public is to be guaranteed that there are effective protections against the potential abuse of human subjects. He also suggests that a system of voluntary guidelines directed by the NCBHR is unacceptable because its composition is "heavily weighted to provider and researcher groups, and its guidelines are purely voluntary".[119] If such an independent body is considered necessary for carrying out the task of monitoring biomedical research projects on an ongoing basis, then a province-wide body would appear to be an appropriate candidate to shoulder this vital responsibility.

The Tri-Council Working Group also addressed the need for REBs to engage in a process of monitoring ongoing research and provides some concrete examples of how this process may be implemented. Indeed, Article 2.7 of the Group's *Code of Ethical Conduct for Research Involving Humans* states that

> All ongoing research must be subject to ethics review. The rigor of this review must follow the principle of a proportionate approach to ethics assessment. The minimal requirement for continuing review is submission to the REB of a brief final report at the conclusion of the project.[120]

At the heart of the Working Group's (draft) Code is the notion of a "proportionate approach to ethics assessment", which the Working Group defines as "a stragegy for REBS to adjust their level of scrutiny to the potential harms apprehended in the research (to be applied both to the review of proposed research and to the review of ongoing research)".[121] In accordance with this strategy, the Working Group suggests that, where research exceeds the "threshold for normally appropriate risk",[122] a continuing ethics review may include:

- review of annual reports;
- formal review of the informed choice process;
- establishment of a safety monitoring committee;

---

[118] *Ibid.* at 23. The need for the development of a methodology for effective monitoring had also been recognized at a national workshop on Ethics Review co-sponsored by the NCBHR and the MRC in April 1989. See J.N. Miller, "Ethics Review in Canada: Highlights from a National Workshop: Part 1" (1989) 22 Annals RCPSC 515 at 517.

[119] See D. Coburn, "Health Sciences Research Ethics: A Critique" (1993) 13 Health Law in Canada 192 at 195–196.

[120] *Ibid.* at 22 (Part 2).

[121] *Ibid.* at 5 (Appendix E).

[122] This is defined in the following manner: "when the possible harms implied by participation in the research are within the range encountered by the participant in everyday life, then the research falls within the range of normally acceptable risks". *Ibid.* at 7 (Appendix E).

- periodic review by a third party of the documents generated by the study;
- review of the impact of the research on a collectivity;
- review of reports of adverse events;
- review of patients' charts; or
- a random audit of the choice process.

On the other hand, where research exposes the participants to a degree of risk that is "below the normally acceptable risk", then only a "minimal review process" is required.[123]

A central issue in considering the efficacy of the REB system of regulating biomedical experimentation is that of compliance. In the Province of Quebec, submission of research protocols involving human subjects is required by provincial legislation. However, in the common law jurisdictions of Canada, the REBs have no statutory authority. The NCBHR report recommended that both the MRC and NCBHR should articulate clear statements as to the consequences of noncompliance with national ethical guidelines or with the recommendations of an ethics committee. These should include the possibility that an investigator may face the loss or suspension of the privilege of conducting biomedical experiments. It was also recommended that the Universities put processes in place to complement national standards with a view to curbing, detecting and sanctioning noncompliance and misconduct in research.[124]

Another method of ensuring compliance with the decisions made by the REBs would be to restrict the publication of research results to those projects in which the REB has certified that the protocol has been followed in an acceptable manner. Such restrictions would clearly affect the conduct of biomedical experimentation, because unpublishable results are worthless both to the medical community and to the researcher producing them.[125] This could be achieved by establishing a final reporting requirement to the REB that approved the protocol as a precondition to publication. In this way, only those protocols receiving prior REB approval would be permitted to proceed and only those results receiving final REB approval would be published.

The increasing use of *multi-centre clinical trials* raises the difficulty that REBs at different sites may make decisions at variance with each other.[126] The NCBHR study indicated that many REBs "requested that there be some central way of considering

---

[123] *Ibid.* at 22 (Part 2).

[124] See NCBHR (1995), *supra* note 27 at 24.

[125] See Law Reform Commission of Canada, *Toward a Canadian Advisory Board on Biomedical Ethics* (Ottawa: Law Reform Commission of Canada, 1990) at 15–16:

> In a performance assessment of the committees . . . the importance of self-discipline by researchers should not be minimized. For the most part, biomedical research culminates in the publication of papers that are widely disseminated in the scientific community. There clearly exists a direct link between compliance with ethical standards, and approval and recognition of the research results by that community. It is not surprising, then, that researchers do comply with the standards, since it is in their own interest to do so.

[126] See Meslin (1993), *supra* note 97 at 185.

multi-centre trials prior to the protocol being fixed".[127] It is clearly unacceptable that any particular REB should be placed in the position where it is required to review a protocol that, in effect, cannot be changed because it involves a multi-centre trial. The Council does not recommend a specific solution to this problem. However, in the Canadian Context, a province-wide Board could give conditional approval to a multi-centre proposal; each local REB would have the opportunity to approve or reject it (but not to modify it). If the research project concerned involves centres in more than one province or territory, it would be necessary to negotiate the protocols between the respective research groups or, alternatively, to refer the project to a national advisory committee.

## REFORMING THE REGULATORY SYSTEM IN CANADA

It is clear that a regulatory system, based on the central role played by REBs, is extremely well-entrenched not only in Canada but also in other countries that share a common legal heritage. There is very little pressure to abandon this system altogether. However, the preceding analysis of current practice in Canada and elsewhere demonstrates that there are many aspects of the present regulatory system that require energetic reform.

### *Establishing a Statutory Basis for the Regulatory System*

As has been emphasized above, in common law Canada, the REBs effectively function as part of a system of self-regulation since there is no statutory framework that underpins their operation and establishes their unquestioned authority over all forms of biomedical experimentation. In the Province of Quebec, in contrast, the *Civil Code* clearly articulates some general legal principles concerning the types of biomedical experimentation that are permitted and requires the approval of an ethics committee, established by the Minister of Health, before children or incompetent adults may participate as research subjects.

In order to establish the unequivocal authority of REBs to approve in advance all protocols for biomedical experimentation with human subjects, it is essential that each province and territory of Canada enact the appropriate legislation. A breach of the requirement to obtain such prior approval should normally result in the imposition of a statutory penalty. In addition, such legislation should impose a duty on researchers to provide the appropriate REB with information about the ongoing progress of research that has been approved in advance. Such information should be sufficient to permit the REB to undertake its function of effectively monitoring the implementation of research protocols to which it has given its stamp of approval. The legislation should also articulate the basic legal requirements for the conduct of biomedical experimentation. For example, it should define (as does the legislation in Quebec) the general circumstances in which it is permissible to enroll children and incompetent adults as research subjects.

---

[127] See NCBHR (1995), *supra* note 27 at 15.

There may well be considerable opposition to enacting such legislation on the part of the medical profession, the universities and researchers who may assert that statutory intervention has no useful role to play in the complex arena of biomedical experimentation and that it may indeed establish formidable obstacles to medical progress. However, the Law Reform Commission of Canada has responded to such contentions by pointing out that,

> ... legislative intervention does not mean that society mistrusts its researchers or that it wishes to stop scientific development. To legislate means something else entirely. Where the integrity of the person can legally be endangered, it seems important that limits and rules be clearly defined. It is up to the law to protect basic values, and it cannot and must not leave this role to ethics. Moreover, and contrary to what one might think, there are many researchers nowadays who would like to have a clear idea of what may legally be done and what should be prohibited.[128]

Nevertheless, the legislation in question should stipulate only general and minimal requirements for the conduct of experimentation. It is critical, for example, that the quasi-criminal sanctions contained in such legislation should only be applicable to the most flagrant abuses that may be committed by biomedical researchers. This is not a field in which researchers should feel that they may be subjected to significant statutory penalties even though they have acted in good faith and in accordance with currently accepted standards of investigation that have been approved by REBs. Therefore, it is necessary that legislative intervention should be limited to the articulation of minimal standards for the conduct of biomedical experimentation. Equally, it is important that legal requirements are formulated in general terms since REBs must be able to respond swiftly to changes in the nature of medicine and medical technology and to act flexibly in the face of varying local conditions across such a geographically vast country as Canada. If legal requirements are excessively detailed in nature, REBs will lose one of their clear advantages over other decision-making mechanisms that might be harnessed to the task of regulating biomedical experimentation; namely, their ability to adapt swiftly to rapidly evolving circumstances and conditions. In general, one would expect the REBs to require research protocols to meet higher standards than those reflected in the minimal requirements articulated in legislation. These higher standards would be derived from the ethical guidelines that are interpreted and applied by the REBs. However, these guidelines would permit the REBs to respond in a flexible and sensitive manner to the demands of an ever-changing research environment.

As the Medical Research Council of Canada noted a decade ago, there is always a danger that detailed legislative activity in the forum of biomedical research will prove to be excessively rigid in practice:

> Legislation and regulation under law prescribe standard responses to anticipated scenarios. Its potential to apply justly consists in its treatment of broadly-defined categories of like cases in like ways, and in the finding of operative facts by the authoritative due process of the courts. The ethical assessment of a research proposal may raise a wide range of interests and values, some novel, and may require risk-to-benefit determinations which

---

[128] See *LRC Working Paper No. 61, supra* note 25 at 58.

cannot easily be prescribed or standardized. The facts of ethical priority cannot always be authoritatively established; particular factors may weigh differently at different times and in different circumstances. A proposal may be rejected as premature at one time and be acceptable only a short time later, for instance, because of evolution in the field of inquiry in question.[129]

Finally, it should be emphasized that establishing a legislative basis for the operation of the REBs in Canada may well prove to be a considerable benefit to those involved in the conduct of biomedical experimentation insofar as it may be argued that a researcher who implements in good faith a research protocol, that has obtained the advance approval of a local REB, should be considered to have met the required standard of care that must be followed in relation to human research subjects. If this is the case, then the approval of the REB could be raised as a defence in response to any negligence action brought against such a researcher. This would provide an element of legal certainty that is currently lacking when a researcher embarks on a course of biomedical experimentation.

### *Enhancing the Membership Requirements of the REBs*

Our survey of the functioning of REBs in Canada, the United Kingdom and the United States manifestly demonstrates that the issue of membership of the boards is central to policy discussions about their future role in the regulation of biomedical experimentation. Coburn, for example, concurs with the general criticism that Canadian REBs are largely drawn from the ranks of the researcher's professional peers or fellow scientists. Although their membership may include non-researchers and "occasionally token membership from the general public", he asserts that "the subjects of the research themselves and representatives of the general public are seldom involved in a significant and substantial manner".[130] Coburn proposes that representation from the subject populations and members of the general public should be balanced, because "potential subjects are sometimes willing to undergo procedures that violate their rights which other members of the public would not be quite so willing to surrender". However, Coburn argues that representation of the general public should not be limited to those individuals who are selected at random from the general population. Rather, there should also be people with experience in the area of patients' rights and those representing organizations such as women's health groups.[131]

On the other hand, McNeill contends that the membership of these committees should reflect a balance solely between the representatives of science and the representatives of research subjects. In this approach, there would be no room for representatives of the public at large, perhaps because the history of lay participation in the work of research ethics committees indicates that such participation has made very little real impact on actual decision-making practices.

---

[129] See *MRC Guidelines, supra* note 6 at 10.
[130] Coburn, *supra* note 128 at 195.
[131] *Ibid.* at 195.

A study by McNeill, Berglund and Webster in Australia, for example, demonstrated that lay members were considered to be "significantly less active and significantly less important" in the deliberations of ethics review committees.[132] This later led McNeill to conclude that "the addition of lay and non-scientific members to review committees may not make an appreciable difference".[133] Lawyers were regarded as playing a more influential role, perhaps because they were considered to be "fellow professionals". However, other lay members, including nurses and ministers of religion, were perceived as having relatively little influence on the actual decisions made by the committees.[134]

Increasingly, the view is emerging that the research enterprise should be conducted as a *partnership* between all those who participate in it.[135] The very use of the word "subject" tends to marginalize those upon whom research is conducted since it encourages the notion that they should be passive in the face of the professional expertise of researchers. McNeill points out that people with HIV and AIDS have rejected such passivity and have

> demanded participation in the process at all levels: including the planning of the research process, the manner in which the research was conducted, and the interpretation of the results. It is my prediction that increasingly research in all fields will require negotiation between those who conduct the research and those who participate as volunteers.[136]

This approach is, to some extent, mirrored in the Code of Conduct for Research Involving Human Beings prepared by the Tri-Council Working Group which exhorts researchers to maintain a "balanced perspective" towards their work and, in particular, to "move beyond their own perspectives to see how their work affects research participants, that is, to see things from the point of view of research participants".[137] At a more specific level, the Working Group states that "it is good ethical practice to conceptualize and realize research with collectivities as a partnership between the researcher and the collectivity".[138] Collectivities are defined by the Working Group as "populations with social structures and common customs" and that are "ordinarily groups in which there is mutual recognition of membership both by those in the groups and those outside it".[139] The Working Group emphasizes that researchers should ensure that the members of a collectivity are granted the opportunity to participate in the design of any research protocol that affects it and to respond to the findings prior to completion of the final report. In addition, the Working Group states that

> Researchers should conduct their research in a manner that ensures that the various (and potentially conflicting) viewpoints held by the collectivity regarding the topics being

---

[132] See McNeill (1993), *supra* note 13 at 91.
[133] *Ibid.* at 91.
[134] *Ibid.* at 92.
[135] See Chapter 13.
[136] *Ibid.*
[137] *Ibid.* at 6 (Part 1).
[138] *Ibid.* at 48 (Part 2).
[139] *Ibid.* at 2 (Appendix E).

researched are acknowledged in publications. Good practice generally necessitates that researchers make their best efforts to ensure that the emphasis of the research and the ways chosen to conduct it, respect the many viewpoints of the group in question.[140]

Article 2.3 of the Code unequivocally reflects the principle that researchers and scientists should *not* constitute the majority of members of REBs:

The minimum acceptable membership of an REB is five members, including both men and women, of whom:

(1) at least two members have broad expertise in the methods or in the areas of research that are covered by the REB;
(2) at least one member who is knowledgeable in the discipline of ethics;
(3) at least one member is a lawyer; and
(4) at least one members has no affiliation with the institution, but is recruited from the community served by the institution and, if possible, from potential participants.[141]

The institution's legal counsel must not be a member of the REB.

The Tri-Council Working Group also indicated that appropriate *ad hoc* members could be added to an REB when it is reviewing a project that "requires particular community or research representation" or a project that "requires specific methodological expertise not available from its regular members".[142]

The Working Group's recommendation clearly focusses on the need to ensure that REBs have demonstrated expertise not only in research but also in law and ethics. However, it also includes provision for a member from the community or from the pool of potential research subjects. The suggestion that the membership of REBs encompass representatives of those who may participate in biomedical experimentation as research subjects is one that is in accordance with the view of an increasing number of commentators to the effect that such representatives are more likely to make a significant contribution to the decision-making of REBs than laypersons drawn from the community at large. Furthermore, the membership structure advocated by the Working Group would allocate a majority of the decision-making votes to those members who are not scientists.

It is to be hoped that the recommendation of the Tri-Council Working Group on the issue of REB membership will ultimately be implemented. However, it is clear that considerable energy should be devoted to solving the eminently practical problems associated with finding, and appointing, members who represent the community at large or the pool of potential research subjects. As we saw in our discussion of the operation of RECs in the United Kingdom, it is important that there be an independent body that assists in identifying appropriate nominees for membership of the various boards. In the Canadian context, this function could be assumed by the provincial (or territorial) ethics review board which would be responsible for identifying suitable nominees from community groups and organizations that maintain an interest in biomedical experimentation as well as from those that represent individuals who may become research subjects.

---

[140] *Ibid.* at 48 (Part 2).
[141] *Ibid.* at 17 (Part 2).
[142] *Ibid.* at 18 (Part 2).

### Supervising and Coordinating the Activities of the Local REBs

In the common law jurisdictions of Canada, the existing system of REBs operates in a context of a process of self-regulation by the various local institutions involved in the conduct of biomedical experimentation. Until relatively recently, the only body that assigned itself a mandate to oversee the activities of local REBs was the Medical Research Council of Canada. The Council recognized, in its *Guidelines* of 1987, that researchers themselves have both an "initial and continuing duty" to ensure that their research is ethical. However, it also stipulated that the *primary* responsibility for maintaining the appropriate ethical standards still rests with the various institutions that conduct research (through the medium of the REBs).[143] The MRC nevertheless perceived itself as playing a key role in this process by articulating ethical guidelines for biomedical research with human beings[144] and by refusing funding to those institutions that did not follow these guidelines in accordance with the decisions made by the appropriate REBs. In addition, the Council expressed the view that it had a special duty to foster awareness of ethical issues concerning this type of research. Significantly, the MRC also suggested that it should assume the responsibility for monitoring the activities of local REBs.[145]

The MRC itself was not able to perform a coordinating or monitoring function owing to a lack of resources. However, in 1988, the MRC and Health and Welfare Canada joined together with the Royal College of Physicians and Surgeons to establish the National Council on Bioethics and Human Research (NCBHR). The NCBHR, a private body, was initially created for a period of nine years, after which its mandate will be reviewed.[146] Its membership consists of five representatives of the Royal College of Physicians and Surgeons, a representative of each of five health care associations, a philosopher-theologian, a lawyer, two community members and the President of the MRC who is *ex officio*.

The NCBHR furnishes advice to, and engages in consultation with, the various local REBs around the country. Furthermore, the National Council attempts to enhance the mutual understanding of ethical issues on the part of the members of the various REBs by organizing special educational seminars and workshops and by publishing a newsletter, called *NCBHR Communiqué*; as well as diverse monographs on relevant topics in the field of bioethics.[147] Most significantly, the NCBHR has the authority to establish teams of experts to conduct site visits at the invitation of REBs. Between 1990 and 1993, the NCBHR designated universities with medical faculties as the "primary clients" for such site visits. The data gleaned from such

---

[143] See *MRC Guidelines, supra* note 6 at 43.

[144] The Code of Ethical Conduct for Research Involving Humans, prepared by the Tri-Council Working Group, if ultimately adopted by the three granting councils in Canada, will supersede both the MRC *Guidelines on Research Involving Human Subjects* (1987) and the SSHRC *Ethics Guidelines for Research with Human Subjects*. It "complements" the 1990 MRC Guidelines for Research on Somatic Cell Gene Therapy in Humans. See Tri-Council Working Group, *supra* note 7 at 1 (Foreword).

[145] *Ibid.*

[146] Law Reform Commission of Canada (1990), *supra* note 125 at 15.

[147] See e.g. National Council on Bioethics in Human Research, Consent Panel Task Force, *Reflections on Research Involving Children* (Ottawa: NCBHR, 1993).

visits constituted an important component of the Council's recent report on the function of REBs in Canadian medical faculties.[148]

The NCBHR has emphasized that its principal role is to provide *advice* on ethical issues and that the ultimate responsibility for maintaining high ethical standards in the conduct of biomedical experimentation rests firmly on the shoulders of individual researchers and the REBs.[149] In essence, therefore, its mandate is advisory, rather than controlling, and members of REBs have the option to decide whether or not to act on any advice that is given by the National Council. It is significant that, in 1990, the Law Reform Commission of Canada questioned the comparatively narrow basis of this mandate:

> ... it is not certain whether the Council's mandate (in essence, it can only monitor researchers who themselves asked to be monitored and those who are funded by the funding organizations) extends to private research centres or the research carried out by entirely private entities (pharmaceutical companies, for example). The Council's activities are therefore not general or comprehensive.[150]

There is a real question, furthermore, as to whether the NCBHR is sufficiently independent from the medical profession and the research community. Its membership consists primarily of representatives who are appointed by professional groups such as the Royal College of Physicians and Surgeons, the Canadian Medical Association and the Canadian Nurses' Association.[151] Similarly, insofar as the NCBHR is accountable at all for its activities, it is answerable solely to professional bodies: more specifically, its formal accountability consists of little more than the duty to submit annual reports to the various professional agencies involved in its creation (the MRC, Health and Welfare Canada and the Royal College of Physicians and Surgeons) as well as to the boards of other organizations that are involved in its operation.

There is little doubt that the NCBHR fulfills an important role in the education of the members of the various REBs in Canada and in the promotion of discussion about ethical issues. However, since it is not visibly "at arm's length" from the medical professions, it may not be the most appropriate body to undertake the critical tasks of monitoring the activities of medical researchers and of ensuring that the members of the REBs perform their duties in accordance with generally accepted standards of ethical decision-making within the field of biomedical experimentation. If REBs are to operate on a statutory basis, as recommended above, it is important that an unequivocally independent agency assume responsibility for coordinating and monitoring their operations. Members of the public are more likely to develop confidence in the whole enterprise of biomedical experimentation if the REBs are held accountable to a body that is distinctly

---

[148] See NCBHR (1995), *supra* note 27 at 7.
[149] National Council on Bioethics in Human Research, *National Council on Bioethics and Human Research* (Ottawa: NCBHR, 1989) at 3.
[150] *Ibid.* at 15.
[151] *Ibid.* at 12–31.

separate from the medical professions whose members may be engaged in the conduct of such experimentation.

What agency should be assigned the responsibility for coordinating and monitoring the decision-making of REBs and for maintaining public confidence in the system by means of which biomedical experimentation is regulated? In Ontario, the *Final Report* of the Enquiry on Research Ethics (1995) recommended that the Government establish a Provincial Ethics Review Board (PERB) to accredit and monitor local REBs in the province.[152] This model is one that is well-suited to the Canadian context and, if adopted across the country, would ensure that there would be an independent agency that would oversee the operations of REBs in each province and territory.

One of the primary tasks of the PERB would be to ensure that the membership of each local REB reflects an appropriate balance between those engaged in biomedical research, those who have some degree of expertise in law and/or ethics, those who represent the interests of those who may become research subjects, and members of the local community. The recommendations of the Tri-Council Working Group, discussed in the previous section, provide a suitable starting point for seeking this balance insofar as they ensure that researchers will be in a minority in each REB (which would consist of 2 scientific experts, 1 expert in ethics, one expert in law, and one member of the local community or a representative of potential research subjects). The PERB would not only be responsible for the accreditation of those REBs which met these membership requirements but would also assist local boards to identify potential members with expertise in ethics and law and to develop lists of members of the community at large who have an interest in biomedical experimentation and of individuals who represent those persons who may become involved as research subjects.

The size of the membership of the PERB would necessarily vary with the population of the particular province or territory concerned. However, the composition of the PERB should reflect the balance of membership that is required in the case of local REBs.

The PERB would also be responsible for providing information to members of local REBs about ongoing developments in the law and ethics that are applicable to biomedical experimentation. Whenever necessary, it could also issue its own interpretations of existing ethical guidelines should there be a lack of clarity or a degree of uncertainty in the latter. The PERB would also organize seminars, workshops, etc. to enhance the knowledge base of members of REBs and to encourage their interaction with their counterparts from other REBs. In this vein, another important function, that may be performed by the PERB, is the dissemination of information about legal and ethical issues relating to biomedical experimentation not only to members of the research community and REBs but also to the members of the public at large. In particular, the collection and dissemination

---

[152] *Enquiry on Research Ethics: Final Report* (Chairman: David N. Weisstub, Submitted to the Hon. Jim Wilson, Minister of Health of Ontario, Aug. 28, 1995) [hereinafter *Enquiry on Research Ethics*].

of information about decisions made by the PERB and by local REBs is a task that is particularly well-suited to a such a body. Hopefully, the various PERBs across Canada would coordinate their educational activities with the NCBHR, which could bring a more explicitly national perspective to bear on its own operations in this area.

As noted above, it is recommended that there be a statutory requirement that all biomedical experimentation with human subjects receive prior approval by an REB. The *Final Report* of the Ontario Enquiry on Research advocated the position that, in certain circumstances, such authorization must also be obtained from the PERB. More specifically, the Report recommended that, where there is some uncertainty as to the *capacity* of the prospective research subjects to give a valid consent (e.g., if they are mentally disordered, developmentally disabled, children, elderly, or prisoners), then the prior approval of a panel of the PERB should be required in relation to any research protocol which indicates that the subjects will be exposed to a substantial risk of harm. In essence, the Enquiry's recommendation implies that, where such a vulnerable group of subjects is involved, then the prior approval of a local REB will only be sufficient where the research protocol concerned indicates that there is a "negligible" or "less-than-substantial" risk of harm. Where the risk is deemed to be substantial, then the prior approval of a panel of the PERB will also be required. Furthermore, whenever its approval is required, the PERB panel should include a representative of the particular group of potential subjects identified in the research protocol. The Report also recommended that a register should be established to maintain a record of all protocols involving groups of subjects whose capacity to give a valid consent is in question and that those experiments that have been approved by local REBs as posing a "less-than-substantial" risk should be periodically reviewed by the PERB with a view to ensuring that the REBs have been applying appropriate techniques of assessing the degree of risk involved in such experiments.

The PERB would also serve a most valuable function where research protocols involve the use of multi-centre clinical trials. As noted earlier, the increasing use of *multi-centre clinical trials* raises the difficulty that REBs at different sites may make decisions at variance with each other.[153] The NCBHR study of REBs in Canada unequivocally demonstrated that there was a demand on the part of members of many REBs that there should be "some central way of considering multi-centre trials prior to the protocol being fixed".[154] It was felt that no individual REB should be placed in the position where it is required to review a protocol that, in effect, cannot be changed because it involves a multi-centre trial. We saw that, in England and Wales, there is increasing support for a national ethics advisory committee to deal with the special problems multi-centre trials raise. Such a national committee could give conditional approval to a multi-centre proposal; however, each local research ethics committee would have the opportunity to approve or reject it (but not

---

[153] See Meslin (1993), *supra* note 97 at 185.
[154] See NCBHR (1995), *supra* note 27 at 15.

to modify it).[155] This potential solution could well be adapted to the Canadian context, although it would be more appropriate to consider allocating this function to a PERB when the research centres are located within one province. If the research project concerned involves centres in more than one province or territory, it would be necessary to negotiate the protocols between the respective research groups or, alternatively, to refer the project to a national advisory committee, which would need to be established by the Government of Canada.[156]

The PERB, as a body that is clearly independent from the medical professions and the research community, could also assume the vital task of monitoring the actual implementation of research protocols in order to safeguard the public interest in protecting the rights of human research subjects. If the integrity of the system for regulating biomedical experimentation in Canada is to be upheld, then it is necessary that an independent body conduct independent monitoring of ongoing research projects. It is clear that local REBs have generally been unable to monitor research projects once they have been approved and this situation constitutes a significant flaw in the regulatory system as a whole. The PERB, if assigned adequate resources to do so, would be a most appropriate agency to undertake the routine monitoring of ongoing experiments after they have been approved by local REBs.

## SUMMARY

Since the conclusion of World War II, there has been a series of determined attempts, at both the national and international levels, to articulate general ethical principles for the conduct of biomedical experimentation involving human beings. While the refinement of these principles constitutes an ongoing process, there is nevertheless considerable debate as to the nature of the most efficacious method of regulating biomedical experimentation at a local level. In most countries that conduct biomedical research, the heart of the regulatory system consists of a process of review by institutional ethics review boards or committees. However, the legal status and authority of these boards or committees are unclear in a number of jurisdictions. In Canada, with the exception of the Province of Quebec, REBs have operated at a local level without any formal legislative underpinning and have generally applied a set of ethical guidelines articulated by the Medical Research Council of Canada: in this sense, a system of self-regulation exists in most of the country. While there does not seem to be a widespread sentiment in favour of abolishing the process of ethical review by REBs, there is an increasing number of voices that are calling for significant reforms to the existing system.

It is recommended that the various provinces and territories of Canada enact legislation that unequivocally establishes the authority of REBs to review all research protocols involving biomedical experimentation with human beings. Such legislation should make it an offence to conduct such experimentation without the

---

[155] *Ibid.* at 29.
[156] See generally Law Reform Commission of Canada (1990), *supra* note 125.

prior approval of a local REB and should articulate general rules governing the conduct of researchers in this field. Furthermore, each province and territory should establish a provincial (or territorial) ethics review board that should be entrusted with the task of coordinating and supervising the activities of local REBs. While the proposed legislation should entrench the minimal legal requirements for the conduct of biomedical experimentation, local REBs would also be responsible for applying the highest ethical standards which, if the recent recommendations of the Tri-Council Working Group are ultimately accepted, will be endorsed by the three major granting councils in Canada (MRC, NSERC and SSHRC). The membership of local REBs should reflect an appropriate balance between researchers, representatives of the pool of potential research subjects, members of the public with a special interest in biomedical experimentation, and experts in law and ethics.

The provincial (and territorial) ethics review boards should be unequivocally placed "at arm's length" from the medical professions and the institutions that conduct biomedical experimentation in Canada. Through a process of accreditation, the PERBs would ensure that the membership of local REBs reflects an appropriate balance of interests and would develop lists of potential members in order to assist local REBs in the task of recruitment. In addition, the PERBs would be responsible for interpreting existing ethical guidelines where there may be uncertainty or a lack of clarity; organizing the ongoing education of REB members in relevant fields; and disseminating information both to the members of REBs and to the public at large. The PERBs would also be responsible for performing the vital task of monitoring ongoing biomedical experiments to ensure compliance with ethical requirements and should be allocated sufficient funds to enable them to discharge this responsibility in an effective manner. Finally the PERBs would be assigned special responsibilities in relation to those research protocols that involve human subjects whose capacity to give a valid consent to participation may be in doubt or that require that research be conducted in a number of different sites.

Hopefully, these proposed reforms would provide more effective protection of the rights of individual research subjects while maintaining the degree of decision-making flexibility that is necessary for the development of medical knowledge in an age of ever-changing technology and constantly evolving challenges.

CHAPTER 18

# ESTABLISHING THE BOUNDARIES OF ETHICALLY PERMISSIBLE RESEARCH WITH VULNERABLE POPULATIONS

DAVID N. WEISSTUB, JULIO ARBOLEDA-FLOREZ AND
GEORGE F. TOMOSSY

## INTRODUCTION

The promulgation of the *Nuremberg Code* initiated a process of inquiry on an international scale that has led to the evolution of various general ethical principles intending to guide participants in the research endeavor. Indeed, the *Code* makes the following statement regarding the justifiability of human experimentation:

> The great weight of evidence before us is to the effect that certain types of medical experiments on human beings, when kept within reasonably well-defined bounds, conform to the ethics of the medical profession generally. The protagonists of the practice of human experimentation justify their views on the basis that such experiments yield results for the good of society that are unprocurable by other methods or means of study. All agree, however, that certain basic principles must be observed in order to satisfy moral, ethical and legal concepts . . . [1]

The *Declaration of Helsinki* further acclaimed society's interest in the conduct of human experimentation, providing the justification that ". . . it is essential that the results of laboratory experiments be applied to human beings to further scientific knowledge and to help suffering humanity".[2] Although the *Nuremberg Code*

---

[1] See the introduction to the *Nuremberg Code*. The *Nuremberg Code* constituted part of the judgment in *U.S.* v. *Karl Brandt et al., Trials of War Criminals Before the Nuremberg Military Tribunals Under Control Council Law No. 10.* (October 1946-April 1949).

[2] See the introduction of the *Declaration of Helsinki*. Adopted by the World Medical Association at the 18th World Medical Assembly in Helsinki in June 1964. Amended at the 29th World Medical Assembly in Tokyo in October 1975; the 35th World Medical Assembly in Venice in October 1983; and the 41st World Medical Assembly in Hong Kong in September 1989. Reprinted in (1991) 19 Law, Medicine & Health Care 264.

explicitly banned experimentation with persons who are themselves unable to provide a valid consent, this stance was subsequently relaxed by the *Declaration of Helsinki*, which allowed for the substitute consent of a legal guardian.[3] It is without doubt, however, that the use of members of vulnerable populations as experimental subjects raises many difficult questions regarding the ethical permissibility of their involvement when the research is non-therapeutic. Certainly, the simplest solution would be to ban such research altogether; however, this position has not been universally acclaimed.[4]

The objective here is to delineate general safeguards for the ethical conduct of non-therapeutic biomedical experimentation with vulnerable populations, which will not include "persons" whose claim to personal respect is itself controversial or purely derivative. Therefore, although it is recognized that there are important decisions to be made regarding research procedures performed on, for example, those who lack brain function and on fetal tissue, these matters have to date been regulated exclusively by a concern for the limits of ethical permissibility and have not been widely viewed as raising issues of consent.

With the exception of prisoners, the discussion will be restricted to populations whose incapacity is the result of some personal characteristic. The law has been reticent to recognize social factors as having bearing on capacity. Arguably, the most significant social factor affecting capacity is that of institutionalization. With the exception of incarceration for penal purposes, institutionalization generally coincides with some other form of incapacity. Other social factors include the relationship between prospective participants and their researchers, caregivers, guardians, or family members, all of which represent situations where the participant's subordinate position may have a coercive influence affecting the voluntariness of a decision to enroll in a research protocol.

It is acknowledged that research in institutions raises important ethical issues and that future legislation may wish to be aware of the impact of the social context as a distinct factor affecting the capacity of an individual to consent to participate in research. Further study should be undertaken to determine whether, in fact, the social context has distinct effects on capacity in this area and whether mechanisms can be formulated to safeguard the welfare of affected individuals so as to ensure the voluntariness of their participation in research.[5]

---

[3] Since the first revision of the *Declaration of Helsinki*, the participation of legally incompetent persons has been deemed permissible provided that the consent of "a legal guardian in accordance with national legislation" was obtained. See *Declaration of Helsinki II-IV, ibid.,* s. I-11.

[4] See e.g. Queensland Law Reform Commission, *Assisted and Substituted Decisions: Decision-making by and for People with a Decision-making Disability* (Brisbane: Q.L.R.C., 1996); The Law Commission, *Mental Incapacity* (London: HMSO, 1995); Council for International Organizations of Medical Sciences, in collaboration with the World Health Organization, *International Ethical Guidelines for Biomedical Research Involving Human Subjects* (Geneva: CIOMS, 1993) [hereinafter *CIOMS Guidelines*].

[5] The social context affecting capacity to consent to non-therapeutic biomedical research may include issues relating to gender, economic status, and membership of a cultural minority. See Office for the Protection of Research Risks, National Institutes of Health, *Protecting Human Subjects: Institutional Review Board Guidebook* (Washington, D.C.: U.S. Government Printing Office, 1993) at 6–11–6–17, 6–49–6–51 [hereinafter *OPRR Guidebook*].

The vulnerable populations to which the general safeguards presented here will apply consist of children, the elderly, the mentally disordered, the developmentally disabled, and prisoners. Each of these populations, to a greater or lesser extent, has been recognized by the international community as "vulnerable" to exploitation and abuse in the context of biomedical experimentation, and not without cause. However, before proceeding, it is important to expand upon the reasons underpinning the classification of members of vulnerable populations as "vulnerable".

## VULNERABILITY

"Vulnerability" has been referred to as ". . . a substantial incapacity to protect one's own interests owing to such impediments as lack of capability to give informed consent, lack of alternative means of obtaining medical care or other expensive necessities, or being a junior or subordinate member of a hierarchical group".[6] The Law Commission (UK) proposed that "[a] person is vulnerable if by reason of old age, infirmity or disability (including mental disorder . . .) he is unable to take care of himself or to protect himself from others".[7] In the context of non-therapeutic experimentation, the key criterion suggesting "vulnerability" would be an inability to protect oneself from exposure to an unreasonable risk of harm. With respect to vulnerable populations, questions of competence and of voluntariness lead to concerns regarding the capability of members of these populations to provide a valid and informed consent to participate in biomedical research.

Children, initially incompetent, represent a population with an increasing capacity to make decisions accompanied by a corresponding decrease in dependence upon their guardians. The elderly, presumably fully competent for most of their lives, can be described as a population with a potentially decreasing decision-making capacity and an increasing reliance upon their caregivers. The mentally disordered can best be characterized as having fluctuating levels of competence, which, in certain cases, may also be decreasing. Developmentally disabled persons may have never possessed, nor are ever likely to possess, sufficient competence to make decisions regarding their own welfare. In addition, the members of these populations, and prisoners in particular, owing to social factors associated with institutionalization, reliance upon caregivers, or being in subordinate relationships, may have problems regarding the voluntariness of a decision to participate in non-therapeutic biomedical experimentation. Therefore, each population generates concerns regarding the ability of its members to protect themselves from exposure to unreasonable risks of harm, and are consequently "vulnerable" as defined above. Although somewhat arbitrary, this determination is nevertheless necessary to provide a coherent framework for discussion. Indeed, a generalization that all the

---

[6] See *CIOMS Guidelines, supra* note 4 at 11.
[7] The Law Commission, *Mentally-Incapacitated and Other Vulnerable Adults: Public Law Protection (Consultation Paper No. 130)* (London: HMSO, 1993) at para. 2.29.

members of a given vulnerable population are incompetent or incapable, and consequently require outside intervention, would be unfair, unrealistic, and in particular, disrespectful. Hence, it is important to stress that where a population is deemed to be "vulnerable", by no means should it be inferred that all members are incapable.[8] Rather, individuals in vulnerable populations may possess particular characteristics, or be exposed to certain conditions, which potentially render them incompetent. Absent such factors, individuals should not be treated any differently from those who do not belong to a vulnerable population.

Along these lines, adults should be presumed to be competent and capable of providing valid consent, unless the opposite is found to be the case. Where the competence of adults is in doubt, they should be allowed to make decisions to the maximum extent that they are able. Even children should be entitled to participate in decisions affecting their personal welfare as much as possible. As such, elaboration of the general safeguards presented here with regard to the various vulnerable populations must focus on the need to assess and maximize the decision-making capacity of individuals, taking into consideration those factors that may influence the voluntariness of a subject's decisions.

## GENERAL SAFEGUARDS

It is important to emphasize that those individuals who are competent and can provide voluntary consent, despite belonging to a vulnerable population, should only be subject to such restrictions as are normally placed on a fully competent person. Therefore, the recommendations that follow apply only to those persons who are unable to provide a valid and informed consent to participate in non-therapeutic biomedical experimentation. However, owing to the specific characteristics of vulnerable populations, further protective measures would be required in addition to those detailed below.

### Overlapping Populations

Different vulnerable populations present distinct challenges to the process of establishing limits on the participation of their members in research. Each has special characteristics that will require separate guidelines specific to each population. However, it should also be noted that persons may belong to more than one vulnerable group. For example, a person may be developmentally disabled as well as a child, or a person may be mentally disordered and elderly. In such situations, subjects should be given the benefit of all protective measures applicable to the different categories. Where the safeguards for each form of incapacity differ

---

[8] The Law Commission (UK) was also careful to make this distinction. See The Law Commission, *Mentally Incapacitated Adults and Decision-Making: An Overview (Consultation Paper No. 119)* (London: HMSO, 1991) at para. 1.2.

in kind, they should be combined; where they differ only in degree, the more stringent should apply.

### Identifying the Nature of an Experiment

It is important to acknowledge that many abuses in the context of human experimentation were often justified on the basis that a given experiment was therapeutic in nature. In order to prevent such abuses, it is important to assess carefully the nature of the research in question; that is, to determine whether an experiment is primarily therapeutic or non-therapeutic in nature. An experiment, defined as ". . . a planned cause-effect study in which the action of a particular maneuver is contrasted with the results of a comparative, or control, maneuver",[9] more often than not, falls on a spectrum between "pure therapy", at one end, and "pure research" on the other.[10] Indeed, this assessment should be made by a research ethics committee primarily on an *objective basis*, rather than on the basis of the stated intent of the researcher.

### Assessing The Scientific Validity of an Experiment

The Medical Research Council of Canada made the following statement with respect to the scientific validity of research protocols:

> Medical research should be a deliberate and careful step into the unknown . . . . Modern scientific research must conform to the requirements of rigorous methodologies. Thus, scientific research has its own integrity. It is accepted that ethical research must be scientifically sound, so that observing the integrity of the scientific method is part of the ethics of research.[11]

Although experimentation, by definition, entails a degree of uncertainty, it must not be based on improper methods or assumptions. Therefore, in order to preserve the "integrity of the scientific method", experiments involving human subjects must be restricted to those based on "good science"; that is, an experiment should not be conducted unless it is scientifically valid. This requires that the following conditions

---

[9] A.D. Feinstein, *Clinical Epidemiology — The Architecture of Clinical Research* (Philadelphia: W. B. Saunders, 1985) at 17.

[10] M.A. Somerville, "Therapeutic and Non-Therapeutic Medical Procedures — What are the Distinctions?" (1981) 2 Health Law in Canada 85 at 88–89. It is worth noting that this opinion is not shared by Annas, who states that "few interventions are in the gray zone and [that] an objective distinction can almost always be made between an experimental intervention and a treatment". See G.J. Annas, "Questing for Grails: Duplicity, Betrayal and Self-Deception in Postmodern Medical Research" (1996) 12 Journal of Contemporary Health Law and Policy 297 at 321.

[11] Medical Research Council of Canada, *Guidelines on Research Involving Human Subjects* (Ottawa: Supply & Services Canada, 1987) [hereinafter *MRC Guidelines*]. It should be noted that the *MRC Guidelines* have been superceded by *Code of Ethical Conduct for Research Involving Humans* (July 1997) by Tri-Council Working Group (comprised of the Medical Research Council, the Natural Sciences and Engineering Research Council, and the Social Sciences and Humanities Research Council).

be satisfied: (1) research must be conducted by "a scientifically qualified person"; (2) it must conform to generally accepted scientific principles; and (3) it must be based on a knowledge of the natural history of the disease or other problem under study, and on a thorough knowledge of the scientific literature.[12]

Without launching into a lengthy discourse on the "scientific method", assessing the scientific validity of an experiment on humans must also include a determination of whether or not the experiment proposes to answer a scientifically valid question and "whether the scientific methods are sound and suitable for the aims of the research".[13] However, it is important to emphasize that while the scientific validity of an experiment is a prime requisite, the fact that an experiment is scientifically valid does not imply that it is ethically permissible. The inter-relationship between science and ethics cannot be ignored. Indeed, a failure to satisfy the requirements of "good science" will render an ethical assessment redundant. Human experimentation based on "bad science" would be clearly unethical, but as stated above, the converse is not automatically true. Therefore, science (based on purely scientific principles) and ethics (based on moral and ethical norms) should be assessed sequentially, and therefore tested separately. An experiment may be based on established scientific principles, but the manner in which it is carried out might not only be ethically impermissible, but morally reprehensible as well. As stated by the *Declaration of Helsinki*, "[i]n research on man, the interest of science and society should never take precedence over considerations related to the well-being of the subject".[14] As such, further restrictions governing the ethical permissibility of experimentation on vulnerable persons are required.

### *Limiting the Scope of an Experiment and the Selection of the Subject*

The first factor limiting the "scope" of research is the requirement that it have significant scientific value. The *Nuremberg Code* emphasized the importance of yielding "fruitful results for the good of society that are unprocurable by other

---

[12] See the *Nuremberg Code, supra* note 1, s. 3 read with the *Declaration of Helsinki, supra* note 2, ss. I-1, I-3 and the *CIOMS Guidelines, supra* note 4 at 12. The MRC stated that "[t]he use of human beings in research must be essential for scientific reasons and . . . research should not repeat work already done unless such work requires further substantiation". See *MRC Guidelines, ibid.* at 15.

[13] The importance of addressing this latter concern was suggested in *CIOMS Guidelines, supra* note 4 at 44, as a special responsibility of external sponsoring countries or international agencies. In other words, a proposed experiment should be able to satisfy the following concerns:
- Does the experiment propose to answer a valid question?
- Would the experiment, as outlined in its methodology, properly answer the proposed question?
- Is the experimental design valid?
- Is the experiment likely to yield scientifically valid results (*i.e.* does it make proper use of placebos and blinding)?

See also K.R. Popper, The Logic of Scientific Discovery, 2d ed. (New York: Harper Torchbooks, 1968).

[14] See *Declaration of Helsinki, supra* note 2, s. III-4.

methods or means of study, and not random and unnecessary in nature".[15] As such, human experimentation must, whenever possible, be preceded by experimentation on animal (or other) models.[16]

The requirement that an experiment have "major scientific importance" is a prerequisite for the approval of research in cases where it is not possible to use capable subjects.[17] This "requirement" can also serve as a "justification" for research. For example, the Department of Health and Human Services (DHHS) regulations, in the context of research involving children, allow an IRB to permit research otherwise impermissible under the regulations in such cases where "the IRB finds that the research presents a reasonable opportunity to further the understanding, prevention or alleviation of a serious problem affecting the health or welfare of children".[18]

The scientific importance of an experiment should be regarded as a benefit for the purpose of evaluating the ethical permissibility of an experiment.[19] Research to be conducted on vulnerable populations must be of significant scientific value, measured by its likelihood of producing knowledge furthering the understanding, prevention, or alleviation of a problem directly related to a condition or circumstance affecting the subject, or the class to which the subject belongs. As such, in the process of balancing benefits and risks, a research ethics committee should take into account, but not be swayed by, the scientific importance of an experiment.[20]

A restriction that flows from the above recommendation is that non-therapeutic experimentation on vulnerable populations should be limited to research bearing some direct relation to a condition or circumstance affecting the subject, or the class

---

[15] This criterion was recognized by the Law Reform Commission in the context of research on children and adults with mental disorders. See Law Reform Commission of Canada, *Working Paper No. 61: Experimentation Involving Human Subjects* (Ottawa: Law Reform Commission, 1989) [hereinafter *LRC Working Paper No. 61*]. It was also recognized in the introduction of the *Nuremberg Code, supra* note 1. It is interesting to note that the Law Reform Commission stated that experimentation should be considered legal, even in the event of the deception of a subject, where, among other requirements, "there are no other means of achieving the research goal" and "the research is of major scientific value".

[16] See the *Nuremberg Code, ibid.*, s. 3; *Declaration of Helsinki, supra* note 2, s. I-1; *MRC Guidelines, supra* note 11 at 15. It could also be argued that the requirement of previous study on animal (or other) models before resorting to human experimentation is a requirement for the scientific validity of an experiment. However, the preferential use of alternate models seems more likely to be an *ethical* decision than a *scientific* one.

[17] See *LRC Working Paper No. 61, supra* note 15 at 42, 45. For example, the 1990 New York State Mental Health Office regulations, 14 N.Y.C.R.R. (1990) §527.10, require that an experiment must be "likely to produce knowledge of overriding therapeutic importance for a condition presented by the patients in question". See Delano & Zucker, "Protecting Mental Health Research Subjects Without Prohibiting Progress" (1994) 45 Hospital & Community Psychiatry 601 at 602.

[18] 45 C.F.R. 46 (1991) § 46.407(a).

[19] Indeed, the Office for the Protection of Research risks states that "[r]isks to research subjects posed by participation in research should be justified by the anticipated benefits to the subjects or society". See *OPRR Guidebook, supra* note 5 at 3–1.

[20] The importance of not being swayed by the interests of society as measured by the "scientific importance" of an experiment is emphasized by the *Declaration of Helsinki* in its statement that "the interest of science and society should never take precedence over considerations related to the well being of the subject". See *Declaration of Helsinki, supra* note 2, s. III-4.

to which the subject belongs.[21] But how narrowly should such a limitation on the scope of experimentation be interpreted? In the context of dementia research, Keyserlingk *et al.* made the following statement:

> The research need not be relevant to a patient's condition of dementia. It is sometimes suggested or implied that persons with dementia should only be enrolled in research projects with a potential benefit for the particular patient's dementia, or at least for the class of patients with dementia. We are not persuaded that such a restriction is justified in all circumstances. In effect, such a restriction applied without qualification would unfairly deny a patient with dementia access to research potentially beneficial to himself or herself for other medical or non-medical reasons, or the opportunity to choose to participate in non-dementia research potentially beneficial to others.[22]

Certainly, members of vulnerable populations should not be arbitrarily restricted solely to participation in research related to the condition which identifies them as vulnerable; however, participation in research not related to their condition should be permitted only if they, like any other competent adult, are capable of providing an informed and voluntary consent. Furthermore, if the research in question could be done with the participation of either competent or incompetent persons, then incompetent persons should not be used. Thus, a competent elderly person could, like any other adult, consent to non-therapeutic experimentation of a condition possessed by that individual, even though the problem may not be specific to the elderly population. However, if that person becomes incompetent by reason of dementia, the above rules would exclude that person from participation in research unrelated to those conditions or circumstances which render that person "vulnerable".

A limitation on the researcher's choice of subject and on the scope of proposed research is also founded upon the principle of justice, discussed by the *Belmont Report*, which requires that subjects be selected for "reasons directly related to the problem being studied" rather than "simply because of their easy availability, their compromised position, or their manipulability".[23] Consequently, where alternative research subjects are available, individuals should not be selected simply for reasons of administrative convenience. This would have a particular application to

---

[21] This requirement is found in article 21 of the *Civil Code of Quebec*, where "[a]n experiment on a group of minor persons or incapable persons of full age shall be carried out . . . only if a benefit to the health of persons of the same age group and having the same illness or handicap as the persons submitted may be expected". The *Code de la santé publique* of France (Art. L. 209–6) also states that research on children or incompetent adults can only be carried out when the research is to be utilized to benefit individuals with a similar illness or handicap. The *CIOMS Guidelines, supra* note 4 at 20 & 22, recommend that research involving children and persons with mental or behavioural disorders have the purpose of obtaining knowledge relevant to the health needs of the respective groups. Similar recommendations were also made by the Law Reform Commission of Canada, whereby research on children must be "in close, direct relation to infantile diseases or pathologies"; and research on the mentally disordered must be "in close, direct relation to the subject's mental illness or deficiency". See *LRC Working Paper No. 61, supra* note 15 at 42, 44.

[22] E.W. Keyserlingk *et al.*, "Proposed Guidelines for the Participation of Persons with Dementia as Research Subjects" (1995) 38 Perspectives in Biology & Medicine 319 at 325 [footnotes omitted].

[23] National Commission for the Protection of Human Subjects of Biomedical and Behavioral Research, *The Belmont Report: Ethical Principles for the Protection of Human Subjects in Research* (Washington, D.C.: U.S. Government Printing Office, 1978) at 10 [hereinafter *The Belmont Report*].

populations who may be vulnerable based on their being institutionalized.[24] In other words, the principle of distributive justice requires "the equitable distribution of both the burdens and the benefits of participation in research",[25] and justifies those safeguards which regulate the selection of subjects for research. Wherever a less compromised or less vulnerable population is available for research purposes, that population should be used.

### Limiting the Level of Risk and Risk Assessment

Perhaps the most important restrictions are those associated with the level of risk of harm to which an individual should be exposed in the course of an experiment. Restrictions in this regard are well-supported on an international and national scale. They require the minimization of harm, the assessment of risk, and the balance of risks and benefits, which is necessary to determine whether the risks can be justified.

#### The Researcher's Obligation to Minimize Harm

The principle of beneficence, as discussed in the *Belmont Report*, requires that potential benefits be maximized and potential harms be minimized.[26] The first step in accomplishing this objective begins with the researcher. Criminal liability aside,[27] researchers have many positive duties associated with their obligation to minimize potential harms. These duties exist before, during and after the experiment.

The first set of duties begin before commencing an experiment. The researcher should not conduct research which causes unnecessary suffering and injury, whether mental or physical.[28] Moreover, he or she should avoid engaging in a research project until the reasonably foreseeable risks have been ascertained.[29] If there is an *a priori* belief that death or disabling injury might occur, the experiment should not be conducted.[30] In addition to readily foreseeable risks, "[p]roper preparations

---

[24] The concern regarding institutionalization applies particularly to prisoners, the civilly committed mentally disordered, and any other mentally disordered, developmentally disabled, or elderly person who also happens to be institutionalized.

[25] See *CIOMS Guidelines, supra* note 4 at 10.

[26] See *The Belmont Report, supra* note 23 at 10.

[27] Researchers must pay careful attention to the level of risk expected from non-therapeutic experiments. In Canada, while s. 45 of the *Criminal Code*, R.S.C. 1985, c. C-46, protects physicians if a procedure is reasonable and performed with reasonable care and skill, "[r]esearchers should be aware that section 45, which affords protection to surgeons, will likely not afford protection to the researcher since the nature of research does not involve benefit to the person involved as required by the section". See H. Sava, P.T. Matlow & M.J. Sole, "Legal Liability of Physicians in Medical Research" (1994) 17 Clinical & Investigative Medicine 148 at 152. As such, careful risk assessment may also be important not only for the sake of the prospective subject, but in order to protect the researcher from possible criminal liability.

[28] See the *Nuremberg Code, supra* note 1, s. 4., which states: "The experiment should be so conducted as to avoid all unnecessary physical and mental suffering and injury". The *Declaration of Helsinki, supra* note 2, s. I-6, goes one step further by stating: "The right of the research subject to safeguard his or her integrity must always be respected. Every precaution should be taken to respect the privacy of the subject and to minimize the impact of the study on the subject's physical and mental integrity and on the personality of the subject".

[29] See *Declaration of Helsinki, ibid.*, s. I-7. The *Declaration* states that researchers should not engage in a research project "unless they are satisfied that the hazards involved are believed to be predictable".

[30] See the *Nuremberg Code, supra* note 1, s. 5.

should be made and adequate facilities provided to protect the experimental subject against even remote possibilities of injury, disability, or death".[31]

The second set of obligations arises during the experiment. The researcher is obliged to "remain the protector of the life and health of the person on whom biomedical research is being carried out".[32] As such, he or she must terminate an experiment if there is probable cause to believe that injury, disability, or death of the experimental subject might arise.[33] As suggested earlier, it is also important to ensure that consent, or in certain cases assent, continues throughout the course of an experiment.[34] Whether this should be the task of the researcher or of a "consent monitor" should depend upon the nature of the study. A researcher might also be obligated to provide ongoing professional support, such as monitoring the subject's condition and keeping a substitute decision-maker informed of the progress of an experiment.

Finally, researchers should, in certain circumstances, have obligations after the conclusion of an experiment. In particular, a researcher should provide professional support to those patients (and their families) who require it.

In sum, a researcher has the obligation to minimize harm regardless of who the patient is; however, when vulnerable persons are concerned, owing to the increased possibility of exploitation or abuse, these safeguards are especially important. Researchers have a positive duty to minimize risks before, during, and after the completion of an experiment. These duties include:

- carefully assessing potential risks and avoiding unnecessary ones;
- not conducting the experiment where there is an *a priori* belief that death or disabling injury might occur;
- making necessary preparations to protect individuals against even remote possibilities of injury, disability, or death;
- providing ongoing support throughout an experiment to monitor unexpected complications, and consent or assent, as the situation requires;

---

[31] *Ibid.*, s. 7.

[32] See the *Declaration of Helsinki, supra* note 2, s. III-1. This article was expressed specifically in the context of non-therapeutic biomedical research involving human subjects and referred to the obligation of a physician carrying out the research. The *CIOMS Guidelines, supra* note 4 at 10, also stated that it is a requirement flowing from the principle of beneficence "that the investigators be competent both to conduct the research and *to safeguard the welfare of the research subjects.*" [Emphasis added]

[33] See the *Nuremberg Code, supra* note 1, s. 4; *Declaration of Helsinki, ibid.*, s. III-3. For a discussion of the obligation of the researcher not to conduct research if there is a risk of serious harm arising out of a subject's pre-exisiting vulnerability, see C. Weijer & A. Fuks, "The Duty to Exclude: Excluding People at Undue Risk From Research" (1994) 17 Clinical & Investigative Medicine 115. This rule can also be based on the principle that the clinical needs of the patient must outweigh all research considerations. Such a statement also leads to the general rule that research should not be conducted where a superior alternative intervention is withheld.

[34] See also Sava, Matlow & Sole, *supra* note 27 at 161.

- terminating an experiment if there is probable cause to believe that injury, disability or death of the subject might arise; and
- providing follow-up support for subjects that require it.

*The Process of Risk Assessment — Defining Levels of Risk*
The accurate assessment of risk is a vital step in limiting the risks to which vulnerable subjects may be exposed.[35] It is important to acknowledge that adequate techniques for the assessment of risk constitute a critical component of the process by which the necessary information for rational decision-making is made available to the participants or their proxies. Therefore, it is fundamental that "[e]very biomedical research project involving human subjects should be preceded by careful assessment of predictable risks in comparison with foreseeable benefits to the subject or to others".[36]

The *MRC Guidelines* stated the following with respect to the assessment of risks in a research protocol:

> The investigator and the REB must determine, as completely as possible, both the known and the potential risks involved in the research regimen. Consideration must be given primarily to the risk of physical or psychological harm to the subjects. Attention should also be paid to the degree of harm that the subjects, as a group, may suffer, as well as to any potential or foreseeable harm to society as a whole if the research is allowed to proceed.[37]

The obligation to assess risks is therefore shared by the researcher and the REC; while the duty of the former is one of disclosure, the responsibility of the latter is to make an objective determination. The researcher's assessment should serve only to guide this process. As Levine states in his commentary on risk assessment,

> ... the IRB [Institutional Review Board] should consider the degree of risk presented by research procedures from at least the following four perspectives: a common-sense estimation of the risk, an estimation based upon the investigator's experience with similar interventions or procedures, any statistical information that is available regarding such interventions or procedures and the situation of the proposed subjects.[38]

Risk assessment should be based on more than just the opinion of the researcher, who may be tempted to downplay the element of risk because of an interest in seeing that a protocol is approved and carried out. Apart from a possible conflict of interest, another reason why the researcher may not be the best person to assess the risks

---

[35] In the United States, the United Kingdom and Canada alike, the notion of "minimal risk" is pivotal to the permissibility of research. See L.M. Kopelman, "When is the Risk Minimal for Children" in L.M. Kopelman & J.C. Moskop, eds., *Children and Health Care: Moral and Social Issues* (New York: Kluwer Academic Publishers, 1989) 89 at 92.
[36] See *Declaration of Helsinki, supra* note 2, s. I-5.
[37] See *MRC Guidelines, supra* note 11 at 16.
[38] R.J. Levine, *Ethics and Regulation of Clinical Research* (Baltimore: Urban & Schwarzenberg, 1986) at 165.

involved is a function of the researcher's perception. Two different studies[39] have concluded that researchers "are unrealistically optimistic about the risks of research".[40] A separate study of the perception of the relative seriousness of a variety of illnesses revealed that there was only a minimal relationship between the perception of seriousness and the actual risk of death in the assessment of both doctors and patients alike.[41]

Although risk assessment is a complex and difficult task, that does not mean it cannot be calculated with greater accuracy in the future. There is no reason why actuarial statistics may not be generated in relation to the nature and frequency of injuries caused by most routine interventions. As a matter of principle, research participants are owed a clear and objective assessment of the possibility of injury. This is all the more true when the subjects are children, and the injuries suffered, if permanent, are likely to affect them for the longest time; even if such injuries are not permanent, they may nevertheless have a variety of long-term effects on a child's future development.

As Thompson noted, children may be more susceptible to the socio-emotional disorganization accompanying stressful experiences and are often unable to make sophisticated psychological inferences about the researchers' reactions to them.[42] It must not be overlooked, then, that children's attitudes towards the medical community are in a formative state; the care provided for them in the context of a research protocol may have profound effects on their faith in the medical care that they seek, and receive, later in life. The possibility exists that the negative impact of a poor research design or the insensitive handling of a young subject may well engender a reticence to seek medical care when the affected individual develops into an adult.

If research involving risk to vulnerable persons is to be permitted, it will be necessary to enshrine the accurate assessment of the risk of harm as an essential requirement of ethical conduct. In this manner, families will be enabled to make decisions about the participation of their children, and substitute decision-makers will be able to make decisions on behalf of their dependents, based on something more than faith in the researcher's personal assessment of the degree of risk involved.

---

[39] See B. Barber *et al., Research on Human Subjects: Problems of Social Control in Medical Experimentation* (New Brunswick, N.J.: Transaction Books, 1979) at 53–57; J. Janofsky & B. Starfield, "Assessment of Risk in Research on Children" (1981) 98 Journal of Pediatrics 842.

[40] R.H. Nicholson, ed., *Medical Research with Children: Ethics, Law, and Practice* (Oxford: Oxford University Press, 1986) at 105.

[41] A.R Wyler, M. Masuda & T.H. Holmes, "Seriousness of Illness Rating Scale" (1968) 11 Journal of Psychosomatic Responses 363.

[42] R.A. Thompson, "Vulnerability in Research: A Developmental Perspective on Research Risk" (1990) 61 Child Development 1 at 8–12. As Goldberg recently noted, there have been very few studies conducted in relation to the special psychological risks faced by child participants in research. Recognition of the particular effects of psychological stress, invasion of privacy and deception on child participants can enable researchers to take positive steps to minimize them. See s. Goldberg, "Some Costs and Benefits of Psychological Research in Pediatric Settings" in G. Koren, ed., *Textbook of Ethics in Pediatric Research* (Malabar, Fla.: Krieger, 1993) 63.

The first step in risk assessment involves identification, which lists the adverse events that might occur from the procedure as well as the nature of the potential harms that may result. These adverse events include the standard hazards from any experimentation, those that result from the particular vulnerabilities of the subjects chosen, and those to which the subjects would not have been exposed but for their participation in the research.[43] These considerations will vary depending on the vulnerable population to which a subject belongs.

The process of identification of risks begins with the researcher, who is responsible for itemizing the procedures involved in the research regimen and their corresponding risks. However, while the personal experience of the researcher will inevitably influence this process, the personal values of the researcher should not bias the assessment. The REC should then assign, according to the system of classification described below, the overall level of risk associated with the experiment. Indeed, risk assessment, as a whole, should be conducted in an objective fashion; however, the "particular circumstances of a subject" may increase the seriousness of an otherwise minor risk. Consequently, any system of risk assessment must be tempered by subjective considerations arising from the unique perspective of the subject; the process must be sufficiently flexible to respect the subject's sensibilities.[44]

*Estimation of Risks*

After all the potential risks associated with a protocol have been identified, the overall risk must be estimated. This involves the quantitative assessment of the combined effect of the probability and extent of the harm associated with any given procedure. Actuarialists have developed a range of effective means for quantifying such losses, but schemes developed for insurance purposes have the somewhat different aim of assessing monetary awards.

Rosser,[45] a psychiatrist, conducted a study of patients' attitudes towards several different states of disability; on this basis, she constructed a table which related the varying extents of risks of the disabilities to one another. Having "scored" each state of disability, she was able to include factors such as the average recovery time required by each type of injury, the average life expectancy of the subject, and the number of subjects necessary for the successful conduct of the study. On this basis it was possible to arrive at a total risk score for any given procedure in a protocol. This score could then be compared directly with a total similarly generated for the estimated degree of risk posed by the disease or dysfunction under investigation.

---

[43] As defined by the OPRR, "risk" is "[t]he probability of harm or injury (physical, psychological, social, or economic) occurring as a result of participation in a research study". See *OPRR Guidebook, supra* note 5 at 3–1 — 3–5.

[44] The Office for the Protection of Research Risks emphasized that "IRBs should be sensitive to the different feelings individuals may have about risks and benefits. . . . An elderly person might consider hair loss or a small scar an insignificant risk, whereas a teenager could well be concerned about it". See *OPRR Guidebook, supra* note 5 at 3–8.

[45] R. Rosser & P. Kind, "A Scale of Valuations of States of Illness: Is There a Social Consensus?" (1978) 7 International Journal of Epidemiology 347.

This method was able to produce a convincing calculation of the potential value of the study.

The score obtained in this way could also be compared with other determinations of the risks associated with everyday life, such as the risks of fatality faced in common occupations, sports, and various modes of travel, as well as with determinations of the risks associated with procedures integral to routine physical examinations. This would enable the researcher to make a more objective determination of the relationship of the risks posed by participation in the study to those faced by individuals in everyday life. Finally, the accuracy of these calculations is adjusted to account for the researcher's experience and facility with the specific procedure in question, and for the record of injuries occurring at the facility concerned or in protocols conducted by that researcher. All in all, while it may be difficult to establish concrete methods of weighing harms and benefits that promise the same degree of precision as, say, the calculation of the ultimate results of the study, it is, as one writer recently noted, "imperative that those responsible for designing and approving such research move beyond an intuitive balancing of harms and benefits of research participation".[46]

The National Commission for the Protection of Human Research Subjects of Biomedical and Behavioral Research delineated a series of categories or levels of risk which, when cross-matched with various categories of benefit, created the basis for the construction of a grid that balanced the permissibility of the research with the measures required to safeguard the welfare of the potential subjects involved. The particular levels of risk identified in this manner included "minimal risk"; "minor increase over minimal risk"; and "more than minor increase over minimal risk". The National Commission did not define these categories, but later, the DHHS condensed them into two: minimal risk and greater than minimal risk and defined the former as follows:

> *Minimal risk* means that the probability and magnitude of harm or discomfort anticipated in the research are not greater in and of themselves than those ordinarily encountered in daily life or during the performance of routine physical or psychological examinations or tests.[47]

DHHS did not define the second category, but in the context of non-therapeutic experimentation research on children, decided to limit it to research that is no more than "a minor increase over minimal risk".[48] The *OPRR Guidebook* explains that

---

[46] F. Baylis, "The Intentional Exposure of Children to Research Risks" (Address presented to the NCBHR Workshop on Research Involving Children, 1 December 1992).

[47] 45 C.F.R. 46 (1991) § 46.102 (i). This definition is somewhat modified when applied to prisoners as "the probability and magnitude of physical and psychological harm that is normally encountered in the daily lives, or in the routine medical, dental, or psychological examination of healthy persons" (§ 46.103). The DHHS lists activities such as collection of hair and nail clippings, excreta, external secretions such as sweat or saliva, blood samples (in specified amounts and frequency), voice recordings, moderate exercise, behavior studies of various sorts, and research on drugs or devices for which an investigational new drug exemption exists or is not required.

[48] Ibid. § 46.406 (a). Even so, the research must satisfy other restrictions. *Ibid.* § 46. 406 (b)–(d).

risks and benefits are understood in the realm of probabilities. "Risk", unfortunately, has two meanings: chances an individual wishes to take, and conditions dangerous *per se*. Only the second meaning is of interest for purposes of research risk assessment.[49]

It should be observed that the definition of "minimal risk" has itself been challenged as being poorly formulated.[50] Kopelman argued that there may be significant differences between: all the risks encountered by ordinary people; the risks all people ordinarily encounter; and the risks ordinary people ordinarily encounter. It was, therefore, argued that an inadequate definition of the term "minimum risk" has been responsible for a certain degree of inconsistency in the approval of research protocols. Furthermore, even if the definition were clear, differences in the context of its application would still have to be taken into account for any real prospect of achieving standardized results. Differences in context are significant because although, for example, "the risks to children living in Beirut and Edinburgh are different . . . we would not want to have this automatically influence the sort of research we think would be 'not too risky' for them".[51] Despite the challenge to the definition, Kopelman acknowledged that the basic "decency" of researchers and IRB members enables them to "do better than the definition".[52]

Indeed, the frame of reference in the DHHS regulations has been described as being "the daily life of the subjects of the proposed research and not of hypothetical 'normal persons.' "[53] If such a standard is applied, then minimal risks could include those involved in a subject's everyday care and treatment. However, such an approach could result in risks for some patients being classified as "minimal", but which could be classified as "more than minimal" for others. A "normal person" standard also fails to take into account the special situation of a vulnerable individual whose subjective perception of risk may be entirely different. However, while researchers and RECs should not ignore the subjective element presented by a subject's unique perspective, risk assessment as a whole should be conducted on an objective basis. That is, researchers and RECs should use criteria external to their own subjective evaluations of risk. This is particularly important in the case of incompetent patients.

Schuck asserts that "communicating effectively about risk is inherently problematic".[54] Assessment of "risk" can only be done from a multidimensional point of view and, consequently, any definition that emphasizes one dimension over another is bound to be incomplete. As no definition will be complete and satisfy

---

[49] See *OPRR Guidebook, supra* note 5 at 3–1.

[50] See Kopelman, *supra* note 35 at 95.

[51] *Ibid.* at 91.

[52] *Ibid.* at 98.

[53] Personal communication from Dr. Charles R. McCarthy, Office of Protection from Research Risks, as cited in J.C. Fletcher, F.W. Dommel, Jr. & D.D. Cowell, "Consent to Research with Impaired Human subjects" (1985) 7:6 IRB 1 at 6. Note that the fundamental aspect of the definition of "minimal risk" in the DHHS regulations at the time the above statement was made has remained unchanged in the 1991 revision of the regulations.

[54] See P.H. Schuck, "Rethinking Informed Consent" (1994) 103 Yale Law Journal 899 at 949.

everybody it would be useful to consider some of the dimensions along which "risk" can be defined.

### Subjective versus Objective

Undeniably, every risk is a deep personal experience. Individuals, in making a decision to accept a risk, undergo three experiences. They measure it against some past reality or experience. They prepare emotionally in case the risk materializes. They undergo an emotional reaction in the form of a "*crescendo-duomo-decrescendo*" curve, where apprehension increases as the vicinity of the risk approaches, peaks as it materializes or fails to materialize, and then decreases as they get used to the materialization of the risk or as the threat of materializing subsides with time. Each of these elements is extremely subjective and each impacts on the subjective assessment of the magnitude of the risk. Hence, no matter how much explanation is given, or in which way, explanations of risk are not accepted the same way by everybody. On the other hand, risk is an objective reality as it stems from the dangerousness of a particular situation. While the subjective interpretation of the risk by every research subject is important, a research protocol cannot be expected to assess the risk posed by the proposed research procedure but in an as objective manner as possible. The researcher should address the danger that the procedure could carry and should assess the risk of that danger's materializing on the research subjects.

### General versus Particularistic

Assessment of risk is a general procedure to be applied to a particular research subject. The risk for each subject, however, could increase or decrease according to their medical condition and its severity. As such, the risk can be described and its magnitude assessed in the research protocol, but the weighing of its impact on each subject has to be left to an individual decision of whether the subject, given a condition at such a level of severity, should or should not participate in the project. The protocol can assist this by itemizing proper inclusion and exclusion criteria.

### Treatment Purposes versus Research Purposes

Certainly, risk assumes a different magnitude depending on whether it is undertaken for treatment or for research purposes. A patient may agree to have a lumbar punction if it is explained that it is needed to diagnose a neurological condition. The same patient, however, may not be so willing, if it is explained that the lumbar punction is needed as part of a research protocol in which it is expected the patient could be enrolled as a control subject.

### Comparativistic versus Categorical

Patients, and people in general, find it difficult to measure a category. Even a relativistic pronouncement like, "this will only hurt a little", (a low level in a category of "discomfort") is difficult to grasp: "a little" by whose standards, the nurse's, the patient's, or other people's? If the patient is highly anxious, "a little" will be felt like "a lot". As such, it is best to give the patient a comparativistic assessment of the risk such as "this will give you as much pain as when you receive a needle". Such statements, however, will not be too useful for an ethics review committee, which would like to be told "categorically" how much risk (danger) there is in the procedure.

*Actuarial versus Clinical:*
The magnitude of a risk could be measured by the proportion of individuals in a particular population on whom the risk materializes. For example, some criminals, who are identified by historical characteristics of their pattern of criminality, tend to commit the great majority of violent crimes in a particular community; they therefore have a larger risk of recidivism and could be considered more dangerous. Based on an actuarial prediction, these individuals could be selected for preventive detention. This type of actuarial risk prediction is different than a clinical prediction such as: persons labelled as having "antisocial personality disorder", or "schizophrenia", have a larger risk of committing a crime and are, therefore, more dangerous simply because of their condition. Another example is that young male drivers are, actuarially, more prone to have serious motor vehicle accidents, but this does not mean that a young man should not drive because he may be injured in an automobile accident. Although clinically irrelevant, or based on an ecological fallacy, this type of explanation may be more readily acceptable to patients than the abstruse actuarial pronouncement. On the other hand, an ethics committee would like to have the risk expressed not so much in clinical possibilities as in some quantitative or qualitative form.

*Quantitative versus Qualitative:*
DHHS has assessed risks as "minimal" and "greater than minimal risk".[55] These qualitative measurements lack precision. Yet, quantitative determinations such as "20% risk of injury" tend to give assurances that are misleading because nothing is so sure in medicine or research. The DHHS categories, however, although intuitively appealing to a scientist, may not be so transparent to a research subject and they are not statistically realizable. What could be "greater than minimal?" If the categories have no mathematical base and do not convey a "sense" to a research subject, why should they not be made as simple as possible in language such as "low, medium, high?"

*Assessing Risk versus Communicating Risk*
These are two different functions. The assessment of risk is an intellectual exercise based on a deep knowledge of the subject matter of the research project and the medical conditions affecting the research subjects. Communicating risk is a human skill. No matter how accurate a risk has been assessed, the researcher will not be doing a good job if the risk is not properly communicated to the subject. Unfortunately, while the REC can measure the risk and approve the protocol, it cannot observe the researcher when an explanation is given to the research subject.

---

[55] 45 C.F.R. 46 (1991).

*Recommendation*

A statutory definition of "negligible risk", and an accompanying system for the classification of risk, will provide a consistent reference point or standard by which RECs can assess the risk inherent in experimentation on vulnerable subjects. However, any system of risk assessment should include sufficient flexibility for RECs to increase or decrease the level of risk inherent in a procedure owing to specific conditions or circumstances particular to the experiment, or the situation of the individual research subject. Risk should be assessed at three levels: *negligible, less than substantial,* and *substantial*; and the DHHS definitions of "minimal risk" should be applied to *negligible* risk, "greater than minimal risk" to *less than substantial risk,* and "research not otherwise approvable" to *substantial* risk. In addition, in the process of assessing the overall risk associated with a given experiment, the REC should assess both the risks associated with individual procedures and the experiment as a whole, since individual components of a protocol may only constitute a *negligible* risk but could become more risky from repetition or prolongation. Also, any new risks arising during the course of an experiment should result in the REC's conducting a thorough reevaluation of the overall risk of the experiment.[56]

*The Balance of Risks and Benefits — What Is an Acceptable Risk?*

The final step in limiting risks in experiments involves the balancing of potential risks and benefits. What is an acceptable risk when vulnerable populations are involved? One answer to this question was expressed by the *Nuremberg Code* as: "The degree of risk to be taken should never exceed that determined by the humanitarian importance of the problem to be solved by the experiment".[57] The *Declaration of Helsinki* was somewhat more specific by requiring that: "Biomedical research involving human subjects cannot legitimately be carried out unless the importance of the objective is in proportion to the inherent risk to the subject".[58] These directives translate into the general requirement that in order for an experiment to be ethically permissible, the potential benefits must outweigh the potential risks.[59]

The benefit to society is certainly an important factor in determining the scientific importance of an experiment. From a coarse utilitarian perspective, a significant benefit to society can always outweigh the harms

---

[56] "An IRB shall conduct continuing review of research covered . . . at intervals appropriate to the degree of risk, but not less than once per year, and shall have authority to observe or have a third party observe the consent process and the research". See 45 C.F.R. 46 (1991) § 46.109(3); *OPRR Guidebook, supra* note 5 at 3–9.

[57] See the *Nuremberg Code, supra* note 1, s. 6.

[58] See the *Declaration of Helsinki, supra* note 2, s. I-4.

[59] See also *MRC Guidelines, supra* note 11 at 15; *LRC Working Paper No. 61, supra* note 15 at 24; *OPRR Guidelines, supra* note 5 at 3–8; 45 Code of Federal Regulations Part 46 (1991) § 46.111(a)(2); and Sava, Matlow & Sole, *supra* note 27 at 159. Indeed, in Quebec, even a fully competent person cannot consent to experimentation if the degree of risk is disproportionate to the expected benefits. See Art 21. C.C.Q.

committed against an individual. However, the *Declaration of Helsinki* clearly rejects this notion:

> In research on man, the interest of science and society should never take precedence over considerations related to the well-being of the subject.[60]

A compromise between the two positions, made in the context of research involving children and persons with mental or behavioural disorders, is the requirement that "the degree of risk attached to interventions that are not intended to benefit the individual subject is low and commensurate with the importance of the knowledge to be gained".[61] Similarly, the DHHS regulations require that risks to subjects "are reasonable in relation to anticipated benefits, if any, to subjects, and the importance of the knowledge that may reasonably be expected to result".[62]

In the context of non-therapeutic experimentation, there can be no direct benefit to the subject. Arguably, subjects may receive a non-tangible benefit flowing from their altruism.[63] However, given the vulnerable nature of vulnerable populations, basing the final determinant for the ethical permissibility of an experiment on such a nebulous standard might create a system rife with abuses. Defining the ethical permissibility of experimentation in such a manner would be based on, not one, but two imprecise concepts: "risk" and "benefit to society". Moreover, the principle of distributive justice demands "the equitable distribution of both the burdens and the benefits of participation in research".[64] The excessive use of individuals for purposes which provide them with no personal benefit, but rather benefit society as a whole, would violate this principle. Therefore, in order to prevent the exploitation of vulnerable persons "in the interests of science", it becomes necessary to define what constitutes an acceptable level of risk. Various solutions have been presented.

There is a significant degree of variance in international descriptions of the level of risk to which vulnerable persons should be exposed. In general, ethics review committees should only approve experimentation which presents a *negligible* or *less than substantial* risk. Needless to say, this standard is only a general one, which could be relaxed in exceptional situations. For example, a person might be allowed, using a research directive, to consent to a procedure involving a *substantial* risk.

### The Requirement of Assent

Investigators should avoid confusing consent with assent. While assent may be sufficient in limited circumstances, such as in research which involves no risk

---

[60] See *Declaration of Helsinki, supra* note 2, s. III-4.
[61] See *CIOMS Guidelines, supra* note 4 at 20, 22.
[62] 45 C.F.R. 46 (1991) § 46.111(2).
[63] For example, children may receive the benefit of developing their moral conscience. Similarly, a person with decreasing competence may receive satisfaction in providing advance consent to participate in non-therapeutic experimentation.
[64] See *CIOMS Guidelines, supra* note 4 at 10.

whatsoever, assent is not a substitute for consent. Any agreement to participate in an experiment with less than a full understanding is simple "assent". "Consent" requires a full understanding of the issue at hand.

Assent should be sought at all times, even in cases where the subject is incapable of providing a valid consent; however, it is important to emphasize that cooperation is not necessarily a sign of assent. In fact the DHHS regulations make this point abundantly clear in their definition of "assent".[65] According to the DHHS regulations, IRBs have a positive duty to solicit the assent of children whenever it is possible to do so.[66] The consistent restatement of this principle strengthens the need to ensure that individuals assent to the maximum extent that they are able.[67]

Assent, however, may be useful in serving another purpose: the absence or withdrawal of assent might be viewed as the most obvious manner in which to determine if a subject no longer wishes to participate in a research experiment. As such, wherever possible, investigators should seek and monitor assent throughout the course of an experiment. Such a role might be served either by an interested third party or by a "consent monitor". The latter has been suggested by Melnick *et al.*, who referred to a "research auditor" as someone "who is able to maintain surveillance over the process and progress of research and to assure that the changing condition and perceptions of the patient with regard to continued participation in the research will receive immediate and sufficient regard and respect".[68]

Therefore, the "assent" standard should serve more as an indicator for dissent rather than consent, and should be monitored closely to ensure that assent is maintained throughout the course of an experiment.

### *Preserving the Right to Withdraw from a Protocol*

The importance of respecting an individual's right to object to participation in research, regardless of that person's decision-making capacity has been consistently recognized. This right can be inferred from both the *Nuremberg Code* and the

---

[65] " 'Assent' means a child's affirmative agreement to participate in research. Mere failure to object should not, absent affirmative agreement, be construed as assent". 45 C.F.R. 46 (1991) § 46.402 (b).

[66] "[T]he IRB shall determine that adequate provisions are made for soliciting the assent of the children, when in the judgment of the IRB the children are capable of providing assent . . . If the IRB determines that the capability of some or all of the children is so limited that they cannot reasonably be consulted . . . the assent of the children is not a necessary condition for proceeding with the research". 45 C.F.R. 46 (1991) § 46.408 (a).

[67] See *LRC Working Paper No. 61, supra* note 15, 44; the *Nuremberg Code, supra* note 1, s. 1; *Declaration of Helsinki, supra* note 2, s. 9; *CIOMS Guidelines, supra* note 4 at 14.

[68] V.L. Melnick *et al.*, "Clinical Research in Senile Dementia of the Alzheimer Type: Suggested Guidelines Addressing the Ethical and Legal Issues" (1984) 32 Journal of the American Geriatric Society 531 at 534. Also, the OPRR suggested the use of a consent auditor "especially if the research involves more than minimal risk and no foreseeable benefit". [Emphasis omitted.] See *OPRR Guidebook, supra* note 5 at 6–30.

*Declaration of Helsinki*, and is stated clearly by the *CIOMS Guidelines*.[69] Subjects must be free to leave a study at any time, and must be informed of this fact.[70]

Consent must not only exist at the start of an experiment, but continue throughout.[71] The right to withdraw from an experiment, at any time, regardless of the level of competence of the subject, must be respected. A person should also be free from any threat of reprisal in the event that he or she chooses to withdraw from a research protocol. An objection made by the subject should supersede previous consent obtained while competent, including consent made through the use of research directives; and the will of the subject, even if incompetent, should also supersede the decision of a substitute decision-maker. In other words, a substitute decision-maker should not be able to override an objection by the subject. Any decision to disregard an objection made by a subject should require judicial or quasi-judicial approval, in accordance with a validly prepared research directive (e.g. a Ulysses Contract consenting to either *negligible* or *less than substantial risk*).[72] Similarly, an objection by a child may be overridden where an innovative therapy, as part of a research protocol, is employed as a life-saving device or for the purpose of significantly improving the long-term quality of life of the subject. In such cases, the consent of the legal guardian must be reviewed.[73]

### Other Considerations

Naturally, other relevant considerations could include: the duration of the experiment, the burden placed on the subject and the subject's family, the provision for or lack of long-term and ongoing monitoring, and the availability of professional support. Including such factors in an experimental protocol and requiring that investigators consider them is justified because, by definition, individuals who participate in non-therapeutic biomedical research do so for the benefit of others.

---

[69] The *Nuremberg Code, supra* note 1, s. 9. states that "[d]uring the course of the experiment the human subject should be at liberty to bring the experiment to an end if he has reached the physical or mental state where continuation of the experiment seems to him to be impossible". Also, the *Declaration of Helsinki, supra* note 2, s. I.-9, states that a subject "should be informed that he or she is at liberty to abstain from participation in the study and that he or she is free to withdraw his or her consent to participation at any time". The *CIOMS Guidelines, supra* note 4 at 20,22, provides the right of refusal in the context of children and persons with mental and behavioural disorders. Also, the right to refuse and discontinue participation at any time can be inferred from the DHHS Regulations' requirements for obtaining informed consent. See 45 C.F.R. 46 (1991) § 46.116(a)(8).

[70] See *MRC Guidelines, supra* note 11 at 25. The Law Reform Commission also recommended that the right of refusal of both children and persons with mental disorders be respected. See *LRC Working Paper No. 61, supra* note 15 at 42, 44.

[71] See Sava, Matlow & Sole, *supra* note 27 at 161.

[72] Ulysses Contracts consenting to a *substantial* risk should be prohibited.

[73] The 1990 New York State Office of Mental Health Regulations, *supra* note 17, allow for a judicial authority to override an objection by a subject after proving, through a third party, the existence of a direct benefit to the patient which is unavailable by any other means. Such a requirement would imply that an experiment has therapeutic value, and as such is not relevant to the present discussion. However, the spirit of the regulations requiring judicial intervention to override an objection by a subject is also applicable in the context of non-therapeutic experimentation, albeit, more difficult to justify. See Delano & Zucker, *supra* note 17 at 603.

Since society as a whole benefits from the actions of these volunteers, that same society has an obligation to provide the minimum safeguards to ensure proper care for and support of vulnerable persons who participate in biomedical experimentation.

Another example of a relevant consideration is whether an experiment will have an adverse effect on a patient's treatment plan. This issue becomes particularly relevant in cases of institutionalized mentally disordered persons. In such cases, an experiment should have the approval of the subject's treatment team.[74] This requirement could be waived where the person has provided a research directive specifying a desire to forego treatment in the course of an experiment, or where the person has authorized a substitute decision-maker to do so on his or her behalf.

Therefore, RECs should also be required to assess whether any other relevant considerations, incidental to the proposed research protocol, have been properly addressed.


## CONCLUSION

It is necessary to establish general safeguards on biomedical experimentation involving vulnerable populations. These safeguards should take the form of legislation or regulations promulgated by a statutory authority.[75] The safeguards should serve to protect research subjects from exploitation, but not be so restrictive as to render the conduct of research on vulnerable populations unfeasible. In order to accomplish this objective, certain recommendations should be considered, which can be applied generally to all vulnerable populations, with respect to: determining the nature, validity and importance of an experiment; defining the permissible scope of a research protocol and the acceptable selection of subjects; establishing the positive duties of researchers and the obligations of research ethics committees; developing functional means of risk assessment; and the balancing of foreseeable risks with expected benefits. With respect to the issue of consent, wherever possible, the "assent" of a subject should be obtained. However, to ensure the validity of subjects' consent to participate in an experiment, both their competence to make decisions and the voluntariness of their decisions must be considered. Also, the individual's right to object to participate and to withdraw from an experiment must be requested. As the relevant considerations will be different in each vulnerable population, additional issues must be addressed. However, it is important to emphasize that where an individual is a member of more than one vulnerable population, the safeguards specific to each population must be applied.

---

[74] For example, the 1990 New York State Office of Mental Health Regulations have this requirement. See Delano & Zucker, *ibid.* at 601.

[75] See Chapter 8.

## SUMMARY

Groups to be treated as vulnerable — on the basis that safeguards are needed to ensure that their consent to participate in non-therapeutic biomedical research is truly voluntary, and that they possess the necessary competence to make such a decision — include children, the elderly, the mentally disordered, the developmentally disabled, and prisoners. The following recommendations are made with respect to all members of these populations :

(1) Further study should be undertaken to determine whether social factors, such as institutionalization, have distinct effects on capacity in the area of consent to experimentation, and that mechanisms be formulated to safeguard the welfare of affected individuals so as to ensure the voluntariness of their participation in research.

(2) All proposals that seek the participation of the members of vulnerable populations whose capacity to give informed and voluntary consent is in doubt must be subjected to review by an approved REC.

(3) A person who falls into more than one category of vulnerable subjects should be given the benefit of all of the protective measures that apply to each particular group concerned.

(4) Any experiment involving human subjects must be scientifically valid. That is, research must satisfy the following conditions:
   - it must be conducted by scientifically qualified persons;
   - it must conform to generally accepted scientific principles (including the scientific method) and be based on a knowledge of the natural history of the disease or other problems under study; and
   - it must be preceded by experimentation on animal or other models (wherever possible) and be accompanied by a thorough knowledge of the scientific literature.

(5) Proposed research to be conducted on vulnerable populations must be of significant scientific value, measured by its likelihood of producing knowledge furthering the understanding, prevention or alleviation of a problem directly related to a condition or circumstance affecting the subject, or the class to which the subject belongs.

(6) Non-therapeutic biomedical experimentation with incompetent members of vulnerable populations must be restricted to those conditions or circumstances directly affecting the subject, or the class of subjects to which the person belongs. Furthermore, incompetent members of vulnerable populations should not be used in non-therapeutic research where other, more competent, persons could be available.

(7) Wherever possible, non-therapeutic research should be conducted by researchers who are not the primary providers of health care to the vulnerable child or adult subject.

(8) A researcher has a positive duty to minimize risks of harm to subjects. The researcher must,

*before* an experiment,
- carefully assess potential risks and avoid unnecessary ones;
- make necessary preparations to protect individuals against even remote possibilities of injury, disability, or death; and

*during* an experiment,
- provide ongoing support to monitor unexpected complications, and consent or assent, as the situation requires;
- immediately alert the REC which approved the protocol if any new risks should arise;
- terminate an experiment if there is probable cause to believe that injury, disability or death of the subject might arise; and

*after* an experiment,
- provide follow-up support for subjects who require it.

(9) Before the approval of an experiment, RECs must conduct a thorough risk assessment. While the specific circumstance of the subject should be a factor used in identifying potential risks, the assessment must be conducted on an objective basis.

(10) RECs must determine the overall risk of an experiment as either *negligible, less than substantial or substantial.*

　　*Negligible risk* means:
- That the probability and magnitude of harm or discomfort anticipated in the research are not greater in and of themselves than those ordinarily encountered in daily life or during the performance of routine physical or psychological examinations or tests.

In the process of assessing the overall risk associated with a given experiment, the REC should not only take into account the risks associated with individual procedures, but rather, it should assess the experiment as a whole, since individual components of a protocol may constitute only a *negligible* risk, though with repetition or prolongation of these procedures, risks may become *less than substantial* or *substantial.*

(11) The ethics review committee that approves the research proposal should be responsible for monitoring the conduct of the research insofar as it is reasonable and practicable for it to do so. If unforeseen risks develop once an experiment has begun, the REC should reevaluate the overall risk of the experiment.

(12) The potential benefits must outweigh the potential risks or harms.

(13) RECs should be able to approve only that experimentation which is of *negligible* or *less than substantial* risk. Experimentation which is of *substantial* risk must receive the approval of a judicial or statutory authority.

(14) Adults should be presumed to be competent and capable of providing valid consent, unless the opposite is found to be the case. Even where competence in adults is in question, the decision-making capacity of such persons should be facilitated so as to allow them to make decisions to the maximum extent

that they are able to do. Moreover, children should be allowed and encouraged to participate in decisions regarding their participation in experimentation to the maximum extent possible.

(15) Wherever possible, a subject's *assent*, to the extent of the subject's capabilities, should be obtained. The fact that a person cooperates is insufficient to be treated as a sign of assent. Continued assent should be monitored throughout the course of an experiment.

(16) Unless an experiment has a *negligible* risk and involves an adult who has provided a valid research directive, the participation of children and adults who are unable to provide a valid consent must be preceded by the approval of a legal substitute decision-maker.

(17) A subject may withdraw at any time, regardless of that subject's level of mental competence. Such objection may only be overruled by a judicial or statutory authority, in writing, in accordance with a validly prepared research directive.

CHAPTER 19

# BIOMEDICAL EXPERIMENTATION WITH CHILDREN: BALANCING THE NEED FOR PROTECTIVE MEASURES WITH THE NEED TO RESPECT CHILDREN'S DEVELOPING ABILITY TO MAKE SIGNIFICANT LIFE DECISIONS FOR THEMSELVES

DAVID N. WEISSTUB, SIMON N. VERDUN-JONES AND JANET WALKER

## INTRODUCTION

The use of children as research subjects can be traced back to various experiments conducted with smallpox vaccines in the 1700s among children in England and in the North American colonies.[1] While children were used, sporadically, as test subjects to develop vaccines, their infrequent selection hindered the development of specific regulations for the protection of their welfare as participants in research. In 1772, for example, Queen Caroline of England had ten orphans vaccinated against smallpox as a precautionary test before consenting to the vaccination of her own children.[2] Edward Jenner and Benjamine Waterhouse vaccinated their own children prior to vaccinating the children of others. Thomas James vaccinated 48 institutionalized children under his care and then inoculated the children with

---

[1] S. Lederer & M.A. Grodin, "Historical Overview: Pediatric Experimentation" in M.A. Grodin & L. Glantz, eds., *Children as Research Subjects: Science, Ethics, and Law* (New York: Oxford University Press, 1994) 3 at 4.

[2] R.G. Mitchell "The Child and Experimental Medicine" (1964) 1 British Medical Journal 721 at 722; A. Holder, "Constraints on Experimentation: Protecting Children to Death" (1988) 6 Yale Law & Policy Review 137 at 155.

smallpox to test the effectiveness of his vaccine.[3] Similar trials were conducted with measles vaccines. During the latter half of the Nineteenth Century and the first half of the Twentieth Century, the common practice of institutionalizing members of special populations rendered children (and other institutionalized populations) attractive subjects for research. The apathy of investigators in safeguarding their subjects' well-being reflects the low esteem in which these populations were held. For example, at the turn of the century, Alfred Hess, Medical Director of the Hebrew Infant Asylum in New York, conducted research on the children in his institution because the conditions there were similar to the "conditions which are insisted on in considering the course of experimental infection among laboratory animals, but which can rarely be controlled in the study of infection in man".[4] Unfortunately, there are many such examples of abuse, with extreme examples occurring as recently as a few decades ago in New York at the Willowbrook State School.[5]

Poor sanitation and acute overcrowding at the Willowbrook facility created a breeding ground for hepatitis. Researchers proposed to test a vaccine by inoculating children upon admission and then infecting them with the disease. Parents wishing to place their children in the school were informed that participation was a condition of admission and that, as research subjects, their children would be isolated from the general population and so would receive superior care. It was pointed out that children who were placed among the general population would be expected to contract hepatitis and/or measles, shigellosis, and other parasitic or respiratory infections that were beyond the supervision and control of the investigators.[6] Research ethicists of the day looked beyond the deplorable conditions at the Institution to raise questions about whether the study should have been conducted in the way that it was and whether the consent so obtained was valid; the questions raised at that time have figured prominently in the subsequent debate on the participation of children in research.

---

[3] See Lederer & Grodin, *supra* note 1 at 5.

[4] *Ibid.* at 1.

[5] D.J. Rothman & S.M. Rothman, *The Willowbrook Wars* (New York: Harper & Row, 1984); S. Krugman, "Experiments at The Willowbrook State School (letter)" (1971) 1 Lancet 966. Another example worth noting is the case of the Fernald School in Massachusetts where experiments were conducted on mentally retarded children in the late 1940s. These studies were reviewed in detail by the Massachusetts Task Force on Human Subject Research and the Presidential Advisory Committee on Human Radiation Experiments. Although the Fernald School radiation studies were not as dangerous as those conducted at Willowbrook, the Fernald researchers were criticized on the basis that they took advantage of a vulnerable pool of subjects (institutionalized children), concealed certain key details from their parents (such as the use of radioisotopes in the studies), and solicited participation unfairly (by offering inducements that were coercive in the circumstances, such as extra milk and special outings for the children). See Task Force on Human Subject Research, *A Report on the Use of Radioactive Materials in Human Subject Research that Involved Residents of State-Operated Facilities within the Commonwealth of Massachusetts from 1943 to 1973* (Submitted to Philip Campbell, Commissioner, Commonwealth of Massachusetts Executive Office of Health and Human Services, Department of Mental Retardation, April 1994); Advisory Committee on Human Radiation Experiments, *Final Report* (Washington, D.C.: U.S. Government Printing Office, 1995).

[6] See Krugman, *ibid.*

## DEFINING "CHILDREN"

Simply stated, children are persons between birth and the age of majority, the latter often being established by statute.[7] Many jurisdictions have also enacted legislation that sets a specific age as the point at which a child may consent to treatment without parental authorization.[8] Under that age, parental permission is required, although there are significant exceptions to this requirement. In those Canadian jurisdictions that have not established a specific age at which a child may consent to treatment without parental authorization, the so-called "mature minor" rule applies.[9] Provincial legislation also provides that the parents' right to refuse treatment for their children may be overridden where a court believes the treatment necessary to preserve life or health.[10]

It follows from the above definition of "children" that their rights and obligations are not comparable to those of adults. Their dependency and statutory incompetence necessitate that parents or guardians act as their representatives. Children therefore have a special, and perhaps even a privileged, place within the legal framework of society. They have been accorded a distinctive status in relation to their health care, entitling them both to receive adequate health care and to rely on others, generally their parents, not only to be vigilant concerning the need for medical intervention

---

[7] In half of Canada's twelve jurisdictions, for example, the age of majority is eighteen years; otherwise it is nineteen.

[8] In New Brunswick, for example, treatment may be administered to a child under the age of 16 where the health care practitioner is of the opinion that the procedure is in the best interests of the child, the child is capable of making a decision, and/or this opinion is supported in writing by another practitioner. See *Medical Consent of Minors Act*, R.S.N.B. 1976, c. M–6.1, ss. 1–3.

[9] The essential elements of this common-law rule were summarized in *Ney* v. *Canada (Attorney General)* (1993), 79 B.C.L.R. 47 (S.C.) at 58–59:

> [A]t common law a child is capable of consenting to medical treatment if he or she has sufficient intelligence and maturity to fully appreciate the nature and consequences of a medical procedure to be performed for his or her benefit. It appears that the medical practitioner is to make this determination. If the child is incapable of meeting this test then the parents' consent will be required for treatment. It is not clear whether parental control yields to the child's independence or whether there are concurrent powers of consent. But it is clear that the parents may not veto treatment to which a capable child consents, and that neither child nor parents can require a medical practitioner to treat.

The "mature minor" rule is upheld by the *Health Care and Consent Act, 1996*, S.O. 1996, c. 2, which presumes that any person can make a treatment decision unless there is a finding of incapacity. See generally B.F. Hoffman, *The Law of Consent to Treatment in Ontario*, 2nd ed. (Toronto: Butterworths, 1997).

[10] See *Re McTavish et al. and Director, Child Welfare Act et al.* (1986), 32 D.L.R. (4th) 394 (Alta.Q.B.); *Re E and Minister of Social Services et al.* (1987), 36 D.L.R. (4th) 683 (N.S.C.A.); *Re K. (R.)* (1987), 79 A.R. 140 (Prov. Ct. Fam. Div.). In the case of *B. (R.)* v. *Children's Aid Society of Metropolitan Toronto*, [1995] 1 S.C.R. 315, the Supreme Court of Canada ruled that the Ontario legislation that permitted the overriding of a parental refusal to permit treatment (*Child Welfare Act*, R.S.O. 1980, c. 66) did not violate the *Canadian Charter of Rights and Freedoms*.

but also to be diligent in making sure that it is actually provided.[11] To the extent that they are not capable of meaningful involvement in decisions regarding their health care, children are also entitled to rely on the fact that the medical treatment to which they are subjected does not pose a risk of serious harm. Yet, to the extent that children are capable of meaningful involvement, they are equally entitled to participate in any decision-making process affecting the nature and quality of the medical treatment that they receive.

### Children as a Vulnerable Population

To say that children enjoy a special status is perhaps not entirely accurate, in that their entitlement to these assurances is no different from that of other vulnerable populations incapable of making independent health-care decisions.[12] Yet, children differ substantially from other vulnerable populations.[13] Four of these differences are worth mentioning.

The first of these differences is the universality of childhood dependency. It is the only inevitable dependence during the course of our lives, and so the need to establish and maintain mechanisms for others to make health care decisions on behalf of children is universally accepted. The operation of these decision-making mechanisms has a direct impact on the lives of all children as well as on those who have the obligation to care for them. Arguably, childhood is the relationship of dependence with the greatest repercussions for the general population.

The second consideration pertains to the mechanisms that permit others to make decisions on behalf of children. As a vulnerable population, children are the only

---

[11] The provision of adequate health care for children is assured through the various duties imposed on their caregivers and through the encouragement and resources provided to their caregivers by their communities. The effectiveness of this care is ensured by the criminal law and by a number of legislative measures that become operative should those who are responsible for a child's care fail in that responsibility. See e.g. *Public Hospitals Act*, R.S.O. 1990, c. P–40; *Family and Child Services Act*, R.S.O. 1990, c. C–11. Under s. 215 of the *Criminal Code*, R.S.C. 1985, c. C–46, parents and guardians are obligated to provide their children, who are younger than 16 years of age, with the "necessaries of life" and they will be found to have committed an offence if, without lawful excuse, they do not secure adequate medical treatment when a child is in danger of dying or suffering permanent injury. In *R. v. Naglik* (1993), 83 C.C.C. (3d) 526, the Supreme Court of Canada ruled that liability under s. 215 is imposed on an *objective* basis and that the relevant standard is that of the reasonably prudent parent or guardian.

[12] Seiber states that children are a vulnerable population because:

> (1) they have a limited psychological, as well as legal, capacity to give informed consent; (2) they may be cognitively, socially, and emotionally immature; (3) there are external constraints on their self-determination and independent decision-making; (4) they have unequal power in relation to authorities, such as parents, teachers, and researchers; (5) parents and institutions, as well as the youngsters themselves, have an interest in their research participation; and (6) national priorities for research on children and adolescents include research on drug users, runaways, pregnant teenagers; and other sensitive topics, compounding the ethical and legal problems surrounding research on minors.

See J. Seiber, *Planning Ethically Responsible Research: A Guide for Students and Internal Review Boards* (Newbury Park, N.J.: Sage, 1992) at 111.

[13] For a discussion of the notion of "vulnerability" in the context of non-therapeutic experimentation, see Chapter 18.

group for whom there is another clearly defined class of persons obligated to shoulder this responsibility. Furthermore, as discussed above, the situation of children is unique insofar as the adults who are legally accountable for providing them with health care are presumed to have the necessary authority to give a valid consent to the administration of treatment.[14]

Thirdly, there are the age-specific limits of the dependency. Most children are engaged in developing their own ability to provide or withhold consent.[15] Concern for their welfare must include balancing the need to obtain substitute consent on their behalf and the need children have to develop their own capacity to give consent.

The fourth concern relates to the fact that children have not yet developed values to which a substitute decision-maker can look in order to assess what the child would do if competent.

### Statutory Incompetence and Consent to Research

The competence of children to consent to *treatment* sheds light upon the question of their capacity to consent to *research*, because competence to consent relies in part on the ability to understand the nature and consequences of the procedure to be performed. However, although a lack of capacity to consent to treatment might suggest an inability to consent properly to participation in research, the converse is not automatically true. For example, the *Civil Code of Quebec* permits children aged fourteen or more to make their own, independent decisions concerning therapeutic interventions and minimal-risk medical care not required by the state of their health; however, only persons of "full age" are entitled to give their consent to participate in an "experiment".[16] Perhaps the most important aspect of the Quebec legislation with respect to non-therapeutic experimentation is the explicit recognition that parents (or tutors) may give substitute consent to the participation of children provided that certain safeguards are respected. The *Civil Code* does not, however, recognize that older children themselves might be competent to consent to participate in an experimental protocol, and that the approval of the parent or tutor would not be required.

In common law jurisdictions, the question remains: Should children be statutorily restricted from consenting to participate in non-therapeutic experimentation before reaching a certain age? As will be discussed below, the answer must take into consideration the modern evolution of mental competency law and general principles which support the empowerment of children to make autonomous decisions concerning their personal welfare whenever they are able.

---

[14] Of course, this presumption may be rebutted where the parent or guardian is unavailable, incapable, or irresponsible.

[15] This situation may be contrasted with that facing the mentally disordered, whose capacity to give consent is *fluctuating*; the developmentally disabled, whose capacity to give consent is *static*; and the elderly whose capacity to give consent is potentially *diminishing*.

[16] For a discussion of the law governing non-therapeutic experimentation in Quebec, see Chapter 10.

## AN INTERNATIONAL SURVEY OF ETHICAL AND LEGAL PRINCIPLES

### Canada

With the exception of Quebec, there is a notable absence of legislation in relation to biomedical experimentation with children in Canada. Furthermore, it is uncertain whether such experimentation is permissible under the common law.[17] The principal consequence of this gap is that guidelines internal to the medical and research professions themselves define the ethical permissibility of research with children.

Until recently, the *Guidelines on Research Involving Human Subjects* of the Medical Research Council of Canada in conjunction with publications by the National Council on Bioethics and Human Research were the most influential directives "regulating" experimentation with children. The *MRC Guidelines*, however, have been criticized for not providing sufficient guidance in areas in which the law is unclear.[18] The MRC has consistently opposed the enactment of legislation regulating research involving children, preferring instead to rely on the continued application of the *Guidelines*. The Law Reform Commission of Canada took an opposing view and recommended that Parliament enact legislation restricting the use of children in experimental protocols. The proposed legislation would have permitted non-therapeutic experimentation involving children provided that:

(1) Nthe research was of major scientific importance and not possible to conduct using adult subjects;

(2) it was directly related to infantile diseases or pathologies;

(3) did not involve serious risks for the child;

(4) consent of the parent or independent third party (a judge, ombudsman, or the child's lawyer) was obtained; and

(5) wherever possible, the consent of the child was obtained, with refusals being honoured whatever the child's age.[19]

Against this backdrop, the Consent Panel Task Force of the NCBHR revised its *Report on Research Involving Children* in 1993. The NCBHR recognized the need for research involving children and offered numerous examples of the benefits such research provided and of the deleterious consequences resulting from its absence. The *Report* was premised on the assumption that "research involving children [is] an activity of valued collaboration between investigators and children (and their parents or guardians), [and] such activity in Canada [should be] encouraged, subject to [certain] limitations".[20] The Consent Panel Task Force established a tripartite

---

[17] See Chapter 8.

[18] See F. Baylis & J. Downie, "An Ethical and Criminal Law Framework for Research Involving Children in Canada" (1993) 1 Health Law Journal 39 at 53.

[19] See Law Reform Commission of Canada. *Working Paper No. 61: Biomedical Experimentation Involving Human Subjects* (Ottawa: Law Reform Commission, 1989) at 61.

[20] See National Council on Bioethics in Human Research, Consent Panel Task Force, *Reflections on Research Involving Children* (Ottawa: NCBHR, 1993).

categorization of authorization required for research involving children. For a very young child (under seven years), the "informed and voluntary authorization" of the parents or guardians would be required. For children between seven and fourteen, the assent of the child would be required in addition to parental consent. Finally, for children over fourteen, only the consent of the child would be required. Furthermore, it was recommended that the dissent of children unable to give consent should be given serious consideration and the refusal of a child capable of consent must be respected.

The *MRC Guidelines* have since been superceded by the *Code of Ethical Conduct for Research Involving Humans*, adopted by the three principal federal funding agencies.[21] The *Code* continues to endorse the conduct of experimentation with children, but only in specific circumstances where: authorization is first obtained from a parent or guardian, risk is within normally acceptable limits, and where research is likely to be of sufficient social benefit. Lastly, the assent (or dissent) of the child must be confirmed wherever appropriate. Of course, consent should be sought directly from the child whenever it is possible to do so.[22]

### International Declarations, Codes, and Guidelines

International covenants that protect children from undue risk of harm as participants in research can best be understood if examined within the context of those covenants that entitle children to receive the benefits of medical advances made possible through research with child participants. These rights are contained in both the *Universal Declaration of Human Rights*[23] and *the International Covenant on Economic, Social and Cultural Rights*.[24] The recent statement that children are entitled to the "highest attainable standard of health" contained in the *Convention on the Rights of the Child*,[25] which Canada has signed, arguably strengthens the obligation to fashion its laws to permit research involving child subjects, particularly when such research would directly benefit the health of children.

The early international documents, by requiring free and informed consent *by the subject*, effectively circumscribed research involving children.[26] This position was

---

[21] See Tri-Council Working Group, *Code of Ethical Conduct for Research Involving Humans* (July, 1997). The three funding agencies consist of the Medical Research Council of Canada, the Natural Sciences and Engineering Research Council of Canada, and the Social Sciences and Humanities Research Council of Canada.

[22] *Ibid.* at Part 2, pp. 3–4; Part 3, pp. 35–36.

[23] *Universal Declaration of Human Rights*, G.A. Res. 217A, UN Doc A/810 (1948), art. 27.

[24] *International Covenant on Economic, Social and Cultural Rights*, 993 U.N.T.S. 3 (1966), art. 15(b).

[25] *Convention on the Rights of the Child*, G.A. Res. 44/25, UN Doc. A/RES/44/25 (1989), art. 24.

[26] See the *Nuremberg Code*, s.10 (2). The *Nuremberg Code* is part of the judgment articulated in the case of *U.S. v. Karl Brandt et al., Trials of War Criminals Before the Nuremberg Military Tribunals Under Control Council Law No. 10* (October 1946-April 1949); *International Covenant on Civil and Political Rights*, 19 December 1966, Can T.S. 1976 No. 47, 999 U.N.T.S. 171, 6 I.L.M. 368, s. 7.

relaxed subsequently.[27] The latest restatement of the international consensus on the permissibility of research involving children was enunciated in the guidelines of the Council for International Organizations of Medical Sciences, prepared in collaboration with the World Health Organization, and promulgated in 1993. This document placed significant restrictions on the research that could be conducted with children by identifying a number of conditions that must be met before children may be involved:

- children will not be involved in research that might equally well be carried out with adults;
- the purpose of the research is to obtain knowledge relevant to the health needs of children;
- a parent or legal guardian of each child has given proxy consent;
- the consent of each child has been obtained to the extent of the child's capabilities;
- the child's refusal to participate in research must always be respected unless according to the research protocol the child would receive therapy for which there is no medically-acceptable alternative;
- the risk presented by interventions not intended to benefit the individual child-subject is low and commensurate with the importance of the knowledge to be gained;
- interventions that are intended to provide therapeutic benefit are likely to be at least as advantageous to the individual child-subject as any available alternative.[28]

### *The United States*

Department of Health and Human Services (DHHS) Regulations contain specific provisions protecting children involved in research. The Regulations divide research with children into four categories: (1) "research not involving greater than minimal risk"; (2) "research involving greater than minimal risk but presenting the prospect of direct benefit to individual subjects"; (3) "research involving greater than minimal risk and no prospect of direct benefit to individual subjects, but likely to yield generalizable knowledge about the subject's disorder or condition"; and (4) "research not otherwise approvable, but that presents an opportunity to understand,

---

[27] In 1975, the *Declaration* was amended explicitly to permit research involving children, if the informed consent of the child's legal guardian was obtained in accordance with national legislation. In 1983, the *Declaration* was further amended to require the child's consent to participate in research if the child was able to provide it. See the World Medical Association, *Declaration of Helsinki*. Adopted at the 18th World Medical Assembly in Helsinki in June 1964. Amended at the 29th World Medical Assembly in Tokyo in October 1975; the 35th World Medical Assembly in Venice in October 1983; and the 41st World Medical Assembly in Hong Kong in September 1989. [Reprinted in (1991) 19 Law, Medicine & Health Care 264.]

[28] See Council for International Organizations of Medical Sciences, in collaboration with the World Health Organization, *International Ethical Guidelines for Biomedical Research Involving Human Subjects* (Geneva: CIOMS, 1993) at 20.

prevent, or alleviate a serious problem affecting the health or welfare of children".[29]

The Regulations require that the investigator obtain the assent of the child subject, when, in the judgment of the Institutional Review Board, the child is capable of providing an assent.[30] Assent is defined in the Regulations as "a child's affirmative agreement to participate in research".[31] The IRB should take into account the age and psychological state of the child in determining whether or not the child is mature enough to provide a meaningful assent. The assent of the child is not required if it is determined that the child is not capable of providing a meaningful assent or if the well-being of the child requires participation in the research project.[32]

For research that involves more than minimal risk but offers the potential of direct benefit for the child subject, the risk must be justified with respect to the potential benefit, and the relation between the risk and benefit must be as favourable as it is for the best alternative treatment.[33] The IRB will also approve research that involves more than a minimal risk to the child but that does not offer the potential of direct benefit for the child. The risk, however, must be a minor increase over minimal risk, the intervention must be consistent with the life experience of the child, and there must be a likelihood that the procedure will generate generalizable knowledge about the subject's disorder that is of vital importance.[34] For research that would not be allowed under the mechanisms just described, the Secretary of Health and Human Services, after consultation with a panel of experts, must determine whether the research presents a "reasonable chance of furthering the understanding, prevention or alleviation of a serious problem affecting the health or welfare of children".[35]

### The United Kingdom

The United Kingdom has not legislated with respect to medical research involving human subjects; however, in 1990, the Royal College of Physicians published guidelines developed from the legislation and jurisprudence in related areas of

---

[29] 45 C.F.R. 46 (1991) § 46. 404–46.407.

[30] *Ibid.* § 46. 408 (a). Grisso notes that the requirement of obtaining the permission of the parent *and* the assent of the child seeks a balance in providing adequate protection in two respects. Firstly, were the decision solely that of the minor, often we would be concerned about whether the minor understood the research procedures to which he or she was making a commitment. Secondly, a parent-only decision places the minor — whose participation is being requested — in the role of one who is "lent out" to the researcher, so to speak, ignoring the fact that the minor is a person who may have his or her own reasons for not wishing to undergo the procedure. See T. Grisso, "Minor's Assent to Behavioural Research Without Parental Consent" in B. Stanley & J. Seiber, eds., *Social Research on Children and Adolescents: Ethical Issues* (Newbury Park, N.J.: Sage, 1992) 109 at 111.

[31] 45 C.F.R. 46 (1991) § 46. 402 (b).

[32] *Ibid.* § 46. 408 (a).

[33] *Ibid.* § 46. 406 (a)(b).

[34] *Ibid.* § 46. 406 (a)–(c).

[35] *Ibid.* § 46. 407 (a)(b).

health law.[36] The Royal College's *Research Involving Patients* and the *Guidelines on the Practice of Ethics Committees in Medical Research* recommend that research with children be limited to that involving minimal risk, an example of a minimal risk procedure being venepuncture.[37] Both reports conceive that research involving higher degrees of risk might be permissible in exceptional circumstances that promise great benefit.[38] On the question of consent, the College recommended that parental consent be required for all children under the age of 18,[39] and indicated that the consent cannot be given if it is against the child's interests.[40] The consent of a child capable of understanding is also required.[41] According to the report, *Research Involving Patients*, the objection of the child who is incapable of understanding must be considered, while the *Guidelines* recommend that such refusal be treated as binding.[42]

The Medical Research Council of the United Kingdom published "The Ethical Conduct of Research on Children" in 1991.[43] The MRC recommended that children be included in research only if the following three conditions are met: (1) the relevant knowledge could not be gained through research with adults, (2) the appropriate Local Research Ethics Committee approves the protocol, and (3) either the subjects have given consent, or a parent or guardian has given consent on their behalf, and those included do not object or appear to object in either words or action.[44] With therapeutic research, the benefits to the child participant must outweigh the possible risk of harm; with non-therapeutic research, participation cannot place a child in more than at a negligible risk of harm.[45]

The British Paediatric Association published guidelines in 1978 which provided:

— that research involving children is important for the benefit of all children and should be supported and encouraged and conducted in an ethical manner.
— that research should never be done on children if the same investigation could be done on adults.

---

[36] Royal College of Physicians of London. *Guidelines on the Practice of Ethics Committees in Medical Research Involving Human Subjects*, 2d. ed. (London: RCP, 1990); Royal College of Physicians of London, *Research Involving Patients* (London: RCP, 1990) [hereinafter *RCP Research Involving Patients*].

[37] See Royal College of Physicians of London (1990), *ibid.* at 27; *RCP Research Involving Patients*, *ibid.* at 19–20.

[38] See *RCP Research Involving Patients*, *ibid.* at 19; Royal College of Physicians of London (1990), *ibid.*

[39] See Royal College of Physicians of London (1990), *ibid.*

[40] See *RCP Research Involving Patients*, *supra* note 42 at 20.

[41] *Ibid.*

[42] *Ibid.*; Royal College of Physicians of London (1990), *supra* note 42 at 27.

[43] See Working Party on Research on Children. *The Ethics of Research on Children* (London: MRC, 1991).

[44] *Ibid.* at para. 6.1.2.

[45] *Ibid.* at paras. 6.2.2 and 6.3.4. The MRC defines "negligible risk" as risks of harm no "greater, considering the probability and magnitude of physiological or psychological harm or discomfort, than those ordinarily encountered in daily life or during the performance of routine physical or psychological examination or tests". *Ibid.* at para. 6.3.3.

— that research which involves a child and is of no benefit to that child (non-therapeutic research), is not necessarily either unethical or illegal.

— that the degree of benefit resulting from a research should be assessed in relation to the risk of disturbance, discomfort, or pain, (the "risk/benefit ratio").[46]

The British Paediatric Association did not establish a threshold of risk that child subjects could be exposed to in non-therapeutic research other than that the risk must be proportionate to the benefit that is expected to result from the research.

### Australia

The National Health and Medical Research Council issued a revised version of its *Statement on Human Experimentation* in 1992.[47] The NHMRC will not fund research involving children that exposes the subject to more than minimal risk unless the research has the potential of benefiting the subject himself.[48] The NHMRC therefore establishes minimal risk as the threshold for non-therapeutic research. The consent of the child's parents or guardians must be obtained, as well as the consent of the child where appropriate.[49]

In 1981, the Australian College of Paediatrics issued a report supporting research with children. Acknowledging that research on children is essential to advance knowledge on childhood disease, it recommended that the research be based on sound scientific concepts; that it be performed only when information could not be sought in practice with other groups; and that it be planned and conducted in such a fashion as to ensure that definite conclusions can be obtained. Informed consent is to be obtained from parents or guardians in all but the most exceptional circumstances, and from the children themselves when they possess sufficient maturity and intelligence. The Australian College of Paediatrics suggested that a risk/benefit ratio be used to determine whether therapeutic research be undertaken. For non-therapeutic research, the College suggested that the risks be "so minimal as to be little more than the risks run in every day life". It also recommended that an ethics committee be established in all centres responsible for research in children. This committee is responsible for: protecting the rights and welfare of children involved in research; determining the acceptable level of risk as weighed against the potential benefit; obtaining informed consent; encouraging the performance of necessary and appropriate research in children; and preventing unscientific research.[50]

---

[46] British Paediatric Association, "Guidelines to Aid Ethical Committees Considering Research Involving Children" in F. Cockburn *et al.*, "Research Ethics" (1980) 55 Archives of Disease in Childhood 75.

[47] See National Health and Medical Research Council, *Statement on Human Experimentation and Supplementary Notes* (Canberra: NHMRC, 1992).

[48] *Ibid.* at 9.

[49] *Ibid.* at 8.

[50] Australian College of Paediatrics, "Report on the Ethics of Research in Children" (1981) 17 Australian Paediatrics Journal 162. See principles 1–5. For a further discussion of research involving children in Australia, see C.A. Berglund, "Children in Medical Research: Australian Ethical Standards" (1995) 21 Child: Care, Health and Development 149.

## France

France enacted legislation regarding human experimentation in 1988[51] under the title, *Loi relative à la protection des personnes qui se prêtent à des recherches biomédicales.*[52] This new law amended *La code de la santé publique.* Article L. 209-6 of the new Code now stipulates three basic requirements for non-therapeutic research involving children: (1) the absence of serious risk, (2) the prospect of a benefit to those of similar age as the subjects, and (3) the impossibility of conducting the research on competent adults.[53] Article L. 209-10 provides that the consent of the parent(s) is required for therapeutic or non-therapeutic research, or, in the case of a child under curatorship, the consent of the tutor is required for therapeutic research that does not pose a serious risk. For all other research, the authorization of the Family Council or a Tutorship Judge is required. This article also requires that assent be obtained from children able to express their will and that their withdrawal of consent, or their refusal to participate, be respected.

## Summary

A review of the scientific literature reveals a well-established consensus regarding the importance and necessity of conducting research with children.[54] The fact that drug safety and efficacy in adults can rarely be extrapolated to children is well-recognized; moreover, developmental influences on psychopathalogic features and pharmacological effects necessitate the use of children as subjects in clinical research.[55] It is not surprising, therefore, that national and international documents consistently endorse non-therapeutic research involving children so long as their participation is subject to strict safeguards. These restrictions can be summarized as follows:

— children should not be exposed to more than "minimal risk";
— children should not participate in non-therapeutic research if the research could be carried out with competent adults;
— the research should be concerned with the health needs of children;
— the consent of the child's parent or guardian should be obtained;

---

[51] See *Loi relative à la protection des personnes qui se prêtent à des recherches biomédicales, Loi No. 88–1138 du 20 décembre 1988*, J.O., 22 December 1988.
[52] For a detailed discussion of the French law on experimentation, see Chapter 9.
[53] Article L. 209.6 Code de la santé publique.
[54] In fact, it has been suggested that children have been *over*-protected in the research context, resulting in the creation of a new class of "therapeutic orphans". See L.E. Arnold *et al.*, "Ethical Issues in Biological Psychiatric Research with Children and Adolescents" (1995) 34 Journal of the American Academy of Child and Adolescent Psychiatry 929 at 931; and C. Levine, "Children in HIV/AIDS Clinical Trials: Still Vulnerable After All These Years" (1991) 19 Law, Medicine and Health Care 231 at 235.
[55] B. Vitiello & S. Jensen, "Medication Development and Testing in Children and Adolescents: Current Problems, Future Directions" (1997) 54 Archives of General Psychiatry 871 at 872. See also J.G. Simeon & D.M. Wiggins, "The Placebo Problem in Child and Adolescent Psychiatry" (1993) 56 Acta Paedopsychiatrica 119.

— the assent of the child should be obtained to the extent that it is possible to do so; and

— the refusal of the child to participate should be respected.

## GENERAL SAFEGUARDS FOR EXPERIMENTATION WITH CHILDREN

It is important to bear in mind that children may belong to one or more of the other vulnerable populations which require the protection of special safeguards in the context of non-therapeutic experimentation. For example, a child may be developmentally disabled and may also be institutionalized. Therefore, it is necessary to underscore the principle that such children should be given the benefit of all the protective measures that apply to the members of the other vulnerable populations concerned.[56] For example, this would require that a potential subject receive the benefit of the safeguards that apply not only to children in general but also to the developmentally disabled and/or those individuals who are institution-alized.

Child subjects are particularly vulnerable to trauma that may be caused by the invasiveness of the procedure itself. They may be profoundly affected by the immediate discomfort or pain of some procedures even though such procedures may be associated with only a low degree of risk. A child may be mature enough to anticipate the unpleasantness of a procedure and may object to it. Accordingly, it is our view that this refusal should be sufficient to halt the procedure and the child should be immediately withdrawn from the study. However, there are many other cases in which the child subject does not have the experience or understanding to object in advance to a particular procedure and, therefore, it is not possible to anticipate their reaction to it. For example, in the case of immunization, a variety of factors may affect the likelihood that a child will object to the procedure, including: the specific character of the child, the reassurance of the parent, and the skill of the medical personnel in administering the injection. It is, therefore, not easy to anticipate whether a child is likely, for example, to develop a strong aversion to injections in general as a result of their experience with routine immunizations. Accordingly, it is necessary for researchers to assess the risks of psychological harm posed by the discomfort or the invasiveness of the procedure, quite separately from the risks of physical injury that may, in fact, be considerably less likely to occur.[57]

---

[56] Kaufmann has noted that "institutionalized children, both handicapped and non-handicapped, are at increased risk for exploitation". Furthermore, he argues that "surrogate consent presents special problems in this poplution. Therefore, these children should not be included in clinical trials unless they directly benefit from participation and/or the subject of the trial pertains to their special circumstance of being institutionalized". See R. Kaufmann, "Drug Trials in Children: Ethical, Legal, and Practical Issues" (1994) 34 Journal of Clinical Pharmacology 296.

[57] As Grodin and Alpert note, "(t)here are very few empirical data on children about the assessment of risk. While it is clear that different scales or standards might exist for infants as opposed to young children or adolescents, how to assess discomfort, pain, or inconvenience in these 'incompetent' vulnerable populations is problematic". See M.A. Grodin & J.J. Alpert, "Children as Participants in Medical Research" (1988) 35 Pediatric Clinics of North America 1389.

Furthermore, the significance or meaningfulness of an adverse reaction by the child subject to a particular procedure, whether this reaction is anticipated or not, may best be gauged by members of the family; they are particularly well-placed to interpret such a reaction and to take the necessary countermeasures on behalf of the child.

Therefore, it is our view that children should not be exposed to a more than a minimal level of risk in the course of biomedical experimentation. This limitation should only be exceeded in exceptional cases, and should require specific authorization by a research ethics review board especially established for this purpose. Furthermore, in all those cases where it is possible to do so, the assent of a child must be obtained before enrolling him or her as a subject in a biomedical experiment.

## THE ROLE OF SOCIETY AND THE FAMILY IN THE CONSENT PROCESS

As discussed above, children form the only vulnerable group which readily lends itself to a presumption of incompetence.[58] As children grow towards adulthood, they develop decision-making capacity, with a correspondingly decreasing reliance on their parents or guardians. The most important consideration when discussing bases for providing consent on behalf of children is the fact that, unlike adults, the former have not led mature lives, and have therefore never expressed wishes or values which could be applied towards the substitute decision-making process. It is only possible to anticipate the future values and/or wishes of a child. As such, it becomes important to examine the role of both society and the family in the consent process.

Society's agents in the research enterprise are the institutional research ethics committees, which are generally comprised of interdisciplinary teams of individuals whose expertise and/or experience enables them to make the complex assessments necessary to determine whether a protocol is scientifically and ethically meritorious and, therefore, whether it is appropriate to ask parents to enroll their children.[59] Although the review process certainly functions as an important "line of first defence" insofar as it prevents parents from being asked to enroll their children in ethically unacceptable research, the final decision to allow a specific procedure whose primary purpose is research should not be made by the researcher, the research ethics committee, or by any other representative of society at large; it must

---

[58] See generally D.N. Weisstub (Chair), *Final Report: Enquiry on Mental Competency* (Toronto: Queen's Printer, 1990) at 122–152. [hereinafter *Enquiry on Mental Competency*]

[59] The correct identification of the potential benefits to the health of children that are made possible by research and the accurate assessment of the risk of harm to child participants posed by such research are both essential elements in the review of protocols by research ethics committees. This harm/benefit ratio, along with other ethical requirements such as scientific validity, scientific merit, and confidentiality, serve as the basis upon which a decision may be made as to whether a research protocol should be implemented. For a review of general ethical principles that should guide research involving vulnerable populations, see Chapter 18. For a discussion of research ethics committees generally, see Chapter 17.

be made by the parents or guardians, or by the child on whom the procedure is to be performed.

### The Family and the Child's Developing Ability to Consent

One problem associated with attempting to establish a climate within which children may be considered free to give or withhold their consent to participate in research is the likely reliance of children on their parents' guidance in making this decision. This guidance may be explicit; it may also be implicit and, as is frequently the case, considerably more subtle in its influence. However, regardless of the impact of parental guidance, considerable benefits may be derived from the dynamic involvement of children in the consent process. Foremost among these benefits is the development of a sense of responsibility in the child and the inculcation of an understanding of what issues are important in the context of making such a critical decision. These benefits may well be of much greater importance to the child's development than the actual final decision to participate in research.

Gaylin has examined the converse situation exemplified by a parent who, hoping to foster a sense of social responsibility in his child by encouraging him to participate in an experiment, was frustrated by the insistence that he respect his child's resistance to participation.[60] Although the situations are very different, in each case the central concerns of those involved are not focused on the ultimate inclusion of the child in the protocol, but rather on the effect on a child's personal development of being asked to volunteer, and on the appropriate degree of parental influence on the decision.

These two situations underscore the sense in which a child's developing ability to give or withhold consent may amount to considerably more than a major stumbling block to be resolved to the satisfaction of ethicists and lawyers. To the child, rather than serving as an obstacle to participation in research, the process of consent may instead present an opportunity to develop the ability to evaluate the potential benefits to others and to medical science against the potential harm to herself, and to place this evaluation in the context of her own personal priorities. Clearly, fostering the ability to respond effectively to this challenge has considerable social merit. For example, in examining the development of this ability as a critical step in the achievement of responsible adulthood, Weithorn has argued that positive psychological functioning and well-being in children is promoted by involving them in decision-making from an early age.[61]

---

[60] W. Gaylin, "Competence No Longer All or None" in W. Gaylin & R. Macklin, eds., *Who Speaks for the Child* (New York: Plenum Press, 1982) 27.

[61] L. Weithorn "Children's Capacities for Participation in Treatment Decision-Making" in D.H. Schetky & E.P. Benedek, eds., *Emerging Issues in Child Psychiatry and Law* (New York: Brunner/Mazel, 1985) 22 at 27. Weithorn contemplates a range of involvement possibilities for children, including autonomous decision-making in the event that the child is capable, consultation with the child about the treatmentoption with the child's preferences taken into account, and lesser decisions within the framework of the regimen selected by the parent, such as deciding the balance between increased medication and increased adverse effects of the medication.

### Family Consensus and the Voluntary Activities of Childhood

Children participate in a broad range of activities and pursue a variety of endeavours unhampered by their legal incapacity. Decisions about their participation in sport-related,[62] cultural, community, and religious activities are all made through some form of familial consensus. While all of these activities have earned the approval of society as being beneficial to the development of the children who participate in them, they are nevertheless recognized as being optional; indeed, any one of them might be regarded as unacceptable by some individuals or groups. Families, for example, may well differ on the importance of competitive sports.[63]

Experience demonstrates that the process of giving consent to the participation of children in optional or voluntary activities evinces a broad range of decision-making patterns. Sometimes parents, after signing children up for an after-school activity, discover that their children are not enjoying it, and elect to withdraw them. At other times, parents agree to enroll children in activities with great reluctance and only after extensive entreaties by them. There will always be dutiful children who persevere with extracurricular activities, in spite of difficulty or boredom, in order to please their parents; similarly, there will always be families who feel so strongly about religious or cultural education that they are unwilling to defer to their children's preference for other activities. In other words, the differing desires and responsibilities of family members and the ways in which conflicts between them are resolved are observed in a variety of circumstances, and are not limited uniquely to the conduct of research with children. To construct a model of decision-making that is indifferent to the importance, and the influence, of this age-old process is, at best, artificial and, at worst, injurious.

The notion that the family should play a significant role in the consent process is not a novel one. Hauerwas, writing for the National Commission on the Protection of Human Subjects of Biomedical and Behavioral Research, contended that "the child ought to be conceptualized as a family member, and because of this vulnerable position the consent and guidance of parents is relevant to the participation of children".[64] These arguments were reviewed but not incorporated into the National Commission's final recommendations, perhaps because they could not be adapted to fit a decision-making model that revolved around the relationship between the autonomous individual, on the one hand, and society, on the other.

---

[62] Because a principal concern related to the decision to participate in research is the assumption of risk of physical harm, Baylis & Downie focused extensively on the freedom of parents to enroll their children in minor league hockey. See F. Baylis, & J. Downie, "An Ethical and Criminal Law Framework for Research Involving Children in Canada" (1993) 1 Health Law Journal 39 at 62.

[63] A. Holder, "Disclosure and Consent Problems in Pediatrics" (1988) 16 Law, Medicine & Health Care 219 at 225. Holder cites R.G. Mitchell, *supra* note 2; A. Holder, "Mental Illness and Parental Rights" (1971) 216 Journal of the American Medical Association 575. See also P. Keith-Spiegel, "Children and Consent to Participate in Research" in G.B. Melton, G.P Koocher & M.J. Saks, eds., *Children's Competence to Consent* (New York: Plenum Press, 1983) 179 at 187.

[64] See National Commission for the Protection of Human Subjects of Biomedical and Behavioral Research, *Research Involving Children: Report and Recommendations* (Washington, D.C.: U.S. Government Printing Office, 1977) at 107.

Evidently, much has changed in the intervening decades with respect to the understanding of childhood and the role of the family. As two Canadian authors recently observed, "the courts and the legislature have increasingly turned back towards respecting the subjective actual familial, social unit of the child and have spurned the more objective social models of the 1970s".[65] A striking illustration of the change can be found in the statement contained in the opening paragraphs of the report in which the NCBHR Task Force said:

> The natural advocates for the children in all such development and in the matter of their "concerns" and "interests" are the child's parents or guardians; they are to act as protectors and nurturers, seeking to protect and enhance the child's growth to adulthood. This familiar relationship between the child and the child's parents or guardians, between the child, siblings and parents, is of serious importance in any therapy or research context - not only insofar as there is parental responsibility to minimize harm and promote benefit to the child, but also in terms of the child's evolving independence of the family.[66]

It is important to note in this passage that parents are identified as the natural "advocates", or spokespersons, for the interests of their children and, in many cases, it will be they who express the familial consensus regarding participation.[67] However, it is implied that this advocacy will only benefit the development of the child's sense of independence if it is the articulation of a consensus that is influenced by the child's wishes (at least, to the extent that the child is capable of forming a judgment on the matter).

In formulating their proposals for an ethical model suited to the needs of children, the NCBHR Consent Panel Task Force struggled to defer to the principles of the *Belmont Report*,[68] notably the first principle of "respect for persons".[69] In so doing, it was observed that this principle advocated not only treating persons as autonomous agents but also protecting those who experience any form of diminished autonomy. After careful reflection, the Task Force recommended replacing the principle of "respect for persons" with "respect for the child", arguing that it is "more applicable and relevant in discussing research involving this group".[70] As a consensus has emerged that it is no longer as imperative to fit every decision-making situation affecting personal rights into the model of contractual autonomy, the opportunity emerges to adjust the model to suit the very particular needs of the child as a child.

---

[65] See C. Bernard & B.M Knoppers, "Legal Aspects of Research Involving Children" in B.M. Knoppers, ed., *Canadian Child Health Law: Health Rights and Risks of Children* (Toronto: Thompson Educational Publishers, 1992) 259 at 318.

[66] See NCBHR (1993), *supra* note 27 at 10.

[67] See Levine, *supra* note 54 at 233.

[68] See National Commission for the Protection of Human Subjects of Biomedical and Behavioral Research. *The Belmont Report: Ethical Principles for the Protection of Human Subjects of Research* (Washington D.C.: U.S. Government Printing Office, 1978).

[69] *Ibid.* at 4–9.

[70] See NCBHR (1993), *supra* note 27 at 18.

### Societal Assistance to the Family in Decision-Making

To say that the child's interests are often best represented through family consensus is not to say that the most vocal family member should be presumed to represent the family as a whole nor that society has no interest in ensuring that the child is content with the final resolution. Indeed, as research is not a requirement for the child's physical health, residual disagreement should generally be identified and respected, by withdrawing the child from the protocol.

Moreover, the one overwhelming reality of medical research with children is that a large part of it must be conducted with children who are seriously ill. Apprehension associated with the illness of a child is liable to affect the judgment of parents and to distort the consent process in a variety of ways. Parents who are confused and concerned about the diagnosis or prognosis of their child's condition may find it very difficult to grasp the fact that they have been asked to authorize a procedure for the purposes of research. In addition, their ability to comprehend the nature and probability of the risks involved may be seriously impaired. Overwhelmed by their own helplessness, and by their gratitude for the efforts of health-care providers, they may be eager to do anything they can to assist.[71] Indeed, in this respect, the sick child is placed in a very difficult position with regard to decision-making — a task that is perhaps rendered virtually impossible by young age or the debilitating effects of illness.

It is unfortunate that a considerable amount of medical research is conducted under circumstances that render the process of obtaining consent difficult and potentially unreliable. Yet the prospect of a family's being placed at risk of making a poor decision with lasting negative consequences for a child was not created by the possibility of conducting medical research with children. Indeed, there is a veritable panoply of rules and regulations establishing minimum ages for everything from the consumption of alcohol and the minimum age to drive a vehicle to the freedom to withdraw from formal education. These age limits have been a staple of legislation in many countries. In addition to ensuring that each child reaches a certain age free from the risks or unfortunate consequences associated with engaging in various dangerous activities, these laws protect families from pressures, both from within and without, that may tend to compromise their judgment in a way that could result in harm to the child.

## THE GROWING INDEPENDENCE OF THE CHILD

In growing up, a child undergoes many radical transformations. It would be fictitious to treat childhood as a unitary state. Accordingly, any formulation of a model for

---

[71] Outwater examined the special ethical problems arising in the context of pediatric care, noting that "although the process of obtaining informed consent from distraught parents may be difficult, the family may benefit from having someone explain their child's disease, the therapy being used, and potential outcomes and problems one more time". However, in view of the limited discretion appropriate to decisions in this context, and the often uncertain efficacy of informed consent under such circumstances, the possibility of designing protocols "to render the process of informed consent unnecessary or inappropriate" was considered. K.M. Outwater, "Ethics of Research in Pediatric Critical Care" in G.D. Koren, ed., *Textbook of Ethics in Pediatric Research* (Malabar, Fla.: Krieger, 1993) 107.

obtaining consent that may be considered appropriate to childhood must inevitably take into account the growing autonomy of the child.

### *Capacity and Disagreement among Family Members*[72]

Differences of opinion between family members in the administration of therapeutic treatment, on the one hand, are likely to have significant consequences because treatment decisions are intended to produce effects on the recipient; on the other hand, procedures performed primarily for research are not intended to have any significant ramifications for the participant. Consequently, it is arguable that differences of opinion in the context of research should simply result in the withdrawal of the child from the protocol. If this is the case, then those responsible for obtaining consent should closely examine any meaningful expression of reticence from the child, however subtle.[73] Furthermore, if there is any indication that the child's willingness to participate appears to be the result of parental pressure, this inquiry should be conducted in the absence of the parent.

The interest of society in protecting the welfare of the child militates against exposing them to unnecessary risk in the research process. Therefore, when there is disagreement between members of the family the question should be whether there are circumstances in which a child's desire to participate should prevail over parental opposition.

Since the family also risks suffering harm through the responsibility of caring for any injury that may be incurred or risks the potential disharmony resulting from the child's defiant participation, the factors in favour of permitting participation in these circumstances would have to be very weighty indeed.[74] In view of the serious implications of dispensing with parental authorization, we believe that the threshold requirements for permitting the child's participation should include: (1) the capacity to make the decision independently, (2) research ethics committee approval of the protocol as a valuable study that could not otherwise be conducted if parental authority were required, and (3) the provision of independent advice for the child.

---

[72] Although this discussion considers disagreement between parents and child subjects, it is acknowledged that disagreement could also arise between parents. It is our view that the permission of one parent should be considered sufficient, provided that this parent attests to the absence of disagreement on the part of any other person legally responsible for the child's welfare and that the researcher has no basis for believing that disagreement exists. This recommendation would, of course, only apply to those research protocols that involve nothing more than a purely minimal risk of harm.

[73] Not only should signs of reticence or objection be actively sought but, if they exist, they should be treated as an indication of a lack of the child's assent, which must be pursued whenever possible.

[74] According to Lynch, the distinction between an emancipated minor and one who is "socially immature" (e.g. living at home) is valid in determining the necessity of parental authorization for research participation. She argued that "social maturity", i.e. living independently from one's parents, is a good indicator of competence to consent to research. See A. Lynch, "Research Involving Adolescents: Are They Ethically Competent to Consent/Refuse on Their Own?" in G.D. Koren, ed., *Textbook of Ethics in Pediatric Research* (Malabar, Fla.: Krieger, 1993) 125. However, so-called "emancipation" may also be a signal for caution in the process of obtaining consent. For example, emancipated children may, for various reasons (including familial abuse), find themselves totally outside the family support system and may be particularly susceptible to peer pressure.

There are, of course, limitations that are currently placed upon the treatment of children without parental knowledge, particularly where it is administered in relation to distinctive conditions such as those related to adolescent sexual activity.[75] It is, therefore, anticipated that approval for research without parental approval would be restricted to those protocols that are designed to bring about an improvement in the treatments that are currently permissible without such approval. In any event, it is necessary for all those involved in the research enterprise to be particularly mindful of the social and practical consequences that may flow from the application of those standards that are ultimately developed as a means of determining the capacity of children to make decisions concerning their own participation.

### The Nature of the Decision to Participate

Capacity should be determined with reference to the specific decision in question rather than on a global basis.[76] Moreover, varying outcomes in the application of procedures designed to determine capacity may well reflect differences in the nature or significance of a proposed treatment. We must be acutely aware of the essential differences between research and therapy when considering the developing capacity of children and the nature of the actual decision-making process that is involved in the process of obtaining consent.[77]

Simply stated, as society has sought to minimize the possibility of negative consequences flowing from biomedical research, the decision to participate in research is both more complex and less consequential than a decision to accept treatment. Given these divergences from other forms of medical decision-making which focus on different requirements and results, it is not surprising that the appropriate standard for determining whether there is capacity to consent to research has not yet been considered in depth. It is also not surprising that, given the very complexity of the question of capacity to consent to research, confused results were obtained by a Consent Survey that asked a range of leading American authorities in

---

[75] For a discussion of the legal and ethical issues surrounding research conducted on adolescents without parental knowledge, see G. Melton, "Ethical and Legal Issues in Research and Intervention" (1989) 10 Journal of Adolescent Health Care 36S.

[76] This is referred to as the "functional" model of capacity assessment. See generally *Enquiry on Mental Competency, supra* note 58 at 67.

[77] The capacity to consent to participation in research differs from the capacity to consent to the administration of treatment in at least two important ways. The first difference is that, when performed for purely research purposes, a biomedical procedure is intended to have little if any practical effect (either beneficial or harmful) on the subject. From the subject's point of view, evaluating the desirability of undergoing the procedure does not involve assessing its potential efficacy in attaining a treatment goal but rather an assessment as to the likelihood that it will cause no significant deleterious effects. Secondly, since the benefits of research are intended to accrue to society as a whole through improvements in medical knowledge, the value of that benefit must be weighed by child participants against the cost to them of the effort or discomfort involved. These two calculations, clearly, are a great deal more complex than those made for the purposes of consenting to treatment and, when considered from the standpoint of child participants, they are significantly more challenging. For a detailed discussion of the distinction between "therapy" and "non-therapy" in the research context, see Chapter 6.

the field of child development to specify the age at which children may be considered as capable as adults to consent to a study that was relatively easy to describe. In fact, the responses ranged from age 2 to age 17 with "no 'modal age agreement' emerging".[78]

Despite the confusion and uncertainty about the capacity of children to make the decision to participate in research,[79] there is notable agreement on two issues. Firstly, there are two points in child development that are particularly significant to the child's involvement in the decision whether or not to participate in research. Secondly, it is possible to identify the chronological ages at which these points of development are usually reached.

A variety of studies have considered these aspects of the issue of children's capacity to give consent, including the work of:

— *Janofsky and Starfield*, on the assessment of risk in research with children, in which three-quarters of the researchers studied made the determination of capacity on the basis of clinical judgment rather than chronological age;[80]

— *Piaget and Kohlberg*, on cognitive development in which children were found to develop the ability to engage in rule-governed behaviour by the age of seven, that the understanding of consensual behaviour began to develop at approximately ten years of age, and that an appreciation of society's perspective concerning one's behaviour began to emerge at age fourteen;[81]

— *Lewis, Lewis and Ifekwunigue*, on consent to participation in the swine flu

---

[78] See Keith-Spiegel, *supra* note 66 at 193. See also G. Koren "Informed Consent in Pediatric Research" in G. Koren, ed., *Textbook of Ethics in Pediatric Research* (Malabar, Fla.: Krieger, 1993) 3, for a discussion of the ages that have been set in other countries for independent consent. See also G. Koren *et al.*, "Maturity of Children to Consent to Medical Research: The Babysitter Test" (1993) 19 Journal of Medical Ethics 142 at 147, for their argument that there are "deep inconsistencies in society's perception of a child's perception of a child's maturity with respect to participation in research, as compared to assuming the role of a babysitter". See also L. Lee, "Ethical Issues Related to Research Involving Children" (1991) 8 Journal of Pediatric Oncology Nursing 24; M.E. Broome & K.A. Stieglitz, "The Consent Process and Children" (1992) 15 Research in Nursing and Health 147.

[79] Keith-Spiegel recommends the articulation, for the purposes of research, of a range of questions concerning the capacity of children and ways in which they might be assisted to make decisions concerning their participation responsibly and independently. See Keith-Spiegel, *ibid.* at 204–207. See also R. Abramovitch *et al.*, "Children's Capacity to Consent to Participation in Psychological Research: Empirical Findings" in G. Koren, ed., *Textbook of Ethics in Pediatric Research* (Malabar, Fla.: Krieger, 1993) 11, for an examination of an interesting series of studies designed to determine the independence of judgment exercised by children of various ages.

[80] See J. Janofsky & B. Starfield, "Assessment of Risk in Research on Children" (1981) 98 Journal of Pediatrics 842.

[81] J. Piaget, *The Moral Judgement of the Child*, trans. M. Gabain (London: Routledge & Kegan Paul, 1932); L. Kohlberg, "Moral Stages and Moralization: The Cognitive-Developmental Approach" in T. Lickona, ed., *Moral Development and Behaviour: Theory, Research and Social issues* (New York: Holt, Rinehart & Winston, 1976) 31.

vaccine trial in which it was determined that all but the six-year-olds were able to elicit the information necessary to make an informed decision;[82]

— *Perrin and Gerrity*;[83] *Millstein, Adler and Irwin*;[84] *Kister and Patterson*;[85] and *Eiser et al.*,[86] on the knowledge children have of their bodies and of health and illness, which showed that children generally have considerably less basic understanding than we tend to realize, and so may be less equipped to make rational decisions than we might otherwise believe; and

— *Susman, Dorn and Fletcher*,[87] on the capacity of children, young adults, and adolescents to assent and consent to research, which demonstrated that the elements of the research protocol relating to concrete experiences in the lives of the subjects were most often understood.

These studies provide a sampling of the mixed research findings concerning those factors relevant to determining a child's capacity to make decisions on participation in research. It is certainly true that researchers currently favour a *functional*, rather than a *chronological*, determination of capacity; however, this surely begs the question as to the criteria that should anchor their assessment of capacity. Although an appreciation of rule-governed conduct would seem critical for participation in the process of consent, we have seen that the substance of the decision to participate in research is far more complex with respect to its costs and consequences than even "ordinary" health-care decisions. In this vein, although Lewis *et al.* determined that children aged seven and above were able, as part of a group, to ask the questions necessary for them to make an informed decision regarding the administration of a vaccine, it should be remembered that a vaccine is a medically advisable and routine procedure; therefore, this finding would not necessarily reflect the ability of the children concerned to make a decision regarding a non-therapeutic procedure. Furthermore, it seems essential to the task of gaining an appreciation of the risks posed by any particular procedure that a potential participant have a clear

---

[82] C.E. Lewis, M.A. Lewis & M. Ifekwunigue, "Informed Consent by Children and Participation in an Influenza Vaccine Trial" (1978) 68 American Journal of Public Health 1079. It should be noted that "(t)he children involved were already involved in a program whereby they were given increased decision making power regarding the school health care system, and were thus more accustomed than average children to making their own health care decisions". It may be the case that children in this sample would be more likely than average to make the decision to participate in the study themselves, without parental involvement. If this assumption is correct, then the number of children wishing for parental involvement would be higher in the general population than the study would indicate.

[83] E.C. Perrin & P.S. Gerrity, "There's a Demon in Your Belly: Children's Understanding of Illness (1981) 67 Pediatrics 841.

[84] S.G. Millstein, N.E. Adler & C.E. Irwin, "Conceptions of Illness in Young Adolescents" (1981) 68 Pediatrics 834.

[85] M.C. Kister & C.J. Patterson, "Children's Conceptions of the Cause of Illness: Understanding of Contagion and Use of Immanent Justice" (1980) 51 Child Development 839.

[86] C. Eiser & D. Patterson, " 'Slugs and Snails and Puppy-Dog Tails': Children's Ideas about the Inside of their Bodies" (1983) 9 Child: Care, Health and Development 233; C. Eiser, D. Patterson & J.R. Eiser, "Children's Knowledge of Health and Illness: Implications for Health Education" (1983) 9 Child: Care, Health and Development 285.

[87] E. Susman, L. Dorn & J. Fletcher, "Participation in Biomedical Research: The Consent Process as Viewed by Children, Adolescents, Young Adults and Physicians" (1992) 121 Journal of Pediatrics 547.

comprehension of the basic functioning, and significance, of the various body parts that might be affected by the procedure in question; without such a basic comprehension, the informational basis for making the decision, whether or not to participate, would be seriously deficient.

Weithorn, and Grisso and Vierling, have found that on "a scale of capacity ranging from the mere ability to manifest a choice to the ability to appreciate the nature of treatment, fourteen-year-olds were capable of the highest standard of mental reasoning".[88] Grodin and Alpert determined that,

> [c]hildren less than seven years of age employ reasoning structures which may not be entirely rational . . . [while] children from age seven to thirteen years . . . will see the world in concrete terms and not employ the magical thinking of the younger child. They suggest that the child of this age will have difficulty anticipating the future, however, thus limiting the ability to provide informed consent.[89]

### *Two Developmental Milestones: Ages Seven and Fourteen*

It is manifestly clear that the whole question of the capacity of children to make decisions about participation in research is an extraordinarily challenging one and that a multitude of cognitive, social and consequential factors are involved. Nevertheless, there is a compelling level of agreement among researchers concerning the existence of two critical points of development and the age at which these points are generally reached. In the view of Nicholson, the following conclusions may be drawn from the research literature concerning the significance of these two developmental milestones:

> Before the [developmental] age of 7 years . . . attempts to obtain a child's assent . . . are likely to be meaningless, and it is more important simply to tell the child, as much as possible using his level of language, what is going to be done . . . . The nearer the child is to 14 years old, the more important does his assent to a research procedure become . . . . (I)f the research procedure is . . . non-therapeutic, it should not in general be carried out if the child refuses assent. From the age of 14 years upwards, the adolescent subject's refusal . . . should be binding . . . . (T)he parents' or guardian's refusal of consent should probably only be binding in the case of non-therapeutic research.[90]

The consistency with which these ages are recognized as being significant in developmental terms extends to the recent NCBHR Report on the subject. In this Report, the Task Force responds to the question "Who are the children to be involved in this research?" by stating that it "agreed . . . to focus its attention on three age groups within the 'child' population": those from birth to seven years of age, those from seven to fourteen years of age and those from fourteen years to the

---

[88] *Ibid.* at 144–45. T. Grisso and L. Vierling, "Minors' Consent to Treatment" (1978) 9 Professional Psychology 412; L. Weithorn, "Developmental Factors and Competence to Make Informed Treatment Decisions" (1982) 5 Child and Youth Services 85; L. Weithorn and S. Campbell, "The Competency of Children and Adolescents to Make Informed Treatment Decisions" (1982) 53 Child Development 1589.

[89] *Ibid.* at 145. *The Enquiry on Mental Competency* cites M.A. Grodin & J.J Alpert, "Informed Consent and Pediatric Care" in G.B. Melton, G.P. Koocher & M.J. Saks, eds., *Children's Competence to Consent* (New York: Plenum Press, 1983) 93 at 96.

[90] See R.H. Nicholson, ed., *Medical Research with Children: Ethics, Law, and Practice* (Oxford: Oxford University Press, 1986) at 150–51.

age of majority.[91] While there is widespread support for recognizing these developmental milestones,[92] the question remains as to how best to harmonize this approach with the mutual interests of society and the family, not only safeguarding the welfare of children but also nurturing their growing independence in relation to their participation in research.

## CONCLUSION

In exploring the type of decision-making structure most appropriate to facilitate the participation of children in research, it was recognized that there is a growing need to cultivate their decision-making capacity in collaboration with their families. There is an increasing degree of respect for the capacity of children to participate initially as partners in the decision-making process and, ultimately, to make such decisions for themselves. Moreover, the family as a whole and the potential research subject in particular should be given effective assistance in this by establishing a system of ethics review to protect children against undue risk and by introducing a regulatory régime to provide concrete support to children who are developing an independent ability to give informed consent to participation in research.

In light of our review of existing practice that has emerged on the issue of children's participation in non-therapeutic research, it is possible to articulate some major goals that should be aimed at by any attempt to establish a statutory framework to regulate such research activity. These goals include: (1) the facilitation of much-needed medical research, (2) the protection of the welfare of child subjects, (3) the promotion of the integrity of the family, and (4) the supportive development of decision-making abilities in children.

## SUMMARY

It is our position that research involving children is necessary for the continued advancement of children's health and welfare. However, non-therapeutic experimentation with children necessitates a compromise between advances in medical knowledge and the exposure of children to risk.[93] Owing to the legal uncertainty surrounding substitute consent for children in procedures that may not be in their best interest, including non-therapeutic experimentation, we further believe that the law with respect to biomedical experimentation involving children requires clarification.

With respect to the participation of children in biomedical experimentation, we make the following recommendations:

---

[91] See NCBHR (1993), *supra* note 27.

[92] See also Group for the Advancement of Psychiatry, Committee on Child Psychiatry, *How Old is Enough? The Ages of Rights and Responsibilities: Report 126* (New York: Brunner/Mazel, 1989), who acknowledge the significance of the ages of seven (as generally coinciding with the development of social cognition) and fourteen (as generally coinciding with the growing adeptness with abstract principles) in the development of decisional capacity.

[93] J. Pearn, "A Classification of Clinical Paediatric Research with Analysis of Related Ethical Themes" (1987) 13 Journal of Medical Ethics 26 at 26.

(1) Non-therapeutic research should never be conducted on children who are not capable of making their own decision to participate, if it is possible to conduct the same research with *adults* or with *competent children*.

(2) Non-therapeutic research should not be conducted with children if the risk is *greater than "minimal"* unless *exceptional* approval is obtained from a research ethics review board especially established for the purpose of reviewing protocols involving biomedical experimentation with members of vulnerable populations. Such approval may be given only after the child, if capable, has given consent, or, where the child is not capable of making an independent decision, then only after a parent has consented.

(3) Protocols should encourage children and their parents to make *joint decisions* concerning the participation of the former in biomedical research. However, children who are capable of making an independent decision should be permitted to give their consent to participation without parental approval.

(4) Children who have reached 14 years should be *presumed* to be capable of making an independent decision to participate in biomedical research. If there are any doubts as to the capacity of a child of 14 or more to make such a decision, an assessment should be conducted to determine whether this capacity exists. In accordance with modern trends, there should be a *functional* determination of capacity.

(5) Children under the age of 14 may be capable of making independent decisions concerning participation in biomedical research. However, the existence of this capacity should be established by an appropriate assessment. Furthermore, before a child under the age of 14 may make an independent decision to participate in research, the child should be given *independent advice* by an individual with special expertise in communicating with children.

(6) Where a child is found to be incapable of making the decision to participate in research, consent may be given by a parent, *provided there is no objection on the part of the child*. Every effort should be made to provide the child with as much information as is possible in the circumstances and the child must be informed of the absolute right to refuse to participate in or to withdraw from the research project in question.

(7) An objection by a child may be overridden where an innovative therapy, as part of a research protocol, is employed as a life-saving device or for the purpose of significantly improving the long-term quality of life of the subject. In such cases, the consent of the legal guardian should be reviewed by a research ethics review board.

(8) In no circumstances may non-therapeutic research be either commenced or continued where there is a clear objection on the part of a child. Current objections must override any previous consent that may have been given to participation in such research.

CHAPTER 20

# BIOMEDICAL EXPERIMENTATION INVOLVING ELDERLY SUBJECTS: THE NEED TO BALANCE LIMITED, BENEVOLENT PROTECTION WITH RECOGNITION OF A LONG HISTORY OF AUTONOMOUS DECISION-MAKING

DAVID N. WEISSTUB, SIMON N. VERDUN-JONES AND JANET WALKER

## INTRODUCTION

The age-related dependence experienced by many of the elderly raises fundamental questions that must be addressed in the context of their consent to participate in biomedical experimentation or other non-therapeutic medical procedures. There are those elderly individuals who have suffered a significant decline in cognitive capacity because of the insidious progression of dementia, particularly that of the Alzheimer type. These individuals will not meet the criteria for competence to consent to research and the obvious question that arises is whether they should be eligible at all for participation in experimentation. Other elderly adults may not have experienced any significant degree of cognitive impairment but they may be nonetheless in a dependent position because of factors related to their age. In their case, the question is how can we ensure that any consent that they may give to submit themselves to experimentation is, in fact, a genuinely free and informed consent?

The protective concern for the elderly as a distinct group is based on a developing bedrock of empirical research. There is little doubt that the elderly are peculiarly susceptible to a number of forms of abuse: in particular, physical abuse, emotional abuse, financial abuse, and neglect. It has been estimated that as many as 100,000 elderly Canadians[1] have recently suffered some form of serious maltreatment within

---

[1] E. Podnieks et al., National Survey on Abuse of the Elderly in Canada (Toronto: Ryerson Polytechnical Institute, 1990).

their own homes. It has also been shown that abuse may be a significant problem in both public institutions and private facilities that provide care for the elderly in Canada.[2] Thus, research has unearthed cases where, in large proportions, those individuals and institutions that have been entrusted with protecting the elderly have in fact become their abusers. With respect to research, it is critical that exploitation and mistreatment by those engaged in biomedical experimentation should not be added to the list of abuses to which the elderly are vulnerable.

While the need for the protection of the elderly is becoming increasingly evident, it is also necessary to acknowledge that they have already enjoyed a great deal of decision-making autonomy during their lives. It has been noted that "an overdose of benevolence can be as harmful as the absence of protection and assistance" and that "*limited, benevolent intervention*" is needed in the case of vulnerable elderly adults to ensure that they "receive the most effective but least restrictive and intrusive form of assistance, support or protection necessary to meet their needs".[3] In the specific context of consent to research, therefore, it is difficult not to agree with Cassel who warned that we should be careful not to permit well-intentioned paternalism to extend to the point where we do not allow the elderly "the choices and freedoms that we would allow other people, constricting their experiences unnecessarily."[4]

## OVERCOMING A HISTORY OF NEGLECT

The elderly have experienced a history that, until recently, was marked by widespread indifference to their special health needs, and accordingly they have lagged behind other sectors of society in terms of medical advances that are of particular relevance to their health care. Indeed, one author has commented that geriatric research "has not been perceived as being as glamorous or rewarding as high-tech research in the acute care setting and has suffered from under-funding and limited visibility in the scientific community".[5] Neglect of the elderly by researchers has also been a feature in the reports of various special commissions or learned bodies.[6]

---

[2] L. Belanger *et al.*, *Les Cahiers de L'Association Québecoise de Gerontologie* (Association Québecoise de Gerontologie, 1981).

[3] R.M. Gordon & S.N. Verdun-Jones, *Adult Guardianship Law in Canada* (Scarborough, Ont: Carswell, 1995) at 1–28.

[4] C.K Cassel, "Informed Consent for Research in Geriatrics: History and Concepts" (1987) 35 Journal of the American Geriatric Society 543 at 543.

[5] C.K. Cassel, "Ethical Issues in the Conduct of Research in Long Term Care" (1988) 28 The Gerontologist Supplement 90 at 91.

[6] None of the leading reports in this area specifically included the elderly in their discussion of the special populations in need of particular protection in the context of the research process. See for example: National Commission for the Protection of Human Subjects of Biomedical and Behavioral Research, *Research Involving Those Institutionalized as Mentally Infirm: Report and Recommendations* (Bethesda, Md.: The Commission, 1977) [hereinafter *National Commission: Research Involving Those Institutionalized as Mentally Infirm*]; Law Reform Commission of Canada, *Working Paper No. 61: Biomedical Experimentation Involving Human Subjects* (Ottawa: Law Reform Commission of Canada, 1989); and A.D. Milliken, "Position Paper: The Need for Research and Ethical Safeguards in Special Populations" (1993) 38 Canadian Journal of Psychiatry 681 at 684.

Traditionally, research with the elderly was regarded as highly unfeasible. There were intractable practical difficulties associated with obtaining their consent for research, and governments were unwilling to address those difficulties because of the prevailing apathy concerning the needs of the elderly in general. Although the National Commission's *Research Involving Those Institutionalized as Mentally Infirm: Report and Recommendations*[7] — much of which might have applied to the elderly — was never officially approved, it effectively stifled much potential research "because investigators and administrators had real concerns about the regulations and about their own risk management concerns to avoid scandals".[8] Therefore, societal indifference to the special concerns of the elderly, their complexity as subjects for research, and the problems associated with gaining their consent all served historically to dampen the enthusiasm of the medical community for research into their needs.

Ironically, during the period when institutionalized populations were widely employed as research subjects to achieve medical advances of general application, universal neglect of the elderly resulted in their protection from much abuse. They were widely regarded as being unsuitable subjects because, often, they were suffering from multiple disorders or dysfunctions that would be likely to complicate or distort the experimental results.[9] Thus, with the exception of the discussions of some notable examples of abuse among institutionalized populations, the elderly were neglected both as the topic of, and as potential participants in, research.[10]

Changing demographics and the elderly's "growing prominence . . . as a market for the therapeutics industry"[11] have sparked increased interest in geriatric research and have fostered a growing recognition of the need to face the problems associated with safeguarding their interests and obtaining consent for their participation. The obvious connection between a developing interest in research with a special population and a rising concern for that population's welfare has rendered further neglect unacceptable. It has been argued that "the exclusion of older people from

---

[7] See *National Commission: Research Involving Those Institutionalized as Mentally Infirm, ibid.*

[8] C.K. Cassel, "Research in Nursing Homes: Ethical Issues" (1985) 33 Journal of the American Geriatric Society 795 at 796.

[9] As Swift has pointed out, "intersubject variability and the coexistence of confounding factors (such as concurrent disease and drugs) increases with age". See C.G. Swift, "Ethical Aspects of Clinical Research in the Elderly" (1988) 40 British Journal of Hospital Medicine 370.

[10] One notorious exception to the neglect of research among the elderly was a protocol from the Sloan-Kettering Cancer Research Center in New York. See E. Langer, "Human Experimentation: Cancer Studies at Sloan Kettering Stir Public Debate on Medical Ethics" (1964) 143 Science 551. Researchers from this Centre injected live cancer cells into long-term-care patients at the Jewish Chronic Disease Hospital to examine the immunologic response to malignant cells in those people likely to have suppressed responses owing to chronic illness. Twenty-two long-term-care patients were enrolled in the experiment but neither they nor their relatives were informed of the experimental nature of the study or of the fact that the injections contained cancerous cells. In reality, since many of the patients were demented, non-English speaking, or extremely deaf, it is likely that they were chosen precisely because they were not in a position to be informed of the nature of the experiment. When asked in a courtroom whether they had explained to the subjects that the injections contained malignant cells, the investigators replied, "[o]f course not; they never would have agreed to the study if we had told them that". See Cassel (1985), *supra* note 8 at 796.

[11] See Swift, *supra* note 9.

clinical research is itself unethical;"[12] it is unethical simply because "we know we can do a better job in the care of the elderly population if we do more research".[13]

In keeping with this growing concern for the welfare of the elderly, recent interest on the part of the medical community in conducting research with this population has focused on areas of particular application to the elderly themselves. Thus, though the elderly may have been spared inclusion in unethical research in the past as a result of the complexity of their medical status, this very complexity is now the subject of ethically sound research into their needs. Research on the elderly has made progress in overcoming such obstacles as societal indifference and scientific complexity, but problems associated with obtaining the consent of the elderly to participate in research still remain.[14]

### *The Special Place of the Elderly and the Urgency of Caring for Their Needs*

Questions of consent that arise in relation to biomedical experimentation involving a vulnerable population flow from a recognition of the particular social characteristics and needs of that population, and a sense of the urgency of attending to those needs, including research whose results may have a potential benefit for them.[15] The recognition of both the urgency of extending the knowledge about their needs through research and the importance of guaranteeing them proper protections acknowledges the special place of these populations in society. Unfortunately, it has not been easy to identify the special needs of the elderly in the context of consent to biomedical experimentation because their special status has not, until relatively recently, been accorded the same degree of recognition as that given to, for example, children.[16]

Is the incentive to conduct research to be understood as being limited to the impact of changing demographics and the discovery of the elderly as a "market for the therapeutics industry"?[17] How do the general public and the caregivers of the elderly view their efforts with respect to those adults who, while they were once full-fledged participants in society, now experience diminishing capacity? In terms of society's pragmatic and varying commitments to different groups, one writer commented that,

> we live in a highly urbanized, industrialized, technological society which emphasizes productivity and technical skills. Those who can contribute productively and skillfully to the economy are highly valued; those who no longer can, or are no longer allowed to, are

---

[12] *Ibid.*

[13] See Cassel (1985), *supra* note 8 at 797.

[14] See e.g. G. Hodge, "Ethics of Research with Dementia Sufferers" (1989) 4 International Journal of Geriatric Psychiatry 239; R.E. Kendell, "Ethics of Research with Dementia Sufferers: Comment" (1989) 4 International Journal of Geriatric Psychiatry 239: G. Langley, "Review of 'The Clonidine Test in Patients with Dementia'" (1989) 4 International Journal of Geriatric Psychiatry 241.

[15] For a general discussion of the notion of "vulnerability" in the research context, see Chapter 18.

[16] See Chapter 19.

[17] See Swift, *supra* note 9.

not valued. Predominant value orientations ... leave the elderly in a vulnerable and devalued position.[18]

Thus, it may actually be that research into the needs of the elderly with the promise of finding solutions for their problems are just the result of a consequentialist approach. Such an approach would be founded on changing demographics that allow the elderly to have a voice like any other lobby group, and on concern that, eventually, the great majority among us will be joining their ranks. To reduce it to its basest level, the possible return on society's investment in the elderly is only now becoming clear.[19] That return is encapsulated in the so-called "golden rule": do unto others as you would have them do unto you.

### *Recognition of the Elderly as a Special Population*

Paradoxically, heterogeneity may be the common element that unites the elderly as a group.[20] Even a definition of age is controversial in relation to the elderly. For example, in establishing regulations for Institutional Review Boards in the U.S., "proposals to develop separate regulations for the elderly were dropped because it was impossible to define the aged as a population sharing any common characteristics".[21] Furthermore, many elderly persons incapable of giving consent are institutionalized and/or may also be physically dependent on others for their daily care. All in all, definitions of the elderly as a class of individuals have traditionally waffled between recognition of their residual vigour and inapt, unfair comparisons with other vulnerable groups.

The emergence of a positive definition of "the elderly" from within the elderly population itself provides a key element for developing a cohesive understanding of the elderly as a population. Political activism among the elderly, such as "senior" and "grey power"[22] movements, has contributed to an improved definition of the elderly as a community. This, in turn, has fostered the commitment of society as a whole to their welfare[23] and to an acknowledgment of their right to give or withhold consent in matters affecting their personal care. This is critical because, as has been

---

[18] E. Ozanne, "Informed Consent and the Elderly: Professional Defence or Consumer Right?" in Law Reform Commission of Victoria, *Medicine, Science and the Law: Informed Consent* (Melbourne: Globe Press, 1987) 50 at 51.

[19] See G.A. Sachs & C.K. Cassel, "Biomedical Research Involving Older Human Subjects" (1990) 18 Law, Medicine & Health Care 234 at 241.

[20] M.B. Kapp & A. Bigot, *Geriatrics and the Law: Patient Rights and Professional Responsibilities* (New York: Springer, 1985) at 177–8.

[21] B. Stanley, ed., *Geriatric Psychiatry: Clinical, Ethical and Legal Issues* (Washington, D.C.: American Psychological Association Press, 1985) at 83.

[22] See e.g. S.G. Rice, "Beyond War: Empowerment for Senior Citizens in a Nuclear Age" (1988) 15 Journal of Sociology and Social Welfare 73.

[23] "The law affecting the elderly is a reflection of how society regards older people, the relationships that exist between and among the generations, the views people have about growing old, and the passage of time in another era". See E.S. Cohen, "Realism, Law and Aging" (1990) 18 Law, Medicine & Health Care 183 at 184.

suggested, "special attitudes and assumptions regarding aging and the aged, whether emanating from a sincere altruistic instinct to protect those we perceive as vulnerable or from a perverse ageist bias akin to other irrational 'isms', shape the ways in which society uses law to define the rights and responsibilities of the elderly".[24]

The more important question remaining, however, is whether we can build on these positive trends in social attitudes towards the elderly and develop a process for obtaining consent that is appropriate to their special needs. How, then, might we begin to characterize a positive and cohesive view of the elderly, and in what ways may the various elements of this view affect the manner in which issues of informed consent are perceived by those involved in the research enterprise?

### The Special Nature of the Elderly Population

Dubler characterized the elderly as being divisible into two categories. The first category covers "those persons who are chronologically advanced in life's predetermined span of years, but who are otherwise vigorous, independent, and autonomous adults". In Dubler's view, "there are no logically compelling reasons for distinguishing these adult persons from those ten or even thirty years their junior" and they should, therefore, be subject to the general research procedures and regulations that are applicable to the adult population at large.[25] The second category, however, consists of "the elderly incapable" whose various cognitive deficits inevitably dictate the need for a degree of special treatment. Dubler, therefore, contends that there is a case for making a clear distinction between the "chronologically advanced" and the "elderly incapable", and for treating the former in the same manner as those who are decades younger and devising special methods for dealing with the latter.

There are serious drawbacks to this bifurcated characterization of the elderly. First, as Dubler herself notes, splitting the elderly into two sharply divided groups does not adequately account for the "gray areas", or the marginal and fluctuating competence that is common in those whose capacity is affected by the aging process.[26] In other words, it does not readily promote the development of "appropriate standards for the treatment of those with some compromised intellectual, emotional and judgmental capacities and some residual capacity . . .

---

[24] M.B. Kapp, "Introduction: Law and Aging" (1990) 18 Law, Medicine & Health Care 181.

[25] See N.N. Dubler, "Legal Judgments and Informed Consent in Geriatric Research" (1987) 35 Journal of the American Geriatric Society 545.

[26] A study by Stanley, Stanley & Pomara revealed that "[t]he elderly demonstrate poorer comprehension [than their younger counterparts] of each of the specific elements of informed consent information, that is, knowledge of risk, benefits, and purpose of the project and procedure. Thus, as a group, geriatric patients may have some impairment in their competency to give informed consent to research. However, this impairment does not appear to have a significant impact on the quality of their decisions". See B. Stanley, M. Stanley & N. Pomara, "Informed Consent in Geriatric Patients" in B. Stanley, ed., *Geriatric Psychiatry: Clinical, Ethical and Legal Issues* (Washington, D.C.: American Psychological Association Press, 1985) 17 at 26–27.

precisely the situation that describes many elderly persons with some cognitive deficits".[27]

In this regard, it is important to underscore the value of a functional approach to competency assessment, rather than on global determinations of competence.[28] This approach seems ideally suited to a class of individuals whose capabilities may gradually diminish in a piecemeal fashion, and who are otherwise entitled to be presumed competent.

Creating legal distinctions within the elderly population on the basis of global assessments of capacity can also have deleterious effects on the community itself. Although heuristically sound, legal distinctions do not promote a community of individuals cognizant of each others' strengths and supportive of each others' needs; instead, they exacerbate the historical situation in which there has been a tendency for the able-bodied and quick-witted to disavow any connection with, or sympathy for, those of similar years who are deemed to be less capable. The bifurcated approach would create a class of individuals identified by their weakness and dependency and who, accordingly, are not challenged to retain their abilities following the precipitous declaration of their incompetence, and another class of elderly persons who, though largely functionally independent, may suffer as a result of masked inabilities that they are afraid to reveal lest they be ·adjudged incompetent.[29]

For the elderly, the rate of decline is a highly individualized process; there is very little correlation between the rates of aging from one individual to another. Many individuals in our society require assistance in managing their personal affairs in their sixties or seventies while others continue into their eighth decade as leaders in government or their chosen professions. Moreover, a rapid decline in functioning, such as that precipitated by the loss of a spouse, has often been observed in certain elderly persons while others experience a much more gradual decline in their abilities.

The marked variation in the observable onset, progress, and effects of the aging process all point to perhaps the central heuristic confusion with respect to the definition of the elderly as a special population: that, although incapacity is age-

---

[27] See Dubler, *supra* note 25 at 546.

[28] See D.N. Weisstub (Chair), *Final Report: Enquiry on Mental Competency* (Toronto: Queen's Printer, 1990) at 74.

[29] But see Dubler, *supra* note 25 at 548:

> Should . . . [specific regulations] be for all elderly persons, as a class, or for the subset of persons of diminished capacity or mental infirmity? The latter option is preferable both for individual determinations and for public policy reasons. Designating all elderly persons as a class in need of special protection may tend to further separate and stigmatize all elderly persons. Ageism need not be provided with any further support.

This view addresses concerns that arise in an environment in which global assessments of (in)competence tend to stigmatize persons, and in which protections established through regulations tend to be inflexible. However, in the context of decisional capacity determinations, recognition of the possibility of diminished capacity among the elderly, sensitively handled, is neither disrespectful, nor does it promote ageism. Rather, it provides protection from the possibility of exploiting a consent obtained from those not yet deemed incapable of giving it.

related in the elderly, there is no clear correlation between chronological age and declining capacity.[30] That is, because we cannot easily determine at what age one becomes "elderly", we have been reticent to acknowledge the elderly as a distinguishable class of persons. In the past, society has tended to make inapt comparisons with children and to regard the elderly as too heterogeneous to be understood as a special population with common needs and interests.[31] In addition, it has tended to doubt that an age-based class of individuals can exist beyond strict chronological parameters. Nevertheless, despite these persisting tendencies, more positive images are gradually emerging of a cohesive community of elderly persons that is striving to retain and maximize their capacity for independence and self-determination — a capacity which society at large possesses and which they themselves once took for granted.

### *Alzheimer's Dementia and its Influence on Models of Consent for the Elderly*

In the preceding material, we have provided a characterization of our historical neglect with regard to the elderly. This neglect has been attached to a resistance to attend to their needs and even to the recognition of this population as deserving a special status.[32] A disease such as Alzheimer's, which especially inflicts the elderly, has become such a common occurrence, due to increased longevity, that there is a strongly-sensed need in our society to generate an effective response through research.

Alzheimer's Dementia (AD) is "a condition marked by continued cognitive deterioration beginning with simple forgetfulness and ending with the inability to eat, to recognize loved ones, and to control one's bodily functions".[33] "The average

---

[30] As Bowsher *et al.* note:

> The diversity that exists among elderly people increases with age, owing to the effects of varied life events, environments and resources. Such effects influence the course of development of frail elderly people. Development is time and change dependent. For these reasons, the degree of change for any elderly person or group of elderly people can only be measured by taking measurements across time.... The false assumption of homogeneity, that is believing that 60-year-olds are like 90-year-olds, can be prevented by the selection of subjects by cohort groups. True differences between the "young-old" and the "old-old" will then be more clear. When subjects are "lumped" into broad age groups, the ability to generalize from the data obtained is seriously jeopardized.

See J. Bowsher *et al.*, "Methodological Considerations in the Study of Frail Elderly People" (1993) 18 Journal of Advanced Nursing 873 at 874.

[31] Glass reaffirms the view that "[i]t is legitimate to consider the elderly as a group for some purposes and to do so without stereotyping them". In so doing, she emphasized the need to respect "the elderly person's altered value system and perception of risk, whether 'accurate' or not". See K.C. Glass, "Informed Decision-making and Vulnerable Persons: Meeting the Needs of the Competent Elderly Patient or Research Subject" (1993) 18 Queen's Law Journal 191 at 204, 224.

[32] R. Ratzan, "Being Old Makes You Different: the Ethics of Research with Elderly Subjects" (1980) 10:5 Hastings Center Report 32.

[33] A. Moorhouse, "Ethical and Legal Issues Associated with Alzheimer's Disease Research and Patient Care: To do Good Without Doing Harm" in S.N. Verdun-Jones & M. Layton, eds., *Mental Health Law and Practice Through the Life Cycle: Proceedings from the XVIIth International Congress on Law and Mental Health* (Burnaby, B.C.: Simon Fraser University, 1994) 43 at 43.

duration of the illness, from first onset to death is 8.1 years, although the time from the diagnosis to death averages 3.4 years. The duration is unpredictable, however, and can last up to 25 years.[34] Among the many tragic consequences of AD is its effect on the ability of its sufferers to consent to participation in research that may yield valuable knowledge in the search for its prevention, treatment, and cure. AD diminishes global cognition and leaves the "capacity [to consent] progressively impaired and ultimately extinguished".[35]

Major ethical and legal problems arise in the context of research into AD because such research is generally of a non-therapeutic nature. Although AD research subjects may contribute to the advancement of scientific knowledge about this disease, they generally will not receive any direct and immediate benefit for themselves as a consequence of their participation.[36] Furthermore, given the short lifespan that may be expected by AD research subjects, any improvements in treatment that may flow from AD research are almost certain to be limited in their impact to future sufferers. This inevitably raises the question of whether incompetent patients should be permitted to participate in AD research at all.

In this respect, researchers find themselves perched uncomfortably on the horns of a dilemma. Since their research will, in many cases, be non-therapeutic in nature, there is, on the one hand, considerable uncertainty as to whether, at common law, a valid third-party consent can be given on behalf of incompetent subjects.[37] On the other hand, there is no doubt that the scourge of AD will only be fought on a more effective basis if we increase our scientific understanding of the disease through systematic research. A further difficulty regarding research on AD is that it has no acceptable animal model, thereby necessitating the use of human subjects.[38]

Campion, among others, has identified the urgent need for AD research to continue notwithstanding the problems associated with obtaining the consent of subjects who are incompetent as a consequence of the disease itself:

> Ultimately, the future care of SDAT patients and the hope of improving that care relies very largely upon research. The costs of the disease are staggering — in hard dollars and in human despair — that we cannot afford to let Alzheimer's research become stalled or, worse, to die the death of a thousand qualifications. No disease affects us more profoundly nor threatens us more tangibly than Alzheimer's. Because it is so common and so highly age-related, the care of the Alzheimer's patient is closely linked to the health care of the elderly in general, and of the seriously impaired elderly in particular.[39]

[34] Congress of the United States, Office of Technology Assessment, *Losing a Million Minds: Confronting the Tragedy of Alzheimer's Disease and other Dementias* (Washington, D.C.: U.S. Government Printing Office, 1987).

[35] See V.L. Melnick *et al.*, "Clinical Research in Senile Dementia of the Alzheimer Type: Suggested Guidelines Addressing the Ethical and Legal Issues" (1984) 32 Journal of the American Geriatric Society 531.

[36] Moorhouse, *supra* note 33 at 43.

[37] See Chapter 8.

[38] See Melnick *et al.*, *supra* note 35 at 531–532; Moorhouse, *supra* note 33 at 44; E.W. Keyserlingk *et al.*, "Proposed Guidelines for the Participation of Persons with Dementia as Research Subjects" (1995) 38 Perspectives in Biology & Medicine 319 at 319.

[39] See E.W. Campion, "Ethical Issues in the Care of the Patient Involved in Alzheimer's Disease Research" in V.L. Melnick & N.N. Dubler, eds., *Alzheimer's Dementia: Dilemmas in Clinical Research* (Clifton, N.J.: Humana Press, 1985) 71 at 76.

The intractable difficulties associated with obtaining informed consent from AD patients to participation in research have considerably limited the range of potential subjects.[40] On the one hand, recruitment of research subjects in the early stages of the disease is inefficient because the early symptoms of AD are common to a variety of other conditions.[41] On the other hand, recruitment in the later stages is problematic because cognition has usually deteriorated to such a marked extent as to render communication virtually impossible. Thus, researchers have largely been required to confine their recruitment efforts to a very narrow window of time between the making of the diagnosis and the onset of incapacity.

As disturbing as AD is, both the incentives that it creates for research and the problems that it presents for the ethical conduct of that research are only the more striking examples of the incentives and problems that are associated with cognitive-aging research in general. Elderly persons may experience a variety of physical dependencies and disabilities in common with members of other vulnerable populations; however, it is cognitive aging, found in the frighteningly unexpected, rapid and relentless shape of Alzheimer's Dementia, that is at the heart of the special needs of the elderly with respect to their participation in biomedical experimentation. Growing awareness of the suffering caused by AD has certainly instilled determination in the medical community to develop ways of treating those suffering from it and it has also created the incentive for legal and ethical experts to search for ways in which research into AD can be conducted with human participants who are actually suffering from it.

Many elderly persons have cognitive abilities that are just as sound as those of younger adults and many others have a range of health requirements that are unrelated to diminishing cognitive capacity. Nevertheless, AD patients are a compelling example of the typical challenges experienced by the elderly in the context of their ability to consent to participate in research that could assist in identifying their own particular needs. Consequently, it will be helpful to draw on examples from the experience of Alzheimer's research when discussing the various practical problems faced by many elderly adults, such as loss of memory and the

---

[40] For a discussion of the ethical justifications for involving elderly persons in research protocols, see generally, B. Brown, "Proxy Consent for Research on the Incompetent Elderly" in J.E. Thornton & E.R. Winkler, eds., *Ethics in Aging: The Right to Live, the Right to Die* (Vancouver: University of British Columbia Press, 1988) 183. See also H. Helmchen, "The Problem of Informed Consent in Dementia Research" (1990) 9 Medicine and Law 1206.

[41] Stanley amplifies the discussion of the problems of recruitment in the early stages of Alzheimer's for research:

> Generally, in the early stages, where the deficits are mild, the diagnosis is made with much less certainty than in its later stages. Therefore, it is possible that some individuals having mild cognitive impairment may not only be wrongly diagnosed as having Alzheimer's disease, but may also be enrolled in studies and subjected to unnecessary risks. Also, in progressive illnesses such as Alzheimer's disease, it may not be possible to detect underlying biochemical changes early in the course of the illness. This, in turn may yield inconclusive findings or falsely negative results.

B. Stanley, "Senile Dementia and Informed Consent" (1982) 1 Behavioral Sciences & the Law 551 at 562.

inability to read small print, as well as the more complex problems, such as those of assessing whether the individuals' dependency on others is a coercive influence on their willingness to participate in research, or whether a family member can be relied on to provide an accurate representation of the participant's competent wishes. In short, the growing understanding by the medical, ethical, and legal communities of the ways in which AD patients may participate in research can be used as a model for research on senility and other types of conditions specific to the elderly.

## THE ELDERLY AND CONSENT TO PARTICIPATE IN RESEARCH

The ethical and legal communities face three principal concerns when searching for ways in which those elderly adults who are experiencing diminishing capacity as a result of cognitive aging[42] may participate in medical research. The first concern relates to the ability of the elderly to consent; this can be compromised by a variety of practical problems, for which there may be equally practical solutions. Surprisingly, ways of making the potential harms and benefits of a protocol understandable to prospective patients who are cognitively impaired are only just now being developed. The second concern involves devising methods of responding to unpredictable, fluctuating and declining cognitive abilities and the presumptions or determinations of capacity to consent that are associated with these abilities. The third concern arises in those situations where there is either a serious doubt concerning the capacity of the elderly individual to consent, or where it is clear that the individual no longer has that capacity; in these circumstances, it becomes necessary to address the whole question of substitute decision-making and the potential role of the care-giver in the consent process.

### *Practical Problems and Solutions*

It is now becoming evident that efforts to satisfy the requirements of informed consent with elderly subjects can teach researchers a great deal about the relationship between the capacity for self-determination and the more mechanical and mundane abilities of those who are asked to provide their consent. For the elderly, the appreciation of the nature of independent decision-making and the resulting capacity to consent may endure despite a decline in other abilities such as perception or memory.[43] Adults tend to consider the mechanics of the physical abilities they once took for granted only when they begin to experience difficulties with them or when they sense that new limitations are being imposed on their freedom of movement; in just the same way, careful attention to the ethical and legal aspects of the consent process in relation to the elderly has only recently been responsible for researchers acquiring more knowledge about the strictly mechanical aspects of obtaining informed consent.

---

[42] Either normal or pathological aging (*i.e.* as a result of Alzheimer's Dementia).
[43] See Weisstub, *supra* note 28 at 28–29.

Several studies assessing the effectiveness of the consent process with respect to informing the patient have suggested that subjects "may have inadequate comprehension and memory of the materials even though they sign the consent document and state that they understand the information presented".[44] However, "even though comprehension is assumed to be necessary for a rational decision about research participation, most evaluations have actually measured memory (e.g. immediate or delayed recall) . . .". despite the fact that "poor memory or recall does not necessarily imply poor comprehension".[45] Moreover, it has been pointed out that "(t)he fact that individuals forget information following a decision does not mean that they did not use it in their decision making".[46] Indeed, the confusion caused by failing memory could be clarified by encouraging participants in research to retain and refer to the information sheets throughout the duration of the study in question.[47]

Another study of consent with the elderly made the simple observation that many elderly persons "have slower reaction time and require more time to process complex information" and are more likely than others to have hearing or vision impairments,[48] and they may be more prone to experiencing comprehension problems arising from the questionable readability of consent forms.[49] The use of large-type forms, written in simplified language and presented in information sessions tailored to suit the time requirements of the prospective participant, provides an eminently practical method of maximizing the ability of individuals to engage meaningfully in the decision to participate in research. Furthermore, proper practical adjustments that enable the involvement of those who are experiencing diminishing capacity in relation to the consent process should no longer be considered optional.

In the last decade, researchers in California extensively examined the health care decision-making process as it affects both elderly and developmentally disabled subjects.[50] Tymchuk argued that the current understanding of the process of informed consent is flawed. Consent is currently conceived of as an "instantaneous

---

[44] H.A. Taub, "Comprehension of Informed Consent for Research: Issues and Directions for Future Study" (1986) 8:6 IRB 7.

[45] See H.A. Taub, G.E. Kline & M.T. Baker, "The Elderly and Informed Consent: Effects of Vocabulary Level and Corrected Feedback" (1981) 7 Experimental Aging Research 137. See also A. Meisel & L.H. Roth, "Toward an Informed Discussion of Informed Consent: A Review and Critique of the Empirical Studies" (1983) 25 Arizona Law Review 388; Weisstub, *supra* note 28 at 167–68.

[46] See Stanley, *supra* note 41 at 555.

[47] See Taub, *supra* note 44 at 8.

[48] R.L. Schwartz, "Informed Consent to Participation in Medical Research Employing Elderly Human Subjects" (1981) 75 Journal of Contemporary Health Law & Policy 15 at 24, 55.

[49] See Taub, *supra* note 44.

[50] A.J. Tymchuk, J.G. Ouslander & N. Rader, "Informing the Elderly: A Comparison of Four Methods" (1986) 34 Journal of the American Geriatric Society 818; A. J. Tymchuk *et al.*, "Medical Decision Making Among Elderly People in Long Term Care" (1988) 28 The Gerontologist Supplement 59; A. Tymchuk, L. Andron & B. Rahbar, "Effective Decision-making/Problem-solving Training with Mothers who have Mental Retardation" (1988) 92 American Journal of Mental Retardation 24; A. Tymchuk, "An Alternative Conceptualization of Informed Consent with People who are Elderly" (1992) 18 Educational Gerontology 135.

event"; this conception fails to consider the underlying "information processing framework"[51] and it presumes that "one method would be satisfactory for all people". Revision of the current conception is urgent because

> ... without such adaptations not only would there be a great likelihood that these populations would not understand what it is they are agreeing or disagreeing to but also limits would erroneously be placed upon our understanding of their capabilities. Both of these results would invariably lead to an abrogation of the principles of self-determination and autonomy on which the concept of informed consent is based.[52]

Tymchuk recommended a model for consent based on a cognitive structure revolving around "three distinct phases: input, assimilation and output".[53] By "operationalizing" each of the phases, it is possible to identify and attend to any difficulties experienced by the individual subject. For example, Tymchuk and his group found that elderly patients in long-term care facilities read on average at a grade-five level and that, by simplifying the information presented and questioning the participants after receiving it, the material was much better understood. In addition, one of these studies supported the contention that a storybook format incorporating large type would contribute positively to the acquisition of an overall understanding of the proposed medical procedure and its possible consequences.[54] One valuable study demonstrated that the exposure of nursing home residents to a simplified patient's Bill of Rights significantly improved the understanding, and application, of the individual's rights in the consent process.[55]

As indicated by these studies and by the ongoing work of specialized teams such as that of the Competency Clinic for the Elderly at Baycrest Centre,[56] there is much to be learned about the effect of aging on cognitive processes and the relationship of cognitive aging to the capacity to consent. In 1990, the *Enquiry on Mental Competency* noted that a precise notion of psychiatric "normality" in the elderly was poorly understood because "cognition is usually measured in reference to the prototypic middle-aged adult". Accordingly, it was recommended that cognitive parameters not be weighed as heavily in assessing decisional capacity, and that self-functioning and sedimented life preferences[57] be accorded increased significance in

---

[51] See Tymchuck, *ibid.* at 138.

[52] *Ibid.* at 136.

[53] This structure was derived from F. Craik & R. Lockhart, "Levels of Processing: A Framework for Memory Research" (1972) 11 Journal of Verbal Learning and Verbal Behaviour 671.

[54] See Tymchuk, Ouslander & Rader, *supra* note 50; Tymchuk *et al.*, *supra* note 50.

[55] The results of which were published in Tymchuk, Ouslander & Rader, *ibid.* See also D.M. High *et al.*, "Guidelines for Addressing Ethical and Legal Issues in Alzheimer Disease Research: A Position Paper" (1994) 8:4 Alzheimer Disease and Associated Disorders 66 at 70–71 for a discussion of the importance of communication in obtaining the informed consent of elderly persons.

[56] See M. Silberfeld *et al.*, "A Competency Clinic at Baycrest Centre" (1989) 10 Advocates Quarterly 23; M. Silberfeld, "The Mentally Incompetent Patient: A Perspective from the Competency Clinic" (1990) 11 Health Law in Canada 33.

[57] See Weisstub, *supra* note 28 at 161, explained the nature and significance of "sedimented life preferences" as follows: "Sedimented life preferences are the established patterns of behaviour and choices of the geriatric person. Choices that comport with an individual's life preferences should be recognized in the absence of serious cognitive or functional deficits".

assessing whether the actions of an elderly person reflected a capacity for self-determination. In sum, further research is likely to yield valuable insight into what is now coming to be regarded as a complex relationship between cognition and self-determination; indubitably, elderly individuals, whose abilities in both areas[58] once passed unchallenged, are likely to be both valuable participants in, and direct beneficiaries of, such research.[59]

### Generational and Cultural Factors

"*Research* can be an emotionally laden term that frequently causes mistrust of the researcher's objectives and arouses concern about the well-being of the patient".[60] In addition, generational differences tend to result in the elderly's being less accustomed to the formal consent procedures applicable to minimally invasive procedures and to the protocols employed in the highly regulated fashion of today's research. Accordingly, images of the high-risk, poorly regulated research of past decades may be conjured up by any reference to "medical research", and the resulting anxiety may be increased by the formality of a written consent procedure. In fact, it is only to be expected that those who were part of a generation that did not have the benefit of the requirement of informed consent, even for highly invasive or risky treatment decisions, would have difficulty grasping the fact that extensive information is being provided to them with a view to gaining their consent to a minimally invasive procedure such as, for example, venepuncture. Although misgivings such as these are likely to change only with time, it is also probable that much assistance may be derived from promoting efforts to educate the general public as to the nature and significance of currently-required safeguards in research and the associated process of obtaining informed consent.[61]

Special attention should also be given to particular cultural backgrounds of certain elderly communities. In her letter to the Editor of the *New England Journal of Medicine*, Barbara Mishkin, in commenting on one study of the feasibility of obtaining informed consent "by proxy", noted that:

> all the foreign-born proxies for residents in the Jewish nursing home were from Eastern Europe, and that of those, only 18% consented to their relatives' participation in the research .... Surprisingly, the authors provided no further analysis of the extent to which

---

[58] And whose autonomous preferences are often remembered by family or otherwise documented.

[59] An elderly person's sense of well-being may be impaired if an opportunity to participate in the consent process is refused. Accordingly, it is crucial for researchers to be sensitive to the differential requirements of the elderly, particularly insofar as the methods by which informed consent is obtained are viewed as a tangible demonstration of respect for their individual autonomy. It is also important to recognize the differences in the nature and extent of the dignitary harms experienced by members of different special populations whenever there is a failure to respect their autonomy.

[60] E. Schiaffino-Purvis, "To the Editor" (1987) 316 New England Journal of Medicine 1029.

[61] M.S. Brod & R.I. Feinbloom, "Feasibility and Efficacy of Verbal Consents" (1990) 12 Research on Aging 364. See Eisch *et al.* who found that "signing a form was more threatening for some of the participants than acutally agreeing to the research". J.S. Eisch *et al.*, "Issues in Implementing Clinical Research in Nursing Home Settings" (1991) 22 Journal of the New York State Nurses Association 18 at 19.

country of origin or sociocultural background affected attitudes toward biochemical research. It is equally surprising that no one commented on the fact that most . . . [were likely to be] holocaust survivors who understandably may have strong negative feelings about human experimentation. Had I been a member of the institutional review board that reviewed this protocol, I would have urged recruitment from a different facility, in part because of the possibility that immigrant Jewish families might be offended by the proposed research.[62]

Similarly, other groups, based either upon a perceived physical vulnerability due to a trauma, or because of an espoused religious belief system which creates a hyper-sensitivity to potential violations, will differ with respect to levels of resistance to research interventions, even when there is a calculable benefit to the group itself or in its relation to a wider population. Owing to the impact of generational and cultural factors, the relationship between informed consent and autonomy is therefore a complex one. While a sense of autonomy may be critical to the well-being of many elderly persons, demonstrating respect for that autonomy may not always require a strict adherence to the current requirements of informed consent: the precise procedures to be followed may vary in light of the particular cultural or generational background of the elderly subjects concerned. This is by no means to argue for dispensing with informed consent in the absence of a participant's objections. Rather, it is to point out that many elderly persons are accustomed to a relationship with health care practitioners that involves a greater degree of trust and a lesser degree of information than is currently desirable; they are more comfortable delegating the sometimes complex assessment of small risks of harm or potential benefit to a family member or their physician than they are in reviewing the information in its entirety. They too should be respected for this choice. In other words, sensitive consideration of the generational and cultural factors affecting the participation of the elderly in the consent process reminds us that informed consent is not an end in itself, but rather a means to promote the sense of autonomy and self-determination in those individuals asked to provide it. The requirements of informed consent should be structured to achieve that end, and not just to meet purely abstract specifications.

### *Capacity: The Problem of Uncertainty and the Use of Presumptions*

Uncertainty with respect to capacity to consent to participate in research is common in elderly persons for at least three reasons. First, the decision-making capacity of an elderly person may be in a process of gradual or rapid decline, or it may be subject to fluctuation as a result of a variety of environmental, emotional or even pharmacological reasons. Accordingly, elderly persons may remain in a state of questionable capacity for an extensive period. Second, a variety of perceptual and cognitive faculties only peripherally related to a sense of self-determination among the elderly may be subject to decline in ways that mask the elderly's underlying

---

[62] B. Mishkin, "To the Editor" (1987) 316 New England Journal of Medicine 1030, commenting on J.H. Warren *et al.*, "Informed Consent by Proxy. An Issue in Research with Elderly Patients" (1986) 315 New England Journal of Medicine 1125.

ability to consent. Finally, as the decision to participate in research is not generally the kind of decision that forms part of an individual's daily routine, it may be difficult to determine the extent to which the elderly are capable of making this kind of decision through observation of their abilities in other areas.[63]

The problem of uncertainty with respect to capacity is common to health care decision-making involving members of several special populations. In the past, this problem has been resolved through the use of presumptions. Both presumptions and determinations are "socio-legal" in nature;[64] in other words, they are very often a function of society's understanding of the nature and difficulty of the decision in question as well as society's appreciation of the consequences of permitting an individual of uncertain capacity to make such a decision.[65]

The interests of society with respect to the treatment decisions of other vulnerable populations may vary as the balance shifts from emphasizing the concern to ensure a high standard of health care — regardless of the effect a presumption of incapacity may have on the individual's autonomy — to promoting a concern to ensure respect for that autonomy. In addition, the current integration of the underlying principles of personal autonomy, "best interests" and community concerns favours individual autonomy.[66] The result of this integration is that the presumption of incapacity, even in its explicitly rebuttable form, has been eliminated, for example, in current Ontario legislation.[67]

In the context of participation in research, where the underlying principles of "best interests" and community concerns may be somewhat at odds with one another, the autonomy principle takes on an even greater significance. It is

---

[63] One significant factor favouring decision-specific capacity noted in the *Enquiry on Mental Competency* is the difference in the "situational parameters" of different kinds of decisions. See Weisstub, *supra* note 28 at 81.

[64] "Although much studied in the biological and behavioral sciences, the term competency is so inextricably linked to social norms that it has become a value-laden concept serving the socio-utilitarian ends of assuring conformity in society".*Ibid.* at 32–33.

[65] Children, for example, have been historically regarded to be incapable of making treatment decisions, regardless of the empirical data either supporting or discounting the incapacity of any individual child in particular. The prime consequence of presuming young children incapable is that treatment decisions will be made by their physicians and those legally responsible for their care.

[66] The *Final Report* argued the point as follows:

> Practical difficulties aside, the integration of personal autonomy, "best interests" and community concerns must reflect the legal and philosophical context in which it is being performed. Since there is not, and cannot be, an objectively "correct" specification of this integration, one must attempt to structure the testing of decision making capacity in such a manner that the applicable social values, as decided through the appropriate democratic processes, are incorporated and reflected in both the substance and procedure of the test.
>
> At present, although this integration favours individual autonomy and, therefore, requires that presumptions, burdens, etc. favour the protection of individual rights, clear evidence of functional incapacity will rebut the presumptions and facilitate an intervention to protect the "best interests" of the individual in question. The entire testing process must reflect the autonomy and "best interests" principles while conforming to social needs for proportionality, administrative simplicity and relevancy of decision making.

*Ibid.* at 54–55.

[67] *Health Care and Consent Act, 1996*, S.O. 1996, c. 2.

increasingly accepted that the involvement of persons in non-therapeutic research cannot be justified on the basis of their "best interests", except insofar as it reflects respect for their autonomy. Accordingly, in some of these situations where capacity may be in question, society may put autonomy aside, and presume individuals to be incompetent so as to protect them from participation in research. This presumption, however, may not be necessary in the current context of comprehensive ethical review of protocols. Indeed, as participation by adults in research is required to be a purely voluntary activity, and since either the lack of a decision or a decision not to participate would exclude them from protocols, a presumption of incapacity would serve only to provide greater assurance of exclusion.[68] This virtually absolute preclusion of participation in research is problematic insofar as it may conflict with the discernible desires of the elderly. Thus, despite the fact that we are caught between "too readily accepting a person's decision to participate and too readily rejecting it (because both can constitute a wrong to the person)",[69] there are strong arguments against the use of presumptions of capacity or incapacity to resolve the problem.

### Capacity Determinations — Distinguishing the Tests of Capacity and Reasonableness

There remain many concerns that the continuing presumption of capacity may result in the participation in experimentation of those who are not capable of consenting. Therefore, it becomes necessary to address the difficult issues surrounding the determination of capacity. Having already stated that capacity determinations may themselves have a deleterious effect on an elderly person's self-esteem, the questions of when a determination should be sought, who should make it, and on what basis it should be made, are all important considerations in this regard.[70]

Accordingly, the nature of decision-making must be examined in a variety of decisional contexts. In the context of research, it must first be acknowledged that the procedure is not proposed for the benefit of the individual but rather for research purposes. Secondly, the question of voluntariness is of particular significance. In considering a method for assessing capacity, the functional abilities of the individual should be the basis for the assessment of capacity, rather than the individual's status or the outcome of the decision-making process.[71]

On this note, academics and jurists have been slow to meet the challenge presented by the need to distinguish the assessment of capacity from the

---

[68] That is, if the elderly were presumed incapable, they would have to both volunteer to participate in research and demonstrate their capacity to do so. It is not clear that risk levels prevalent in research currently justify a level of caution that would require challenging their capacity, and hence their autonomy, in order to permit them to be enrolled in a protocol.

[69] See K.C. Glass & M.A. Somerville, "Informed Consent to Medical Research on Persons with Alzheimer's Disease: Ethical and Legal Parameters" in J.M. Berg, H. Karlinsky, & F.H. Lowy, eds., *Alzheimer's Disease Research* (Toronto: Thomson Professional Publishing, 1991) at 38–9.

[70] See generally Weisstub at 28.

[71] *Ibid.* at 83.

reasonableness of the decision. For example, in a recent consideration of competence to consent to research with Alzheimer's patients, one writer contrasted the level of difficulty of health care decision-making in the context of treatment with that of research, on the basis of the kind of critical thinking required to engage in an independent assessment of the recommendations of a practitioner. It was noted that "only modest competence is needed to decide to follow one's doctor's advice" since "a prudent person would" do so; and that it would be wise to set a low "threshold of entry to indicated medical care" because to do otherwise "might compromise the well-being of people of borderline or uncertain competence".[72] However, "as the recommended treatment becomes more speculative and invasive, more prone to adverse side-effects and liable to prove unsuccessful, the balance of advantage between receiving and rejecting the treatment shifts, and deciding not to undertake it may require no more capacity than accepting it".[73]

Nevertheless, acquiescing to the course of action proposed by one's physician is generally less demanding on the capacity and competence of the patient than refusing to follow such advice. It is, therefore, critical to find ways to construct an environment in which the prospective participant in research will feel equally comfortable whether accepting or declining the request to enroll in a protocol.[74] With respect to the ease of doing "what a prudent person would," it is acknowledged on an evidential level that the more reasonable a decision appears, the easier it is for others to recognize it as a capable and voluntary decision. However, neither the difficulty of expressing one's desire to refuse the request of the researcher to participate, nor the difficulty of recognizing as competent a wish to do something other than that which a reasonable person would do, shed any light on the actual difficulty of the decision itself. In other words, conflating the nature and difficulty of the decision with the question of whether a socially acceptable outcome is achieved by acquiescing to the proposal of the researcher creates a problem; it glosses over the critical issues raised by the difficulty of making the actual decision involved and the capacity required to make it.

Capacity assessments, based on the reasonableness of the outcome and on the difficulty of distinguishing a decision that is idiosyncratic from one that is the product of incapacity or coercion, have been common, for example, in Canadian jurisprudence.[75] Case law has held that the information necessary to consent to research "may considerably exceed what needs to be given for therapeutic care

---

[72] See B.M. Dickens, "Substitute Consent to Participation of Persons with Alzheimer's Disease in Medical Research: Legal Issues" in J.M. Berg, H. Karlinsky & F.H. Lowy, eds., *Alzheimer's Disease Research* (Toronto: Thomson Professional Publishing, 1991) 60 at 62.

[73] *Ibid.* at 63.

[74] Morris, in his discussion of conflicts of interest, notes that "[t]he method of presentation, type of data chosen for discussion, level of enthusiasm in describing one alternative versus another, and other factors all may be used, knowingly or unknowingly, by the clinician to influence or coerce patients during recruitment for clinical trials". See J.C. Morris, "Conflicts of Interest: Research and Clinical Care" (1994) 8:4 Alzheimer Disease and Associated Disorders 49 at 52.

[75] See *Halushka* v. *University of Saskatchewan* (1965), 53 D.L.R. (2d) 436 (Sask.C.A.) [hereinafter *Halushka*]; *Weiss* v. *Solomon*, [1989] R.J.Q. 731 (C.S.).

alone, and it needs to be better understood . . .";[76] that no therapeutic privilege applies to relax the standard of fully informed consent;[77] and, consequently, that "a patient sufficiently competent to consent to therapy may not be sufficiently competent to decide on entering a research study . . .".[78] The question that needs to be asked, however, is whether it is more difficult *per se* to make the decision to participate in research than it is to make the decision to consent to treatment, or whether the possibility of undue influence simply makes it more difficult to recognize the voluntariness of the decision.

In distinguishing the considerations underlying the need to determine the capacity of an individual consenting to participate in research from one consenting to treatment, one writer discussed the problem of undue influence as follows:

> We must maintain a constant vigilance (or index of suspicion) that the subject is possibly not acting in the truest sense of autonomy and needs therefore our most careful protection. It may be easy enough to allow cognitively impaired subjects to agree to investigative procedures in the name of respecting their own autonomy, if that agreement furthers our own projects. [However,] "[r]easonableness" criteria have been used in situations of evaluating competence to consent to treatment, not to experimentation. We might, in certain cases, support a measure of competence when the patient's well-being is at stake but participation in research is more optional, and therefore we have no overriding paternalistic duty pressing us to question the "reasonableness" of the patient's decision.[79]

In other words, setting aside the question of the actual difficulty of the decision in both cases, the concern with respect to individual subjects' capacity to consent to treatment is ensuring that they receive adequate health care, whereas the concern in the context of participation in research is that they be protected from having their possible susceptibility to undue influence exploited.[80]

In the treatment context, there is significant social concern with respect to the consequences for the individual of an outright rejection of recommended treatment or even of a failure to make a decision to consent to that treatment. However, in the research context, where no direct benefit is intended and where the ethical review of protocols is, in part, intended to reduce the risk of harm, the social concerns would be twofold: first, with establishing a context in which the individual is genuinely free to consent or to refuse to participate; and second, with establishing a viable method for ensuring that the individual's participation in this voluntary activity represents an *affirmation* rather than a *violation* of their autonomy. Once again, the reasonableness of the patient's decision does not count so much as its voluntariness.

Jurists have recently begun to require direct evidence of functional incapacity and not just mere evidence of an "unreasonable" decision before they will consider displacing the presumption of capacity in a variety of contexts, including that of

---

[76] See Dickens, *supra* note 72 at 64.
[77] See *Halushka*, *supra* note 75.
[78] See Dickens, *supra* note 72.
[79] See C.K. Cassel, "Research on Senile Dementia of the Alzheimer's Type: Ethical Issues in Informed Consent" in V.L. Melnick & N.N. Dubler, eds., *Alzheimer's Dementia: Dilemmas in Clinical Research* (Clifton, N.J.: Humana Press, 1985) 99 at 103–105.
[80] The need to guard against the possibility of undue influence in this context is very similar to that which exists in the capacity to make a gift or a will.

treatment. The right to refuse treatment has not been held to be dependent on the ability to demonstrate the reasonableness of the decision; rather, it has been viewed as the logical correlative of the requirement of informed consent.

### *Evidencing Voluntariness*

The problem of *formally* evidencing voluntariness in a context in which there is the possibility of undue influence may be novel in the field of medical research, but it has arisen in several areas of law, such as the law of wills and estates. The challenge of evidencing voluntariness in the course of testamentary disposition is similar to that which arises in the process of obtaining consent to participate in research on the part of those who are considered to be (or seem to be) of uncertain capacity; one party receives a gratuitous benefit (*viz.* the subject's participation) in a situation in which it may not be possible to verify the donor's actual intentions directly.

Evidencing voluntariness in testamentary disposition is achieved most commonly through the use of witnesses to the signing of a will. It is worthwhile to employ this device in the context of obtaining consent for research from individuals who are of uncertain capacity. Family members frequently are able to assist in the consent process not only by ensuring that the prospective participant understands the procedure as completely as possible, but also by assisting them to articulate their wishes. It is not clear to what extent the understanding of the elderly individual's wishes by family members is based on their comprehension of the individual's contemporaneous expressions or their familiarity with the "sedimented life preferences"[81] of the individual; however, this is probably not important because the contemporaneous willingness to participate is a prerequisite for continued enrollment in experimentation regardless of any signed consent form or the attestation of any third party.

## SUBSTITUTE DECISION-MAKING AND THE ROLE OF THE FAMILY IN THE CONSENT PROCESS

Undoubtedly, there remains a great deal to learn about how we may best provide practical assistance to potential research subjects whose capacity may have been diminished by the aging process. Suitable adjustments in everything from the design of packages of information materials, through the conduct and approach of the individual seeking consent, to the nature of the setting and the role of those present, can assist potential participants whose interest in self-determination continues despite the loss of their ability to exercise it easily. Further, the practice of encouraging family members to witness the consent of elderly persons may achieve a great deal by enabling those of uncertain capacity to have their wishes respected without harming their dignity either through questioning their capacity or appearing to transfer their legal right to consent to someone else.

---

[81] See Weisstub, *supra* note 28.

Nevertheless, there are occasions upon which research is proposed in relation to subjects who are no longer able to articulate their attitudes toward medical research and participation. As much as researchers have been encouraged, and should continue to be encouraged, to devise alternatives to the involvement of subjects who cannot consent, there will remain occasions on which valuable research cannot be conducted without the participation of such subjects. On these occasions, if the research is to be conducted at all, and if respect for the autonomy of these individuals is to continue to be a requirement satisfied through the process of informed consent, the question must be posed as to how the wishes of these individuals may be represented in that process.

Research directives provide individuals with a way to document their wishes in advance.[82] Further, to the extent that they may stipulate or require the interpretive support of family members or other nominated persons, they encourage participants to discuss their wishes with others while they retain capacity to consent. Finally, they may provide comfort to those anticipating diminished capacity and increased dependence by giving them the opportunity to take positive steps for the future, and possibly to make a valuable contribution to medical advances in areas of significance to them.

However, in the absence of an advance research directive, there is bound to be considerable debate concerning the desirability of permitting family members to represent the wishes of an incapable person in the context of a request to enroll that person in a research project. For example, in the context of Alzheimer's Dementia and other illnesses in which it is suspected that there is a genetic element,[83] family members may have a keen interest in the advancement of medical understanding.[84]

In conferring a gratuitous benefit, it is essential that those who witness the voluntariness of the consent be disinterested parties (i.e. those who are not potential beneficiaries of the decision). The potential for a conflict of interest should disqualify one as a witness to the consent of the individual.[85] In response to this concern, the Task Force established by the U.S. National Institute on Aging included the following recommendation in their suggested guidelines for research in dementia:

> 8. Authorization by a "legally authorized representative"
>
> ... substitute decision-makers should possess the following characteristics:

---

[82] See also Chapter 11.

[83] For a discussion of the genetic basis of Alzheimer Disease and some related ethical issues, see S.G. Post, "Geriatrics, Ethics and Alzheimer Disease" (1994) 42 Journal of the American Geriatric Society 782.

[84] Moreover, there may be other sources of conflicts of interest on the part of a family member. For example, an overburdened caregiver may find it difficult to make an impartial decision about the participation of an incompetent family member in an experiment that involves keeping the latter in a medical facility for a fixed period; clearly, the prospect of a degree of welcome release from the incessant pressures associated with looking after the incompetent family member would inevitably place the caregiver member in a situation of conflict of interest.

[85] See Melnick *et al.*, *supra* note 35.

a. No evident or substantial conflict of interest that would be likely to lead to a decision contrary to the best interests of the patient.
b. An ability to participate in a vigorous, informed and conscientious manner in the decision.
c. An ability to remain a vigorous advocate of the incompetent's interest in maintaining control of decision-making throughout the course of the patient's participation in research.

Clearly, it may be difficult for devoted caregivers to avoid being offended by potential disqualification, and it is equally true that recognition of the potential benefit to a caregiver might be a strong motivation for competent persons to consent to participation in non-therapeutic research. However, the extension of the limits of permissible research to include those who cannot provide contemporaneous informed and voluntary consent can only be justified if it can be shown not to violate the interest in autonomy served by the requirements of informed consent. Since the person providing consent on behalf of the incapable individual must do so because of the difficulty of evidencing one's wishes created by incapacity, any factor that would cast doubt on the apparent disinterest of the witness would defeat the purpose of their participation in the consent process.

Despite the fact that questions have been raised about the reliability of family members accurately to represent the wishes of their incapable relatives, it is nevertheless not recommended that caregivers and family members be excluded from the consent process. Indeed, much research would be impractical in the absence of family support.[86] Rather, a method must be established for recognizing potential conflicts of interest and preventing family members from permitting participation in research that would be contrary to the interests of incapable subjects. The underlying societal considerations in research seem to favour retaining a duty of the researcher, or any other person participating in the consent process, to refuse to accept the consent of someone whom they believe has a conflict.[87]

## ETHICAL AND LEGAL PRINCIPLES

Very few of the international and national documents concerned with the regulation of biomedical experimentation contain provisions that are specifically concerned with ensuring the ethical conduct of experimentation with elderly subjects. However, some recent domestic and international ethical guidelines have given the

---

[86] American Psychiatric Association Task Force on Alzheimer's Disease, "Editorial: The Alzheimer's Disease Imperative — The Challenge for Psychiatry" (1988) 145 American Journal of Psychiatry 1550 at 1550–1551.

[87] A requirement that a family member representing the wishes of an incapable person make a declaration that there is no conflict of interest assists in two ways: first, by alerting them to the requirements for their eligibility to represent another's wishes, and, second, by ensuring that a potential conflict of interest does not lead to a decision to enroll a subject in a protocol whose interests would not be adequately served.

elderly some recognition as a distinct population.[88] For example, the World Medical Assembly in Hong Kong in 1989, with the publication of its *Declaration of Abuse of the Elderly*, contributed towards the recognition of the elderly as a population with special needs.[89] This declaration was a signal of a growing sensitivity both to the importance of conducting research with the elderly and to the principle that failure to respect the wishes of elderly participants should be considered a form of abuse.

In the United States, the guidelines from the Office for the Protection of Research Risks, although acknowledging that the DHHS regulations contain no specific provisions regarding research with elderly subjects state that:

> it is generally agreed . . . that the elderly are, as a group, heterogeneous and not usually in need of special protections, except in two circumstances: cognitive impairment and institutionalization. Under those conditions, the same considerations are applicable as with any other, nonelderly subject in the same circumstances.[90]

The OPRR also recognizes that the presence of elderly persons in institutions such as nursing homes or hospitals increases the chances of coercion and undue influence because of a lack of freedom. The *Guidebook* therefore recommends that research in these settings be avoided unless the involvement of the institutional population is necessary to the conduct of the research.[91]

---

[88] Specific mention of the elderly as a special population was made in the *MRC Guidelines*, which state that: "Research into disorders of the elderly pose special problems. Research into Alzheimer's disease, for instance, may require affected subjects to be exposed to uncomfortable and above-minimum-risk procedures. Subjects may not themselves benefit from results of individual studies, and they may lack competence to give consent". See Medical Research Council of Canada, *Guidelines on Research Involving Human Subjects* (Ottawa: Supply & Services Canada, 1987) at 31 [hereinafter *MRC Guidelines*]. Similarly, the *CIOMS Guidelines* recognize the special nature of Alzheimer's Dementia, which, of course, targets the elderly. See Council for International Organizations of Medical Sciences, in collaboration with the World Health Organization, *International Ethical Guidelines for Biomedical Research Involving Human Subjects* (Geneva: CIOMS, 1993) at 22. However, the *Guidelines* do not provide any additional safeguards aside from those which apply to persons suffering from behavioural or mental disorders. The Law Commission (UK), however, throughout its analysis of legal reform with respect to mentally incapacitated adults, included "elderly people with mental infirmity" among the four categories of adults identified amongst whom incapacity may occur. See The Law Commission, *Mentally Incapacitated Adults and Decision-Making: An Overview (Consultation Paper No. 119)* (London: HMSO, 1991); The Law Commission, *Mentally Incapacitated Adults and Decision-Making: A New Jurisdiction (Consultation Paper No. 128)* (London: HMSO, 1992); The Law Commission, *Mentally Incapacitated Adults and Decision-Making: Medical Treatment and Research (Consultation Paper No. 129)* (London: HMSO, 1993); The Law Commission, *Mentally Incapacitated and Other Vulnerable Adults: Public Law Protection (Consultation Paper No. 130)* (London: HMSO, 1993); The Law Commission, *Mental Incapacity* (London: HMSO, 1995).

[89] See "Hong Kong World Medical Assembly — Full Report" (1990) 37 World Medical Journal 4 at 13. For further discussion of the growing concern over the problem of elder abuse, see Council on Scientific Affairs, "Elder Abuse and Neglect" (1987) 257 Journal of the American Medical Association 966.

[90] See Office for Protection from Research Risks, National Institutes of Health, *Protecting Human Subjects: Institutional Review Board Guidebook* (Washington, D.C.: U.S. Government Printing Office, 1993) at 6–47.

[91] *Ibid.* at 6–48. The problem of obtaining a voluntary consent from elderly persons who are also institutionalized was also recognized by the National Health and Medical Research Council in Australia. The Council clearly states that the elderly constitute a special group in a dependent relationship. See National Health and Medical Research Council, *Statement on Human Experimentation and Supplementary Notes* (Canberra: NHMRC, 1992) at 23.

The *Guidebook* also states that:

> Despite [certain] difficulties, the inclusion of older persons in the research enterprise is important. IRBs should ensure that where they are excluded or treated specially, older subjects are in need of protection and are not the object of disdain, stereotyping or paternalism. Together, researchers and the IRB should enable older persons to share in the benefits and burdens of research.
>
> The use of age as the criterion of ability to consent and therefore participate in research is not valid. Studies have shown that education, health status, and inadequate communication about the research rather than age contribute to the lack of comprehension and recall. While it is recognized that memory may be a problem for some elderly subjects (thus putting into question their ability to provide continuing consent), the question for the IRB is whether, despite some impairment to competence, subjects can make reasonable choices.[92]

The conclusion to be drawn is that the elderly may be a population which merits special attention before participating in research to ensure that their consent is informed and free.

Finally, provisions concerned with whether the values of a competent adult regarding participation in research are respected when the adult becomes incompetent, are of particular relevance to the elderly, who may suffer from degenerative diseases, and therefore have a potentially decreasing capacity to give consent to participate in research. The MRC recognized that treatment involving mentally-incompetent adults must reflect their:

> mature personalities and, in particular, avoid any procedure which the subjects would probably have refused, were they still fully competent. The ability of all incompetent potential subjects to exercise choice must be maximized, and their dignity must not be compromised by exposing them to procedures which demean them or exacerbate their dependency.[93]

The above observation is especially applicable to the elderly, given the fact that, as mature adults, they have had ample time to develop their personal values and beliefs upon which a substitute decision, if necessary, could be based. Hence, various guidelines, including the present legislation in Ontario,[94] emphasize the need for substitute decision-makers to base their decisions on the previously expressed values and beliefs, thereby preserving the right of individuals to project autonomous

---

[92] See *OPRR Guidebook, supra* note 90.

[93] *MRC Guidelines, supra* note 88 at 30. The revised *Code of Ethical Conduct for Research Involving Humans* reaffirmed this principle, namely in its assertion that although there might be concerns with respect to competence in the case of elderly persons, these concerns should be addressed in their own right and not on the basis of age. See Tri-Council Working Group, *Code of Ethical Conduct for Research Involving Humans* (July, 1997), at Part 2, p. 36. The Tri-Council consists of the Medical Research Council of Canada, the Natural Sciences and Engineering Research Council of Canada, and the Social Sciences and Humanities Research Council of Canada.

[94] *Substitute Decisions Act, 1992*, S.O. 1992, c. 30, s. 66.

decisions into a future time where they may be mentally incompetent. Recommendations often suggest that this be accomplished through research directives.[95]

## CONCLUSION

The elderly are difficult to categorize as a distinct population — possibly because society has not until recently recognized them as an identifiable group. It is apparent, however, that a growing recognition of the needs of the elderly has emerged in response to changing demographics, an increasingly outspoken elderly population, and an improved awareness of the tragedy of Alzheimer's Dementia. These factors have all served to focus attention on the importance of medical research concerned with the health needs of the elderly.

One significant characteristic of the elderly in the context of health care decision-making is that the elderly might possess *uncertain and fluctuating capacity caused by diminishing cognitive abilities*. Because declining capacity in the elderly generally follows several decades of autonomous functioning, cognitive deficits often mask both a residual desire, and a continued capacity for, self-determination that should be respected. Accordingly, every effort should be made to eliminate, or at least reduce, practical impediments to the participation of the elderly in the process of providing informed consent. Similarly, since many members of the current generation of elderly persons have developed their perspectives as to the nature of an appropriate relationship between individuals and health-care practitioners in contexts that differ from that of the current concern for informed consent, deference should be shown to their standards of respect for persons in health-care decision-making.

While presumptions of capacity to consent to participation in research may place some elderly participants at risk, presumptions of incapacity nevertheless provide protection that is largely unwarranted and often offensive. Formal determinations of capacity are, in the same way, unduly intrusive. By distinguishing measures of capacity that focus on the decision-making process from those concerned with the reasonableness of the outcome, capacity to consent to research participation[96] emerges as the ability to appreciate that the proposed procedure is not intended to benefit oneself directly and that the giving or withholding of consent to participate should have no effect on one's current or future entitlement to health care. Thus, the capacity to consent to research essentially consists of the ability to appreciate it as an entirely voluntary and optional activity. Because the decision to participate in

---

[95] For example, the CIOMS Guidelines, *supra* note 88 at 23, suggest that persons who foresee cognitive impairment might make use of research directives — either by stating acceptable conditions for participation in research, or by designating a substitute decision-maker. Similarly, the guidelines presented by the task force sponsored by the National Institute on Aging encouraged the use of documentary methods of establishing a person's wishes, such as the use of "durable powers of attorney", and the conduct of long-range and long-term protocols so as to permit the enrollment of persons at a time when they were competent. See Melnick *et al.*, *supra* note 35.

[96] As distinct from the capacity to consent to treatment with which it shares the requirement that one be able to understand the nature of the proposed procedure and its relationship to one's physical condition.

research involves the conferring of a gratuitous benefit by one whose capacity to do so may be uncertain, the inherent challenges are to eliminate the possibility of undue influences and to ensure the provision of reliable evidence of the voluntariness of the decision. In the case of a participant of uncertain or borderline capacity, both can be accomplished through the witnessing of the consent of the participant by a person familiar with the individual and independent of the researcher. The involvement of a witness in the consent process in the event of any uncertainty with regard to capacity obviates the need for either presumptions or determinations of capacity and the threat they may pose to the dignity of elderly persons.

Although many persons retain capacity or borderline capacity throughout their lives, many others experience a period of incapacity at the end of their lives. In some cases, especially when this incapacity is a result of a degenerative condition such as Alzheimer's Dementia, these individuals or subsequent generations of elderly persons are entitled to receive the benefits of medical advances that may arise in the future as a consequence of their participation in research. If the commitment to informed consent is to be sustained, this research can only go forward if provision is made for substitute consent.

This should be accomplished through a legislative framework for substitute consent that may be applied in a variety of areas including that of consent to treatment. Such legislation should be, by and large, applicable to research participation and provisions made for encouraging the use of documentary methods, such as continuing powers of attorney, as a method of establishing the prior capable wishes of incapable persons, would be most helpful in determining the continuing wishes of those incapable of articulating them. Also, mechanisms should be enunciated for determining the substance of the incapable person's wishes and for selecting a person to represent those wishes should also be applicable to the decision whether or not to participate in research.[97]

However, because research participation is a voluntary activity, unrelated to the therapeutic best interests of the individual, the legislative provisions relating to the therapeutic need for, or likely effect of, a proposed procedure are clearly inappropriate when considering the incapable person's wishes regarding their participation in a research protocol. Furthermore, because the possibility of conflicts of interest can extend to those who might be called on to consent on behalf of the incapable person, those with potential conflicts should be regarded as being ineligible to provide consent unless the factors that might affect their decision will not lead them to consent to research participation on the part of the incapable person against that person's wishes. Researchers should have a continuing obligation to decline the consent of those they feel ought be disqualified on this basis. A declaration to be signed by family members acting as substitute decision-makers, for example, should include an appropriate statement declaring that they are free from any conflict of interest.

---

[97] See also Chapter 8.

The elderly are coming to be recognized as a special population with health care needs requiring research with human participants. There is growing concern for the specific health care needs of the elderly and an increasing degree of recognition that respect for their dignity requires the establishment of mechanisms for determining their continuing wishes with respect to research participation.

## SUMMARY

Geriatrics is, comparatively speaking, not a very old medical specialty. In the past, illnesses suffered by older persons gained the attention of health professionals for purposes of cure or maintenance of comfort; it was rare for individuals to live to a "ripe old age". The comparative rarity of old age delayed the recognition of the existence of a group of older persons with physiological characteristics and health requirements that were distinct from those of the general population. Longevity is a new social reality that has obliged us to regard old age as a second likely period of dependency. Indeed, it is now becoming increasingly apparent to society as a whole, including the medical and legal communities, that age-based incapacity is likely to affect many people not only at the beginning, but also at the end, of their lives. In the light of these major considerations, we make the following recommendations:

(1) Non-therapeutic research should never be conducted on incompetent elderly subjects if it is possible to conduct the same research with competent subjects.

(2) Although treated as a special population, the elderly must nevertheless be treated as adults whose autonomy of decision-making is strictly respected *unless* there is clear and convincing evidence of incapacity to give or withhold consent to participation in non-therapeutic experimentation. For this reason, there should be a presumption of capacity.

(3) Where an elderly person is institutionalized in a hospital or nursing home, or where there is a reasonable doubt as to the person's ability to voluntarily consent to participate in research, the giving of consent should be witnessed by a family member or, if appropriate, a close friend. Where research is conducted in an institutional setting, the prospective subjects should be given the opportunity to discuss the relevant issues with an independent third party. There should also be a reduction of incentives that might cause a prospective subject to become involved in research where others would not.

(4) Where an elderly person has been placed under the guardianship of an attorney for personal care, or where there is clear evidence of incapacity to give consent to participate in research, the consent must be obtained from the appropriate substitute decision-maker.

(5) Those individuals qualified by their relationship to an incapable person to give a substituted consent to participate in research should be required to provide a statement that certifies the absence of any conflict that would be likely to lead to a decision contrary to the wishes of the incapable person, and

that they will assume responsibility for decision-making regarding continued participation throughout the protocol. A person unable to meet this requirement must be disqualified.

(6) Where substitute consent is given, it must be based, wherever possible, on the *previously expressed wishes, values or beliefs of the elderly person concerned* — assuming that these were expressed at a time when the latter was competent to do so.

(7) In no circumstances may non-therapeutic research be either commenced or continued where there is a clear objection on the part of the incompetent subject. Current objections must override any previous consent that may have been given to participation in such research.

(8) Elderly persons should be encouraged to give advance research directives, especially when a person has been diagnosed as being in the early stages of a disease that is known to cause cognitive impairment. These directives should express the wishes of the person concerned and should appoint a substitute decision-maker of their own choice.

(9) An objection on the part of the subject, or previously expressed wishes (such as an advance directive stating an unwillingness to participate in non-therapeutic experimentation), should override consent provided by a substitute decision-maker.

CHAPTER 21

# ETHICAL RESEARCH WITH VULNERABLE POPULATIONS: THE MENTALLY DISORDERED

JULIO ARBOLEDA-FLOREZ AND DAVID N. WEISSTUB

## THE MENTALLY DISORDERED AS A VULNERABLE POPULATION

The mentally disordered stand at the intersection of several social systems, particularly health and justice. Health care systems claim control over the mentally disordered on the premise that they suffer from an illness that deprives them of the ability to make proper decisions, entitling them to specialized care, treatment, and protection. The justice system, on the other hand, claims that the mentally disordered are in need of protection from themselves and even from those persons whose job it is in other systems to protect them. Even when the mentally disordered commit offences, the justice system considers that they should not be punished with the full force of the law. Health and justice are both geared to protect the mentally disordered; their aims are the same. Their methods, however, differ; for while health care can be understood as trying to protect the mentally disordered at the expense of their rights, justice can be seen as protecting their rights at the expense of their autonomy. In the quest to develop protections for mentally disordered persons, the legal system has introduced a variety of structures, including guardianship laws, regulation and liability of caregivers, and exemptions from criminal liability. However, these régimes are designed not only to protect the mentally disordered, but also, in certain cases, to protect society. Unfortunately, and as might be expected, these two antithetical objectives can lead to extremely difficult decisions. As a consequence of institutionalization, whether for short or long periods, the mentally disordered become highly vulnerable to exploitation and abuse. Both their dependency upon others and their general lack of freedom to use independent judgment raise doubts about the voluntariness of their decisions. Moreover, owing to the inherent nature of mental disorders, the cognitive capacity of affected individuals is frequently compromised, which affects their competence to make decisions in their best interests. These two factors place the mentally disordered among special populations who present specific problems in the context of human experimentation.

### *Defining "Mental Disorder"*

Before elaborating upon the legal and ethical issues raised by the conduct of non-therapeutic research with the mentally disordered, it is important first to identify what is meant by the term "mental disorder".

Mental disorders are not homogeneous. Indeed, many questions have been raised about the ways in which they could be diagnosed.[1] While the definition of this term has been an evolutionary process that has depended upon medical, social and legal developments,[2] it is also a contextual enterprise, varying according to the given circumstances. Traditionally, there has been a difference between legal and clinical definitions of the term.[3] While clinical definitions emphasize symptoms, disability or distress, and tend to consist of ambiguous generalities,[4] legal definitions tend to emphasize constructs such as incapacity and incompetence. Finally, although the presence of a "mental disorder" is the threshold requirement for any further legal decisions, it lacks a precise operational definition.[5]

---

[1] T. Szasz, *The Myth of Mental Illness: Foundations of a Theory of Personal Conduct* (New York: Hoeber-Harper, 1961).

[2] For more on this subject, refer to the English Court of Appeal decision *W.* v. *L.*, [1973] 3 All E.R. 884, and *Winterwerp* v. *The Netherlands* (1979), 2 E.H.R.R. 387; and *X.* v. *The United Kingdom* (1981), 4 E.H.R.R. 188; *Luberti* v. *Italy* (1984), 6 E.H.R.R. 440.

[3] J. Arboleda-Florez & M. Copithorne, *Mental Health Law and Practice* (Toronto: Carswell, 1994) at 1–42.

[4] J.M. Livermore, C.P. Malmquist & P.E. Meehl, "On the Justification for Civil Commitment" (1968) 117 University of Pennsylvania Law Review 75 at 80.

[5] For example, the Diagnostic and Statistical Manual of Mental Disorders defines "mental disorder" as:

> . . . a clinically significant behavioral or psychological syndrome or pattern that occurs in a person and that is associated with present distress (e.g., a painful symptom) or disability (i.e., impairment in one or more important areas of functioning) or with a significantly increased risk of suffering death, pain, disability, or an important loss of freedom. In addition, this syndrome or pattern must not be merely an expectable and culturally sanctioned response to a particular event, for example, the death of a loved one. Whatever its psychological cause, it must currently be considered a manifestation of a behavioral, psychological, or biological dysfunction in the individual. Neither deviant behavior (e.g., political, religious, or sexual) nor conflicts that are primarily between the individual and society are mental disorders unless the deviance or conflict is a symptom of a dysfunction in the individual, as described above.

See American Psychiatric Association, *Diagnostic and Statistical Manual of Mental Disorders*, 4th ed. (Washington, D.C.: A.P.A., 1994) [hereinafter *DSM IV*] at xxi–xxii.

This definition clearly covers the broad range of "developmental disorders", "organic mental disorders" (e.g. intoxications, "primary degenerative dementia of the Alzheimer type"), disorders resulting from injury (e.g. in children, for example, receiving a severe injury), "functional" mental conditions (e.g. schizophrenia, affective disorders), "neurosis" (e.g. anxiety, panic disorders, obsessive compulsive disorders), and personality disorders of every kind. The heterogeneity of individuals included in this definition translates into an array of management options including ambulatory care, short term acute admission, or long term institutionalization, voluntarily or otherwise. While for some chronically mentally disordered persons, permanent institutionalization may be an option, for others, suffering from seriously acute and severe conditions such as irreversible coma, permanent hospitalization is the only option.

Clinical definitions of "mental disorder" suffer from ambiguities. Because of overlapping categories and the absence of defined borders among conditions that affect different systems simultaneously, these definitions are useful for clinical purposes, but they lack legal functionality. On the other hand, legal definitions tend to differ according to whether they are for purposes of criminal proceedings or civil commitment.[6] In Canada, the *Uniform Mental Health Act*[7] (adopted by Uniform Law Conference representatives in 1987, but never implemented provincially as such), the term is defined as "a substantial disorder of thought, mood, perception, orientation or memory that grossly impairs judgment, behaviour, or capacity to recognize reality or ability to meet the ordinary demands of life".[8] This definition, apart from being more functional, and possibly broader than the *DSM IV* definition, is also more helpful to conceive legal standards that could be used to establish legal parameters for research involving the mentally disordered.

### *Capacity and the Mentally Disordered Person*

The key component of a functional definition of "mental disorder" is the presence of a mental or physical impairment which often seriously reduces the subject's capacity to provide consent.[9] In this sense, *capacity* refers to whether a person possesses the psychological and physiological foundations that constitute the wherewithal required to make decisions.[10] A determination of *capacity* is a *medical*

---

[6] The *Criminal Code*, while providing a circuitous definition of "mental disorder" as a "disease of the mind", uses the term in s. 16 (the "not-criminally responsible" clause) not as a definition but as a way to describe a person who may not be criminally responsible. See the *Criminal Code*, R.S.C. 1985, c. C–46. In Canadian mental health legislation, the definition of "mental disorder" is varied. For example, a leading Canadian legal text uses the term "mental disability" in a generic sense to include "any mental disorder or developmental disability, including mental illness, mental retardation and related developmental disabilities". See G.B. Robertson, *Mental Disability and the Law in Canada*, 2d ed. (Toronto: Carswell, 1994) at 2.

[7] See M.A. Gaudet, *Overview of Mental Health Legislation in Canada 1994* (Ottawa: Supply and Services Canada, 1994) at 19–20.

[8] The definition contained in the *Uniform Mental Health Act* (1987) has been adopted in several provinces, *in toto*, as in Alberta, or with modifications, as in Manitoba, Saskatchewan, New Brunswick, P.E.I., the Yukon and the Northwest Territories. In Ontario and Newfoundland, however, mental disorder is defined as "any disease or disability of the mind", while in British Columbia a "mentally ill person" is defined as "[a] person who is suffering from a disorder of the mind that seriously impairs his ability to react appropriately to his environment or to associate with others; and that requires medical treatment or makes care, supervision and control of the person necessary for his protection or for the protection of others". In Nova Scotia, a person suffering from a mental disorder is not referred to as having a mental disorder, but is rather defined as " 'an adult in need of protection' who is not receiving adequate care and attention in the premises where he resides and 'is incapable of caring adequately for himself by reason of . . . mental infirmity, and refuses, delays or is unable to make provision for his adequate care and attention' ". Finally, the legislation in Quebec does not provide a definition of "mental disorder". *Ibid.*

[9] "In a wide spectrum of disorders and conditions, mental or physical impairment seriously reduces the subject's capacity to give informed consent. Among these conditions are Alzheimer's disease, schizophrenia, manias with depression and suicidal behavior, types of aphasia, and states of partial or total coma". See J.C. Fletcher, F.W. Dommel Jr. & D.D. Cowell, "Consent to Research with Impaired Human Subjects" (1985) 7:6 IRB 1 at 1.

[10] "In contrast to competency, decision making capacity is defined as the ability to make an acceptable choice with respect to a specific decision". See D.N. Weisstub (Chair), *Enquiry on Mental Competency: Final Report* (Toronto: Queen's Printer, 1990) at 31.

act.[11] However, the factor taken into account most often when considering the mentally disordered is *competency*, which involves the ability to understand and appreciate the nature and consequences of one's decisions. *Competence* is a *legal categorization*.[12]

*Cognitive impairment*, in the context of experimentation, "... renders persons incompetent to make their own decisions to participate in research if it eliminates the person's ability to understand, make choices about, or communicate a decision regarding particular research".[13] Such a characterization, although often convenient, is overly broad, including the mentally ill, mentally handicapped, the demented, and even the unconscious. While it may often be convenient to use a broad all-inclusive description of "mental disorder", it is preferable, for the purposes of the present discussion, to use a more restrictive term that would only include, for example, dementias and psychotic disorders, distinguishing those persons afflicted with mental disorders from individuals with developmental disabilities. This distinction is made primarily on the basis that the latter have never possessed, nor are they ever likely to possess, to varying degrees, sufficient competence to make decisions regarding their own welfare. As a result, the developmentally disabled will be discussed separately.[14]

It is possible that mentally disordered persons may exhibit increasing or decreasing levels of competence, and so, they could best be characterized as having fluctuating periods of lucidity, during which times they may be competent to a certain degree.[15] The frequency of such periods and the corresponding quality of decision-making capacity may vary depending upon the type and severity of the mental disorder. For example, specific delusions could affect the decision at hand, and seriously depressed patients may not have the energy or mental focus required to understand the issues and may simply waive their rights. Under such

---

[11] "Medical" is understood as covering all medicine-related professions.

[12] "Competency" has been defined as "the capacity to function in a particular way, the ability to process and understand information and to make relevant, well-circumscribed decisions based on that understanding ... [but] in plain language, however, individuals are 'competent' if they are recognized as having achieved a certain level of expertise or ability in a particular area". See Weisstub (1990), *supra* note 10 at 26.

[13] American College of Physicians, "Position Paper on Cognitively Impaired Subjects" (1989) 111 Annals of Internal Medicine 842 at 843.

[14] See Chapter 24. In addition to the basic definitional problem associated with the mentally disordered, there is also the problem of overlap with other vulnerable groups. For example, mental disorders can be found among children, the elderly, prisoners, and the developmentally disabled. In such cases, the protective principles applicable to each population must be observed. There is also the practical problem of identifying an individual as "mentally disordered", especially in the case of the developmentally disabled because "[i]ndividuals with Mental Retardation have a prevalence of comorbid mental disorders that is estimated to be three to four times greater than in the general population". See *DSM–IV, supra* note 5 at 42. Owing to an individual's inability to communicate effectively, mental disorders may go undiagnosed in such persons, or they may be misdiagnosed.

[15] The problem of fluctuating periods of lucidity, accompanied by a general decline in decision-making capacity as a whole can be found in patients suffering from, for example, Alzheimer's disease. This can be a serious problem in lengthy research proposals where a patient may be competent at the commencement of the experiment, only to be incompetent by its conclusion. See E.W. Keyserlingk *et al.*, "Proposed Guidelines for the Participation of Persons with Dementia as Research Subjects" (1995) 38 Perspectives in Biology & Medicine 319 at 320, 326.

circumstances, it would be unwise for a researcher to assume that the patient had the proper capacity to make a decision.[16] Fluctuating degrees of competence will therefore have special consequences on the means by which the mental competency of such individuals is assessed and on the question of how consent to participation in research should be obtained.[17]

It should be noted that, apart from acute and serious conditions such as strokes, a person does not become totally incapacitated overnight; incapacities tend to develop gradually. A person may also be incapable of making decisions in certain situations. Therefore, although mentally disordered persons are, in general, capable of making decisions on their own behalf, because of a potentially fluctuating capacity that may affect them from time to time, their capacity to act autonomously may be unclear. These concerns raise the important question of whether, because of their restricted autonomy, the participation of the mentally disordered in non-therapeutic research should be circumscribed. The dilemma is complicated further by the potential problem of voluntariness.

### *Voluntariness*

Voluntariness invokes the question of whether a decision is made freely and whether it is truly representative of the individual's will. This issue is especially of concern where the subject has been civilly committed. Like prisoners, civilly committed individuals may be highly susceptible to inducements in exchange for participation in research.[18] Indeed, it has been observed that "... institutionalization can contribute to a person's vulnerability, increase dependency, foster peer pressure and a desire to please caregivers, and put in doubt the voluntariness of a resident's consent".[19]

Threats against the voluntariness of decision-making among the institutionalized mentally ill increase the importance of the question of whether civilly committed patients should be excluded automatically from participation in non-therapeutic research protocols.[20] Further, an individual may meet the criteria for commitment

---

[16] J. Arboleda-Florez, "*Reibl* v. *Hughes*: The Consent Issue" (1987) 32 Canadian Journal of Psychiatry 66 at 69.

[17] The powers given to guardians are also relevant in this context. See Chapter 8.

[18] See Chapter 6.

[19] See Keyserlingk *et al.*, *supra* note 15 at 322.

[20] From time to time, mentally disordered persons may be placed in institutions, and consequently present themselves as a convenient group for research purposes. While cases of flagrant abuse and exploitation of the mentally disordered have occurred in the experimental context, known cases are fortunately uncommon. However, since cognitively impaired individuals may not be in the best position to vocalize their objections, it is possible that abuses of the mentally disordered may be far more common than believed. Unlike the case of prisoners, where experimental abuses were more frequent, part of the reason for a decreased number of incidents of abuse of the mentally disordered in the experimental context may be due to the population's general unsuitability for experimentation. There are many problems with aspects of the standard experimental design (such as randomisation, the use of control groups and blinding) that arise when conducting research on the mentally disordered. For a discussion of such problems, see R.J. Dworkin, *Researching Persons with Mental Illness* (Newbury Park, N.Y.: Sage, 1992) at 50–55. Consequently, research utilizing the mentally disordered as subjects will tend to be limited to the study of mental disorders, rather than general disorders which can be tested on the general population.

and yet may still be competent to make decisions about treatment and participation in research.

The concern for voluntariness relates to the notion that the legal requirement of informed consent may not be an adequate safeguard to ensure voluntariness when applied to institutionalized populations. Even if the criteria for providing informed consent are met and a seemingly valid consent is obtained, mild inducements (from an "outside" perspective) may seem significant, almost coercive, in an institutional setting.[21] Consent, therefore, while it may appear informed, may not be voluntary.[22]

However, one should not exaggerate the impact of institutionalization. Institutions differ widely in quality of care and degree of independence, and may have been chosen by the patient before a serious mental condition, such as dementia, sets in. Institutionalization, in and of itself, should not be a limiting factor.[23] "People do not automatically become incapable of competent and voluntary consent the moment they enter a mental institution".[24] Rather, institutionalization should be viewed as a factor in assessing the legitimacy of a potential subject's consent to participate in an experiment.

## APPROACHING THE PROBLEM OF PROTECTION *VERSUS* AUTONOMY

The primary goals of the New York State Office of Mental Health regulations (1990) in relation to medical research on the mentally disordered have been described as follows:

(1) maximization of patient autonomy and control over participation in research;

(2) protection of the rights and welfare of this potentially vulnerable population; and,

(3) creation of an environment in which appropriate research may be conducted.[25]

---

[21] For example, an unspoken promise of better treatment in exchange for participation in research may serve as an unbalancing factor in the decision-making process of an institutionalized person.

[22] It is not only the physician who may affect voluntariness; close family members may have an influence as well, placing additional pressures on the patient to participate in research. The effects of family members on decision-making was pointedly described by Lord Donaldson as "subtle, insidious, pervasive and where especially powerful religious beliefs may be involved". His Lordship concluded that two factors were of crucial importance in assessing the effect of outside influence: the strength of the will of the patient ("one who is very tired, in pain or depressed will be much less able to resist having his or her will overborne than one who is rested, free from pain and cheerful"); and the relationship of the "persuader" to the patient. See *In Re T.*, [1992] 4 All E.R. 649 (C.A.) at 667.

[23] Also, by entirely banning research on institutionalized populations, one may preclude the study of phenomena which are *particular* to institutionalized settings, for example, the study of the effects of institutionalization, or the spread of contagious diseases in a mental health facility.

[24] Office for Protection from Research Risks, National Institutes of Health, *Protecting Human Subjects: Institutional Review Board Guidebook* (Washington, D.C.: U.S. Government Printing Office, 1993) at 6–28.

[25] See S.J. Delano & J.L. Zucker, "Protecting Mental Health Research Subjects Without Prohibiting Progress" (1994) 45 Hospital & Community Psychiatry 601 at 602. The Regulations referred to are: *New York State Office of Mental Health Regulations Governing Research*, 14 N.Y.C.R.R. (1990) § 527.10.

These statements recognize the need to balance two major ethical considerations: protecting individuals while respecting their autonomy. Enabling the mentally disordered to make decisions on their own behalf is but one consideration. Adequate safeguards designed to protect them from exploitation and abuse must also be enforced. Therefore, in addressing issues of competence and voluntariness as they relate to the problem of consent to participation of the mentally disordered in non-therapeutic research, any ethical, moral and legal discussion must also engage the conflicting ethical objectives of protecting individuals through interventions while respecting their autonomous rights.

### The Presumptions of Competency and Capacity

The common law presumption of competency extends to the mentally ill.[26] This decision places a responsibility on physicians to test the competency of a mentally disordered person prior to treatment.[27] The same considerations would apply, *mutatis mutandis*, to researchers, especially if it is assumed that the competency of mentally disordered persons may be compromised according to the fluctuating characteristics of their mental condition. Any system of enabling the mentally disordered must be geared toward capitalizing on those moments when they are lucid, and toward maximizing their decision-making capacity when not fully competent. They should be empowered to make decisions on their own behalf insofar as they are able to do so, as determined by a mental competency assessment, and they should be able to exercise their right to self-determination through the use of research directives and durable powers of attorney while competent. Indeed, direct consent by an individual while competent, including through the use of research directives, may provide the most suitable means by which mentally disordered persons would be allowed to participate in biomedical research, and should be sought whenever possible.

Mental incapacity can also occur along a gradient. A declaration of incapacity for a particular legal event does not necessarily imply that the person cannot have capacity for other events. For example:

> The more complicated cases with regard to assessing capacity are those in which the individual appears capable of making some decisions but not others . . . . [Thus,] [b]earing in mind that in all other cases individuals' capacities are marginal, the desirability of allowing individuals to exercise the greatest possible measure of control of their bodies and their lives demands that tests of decisional capacity accurately reflect the requirements of a particular decision.[28]

The importance of avoiding false presumptions regarding the mentally disordered was also noted as follows:

> It is commonly assumed that persons who are mentally ill have a significantly impaired ability to provide meaningful informed consent and, in the case of those who are

---

[26] For the sake of the present discussion, it is presumed that those individuals who are affected by mental disorders that are so severe as to render them entirely incompetent are in the minority.

[27] See Arboleda-Florez & Copithorne, *supra* note 3 at 5–15.

[28] See Weisstub (1990), *supra* note 10 at 47.

institutionalized, that they are further restricted in their ability to consent by the inherently coercive nature of their setting . . . neither of these propositions is necessarily true.[29]

The purpose of a presumption of capacity, rather than incapacity, to consent is to arrest the fallacy that persons with mental illness are automatically incapable of making life decisions.[30] Therefore, mentally disordered persons should be presumed competent to consent to participate in biomedical experimentation unless proven otherwise. By enabling the mentally disordered to make decisions on their own behalf, the effect of stigmatization on a population, already distinguished as "vulnerable", can be offset.

In addition to the right to be informed, the most important aspect of the principle of respect for persons is the respect for autonomy. This requires ". . . that those who are capable of deliberation about their personal choices should be treated with respect for their capacity for self-determination".[31] Naturally, enabling the mentally disordered to make their own decisions can only extend so far as those persons are capable of making those decisions. Limitations should be determined on a case-by-case basis, grounded in the assessment of an individual's capacity (or competence) to make decisions.

Enabling an individual requires the presumption that a person is both capable and competent to make decisions. Where these assumptions fail, as in the case of an assessment resulting in a declaration of incompetence, the person would be precluded from making the decision of whether or not to participate in an experiment. In such a situation, it may be necessary to consult a substitute decision-maker or to produce a research directive. It is important, therefore, in anticipation of future mental incapacity, to prepare advanced directives for research, and if the law permits, to appoint in advance a substitute decision-maker.[32] In addition, the guarantees extended to competent research subjects, including the rights to refuse to participate or to withdraw from a research protocol without jeopardizing their treatment or care, should also apply to incompetent subjects. Thus, while mentally disordered persons may be inducted into a research protocol through the consent of guardians or substitute decision-makers, the power of the surrogate should not be extended to override the incompetent patient's decision to refuse, or to withdraw at a later date.

In conclusion, there are two ethical considerations encapsulated in the fundamental principle of respect for persons: a respect for autonomy in competent individuals, and the protection of persons with diminished autonomy.[33] A protectionist solution, such as completely banning experimentation on the mentally

---

[29] See Delano & Zucker, *supra* note 25 at 601.

[30] In Canada, a presumption of incapacity could amount to discrimination under the *Canadian Charter of Rights and Freedoms*.

[31] Council for International Organizations of Medical Sciences, in collaboration with the World Health Organization, *International Ethical Guidelines for Biomedical Research Involving Human Subjects* (Geneva: CIOMS, 1993) at 10.

[32] For a detailed discussion of the ethical and legal considerations raised by advance directives for research and durable powers of attorney, see Chapter 11.

[33] See *CIOMS Guidelines*, *supra* note 31 at 10.

disordered would be incompatible with the primary consideration that individuals should be treated as autonomous agents. Rather, a flexible position that protects the incapacitated persons while at the same time permitting and promoting participation in research should be pursued. Such an approach is supported in the recent recommendations of the Law Commission of the United Kingdom,[34] and indeed should be adopted as the course for reform of the law on experimentation in other jurisdictions.

## ETHICAL AND LEGAL PRINCIPLES: INTERNATIONAL AND CANADIAN SOURCES

Developing a set of guidelines for the conduct of ethical research involving the mentally disordered requires a consideration of the various international and national declarations, codes, guidelines, and laws that address this issue. Upon completing such a survey, it becomes evident that a modern consensus exists which holds that experimentation on the mentally disordered should not be banned, but that adequate safeguards are essential.

With the *Nuremberg Code*,[35] the *International Covenant on Civil and Political Rights*[36] and the first incarnation of the *Declaration of Helsinki*,[37] there was a ban on research with individuals unable to provide consent. Over the years, however, this restrictive stance was relaxed, as can be observed in the evolution of the *Declaration of Helsinki* into its present form, and this relaxed stance is supported by the World Psychiatric Association,[38] the United Nations General Assembly's Resolution on *Principles for the Protection of Persons with Mental Illness and the Improvement of Health Care*,[39] the *Draft Bioethics Convention* presently being considered by the Council of Europe,[40] and the *International Ethical Guidelines for Biomedical Research Involving Human Subjects*, promulgated in 1993 by the Council for

---

[34] The Law Commission, *Mental Incapacity* (London: HMSO, 1995) at 100.

[35] The *Nuremberg Code* constituted part of the judgment resulting from *U.S.* v. *Karl Brandt et al., Trials of War Criminals Before the Nuremberg Military Tribunal Under Control Council Law No. 10.* (October 1946–April 1949).

[36] *International Covenant on Civil and Political Rights*, 19 December 1966, Can T.S. 1976 No. 47, 999 U.N.T.S. 171, 6 I.L.M. 368, art. 7.

[37] World Medical Association, *Declaration of Helsinki*. Adopted at the 18th World Medical Assembly in Helsinki in June 1964. Amended at the 29th World Medical Assembly in Tokyo in October 1975; the 35th World Medical Assembly in Venice in October 1983; and the 41st World Medical Assembly in Hong Kong in September 1989. [Reprinted in (1991) 19 Law, Medicine & Health Care 264.]

[38] See World Psychiatric Association, *Declaration of Hawaii* (1977), s. 4. The full text of the *Declaration of Hawaii* appears in D. Clarence & D. Blomquist, "From the *Oath of Hippocrates* to the *Declaration of Hawaii*" (1977) 4 Ethics in Science & Medicine 139. The *Declaration* has now been superseded by the *Declaration of Madrid*, ratified at the Congress of the World Psychiatric Association in Madrid, on August 21, 1996, which endorses a relaxed stance on the matter of consent.

[39] GA Res. 46/119 (1991).

[40] Steering Committee on Bioethics, *Draft Convention for the Protection of Human Rights and Dignity of the Human Being with regard to the Application of Biology and Medicine: Bioethics Convention and Explanatory Memorandum* (Strasbourg: Directorate of Legal Affairs, 1994); Council of Europe, Committee of Ministers, Recommendation No. R. 90 (3) (1990); and Parliamentary Assembly, Council of Europe, *Opinion No. 184 of the Parliamentary Assembly of the Council of Europe on the Draft Bioethics Convention* (Strasbourg: Steering Committee on Bioethics, 1995).

International Organizations of Medical Sciences (CIOMS) in collaboration with the World Health Organization.[41] However, despite the elaboration of ethical principles on an international level, surprisingly few jurisdictions, with France being a notable exception,[42] have legislation or regulations addressing the problem of non-therapeutic research involving the mentally disordered.

In the United States, the Belmont Commission examined the problem of experimentation on vulnerable populations between 1974 and 1978. These studies led to the publication of recommendations for research involving the mentally disordered,[43] but were criticized as failing to "specifically address research involving persons with mental illness", and thus "were never successfully incorporated in federal regulations, despite attempts to do so by the Department of Health and Human Services (DHHS)".[44] The DHHS regulations only allude to research involving persons with mental illness, and fail to provide specific guidelines.[45] Consequently, the American College of Physicians developed its own guidelines by analogy, based on the federal stance on research with children.[46] Nevertheless, although federal regulations do not address research with the mentally disordered explicitly, various states have enacted legislation or issued regulations to supplement those promulgated by the DHHS.[47]

In Canada, with the exception of Quebec,[48] there is a complete absence of statutory régimes regulating research with the mentally disordered. Rules for ethical conduct exist only in guidelines promulgated by professional associations.[49]

[41] See CIOMS, *supra* note 31 at 22.

[42] Arts. L.209–1 to L.209–18 *Code de la santé publique.* An Experimentation Bill was also recently under consideration in the Netherlands.

[43] This report stated that experimentation should be performed on the mentally infirm only upon satisfying two conditions: that there is minimal risk; and that the subject's approval, if at all possible, should be obtained, but the approval of a legal representative should be necessary in all cases. See National Commission for the Protection of Human Subjects of Biomedical and Behavioral Research, *Research Involving Those Institutionalized as Mentally Infirm: Report and Recommendations* (Bethesda, Md.: The Commission, 1977) at 44.

[44] See Delano & Zucker, *supra* note 25 at 601.

[45] See 45 C.F.R. 46 (1991) § 46.111 (7)(b).

[46] See American College of Physicians (1989), *supra* note 13 at 846.

[47] For example, in 1990, the New York State Office of Mental Health issued regulations to govern research involving patients in psychiatric facilities operated or licensed by the State. See *New York State Office of Mental Health Regulations Governing Research,* 14 N.Y.C.R.R. (1990) §527.10. However, it is important to note that, at the time of writing, these regulations do not have the force of law. On February 28, 1995, Judge Edward J. Greenfield of the Supreme Court of New York, New York County, held that the "OMH regulations for the conduct of human subject research were promulgated by the Commissioner of OMH beyond his authority and are thus invalid. Accordingly, . . . the regulations codified at 14 NYCRR 527.10 [are] invalid and unenforceable". See *T.D.* v. *New York State Office of Mental Health* 626 N.Y.S. (2d) 1015 (S.C. 1995). While the judgment was stayed upon the application by the OMH for appeal, the Court of Appeal, on August 3, 1995, vacated the stay. See *T.D.* v. *New York State Office of Mental Health,* M–3213, Index No. 5136/91 (N.Y. Ct. App. 1995) [unreported].

[48] Arts. 20–25 C.C.Q.

[49] See Medical Research Council of Canada, *Guidelines on Research Involving Human Subjects* (Ottawa: Supply and Services Canada, 1987) at 30–32. This document is currently under revision: see Tri-Council Working Group, *Code of Conduct for Research Involving Humans* (July, 1996). The *Code* superceded the *MRC Guidelines.* It emphasizes the need to designate an appropriate substitute decision-maker, but, like its predecessor, fails to provide a legal solution for the problems raised by such an alternative form of consent. Thus, the revised *Code* falls short in providing any impetus toward much needed legal reform.

Although professional guidelines play a significant role in education and encouraging ethical behaviour, the mere availability of ethical codes, from a legal point of view, is of limited worth as a regulatory authority. Such sources, even if adhered to, may be insufficient to establish a standard of professional conduct that would absolve researchers and their affiliated institutions from liability in the event of an injury or death resulting from research activity.[50] Therefore, in Canada, there exists a legal necessity to clarify the obligations of researchers and the rights of subjects in non-therapeutic experimentation, especially in cases involving vulnerable populations such as the mentally disordered. It is proposed that the solution lies in the establishment of a statutory régime in Canada's common law provinces, and in the expansion of the existing régime in Quebec. Indeed, there is a trend favouring external regulation, characterized by the enactment of statutes or regulations, rather than internal controls, such as through professional codes. Advocates for legislation include the Law Reform Commissions of Canada,[51] the United Kingdom[52] and Queensland, Australia,[53] along with, most recently, the *Enquiry on Research Ethics* submitted to the Government of Ontario.[54]

## ESTABLISHING SAFEGUARDS FOR RESEARCH ON THE MENTALLY DISORDERED

All experiments involving human subjects should be required to undergo review by an independent body, such as a Research Ethics Committee (REC).[55] The REC must ensure that the research itself is of significant value and scientifically valid, that is, conducted by scientifically qualified persons, conforms to generally accepted scientific principles, has been preceded by experimentation on animal or other models, and is accompanied by a thorough knowledge of the scientific literature. Members of vulnerable populations should not be employed as subjects where other, more competent, persons could be available. Hence, research with the mentally disordered must be restricted to that which furthers the understanding, prevention or alleviation of a problem directly related to a condition or circumstance affecting the subject, or the class to which the subject belongs.

Researchers have a positive duty to minimize risks of harm to subjects. This duty exists before an experiment, through a careful risk assessment and preparations to prevent injury; during an experiment, with ongoing support and monitoring, including the immediate termination of the protocol if there is probable cause to

---

[50] See Chapter 8.
[51] See Law Reform Commission of Canada, *Working Paper No. 61: Biomedical Experimentation Involving Human Subjects* (Ottawa: Law Reform Commission, 1989). at 45
[52] See The Law Commission, *supra* note 51 at para. 6.3.2.
[53] See Queensland Law Reform Commission. See Queensland Law Reform Commission, *Assisted and Substituted Decisions: Decision-making by and for People with a Decision-Making Disability, Vol. 1* (Brisbane: Queensland Law Reform Commission, 1996).
[54] *Enquiry on Research Ethics: Final Report* (David N. Weisstub: Chairman. Submitted to the Hon. Jim Wilson, Minister of Health of Ontario, Aug 28, 1995).
[55] P.M. McNeill, *The Ethics and Politics of Human Experimentation* (Cambridge: Cambridge University Press, 1993) at 138.

believe that injury, disability or death of the subject might arise; and after an experiment, by providing follow-up support for those subjects who require it. Generally, research involving the mentally disordered should not raise more than minimal risks. Wherever possible, non-therapeutic research should be conducted by researchers who are not the primary providers of health care to the subject.[56]

In addition to such general requirements, ethical research with the mentally disordered requires the elaboration of further safeguards taking into consideration the special characteristics of the population, specifically with regard to questions of voluntariness and competence, and the approach described above.

Research ethics committees must conduct a careful assessment of both the level of risk inherent in a research protocol, and also the level of mental competency possessed by a prospective subject. These two factors will interact to suggest what sort of experiments may be permitted on a particular mentally disordered person. A non-therapeutic experiment with minimal risk raises minor concerns regarding the consent of a subject, and only demands general protective measures. A procedure with no foreseeable risk is even less problematic. However, once an experiment exposes the subject to more than a minimal level of risk, additional protections become important, especially where the subject is a vulnerable person.

If primacy is given to the principle of autonomy, a mentally competent person should be allowed to assume a high level of risk even without a direct personal benefit. However, by virtue of the principle of personal inviolability, this right is often circumscribed by law.[57] As such, the determination of the level of risk of a research proposal must be conducted prior to any assessment of the mental competency of prospective subjects. "The criteria for determining competence might vary according to the degree of risk or discomfort presented by research procedures".[58] Consequently, competency assessments can serve to determine whether an experiment, classified according to level of risk, is permissible on a particular subject or class of subjects.

As a general rule, protection should be proportionate to the level of risk involved. As stated by Melnick *et al.*:

> The greater the risks posed by the research, the lesser the direct benefits likely to accrue to the subject from participation in the research, and the more complex the procedures involved in the research, the greater the need for careful scrutiny of the potential subject's capacity to provide consent.[59]

In order to properly determine the validity of an individual's decision to participate in a research experiment, it is necessary to assess that individual's level

---

[56] For a detailed discussion of general safeguards for research involving vulnerable populations, including the notions of risk and risk assessment, see Chapter 18.

[57] See Chapter 10.

[58] See Office for the Protection of Research Risks, National Institutes of Health, *Protecting Human Subjects: Institutional Review Board Guidebook* (Washington, D.C.: U.S. Government Printing Office, 1993) at 6–30 [hereinafter *OPRR Guidebook*].

[59] V.L. Melnick *et al.*, "Clinical Research in Senile Dementia of the Alzheimer Type: Suggested Guidelines Addressing the Ethical and Legal Issues" (1984) 32 Journal of the American Geriatric Society 531 at 533.

of mental competence. The topic of mental competency assessment has been discussed at length by the *Enquiry on Mental Competency.*[60] Various factors are taken into account in any such process, and a variety of techniques and preferred methods exist.[61] Unfortunately, assessment protocols tend to be concerned mainly with the issue of consent to treatment, rather than with research. The objectives behind consent to treatment and consent to participation in biomedical research are very different. Therefore, when assessing mental competency for research purposes, assessors must modify their approach.

It is important to acknowledge the fact that among the mentally disordered, a wide spectrum of capacities to provide consent can be found. The issue, therefore, is not the particular diagnosis or mental condition, but the incapacity that such a condition or diagnosis could cause from time to time. Competence can vary in terms of degree and in duration. The following classification scheme, for patients suffering from dementia, for example, demonstrates the fluctuating nature of a mental disorder's effect on an individual's competence, and consequently on the capacity to provide consent:

(1) Patients at a pre-dementia or early dementia stage who are presently competent and who have no diminishment or fluctuation in cognitive and reasoning skills, short-term memory, or expressive capacities.

(2) Patients at an early dementia stage, who are presently competent despite a minor diminishment or fluctuation in cognitive and reasoning skills, short-term memory, or expressive capacities.

(3) Patients of uncertain competence and those who are clearly incompetent in view of a more serious diminishment or fluctuation in cognitive or reasoning skills, short-term memory, or expressive capacities . . .[62]

This classification also illustrates the importance of competence as a factor used to assign a patient to a category that will have consequences for determining the type of experiment in which such a person may participate. "Competence" generally refers to an individual's decision-making capacity, and an assessment of this ability might be justified by a mere doubt on the part of the clinician or researcher. Fletcher,

---

[60] See Weisstub, *supra* note 10.

[61] For example, among the various sorts of information used when assessing competence, B. Quarrington, in "Approaches to the Assessment of Mental Competency" (1994) 15:2 Health Law in Canada 35, states his preference for information provided by a person familiar with the potential subject. That is, a person who,

> has had the opportunity to observe the person spontaneously engaging in the activities of everyday life in which incompetence is alleged or suspected. If this familiar individual is governed by what is in the best interests of the person, and if the observations are recent and reasonably extensive, we accept his/her judgments of competency since they have an obvious validity. No additional data or information is needed. Given the good will of the familiar, no special qualifications are needed beyond those of understanding the life situation of the person, and being able to recognize success and failure in the coping behaviour of that person.

[62] See Keyserlingk *et al.*, *supra* note 15 at 322. Note that similar concerns can be raised about the competence of individuals with psychotic disorders, who would similarly have fluctuating cognitive capacities.

*et al.*, citing a 1984 NIH working committee report, stated three key assumptions for the assessment of decision-making capacity:

(a) In determining capacity, the assessment of a patient's *actual functioning* in decision-making situations is given precedence over the outcome of a patient's decision . . .

(b) Except as noted, individuals should be assumed to possess decisional capacity unless otherwise demonstrated; *incapacity* should be found to exist only when the individual lacks the ability to make decisions that promote his/her well-being in conformity with his/her previously expressed values and preferences. Rarely should incapacity be considered to be absolute; even when ultimate decisional authority is not left with the individual, a reasonable effort should be made to provide to the subject relevant information about the protocol and the available options and to accommodate his/her preferences.

(c) At least three criteria are relevant to determining level of impairment:
  (1) how independently the subject has been functioning in making choices in daily life;
  (2) evaluations of mental status requested from . . . consultants in psychiatry or neurology; and
  (3) results of the physician's initial efforts to inform the subject in the presence of a family member or significant other about the protocol and the available options.[63]

The criteria suggested as relevant to determining the level of impairment of decision-making capacity no doubt form the basis of many methods for competency assessment. However, in addition to the stigmatizing nature of mental incompetency, many deficiencies in relation to current practices in assessment procedures have been pointed out. These criticisms include: a failure in many evaluations to proceed with the fundamental assumption that an adult is competent rather than incompetent; the futility of attempting to establish a single criterion of incompetency; the fact that a medical diagnosis of mental disorder does not immediately imply incompetency; and the fact that incompetency will often fluctuate and vary in relation with different areas of an adult's life.

As an assessment of competency is important to any consent process, several useful guiding principles should be recognized. A mental disorder should not be presumed to diminish a person's capacity to consent to participation in research. Even where a court has declared a person to be incompetent, unless the court specifically states that this declaration includes the capacity to consent to participation in research, such incapacity should not be presumed, especially since certain mental disorders may result in fluctuating levels and periods of incompetence. A competency assessment should never be influenced by the outcome of a patient's decision, should not be prejudged, and should only occur where justified by the circumstances.

## *The Doctrine of Informed Consent*

Whether a mentally disordered person is competent (and decides personally) or is found to be incompetent (and a substitute decides), the first and most important safeguards are to ensure that the consent to participate in research is both free and

---

[63] See Fletcher, Dommel Jr. & Cowell, *supra* note 9 at 4.

informed. This goal can be difficult to achieve with the mentally disordered owing to difficulties in establishing competence and voluntariness.

To obtain an informed consent, the researcher must provide proper disclosure of all information relevant to the subject's decision, and the prospective subject must be capable of fully understanding the information presented. In no case should research ethics committees and investigators assume that mentally disordered subjects "are unable to grasp any element of a protocol. While they may not be legally capable of making a valid choice to participate, appropriate respect for these prospective subjects requires that they be given the fullest possible explanations consistent with their abilities and in terms most easily understandable to them".[64]

The Medical Research Council (UK) emphasized both the need to seek consent and the elements of such consent:

> . . . [s]eeking the consent of an individual to participation in research reflects the right of that individual to self-determination and also his fundamental right to be free from bodily interference whether physical or psychological. These are ethical principles recognized by English law as legal rights. We identified three elements to consent in its broadest sense — the information given, the capacity to understand it and the voluntariness of any decision taken.[65]

"Understanding", or "comprehension", is certainly the most important element, and depends on a number of factors. The Medical Research Council (UK) stated that understanding consists of comprehension of the nature and purpose of a course of action and of the long-term risks and benefits.[66] The definition of "cognitive incapacity" provided by the Law Commission (UK) demonstrates this point:

> A person should be considered unable to take the decision in question (or decisions of the type in question) if he or she is unable to understand an explanation in broad terms and simple language of the basic information relevant to taking it, including information about the reasonably foreseeable consequences of taking or failing to take it, or to retain the information for long enough to take an effective decision.[67]

Nevertheless, too much reliance on the doctrine of informed consent as a safeguard is not advisable. A test for informed consent can often serve as a test of comprehension, not of consent. For example, Irwin *et al.* noted in their study on psychotic patients' understanding of informed consent that "most patients stated they had understood the informed consent information. Objective assessments of their understanding did not confirm this".[68] Therefore, the doctrine of informed

---

[64] See Keyserlingk *et al.*, *supra* note 15 at 339.

[65] See Working Party on Research on the Mentally Incapacitated, *The Ethical Conduct of Research on the Mentally Incapacitated* (London: Medical Research Council, 1991) at para. 3.2.1. This appears to be a standard categorization of the three required elements of consent. The *Belmont Report*, *supra* note 43 at 4, specified the same three elements to consent.

[66] See Working Party on Research on the Mentally Incapacitated, *ibid.* at para. 3.2.3.

[67] See the Law Commission (1992), *supra* note 34 at 28

[68] M. Irwin *et al.*, "Psychotic Patients' Understanding of Informed Consent" (1985) 142 American Journal of Psychiatry 1351 at 1352. The authors went on to say that, "[t]he patients, all of whom were acutely psychotic were able to read the informed consent information, and most reported that their understanding of the information about antipsychotic medication was good. Objective measures, however, did not confirm their self-reports. Many simply affirmed understanding to mask confusion while reading the information about antipsychotic medication".

consent only serves as a useful safeguard for valid consent with mentally disordered persons when they are competent (assuming that competence is not confused with comprehension). In fact, a patient may comprehend (cognitively) the elements of the information presented, and yet not appreciate the impact of the same information at an emotional level. Although the information may be comprehended, the patient may not be able to apply it. Finally, it should also be noted that the scope of consent that a subject provides may be different in length and complexity than what the researcher has in mind, or that the subject may have doubts about ongoing participation as the project goes on. Therefore, it is essential that the researcher questions the subject on this issue from time to time throughout the duration of the project.

### Voluntary Consent

As with prisoners, other institutionalized individuals may have a different set of values and a different notion of self-worth than the general public.[69] For example, an institutionalized individual may consent to a far greater level of risk for a relatively minor reward (which may nevertheless seem significant to the subject) than would a person whose liberty is not restricted. Civilly committed individuals would be prone to similar exploitation and other forms of coercion possible in an institutional setting. As such, there should be a cap therefore on the level of risk to which institutionalized mentally disordered persons may be subjected.

Thus, in addition to an assessment of mental competence, ethics review committees should include special safeguards to ensure that consent is truly voluntary. These safeguards might include:

- ensuring that representatives of the mentally disordered are granted meaningful roles as members of ethics review committees;
- ensuring the presence of an independent third party throughout the consent process (a role that might best be served by a family member); and
- reducing incentives that might cause a prospective subject to become involved in research where others would not. This is especially relevant in the case of civilly committed patients.

### Advance Directives for Research and Substituted Consent

As the mentally disordered represent a population that can have fluctuating periods of lucidity or have been competent at some time in the past, the use of research directives is clearly relevant to their situation. Research directives should not be considered to be an exception to the general rule that research should only be conducted with the subject's consent, but rather as an alternative means of indicating and ensuring consent. A research directive would provide a specific direction in place of future consent. Ethical and legal issues pertaining to research directives are discussed elsewhere in this Volume.[70]

---

[69] See e.g. J.L. Hill, "Exploitation" (1994) 79 Cornell Law Review 631 at 697.
[70] See Chapter 11.

As a general rule, the consent of participants is preferable to that of surrogates. However, in the absence of a competent contemporaneous or advance decision to participate in an experiment, experimenters must rely on substitute decision-makers.

There is an important distinction between a decision made by a competent individual and that of an incapable person's guardian. The former can be characterized as a direct expression of an individual's will; the latter represents a subjective evaluation of what the incapable individual would have decided if competent to do so, or a decision made in the incapable person's best interests.[71] One can safely say that all modern legal systems have made provisions for guardianship of incapable adults.[72] For example, the Law Commission (UK) report *Mental Incapacity* recommends that local social service authorities be named as guardians.[73] As discussed elsewhere in this Volume, similar dispositions for the protection of research subjects should be made.[74]

The participation of mentally disordered subjects should be permitted only in a very limited number of circumstances. These might include the following situations:

(1) the individual has verbally consented to an experiment while lucid;
(2) the individual has prepared an advance directive in which he or she specifically consents to participate in the experiment or has named a substitute decision-maker in a durable power of attorney specifically for the purpose of consent to participation in research; or
(3) a valid and informed consent has been obtained by the legal guardian of the individual.

Even if an individual has consented in one of the ways listed, the level of risk to which a substitute decision-maker could consent must be limited. Various authors have stated that, absent a research directive to the contrary, a substitute decision-maker must not be allowed to consent to research involving more than a low risk.[75] Substitute consent to a procedure involving a *substantial* risk should be approved by a judicial or quasi-judicial authority, and should only be allowed in exceptional cases.

Another consideration is the selection of the SDM, a role that can only be fulfilled by those with sufficient knowledge about both the patient and the research in question. Precedence should be given for the surrogate chosen by the patient. Lacking such an indication, a spouse, or parent or guardian (if a child), should be

---

[71] That is, it is important to distinguish between a decision made "in the shoes of" or "in the best interests of" an incapable person. See e.g. S.A. Kline, "Substitute Decision-making and the Substitute Decision-maker" (1992) 13 Health Law in Canada 125 at 127.

[72] In the United States, the DHHS regulations clearly allow for the use of substitute consent by stating that "I[i]nformed consent will be sought from each prospective subject or the subject's legally authorized representative". 45 C.F.R. 46 (1991) § 46.111 (a)(4).

[73] See Law Commission (1995), *supra* note 34.

[74] See Chapter 8.

[75] See the American College of Physicians, "Position Paper on Cognitively Impaired Subjects (1989) 111 Annals of Internal Medicine 843 at 845; and Keyserlingk *et al.*, *supra* note 15 at 348, both of whom also place the limit at minimal risk.

preferred unless a patient objects. Any such objection should be overruled only by a court order. The court may then appoint a close friend who must not be connected to the facility conducting the study.[76]

Regardless of who is chosen as the surrogate decision-maker, decisions continue to be in the mentally disordered person's best interests.[77] Attention should be paid to any bias which the surrogate may possess. Family members may normally be in the best position to interpret an incompetent person's cognitive status and anxieties; however, because of concerns for the inheritability of a disease, family members may have a vested interest in the results of the research, and may place undue pressure on the patient to participate.

## CONCLUSION

The preferred method for enrolling mentally disordered persons in research protocols should be by obtaining contemporaneous, or in the alternative, advance consent, including by way of a research directive. In the case where it is impossible to obtain either, substituted consent by a legal guardian should be permitted, but only in cases of negligible to minimal risk. Exceeding this level should require judicial or quasi-judicial approval. It is recommended further that those jurisdictions lacking regulatory régimes enact legislation to regulate non-therapeutic research involving the mentally disordered, including provision for substituted and advance decisions. Such legislation must serve to protect this population, while at the same time strive to maximize and facilitate individual autonomy. All protocols must receive approval from established RECs, with particular care being given to ensuring voluntary and informed consent to participation. Sanctions should be imposed on researchers who do not adhere to the suggested guidelines.

---

[76] See Kline, *supra* note 71 at 126.

[77] Determining a subject's "best-interests" includes, of course, consideration of any prior wishes, values or beliefs known to have been held by the subject.

CHAPTER 22

# ETHICS IN PSYCHIATRIC RESEARCH WITH INCOMPETENT PATIENTS

HANFRIED HELMCHEN

Psychiatric research with incompetent[1] patients has become a focus of interest in medicine as well as in the public sphere. The major questions are:

- Why is it a *current topic* at all?
- Is such research *really needed*?
- Is such research *ethically justifiable*?
- Is such research *legally permitted*?

The following answers imply a European perspective[2] and, with regard to specific details, will refer mostly to German regulations.[3]

## ON THE CURRENT RELEVANCE OF THE TOPIC

Ethics committees have to deal increasingly with proposed research projects involving patients who are not competent to give informed consent. This is not only valid for research projects in pediatrics and psychiatry, but likewise for research projects in the field of neurology and neurosurgery, and above all in anesthesiology and intensive care medicine. Ethics committees occasionally appear to be uncertain in their decisions when they ask for additional legal advice or when their judgments contradict those of other ethics committees involved in the same (multi-centric) project. This is especially true for research projects which do not deal with

---

[1] In the following context this term means the same as "not able to consent" or "without the capacity to consent" or "incapacitated" or "not capable of giving consent".

[2] H.G. Koch, S. Reiter-Theil & H. Helmchen, eds., *Informed Consent in Psychiatry* (Baden-Baden: Nomos, 1996).

[3] H. Helmchen, "Summary and Proposals" in Koch, Reiter-Theil & Helmchen, *ibid.*; H. Helmchen & H. Lauter, eds., *Dürfen Ärzte mit Demenzkranken forschen?* (Stuttgart: Thieme, 1995).

therapeutic research in the narrow sense of the word. In Germany the Central Ethics Committee (with the Federal Board of Physicians) as well as the national working group of regional institutional review boards ("Arbeitskreis Medizinischer Ethik-Kommissionen in der Bundesrepublik Deutschland")[4] currently are intensively engaged in this field. Obviously there is a need for clear and binding rules of judgment.

The subject is also highly topical because the European Council is presently discussing the draft of a so-called "European Bioethics Convention", which was made public for the first time in the late summer of 1994 as a common legal framework for biomedical research with human beings. Above all, Article 17.2 of the final draft[5] was under debate and is considered controversial in this latest version as well. It attempts under the provision of a number of prerequisites, formulated in the preceding Articles 17.1, 16, and 5, to establish the criteria under which research that "does not have the potential to produce results of direct benefit to the health of the person concerned" could be admitted as an exception.

Supporting arguments for conducting psychiatric research on incompetent[6] patients are, according to Roscam Abbing,[7] primarily:

> "that without this non-therapeutic research, the very conditions which make the *individual* incompetent would be hard to study and treat";
> the contribution of the research to the health of groups of *others*;
> great potential interest of the research for *society* with regard to the burden placed on the next of kin and health care expenditure;
> "*solidarity* which one may expect from every citizen".

Whereas medical research yielding no direct benefit for incapacitated persons, as formulated as a defined exception in Article 17.2, is regarded as permissible in other European countries — as e.g. under the French Research Law of 1988[8] or as proposed by the British Law Commission of 1993/1995[9] — this is not permitted under German Law and is judged more critically by the German public, or is even rejected vociferously, as shown by a hearing of the German parliamentary committees of Law, Health, and Education, Science and Research.[10] The main

---

[4] R. Toellner & U. Wiesing, eds., *Wissen — Handeln — Ethik: Strukturen ärztlichen Handelns und ihre ethische Relevanz* (Stuttgart: G Fischer, 1995) at 82ff.

[5] Council of Europe, *Draft Convention for the Protection of Human Rights and Dignity of the Human Being with Regard to the Application of Biology and Medicine: Convention of Human Rights and Biomedicine* (Strasbourg: Council of Europe, 1996).

[6] In the following context this term means the same as "not able to consent" or "without the capacity to consent" or "incapacitated" or "not capable of giving consent".

[7] H.D.C. Roscam Abbing, "Medical Research Involving Incapacitated Persons: What is Legally Permissible?" (University of Utrecht, Institute for Private Law, Department of Health Law, Faculty of Law, 1994) [unpublished].

[8] A.M. Fagot-Largeault, "National Report: France" in Koch, Reiter-Theil & Helmchen, *supra* note 2.

[9] The Law Commission, *Mentally Incapacitated Adults and Decision Making: Medical Treatment and Research (Consultation Paper No. 129)* (London: HMSO, 1993); The Law Commission, *Mental Incapacity* (London: HMSO,1995). Abridged version of the summary in (March 1995) Bulletin of Medical Ethics 13.

[10] Deutscher Bundestag, Rechtsausschuß: Zusammenstellung der Stellungnahmen zur gemeinsamen Anhörung des Rechtsausschusses, des Ausschusses für Gesundheit und des Ausschusses für Bildung, Wissenschaft, Forschung, Technologie und Technikfolgenabschätzung, 17 May 1995.

arguments of critics of research without direct benefit for the patient who is not able to consent are that:

> there is *no real need* for such research;
> such research is *not in conformity with* the respect for human dignity guaranteed by the *Constitution*;
> this kind of research — not least on account of the German experiences with criminal experiments on humans during the period of National Socialism and especially with a view to the crimes committed on the mentally ill[11] — would not be controllable and thus could signal the *bursting of the dam*;

As a consequence of this public debate Germany was the only European country in the Council of Europe which voted in 1996 against the above-mentioned European Draft Convention on Human Rights and Biomedicine.

## LEGAL FRAMEWORK AND PROFESSIONAL GUIDELINES

Opponents of any research with persons incompetent to give informed consent argue that such research is prohibited by the *Nuremberg Code*[12] and subsequent regulations, such as the UN Covenant on Civil and Political Rights (1966).[13]

And, indeed, it could be concluded "to one's best knowledge" and usually from the concept "his free consent" in the context of Article 7 of the above-mentioned UN Covenant that the representation of incompetent patients as participants in medical-scientific studies is excluded. However, if one interprets the words, "his free consent", in Article 7, arguing as does the Meijers Commission:[14]

> ... taking into account the development of views in the international community it is justifiable to interpret the word "his" in "his free consent" such that the consent of a guardian is not strictly excluded. The Commission is of the opinion that neither the function of law of ensuring order nor that of safeguarding the individual makes such a narrow interpretation mandatory. It is obvious that both functions are always influenced to a considerable degree by ethical and social views. In addition, it can be seen from the preparatory work of the Commission that the problem of non-therapeutic medical-scientific studies was not precisely taken into consideration at the time ... [and, therefore,] this interpretation of Article 7 is justified by the development of views within the international community ...
>
> According to the conditions formulated by the Commission and taking into consideration the fact that the medical necessity of the scientific study has been established, carrying out this study with incompetent patients whose legal guardians have given consent does not in the opinion of the Commission constitute a violation of Article 7 of the Covenant ...

---

[11] A. Mitscherlich & F. Mielke, *Medizin ohne Menschlichkeit: Dokumente des Nünberger Ärzteprozesses* (Frankfurt: Fischer, 1960); E. Klee, "Euthanasie" in NS-Staat, *Die "Vernichtung lebensunwerten Lebens"*. (Frankfurt/Main: Fischer, 1983).
[12] The *Nuremberg Code* formed part of the judgment in *U.S. v. Karl Brandt et al., Trials of War Criminals Before the Nuremberg Military Tribunals Under Control Council Law No. 10* (October 1946-April 1949).
[13] *International Covenant on Civil and Political Rights*, 19 December 1966, Can. T.S. 1976 No. 47, 999 U.N.T.S. 171, 6 I.L.M. 368.
[14] L.C.M. Meijers *et al.*, "Medical Experiments With Incapacitated Persons" report to the Ministry for Health, Welfare, and Sport and the Ministry of Justice. Den Haag, 1995 [hereinafter Meijers Committee].

The Commission draws the following conclusion from its extensive analysis:

> In all, the Commission concludes that significant practice in application in other countries as well as views on scientific studies with incompetent patients in the international community — both of which are vital for the interpretation of Article 7, Sentence 2 — give no occasion to assume that Article 7, Sentence 2 tends to make non-therapeutic medical-scientific studies impossible. The intention of Article 7, Sentence 2 is to offer protection from excesses. If the preparatory work of the Covenant as a supplementary means of interpretation is valid at all, this provides no reason to make assumptions to the contrary.
>
> However, in dealing with scientific studies with incompetent patients, additional security measures must be present.[15]

Opponents of research with incompetent persons also refer to the first point of the *Nuremberg Code*, which states that only persons who have voluntarily given informed consent can be included in research. The aim of this statement, made 50 years ago, was primarily to condemn "research" by the National Socialists in Germany, performed on uninformed, involuntary and deceived persons. However, the *Nuremberg Code* does not mention incompetent patients, and its authors very likely did not have in mind such patients. In any case, only by virtue of this interpretation does the *Nuremberg Code* not stand in contradiction to the *Declaration of Helsinki*,[16] which in Section I, Paragraph 11 states that consent to research in the case of incompetent patients may be substituted by the consent of the informed legal guardian. The background of this regulation was apparently the understanding of the World Medical Association that clinical research, including with incompetent patients, cannot be renounced, since research is a prerequisite for improving diagnostic procedures and patient treatment.

Physicians are motivated to conduct such research in view of the immediate experience of the patient's suffering, for example, the demented patient and his/her relatives and caregivers. They wish to treat such illnesses more effectively than has been possible to date, and therefore to include incompetent patients in research in defined cases where there is no other way to achieve this objective. Incompetent patients are of course particularly vulnerable because their illness — not the physician! — has deprived them of the possibility of advocating their own rights. Therefore, physicians attempt to find improvements which eliminate or at least alleviate the illness to the degree that the patient regains his capacity to exercise his own rights. Physicians are not motivated by a desire to freely dispose over such patients, weak and defenseless as they are, including them in research because of their inability to resist. On the contrary, physicians are conscious of an obligation to take measures to alleviate the dire condition of such patients, and wish to include them in research under defined safeguards precisely because their illness is so severe and so insufficiently treatable. In addition, these concepts are legitimized by

---

[15] *Ibid.*

[16] World Medical Association, *Declaration of Helsinki*: Adopted at the 18th World Medical Assembly in Helsinki in June1964. Amended at the 29th World Medical Assembly in Tokyo in October 1975; at the 35th World Medical Assembly in Venice in October 1983; and at the 41st World Medical Assembly in Hong Kong in September 1989.

professional law which states that the physician must serve "the population" (e.g. *Berufsordnung der Ärztekammer Berlin*, 1990)[17] and by the German constitution (*Grundgesetz*: GG) which guarantees the freedom of research (Article 5, GG). Nevertheless, it must be clear that these obligations and rights are secondary to and limited by the physician's primary obligation "to serve the health of the individual human being".[18]

These ethical norms are expressed both internationally and nationally in more or less specified guidelines for research with incompetent persons, published primarily by professional bodies such as the World Medical Association,[19] the Council of Medical Sciences[20] and National Research Councils (for an overview cf. Roscam Abbing).[21] These guidelines also find their expression in general as well as in more specific laws and by the fact that a number of legally binding regulations refer to them. However, specific legal regulations for research with patients not competent to give consent are currently being outlined only in a few countries.

This may be illustrated by the German legal situation. In this context the most important general laws are the Constitution (*Grundgesetz*: GG), which guarantees the dignity and autonomy of the human being, and the penal code (*Strafgesetzbuch*: StGB), with its safeguards of body and life. These basic human rights (*Grundrechte*) are safeguarded more specifically by the drug law (*Arzneimittelgesetz*: AMG), particularly its Paragraphs 40–42, regulating the scientific investigation of potential new drugs by means of clinical trials involving patients. According to Paragraph 41 AMG, clinical trials (i.e. therapeutic research in the proper sense) are also permissible with incompetent patients for whom a legal guardian has given informed consent. On the basis of the new law regulating legal care (*Betreuungsgesetz*: BtG), which in 1992 replaced the former guardianship law (*Vormundschafts- und Pflegschaftsrecht*), the legal guardian is allowed to give consent only corresponding to the wishes and interests of the patient, i.e. he is not permitted to consent to research lacking potential individual benefit.

Additionally, both the recently passed fifth revision of the drug law and the physician's codex (*Berufsordnung*) oblige the researching physician to consult the Ethics Committee of a university medical faculty or state Board of Physicians (*Landesärztekammer*). These Ethics Committees advise the researchers corresponding to the revised *Declaration of Helsinki* on the regulations which he must be aware of and adhere to.

Thus, the drug law is the only law that deals specifically with research, and particularly with the safeguarding of human beings in research. However, it regulates only drug research and no other medical research involving human beings.

---

[17] Ärztekammet Berlin, "Berufsordnung: Amtsblatt für Berlin" (14 September 1990) at 1884.
[18] *Ibid.*
[19] *Declaration of Helsinki*, *supra* note 16.
[20] Council for International Organizations of Medical Sciences, in collaboration with the World Health Organization, *International Ethical Guidelines for Biomedical Research Involving Human Subjects* (Geneva: CIOMS, 1993).
[21] See Roscam Abbing, *supra* note 7.

The legal situation is for this reason quite unclear. Furthermore, the medical concern of law in general provides little help for the legal assessment of medical research. It has so many gaps that it does not even sufficiently regulate the prerequisites for the patient's valid informed consent. This gap has been filled by court decisions establishing precedents, but only very rarely with regard to research. In view of this deficit the question must be asked whether and how far Paragraphs 40–42 of the drug law can be used analogously for the legal assessment of medical research not involving drugs.[22]

Some other countries have outlined specific legal regulations for non-therapeutic research with patients not competent to give informed consent. This is true for France, with the French Research Law of 1988[23] and Canada with the *Civil Code of Quebec* of 1993,[24] as well as for the United Kingdom with proposals by the British Law Commission for corresponding legal regulations[25] and the Netherlands, where the Committee of the Ministry of Justice and Health Care has advanced a draft Bill on Medical Experimentation (1992). This committee in the Netherlands made thorough "recommendations with regard to the regulation of medical-scientific studies with minors and incapacitated adults".[26] Simultaneously and independently in Germany a group of psychiatrists, jurists, theologians and lay people concerned with these questions has met since 1991 to determine and publish:

> arguments for the *need for research* with incompetent patients;
> criteria and measures for the *assessment of the incapacity* to give consent, as well as
> recommendations for *criteria for the safeguarding* of the patient and
> recommendations for *modifications/amendments* to the Drug Law as the only specific
> legal regulation of such research in Germany, e.g. for dementia.[27]

In conclusion it must be noted that the international guidelines and laws published to date are

- not very *specific*;
- lacking in the precision of some definitions such as *minimal risk*;
- ambiguous with regard to some concepts such as
  - *benefit* (direct versus no benefit)
  - *research* (therapeutic versus non-therapeutic).

## THE BASIC ETHICAL PROBLEM

The ethical principle of respecting the patient's autonomy has become most important in modern medicine. This means accepting that every individual has the right to decide himself, according to his values and beliefs, whether to take action

---

[22] See Helmchen and Lauter, *supra* note 3.
[23] See Fagot-Largeault, *supra* note 8.
[24] See Roscam Abbing, *supra* note 7.
[25] See Law Commission (1993), *supra* note 9; Law Commission (1995), *supra* note 9.
[26] See Meijers Committee, *supra* note 14.
[27] See Helmchen and Lauter, *supra* note 3.

and what action to take, with regard to his health.[28] Corresponding aspects, constituting the individual's dignity, are respect for his/her personal attitudes, for the integrity of his/her body and for his/her privacy. Respect for human dignity manifests itself primarily in direct interpersonal contact, i.e. in this case contact between physician and patient, and particularly in the comprehensive recognition of the patient's wishes, interests, views, social and cultural values. This premise attains special significance if the aim of an intervention is not restricted to the individual but, through the acquisition of knowledge by means of research, also pertains to a purpose beyond the individual. Thus it is essential to safeguard human dignity in all research with human subjects. This is particularly true in research with "vulnerable" patients, i.e. those whose human rights could be violated because they are incompetent to give informed consent due to mental disorders or somatic conditions involving impairment of conscience, or in the cases of minors. (Within the framework of this chapter the specific problems of the two latter groups will not be dealt with specifically.) Thus the recommendations follow[29] that in such cases, and particularly with regard to non-therapeutic research (see below):

- besides the necessary informed consent by the legal *guardian* or next of kin,
- the *risks and inconveniences* of the investigation must be only minimal;
- the patient must be *informed* corresponding to his capacity to comprehend; and
- the patient's *refusal* must be accepted.

If non-therapeutic research is defined as that with no direct potential individual benefit, the question is whether such research in any case contradicts the human dignity of the participating incapacitated person. The Meijers Commission argues that in society the significance of an individual contribution to the common welfare is generally accepted, and thus it can be assumed that a certain willingness to participate in medical-scientific investigations exists in society. Thus, incapacitated persons must not be categorically excluded and therefore could under strict conditions be included in non-therapeutic medical-scientific studies.[30] A related argument is that individuals may more likely be willing to participate in such research in solidarity with other persons with the same disorder.[31] This bond may justify the inclusion of incapacitated persons in non-therapeutic research under the following conditions (among others).

Such research

- *cannot otherwise be performed* without incapacitated persons;

---

[28] T.L. Beauchamp & J.F. Childress, *Principles of Biomedical Ethics*, 3rd ed. (New York and Oxford: Oxford University Press, 1989).

[29] See Meijers Committee, *supra* note 14.

[30] *Ibid.*

[31] See Helmchen & Lauter, *supra* note 3.

- must be relevant for the group of persons with the *same disorder*; and
- involves no more than *minimal risks* and inconveniences.

## TYPES OF MEDICAL RESEARCH

### *Therapeutic versus Non-Therapeutic Research*

The *Declaration of Helsinki* distinguishes between clinical research and non-clinical research: the former is understood as having the potential of individual benefit, whereas the latter provides no individual benefit for the participating patient.[32] Relevant specific national proposals mainly differentiate between therapeutic and non-therapeutic research. However, although the *Declaration of Helsinki* equates non-therapeutic and non-clinical research, there is no identity of clinical and therapeutic research or non-clinical and non-therapeutic research with regard to individual benefit. This has consequences for the question under debate as to whether diagnostic, preventive, palliative, and (institutional) care research is classified as therapeutic or non-therapeutic research. The former is suggested by the *Declaration of Helsinki* insofar as in non-clinical research "the experimental design is not related to the patient's illness", from which the inference can be made that clinical research includes all research related to the individual patient's illness and thus also research projects from the above-mentioned fields. Furthermore, at least some of these research projects will have direct potential individual benefit.

Apparently the range of definition of "therapeutic research" is of significance, because therapeutic research is identified with potential individual benefit by offering "hope of saving life, restoring health or alleviating suffering",[33] and because potential individual benefit is a decisive criterion in research with persons who are incompetent to give consent. However, only the narrow definition of therapeutic research implies categorically and by definition potential individual benefit in the sense of an immediate or direct benefit, whereas such a narrow definition of therapeutic research excludes diagnostic research projects or those in the other above-mentioned fields, all of which may also have potential individual benefit. The consequence of a narrow definition of therapeutic research is that in widespread opinion those other projects mentioned are classified as non-therapeutic research, with the implication that they have no potential individual benefit.

Therefore, it must be made clear that the borderline between therapeutic and non-therapeutic research is not identical with that between research with and without potential individual benefit. In other words: between therapeutic research in the narrow sense with immediate potential individual benefit on the one hand and non-clinical research with no individual benefit on the other, there exists a group of clinical research fields in which potential individual benefit must be proved as being either immediate or only mediate, or as not at all given.

---

[32] See *Declaration of Helsinki, supra* note 16.
[33] *Ibid.*

### Benefit versus No Benefit[34]

Some ambiguity also exists with regard to the directness or immediacy of the potential individual benefit, particularly its probability (not to be excluded; potential; probable), its extent and intensity (questionable, demonstrable, evident) and, not least of all, its relation in time to the intervention. Furthermore, the term "direct" benefit implies, or at least suggests, some form of "indirect" or "mediate" benefit, since the specification "direct" would not be necessary if there existed solely the categories "benefit" and "no benefit". Support for this interpretation can be found in the above-mentioned Article 17.2 of the European Convention on Human Rights and Biomedicine. Here it is stated that research which "does not have the potential to produce results of direct benefit to the health of the person concerned" is permissible under exceptional circumstances, if it "has the aim of contributing . . . to the ultimate attainment of results capable of conferring *benefit to the person concerned* or to other persons . . .". [our emphasis]. However, the specification of a benefit as "direct" could express the wish that only a high degree of "directness" of benefit justifies research with incapacitated persons.

These points of clarification and differentiation between therapeutic versus non-therapeutic research and between benefit versus no benefit are related to the empirical reality of medicine. However, there are strong opinions that such inductive argumentation on the part of the medical profession must be secondary to deontological or other deductive prescriptions set forth by society. Furthermore, the opinion has been expressed in a more political sense that such phrases are merely a strategy used to blur the fundamentally clear borderline between benefit and no benefit, with the aim of expanding medical research into dangerous areas or even of concealing illegitimate research.[35]

---

[34] Here the "best interest standard" cannot be discussed in its various aspects, such as well-being, welfare, good, benefit, etc., or its relation to various theories of the good, as in e.g. D. Parfit, *Reasons and Persons* (Oxford: Clarendon Press, 1984) or in the position of e.g R.M. Veatch, "Abandoning Informed Consent" (1995) 25:2 Hastings Center Report 5, that it is impossible for a physician to determine what constitutes the good for another being. However, one relevant relationship of "benefit" to "the best interest" of the individual should be mentioned at least descriptively. For example, the definition of "best interest" given by the British Law Commission (1995) includes "past and present wishes and feelings" of the incapacitated person as well as this "person's welfare". Due to the fact that there may be a conflict between the "subjective" and the "objective" best interest of the person concerned, the question may arise as to which of these two aspects of best interest should prevail as benefit. In the case of a research project with no direct benefit the intervention is presumably not in the "objective" best interest of the incapacitated person. However, it may and should not be against the "objective" best interest of the incapacitated person and must not be against his/her "subjective" best interest. In other words, with regard to the best interest of the incapacitated person included in research with no direct benefit, no more than minimal risks or negligible discomforts are acceptable, and any refusal of the individual must be accepted.

[35] U. Fuchs, *Stellungnahme zum Entwurf einer Bioethik-Konvention des Europarates*, Deutscher Bundestag, Rechtsausschuß, Zusammenstellung der Stellungnahmen zur gemeinsamen Anhörung des Rechtsausschusses, des Ausschusses für Gesundheit und des Auschusses für Bildung, Wissenschaft, Forschung, Technologie und Technikfolgenabschätzung, 17 May 1995; J. Paul, *Stellungnahme zum Entwurf einer Bioethik-Konvention des Europarates*, Deutscher Bundestag, Rechtsausschuß, Zusammenstellung der Stellungnahmen zur gemeinsamen Anhörung des Rechtsausschusses, des Ausschusses für Gesundheit und des Auschusses für Bildung, Wissenschaft, Forschung, Technologie und Technikfolgenabschätzung, 17 May 1995.

*Therapeutic Research*

In any case, therapeutic research in the proper sense (i.e. primarily controlled clinical trials) even involving incompetent patients, is permissible according to Section I.11 of the *Declaration of Helsinki*[36] as well as according to national law, e.g. Paragraphs 41.2–5 of the German Drug Law (*Arzneimittelgesetz*: AMG), provided that the legal guardian, after receiving the appropriate information, gives his consent under the general prerequisite codified in paragraph 41.1 of the German Drug Law as follows: "The application of the drug to be tested is indicated by the state of knowledge of medical science to save the patient's life, restore his health or relieve his suffering". This is in conformity with the duty of the legal guardian codified in paragraph 1901 of the German Civil Law Code, which states that the guardian must orient his decisions according to the benefit and the will of the patient, insofar as the latter is known and is not opposed to the patient's best interest.

Thus, therapeutic research with immediate benefit for the patient is correspondingly allowed by law with incompetent patients. But in practice, this frequently raises difficulties. Especially in an acute state of illness, as e.g. in stroke therapy research, a legal guardian does not exist, and the possibility of establishing guardianship for the sole purpose of research will be doubtful. If in such an urgent case legal proceedings are begun to appoint a guardian, this is often not granted as quickly as would be necessary in connection with acute therapy requirements. Paragraph 41.5 of the German Drug Law (AMG) provides for the possibility of carrying out necessary acute therapeutic treatment without the legal guardian's consent if it is expected that the guardian will subsequently supply this consent. In a case where no guardianship exists, the possibility of following such a procedure is legally questionable. As a subsidiary point, according to Paragraph 40.3 of the AMG, patients are excluded from therapeutic research if they are "kept in custody in an institution by court order or a public authority", even if their legal guardian would consent to such therapeutic treatment, as e.g. with demented patients who urgently need treatment and who are confined in an institution as a consequence of their state of agitation or confusion.[37] This is primarily to exclude any form of coercion. However, it might be questioned whether these individuals should in any case be excluded from research with potential specific benefit for them.[38] In this context the ethical principle of "*primum nil nocere*" can also be interpreted as "do not deprive a patient of a potential benefit". This argument may also be of value in cases such as acute (e.g. delirium or stroke) or chronic (e.g. dementia) organic mental disorders in which an excessively restrictive attitude towards the obtaining of informed consent might harm psychiatric patients.[39]

On the whole, therapeutic research as outlined here is legally permissible but not

---

[36] See *Declaration of Helsinki*, *supra* note 16.
[37] K. Amelung, "National Report: Germany" in Koch, Reiter-Theil & Helmchen, *supra* note 2; R. Rosenberg, "National Report: Denmark" in Koch, Reiter-Theil & Helmchen, *supra* note 2.
[38] See Amelung, *ibid.*
[39] See Rosenberg, *supra* note 36.

feasible without considerable difficulties in practice.

*Non-Therapeutic Research*

In general non-therapeutic research is not permissible with incapacitated patients. However, according to Section II.1 of the *Declaration of Helsinki* the development of a "novel *diagnostic* procedure" may be considered permissible and ethically justifiable if the patient participating in a research project may have "hope for saving his life, restoring his health or alleviating his suffering".[40] In this sense diagnostic research may be subsumed under therapeutic research as mentioned above.

However, it becomes problematic if such recognition of the disease is interpreted in a broader sense, i.e. in terms of the identification of its conditions and causes. Certainly any such research in exceptional and precisely defined and controlled cases may be well-founded as ethically justifiable if benefits are likely to be expected for the patient. However, the discussion becomes very controversial if this potential personal benefit is considered as being only mediate and less probable, i.e. if the acquisition of knowledge and thus benefit for others will dominate the intervention. A potential, at least mediate benefit may be assumed if e.g. the findings on the pathogenesis of the disease appear to enable the improvement of the therapy, from which the patient included in the research project could also benefit.

Such an assumption must be specified in the interest and for the safeguarding of the patient, since the safeguarding of the patient must be all the more stringent the less the patient is likely to benefit personally from it. Therefore, in Germany a number of criteria are under discussion which must be observed if this kind of research is to qualify as permissible in exceptional cases defined by law. These safeguarding criteria are:

(1) The research project *must be related to the illness resulting in the patient's incapacity* to give informed consent. In adults this criterion is of decisive importance for the exclusion of patients from research which promises benefit exclusively for others, as defined in Section III of the *Declaration of Helsinki* as non-clinical biomedical research. According to this criterion, demented patients are excluded from research projects which have nothing to do with dementia. However, this criterion is not valid for minors who, independent of a given disorder, are incompetent to consent due to their developmental state.

Correspondingly, the French version of this criterion states that the research study must be of value to persons of the same age, suffering from the same illness or handicap characteristics.[41] An Irish version includes an apparently broader definition, i.e. that benefits must be expected at least for other patients.[42]

(2) The research project *can only be performed with incapacitated patients*; i.e. there is no alternative research strategy to answer the research question.[43] If, however, the question could also be clarified with patients competent to consent, any

---

[40] See *Declaration of Helsinki*, *supra* note 16.
[41] See Fagot-Largeault, *supra* note 8.
[42] P.R. Casey, "National Report: Ireland" in Koch, Reiter-Theil & Helmchen, *supra* note 2.
[43] See Fagot-Largeault, *supra* note 8.

such research with incompetent patients would not be permissible. Examples of this kind of research would be defined pediatric projects or those projects with incapacitated demented persons which cannot be carried out with less demented patients still able to consent, because a precise diagnosis cannot be provided early enough in the course of the disease with the necessary degree of certainty, or because the factors determining the course of the disease may differ completely in the late stage of the disease compared to its early stages.

Such a research question would be, for example, whether the assumption could be verified by an epidemiological observation and inquiry that there are differences with regard to the degree of the need for care between patients in the late stage of dementia still living at home, and others who are institutionalized, with the consequence that care could be specified and improved on account of these findings. Or, an answer to the question as to whether the pattern of disturbances and the duration of progressing late stages of dementia differ between young-old and old-old patients would be of importance in solving the problem with old-age dementia in distinguishing between accentuated aging and brain diseases.[44] Another question in research would be, for example, whether there exists — possibly by charting the profile of neurotrophines in venous blood or using data on the metabolism of certain cerebral proteins obtained by magnetic resonance spectroscopy — a regenerative potential of the brain even in the late stages of dementia, which could then serve as a basis for a specific therapy.

The Meijers Commission cites the following example: "Because in current society individuals with Down's syndrome reach an older age than in the past it can be observed that Alzheimer's disease is relatively frequent among these individuals. Cognitive regression in these individuals is more pronounced than it normally is in other persons with Alzheimer's disease without Down's syndrome. Examinations are now conducted with reliable diagnostic predictability so that care can be adapted to the patient's needs in due time. In 18 persons with Down's syndrome and a large group of volunteers with no mental handicap a CT scan and an MRI were taken to measure atrophy in a defined area of the brain. The study shows that the degree of cognitive regression resulting from Alzheimer dementia in individuals with Down's syndrome can be predicted on the basis of radiological data. This study represents a significant step in the development of non-invasive diagnostic procedures for these patients".[45]

(3) The research project is expected to *contribute* not only incidentally but *considerably* to the diagnosis or treatment of the patient's disease.[46] This criterion also is valid for minors, but without a relationship to the disease. "Me too" research and research exclusively for the purpose of generating hypotheses will not be permitted. However, in making its judgment, the ethics committee should consider

---

[44] F.M. Reischies & U. Lindenberger, "Grenzen und Potentiale kognitiver Leistungsfähigkeit" in K.U. Mayer & P.B. Balters, eds., *Alter: Die Berliner Altersstudie* (Berlin: Akademie-Verlag, 1996) 351.
[45] G.D. Pearlson *et al.*, "Brain Atrophy in 18 Patients with Down's Syndrome: A CT-Study" (1990) 11 American Journal of Neuroradiology 811.
[46] See Amelung, *supra* note 37.

the need for research which, e.g. for dementia, is compulsory because dementia, progressing over many years, entails continuous suffering for the patient and his next of kin, and because its causes cannot yet be treated, and, finally, because dementia is becoming an increasing burden on public health on account of its present and steadily increasing frequency. This may be seen in the public discussion about the Law of Old-Age Nursing Insurance in Germany.

(4) The research project involves *only minimal foreseeable risks and no significant burden* for the patient.[47] The formulation of the question of risks must be stringent since no standardized and generally accepted definition of the minimal risk exists.

Whereas it is standard in therapeutic research that an acceptable relationship between unavoidable risks and intended benefits must be given, there are differences in non-therapeutic research as to whether minimal risks[48] or no risks at all are permissible.[49] Minimal risk is not clearly defined: as "small",[50] "not serious",[51] "the risk of a routine medical treatment"[52] or as in only "observational" ("non-interventional") studies.[53] It may be added that research without potential individual benefit is not necessarily an intervention with risks or harm.

It is safe to assume that standardized observations and inquiries, examinations by psychological tests, EEG and magnetic resonance tomography (MRT), as well as taking venous blood, are procedures with only minimal risk or "with no kind of risk".[54] However, in relation to the individual's disposition some of these procedures, as e.g. MRT, may provoke considerable discomfort.

(5) The patient's *refusal must be accepted*.[55] The burden placed on the patient by one of the examination procedures mentioned above may differ according to the individual. In any case, the examination must be stopped when the patient, and likewise the incapacitated patient, indicates that the intervention is so unpleasant for him that he wishes it to be discontinued.

A specification was given in the Netherlands: the research will not continue if the incompetent subject protests, and his protest can be considered to deviate from the normal behavior encountered in the concerned group of incompetent human beings.[56]

---

[47] See Fagot-Largeault, *supra* note 8.
[48] See Amelung, *supra* note 37; Fagot-Largeault, *supra* note 8; and R.L.P. Berghmans, "National Report: The Netherlands" in Koch, Reiter-Theil & Helmchen, *supra* note 2.
[49] T. Hope & K.W.M. Fulford, "National Report United Kingdom" in Koch, Reiter-Theil & Helmchen, *supra* note 2; E.L. Mordini,"National Report: Italy" in Koch, Reiter-Theil & Helmchen, *supra* note 2; Casey, *supra* note 42.
[50] See Berghmans, *supra* note 48.
[51] See Fagot-Largeault, *supra* note 8.
[52] See Mordini, *supra* note 49.
[53] See Berghmans, *supra* note 48.
[54] See Mordini, *supra* note 49.
[55] See Fagot-Largeault, *supra* note 8.
[56] See Berghmans, *supra* note 48

(6) The *informed consent granted by the legal representative or guardian*[57] or by the judge responsible for guardianship[58] must be on hand. In many cases it may be doubtful whether merely expected mediate benefit will be sufficient ground for the guardian's consent in the interest of the patient. Therefore, a so-called "living will" or an authorization in matters of health as a significant indication of the patient's will would be helpful. Another possibility is the demand of the proposed law in the Netherlands that the participant must be informed in conformity with his comprehension.[59]

(7) The regional *ethics committee* (or Institutional Review Board) *must have decided positively* on the research project. The ethics committee must explain its positive vote with regard to the criteria mentioned above[60] and, on the basis of this vote, recommend a judicial decision — either directly or for purposes of appointing a guardian.

(8) In view of the general interest in public *safety* it is worth considering the extent to which there would be a need to call in the National Ethics Committee additionally. Likewise, specific research projects could be exclusively admitted at specific institutions, which would then be subject to on-the-spot control.[61]

To sum up, in Germany non-therapeutic research with incompetent adult patients is not legally permitted at present. Non-therapeutic research with potential mediate benefits for the patient participating in such a research project, as expected by diagnostic research in the true sense of the word, could be considered as ethically justifiable and should become legally permissible for incapacitated adult patients, as is the case for minor patients. Non-therapeutic research without immediate, but at least mediate, benefit for the patient seems to be ethically justifiable as a legally defined exception, provided that stringently formulated safeguarding criteria as mentioned above are observed. Therefore, in such exceptional cases the legal guardian must be convinced to his own satisfaction that the research procedure is not against the patient's best interest, or — at most — involves no more than minimal risks and negligible discomfort, and that the patient's refusal must be accepted.

## CONCLUSION

Today in most countries informed consent seems to be legally codified, ethically recognized and increasingly in use as a necessary prerequisite for all medical interventions. This is grounded in respect of autonomy as a basic right of a human being to decide him/herself what is in his/her best interest. This principle had been developed particularly with regard to medical research because research aims

---

[57] See Amelung, *supra* note 37.
[58] O. Benkert, "Klinische Prüfungen bei einwilligungsunfähigen Patienten. Ein Mainzer Vorschlag zur Korrektur des Arzneimittelgesetzes" (1995) 66 Nervenarzt 864;
Fagot-Largeault, *supra* note 8.
[59] See Berghmans, *supra* note 48.
[60] *Ibid.*
[61] See Mordini, *supra* note 49.

primarily at furthering scientifically-proven knowledge as a benefit, the scope of which goes beyond the individual. Therefore, medical research with persons incompetent to consent seems to be accepted as ethically justifiable and legally permissible only if the participating person can also expect an individual benefit and if specific safeguards such as substitute informed consent by a legal guardian are fulfilled.

A real problem remains, namely, medical research with incompetent patients from which only questionable or no individual benefit (often incorrectly equated with non-therapeutic research) may be expected. This is reflected in the fact that such research is legally permissible under defined conditions in some countries, like France and presumably the United Kingdom, but not in other countries like Germany. From a medical point of view such research seems ethically justifiable if its need is proven specifically according to defined criteria and if defined safeguards are fulfilled, particularly no more than minimal risks and negligible inconveniences and acceptance of a refusal of the participating person as explained in this paper. Furthermore, concepts and definitions of benefit and minimal risk, of criteria, rules, and procedures for weighing them against each other and for weighing individual benefit versus societal benefit, and the question of who is to conduct this assessment must be elaborated further.

The discussion of so-called non-therapeutic psychiatric research with persons incompetent to consent reinforces fears of medical science with its sometimes dehumanizing aspects. Because such research more than other medical interventions implies the risk of instrumentalizing human beings, it touches deeply the basic human right of respect of the person's dignity. Therefore, a public debate on this issue is necessary, not only because respect for dignity is bound to openness and an effort to understand the other, but also because in an open society an understanding of public opinion is a prerequisite for the representative societal decision-makers to set the framework for such research. In this debate philosophical researchers may explain the relationship of utilitarian to deontological ethics and its practical consequences, whereas medical researchers have the difficult task of making convincingly clear: both the individual and societal need for, and consequences of, such research by means of specific examples; the differentiated reality of its benefits, risks, burdens; and the ethical justification of such research.

Although in a rapidly changing world it is understandable if people try to hold onto what they take to be undisputable, seemingly clear principles, it is also a serious question what the empirical consequences of these principles are, particularly with regard to successful and humane adaptation to the mentioned changes. Thus, the burdens of such a thorough public discussion must be borne.*

---

* Reference follow-up: H. Helmchen, "Research with patients incompetent to give informent consent", Current Opinion in Psychiatry 1998, 11: 295–297; H. Helmchen, "Research with incompetent detained patients", Eur Psychiatry (submitted).

CHAPTER 23

# THE CONDITIONS OF PERSONHOOD AS APPLIED TO INCOMPETENT PERSONS

MICHEL SILBERFELD

## INTRODUCTION

The assessment of competence has become the standard way to triage the population in order to preserve the autonomy of capable subjects and to protect the vulnerable. The status accorded to incompetent subjects is of paramount concern if this triage is to provide an ethically acceptable result. This is even more apparent given the lack of uniformly accepted reliable and valid criteria for assessing mental capacity.

With the application of more restricted notions of competence, such as decision-specific capacity, many incapable subjects will still be able to voice wishes.[1] The weight given to wishes of the incompetent is critical in defining the conditions of personhood accorded to incompetent subjects. Incompetence may justify beneficent intervention, and some research interventions sanctioned by a suitable authority. However, a finding of incompetence is a weak justification for non-protective, non-healing interventions, because incompetence does not invalidate personhood. Some conditions of personhood may actively support research on incompetent subjects. Such support is based on the general conditions of being a person, rather than the subject being incompetent. Therefore, consent to research is sufficient for the capable subject's participation; but a lack of consent is insufficient to preclude research.

In the realm of research on incompetent subjects, there is an advantage to be gained by introducing wider considerations that reflect a fuller account of persons in general and vulnerable persons in particular. The narrow account is the legal one that defines persons as having a primarily negative right, such as freedom from unwanted

---

[1] M. Silberfeld, K.V. Madigan & B. Dickens, "Liability Concerns about the Implementation of Advance Directives" (1994) 15:9 Canadian Medical Association Journal 285.

466

interference. This narrow legal account has dominated the debate on consent to treatment.[2] The right to freedom from unwanted interference has been advocated as trumping all other rights, with very few exceptions, having to do with the preservation of life. But even when life is at stake, the right to self-determination has been used to argue in favor of a person's wish to die. That right has been supported by the courts.

While conceding the grave importance of the right to freedom from unwanted intrusion, there is room for other rights to gain primacy depending on the circumstances. The right of self-determination construed in a more positive sense may bring some persons into conflict with their concurrent right to be free from unwanted intrusion. For example, persons may wish to forgo the right to be free in favor of receiving treatment, or preserving important relationships, or safeguarding their work. This limited view is uniquely adversarial and reflects the needs of the law to sort things out in the way that it does, in the courtroom context. From a different vantage point, competing rights are an expression of competing values within the individual. Within the individual these competing values coexist, and may not be resolved in a mutually exclusive way.

A fuller account of persons and vulnerable persons (diagnosed with Alzheimer's disease for the purpose of this paper) will be provided without, by necessity, being fully extensive. The purpose of this account is to illustrate the characteristics given to persons, and vulnerable persons in particular, that augment their status as persons. These "conditions of personhood" are used to show the limitations of the use of consent as a vehicle for fully resolving the issue of participation in research at the time that research is undertaken. By way of discussion, the use of the Ulysses contract is proposed for the participation in research of those diagnosed with Alzheimer's disease. The Ulysses contract is a legal contractual device through which an individual anticipates future incapacity and requests to be treated notwithstanding objections that the individual foresees he or she may raise while incapable. The Ulysses contract permits us to see the confinement of personal wishes, and the diminished view of persons created by an assertion of negative rights alone.

I will conclude from my argument that there are conditions of personhood that are of paramount significance, some of which would lead persons to participate in research in violation of an absolute right to freedom from intrusion. This conclusion can be reached without exhaustively reviewing the literature on the Ulysses contract. Whether such conditions are widespread in the population is a question in answer to which an empirical solution should not be neglected. The law of contracts can accommodate a broader account of persons as complex contradictory beings who find solutions to life's difficulties in different ways not suitable to the prevalent legal trends with respect to consent. Being afflicted with an irrevocable, eventually fatal, illness, may at times lead some persons to forgo freedom from unwanted intrusion

---

[2] L.B. McCullough & S. Wear, "Respect for Autonomy and Medical Paternalism Reconsidered" (1985) 6 Theoretical Medicine 295.

for a more positive assertion of self-determination, personal integrity, and privacy. The latter conditions may be fulfilled by the desire to participate in research.

## CONDITIONS OF PERSONHOOD

People retain their membership in the class of persons through a variety of affiliations. It is not my purpose to enumerate these conditions, even if this could be done exhaustively. I will raise only some of the conditions of personhood necessary for this paper, and make sure to distinguish conditions not fully or well encompassed within the legal notion of personhood. Legal persons (individuals) in the law are recognized primarily by their possession and exercise of rights and duties before the law. However, people who do not exercise their legal rights, and who do not have outstanding obligations toward the law are still persons. Clearly the various ways in which persons are not legal persons will have legal implications because the law encompasses so much of our lives, addressing almost every aspect of life if only in a limited way. Three conditions of personhood will be considered as they apply to the circumstances of the newly diagnosed Alzheimer's patient: (1) fostering personal integrity; (2) promoting interpersonal connection; and, (3) the search for personal meaning.

To paraphrase Freud, people live to love and to work but they can do neither successfully without hope of fulfillment. The ability to anticipate the future and to shape the future to one's aspirations is a distinctly human quality. The continuity of persons is recognized by themselves and others according to the trajectory along which enduring values have led them. People stand for whom they are and hold themselves accountable for their lives. People change and yet retain their identity no matter how poorly we can explain that. The expectedness of the change is crucial; it is change that is absorbable into the whole. When unexpected illness and the reality of death intrude into ordinary expectations, these new certainties have to be absorbed to maintain personal integrity.

Personal integrity requires effort, struggle, self-scrutiny, etc., and it is not easily achieved if it ever is. At any one time, persons are generally divided within themselves. Personal coherence, such as it exists, is not brought about by rationality, but by the person having a sense of themselves imbedded in their life history. Nor is personal coherence a result of consistency over time, for there are many occasions when persons are not consistent; for example, they may fail in their prior commitments, and yet they maintain their integrity in other ways. The relevance of these considerations will become apparent in discussing the reactions of persons to illness and impending death.

Illness and impending death are radical intrusions into the life of persons. Common reactions to both include a struggle with hopelessness, helplessness, a search for comfort from others, a search to make the changes in life expectations meaningful. People's expectations for their present and future constitute, psychologically, an ordinary entitlement. With the loss of that entitlement that ensues upon a diagnosis of Alzheimer's disease, the alterations are felt as losses. In the face of losses, hope has to be restored and maintained. This can only occur with some alteration of the person's identity, expectations, and often a profound realignment of values.[3] It is in this way that a person may not remain consistent with themselves in the face of death, and yet sustain their personal integrity.

With Alzheimer's disease and its profound deterioration of memory and other brain fragments/functions, a point is reached where personal integrity breaks down. The center cannot hold, the traces of the person's own life history have been eroded, the capacity to integrate the disparities is greatly diminished, personal meaning cannot be mustered, and significant others are not recognized. At this point, depending on the practical problems to be solved, persons with Alzheimer's disease may or may not be granted status as persons. Legal personhood may be withdrawn when the person fails to pass the test of mental capacity for certain specified legal tasks. Policy makers may choose not to allocate resources to Alzheimer's patients who have lost their sense of personal identity, and refuse life-sustaining treatment for them.[4] In contrast, those who provide continuing care may have a clinical focus that permits them to see subtle signs of integrity and identity. There are ascriptions of personhood that may go unnoticed even in the most profoundly demented: demented persons retain their status as married in the eyes of the law and others despite their failure to be able to live up to the commitment. All of which is to say that incompetent patients retain some of the conditions of personhood even if they do not retain all of their legal rights.

## RESTITUTION FOR LOSSES INCURRED BY ILLNESS AND INEVITABLE DEATH

When people suffer loss, they try out various maneuvers to attempt to make good what they have lost, physically or symbolically. With untreatable illness invariably leading to death, restoring the loss is not possible, nor is there a substitute restoration that would preserve the life of the afflicted. All attempts to make good the losses will be a compromise.

Compromise ways of compensating for losses are ubiquitous.[5] This is because the means for restoration of losses are not always present or thought of. Compromise is second best only if there really is a better way, otherwise compromise is the way to restore a sense of personal integrity in the face of inevitable loss, even if it does not fully succeed.

Self-sacrifice is one way to restore a sense of integrity, and to create meaning in the face of inevitable loss, and, since sacrifice is for another, it is a way to affirm interpersonal connections when they are just about to be lost. Sacrifice of life on the

---

[3] M. Silberfeld, "The Psychology of Hope and the Modification of Entitlement Near the End of Life" in V.D. Volkan & T.C. Rodgers, eds., *Attitudes of Entitlement: Theoretical and Clinical Issues* (Charlottesville: University of Virginia Press, 1989) 41.
[4] D.W. Brock, "Justice and the Severely Demented" (1988) 13:1 Journal of Medicine and Philosophy 73.
[5] C. Brenner, *The Mind in Conflict* (New York: International Universities Press, 1982).

battlefield, in aid of medical research, and in exchange for the life of a loved one are but a few examples where the sacrifice is accorded great merit. However, sacrifice does not have to involve a life.

For a long time, psychoanalysts have identified the inclination of people to sacrifice a part to preserve the whole. This is very evident in illness when the afflicted person promises to be virtuous in exchange for immortality. Yet, sacrifices do not have to be based in illusion. Sacrifices that have a potential for fruitful results can enhance personal integrity in the face of illness by giving meaning to one's actions near the end of life, and promoting interpersonal connections by offering hope to those that remain.

I believe that some persons afflicted with Alzheimer's disease, perhaps many, will want to make the sacrifice that is required for participating in a well conducted research. I believe that these people will accept the burden of discomfort, possible pain, and unperceived risk in favor of doing something of benefit to others (and perhaps themselves, if fortunate). On the benefit side of the compromise, the sacrifice offers a great deal. It offers a chance to participate in overcoming the disease that claims them. Meaningless pain is worse. And pain remains private until it is made meaningful.[6]

## GIVING AND RECEIVING

In recent times, our attention and that of the law has been on the entitlement rights of citizenship. During this time of relative abundance, people examined the claims they could make on each other, and on the state. There was a strong emphasis on what individuals should receive as a matter of right, including services from the state. Now with abundance perceived as receding, there is a renewed interest in philanthropy and in giving in a more personal way than through taxes.

Philanthropy, however, remains suspect today. The motives for giving appear more dubious than the motives for receiving. Nevertheless, it is good to remember that philanthropy is well established in the history of humankind. Throughout the ages the motives for giving have been varied, and have changed over time (for example, Veyne[7] gives an account of public giving; Titmuss[8] gives a sociopolitical account of blood donation across nations; Silberfeld and Fish[9] describe giving to incompetent persons; and Weisstub[10] reviews the issue for research). Regardless of the perceived merits of the motives for giving, people want to give. Giving is a social act that brings people into connection with others. But it has more personal rewards as well. Some of those rewards may have as an indirect consequence compromising

---

[6] D. Bakan, *Disease, Pain, and Sacrifice* (Chicago: University of Chicago Press, 1971).

[7] P. Veyne, *Bread and Circuses* (London: Penguin Books, 1992).

[8] R. M. Titmuss, *The Gift Relationship* (London: George Allen & Unwin, 1990).

[9] M. Silberfeld & A. Fish, *When the Mind Fails: A Guide to Dealing with Incompetency* (Toronto: University of Toronto Press, 1994).

[10] *Enquiry on Research Ethics: Final Report* (David N. Weisstub, Chairman. Submitted to the Hon. Jim Wilson, Minister of Health of Ontario, Aug. 28, 1995) [hereinafter *Enquiry on Research Ethics*].

oneself to some degree. Even so, people want to give. Not all of them can be discredited on the basis of "bad faith".[11]

I will assume that giving is an integral part of social life. Today, giving includes the donation of one's body after death to medical schools, the donation of body parts to transplant recipients (bone marrow, corneas, etc.), and the donation of blood products. This has become commonplace. Often, these gifts are given to strangers (unknown to the donor) who are part of the community in which we live.

Giving reinforces the sense of membership in the community in which we live, and strengthens our identity as contributing members of that community. What is perceived as our community can be altered by illness. The recent appearance of support groups and advocacy groups for people afflicted with a particular illness is not surprising. People afflicted with an illness have always felt an affinity, and they do indeed share a common good, with others similarly afflicted.[12]

There is a large community of people afflicted with Alzheimer's disease. This community is expected to become larger. Many members of this community will feel the desire to participate in finding a cure, or some amelioration, for Alzheimer's disease. Some will be willing to give of themselves by participating in research. They will want to feel that they can continue to give by contributing to research if and when they become mentally incompetent. Incompetence is one of the most distressing consequences of Alzheimer's disease because the period of mental incapacity can last for so long. For many, it will seem rewarding to be able to give in some fashion during this time of incapacity when their opportunities to do so are very restricted.

## THE ULYSSES CONTRACT

The Ulysses contract permits capable persons to plan in advance of impending incapacity to safeguard their wishes. The Ulysses contract sets down the conditions whereby a person contracts to be treated in the face of their own opposition when incapable. The objective is to give permission to the physician to disregard objections to treatment uttered by an incapable person, and thereby allow a full course of treatment to proceed. Freedom from interference while incapable is given second place to the desire to maintain relationships — therapeutic as well as others — favoring a course of treatment. Positive rights (right to life, security of the person, and privacy) are being asserted; a negative right (freedom from interference) is being forsaken. Furthermore, the opportunity becomes available to give through the sacrifice of the arresting power of later incapable protestations.

Could a variant of the Ulysses contract have a role in solving the dilemma of research on persons likely to become incapable by virtue of Alzheimer's disease? At present in Ontario, as well as many other jurisdictions, the use of advance directives

---

[11] J. Elster, *Sour Grapes* (Paris: Cambridge University Press, 1985).
[12] B. Brown, "Proxy Consent for Research on the Incompetent Elderly" in J.E. Thornton & E.R. Winkler, eds., *Ethics in Aging: The Right to Live, The Right to Die* (Vancouver: University of British Columbia Press, 1988).

of various types has been encouraged by policy makers, and enabled by legislation. As a large segment of the population ages, more and more people will become vulnerable by virtue of mental incapacity. As well, we have gone through a period in our history in which the right to freedom from interference by medical intervention has been strongly advocated and supported by changes in the law. Advance directives are promoted as the vehicle whereby the wishes of the person can be asserted against the interference of forcible treatment. The Ulysses contract is an advance directive as well, but unlike a living will, it does not prohibit interference.

The Ulysses contract is more like a testament or last will.[13] The last will of a person is a vehicle for asserting wishes in face of the inevitability of death. It assures the person while living the likelihood that their wishes will survive them. The law treats these last wishes seriously and provides the administrative and procedural apparatus to enforce them. One can see why wills and Ulysses contracts are analogous. The latter, made by a competent person with an early diagnosis of Alzheimer's disease, would express wishes in the face of an impending lengthy period of mental incapacity that is irreversible, and ends only with death. A will can have some limitations imposed by other laws, such as matrimonial laws, which would restrict the enforcement usually accorded by the law.[14] Similarly, limits can be imposed on the Ulysses contract and the analogy with a testament can still hold.

Recent (1995) changes in the Ontario health laws enabled such contracts. Nevertheless, possible objections are worth examining to see their potential applicability to the circumstances of a person recently diagnosed with Alzheimer's disease who remains capable and wishes to make provisions to participate in research when incapable. The motives for participation will be germane to, and encompass, the considerations above. The following will also make clear the relevance of making advance directives disease-specific, and specific (in this case) to the context in which research is carried out.

The Ulysses contract should fulfill certain conditions. It must be made by a competent person. It must be clear on what is to be disregarded. For the persons diagnosed with Alzheimer's disease desirous of participating in research even if they become incapable, it would be possible to clarify for researchers what objections to disregard. Furthermore, it would also be possible to make clear what objections would have to be sustained. The person making the Ulysses contract would be helped in his or her task by the research protocol, and the process of obtaining informed consent for the particular research protocol. Where the research protocol has not been examined in advance of incapacity, the task of doing so could be knowingly assigned to a proxy.

I am assuming that there will be some local scrutiny in place that would govern the conduct of research. Local circumstances would determine the limitations that

---

[13] M. Silberfeld, C. Nash & P. Singer, "Capacity to Complete an Advance Directive" (1993) 41 Journal of the American Geriatrics Society 1141.

[14] M. Silberfeld & W. Corber, "Permissible Errors for Will-Making Capacity" (1996) Estates and Trusts Journal (In Press).

are set on the Ulysses contract. This may vary from one regulated environment to another. Helmchen[15] shows that there are differences amongst European countries with respect to the conduct of therapeutic and non-therapeutic research. Eventually, some international standard may be achieved along the lines of the Helsinki Declarations.[16] So long as there are local standards, regulatory bodies, and the potentially vulnerable person is made aware of them, a competent person could come up with directions as to what is to be disregarded, and what is not to be disregarded. My argument is not contingent upon a single set of limitations that should be set on the Ulysses contract.

I am in favor of regarding competent aged persons who are recently diagnosed with Alzheimer's disease as full persons. They should be entitled to make pledges about their future and to participate in research if it is a meaningful personal reaction to being afflicted. That entitlement is grounded in the individual and personal struggle in the face of illness.

Unlike state-based arguments, or cooperativist ethics, the wellspring for participating in research has been made out to be an act of restitution for the sense of loss and an attempt to retrieve some benefit from lost time due to mental incapacity. These motivations might persuade persons to waive the right to personal inviolability and to accept some degree of unforeseen risk (within regulatory safeguards as recommended by the Weisstub Enquiry on Research Ethics).[17]

It is worth remembering that the elderly person making such a contract has a long history of being capable, and has a wealth of accumulated life experience. This is no moral fiction. The state should provide the regulatory framework for assuring that the research is of merit, but I question whether the state has a role in "protecting" people against a voluntary, informed, and capable choice to participate in research.

## POTENTIAL OBJECTIONS TO THE ULYSSES CONTRACT

A person may not be able to anticipate sufficiently the impact of research on themselves when incapable. Most persons will not have had a prior experience of being the subject of research. A research intervention may not be very different from any proposed treatment in its potential harms, though it may be if there are not sufficient procedural safeguards on the quality of research and its execution. For this paper, I will assume that the procedural safeguards with respect to research will be sufficient, given the deserved attention they have received. Of course, the viability of the contract — or contracts, should there be more than one — would have to be examined for every research project separately. Though consent to treatment is specific to the treatment offered, an advance directive can be more general. It can authorize treatment consistent with sound medical judgment. Similarly, the Ulysses

---

[15] G. Helmchen, " Ethics in Psychiatric Research with Incompetent Patients" (1996) (In this Volume).

[16] G. J. Annas & M.A. Grodin, eds., *The Nazi Doctors and the Nuremberg Code.* (New York: Oxford University Press, 1992) 331.

[17] See *Enquiry on Research Ethics, supra* note 10.

contract could authorize research consistent with the highest principles of ethics which allow for the conduct of research.

Having been diagnosed with Alzheimer's disease may lead a person to sacrifice freedom from interference and accept potential harms resulting from research because of the shock of the diagnosis and the helplessness resulting from a lack of effective treatment. It is true that the motives for entering into a Ulysses contract could vary depending on personal circumstances. For a recently diagnosed person, entering into such a contract could be a poorly considered reaction to being ill, or not. Furthermore, nondelusional reasons are generally accepted in the area of consent, even if they are perceived to be ill-advised or eccentric, because they are considered to be meaningful to the person. Preferably, such a contract would be drawn up at a time when the person has had some time to adjust to the presence of illness sufficiently to appreciate its future consequences.

Could it be that the bargaining power of a person afflicted with Alzheimer's disease, though capable, puts them invariably in a disadvantageous position? Only capable persons can enter into a contract, including the proposed Ulysses contract. This danger can be circumvented. The person may make a Ulysses contract without direct involvement of a researcher. When the researcher is involved, independent advice may be offered or even required. Research Ethics Boards may be consulted in the institution where research is likely to be carried out (teaching hospitals), or in some other umbrella institution (University). This may compensate for the individual's lack of personal experience with research, and for the antecedent nature of the choice.

The determination of mental capacity is crucial to the implementation of the Ulysses contract. It is only at the turning point when capacity is lost that permission is given to disregard current wishes. The definition of mental capacity has been evolving. There has been a retreat from viewing capacity as global in nature toward viewing it as task-specific. The particular task to be performed in most cases is decisional, and this explains the ability or lack thereof to undertake any dependent, related functional task. The criteria for assessing mental capacity have yet to be widely agreed upon. However, some general direction is available. This general direction has yet to be shown to be sufficient to ensure fairness in the application of the criteria. Nevertheless, this problem is surmountable, providing one is prepared to accept that in this area a judgment has to be passed on another person in a mode that is common in many other social judgments, especially in the law. Today, more and more quantitative tests are being developed with a view to meeting some of the objections about reliability and uniformity of application of mental capacity criteria. If the Ulysses contract for research were to become acceptable, no doubt standards could be tightened further in the determination of capacity to consent to the research procedure in question at the time. One might want to set a lower threshold for refusal in such circumstances than might exist for consent to treatment.

With a more refined assessment of mental capacity being required, there is no doubt persons will be found who are incapable, and are yet able to voice wishes refusing participation in research. Some will be able to acknowledge their instructions to disregard them as contained in their Ulysses contract, while others may not be able to acknowledge their advance directive. Should this vitiate the contract, and the whole endeavor of entering into Ulysses contracts? For the most part, advance directives assume that persons would use this device to overcome ill-considered decisions and actions resulting from incapacity. It is also a common practice in everyday life to use devices to protect oneself from oneself. Human development requires people to gain control over themselves when they anticipate losing control in their decisions and actions. Restraint over one's emotions is required to avoid irrational choices issuing from "weakness of will"[18] and yielding to impulse rather than due deliberation. Given that the person is truly incapable, and procedural safeguards are in place (for example, recourse to an advocate), it is in keeping with other advance directives to permit the contract to stand in the face of opposition by an incapable person. To give an advance directive capably a person must appreciate that at the point of incapacity they will forgo the opportunity to withdraw the directive even if the full implications of the directive are not available at the time of giving.[19]

Can we recognize circumstances where a person can forgo the right of self-determination even when they voice a well established right to refuse interference? If the person is currently incapable, this thorny dilemma can also exist with respect to consent to treatment when an advance health care directive exists.[20] Where life-sustaining treatment is the issue, the problem appears unresolvable at present. But, in the context where the Ulysses contract has been prepared with all the safeguards discussed above, what is at stake here? Some may believe that there are no "incompetent" wishes when refusal is the issue. A refusal should be taken at face value no matter the circumstances, Ulysses contracts notwithstanding.

The Ulysses contract as an entity poses a challenge to the prevalent view that legal rights are primarily negative, and are only used to assert freedom from interference from unwanted intrusion. With the Ulysses contract persons asserting other rights may trump their right to freedom from unwanted intrusion expressed when they are incapable. The rights asserted may include self-determination (control over oneself, and one's future), privacy, and integrity (the fulfillment of one's better wishes, as in a last will and testament). Persons afflicted with Alzheimer's disease will know if the Ulysses contract is properly executed as well as the rights they are waiving (along with the other considerations above). Good reasons will have to be provided by them to do so, and these should appear in the contract.

Patients (persons afflicted with Alzheimer's disease) are more likely than some legal advocates, and some politically motivated persons, to view research as a benefit to mankind which does not inherently conflict with respect for autonomy. For these persons who are willing to enter into a Ulysses contract permitting research, they share (more or less) common values and interests with the community

---

[18] See Elster, *supra* note 11.
[19] See Silberfeld, Nash & Singer, *supra* note 13.
[20] See Silberfeld, Madigan & Dickens, *supra* note 1.

of researchers, and the community of others so afflicted who seek a remedy both for themselves and/or others.

With chronic deteriorating illnesses such as Alzheimer's disease, which invariably result in eventual death, what right will a person recover if their incapable wishes are asserted against their previously made directive? I have tried to show that little is lost and much is gained by permitting some persons with Alzheimer's disease to enter into a Ulysses contract for research into this condition which as yet has no effective treatment. My considerations and conclusions may be different for other illnesses,[21] such as acute relapsing psychotic or non-psychotic conditions.[22]

We don't know the likelihood that Ulysses contracts will be employed, or challenged when they are present. Macklin has argued, for mental patients, that the law should have no interest in upholding the contract where it is challenged, nor should there be a social apparatus to enforce it.[23] Should this be the same for Alzheimer's Disease? A challenge would likely only arise if the contract was being adhered to in violation of local restrictions imposed by regulatory bodies on such contracts in general, and therefore in opposition to what the person giving the Ulysses contract had anticipated. Such a contract should not be enforced. Where the "challenge" is the person's anticipated incompetent objection, I would hope that local regulations would create a process for the researcher to seek independent confirmation before proceeding, especially in those circumstances where the research allowed has a permissible level of risk and associated pain that are on the "higher" end.

It has been argued that persons cannot make choices for themselves in anticipation of the future because they will not be the same person in the future, and therefore this is equivalent to making a choice for another. Taking this argument to its logical conclusion, any previous decision (separated in time) can be discredited. It is true that for observers it is hard to ascribe an identity to persons in the late stages of Alzheimer's disease. However, we know so little about the subjective state of these patients that I would be loath to concede that they should be deprived of their customary identity to the extent of not being permitted to plan for future incapacity.

Another argument against making decisions in anticipation of the future is the inability to fully appreciate future risk, and to properly allow for unforeseen risks. I have argued in another context that there is no justification for assessing risks in any manner that is not consistent with everyday life. Setting a greater standard of prudence is a covert way of depriving people of opportunities that are available to all (whose risks are not being assessed). Almost all decisions made about the future are made with limited knowledge; the limits on knowledge are due both to the fact that we don't know the future with any certainty and the fact that we cannot in any circumstance anticipate *unforeseen* risk. Nevertheless, we carry on, and the

---

[21] P.A. Singer, "Disease Specific Advance Directives" (1994) 344 Lancet 594.
[22] A. Macklin, "Bound to Freedom: The Ulysses Contract and the Psychiatric Will" (1987) 45:1 University of Toronto Faculty of Law Review 37.
[23] *Ibid.*

assumption of risk is part of what makes our lives meaningful. There is no decision-making body that can circumvent these limits on our knowledge of the future.

A further argument against making the choice to participate in future research is that the decision to do so is an unprecedented one, in the sense of being unfamiliar: that is, it is unlikely that a person would have had prior experience making decisions of this sort. It is likely that few people would have had the experience of participating in research protocols prior to being afflicted. In the future, it may be more likely that people will have had such experience. Nevertheless, people make decisions in unfamiliar circumstances all the time, and they do so competently, based on their customary problem-solving styles. The decision to accept treatment for an illness the first time around is an unfamiliar one. It may remain unfamiliar for another type of illness, or treatment at a subsequent time. Making a will is not a familiar experience (except for those who wield their wills as a weapon against their beneficiaries). The lack of prior experience speaks to the requisites for making a good decision, but does not speak against making a decision *per se*.

One thorny issue has to do with the right to withdraw at any time regardless of competence. It is well known that persons afflicted with Alzheimer's disease will in the late stages of the illness object to a great many things that change their circumstances or otherwise challenge them. Routine care such as bathing is one of many common examples. This tendency to object, even to innocuous things, is a part of the illness profile. It is difficult then to take the withdrawal of assent as sufficient to overcome a prior direction. However, it is also difficult to allow objections on the part of the patient to be disregarded without approval of someone, or some authority that places the patient's best interests above the research objectives. It is clear that in these circumstances, the patient's best interests to be determined include their pre-stated desire to be a research subject. This is why a withdrawal of assent, or even an objection while incapable, should not necessarily lead to a withdrawal from research.

In sum, the consent-oriented approach to resolving the question of participation in research by patients with Alzheimer's disease is misplaced. The consent/contract approach creates a representation of people for a particular purpose(s). In the case of research it overreaches its grasp. This is not to say that this approach may not play a part in deciding who can participate in research and in what circumstances. Rather, what is asserted is that the conception of persons created to deal with consent to treatment is not adequate nor sufficient to deal with their participation in research (particularly non-therapeutic research), especially after the point of incompetence. Trying to make it "fit" doesn't work well. It leads to confused and confusing arguments. One such argument is that participation in Ulysses contracts leads to "self-paternalism".[24] Once we invent such notions there is nothing left by which to identify authentic actions: even looking after your best interests when capable, as you see them, is suspect.

---

[24] *Ibid.*

## CONCLUSIONS

I have argued that personal aspirations and the need for psychological restitution and restoration in the face of illness are part of the conditions of personhood. The fulfillment of such needs is justifiably attributed to the vulnerable, and constitutes another form of self-determination.

These attributes are applicable to persons made vulnerable by Alzheimer's disease. They may wish to be subjects for research on the illness that afflicts them. Recent developments in the genetics of Alzheimer's disease pointing to a common genetic inheritance may multiply their motives for wanting to make such a commitment to assist their relatives.

The consent/contractual approach to safeguarding research subjects is of limited effectiveness. Society must determine through other means the merits of permitting the research to take place. Participation is a secondary consideration. Participation in non-therapeutic research cannot be properly modeled on consent to treatment. Participation in research is more akin to making a will, more akin to an expression of a binding wish. Unlike consent to treatment, participation in research does not require a reciprocal exchange. It is an act of giving. There is a hope of receiving, but not in a contractually binding fashion.

Not all sacrifices are the same. Given a framework for the proper monitoring of research, as contained in the *Enquiry on Research Ethics* (1995), the sacrifice of participating as a subject in Alzheimer's research may be acceptable to the point of allowing people to waive the right of inviolability of the person in favor of some personally meaningful purpose. We need to have data on how many people in the population would have a general willingness to participate in research if afflicted by an incurable illness leading to inevitable death. I believe people have been inspired and made hopeful by the recent discoveries about the genetics of Alzheimer's disease. There is a desire to find hope, and though it can be misled, it should not be suppressed nor thwarted, lest hopelessness itself be added to our afflictions.

CHAPTER 24

# ETHICAL RESEARCH WITH VULNERABLE POPULATIONS: THE DEVELOPMENTALLY DISABLED

DAVID N. WEISSTUB AND JULIO ARBOLEDA-FLOREZ

## INTRODUCTION

Although developmental disorders have often been included in a broad notion of "mental disorder"[1] based on cognitive impairment, general definitions of developmental disability neglect important and unique characteristics of those afflicted that affect both their capacity to make autonomous choices and the ability of others to help them improve their lives. Whereas all mentally retarded individuals have a level of intellectual functioning that places them in the lowest 2.5% of the population, developmental disabilities impair intellectual activity to a much more varying extent.[2] This fact distinguishes the developmentally disabled from the other

---

[1] See e.g. the definition for mental disorder in *DSM IV*. See American Psychiatric Association, *Diagnostic and Statistical Manual of Mental Disorders*, 4th ed. (Washington, D.C.: A.P.A., 1994) [hereinafter *DSM IV*] at xxi-xxii. The Office for Protection from Research Risks defines the "cognitively impaired" as those persons:

> [h]aving either a psychiatric disorder (e.g. psychosis, neurosis, personality or behavior disorders), an organic impairment (e.g. dementia), or a developmental disorder (e.g. mental retardation) that affects cognitive or emotional functions to the extent that capacity for judgment and reasoning is significantly diminished. Others, including persons under the influence of or dependent on drugs or alcohol, those suffering from degenerative diseases affecting the brain, terminally ill patients, and persons with severely disabling physical handicaps, may also be compromised in their ability to make decisions in their best interests.

See Office for Protection from Research Risks, National Institutes of Health, *Protecting Human Subjects: Institutional Review Board Guidebook* (Washington, D.C.: U.S. Government Printing Office, 1993) at 6–26.

[2] J.W. Ellis "Decisions By and For People with Mental Retardation: Balancing Considerations of Autonomy and Protection" (1992) 37 Villanova Law Review 1779.

479

vulnerable populations. Children show an increasing capacity to make decisions. The elderly, while presumably fully competent for most of their lives, may potentially exhibit the opposite tendency of decreasing decision-making capacity. The mentally disordered have been characterized as having fluctuating periods of lucidity, and may have been mentally competent at some point in their lives. The developmentally disabled, however, have never possessed and are never likely to possess sufficient competence to make all decisions regarding their own welfare. And in cases where such persons are institutionalized, the voluntariness of their decisions to participate in an experiment may also be cast into doubt.

Any discussion with regard to the participation of developmentally disabled persons in non-therapeutic biomedical research must therefore, address issues of competence and voluntariness. A policy governing experimentation on this population must also aspire to maximize and assist the decision-making capacity of such individuals. But before addressing these issues, it is important to define this population in light of its special characteristics.

## DEFINITIONS

Members of special populations are not all equally vulnerable. This is no less true of the developmentally disabled where disorders include a broad range of conditions such as Mental Retardation, "Pervasive Developmental Disorders" (e.g. Autistic Disorder),[3] Learning Disorders, Motor Skills Disorders,[4] and Communication Disorders. Each disability has characteristics that will require the researcher to adjust the approach taken to assess the person's ability to make decisions. For the purposes of this discussion, the category of Mental Retardation is the one which is most likely to generate concerns about the capability of a potential subject to give a valid consent to participation in biomedical research. According to the *DSM IV*, the criteria for this condition are the following:

> ... significantly subaverage general intellectual functioning (Criterion A) that is accompanied by significant limitations in adaptive functioning in at least two of the following skill areas: communication, self-care, home living, social/interpersonal skills, use of community resources, self-direction, functional academic skills, work, leisure, health and safety (Criterion B). The onset must occur before age 18 years (Criterion C).[5]

This category covers a broad range of disabilities which, according to the *DSM IV*, can exist in four degrees of severity, reflecting the level of intellectual impairment:

---

[3] Pervasive Developmental Disorders are characterized by "severe deficits and pervasive impairments in multiple areas of development. These include impairment in reciprocal social interaction, impairment in communication skills, and the presence of stereotyped behavior, interests, and activities". See *DSM IV*, *supra* note 1 at 38.
[4] *Ibid.* at 37–39.
[5] *Ibid.* at 39.

- *Mild Mental Retardation*, which reflects an IQ level from 50–55 to approximately 70; approximately 85% of all those persons with Mental Retardation fall within this particular group;
- *Moderate Mental Retardation*, which reflects an IQ level from 35–40 to 50–55; individuals in this group constitute about 10% of all those affected by Mental Retardation;
- *Severe Mental Retardation*, which refers to those individuals with an IQ level from 20–25 to 35–40: this group constitutes some 3% to 4% of all individuals with Mental Retardation; and
- *Profound Mental Retardation*, which reflects an IQ level that is below 20 or 25; this group comprises only 1% to 2% of people with Mental Retardation.[6]

Another widely accepted definition of mental retardation is that of the American Association on Mental Retardation ("AAMR"):

> Mental retardation refers to substantial limitations in present functioning. It is characterized by significantly sub-average intellectual functioning, existing concurrently with related limitations in two or more of the following applicable adaptive skill areas: communication, self care, home living, social skills, community use, self-direction, health and safety, functional academics, leisure, and work. Mental retardation manifests before age 18.[7]

This definition focuses on a functional construct requiring that the intellectual impairment be accompanied by related limitations in particular skill areas.[8] This requires researchers to consider the individual in the context of a specific experiment.

Although there is considerable variation in their levels of intellectual impairment, developmentally disabled persons share certain characteristics, with corresponding implications for their participation in research:[9]

(1) There is a deficit in basic knowledge; they often lack basic information that is relevant to making a decision. This means enhanced duties at the information stage of the consent process.

(2) Communication skills are greatly impaired. Researchers cannot assume silence is either a lack of comprehension or an assent.

(3) The individual often denies the disability preventing him or her from seeking help in making a decision when such help may be badly needed.

(4) There is a reduced ability to make decisions. For example, the individual may impulsively seize on the first solution regardless of consequences when faced

---

[6] *Ibid.* at 40–42.

[7] This is a recent revision of the AAMR's definition of mental retardation. Mental retardation previously referred to a "significant sub-average general intellectual functioning existing concurrently with deficit inadaptive behavior and manifested during the developmental period". See American Association on Mental Retardation, *Mental Retardation: Definition, Classification and Systems of Supports*, 9th ed. (Washington, D.C.: American Association on Mental Retardation, 1992).

[8] See Ellis, *supra* note 2 at 1782.

[9] *Ibid.* at 1784–6.

with decisions which require a greater degree of assertiveness, or which could have major consequences, or in which options for action are not clear.[10]

(5) The settings in which the individual is asked to make a decision may be so coercive as to call into the question the legal adequacies of the decision the individual is asked to make.

(6) The disability is permanent in the sense that the intellectual impairment is not "curable" or "changeable" in the ordinary sense of those terms, although significant changes may occur over the life span of the person and the ability to make choices may be affected by successful special education programs or environmental changes.[11]

There is a significant likelihood that individuals with Mental Retardation will also be affected by some other form of mental disorder. According to *DSM IV,*

> Individuals with Mental Retardation have a prevalence of comorbid mental disorders that is estimated to be three to four times greater than in the general population. In some cases, this may result from a shared etiology that is common to Mental Retardation and the associated mental disorder (e.g., head trauma may result in Mental Retardation and in Personality Change Due to Head Trauma). All types of mental disorder may be seen and there is no evidence that the nature of a given mental disorder is different in individuals who have Mental Retardation.[12]

The *DSM IV* also points out that diagnosing comorbid mental disorders may be complicated by "the fact that the clinical presentation may be modified by the severity of the Mental Retardation and associated handicaps".[13] For example, poor communication skills manifested by a person with Mental Retardation may render it impossible to acquire the personal history adequate for an accurate diagnosis. Indubitably, it is vital that any assessment of an individual's capacity to participate in biomedical experimentation should be extremely sensitive to the possibility that the individual may belong to more than one vulnerable population. In such cases, the individual is entitled to the protections applicable to all of the groups to which he or she belongs.

## ISSUES CONCERNING THE DEVELOPMENTALLY DISABLED

### *Historical Influences*

It is not necessary to reach far back into history to encounter a period during which the central question raised by biomedical research was not whether it was appropriate for the developmentally disabled to participate in such research but

---

[10] J.C. Jenkinson, "Who Shall Decide? The Relevance of Theory and Research to Decision-Making by People with an Intellectual Disability" (1993) 8 Disability, Handicap & Society 361 at 368.

[11] See Ellis, *supra* note 2 at 1784–6.

[12] See 2DSM-IV1, *supra* note 1 at 42.

[13] *Ibid*. at 42–43.

whether such persons were suitable subjects for experimentation.[14] Unfortunately, this suitability was greatly enhanced by the widespread practice of institutionalization, which itself was a product of common prejudices supported by crude scientific theories.

### The Effect of Institutionalization

Physical segregation and the gathering of the developmentally disabled in institutions made them particularly convenient subjects for a broad range of protocols unrelated to their particular disabilities. There are several notorious examples of abusive research that was undertaken with the developmentally disabled, including the Fernald School radiation experiments.[15] Institutionalization was considered to be such a significant factor in the creation of conditions permitting abuse that the National Commission for the Protection of Human Subjects of Biomedical and Behavioral Research identified it as a necessary condition for inclusion of potential subjects in the vulnerable populations that were being reviewed. Yet, the Commission did not investigate the particular needs and vulnerabilities of either the developmentally disabled or psychiatric patients except to the extent that they were "Institutionalized as Mentally Infirm".[16]

The developmentally disabled represent a vulnerable population whose interests and needs cannot be assimilated to those of the majority at some supposedly analogous developmental point in their lives. Decisions about their involvement in research must therefore be based on something other than a pragmatic or consequentialist rationale. Unlike "normal" children, the developmentally disabled will not become the providers for today's adult generation in their old age and, unlike the elderly, they at present do not motivate the majority to develop techniques and patterns of care from which ultimately they themselves hope to benefit. Indeed, it is a sad comment that, in the past, this lack of a pragmatic or consequentialist interest in the needs of the developmentally disabled has often removed them from

---

[14] On the question of the suitability of the institutionalized disabled for research, it is apparent from a summary of conference proceedings of the National Association for Retarded Children that, as recently as 1964, there was a consensus that the challenge in this area, especially in the case of isolated rural institutions not directly affiliated with a major university or general hospital, was to attract researchers and the financial assistance provided by research grants. It was further agreed that this could be done by encouraging the research community to regard such a facility as a "ready-made laboratory" in which environmental controls and the establishment of control groups was more easily accomplished than in other settings. See E. Hart, *The Role of The Residential Institution in Mental Retardation Research* (Report of the Conference Sponsored by The National Association for Retarded Children, 23–25 May 1964) at 8–11.

[15] Advisory Committee on Human Radiation Experiments, *Final Report* (Washington, D.C.: U.S. Government Printing Office, 1995) at 342-346; Task Force on Human Subject Research, *A Report on the Use of Radioactive Materials in Human Subject Research that Involved Residents of State-Operated Facilities within the Commonwealth of Massachusetts from 1943 to 1973* (submitted to Philip Campbell, Commissioner, Commonwealth of Massachusetts Executive Office of Health and Human Services, Department of Mental Retardation, April 1994).

[16] See National Commission for the Protection of Human Subjects of Biomedical and Behavioral Research, *Research Involving Those Institutionalized as Mentally Infirm: Report and Recommendations* (Bethesda, Md.: The Commission, 1977).

human reference points[17] and reduced the evaluation of research into their health needs to a mere cost-benefit analysis.[18]

Those who set the objectives and standards for research involving the developmentally disabled have done so, historically, on the basis of limited personal experience of their particular fears, aspirations, dislikes, and desires. As perceived differences from the majority of the population once justified institutionalizing the developmentally disabled in remote rural facilities, it was rare for those who had not themselves experienced a form of developmental disability to have encountered a disabled individual in "normal" social interactions and to have acquired insight into their special interests and needs.

### The Impact of Deinstitutionalization

Much has changed over the past few decades. The trend towards deinstitutionalization[19] and integration within the community has meant that developmentally disabled persons are much more likely to be known and understood as individual members of families in private households, or as residents of group homes, than as inmates of distant institutions. This has greatly increased the hope that the interests of this vulnerable population will be addressed not only in terms of their need for protection as research subjects, but also in terms of the value and quality of the research conducted, as well as the global allocation of research resources.

If we are incensed by inmates being subjected at institutions like Willowbrook to non-consensual participation in biomedical research, then we should be similarly troubled by the research design and the use subsequently made of studies conducted in institutions such as the Vineland Training School. These studies involved the ill-treatment of those deemed "retarded", and in turn were actually used to justify that treatment. The developmentally disabled were abused not only as a consequence of their involuntary exposure to an unacceptable degree of risk in the research of this era, but also by virtue of the nature of the treatment that was "indicated" by the results of these studies.[20]

In marked contrast, during more recent times, the reintegration of the developmentally disabled into the community has helped to create a group of competent lay persons who possess a special insight into the health care needs of this vulnerable population and who have a particular interest in influencing the direction of future research. They can point out the political short-sightedness that

---

[17] During the conference of the National Association for Retarded Children, referred to *supra* note 14, "the question was raised whether it is appropriate to establish an animal laboratory within an institution [for disabled children]. On the positive side, it was pointed out that ideas gained from the behavior of children may be explored and studied in animals and that juxtaposition has heuristic virtues". See Hart, *supra* note 14 at 11.

[18] Conference participants recognized that continued delay in developing the full potential for collaborative research activities within institutions means that many individuals are bearing the impact of retardation, and society is bearing the cost of services for such individuals, unnecessarily. *Ibid.*

[19] See The Law Commission, *Mentally Incapacitated Adults and Decision-Making: A New Jurisdiction (Consultation Paper No. 128)* (London: HMSO, 1992). at para. 2.2.

[20] M.S. Crissey, "Vignettes in Mental Retardation" (1983) 18 Education and Training of the Mentally Retarded 117.

neglects public health concepts of prevention and early intervention in favour of crisis management.[21] Most importantly, they advocate habilitation within a developmental model that stresses learning life skills, although much remains to be learned about how this can be effected. When members of the National Commission for the Protection of Human Subjects of Biomedical and Behaviorial Research visited the Eunice Kennedy Shriver Mental Retardation Research Centre, they were particularly distressed by three facts:

(1) There was a large number of individuals with retardation over age 21 who were in a consent limbo: they were clearly incompetent to make decisions, but no guardian had been appointed for them by a court and often there was not an involved parent;

(2) There was a lack of a clear transitional line from innovative research or training to standard practice; and

(3) The patients they saw made them question whether it was ethical not to conduct research when there is a clear need to improve the care and training of persons with retardation, not only to prevent mental retardation but to avoid placement in an institutional setting.[22]

Ironically, the global improvement in the lot of this vulnerable population has transformed a concern for the financial burden they place on society for their care into a complaint about the increased cost of undertaking research. It was once asserted that medical research should be considered economically prudent insofar as it reduced the need to maintain the developmentally disabled in rural institutions. Today, the complaint is increasingly being expressed that deinstitutionalization and enhanced requirements for consent have made recruiting research subjects a much more onerous task and, in some cases, have actually cast doubt on the perceived cost-efficiency of conducting any research whatsoever into the specific needs of the developmentally disabled.

While it may provide satisfaction to those concerned with preventing abuse to discover that both the challenge to find willing subjects and the enhanced procedural protections now entrenched in enrolling them may deter abuse, it is nevertheless sobering to bear in mind the other, perhaps more subtle and damaging, form of abuse found in the neglect of and indifference to the particular needs of this population.[23]

---

[21] See Brakel *et al., The Mentally Disabled and the Law*, 3rd ed. (American Law Foundation, 1985) at 617.

[22] D. Alexander, "Decision Making for Research Involving Persons with Severe Retardation: Guidance from the National Commission for the Protection of Human Subjects of Biomedical and Behavioral Research" in P. Dokecki & M. Zaner, eds., *Ethics of Dealing with Persons with Severe Handicaps: Toward a Research Agenda* (Baltimore: Paul H. Brooks, 1986) 39 at 44.

[23] The National Commission for the Protection of Human Subjects of Biomedical and Behavioral Research was given the mandate to examine a variety of populations that were potentially in need of regulatory protection against abuse in research. However, it was impressed in its consideration of mentally retarded persons by the ignorance relating to the care and treatment of these persons and it was anxious to make recommendations that did not unnecessarily impede the research it believed critical to improved diagnosis and treatment of mental disabilities. *Ibid.* at 48-50.

A quarter of a century ago, Haywood spoke of "the right of mentally retarded persons, as well as of other persons, to the best methods of care, treatment, education, and habilitation that we have the power to give them", adding that the right not to participate in research may come into direct conflict with the impetus to achieve concrete improvements in these areas.[24] It may seem axiomatic, in light of current ethical principles, that members of a vulnerable population should enjoy the benefits of high standards of care regardless of their participation in research. Nevertheless, the potential of enhanced procedural safeguards in recruiting subjects to restrain the pace at which medical advances are accomplished should spur those concerned with the welfare of the developmentally disabled to ensure that these safeguards are limited to those measures truly necessary to preserve the safety and dignity of the individuals involved.

The inherent magnitude of this dilemma is reflected in the fact that it was one of the underlying reasons for the failure of the National Commission for the Protection of Human Subjects of Biomedical and Behavioral Research to have its *Recommendations for those Institutionalized as Mentally Infirm* implemented as concrete regulations. The Commission's proposed regulations, influenced by its concern that too little research was being undertaken into the needs of the developmentally disabled, were subsequently rendered far more stringent by the Department of Health, Education & Welfare (DHEW) and, in turn, the resulting popular protest on the part of researchers and advocacy groups effectively prevented regulation in this area.[25] A general requirement that Institutional Review Boards (IRBs) should ensure that appropriate additional safeguards have been included in the research protocol in order to "protect the rights and welfare" of research subjects[26] has encouraged these IRBs to consider both the National Commission's original Proposal and the DHEW's rejected regulations; however, that this issue is unresolved serves as a powerful reminder of how profound the controversy involved is.

### *Specific Concerns*

Today, researchers who wish to conduct non-therapeutic experiments involving the developmentally disabled confront two main questions:

(1) Can a person from this population give a valid consent to the proposed experiment?
(2) If not, can a substitute decision-maker (SDM) ever consent to non-therapeutic experimentation?

The very expansive range of disabilities that may be present among those with differing levels of Mental Retardation requires a close examination of the extent to

---

[24] H.C. Haywood, "The Ethics of Doing Research . . . and of Not Doing It" (1976) 81 American Journal of Mental Deficiency 311 at 312.
[25] See Alexander, *supra* note 23 at 51.
[26] 45 C.F.R. 46 (1991) § 46.111 (7)(b).

which a potential subject is capable of making an independent decision whether or not to participate in biomedical experimentation. This question can be satisfied by utilizing a mental competency assessment.[27] For example, it may well be the case that many of those individuals with only a mild degree of Mental Retardation are capable of making such a decision on their own, particularly if they receive help and support from relatives. On the other hand, those with severe or profound levels of Mental Retardation will clearly not be able to engage in such decision-making and, if they are to participate at all, consent must first be given by a third party.[28]

Consequently, the focus of inquiry in respect of this population will be on competence and voluntariness of consent, and recommendations will emphasize the need to assess and maximize the decision-making capacity of the developmentally disabled, as well as the ambit and influence of SDMs.

### Mental Competency

The presumption of competency is based in the principle of respect for persons, and has as its correlate the requirement that incompetency *not* be presumed without substantiating evidence. This includes recognition of the fact that incompetence in one area of decision-making does not imply the inability to make decisions in other aspects. However, these concepts must be considered in conjunction with the potential for abuse made possible by presuming competence in one who is incapable of understanding and does not actively object to the performance of a procedure for research. Although people with developmental disabilities have every right to take risks and accept both benefits and burdens as they choose, this right should be exercised only in accordance with a requisite level of competence.

It is also recognized that, notwithstanding the shared diagnosis of developmental disability, individuals within this vulnerable population span the entire spectrum from those capable of making independent decisions through those who are capable of making such decisions with assistance to those incapable of involvement in any decision-making of this sort. The common law presumption of competency should be respected, but the rights of the developmentally disabled person must be protected by a competency assessment whenever there is a reasonable doubt as to the individual's capability to make a specific decision.

For the developmentally disabled, reduced cognitive capacity calls into question their ability to make or effectively communicate independently, an informed and voluntary choice to become involved in biomedical research. Effective consent requires the ability to understand the nature of the various research procedures concerned, appreciate the consequences of a decision to participate in research, and communicate the nature of such a decision. Neither legal nor functional competence

---

[27] See D.N. Weisstub (Chair), *Enquiry on Mental Competency: Final Report* (Toronto: Queen's Printer, 1990) for a detailed discussion of competency assessments.

[28] Also, the "markedly abnormal or impaired development in social interaction and communication" that is characteristic of Autistic Disorder clearly implies that individuals with this disorder would almost always be considered incapable of consenting to participation in research. For discussion of Autistic Disorder, see *DSM IV, supra* note 1 at 66-71.

is a general, static concept, but should be evaluated in relation to required specific decisions. Whether an individual will be able to consent to participate in an experiment depends on the abilities of the person, the nature of the decision to be made, and the likely consequences of the decision for the person".[29] The same person may be able to function competently in a specific environment, but be unable to consent to an experiment. Conversely, he or she may need a great deal of assistance physically, but be competent to consent to a specific procedure. It is important to realize that competency can vary among the developmentally disabled. A particular individual may have capacity to consent to some proposed experiments, but not to others, and often, it is not clear whether that individual is capable of making a particular decision.[30]

One obvious challenge presented by this range is that of finding an accurate method of assessing competence so that individual procedural safeguards may be fashioned. Furthermore, the method of assessment ultimately employed must be directed toward identifying the specific ability of the individual to make decisions concerning participation in the research process. With the elderly, the principal uncertainty with regard to their ability to understand basic issues is not likely to relate to the nature of the procedure as much as to its purpose (i.e. whether it can be considered as therapy or research). However, in the case of a developmentally disabled person, the capacity to make any medical decisions whether in relation to research or treatment may be ambiguous at best. Also, as with the mentally disordered, individuals may claim to understand when objectively they do not.[31] Therefore, competence must not be confused with alleged comprehension.

A considerable literature exists on methods of assessing competency.[32] Advances have been made in the accuracy of methods of assessing consent.[33] For many developmentally disabled individuals, the reduction of the competency threshold to a level commensurate with the risk presented by the research project will mean these persons will have capacity to make this decision. Where the research is not invasive and involves low risk, individuals who normally would be unable to make medical or therapeutically beneficial decisions may have capacity to participate in a project.[34]

---

[29] See Jenkinson, *supra* note 10 at 372.

[30] See Ellis, *supra* note 2 at 1802.

[31] M. Irwin *et al.*, "Psychotic Patients' Understanding of Informed Consent" (1985) 142 American Journal of Psychiatry 1351.

[32] See Weisstub (1990), *supra* note 27.

[33] See e.g. R.D. Tustin & M.J. Bond, "Assessing the Ability to Give Informed Consent to Medical and Dental Procedures" (1991) 17 Australia and New Zealand Journal of Developmental Disabilities 35.

[34] Among the many requirements for ethically permissible research is the critical stipulation that the involvement of incapable persons be considered only in relation to research that cannot be conducted with persons capable of giving informed consent. It is recommended that this requirement be vigorously sustained within this population as well. To be more specific, not only can research with incapable persons be considered only when that research is not possible with capable persons, but subjects must also be selected in a hierarchical fashion, with those considered least capable of ever giving consent being selected last for the study. In this respect, it should be required that a research protocol explicitly indicate how this assessment is made and how it will be observed in recruiting developmentally disabled subjects.

Once a developmentally disabled individual has been determined incapable of providing independent consent and the caregiver has expressed willingness to support participation in research presenting only a minimal degree of risk, researchers should be prevented, however, from dealing on future occasions (for reasons of expediency) with the caregiver alone. An individual's capacity to make decisions should be reassessed whenever a reasonable doubt is raised as to the existence of this capacity, whether because of the passage of time, variations in the protocol, or other significant changes in circumstances. In all cases, the researcher should proceed on the basis of the most recent appraisal of capacity.

Researchers also have an obligation to their developmentally disabled subjects to maximize capacity itself. Although legal competence may be absent, researchers should recognize an ethical responsibility to foster *functional* competence where possible, because of its significance in habilitation. The developmentally disabled individual may lack the legal ability to provide a valid consent to participate in research, but retain functional capacity to make certain decisions. As Jenkinson noted:

> Decision-making by people with an intellectual disability is not only in accord with principles of human dignity and autonomy, but also conveys benefits such as increased motivation and improved task performance. Despite this, and despite an awareness by service providers and care-givers of the need to encourage and create opportunities for decision-making, there is still considerable evidence that even in many routine activities of daily life opportunities for self-determination are not occurring to the same extent as for others in the community. Environments should be structured to allow the individual maximum control over options, to ensure that options are clear, and to help the decision-maker become aware of his or her preferences without feeling under pressure or influenced.[35]

A more positive environment is created when researchers seeking the participation of a developmentally disabled individual in a protocol presume the individual to be capable of participating in the decision, devise methods of presenting the information to maximize the contribution of the individual, and reconsider the capacity of the individual on an ongoing basis.[36] Participation in the decision-making process may improve the skills of the individual so that competency may increase as the individual's knowledge of the research project increases and as his or her comfort level with any changes in the environment caused by the project increases. As well, as the project progresses, the level of risk may decrease, lowering the threshold for legal competency.

*Voluntariness*

Problems in discerning capacity are exacerbated by the fact that the developmentally disabled generally live in a coercive environment created by:

(1) pthe power of a legal guardian, or the authority of institutional staff; and

---

[35] See Jenkinson, *supra* note 10 at 372.
[36] See Medical Research Council of Canada, *Guidelines on Research Involving Human Subjects* (Ottawa: Supply & Services Canada, 1987). These guidelines have been superseded by the Tri-Council Working Group, *Code of Ethical Conduct for Research Involving Humans* (July, 1997).

(2) the belief (usually valid) that: "as a practical matter, they must obtain 'permission' from non-disabled individuals to do things that no other adults in society must obtain permission to do. Both people with Mental Retardation and non-disabled individuals who deal with them on a regular basis assume that such authority is natural, necessary and appropriate".[37]

Thus, voluntariness is severely compromised. Individuals will assent to participation in an experiment because they feel they are expected to by others. This will not be a valid consent in the technical meaning of the term. A person may not have capacity to give a valid consent, but may assent to the procedure by not objecting. In a coercive atmosphere, even this assent (which is really a lack of objection) may not be truly voluntary. It is for this very reason that, wherever possible, the subject's assent, to the extent of the subject's capabilities, must be sought. Failure to provide assent or mere cooperation should not be treated as assent, and continued assent should be monitored throughout an experiment.[38]

*Substitute Decision-Making*
Where decisions are being made about the participation of children, the elderly, or the mentally disordered, it is usually feasible to create an external reference point based on the anticipated or remembered personality of the specific individual concerned. In the case of the developmentally disabled, however, it is manifestly impossible to establish such a standard. For this vulnerable population, there is no opportunity for SDMs to take into account a developing, diminishing, or previous capacity for independent decision-making.[39]

A legally appointed decision-maker should have the duty to ascertain current wishes and to encourage the participation of the individual in the decision.[40] It is recommended that similar principles apply to decisions made on behalf of participants in non-therapeutic research, because there can be a therapeutic benefit conferred by the opportunity to participate in the decision-making process. All caregivers have a particular responsibility to encourage control by the developmentally disabled over values, decisions and choices. This means SDMs should ensure that their own values and preferences do not have undue influence on the choices and options provided to the developmentally disabled person.[41] This process should not be confused with obtaining informed consent in the legal sense from the individual.[42]

It can be argued that the term "substitute consent" is a misnomer when applied, in the context of very severe Mental Retardation, to the condition of a person who

---

[37] *Ibid.* at 1802.
[38] See Chapter 18.
[39] Therefore, the focus of inquiry for developmentally disabled persons will often not be on the adequacy of consent, but on what areas of protection are needed and when. A person who is severely retarded may have no preference, or may not be committed to preferences or values or to anything that would determine what the person's preference would be if not disabled.
[40] *Substitute Decision Act, ibid.*, ss. 66(4) & (5).
[41] See Jenkinson, *supra* note 10 at 370.
[42] See Ellis, *supra* note 2 at 1809.

may have never possessed the capacity to express a value, belief or desire.[43] In these circumstances, substitute consent has been described as a legal fiction.[44] In the context of treatment decisions, Dresser has pointed out:

> the effort to force these patients into the model of ourselves as autonomous decision-makers (seeking desperately to avoid their dire situations) distracts us from the real people before us. In consequence, we miss seeing who they are. It is all too true that these patients are difficult to find. Many cannot speak with us; those who can tend to speak in tongues that seem impossible to decipher. If we want to know them, to understand the value life has for them, we must depart from the customary, comfortable methods we have for exploring the subjective world of another human being. We must undertake a different approach, since these patients typically cannot talk with us about "what it is like" to be in their situation. Yet, the existing legal doctrine barely recognizes this need and consequently creates little incentive for decision-makers to do so.[45]

Harmon concludes that the substituted judgment test allows the state to invade the bodily integrity of the incompetent person without having to justify the invasion.[46] As an alternative, Dresser suggests that it is preferable to drop, or at a minimum, seriously reroute the effort to identify the individual's hypothetical choice. Instead, decision-makers should focus on the person's current conditions, the concerns of those who love and care for that person, and the concerns of the larger community to which she belongs.[47]

Legislation should recognize the unique qualities required by a SDM of a developmentally disabled person, particularly in the context of consent to research.[48] As such, the following would apply:

(1) The SDM's willingness and reasons for serving as SDM must be established.[49]

(2) The SDM must be competent to make the decision to consent to research. The SDM must be capable of assessing possible risk to a particular individual and be aware of any specific personal characteristics that would put the person at increased risk. The decision-maker should be sensitive to various types of harms (social or moral, not only medical) to which this particular person is vulnerable. Only then can a proper balance of risks and benefits be made.

(3) The SDM should disclose reasons for the decision. This should indicate that the decision-maker recognizes the complexities of making these decisions for

---

[43] See R.M. Veatch, "Persons with Severe Mental Retardation and the Limits of Guardian Decision Making" in P.R. Dodecki & R.M. Zaner, eds., *Ethics of Dealing with Persons with Severe Handicaps* (Baltimore: Paul H. Brooks, 1986) 239 at 244 in which it is said that in the case of all patients who have never been competent, the moral mandate for others is to promote the patient's best interest.

[44] L. Harmon, "Falling off the Vine: Legal Fictions in the Doctrine of Substituted Judgment" (1990) 100 Yale Law Journal 1.

[45] R. Dresser, "Missing Persons: Legal Perceptions of Incompetent Patients" (1994) 46 Rutgers Law Review 609 at 612.

[46] See Harmon, *supra* note 44.

[47] See Dressser, *supra* note 45 at 612.

[48] See Chapter 8.

[49] See S.A. Kline, "Substitute Decision-making and the Substitute Decision-maker" (1992) 13 Health Law in Canada 125.

another person, and, in particular, understands the standards to be applied. Any connections between the researcher and the SDM should be revealed. It may be the case, for example, that pressures be exerted on the decision-maker to appear cooperative with the staff in order to assure good treatment for the developmentally disabled person.

(4) Conflicts of interest must be revealed. For example, if a proposed project would relieve the decision-maker of some of his or her caregiving responsibilities, and thus confer a benefit, this should be revealed. This would not necessarily constitute a ground for refusing to give the SDM powers, but the information should be revealed. If the conflict of interest is serious, another decision-maker should be sought.

(5) The SDM should be able to describe steps taken to engage the participation of the developmentally disabled individual in the ongoing consent process.

If the SDM first named fails to meet these guidelines, a person who is able to meet the following criteria should be named:

(1) competency to make decisions;
(2) willingness;
(3) no serious conflict of interest;
(4) the ability to ascertain the wishes or feelings of the developmentally disabled person; and
(5) the ability to engage the developmentally disabled person in the decision-making process.

Whenever the condition of the developmentally disabled person permits, the opinion and approval of that person should be obtained in the selection of a SDM.

## SOME GENERAL SAFEGUARDS FOR RESEARCH ON THE DEVELOPMENTALLY DISABLED

Most international and domestic guidelines and codes do not refer specifically to the developmentally disabled, but include this population in their recommendations concerning experimentation with the mentally disordered. Although most of the recommendations made in respect of experimentation with mentally disordered persons are equally applicable to the developmentally disabled, given that the members of both populations possess varying degrees of cognitive impairment, the unique situation of the developmentally disabled requires that application of these principles take into consideration the special characteristics of this population.

In addition to the specific recommendations stated above, general safeguards should be adopted.[50] These include: the requirement of the scientific validity and importance of an experiment; the requirement on the part of all participants in the

---

[50] See Chapter 18.

research process to minimize risks and to balance possible risks with potential benefits; a general limitation of the level of risk to which a vulnerable person may be exposed; the requirement of assent; and the importance of respecting an individual's right to object to and withdraw from (or be withdrawn from) an experiment. Any objection by the subject, even if he or she is legally incompetent, should be respected and should supersede the decision of a SDM.[51]

When developmentally disabled persons are research subjects, it is therefore recommended that:

(1) the assessment of competency include the assessment of the ability of the developmentally disabled person to express his refusal. Severely and profoundly retarded persons may have difficulty communicating their refusal. If this is the case, a record should be made to alert researchers and SDMs that extra care should be taken in monitoring the experiment.

(2) If a SDM has been appointed for the person, the SDM should monitor the progress of the experiment and be sensitive to expressions of disapproval by the person. A researcher may not know the individual well enough to recognize signs of objection. The SDM must recognize that in addition to providing consent of behalf of the individual he or she must also honour an objection on behalf of the developmentally disabled person.

## CONCLUSION

Owing to the unique factors that affect the capacity to provide consent in developmentally disabled persons, it should be recognized that there is a substantial difference between this population and other vulnerable groups. There are substantial mental and communication disabilities that may not be apparent to outsiders, the individuals may live in a uniquely coercive environment, and many of the decisions involve possible deprivation of fundamental rights.[52] In this regard, the issues involved depend on the incapacity of the developmentally disordered person to give an informed consent, and on the potential involuntariness brought about by either institutionalization or complete dependence on SDMs. Every attempt should be made to maximize and facilitate the participation of developmentally disordered persons in the decision-making process, taking into consideration the variance in levels of capacity that exists in this population. Competency assessments should therefore be reevaluated during the course of an experiment if required, owing to the likelihood of increased familiarity and understanding of the nature and conse-quences of participation on the part of the subject. Finally, the point is often made

---

[51] The UK Law Commission Paper #128 stated that it is not reasonable for a researcher to force an incapacitated person to act in accordance with a decision to which the incapacitated person objects, unless such action is essential to prevent an immediate risk of serious harm to that person or others. This principle should be applied when considering whether to overrule the objection of a subject.

[52] See Ellis, *supra* note 2 at 1804.

that to establish the parameters and the internal structure for the process of obtaining a valid consent to participate in non-therapeutic experimentation, "diagnoses are not always accurate and the levels of mental acuity may vary considerably within the category and over time".[53] Therefore, assessments for competency should be made on the issue of competence independent from the issue of diagnosis or the level of intellectual capacity.

---

[53] D. Wikler, "Reflections on Research on Mentally Disabled Human Subjects" (1987) 23 Psychopharmacology Bulletin 372 at 373.

CHAPTER 25

# ETHICS IN RESEARCH WITH DETAINED INDIVIDUALS

NORBERT NEDOPIL

## INTRODUCTION

It is a difficult and delicate task for a German scientist to talk or write about ethical questions on medical research with detained individuals 50 years after the declaration of the *Nuremberg Code*. This Code was created as a reaction against the inhumane and cruel misbehavior of medical researchers during the Nazi Régime in Germany. Although its ideas were provoked by the massive violations of the rights and dignity of captives, the *Code* established important general principles for research on human subjects. The awareness of the inhumane practices of medical research during the Third Reich has not only increased awareness about ethical questions concerning scientific experiments in humans, but has also led to a general distrust of research on humans among many people — especially in Germany. Psychiatric research is particularly distrusted, especially if specific populations that are regarded as vulnerable are concerned. This is true for research with old and/or demented individuals, with prisoners, as well as with patients involuntarily committed to hospitals.

The ethical principles of research, which were agreed upon and published in several international declarations and which regulate — and render possible — research in several comparable countries are interpreted rather narrowly in Germany.[1] § 40,I,3 of the *German federal drug law* states that no clinical trial of

[1] See e.g. World Medical Association, *Declaration of Helsinki*. Adopted at the 18th World Medical Assembly in Helsinki in June 1964. Amended at the 29th World Medical Assembly in Tokyo in October 1975; the 35th World Medical Assembly in Venice in October 1983; and the 41st World Medical Assembly in Hong Kong in September 1989. [Reprinted in (1991) 19 Law, Medicine & Health Care 264.] See also *Good Clinical Practice for Trials on Medicinal Products in the European Community*; Council for International Organizations of Medical Sciences, in collaboration with the World Health Organization, *International Ethical Guidelines for Biomedical Research Involving Human Subjects* (Geneva: CIOMS, 1993); or World Health Organization, *Guidelines for Good Clinical Practice (GCP) for Trials on Pharmaceutical Products* (Geneva: WHO, 1995).

drugs may be conducted if and as long as the individuals are detained in an institution by court or administrative order. Although these laws regulate research pertaining to the efficacy of drugs and although there are no specific laws for other biomedical research, almost the same principles are applied for all biomedical experiments on humans.

When biochemical examinations of violent offenders that demanded blood and CSF taking were proposed to the deciding authorities, the authorities from a university suggested this kind of research be given up, prisons refused to cooperate, and special hospitals were very reluctant in their answers. The resistance against examinations that could cause even minimal physical harm to the prisoners or the committed patients was too great to pursue this research in Germany. Only psychopathometric examinations were regarded as ethically acceptable and even found some support among the prison authorities. When drug companies were asked to give support to a research symposium on violent offenders they refrained from doing so because they did not want to be connected with research on vulnerable populations, especially with detained individuals. So in Germany research with detained individuals is possible only on a very restricted scale and taking into account many ethical and legal precautions.

Although biomedical and medicolegal research with detained individuals do not have the same juridical limitations in other countries, the ethical principles are discussed with much controversy, and guidelines for research protocols with this vulnerable population are more often demanded than really proposed.[2] All authors agree that experiments that can be done on free individuals should not be carried out on imprisoned or committed persons. However, certain deviations of behavior can only be studied in forensic settings.

Different behavioral sciences, like biological psychiatry, psychopharmacology, psychology and psychoanalysis have increasingly turned their attention towards aggression and dissocial or antisocial behavior during the last decades. While psychoanalysis has studied behavior that deviates from what is accepted as normal and speculated on its causes for a long time, the other disciplines only lately have developed new methods of experimentation by which new results have been obtained, and knowledge could be extended: low 5-hydroxyindolacetic acid levels have been found in the cerebrospinal fluid (CSF) of habitually impulsive aggressive criminals as well as in the CSF of suicidal patients.[3] Psychopharmacological research developed specific serotonin reuptake inhibitors, so-called serenics, which

---

[2] A.D. Milliken, "The Need for Research and Ethical Safeguards in Special Populations" (1993) 38:10 Canadian Journal of Psychiatry 681; C. Potler, V.L. Sharp & S. Remick, "Prisoners' Access to HIV Experimental Trials: Legal, Ethical, and Practical Considerations" (1994) 7:10 Journal of Acquired Immune Deficiency Syndrome 1086; S. Ayer, "Submitting a Research Proposal for Ethical Approval" (1994) 9:12 Professional Nursing 805.

[3] M. Asberg *et al.*, "Psychobiology of Suicide, Impulsivity and Related Phenomena" in H. Y. Meltzer, ed., *PsychoPharmacology: The Third Generation of Progress* (New York: Raven Press, 1987) 655; M. Virkunnen *et al.*, "CSF Biochemistries, Glucose Metabolism, and Diurnal Activity Rhythms in Alcoholic, Violent Offenders, Fire Setters and Healthy Volunteers" (1994) 51 Archives of General Psychiatry 20.

prevent aggressive behavior in animal experiments.[4] Psychology elaborated on checklists that should be able to predict aggressive and antisocial behavior of humans more precisely.[5] The series of new results of research on deviant behavior is quite long. Many of the results were presented at a much publicized congress of the American Neuroscience Society in San Diego, Cal. in 1995. It is a current trend to put together knowledge acquired with different methods and to investigate the interactions of the individual parameters. The subjects, from which this knowledge is derived and on which the upcoming hypotheses are to be tested, are mostly imprisoned violent criminals or serial offenders.

## GENERAL ETHICAL PRINCIPLES APPLIED TO DETAINED INDIVIDUALS

The ambivalence between the necessity to do research in this special area and the restrictions based on ethical, legal or political reasons or simply based on distrust render it necessary to summarize the special ethical principles that should guide examinations and experiments with detained individuals. At the same time one should examine how research can be justified and how the conditions for research can be improved while keeping in mind the rights and privileges of those under examination. These include not only the principles of beneficence, autonomy and justice as required by the Belmont Commission in the U.S. and adopted in other countries[6] but extend to other requirements due to the special situation under which these individuals live and are cared for. All these principles can be seen as problematic with regard to research on detained individuals:

### *Beneficence*

The first ethical principle of research on humans requires that the knowledge gained from the experiment serves the individual under study or at least other individuals, that are at the same time or will be later in a similar condition. One can, however, question whom research on detained individuals really serves. Who will profit, if we can identify markers of aggression? What would be the real consequences if we developed a reliable method of identifying predictors of future violence? Dr. Hare expressed considerable reservations about the use of the psychopathy checklist currently the most valid single instrument for the prediction of future delinquency[7] in Texas as a tool to support an unfavorable prediction of violence.[8] A prediction of

---

[4] B. Olivier, J. Mos & D. Rasmussen, "Behavioral Pharmacology of the Serenic, Eltoprazine" (1990) 8 Drug Metabol. Drug Interact. 31.

[5] R.D. Hare, *The Hare Psychopathy Checklist — Revised.* (Niagara Falls and Toronto: Multi-Health Systems, 1990); C.D. Webster & D. Eaves, *The HCR-20 Scheme The Assessment of Dangerousness and Risk* (Vancouver: Simon Fraser University and Forensic Psychiatric Services Commission of British Columbia, 1995).

[6] F. Baylis *et al.*, *Health Care Ethics in Canada* (Toronto: Harcourt Brace, 1995).

[7] See Hare (1990), *supra.*, note 5.

[8] R.D. Hare, *Without Conscience* (Toronto: Pocket Books, 1993).

future delinquency is one of prerequisites used to justify capital punishment in that state.

### *Autonomy*

A second ethical principle for research on humans is that the individuals consent voluntarily to the experiments. This autonomy does not only mean that no pressure or coercion is used on the potential participants of research, but also that these may not be promised any advantages other than those granted by the project itself. In prisons, where the expectations of privileges, the manipulations of the system, the breaking of rules and the exploitation of loopholes in the regulations are daily routine, two reasons are given to object to research projects:

(a) The prison is afraid that loopholes are created in the system of rules, and that prisoners might demand privileges for participating in an experiment. Loopholes and privileges then could make it difficult to maintain a regular routine in the prison. Under these conditions the participation in a research project would not be voluntary but in direct expectation of privileges to be obtained from the justice system.

(b) A more important argument is used even if participation is truly voluntary and can be so documented in the research protocol: prison authorities fear that misinformation, rumors and publications that are based on them might induce the impression that research is done on involuntarily committed individuals. The wrong impression of involuntary participation in experiments has led to the premature termination of two projects in prisons neighboring the author's institution.

The individual prerequisites of autonomy, the mental competence to consent after being fully informed about the experiments, their risks and benefits are hardly disputed for prisoners in Germany. Individuals who lack the capacity to consent are mentally so disturbed that their criminal responsibility has to be questioned. These individuals are normally regarded as irresponsible or having diminished responsibility for their criminal acts and, if criminal recidivism is predicted, are sent to special hospitals; they are not found in prisons. For these patients, not only their competency to understand the technical procedures of the experiment, the individual consequences of the participation, and the consequences of non-participation have to be examined,[9] but also their ability to choose between alternatives, to act according

---

[9] L.H. Roth, A. Meisel & C.W. Lidz, "Tests of Competency to Consent to Treatment" (1977) 134:3 American Journal of Psychiatry 279; C.H. Cahn, "Consent in Psychiatry — The Position of the Canadian Psychiatric Association" (1980) 25 Canadian Journal of Psychiatry 78; P.S. Appelbaum & T. Grisso, "Assessing Patients' Capacities to Consent to Treatment" (1988) 319:25 New England Journal of Medicine 1635; D. N. Weisstub (Chair), *Final Report: Enquiry on Mental Competency* (Toronto: Queen's Printer, 1990); J.S. Janofsky, R.J. McCarthy & M.F. Folstein, "The Hopkins Competency Assessment Test: A Brief Method for Evaluating Patients' Capacity to Give Informed Consent" (1992) 43 Hospital and Community Psychiatry 132.

to their own system of values and to defend their own decision logically and understandably.[10]

But even if that is the case, biomedical research is not possible on this population for the above-mentioned legal reasons.

### *Justice*

Justice of research with detained individuals refers to the above mentioned principle, that the expected results can only be obtained within that population and that they conform with standards and beliefs not only of society in general, but also of the individuals that are approached for participation in that specific research. Difficulties may arise, if the questions of the experiment may be of interest to scientists and to prisoners but of little or no interest to prison authorities. Sometimes it can be more difficult to convince these authorities to collaborate than the prisoners.

## SPECIAL ETHICAL PROBLEMS FOR RESEARCH WITH DETAINED INDIVIDUALS

The additional ethical issues arise from the special legal and social status of detained individuals:

### *Health Care and Health Risks in State-Controlled Institutions*

Medical treatment and health care for most detained individuals are provided by the institutions that are responsible for them and which in turn are controlled by state authorities. In many instances the detained individuals have to tolerate necessary treatment. This is especially true if their hospital order is the consequence of a treatable disorder. The almost total health care in institutions also means, that possible health-risks have not only to be carried by those who participate in research but also by the institutions. So benefit-risk analyses have not only to be considered by the participants but also by the institutions that house them. The permission to do research depends in most institutions on the guarantee that the personnel will not be burdened by additional work or strain. If one considers — for example — the clinical trial of an antiaggressive drug on prisoners, the burden for the staff would not only be that the participating prisoners have to be observed intensively, and that the desired effects and the unwanted effects have to monitored continuously, but also that possible health risks are recognized by the institution in time. Long-term consequences of these trials may cause a longlasting strain on the institutions. Research projects that might pose an even minimal risk to the health of prisoners have not been permitted in penitentiaries in the state of Bavaria.

### *Critical Information Gathered during a Research Project*

Research on prisoners is predominantly aimed at behavior that is not as readily observable in people outside prisons, like aggression, violence or sexual perver-

---

[10] H. Helmchen & H. Lauter, *Dürfen Ärzte an Demenzkranken forschen?* (Stuttgart: Thieme, 1995).

sions. Information gathered during the research project might not only be of value for scientific knowledge but also for the staff and the authorities of the institution, e.g. if the information pertains to the security of the institution, if a potential of threatening violence or if sexual abuse of other inmates are detected during the project. This information might lead to a conflict of interests, when the immediate wishes and needs of those under research are in disagreement with the interest of the institution. These conflicts of interest might overburden a study and wreck the research with detained individuals. The anticipation of conflicts of interest is one of the reasons for the limited value of research among prisoners, since they are often inclined not to disclose their real motives and behavioral attitudes to the researcher. The studies of Toch[11] who employed fellow inmates and not academics as interviewers demonstrated the masking tendencies that the participants employed towards academics, who did not speak their language and were associated with the authorities.

So there are very important ethical reservations against research on detained individuals, reservations that can also lead to severe practical problems.

## NEED FOR RESEARCH AND GUARANTEE FOR RESEARCH

On the other hand there exists a genuine necessity, and finally even an ethically founded demand, to carry out research on detained individuals.

The *Constitution of the Federal Republic of Germany* and those of some other countries guarantee the freedom of science and also guarantee research itself. In medicine, research is essential, since giving up research would mean a resignation to avoidable diseases, disorders and harm. Such a resignation would then lead to disadvantages and handicaps for every person who suffers from the disorder and for the general public, which has to compensate for these handicaps.[12]

According to modern thinking health is not only a quality that concerns the individual but has definite impact on society as a whole. We all live in a society built on solidarity, a society that grants many advantages and rights to the individual. Especially in the health sector in Western Europe the individual profits from almost complete social welfare from the society. In the perspective of the social contract the individual cannot only profit from the benefits of social welfare but has to meet his obligations toward this society. Ethics of medical conduct do not have to limit themselves on the basis of the autonomy of the individual but should also consider the interests of solidarity within the society.[13] Insofar as it seems appropriate, medical research can depart from the limits set by medical ethics, which considers only the benefit of individuals as the justification for diagnosis, treatment or

---

[11] H. Toch, *Violent men* (Washington, D.C.: American Psychological Association, 1992).

[12] E. Deutsch, "Der Beitrag des Rechts zur klinischen Forschung in der Medizin" (1995) 48:46 Neue Juristische Wochenschrift 3019.

[13] N. Nedopil, "Zur Instrumentalisierung von Psychiatrie und Psychologie durch Justiz und Öffentlichkeit — aus forensisch psychiatrischer Sicht" (Vortrag bei der 10. Mnnchner Herbsttagung der AGFP am 23 October 1995) [In Press].

experiment. At least the question should be allowed, whether those who severely offended the solidarity of society and transgressed the social contract of society should not be asked for a special sacrifice for society.

The request for such a sacrifice could make sense, if these individuals do not suffer subjectively from their disorder, but posed undue burden on the society or on other individuals of this society. Research on recidivists with aggressive or sadistic sexual delinquency could be approached by way of these considerations.

The idea of responsibility of the individual for society and its solidarity and of sacrifice of those having burdened this society should not be accepted without serious reflection or be interpreted as a necessary duty for the individual. The history of Germany — but also that of other countries — demonstrates to what kind of misuse with minorities or socially underprivileged persons scientists have been tempted in the name of common welfare.[14] The legal system — according to its origins — has to protect the weak against the trespasses of the powerful in these cases. Therefore very careful regulation is necessary to make research possible for the sake of common welfare and for the sake of freedom of research and at the same time to protect the legitimate interests of all participants in research.[15] Review boards have been established in medical institutions to guard the ethical and legal principles of research on patients and healthy volunteers and to guarantee the rights of participants in research. Expertise and competence of medical review boards are however sometimes not quite sufficient when it comes to research on vulnerable populations. This limited expertise becomes evident when incompetent individuals, e.g. children, are to be tested for a scientific question. Children are rarely able to accomplish a realistic benefit-risk analysis, as they do not dispose of a binding set of values as the basis for — individual — ethical decisions. Therefore scientific and individual benefit and health risks have to be decided upon beforehand. Medical review boards are, however, overburdened with this task. The decision has to be left to scientists and practitioners who are very familiar with the questions which are posed by the specific kind of research and the situation the participants are in. In Germany, the question is intensely debated, whether a scientific board of advisers or a central review board in addition to the local review board should exercise an additional control, when research is being planned with incompetent individuals.[16]

## PROPOSAL FOR A PRELIMINARY SOLUTION

Although prisoners are able to accomplish a benefit-risk analysis, their decision will never be free from pressures and hopes that aim beyond the scope of the research project. Another dilemma results from the fact that information gathered by the

---

[14] See e.g. T. Benedek, "The 'Tuskegee Study' Of Untreated Syphilis: Analysis Of Moral Aspects Versus Methodological Aspects" (1978) 31 Journal of Chronic Diseases 35.

[15] J. Robertson, "The Law of Institutional Review Boards" (1978) 79 UCLA Law Review 79; E. Deutsch, "Verkehrssicherheitspflicht bei klinischer Forschung — Aufgabe der universitüren Ethik-Kommissionen" (1995) 12 Medizin und Recht 483.

[16] O. Benkert, "Klinische Prüfung bei einwilligungsunfähigen Patienten" (1995) 66 Nervenarzt 864.

scientist can be of special value to those in charge of the prisons. This may lead to manipulations on the side of the participants in research or to premature interventions of controlling agencies in the institutions. The problems become quite evident when one considers what consequences it would have, if an epidemiological study on the use of illegal drugs by interview and by blood, urine and hair analysis detected the channels of commerce. For these studies confidentiality has to be granted to the participants; at the same time scientists have to agree beforehand to terminate the study prematurely to allow interventions in order to avoid greater harm. A proposal to solve this problem would be to establish a council consisting of scientists, ethics professionals and authorities from the institution. This council would examine not only the ethical justification of proposed studies and consents at the beginning of it, but receive and examine intermediate anonymous reports in order to decide on the continuation or the termination of the study. This council has on the one hand to guarantee professional confidentiality[17] to the participants and it has on the other hand the duty to minimize the risks to the participants and to the institutions in order to exclude, as best as possible, any permanent damage. If the intentions of research, the conditions for participation, and the guarantees for participants and institutions are made public, the distrust that biomedical research on detained individuals is currently confronted with in Germany could be reduced. The public disclosure of aims and design of research projects in prisons has prompted an encouraging cooperation from the prisoners, the institutions and the control authorities of the state. The establishment of a council guaranteeing confidentiality and safeguarding the interests of participants and institution could contribute to the appreciation that this kind of research deserves and avoid the danger that this very sensitive area of science will be discredited by partial interests and thus become impracticable.

---

[17] H.J. Bochnik, "Ein 'Medizinisches Forschungsgeheimnis' konnte Forschung fördern und Persönlichkeitsrechte schützen" (1995) 47 Versicherungsmedizin 151.

CHAPTER 26

# PRISONERS AS SUBJECTS OF BIOMEDICAL EXPERIMENTATION: EXAMINING THE ARGUMENTS FOR AND AGAINST A TOTAL BAN

SIMON N. VERDUN-JONES, DAVID N. WEISSTUB AND
JULIO ARBOLEDA-FLOREZ

## INTRODUCTION

Research on prisoners is a major topic of contention and polarization. It could not be otherwise. The history books are replete with horror stories concerning the abuse of convicted prisoners by those engaged in various forms of biomedical experimentation. For example, in Ptolemaic Alexandria, vivisection is said to have been conducted on prisoners[1] and, as Louis Lasagna notes,

> The ancient Persian kings and the Egyptian pharaohs are said to have treated criminals as expendable material, much as a modern laboratory researcher might order a supply of rats or rabbits. The practice was apparently still in vogue in eighteenth-century England, since Caroline, Princess of Wales, "begged the lives" of six condemned criminals for experimental smallpox vaccination before submitting her own children to the procedure.[2]

In more recent times, revulsion toward medical experiments on prisoners of concentration camps in Europe, during World War II, has led to an awakening of conscience about the use of humans as subjects of research and about the ethical foundations of medical experimentation. Even now, fifty years after the war, the

---

[1] J. Scarborough, "Celsus on Human Vivisection at Ptolemaic Alexandria". (1976) 11:1 Clio Medica 25. This practice is now attributed to the Japanese during World War II in experiments among the Chinese to determine the spread of bubonic plague in the viscera of affected prisoners. *Ibid.*

[2] L. Lasagna, "Special Subjects in Human Experimentation" in P.A. Freund, ed., *Experimentation with Human Subjects* (New York: George Braziller, 1970) 262 at 262. See also N. Howard-Jones, "Human Experimentation in Historical and Ethical Perspectives" (1982) 16 Social Science & Medicine 1429; C. Bernard, *Introduction à L'étude de la médecine expérimentale* (Paris: Librairie Joseph Gilbert, 1865).

popular press tells us how prisoners were abused at the Pacific front.[3] Similar abuses in the prisons of America have left feelings of horror and futility about the success of guidelines, or the appropriateness of legislation, to guarantee the protection and stem the abuse of prisoners as subjects of medical research. A strong reaction leading to a ban of research on prisoners is understandable. Strong visceral reactions, however, usually tend to produce problems of another nature, or create unexpected injustices against the very group the reaction was intended to protect. Consequently, voices have been raised denouncing the ban. This Chapter will review the ethical issues involved when conducting research among prisoners and will propose a balanced position supporting such research.

## DEFINITIONS

Prisoners constitute a heterogeneous group, and "prisons" is a common term encompassing a broad range of different kinds of institutions and forms of custody. For example, the Department of Health and Human Services in the United States defines the term "prisoner" as follows:

> "Prisoner" means any individual involuntarily confined or detained in a penal institution. The term is intended to encompass individuals sentenced to such an institution under a criminal or civil statute, individuals detained in other facilities by virtue of statutes or commitment procedures which provide alternatives to criminal prosecution or incarceration in a penal institution, and individuals detained pending arraignment, trial, or sentencing.[4]

Similarly, in Canada, prisons are either federal or provincial; the sentence rather than the seriousness of the charge determines whether an individual spends time in a federal penitentiary or in a provincial correctional facility.[5] The *Criminal Code* defines "prison" as including:

> [a] penitentiary, common jail, public or reformatory prison, lock-up, guard-room or place in which persons who are charged with or convicted of offences are usually kept in custody.[6]

Canadian prisons may differ substantially not only in category (i.e. federal or provincial), but also in average lengths of sentences and, more importantly, in perceived openness to outside systems. Most prisons allow outside organizations to carry on activities within the prison on behalf, or for the benefit, of prisoners. Other prisons allow conjugal visits and, depending on the behaviour of the prisoner and level of security, prisoners are allowed passes for compassionate, socialization, or

---

[3] "Chinese remember Japanese Atrocities" *The [Toronto] Globe and Mail* (7 August 1995) A1; "Ghoulish Experiments still Fresh in Chinese Minds" *The [Toronto] Globe and Mail* (7 August 1995) A5.
[4] 45. C.F.R. 46 (1991) § 46.303 (c).
[5] *The Criminal Code*, R.S.C. 1985, c. C-46, s. 743.1. This provision states that inmates sentenced to two years or more must serve their sentences in a federal penitentiary, while those who are sentenced to less than two years must serve their sentences in a provincial correctional facility.
[6] *Ibid.*, s. 2. The term "penitentiary" is defined in s. 2 of the *Corrections and Conditional Release Act*, S.C. 1992, c. 20.

rehabilitation purposes. With some exceptions, therefore, prisons are not the isolated, "total institutions"[7] of years gone by. Significantly, remand centres are the prisons that are most frequently found throughout Canada. They are the most commonly used and the most crowded institutions in the justice system; however, they are also the facilities where prisoners receive the fewest amenities and protections in spite of the fact that they may not have been found guilty of an offence, let alone sentenced to a prison term.

A definition of prisoners and prisons cannot be complete without an understanding of the social functions of prison. Prison populations have exploded in many countries — a phenomenon that testifies to a failure of many social systems to accommodate the needs of their populations. In several countries, prisons may be the repositories of political dissenters, not just offenders convicted of conventional crimes. Moreover, in many places in Canada and the United States, policies of deinstitutionalization and an unarticulated policy of transmigration[8] have, to some extent, converted prisons into extensions of the mental health system, so that they should more appropriately be considered asylums rather than prisons.[9] Simply put, prisons and prisoners are not all the same and the marked diversity of the conditions under which prisoners live is a factor that must always be taken into account when developing guidelines about research on prisoners and the types of research to be allowed in prisons.[10]

## PRISONERS AS RESEARCH SUBJECTS

Prisoners constitute a markedly different group from the other "special populations" that may be identified in the context of biomedical experimentation. As far as children, the elderly, the mentally disordered, and the developmentally disabled are concerned, the major focus of inquiry is whether the prospective research subject is *competent* to give an informed consent to participation in such experimentation. In the case of prisoners who do not fall into one or more of these other groups, the

---

[7] E. Goffman, *Asylums* (New York: Doubleday Anchor, 1961).

[8] Transmigration refers to the movement, by policy or by default, of mental patients from one social system to another, especially between the mental and the justice/correctional systems. See J. Arboleda-Florez & H. Holley, "Criminalization of the Mentally Ill: Part II. Initial Detention" (1988) 33:2 Canadian Journal of Psychiatry 87 at 88.

[9] J. Arboleda-Florez, *The Prevalence of Mental Illlness in a Remanded Population and the Relationship between Mental Illness and Criminality* (Ph.D. Dissertation, University of Calgary, 1994).

[10] The Tri-Council Working Group's *Code of Ethical Conduct for Research Involving Humans* clearly recognizes the need to take account of the specific institutional context in which potential research subjects may be recruited whenever an assessment must be made as to whether their agreement to participate is truly voluntary. The Working Group states, for example, that "the pervasiveness of authority relationships on voluntary choice must be judged according to the particular context of prospective participants (e.g., the difference between a maximum and minimum security institution)" and notes that authority figures may have a considerably less significant impact on inmates in a short-term, as opposed to a long-term, institution. See Tri-Council Working Group, *Code of Ethical Conduct for Research Involving Humans, Part 2* (July 1997) at 11. The Working Group represents the Medical Research Council of Canada, the Natural Sciences and Engineering Council of Canada and the Social Sciences and Humanities Research Council of Canada.

primary emphasis is on whether potential recruits for involvement in research are capable of acting *voluntarily* in light of the institutional environment in which they are held.

### Competence

Some commentators have suggested that the competence of prisoners to make decisions, such as participating in research, should be vigorously questioned. For example, Bach-y-Rita indicates that prisoners may not be able to make a competent assessment of the degree of risk involved in participation in biomedical research:

> judgment about an acceptable degree of risk requires contact with the free world as opposed to the prison environment. What may be perceived as an acceptable risk for a person inside prison may be totally unacceptable for the same person outside.[11]

It has also been suggested that institutionalization strips inmates of their self-esteem and this effectively diminishes their capacity for making decisions about whether to participate in biomedical research.[12]

These views seem to be based on a lack of appreciation of the nature of the conditions under which most prisoners are actually held and seriously underestimate what they are capable of doing as humans who have been deprived of their freedom. Furthermore, such views fail to recognize that there are many gradients and lengths of institutionalization. For example, prisoners are rarely, if ever, totally cut off from the outside world. Apart from the regular media to which they have access (radio, TV), prisoners maintain informal communications pipelines that, in many instances, are more accurate and rapid than official channels. In addition, for many prisoners, being in prison is perceived as being nothing more than an inescapable, albeit, unfortunate trade off against their activities in the outside world and, for these individuals, prison becomes a place to meet new acquaintances and renew old friendships rather than a place of isolation.

On the other hand, commentators have not given enough attention to the fact that many prisoners could be incompetent, not because they are in an institution, but simply because many of them are affected by mental disorders and/or developmental disabilities. Thus, in addition to those suffering from functional mental illnesses and/or significant developmental disabilities, inmates affected by serious antisocial personality disorders or drug and alcohol dependencies are quite prevalent in the prison population[13] and they may have considerable difficulty understanding the nuances of research protocols; where this occurs, such prisoners should be considered incompetent to participate in experimentation. In these cases, the

---

[11] G. Bach-y-Rita, "The Prisoner as an Experimental Subject" (1974) 229 Journal of the American Medical Association 45.

[12] See P.S. Appelbaum, C.W. Lidz & A.J. Meisel, *Informed Consent: Legal Theory and Clinical Practice* (New York: Oxford University Press, 1987) at 230—231. See also R.A. Burt, "Why We Should Keep Prisoners from the Doctors" (1975) 5:1 Hastings Center Report 25.

[13] See Goffmann, *supra* note 7. See also R.C. Bland *et al.*, "Prevalence of Psychiatric Disorder and Suicide Attempts in a Prison Population". (1990) 35 Canadian Journal of Psychiatry 407.

concern about competency has nothing to do with the fact of institutionalization *per se* but rather with the presence of a mental condition. Indubitably, where an individual lacks the capacity to engage in autonomous decision-making, protective intervention is fully warranted. Similarly, if drastic medical intervention (e.g. psychosurgery) has the potential to destroy an individual's future capacity to make autonomous decisions, then intervention is necessary to preserve the existing autonomy of the subject. In both of these examples, prohibiting a prisoner from participating in experimentation can legitimately be viewed as preserving the autonomy of the individual at risk.

### *Voluntariness*

The main thrust of the debate concerning the involvement of prisoners in research has turned on the question of the voluntariness of their decision to participate. The issue of voluntariness is at the heart of the pressing moral, ethical, and legal dilemmas surrounding the participation of prisoners in biomedical experimentation.

Both competence and voluntariness are essential components of the doctrine of informed consent, although it is significant that there is a relative paucity of analysis of the latter concept.[14] A competent person can, of course, make a voluntary or an involuntary decision. "Voluntariness" refers to freedom to decide, unencumbered by constraints or undue enticements of any sort. In terms of informed consent, therefore, it is important to recognize that some individuals may be perfectly competent to make decisions about participation in biomedical research but these decisions may be defective because they are involuntary.[15] Many would argue that this is precisely the situation that pertains in the case of prisoners who are recruited as subjects of biomedical research.

> An important issue is whether prisoners can ever provide a true and voluntary consent to participation in research. This question is prompted by the belief, forcefully expressed by a U.S. judge more than two decades ago, that a prisoner is in "an inherently coercive atmosphere even though no direct pressures may be placed on him".[16] According to this view, the very role ascribed to prisoners in our society may effectively preclude them from making decisions that may be considered genuinely voluntary on their part.[17]

The truth is that much of the frequently acrimonious debate concerning the participation of prisoners in biomedical experimentation is really fueled by

---

[14] See P.S. Appelbaum, "Informed Consent" in D.N. Weisstub, ed., *Law and Mental Health: International Perspectives*, vol. 1 (New York, Pergamon Press, 1984) 45 at 60—62.

[15] *Ibid.* at 197—198.

[16] See *Kaimowitz* v. *Michigan Department of Mental Health*, Civil Action No. 73—19434—AW (Wayne County, Michgan Circuit Court 1973) [unreported]. See also B.M. Dickens, "Coercion and Inducement in Medical Experimentation" in Law Reform Commission of Victoria, *Medicine, Science and Law: Informed Consent* (Melbourne: Globe Press, 1987) 92 at 97.

[17] See G.J. Annas, L.H. Glantz & B.F. Katz, *Informed Consent to Human Experimentation: The Subject's Dilemma* (Cambridge, Mass.: Ballinger, 1977) at 106.

competing factual assumptions as to whether the nature of the prison environment invariably precludes the giving of a voluntary consent.[18]

There is an intimate connection between the concepts of voluntariness and autonomy. In decision-making, voluntariness can be regarded as a marker of, or test for, self-determination.[19] If a decision is not voluntary, then it does not seem appropriate to call it truly self-determined. In this sense, therefore, requiring voluntariness preserves a prisoner's autonomy.

However, autonomy can be brought to bear in more than one direction. When the prospective research subject is fully competent, healthy, and possesses the capacity to give an informed, voluntary consent, then the value of autonomy is clearly diminished if he or she is prevented from participating in biomedical experimentation.[20] While there are undoubtedly strong arguments for preventing prisoners from volunteering for non-therapeutic research that involves a high degree of risk to their health,[21] the question arises as to whether it is legitimate to prevent competent prisoners from exercising their autonomy to participate even in low-risk biomedical research. The legitimacy of such a prohibition is particularly acute when based on a blanket presumption that not a single inmate can ever give a truly voluntary consent to such participation.

Viewed in this context, it can be seen that the use of prisoners in biomedical experimentation throws into focus the underlying conflict between enabling individuals to make autonomous choices (free from state control) and paternalism (the protective side of the autonomy principle).[22] Autonomy in its guise as a champion of the individual's freedom from the state control has been described as "one of the central values that shapes law generally".[23] It is such an important value in our legal system that a desperately ill, but competent, adult may refuse life-saving treatment even though his or her physicians know that such refusal will lead to death.[24] Indeed, it has been suggested that permitting individuals to make their own "voluntary choices in matters vitally affecting them is developmentally beneficial

---

[18] See R. Macklin & S. Sherwin, "Experimenting on Human Subjects: Philosophical Perspectives" (1975) 25 Case Western Law Review 434 at 450.

[19] See M.A. Somerville, "Labels versus Contents: Variance between Philosophy, Psychiatry and Law in Concepts Governing Decision-Making" (1994) 39 McGill Law Journal 179 at 196—197.

[20] K. Schroeder, "A Recommendation to the FDA Concerning Drug Research on Prisoners" (1983) 56 Southern California Law Review 969 at 982. Indeed, one philosopher has argued that even inmates in "total institutions" should not necessarily be pre-empted from giving consent to such highly controversial therapies as psychosurgery. See J.G. Murphy, "Therapy and the Problem of Autonomous Consent" (1979) 2 International Journal of Law and Psychiatry 415. The general arguments used to de-mythologize autonomy can be applied equally to the research context. In addition, see J.G. Murphy, "Incompetence and Paternalism" (1974) 60 Archiv Für Rechts und Sozialphilosophie 465; and J.G. Murphy, "Total Institutions and the Possibility of Consent to Organic Therapies" (1975) 5 Human Rights 25.

[21] A. Capron, "Medical Research in Prisons: Should a Moratorium Be Called?" in T.L. Beauchamp & L.Walters, eds., *Contemporary Issues in Bioethics* (Belmont, Ca.: Wadsworth Publishing, 1978) 497.

[22] See J. Katz, "Informed Consent — A Fairy Tale? Law's Vision" (1977) 39 University of Pittsburgh Law Review 137 at 139.

[23] See B.J. Winick, "On Autonomy: Legal and Psychological Perspectives" (1992) 37 Villanova Law Review 1705 at 1706.

[24] See *Malette* v. *Shulman* (1990), 72 O.R. (2d) 417 (C.A.); *Nancy B.* v. *Hotel-Dieu de Québec et al.* (1992), 69 C.C.C. (3d) 450 (Qué. S.C.).

and may be essential to their psychological well-being".[25] In this particular respect, Winick has asserted that,

> treating individuals as competent adults able to make choices and exercise a degree of control over their lives rather than as incompetent subjects of government paternalism and control will predictably have a beneficial effect. Denying people a sense of control over their lives can have strongly negative consequences. Indeed, when people feel that they have no influence over matters that vitally affect them, they may develop "learned helplessness".[26]

Arguably, these considerations apply with particular force to prison inmates, given their need to cope with the many disabling aspects of forced institutionalization. However, it would be a fundamental mistake to assume that most prisoners are "helpless" because of ingrained patterns of learning established during cycles of institutional abuse. Indeed, the great majority of prisoners are not helpless and unable; in fact, the hypernomic[27] aspects of imprisonment, rather than stunting creative capacities, could lead to unusual problem-solving, such as outright rebellion or subtle ways of manipulating the environment.[28]

It would be quite possible to apply both the protective and self-determining aspects of the principle of autonomy to the question of whether prisoners should be permitted to participate in biomedical research. The protective aspect of the principle would come into play through the establishment of review mechanisms to ensure that a prisoner's decision to participate in a particular project is truly voluntary and would operate to exclude any prisoner whose voluntariness is in doubt. The self-determining aspect of the principle, on the other hand, would permit those prisoners, who have given a genuinely voluntary and informed consent to such participation, the right to do so. However, it appears that biomedical research is not conducted with prisoners in Canada because there is an assumption that it cannot be established that their involvement could ever be considered truly voluntary. If such were indeed the case, then the existing *de facto* ban on prisoner involvement in research should be continued. On the other hand, if it is possible to establish review mechanisms that ensure that competent prisoners' participation is indeed voluntary, then it would surely be wrong to deprive them of the opportunity to reaffirm their autonomy. This point assumes considerable force when it is borne in mind that, in a non-therapeutic situation, a prisoner's decision to become involved in biomedical research may be viewed as an altruistic act, since the community at large, rather than the prisoner him or herself, stands to gain from such research.

### Summary

This Chapter will assume that prisoners are, in fact, competent to make decisions about involvement in biomedical rexperimentation. The major emphasis will,

---

[25] See Winick (1992), *supra* note 23 at 1706—1707.

[26] *Ibid.* at 1765.

[27] H. Holley & J. Arboleda-Florez, "Hypernomia and Self-Destructiveness in Penal Settings" (1988) 11 International Journal of Law and Psychiatry 167.

[28] J. Arboleda-Florez & H. Holley, "Predicting Suicide Behaviour among Incarcerated Offenders Awaiting Trial" (1989) 34 Canadian Journal of Psychiatry 668.

therefore, be on the issue of voluntariness. In approaching this issue, it is important to recognize that: (1) not all prisons are the same; (2) there are prisons where inmates cannot be considered "institutionalized", given short sentences or openness to outside systems; (3) depending on the definition of "prison", not every individual housed in a prison has been found guilty; (4) not all prisoners are the same in terms of their cognitive and intellectual capacities; and (5) prisoners may be members of one or more of various other "special populations" (e.g. the mentally disordered or the developmentally disabled) and, as a consequence, the competence among members of these subgroups may be called into question.[29]

## WHAT IS THE EXPERIENCE WITH PRISONER PARTICIPATION IN BIOMEDICAL EXPERIMENTATION?

In recent history, the use of prisoners in non-therapeutic medical experimentation has been confined primarily to the United States,[30] whereas it is expressly prohibited in European countries such as France and Germany, and does not take place in England and most other countries even though there is no express prohibition against it. As far as Canada is concerned, such research is not permitted in either federal or provincial correctional facilities.[31]

In reviewing the literature on this matter it is significant that, despite the fact that health care has long been considered a major issue in prison law within the United States, comparatively little attention has been paid in the prison law literature to the specific question of biomedical experimentation with inmates.[32] Indeed, even in the mid-1970s when biomedical experimentation with prisoners was at its zenith, the leading practitioner's sourcebook on prisoners' rights did not so much as mention such activity.[33]

---

[29] In these cases, inmates' participation in biomedical research should be governed not only by the special safeguards that may be applicable to prison volunteers, but also by the various mechanisms that are in place in order to protect the interests of members of the special population concerned.

[30] See Law Reform Commission of Canada, *Working Paper No. 61. Biomedical Experimentation Involving Human Subjects* (Ottawa: Law Reform Commission of Canada, 1989) at 38 [hereinafter *LRC Working Paper No. 61*]. In making this assertion, the Commission refers to the work of the Belmont Commission. See National Commission for the Protection of Human Subjects of Biomedical and Behavioral Research, *Research Involving Prisoners* (Bethesda, Md.: The Commission, 1976) [hereinafter *National Commission: Research Involving Prisoners*].

[31] See *LRC Working Paper No. 61, Ibid.* at 38. The Commission's reference to the ban on experimentation with prisoners in Germany has been confirmed more recently in Chapter 25. In the case of the German federal drug laws, the ban on experimenting with drugs on prisoners is explicit: in other types of biomedical experimentation, there is a *de facto* ban, except in very limited circumstances.

[32] See A.J. Fowles, *Prisoners' Rights in England and the United States* (Brookfield: Gower Publishing Company, 1989); R. Hawkins & G.P. Alpert, *American Prison Systems — Punishment and Justice* (Englewood Cliffs, N.J.: Prentice Hall, 1989); B.B. Knight & S.T. Early, *Prisoners' Rights in America* (Chicago: Nelson-Hall Publishers, 1986). The same situation pertains in Britain. See S. Livingstone & T. Owen, *Prison Law: Text and Materials* (Oxford: Clarendon Press, 1993); G. Richardson, *Law, Process and Custody: Prisoners and Patients* (London: Weidenfeld and Nicolson, 1993).

[33] M.G. Hermann & M. Haft, *Prisoners' Rights Sourcebook: Theory, Litigation, Practice* (New York: Clark Boardman, 1973).

### Research with Prisoners in the United States

Biomedical experimentation on prisoners in the United States burgeoned during the years of World War II, primarily in relation to the testing of drugs designed to combat the infectious diseases that might be contracted by members of the various armed services.[34] Prison furnished the ideal environment for conducting controlled experiments of drugs.[35]

In addition, the participation of prisoners in experimentation was viewed as a form of "patriotic" contribution to the war effort.[36] Just as soldiers, sailors and aircrew put their lives on the line for their country, so did prisoners at home put their health at risk to contribute to research that might save thousands (or possibly hundreds of thousands) of service personnel at the front. During the 1950s and 1960s, there was a "huge expansion" of medical experimentation in the prisons, although the rationale for this expansion was primarily based on crass considerations of profit and cost-benefit rather than on ringing appeals to patriotism.[37] In 1962, for example, the Kefauwer-Harris amendments to the Food and Drug Act "encouraged the use of prisoners as subjects in the testing of drugs".[38] Indeed, in 1973, Mitford recorded the view of one British scholar who recounted that "one of the nicest American scientists" had told him that, "criminals in our penitentiaries are fine experimental material — and much cheaper than chimpanzees".[39]

One important factor in the continued practice of involving prisoners in drug testing was the absence of any restrictive federal or state regulations that specifically protected this group of research subjects.[40] On the other hand, the contemporary FDA regulations concerning new drugs required that they be tested on human beings before they were marketed, and dictated that the first phase of such testing must take place in an institutional setting. As one physician put it, "the only place available for such large-scale toxicity studies is prison".[41]

As the 1960s drew to a close, 85% of all new drugs were tested on prisoners in forty-two U.S. prisons.[42] However, the 1970s witnessed a dramatic reduction in the

---

[34] See *LRC Working Paper No. 61, supra* note 30 at 38.

[35] As the Law Reform Commission of Canada noted,

> Why are prisoners so attractive to researchers? The answer is simple. Prisoners are a captive population leading a routine existence. Their lifestyle greatly facilitates the administration of research protocols and the collection of data, especially where the research is related to new medications and pharmaceutical products.

See *Ibid.* at 38. This, by the way, is a false representation of life in a prison. If anything, life in a prison is only routine on the surface. Away from the gaze of the guards, a prisoner is constantly cutting deals, always on the alert for new possibilities, always suspicious of any new developments. In reality, a prison is a constant beehive of activity.

[36] See Schroeder, *supra* note 20 at 971.

[37] See J. Mitford, *Kind and Unusual Punishment — The Prison Business* (New York: Alfred A. Knopf, 1973) at 139—140.

[38] See J.L. Hill, "Exploitation" (1994) 79 Cornell Law Review 631 at 646.

[39] See Mitford, *supra* note 37 at 139—140.

[40] C.M. McCarthy, "Experimentation on Prisoners" (1989) 15 New England Journal on Criminal and Civil Commitment 55 at 58.

[41] See Mitford, *supra* note 37 at 140.

[42] See Schroeder, *supra* note 20 at 971.

use of prisoners for drug testing. During this decade, the general practice of using human subjects in biomedical and behavioural research was brought under close public scrutiny as a consequence of various scandalous examples of abusive research practices coming to light.[43] This scrutiny gave birth to significant developments, including the *National Research Act*.[44] The legislation created the National Commission for the Protection of Human Subjects of Biomedical and Behavioral Research[45] and directed the Secretary of the Department of Health, Education, and Welfare (DHEW) to require that all those institutions that received research funds from DHEW establish IRBs (institutional review boards or research ethics committees); and the introduction of a series of increasingly restrictive regulations concerning prison research.[46] These developments clearly signaled a consensus that the pace of medical progress must be subordinate to the need to protect human subjects who are involved in biomedical research. Indeed, by 1980, only 15% of all drug testing was carried out in prisons and, then, only in two facilities (in Michigan and Montana).[47]

During the years 1973—1981, the federal government, through DHEW and its successor the Department of Health and Human Services (DHHS) and the Food and Drug Administration (FDA), introduced a series of restrictive regulations that specifically addressed the need to protect the interests of prisoners involved in biomedical research.[48] The DHHS regulations applied to all research that was carried out or funded by DHHS.[49] The type of research permitted by DHHS was very restricted in nature. Originally, the research had to meet two requirements: it had to be intended to ameliorate the health of the individual inmate and had to pose no more than a minimal risk.[50] Later, the rules were modified to permit "research where prisoners served as control subjects (and therefore could not expect to receive any direct benefit from the research) or where the general conditions of life in prison

---

[43] See H.K. Beecher, *Research and the Individual: Human Studies* (Boston: Little, Brown, 1970); G. Edsall, "A Positive Approach to the Problem of Human Experimentation", in P.A. Freund, ed., *Experimentation with Human Subjects* (New York: George Braziller, 1970) 276. (discussing the Willowbrook studies); M.H. Pappworth, *Human Guinea Pigs: Experimentation on Man* (London: Routledge & Kegan Paul, 1967); E. Langer, "Human Experimentation: New York Affirms Patient's Rights" (1966) 151 Science 663 (discussing the Sloan-Kettering episode); "Unethical Studies" (1974) 33 Massachusetts Physician 19 (criticizing the Tuskegee syphilis study).

[44] See *The National Research Act*, Pub.L. No. 93—348 (1974).

[45] See *National Commission: Research Involving Prisoners*, *supra* note 30.

[46] DHEW issued regulations for the protection of prisoners involved in human research in 1981. See 45 C.F.R. 46 (1981). The Food and Drug Administration (FDA) issued similar regulations in 1981. See 46 Federal Register 61 (1981) § 666. Some states also introduced regulations at around the same time (Oregon and Iowa). See McCarthy, *supra* note 40.

[47] See Schroeder, *supra* note 20 at 971.

[48] See D.M. Maloney, *Protection of Human Research Subjects: A Practical Guide to Federal Laws and Regulations* (New York: Plenum Press, 1984) at 347—360. See also President's Commission for the Study of Ethical Problems in Medicine and Biomedical and Behavioral Research, *Protecting Human Subjects: The Adequacy of Federal Rules and their Implementation* (Washington, D.C: U.S. Government Printing Office, 1981).

[49] See 45 C.F.R. 46 (1991) § 46.101(a).

[50] *Ibid* § 46.306(b).

were investigated".[51] In addition, an IRB was required to document the subject's informed and voluntary consent. Prisoners had to be told beforehand that their participation in medical research would have no impact on their parole status, and that their living conditions would not be significantly improved to induce their consent.[52] These requirements have led some to suggest that DHHS had "essentially forbidden most research in prisons" because minimal risk research "often does not actually improve the health of the subjects" concerned and therefore does not meet the basic requirements of DHHS regulations.[53]

In May of 1980, the FDA proposed regulations that would have effectively placed a complete ban on non-therapeutic, experimental drug research involving prisoners. However, these proposed new regulations were subsequently challenged by four inmates of the State Prison of Southern Michigan at Jackson and the pharmaceutical company, Upjohn. It was alleged that closing down non-therapeutic research in prisons would, *inter alia*, deprive inmates of their right to decide freely whether or not to participate in such experimentation and would cause harm to the public by depriving the drug companies of the most suitable research populations for certain types of studies.[54] In response to this lawsuit, the FDA issued a so-called "reproposal" in December of 1981.[55]

The guidelines set out in the reproposal did not impose a total ban on non-therapeutic drug research in prisons. However, they did require that there be "compelling reasons" (not defined in the guidelines) for involving prisoners in medical research, and that all efforts be made to ensure that their consent was obtained on a demonstrably voluntary basis. In addition, the reproposal called for prisoner representation on research review boards and a requirement that research only be approved when nonprisoner volunteers would be willing to accept the inherent risks of the experiment concerned.[56] In spite of the changes to the earlier proposal, it has been suggested that the stringency of the provisions of the proposals nevertheless imposed a *"de facto* ban" on non-therapeutic experimentation.[57] It is significant that, in 1987, it was noted that "due to the established regulations and the ethical standards no experimental testing with prisoners as subjects would be allowed in federal prisons" because of the belief that "the existence of undue influence and coercion on inmates to participate in experimentation outweighed the advantages of using a segment of the population in a controlled setting".[58] No matter what the substance of the federal regulations may have been, the development of stringent ethical standards appears to have been the critical factor in the demise of medical experimentation involving federal prisoners.

---

[51] See Maloney, *supra* note 48 at 349.
[52] See 45 C.F.R. 46 (1991) § 46.109.
[53] See McCarthy, *supra* note 40 at 68.
[54] See Schroeder, *supra* note 20 at 986—987.
[55] See 46 Federal Register 61, *supra* note 46 § 666.
[56] See Schroeder, *supra* note 20 at 988—999.
[57] *Ibid.* at 1000.
[58] See McCarthy, *supra* note 40 at 70.

Another important development, at this time, was the establishment of the President's Commission for the Study of Ethical Problems in Medicine and Biomedical and Behavioral Research in 1980. Over a three-year period, the Commission produced ten reports, many of which dealt directly with the adequacy of the federal regulatory framework established to protect human research subjects.[59] An important issue raised by the Commission was the whole question of the adequacy and uniformity of the federal research regulations and their implementation in the various federal agencies affected.[60] One recommendation was to bring all research sponsored by federal departments into compliance with the existing DHHS regulations. To achieve this end, a Model Federal Policy was published in the *Federal Register* in June of 1986, although some changes were made to the existing DHHS regulations at that time. However, this Model Federal Policy has now been updated.[61]

In 1991, the Federal Government issued the *Federal Policy for the Protection of Human Subjects*.[62] This policy is to be applied by most federal departments that engage in biomedical research.[63] In articulating a detailed set of procedures for the establishment and conduct of IRBs, the *Federal Policy* states that if "an IRB regularly reviews research that involves a vulnerable category of subjects, such as prisoners, consideration shall be given to the inclusion of one or more individuals who are knowledgeable about and experienced in working with these subjects".[64] Clearly, the new *Federal Policy* does not impose an absolute ban on biomedical experimentation involving prisoners.

The DHHS regulations were modified in 1991 in accordance with the requirements of the *Model Federal Policy*. The revised Regulations provide that research can be conducted on prisoners if, in the judgment of the Secretary of the Department of Health and Human Services, the proposal was concerned with one or more of the following topics: the study of possible causes, effects, and processes of incarceration and criminal behaviour, and the study of prisons as institutional structures, or of prisoners as incarcerated subjects. Furthermore these limited topics could only be studied if the research protocol in question presents no more than minimal risk and no more than inconvenience to the prisoner-subjects. Research on

---

[59] President's Commission for the Study of Ethical Problems in Medicine and Biomedical and Behavioral Research, *Summing Up: Final Report on Studies of the Ethical and Legal Problems in Medicine and Biomedical and Behavioral Research* (Washington, D.C.: U.S. Government Printing Office, 1983).

[60] See Maloney, *supra* note 48. See also President's Commission for the Study of Ethical Problems in Medicine and Biomedical and Behavioral Research, *Implementing Human Research Regulation: Second Biennial Report on the Adequacy and Uniformity of Federal Rules and Policies, and their Implementation, for the Protection of Human Subjects* (Washington, D.C.: U.S. Government Printing Office, 1983).

[61] See P.R. Benson & L.H. Roth, "Trends in the Social Control of Medical and Psychiatric Research" in D.N. Weisstub, ed., *Law and Mental Health: International Perspectives, Volume 4* (New York: Pergamon Press, 1988) 1 at 31—34.

[62] See 56 Federal Register 117 (1991).

[63] See R.J. Kelly, S.S. Fluss & F. Gutteridge, "The Regulation of Human Subjects: A Decade of Progress" (Paper presented to the XXVIth CIOMS Conference on Ethics and Research on Human Subjects: International Guidelines, 1992) [unpublished] at 12.

[64] See 56 Federal Register 117, *supra* note 62 at 28015.

conditions particularly affecting prisoners as a class (e.g. hepatitis), or social and psychological problems (e.g. alcoholism), and research on practices with a good probability of improving the health or well-being of the subject, could only be conducted provided that the study obtained the prior approval of the Secretary after extensive consultations and that a notice were published in the *Federal Register* of intent to approve.

At the state level, during the 1980s, a number of state governments introduced legislation regulating the participation of prisoners in medical research, although the substance of the statutory schemes established was by no means uniform in nature. For example, Oregon banned such research completely,[65] while Iowa introduced only a series of limited restrictions upon such activity.[66] However, even though the Iowa state law did not prohibit the involvement of inmates in experimentation, the state prison nevertheless adopted ethical standards that, in effect, precluded such activity, primarily on the ground that the existence of various forms of potential coercion within the institution cast doubt upon the voluntariness of any inmate consent to participation in research.[67]

During this era, California also enacted provisions that imposed restrictions on prisoner involvement in biomedical research. These provisions focused on the need to obtain informed consent and establish a complex list of conditions designed to ensure that such consent had indeed been obtained without, for example, undue influence or coercion.[68] However, in practice, such experimentation did not take place because of the application of ethical standards within the state prison system.[69]

In contrast, in Michigan, where a major U.S. pharmaceutical manufacturer is located, drug testing on prisoners apparently continued in at least one prison facility, although the State Department of Corrections issued directives concerning the need to obtain informed consent from inmates who decided to participate in such research.[70]

The question of prisoner involvement in medical research was addressed in *Bailey v. Lally* (1979).[71] State prisoners in the Maryland House of Correction alleged that their constitutional rights had been violated by virtue of their involuntary participation in such experimentation. Their contention was that they had been coerced into participation as a consequence both of the comparison between the very poor living conditions in the main prison and the more advantaged living

---

[65] Or. Rev. Stat. § 421.085(b)(2) (West 1985).

[66] Iowa Code Ann. § 246.47 (West 1967).

[67] See McCarthy, *supra* note 40 at 71.

[68] Cal. Penal Code § 3502, 3505 & 3521. In September 1989, the Governor of California approved certain amendments to those provisions of the Penal Code that relate to medical experimentation with prisoners. The amendments were designed to ensure that prisoners are not denied access to drugs and treatment that may be required for good medical care (e.g. experimental therapy). The amendments are to § 3502.5.

[69] See McCarthy, *supra* note 40 at 72.

[70] *Ibid.* at 72—73.

[71] 481 F. Supp. 203 (D. Md. 1979).

environment in the research unit and of the subtle effects of the various other inducements offered to research subjects. The Court denied the plaintiffs relief. The Court noted that the pay of $2 per day was not so attractive as to induce participation by those who would not otherwise participate; that the inmates were free to withdraw at any time; that both oral and written statements of consent had been obtained from each inmate who had participated in research and that the consent process had been scrutinized by review committees; and that only temporary illness or discomfort was involved. In the Court's view, there was no evidence of actual coercion; rather, the situation was that each prisoner had been offered a choice to "participate in a worthwhile but unpleasant activity which may be attractive to him because of his environment".[72] Schroeder notes that "by negative inference, *Bailey* established that prison conditions do not necessarily inhibit a prisoner's ability to give informed and voluntary consent".[73] On the other hand, other commentators have faulted the Court for having ignored the very subtle forms of coercion that operate in the prison environment to negate the voluntariness of consent to participation in research.[74]

Subsequent to this case, however, a 1986 report by a Subcommittee of the House Committee on Energy and Commerce raised fundamental concerns about the potential for serious abuse when biomedical experiments are carried out on prisoners.[75] This report addressed a series of radiation experiments carried out in the United States between the mid-1940s and the early 1970s. One of the experiments revealed by the report involved the use of prisoners in a form of biomedical experimentation that clearly posed a considerable risk to their health. In the words of the report,

> From 1963 to 1971, 67 inmates at Oregon State Prison and 64 inmates at the Washington State Prison received x-rays to their testes to examine the effects of ionizing radiation on human fertility and testicular function. These experiments were conducted by the Pacific Northwest Research Foundation and the University of Washington. Subjects had to agree to receive vasectomies after completion of the experiments. The Energy Research and Development Administration planned to begin medical follow up of the irradiated prisoners, but these plans were dropped in 1976 at the request of the U.S. Attorney in Portland after several irradiated inmates filed suits against the state and federal governments.[76]

In the Oregon experiments, prisoners were generally given radiation doses ranging from 8 to 600 roentgen in a single exposure. The *Report* notes that, in the mid-1980s, the occupational limit that had been set for X-ray exposure to reproductive organs was only 5 roentgen per year.[77] The follow-up studies in Oregon

---

[72] *Ibid.* at 220.
[73] See Schroeder, *supra* note 20 at 978.
[74] See McCarthy, *supra* note 40 at 77.
[75] See E.J. Markey (Chair), *Report Prepared by the Subcommittee on Energy Conservation and Power of the Committee on Energy and Commerce, U.S. House of Representatives: American Nuclear Guinea Pigs: Three Decades of Radiation Experiments on U.S. Citizens* (Washington, D.C.: U.S. Government Printing Office, 1986).
[76] *Ibid.* at 2—3.
[77] *Ibid.* at 15.

were halted in 1976 because of lawsuits brought by a number of irradiated inmates.[78]

It is significant that the Washington experiments, which involved single doses of radiation ranging from 7 to 400 roentgen, were terminated in 1969 because the Human Subjects Review Board at the University of Washington refused to authorize further irradiation.[79] This suggests that the type of experiment conducted in the Oregon and Washington prisons would never be approved by an IRB today. However, it is profoundly disturbing that medical researchers decided to participate in non-therapeutic research that, in fact, necessitated deliberately inflicting varying degrees of potentially permanent damage on any human subjects, let alone prisoners. The uncovering of these experiments in the mid-1980s has indubitably contributed to the maintenance of a negative attitude towards the recruitment of prisoners as subjects in any form of biomedical experimentation in the United States.

### Canada

The Law Reform Commission of Canada has noted that non-therapeutic experimentation on prisoners is apparently prohibited in both federal and provincial institutions.[80] As an example, the Commission points to the applicable provision in Quebec, which states that, "an imprisoned person may not be subjected to medical and scientific experiments that may be detrimental to his mental or physical integrity".[81]

The situation in federal correctional institutions is governed by the Directive of the Commissioner of Corrections of Canada that was issued in 1987.[82] The Directive states that medical research focuses on improving health and the treatment of disease, while "behavioral research focuses on the understanding, treatment, and management of offenders".[83] The Directive stipulates that informed consent must be obtained in writing from all participants in experimental research. At a minimum, the consent form must:

(1) provide a written explanation of, and a justification for, the study, including potential risks and tasks to be performed;

(2) state that participation is strictly voluntary and may be terminated at any time; and

(3) in the case of offenders, shall make it clear that participation will in no way affect the terms or length of their sentence.[84]

---

[78] *Ibid.* at 16. However, in September 1976, the District Court for the District of Oregon dismissed the case brought against the federal defendants.

[79] *Ibid.* at 17.

[80] See *LRC Working Paper No. 61, supra* note 30 at 38.

[81] *Regulation respecting Houses of Detention,* R.R.Q. 1981, c. P-26, r.1, s. 21. The regulation was made under *An Act respecting Probation and Houses of Detention,* R.S.Q. 1977, c. P-26.

[82] Number 009 (1987—01—01).

[83] *Ibid.,* ss. 2—3.

[84] *Ibid.,* s. 14.

To ensure that there should be no subtle coercion in obtaining inmate consent to research, the Directive provides that no inducements may be given to participate:

15. No inmate shall be offered privileges or earlier release in return for participation in a research project.

16. Participation or lack of it in an approved research project shall not affect an inmate status or pay under the inmate pay system.

Any research proposal that involves a "new medical treatment program or experimental methodology" must obtain prior review and approval at the institutional, regional and national levels, including the imprimatur of the Regional Psychiatric Centre/Regional Treatment Centre Research Review Committee and Regional Research Committee. The Director General of Health Care Services is also required to "seek the guidance of the appropriate professional and scientific bodies, for any major project in relation to ethical considerations, methodology and scientific rigour".[85] Section 19 clearly indicates that inmates may only participate in research that has the potential of offering them some therapeutic benefit:

An inmate may volunteer to participate in a medical research study, only if he:

(1) has been diagnosed as having a condition which the study addresses;

(2) clearly understands the objective of the study;

(3) understands and accepts the methods to be used including the use of controls, placebos, and randomization;

(4) is aware of the anticipated benefits and risks, in comparison with the current best treatment or no treatment; and

(5) has signed a consent form which clearly described the objective of the study, the inmate's understanding of his or her involvement and the impact of his or her consent.[86]

Section 20 also implicitly prohibits non-therapeutic medical research: "the inmate involvement shall be discontinued if requested and the current best treatment shall be re-instituted where applicable".[87]

The Canadian Psychiatric Association identified prisoners as constituting a special population who are in need of "special protection" in the context of research. The Association does not recommend a ban on experimentation with prisoners but rather emphasizes the need to apply safeguards to ensure that they are protected against abuse. The need for informed consent requires specific attention to voluntariness and capacity with patients who fall into categories of special populations (children, the mentally infirm of any age, and prisoners). Not to permit such groups to volunteer can discriminate against them, yet their vulnerability must be protected.[88]

---

[85] *Ibid.*, s. 17.

[86] *Ibid.*, s. 19.

[87] *Ibid.*, s. 20.

[88] See A.D. Milliken, "Position Paper: The Need for Research and Ethical Safeguards in Special Populations" (1993) 38 Canadian Journal of Psychiatry 681. This paper was approved by the Board of the Canadian Psychiatric Association on 3 April 1993. See also J. Arboleda-Florez, "Ethical Issues Regarding Research on Prisoners" (1991) 35 International Journal of Offender Therapy and Comparative Criminology 1.

### England and Wales

Non-therapeutic research with prisoners is not explicitly prohibited in England and Wales, but in practice it is not undertaken.[89] Research conducted internally or sponsored by the Prison Service or the Home Office is coordinated by the Prison Service's Planning Unit and is subject to review by the Prison Board's Executive and the Prison Service's Health Research Ethics Committee. Research conducted by external researchers is the responsibility of the local governors of the various institutions. External research must be approved by the "appropriate ethics committee of an established academic institution or a national health service body" and the researchers themselves must have "at least graduate standing" and should be supervised by a "well known academic".[90]

In a document describing the role and functions of the Prison Service Health Research Ethics Committee,[91] Oliver and Joyce note that all medical research funders in the United Kingdom now require independent ethical approval, usually obtained from the Local Research Ethics Committees (LREC) established under the guidance of the Department of Health.[92] The Prison Service established its own Health Research Ethics Committee (HREC), under the auspices of the independent Health Advisory Committee (HAC), because "much prison research is multicentre" which might otherwise involve "a costly and unwieldy approach to several LRECs, sometimes resulting in conflicting advice" and, in addition, "many of the existing research ethics committees lack the necessary expertise or familiarity with the special problems of research in prisons".[93]

The membership of the HREC consists of an independent professional chairperson and other individuals who are drawn from the HAC or externally (but, in any event, not from the Prison Service itself). Where special expertise is required, additional members may be co-opted on an *ad hoc* basis. Membership must include at least one woman, a nurse, and two lay members. The objectives and purpose of the Committee are as follows:

> In conformity with accepted principles, the objectives of the Committee are to promote the ethical standards of practice of health research, to protect the subjects of such research from harm, to protect the subject's rights and to provide reassurance to the public and the prison authorities that this is being done. To these ends the Committee will consider the ethics of health research projects involving human subjects in prisons in England and

---

[89] See *LRC Working Paper No. 61, supra* note 30 at 38.

[90] Personal Communication (to Mr. Vincent Cheng Yang) from Mr. Danny Clark, Research Manager, Planning Unit, HM Prison Service (28 February 1995).

[91] R. Oliver & L. Joyce, *Prison Service Health Research Ethics Committee* (17 November 1994) [unpublished].

[92] See Department of Health, *Local Research Ethics Committees* (London: HMSO, 1991). In 1990, the Royal College of Physicians of London published guidelines for the conduct of ethics committees in medical research involving human subjects: see Royal College of Physicians of London, *Guidelines on the Practice of Ethics Committees in Medical Research Involving Human Subjects*, 2nd ed. (London: RCP, 1990). The RCP also issued Research on Healthy Volunteers in 1986. See Royal College of Physicians of London, *Research on Healthy Volunteeers* (London: RCP, 1986) Both publications contain sections on the use of prisoners.

[93] See Department of Health, *Ibid.* at 1.

Wales and recommend to the Directorate of Health Care whether ethical approval should be given.[94]

The document also notes that research involving prisoners raises special ethical issues because of the custodial setting and the character of the prison population. In particular, it is observed that "free and informed consent" is a critical issue and that, in this respect, "there is always the real or perceived risk that undue pressure may be brought to bear on prisoners to participate in research or not to withdraw from it". In addition, it is underscored that special care needs to be taken in relation to the "mental capacity of some prisoners" as well as "language or literacy problems".[95] No research involving prisoners may proceed without their written consent, assuming they are adults. It is explicitly stipulated that, "no significant reward or inducement should be offered to encourage participation in research" although "trivial and appropriate gifts in recognition of inconvenience" are considered acceptable.[96]

### The International Context

International concern in relation to the participation of prisoners in non-therapeutic experimentation was first aroused during the Nuremberg trials. The critical element in the case against the Nazi physicians was that their victims had obviously not given their consent to participation in biomedical experimentation.[97]

The *Nuremberg Code* did not explicitly ban non-therapeutic research involving prisoners; however, it did emphasize the basic requirement of informed consent. Similarly, the World Medical Association's *Helsinki Declaration* of 1964 did not deal directly with the issue of prisoner participation in research but it clearly stipulated that "clinical research on a human being cannot be undertaken without his free consent after he has been fully informed".[98] Similarly, the General Assembly of the United Nations has adopted a number of resolutions that have a very clear impact on the question of prisoner participation in medical experimentation.[99]

---

[94] *Ibid.* at 2.

[95] *Ibid.* at 3.

[96] *Ibid.* at 4.

[97] The legal defence team for the Nazi doctors tried to establish that the concentration camp experiments were similar to those carried out in American prisons during the war years. In particular, prisoners in Joliet Prison, Illinois, had been deliberately infected with malaria to find suitable treatments for this disease. This attempt to draw an analogy failed because the Court took the view that the American research involved volunteer subjects whereas the Nazi physicians used non-volunteer subjects. Also, the Nuremberg trials did not allow for the accused to plead *tu quoque* ("you did it too") defences. The Court did not address the question of exactly how voluntary the participation of the American prisoners really was, given the circumstances of their incarceration and the pressure of the war effort. See McCarthy, *supra* note 40 at 57n.

[98] See World Medical Association, *Declaration of Helsinki*. Adopted at the 18th World Medical Assembly in Helsinki, June 1964. Amended at the 29th World Medical Assembly in Tokyo in October 1975; the 35th World Medical Assembly in Venice in October 1983; and the 41st World Medical Assembly in Hong Kong in September 1989. [Reprinted in (1991) 19 Law, Medicine & Health Care 264].

[99] See *International Covenant of Civil and Political Rights*, 19 December 1966, Can. T.S. 1976 No. 47, 999 U.N.T.S. 171, art. 7.

The 1993 CIOMS *Guidelines* note that "the involvement of volunteer prisoners in biomedical research is permitted in very few countries, and even in those is controversial". However, it is also pointed out that contradictory arguments about the propriety of using prisoners in biomedical research have "precluded an internationally agreed recommendation" and that, where the practice is permitted, there should be "independent monitoring of the research projects".[100]

The CIOMS *Guidelines* also states that, as there are no diseases that afflict only prisoners, one cannot justify their involvement in biomedical research by drawing analogies with the situation of, for example, the developmentally disabled and children. It may be justified to involve developmentally disabled individuals and children in biomedical research that specifically investigates conditions peculiar to members of these two groups, even though it may not be possible to obtain an informed consent from them. The rationale for this approach rests on the premise that there is no alternative method of studying the conditions in question than to involve members of these special populations. In the case of prison volunteers, this rationale does not apply, since the researchers may recruit subjects from "outside" without impinging on the validity of their studies.[101]

## THE ARGUMENTS AGAINST THE INVOLVEMENT OF PRISONERS IN BIOMEDICAL EXPERIMENTATION

The major argument propounded by those who oppose the participation of prisoners in biomedical experimentation rests on the difficulty of ensuring that their consent is in fact truly voluntary. For example, it has been contended that, owing to the very fact of their incarceration, "there are motivations which induce them to participate in medical experiments and research that other people would not choose to do themselves".[102] Among the various motivations that may coerce or unduly influence a prisoner to volunteer are the following: relief from monotony and oppressiveness of prison routine and escape from a lonely and tedious existence; the ability to obtain compensation; the concept of altruism; hope of receiving favorable treatment in the future by prison authorities and parole officials; direct and indirect seeking of medical or psychiatric help; safety from violence in prison; and the risk incentive correlated with volunteering.[103]

Altruism has been identified as a key factor in inducing participation insofar as the research can serve as a method of releasing guilt through "making it up to the community by volunteering in what (prisoners) may believe to be a noble cause".[104] Furthermore, insofar as medical or psychiatric help is concerned, the suggestion has

---

[100] See Council for International Organizations of Medical Sciences (CIOMS) in collaboration with the World Health Organization (WHO), *International Ethical Guidelines for Biomedical Research Involving Human Subjects* (Geneva: CIOMS, 1993) [hereinafter *CIOMS Guidelines*].
[101] *Ibid.* at 24.
[102] See McCarthy, *supra* note 40 at 60.
[103] *Ibid.* at 61.
[104] *Ibid.* at 62—63.

been made that professional help may be freely available to research volunteers, but not to other inmates in a prison setting.[105]

Personal safety may also be a vital consideration in the prisoner's decision whether or not to participate in research. If volunteers are housed in a separate facility, the opportunity to escape the violence in the regular prison environment coupled with the probability of enjoying markedly improved living conditions may constitute a potent inducement to prisoner involvement in research.[106] Of course, the strength of this particular inducement depends on the conditions that pertain in the regular prison environment. For example, it has been suggested that "if prison conditions should deteriorate below prevailing standards of human rights, the prisoners would clearly be subject to coercion. Coercion deprives victims of something to which they are entitled".[107] However, if prisoners' living conditions do meet the necessary standards of human decency, then the inducement to participate in research could not be viewed as being inherently coercive.

A related argument is that brutal prison conditions can produce a situation in which prisoners are exploited, in the sense that researchers take unfair advantage of the psychological and physical vulnerability of their subjects.[108] If the prison environment involves living conditions so grim that prisoners are physically deprived or rendered psychologically unstable by their appalling ordeal, then it could be argued that their ability to reason and to make decisions has been impaired to the point where they lack the competence to participate in biomedical experimentation. However, this line of argument would only be applicable where prison conditions are, in fact, perceived as being oppressive in nature. With respect to the so-called "risk incentive", it has been noted that:

> The risk accepted by the prisoners in these experiments is rarely recognized in society. But the risk is present, and unfortunately for inmates, the greater the amount of risk involved, the more status is attached to their participation. Therefore, prisoners find volunteering for experimentation to be a daring attempt or an enticing risk to chance rather than a possible hazard to their lives.[109]

A somewhat different version of this argument is that the nature of the prison environment may be such as to "dampen an inmate's usual self-interest". Some prisoners, particularly those who are serving very long sentences, may arrive at a point where they "simply may not care" about the level of risk involved in a research project. In such circumstances, the psychological effects of living in prison may

---

[105] *Ibid.* at 63. Certainly, McCarthy may not be entirely correct on this point. Prison systems have an obligation to provide health care to prisoners equal to that found in the community. Most prisons, if not in USA, at least in Canada (given the virtues of the Canadian health system), are equipped with proper medical, psychiatric, and dental care. As such, this objection may be unwarranted as far as certain jurisdictions are concerned.

[106] *Ibid.* at 64.

[107] See Hill, *supra* note 38 at 697.

[108] *Ibid.*

[109] See McCarthy, *supra* note 40 at 65.

"have a foreseeably detrimental or debilitating influence on the prisoner's capacity to evaluate proposals involving some risk".[110]

The National Commission for the Protection of Human Subjects of Biomedical and Behavioral Research concluded that inmates were not able to give voluntary consent to participate in drug testing in American prisons. The Commission noted that, "although prisoners who participate in research affirm that they do so freely, the conditions of social and economic deprivation in which they live compromise their freedom".[111] Another argument that weighed heavily with the Commission was the notion that principles of equity might be infringed by the involvement of prisoners in biomedical research.[112] This issue turns on the question of whether prisoners receive their fair share of any benefits and any burdens that flow from biomedical experimentation. For example, one of the Commission's concerns was that members of minority groups within prison may find themselves participating in the research that poses the highest risk to their health, while white prisoners might be perceived as reaping most of the benefits ("rewards" or "inducements") that may be made available to "volunteers". Similarly, there was some anxiety that the principle of equity would be seriously infringed if prisoners were induced to accept a higher degree of risk in experimentation than would volunteers "on the outside".

The Law Reform Commission of Canada has stated that "no one would seriously dispute that the voluntary nature of a prisoner's consent must be considered more critically than in the case of a person having the exercise of his freedom".[113] In non-therapeutic research, where by definition, the subject does not stand to gain anything personally in terms of improvement in his or her health, it is especially important that consent should be demonstrably voluntary in nature. However, the impact of the prison environment and the deprivation of freedom inevitably render a prisoner's consent highly suspect. In the words of the Commission, the prisoner "is more susceptible to undue influence; his motive for participating is often not disinterested, but rather related to his hopes for improving his lot (breaking the monotony, earning money or privileges, making a good impression on authorities, obtaining early release, having his sentence reduced or being granted parole)".[114]

The Commission remarks that, except in the situation where research is focused specifically on the impact of prison conditions on the health of inmates, there is no compelling reason why prisoners should be used as subjects for biomedical experimentation. In fact, they tend to be utilized on grounds of convenience rather than on the basis of any scientific imperative; exactly the same type of research could be undertaken using subjects who are not incarcerated. The Commission took

---

[110] See Hill, *supra* note 38 at 697. Appelbaum, Lidz & Meisel, *supra* note 11 at 231, observed that "[p]reliminary empirical research has suggested that prisoners are likely to consent to research for emotional rather than rational reasons and to ignore risk-benefit ratios in making their decision". [footnote omitted] But they go on to add that "[i]t is unclear whether these characteristics of prison populations distinguish them from the population at large".

[111] See *National Commission: Research Involving Prisoners*, *supra* note 30 at 6—7.

[112] *Ibid.* at 10—11.

[113] See *LRC Working Paper No. 61*, *supra* note 30 at 39.

[114] *Ibid.* at 39.

the view that mere convenience could not justify involving prisoners in biomedical experimentation when the risks of abuse are so immense. Moreover, unlike children or the mentally disordered, prisoners do not constitute a "distinct biomedical group" and, therefore, there is no compelling need to recruit them as research subjects. However, on this point, the Commission appears to be in error. By virtue of selection factors, certain types of individuals that present or suffer from particular medical conditions seem to accumulate in prison environments. To obtain proper samples of these individuals from the community is practically impossible, if only because of the fact that admitting to having the condition in question is tantamount to confessing to a punishable crime, or of being highly susceptible to commit one. Thus, highly violent individuals, psychopaths, and serious and violent sexual offenders, could only be found in prisons in sufficiently large numbers to conduct studies with the proper sample sizes to allow for statistical analysis. Similarly, serious physical conditions such as AIDS and hepatitis run rampant in prison environments. Clearly, although prisoners are not a discernible "biological population", the large number of them with peculiar conditions makes them discernible "selected epidemiological populations". Surely conducting research on prisoners for these types of conditions has inherent values accruing to prisoners as a class, and to society in general?

The Commission has also stated that there will always be some degree of uncertainty surrounding the question of whether a prisoner's consent is truly free. The Commission, therefore, urged that, for the moment, the *status quo* in Canada should be preserved; that is, a formal ban should be maintained on the use of prisoners in medical experimentation. However, the Commission undertook to consider this position further before taking a final position on whether the ban should be permanent or whether there might not be some conditions under which experimentation on prisoners might be permitted.[115]

One ironic aspect of this debate is the possibility that prisoners may not constitute the most appropriate population for the purposes of such research as drug testing. Indeed, it has been contended that they "probably represent a special subgroup of research volunteers whose health status may not be representative of the total 'healthy' population".[116] For example, it has been suggested that there may be a higher incidence of subclinical liver disease (a condition that may well affect many drug trials) in a prison population and that prisoners are "unlikely to give accurate or reliable responses in testing situations which rely upon reporting of the subjective effects of drugs with regard to tolerance or pharmacologic effect".[117] In light of the latter observation, Schrogie *et al.* advise that, unless the results of evaluating compliance in prison populations are favourable, then "prisoners should be used only in studies with minimal risk and requiring minimal active cooperation such as

---

[115] *Ibid.* at 39.
[116] J.J. Schrogie *et al.*, "Evaluation of the Prison Inmate as a Subject in Drug Assessment" (1977) 21 Clinical Pharmacology and Therapeutics 1.
[117] *Ibid.* at 1.

bioavailability or pharmacokinetic studies of either marketed drugs or those already well advanced in clinical investigation".[118]

## ARGUMENTS IN FAVOR OF PERMITTING PRISONER PARTICIPATION IN BIOMEDICAL EXPERIMENTATION

Those who support the involvement of prisoners in biomedical experimentation contend that they constitute a particularly suitable population because they are "living in a standard physical and psychological environment" and that "unlike fully-employed or mobile populations they have time to participate in long-term experiments".[119] In addition, it might be emphasized that the participation of prisoners may be absolutely necessary and unavoidable in the type of research that examines health conditions in the prison setting itself (e.g. the spread of hepatitis or HIV/AIDS).

A prisoner's participation in research should be "truly voluntary". However, while there is always the potential for coercion in the prison environment, it does not automatically follow that a prisoner can never give a truly free and informed consent to participation in biomedical research.[120]

But exactly what is meant by "voluntariness" in this context? More specifically, in precisely what circumstances should consent to participation in experimentation be vitiated on the basis of a *lack of voluntariness*? In this respect, The *Nuremberg Code* stipulated that the person purporting to give such consent must be

> so situated as to be able to exercise free power of choice, without the intervention of any element of force, fraud, deceit, duress, overreaching, or other ulterior form of constraint or coercion.[121]

It has already been established that many prisoners possess capacity and are competent to give consent. Of the other elements required "to exercise free power", those of fraud, deceit, or overreaching should be matters for a competent and ethical investigator to address, or for a research ethics committee (REC) to prevent. Equally, elements of force, duress, or other forms of constraint could be easily addressed by the REC or other intramural or extramural research structures. There remains only the element of coercion, often stressed as the intractable stumbling block to such research.

Coercion implies undue influence and subtle pressures. Negative, subtle, coercion (e.g. intimidation) could be dealt with in exactly the same manner as duress. Positive coercion is understood as consisting of perks that would not be available to non-participating prisoners. As listed by Ayd[122], these perks are either external to the

---

[118] *Ibid.* at 7.

[119] See *CIOMS Guidelines*, *supra* note 100 at 24.

[120] See Schroeder, *supra* note 20 at 970.

[121] See the *Nuremberg Code*, s. 1. The Code is contained in *U.S.* v. *Karl Brandt et al., Trials of War Criminals Before the Nuremberg Military Tribunals Under Control Council Law No. 10* (October 1946—April 1949).

[122] F.J. Ayd, Jr., "Drug Studies in Prisoner Volunteers" in T.L. Beauchamp & L.Walters, eds., *Contemporary Issues in Bioethics* (Belmont, Ca.: Wadsworth, 1978) 492.

individual (e.g. financial, or sentencing gains), or internal (e.g. altruism, escaping loneliness, seeking respect). The question in this regard is whether, apart from seeking a sentencing gain, the other perks that the volunteer prisoner seeks are different than those sought by outside volunteers.

Jonas makes it a prerequisite to "authentic" consent that the research subject identifies with the project and develops "[an] appropriation of the research purpose into the person's own scheme of ends".[123] This "ownership" of the project becomes a matter of identification with the project that could only develop from internal sources similar to those expounded by Ayd.[124] But the argument seems to be that these internal motivations should be acceptable in the case of outside volunteers, but in the case of prisoners they should be rejected because they are re-interpreted as "coercive".

Certainly, once appropriate safeguards to prevent abuses are in place, the argument of coercion to deny all prisoners involvement in research activities is spurious and suffers from a double standard at three different levels:

(1) it sends negative messages of social hypocrisy to a population that many believe should identify more strongly with the aims and values of the broader society;

(2) it closes doors to elements of rehabilitation, one of the aims of the criminal justice system[125] and one that, with the explosion of prison populations, should be fostered in any way possible; and

(3) it does not go far in advancing concerns for justice in relation to the equitable distribution of the risks and benefits of research. By refusing to allow prisoners to take some of the burden of research, those upholding a ban are imposing an extra burden on the general public.

Most proponents of the use of prisoners in biomedical experimentation claim that it is discriminatory and unfair to exclude competent prisoners from involvement in such research activity.[126] At least some prisoners are highly motivated to demonstrate their willingness to contribute something of value to the larger society, and there are positive advantages to be gained from enhancing their sense of self-worth through their involvement in an act of genuine altruism. In other words, biomedical experimentation "provides an opportunity for the prisoner to maintain, or develop, a sense of personal value":

---

[123] See H. Jonas, *Philosophical Reflections on Experimenting with Human Subjects* (New York: George Braziller, 1970), at 122.

[124] See Ayd Jr., *supra* note 129.

[125] Ironically, great attention is paid to ensuring that those convicted of crimes are held accountable and that they accept responsibility for their conduct. It is certainly paradoxical that punishment by way of incarceration is increasingly viewed as a method of affirming individual responsibility while participation in biomedical research is denied on the basis that the prisoner's very status automatically strips him or her of the power to act voluntarily. See S.J.M. Donnelly, "The Goals of Criminal Punishment: A Rawlsian Theory (Ultimately Grounded in Multiple Views Concerned with Human Dignity)" (1990) 41 Syracuse Law Review 741.

[126] See *LRC Working Paper No. 61, supra* note 30 at 39.

> Participation as a subject may involve sacrifice, perseverance, altruism — a chance for the inmate to prove to himself or to friends and relatives that he (*sic*) can do something worthwhile.[127]

Furthermore, a prisoner's desire to obtain a respite from the boredom and monotony of prison life or to obtain a temporary improvement in living does not imply coercion. The prisoner is under no threat, and the implicit offer of a reward in no way destroys the capacity to make voluntary decisions.[128] Similarly, financial inducements do not automatically vitiate the voluntariness of consent; it is, after all, a major element in persuading volunteers outside the prison walls to become involved in experiments.[129] Of course, excessive remuneration would constitute an unethical inducement to participate in an experiment whether this occurs inside or outside of a prison setting.[130]

It would be misguided to impose an explicit or disguised ban on all prisoner involvement in non-therapeutic research.[131] Instead, appropriate procedural safeguards should be developed in order to address the issue of voluntariness within a prison environment. For example, procedures should be developed to establish a fair method of selecting inmates to serve as representatives on the relevant institutional research ethics committees (RECs) with a view to ensuring that proper account is taken of inmate perspectives and that there is some effective counterbalance to the views of professional researchers and prison officials.[132]

Another protective mechanism that might operate to ensure that prisoner participation in research is genuinely voluntary is the development of more informative and understandable consent forms and a more efficacious process for obtaining consent.[133] In both respects, it would be necessary to remember that there is a greater incidence of educational and literacy deficits in the prison population than in the greater population at large. As far as the actual process of obtaining consent is concerned, a number of options have been suggested, including (1) administering to each potential subject, before participation proceeds, a multiple choice or "true-false" test covering the critical information that the subject needs to know to make an informed decision; or (2) developing a two-part consent form with

---

[127] See Lasagna, *supra* note 29 at 265.

[128] See LRC *Working Paper No. 61, supra* note 30 at 39.

[129] See C.E.M. van Gelderen, *et al.*, "Motives and Perceptions of Healthy Volunteers who Participate in Experiments" (1993) 45 European Journal of Clinical Pharmacology 15.

[130] See Medical Research Council of Canada, *Guidelines on Research Involving Human Subjects* (Ottawa: Supply and Services Canada, 1987) at 24—25. The Council states that remuneration should be limited to "compensation for expenses actually incurred and losses reasonably assessed, including loss of wages". "Nominal payments" for "time and inconvenience" are also considered to be "acceptable".

[131] See Schroeder, *supra* note 26.

[132] Coburn has pointed out the need to ensure that members of the general public should also be involved in approving research (including, perhaps the power to cast a veto) because "potential subjects of research are sometimes willing to undergo procedures that violate their rights which other members of the public would not be quite so willing to surrender". See D. Coburn, "Health Science Research Ethics: A Critique" (1993) 13 Health Law in Canada 192 at 195.

[133] See Schroeder, *supra* note 20 at 999.

a combination of a traditional consent section and a division with questions for the subject to answer.[134]

Attention may also be paid to reducing the impact on prisoners of any incentives that may distort the voluntariness of the consent process. For example, drug companies could be required to pay market rates for the participation of prisoners in research. This would reduce their incentive to use prisoners as opposed to volunteers from the population at large. On the other hand, prisoner volunteers could be paid at a rate approximating the prison pay scale and the difference between the outside market rate and the prison rate could be placed in a fund that benefits all members of the prison population (for example, a fund to subsidize job training programs or post-release medical care); this would ensure that there is no direct financial incentive for inmates to enroll as volunteers in biomedical research.[135] Similarly, the impact of other elements of potential coercion can be reduced by making it clear to inmates that involvement in research will not affect their chances of obtaining parole or other forms of early release.[136]

In light of such arguments, it has been suggested that the process of obtaining informed consent does not in fact pose any special difficulties that are uniquely applicable to prisoners and that, therefore, the healthy prisoner should be treated in exactly the same manner as a healthy volunteer on the outside.[137]

## CONCLUSIONS

Recent practice in most countries has been to prohibit prisoner involvement in non-therapeutic research on the empirical basis that prison inmates are never in a position to give a truly voluntary consent. In light of the abuses of experimentation that have taken place in the past, it is scarcely surprising that this view prevails. However, it is legitimate to ask whether the maintenance of a complete ban on prisoner participation in research constitutes an overreaction to a sorry legacy which present control mechanisms have now addressed. Since the hideous medical experiments of World War II, and since the Nuremberg trials, there have been major changes in the law and ethics pertaining to biomedical experimentation. The evolution of institutional frameworks for undertaking ethical review of all biomedical research proposals and the emergence of more effective legal remedies enforcing the fundamental principle of informed consent are evidenced in many countries, including Canada.

In addition to the existing safeguards that apply to biomedical research with non-prisoner volunteers (such as mandatory screening by RECs), there are a number of

---

[134] *Ibid.* at 997.
[135] *Ibid.* at 990—991. Note that the payment of researchers by pharmaceutical manufacturers raises an entirely separate series of ethical issues. See e.g. D.S. Shimm & R.G. Spece, "Industry Reimbursement for Entering Patients into Clinical Trials: Legal and Ethical Issues" (1991) 115 Annals of Internal Medicine 148.
[136] See Schroeder, *supra* note 20 at 998—999.
[137] See Lasagna, *supra* note 29 at 267.

special protective steps that might reasonably be taken in the context of the prison setting. These might include:

- Establishing special procedures for obtaining consent in light of the literacy and educational deficits that may exist in a prison setting. Consent forms may be specially designed to be readily understandable and ethical review boards may require that there be evidence of actual understanding on the part of potential subjects (through tests that require specific answers to questions about the critical information presented in relation to the research).
- Establishing a meaningful, participatory role for prisoner representatives on the ethical review committees that screen proposals for research in the prison setting.
- Making sure that research proposals also undergo a thorough review by an external scientific and ethics committee (e.g. university or hospital-based).
- Ensuring that prisoners are informed that their participation in biomedical research will have no impact on their parole or release status.
- Reducing the impact of incentives that might cause a prisoner to become involved in research when a non-prisoner would not do so (e.g. by ensuring that financial and other rewards are modest in nature and that there is not a marked discrepancy between the living conditions of research participants and members of the general prison population).[138]

The dangers of over-protectionism are becoming increasingly clear. For example, a heightened emphasis on protecting prisoners and members of other vulnerable populations has, in some jurisdictions, led to their being excluded even from therapeutic research and has denied them legitimate access to new drug therapies that might improve their condition.[139] In September 1989, the State of California passed legislation which included the following note:

---

[138] The Tri-Council Working Group, *supra* note 10, emphasizes some of the considerations that should be addressed when assessing the extent to which prisoners' participation in research may be considered truly voluntary. For example, the Working Group stresses the importance of privacy and confidentiality in the prison setting and the need to ensure that the "permission of authorities" is not used to "induce or compel participation in research". The Working Group also indicates that the voluntariness of research will inevitably be seriously compromised if participation is secured by the "order of authorities or as a result of coercion or manipulation on the part of others". In particular, researchers should obtain the consent of individual subjects rather than relying on the consent of authorities and undue inducements, such as offering a sentence remission to prisoners, should be scrupulously avoided. In a general sense, the Working Group states that,

> Restricted or dependent participants must be granted their rights to make informed choices. Researchers working with such populations must pay attention to conditions which might adversely affect the voluntariness of their decisions. In the past, there has been abuse of participants under the authority or in the care of others because they could be ordered to participate (e.g., human radiation experiments). There is an overwhelming consensus that such abuses must never occur again.

See Tri-Council Working Group, *supra* note 10 at 11.

[139] See M.L. Elks, "The Right to Participate in Research Studies" (1993) Journal of Laboratory and Clinical Medicine 130, which discusses the need to include patients with AIDS, women, minorities, children and the elderly in research studies so that they might obtain some of the personal benefits of such research or contribute to the societal benefits that may flow from advances in scientific knowledge.

> The Legislature finds that state law designed to protect prisoners from inappropriate medical experimentation has had the unintended effect of preventing prisoners from having access to drugs or treatments which might be required for good medical care.

Consequently, the *Penal Code* was amended so as to "provide prisoners access to certain investigational drugs to treatments on the same basis that they are made available to patients outside the prison setting".[140]

Similarly, in 1993, the Council for International Organizations of Medical Sciences felt constrained to articulate the following principle:

> Prisoners with serious illness or at risk of serious illness should not arbitrarily be denied access to investigational drugs, vaccines or other agents that show promise of therapeutic or preventive benefit.[141]

As noted earlier, it has been asserted that, given the nature of the prison subculture, some prisoners may actually seek risk as a means of enhancing their status within the institution.[142] Similarly, it is possible that some prisoners may seek to expiate their guilt by accepting an unusually high level of risk in the interests of society as a whole.

While these "risk as excitement" considerations may seem at first blush to call for special restrictions on the level of risk that might be accepted by a prisoner volunteer, there remains the basic question of whether a competent adult acting in a genuinely voluntary manner should be treated any differently from those outside the prison walls who wish to participate in biomedical experimentation. Assuming that the general principle requiring an acceptable ratio between the level of risk and the anticipated benefits is routinely applied by RECs to all potential research projects, it is reasonable to assume that there will always be a firm limit to the kinds of risk that may be accepted by any volunteer (prisoner or otherwise) in a non-therapeutic project. It is submitted that it makes no sense to treat prisoners as fully accountable for their actions when punishing them and as less than fully autonomous when letting them decide whether or not to accept a level of risk that is considered to be perfectly ethical in the case of non-prisoner volunteers. By denying them their autonomy, we rob prisoners of what makes them persons and deprive them of a belief in themselves as capable of undertaking non-criminal, positive, social actions.

---

[140] California Penal Code § 3502.5. Similarly, the Tri-Council Working Group, *supra* note 10 at 41, underscores the need to avoid the exclusion of prisoners from the benefits of participation in research:

> Restricted or dependent participants, that is, those in the care, or under the authority, power, or control of others (e.g., students, employees, incarcerated populations, persons in care) should neither bear an unfair share of the burden of participating in research nor should they be excluded from its benefits.

[141] See *CIOMS Guidelines, supra* note 106 at 24.

[142] See McCarthy, *supra* note 40 at 65.

CHAPTER 27

# ETHICAL QUESTIONS PERTAINING TO THE USE OF PLACEBOS

CHRISTIAN MORMONT

## INTRODUCTION: WHERE DO THE PROBLEMS LIE?

It is generally understood that the placebo effect remains an enigma that merits scientific analysis. It is also understood that interest in the role of placebos in experimentation gives rise to methodological reflection. Finally, it is understood that prescribing a placebo in the framework of actual therapeutic activity poses problems, some of which are of an ethical nature. But what objection could there be to administering a placebo to a subject participating in an experiment, since a placebo is, by definition, an inoperative substance? In concrete terms, what could possibly be wrong in giving someone a bit of bread dough or a capsule of glucose? Given that the absorption of such substances can have no effect, nor, *a fortiori*, any direct biological toxicity, the Hippocratic stricture of *primum non nocere* is clearly respected. Moreover, let us take the legitimacy of the research's objectives (increasing knowledge and power) as a given. In what way would placebos violate the Augustinian principle that prohibits against obtaining a good by way of an evil (the end does not justify the means) — since, as a means, a placebo causes no harm?

The unexpected complexity of these questions can only be grasped by situating the use of placebos in its context. Let us recall, with Kissel and Barrucand,[1] that the word "placebo" (I will please), understood substantively, originally designated a flatterer or a flattery. From 1811 on, Hooper's medical dictionary specified that what was designated by the term was "any medicine adapted more to please than benefit the patient". While, ethically speaking, such a prescription may have been debatable,

---

[1] P. Kissel & D. Barrucand, *Placebos et effet placebo en médecine* (Paris: Masson, 1964).

531

it was not to provoke either a general outcry nor in-depth inquiry until the middle of the Twentieth Century. It was then that the placebo began to impose itself in clinical trials, for several reasons, among them:

(1) the experimenter's suggestibility engendering systematic error of evaluation;
(2) the suggestibility of subjects as parasitical origin of behaviours and perceptions;
(3) the possible confusion between spontaneous evolution of the target semiology and evolution induced by the substance under trial.

The placebo neutralises these sources of distortion by guaranteeing all subjects, from the experimental group as well as the control group (and, *mutatis mutandis*, the experimenter), psychological conditions that are strictly identical: they are given the same orders, follow the same procedures, hear the same discourse, absorb *apparently* the same substance and are thus all equally exposed to the phenomena of inducement, suggestion, and resistance. The identity of conditions cannot be respected, it goes without saying, unless the taking of the placebo is unknown to the subject (simple unawareness; we will speak of double unawareness when the experimenter shares in the ignorance), which implies that the placebo is not an inert substance: it is that substance *and* the discourse that prescribes it. A discourse that silences the product's real nature while imposing its absorption. The dough pellet is not a placebo until it is prescribed as medication. "Deception" is thus necessary. But does the deception's necessity render it ethically acceptable? The question has been central among prescribing therapists for almost two centuries; it cannot be ignored simply by virtue of the fact that we view the use of placebos within the experimental framework.

Experimental research often has more than one goal: scientific progress, therapeutic progress, economic benefit, or personal profit. At times these goals compete with one another, and the conflicts then call for ethical consideration. In any case, these ends have no concrete legitimacy unless the means they implement are in accord with Human Rights.

The question that is then raised is whether or not an experimental procedure involving a placebo (conceived not as an inert substance but as a method) and its obligatory corollary, unawareness, strikes a blow at those Rights and more particularly at the essential right to self-determination. The exercise of that right presupposes the best possible knowledge of the situation and, *a priori*, it does not easily tolerate any alteration or restriction of truth. The subject who decides to participate in an experiment can only do so if informed of all the aspects of the experiment, its risks, its goals, and its methods. The placebo, however, cannot fulfil its methodological function unless it is administered without the knowledge of the subject absorbing it. From this, one could conclude that the experimental model including the use of placebos is ethically unacceptable.

But one could also consider that the right to self-determination is fully exercised by the subject who decides to participate in an experiment without deeming it necessary to know all the details. He may consider that they don't concern him, that he would be incapable of understanding them, or that he can trust the experimenter and/or the ethical committees charged with ensuring the legality of the experiment. In effect,

(1) The individual right to self-determination can be exercised in deciding to renounce the right; in other words, the subject who has freely accepted to participate in an experiment can decide not to be party to subsequent decisions; the personal motives driving him to not assert his right to self-determination do not influence the legitimacy of his decision; for example, if the motive behind the choice is financial, then profitability will be the criterion of participation, and the subject will leave the rest (goals, methods, risks) up to the experimenter.

(2) Having a right does not mean one has the duty to exercise it. In the context of an experiment, while the experimenter cannot refuse to disclose the truth to the subject who demands it (although it may mean that the subject becomes ineligible to participate in the experiment), he is not obliged to impose it upon a subject who does not ask. Furthermore, and this goes beyond the framework of experimentation, the subject can exercise the right not to know, and truth cannot be imposed on him. Would the exercise of the right not to know be another condition of ineligibility?

Although this debate highlights truth as a major element in self-determination, the question remains as to the status of truth in experimentation. If one regards truth and reality as synonymous, then saying the truth consists of saying that which is, including — when setting out the outlines of an experiment — the use of placebos. Nonetheless, research has no meaning except in relation to a "that which is not yet" (and which may never be), with proof of the prediction's validity being left up to the experiment. Does setting forth hypotheses and probabilities still amount to stating the truth, even if it is admitted that the truth may be merely virtual or probable? At what point are we affirming a new synonymy that causes a confusion between honesty of information and truth? In that case, honesty is a moral pledge that offers no guarantee of scientific exactitude, or of security. It will be said that what is sought is an increase in knowledge by means believed to be harmless.

One could also consider that the question of truth concerns the experimenter made "blind" by double unawareness. In the experimental model providing for this particular procedure, the difference between the experimenter and the subject lies in the absorption of the placebo. Does this difference have ethical implications concerning the notion of truth? When is ignorance of the absorbed product, an ignorance that is asymmetrical but shared, assimilable to a deception not only of the subject but also of the experimenter? Let us mention in passing the curious problem raised by double unawareness, since in this situation the "liar" also lies to himself, in order to arrive at a greater truth in his assessment of the observed facts. The approach to these problems calls for some reminders and clarifications.

## PLACEBO AND ETHICS

### *The Concept of the Placebo*

The concept of the placebo is complex and can only yield its meaning after several of its aspects have been examined. We will begin thus from an imprecise and empirical meaning of the term (medication prescribed in order to please) and seek to better delimit the field making up the object of our ethical inquiry, notably by distinguishing the placebo (in its substantial aspect) from the reactions it provokes. This distinction will be developed by analysing the concrete or psychic matter of the placebo, looking to its utility, its conditions of use and the individuals to whom it is administered.

### *The Substance Of The Placebo*

Most often, and clearly so in pharmacological experiments, the placebo is presented as medication, which leads to the following questions: firstly, does the placebo exist other than in this medicinal form? Secondly, when presented in this form, what is its substance?

*Does the Placebo Only Exist in Medicinal Form?*
This question receives no clear-cut answer insofar as: suggestibility is a major factor in the placebo reaction; and, other agents (verbal or behavioural) have a suggestive effect, not just pseudo-medicinal ones. By admitting that the placebo could take on other forms, one renders almost impossible the restriction of the placebo domain, which can merge with the infinite territory of suggestibility. One probably does, however, thereby do justice to the "placebogenic" psychic reality that grasps and deals with information (the placebo, in this instance) before converting it into physiological messages that in turn give rise to bodily reactions.

Within this perspective, the chemical, verbal, or behavioural substance of a placebo merely has an incidental effect, that of supporting complex information. The study of the placebo would thus be the study of a particular but not unique case of the processing of information. This point of view is as valid for the patient receiving the placebo, who may be altered at the physical level and the cognitive level (estimate of the placebo's effects), as it is for the administering doctor who risks distorting his evaluation of reality as a function of the meaning he has attributed to the placebo (cf., the Rosenthal effect). The role of beliefs and cognitions, not to mention relational, emotional, and behavioural aspects, is well proven by the establishment, deemed necessary in clinical trials, of the double blind (i.e., double unawareness), which consists of making both experimenter and subject blind, or unaware, in order to impede their impulsive turning to cognitive models and processes that generate distortions. Where this blindness that deprives expectation of a definable object is concerned, there is a tendency to discount a kind of virginity favourable to the emergence of honest interoceptive excitations susceptible of being submitted to an unprejudiced auto- and/or hetero-evaluation.

*What Is the Substance of a Placebo?*

Debate essentially bears on the inert pharmacological nature of the substance: is a placebo necessarily inert (pure placebo), or can it possess active principles (impure placebo) so long as they do not attack the pathology being studied? One might ask what possible reason there could be for administering an impure placebo, given that, either the impure placebo has no effect and does not differ in any way from a pure placebo, or rather it has some sort of effect, thereby giving rise to the problem of allowing the researcher to identify the subjects who have received the placebo. That is not to say that the administration of an impure placebo may not act as a salve on the conscience of the doctor, who finds himself even more in conflict with his professional ideal when prescribing an inert substance than when prescribing one that is "incongruous" but not completely useless. It remains the case that the term placebo is used for both types of products (neutral product; inappropriate and thus inefficient product), even if the first sense is the more common one.

### Notions of Inefficiency and Belief

The notion that the placebo is a substance that is pharmacologically or biologically inactive is not without its consequences. If, as contemporary medicine tends to think, the majority of medications from the Eighteenth Century (and beyond) were objectively inefficient for treating the pathologies they were supposed to cure, one would have to infer that it was a matter of placebos or that placebos designate less a "medication prescribed more to please" (than to be useful), than the conjunction of the belief of the doctor and the desire of the patient. The latter demands that something be done for him and the former responds by means in which he either believes (real medication) or does not (placebo). It is thus, in large measure, the conviction of the doctor that defines the placebo. It is well known that this conviction was not and is still not founded upon established scientific facts. It seems therefore that the placebo is a sort of virtual image born in the interstice between two beliefs, that of the doctor who thinks he's giving an ineffective medicine and that of the patient who thinks he's receiving medicine that is effective.

Underlining the role of beliefs in this manner draws attention to the psychological dimension, with the cognitive, emotional and functional aspects it implies. That dimension cannot be approached without briefly evoking the problems in defining the placebo effect.

### The Placebo Effect

If we wish to refrain from speculations regarding the aetiology of the placebo effect we'll agree on the fact that we are dealing with alterations observed after a prescription but which cannot be attributed to it. This does not imply that the placebo effect arises only following the administration of a placebo. On the contrary, as Pichot suggests,[2] the placebo effect also exists after the administration of an

---

[2] P. Pichot, "A propos de l'effet placebo"(1961) 3 Rev. med. psychosom. at 37–40.

active product; the effect is, thus, "the difference between the alteration that is observed and that which can be imputed to the pharmacodynamic effect of the drug". We formulated this conception in an arithmetically simple manner, at the time of an experiment in double unawareness and the mixed administration of benzodiazepine and a placebo,[3] by subtracting the levels of anxiety obtained under the placebo from those obtained under the benzodiazepine: the remainder can be considered as a measure of the specific effect of the drug, while the change common to both is the measure of the placebo effect.[4]

The placebo effect may turn out to be better than that of the substance that is active or believed to be so, or it may be negative. The use of the term "nocebo" to designate the unpleasant effects of the placebo has not met with great acclaim. In 1969, Herzhaft undertook a fairly exhaustive review of the negative effects provoked by a placebo,[5] and Heusghem and Lechat[6] saw fit to devote a chapter of their work on *The Undesirable Effects of Drugs* to the importance of psychological factors in the evaluation of these effects.[7]

But whether the effect is placebo or nocebo, many believe it to be a matter of an imaginary alteration, in the same sense as one speaks of an imaginary illness. This notion of an illusory effect is reinforced by the fact that, often, placebos take effect rapidly but also become exhausted more rapidly than many active molecules. However, the increasingly sophisticated understanding of the influence of life events on the organism and especially on the immune system leads to the admission that information (in the form of an internal or external event) is capable of provoking objective and enduring alterations of the organism. The scope of these changes surpasses that of changes set off in us by affective signals. It is thus that there are countless examples in history of people who have resisted horrifying conditions because their will to live, their enthusiasm, had mobilised inner resources to extreme levels of performance. Inversely, many are the politicians, business leaders or workers who wither away as soon as their career is over. The placebo effect thus cannot merely be an illusory flash in the pan, triggering as it does complex processes that affect — in positive and negative ways — the overall defence of the organism.

In the relationship between doctor and patient, or between experimenter and subject, recognising that the placebo effect is something more than a mere illusion can have as an important consequence a change in the way the placebo reactor is perceived by the person prescribing or evaluating. Even today, and in spite of all the contrary findings in the research, there is a tendency to view with irony and

---

[3] D. Bobon, J. Fanielle & C. Mormont *et al.*, "Time-blend Videotaped Evolution of Injectable Diazepam, Corazepam and Placebo" (1978) 78 Acta psychiat. belg. 619.

[4] C. Mormont, D. Bobon, & R. Von Frenckell, "Quantitative Evaluation of Placebo Effect as a Corrector of the True Drug Activity" (Paper presented to the 12th CINP congress, Göteborg, 1980).

[5] G. Herzhaft, "L'effet nocebo" (1969) 58 Encéphale 486.

[6] C. Heusghem & P. Lechat, eds., *Les effets indésirables des médicaments* (Paris : Masson, 1973).

[7] C. Mormont, J. Bobon, & C. Heusghem, "Importance des facteurs psychologiques dans l'évaluation des effets indésirables des médicaments" in C. Heusghem & P. Lechat, eds., *Les effets indésirables des médicaments* (Paris : Masson, 1973).

commiseration the suggestible, dependent, and deceivable being that is the placebo reactor.

Moreover, it is appropriate to recall that the hypothesis of a personality-type specific to the placebo reactor, the nocebo reactor, the placebo-resister has become outdated; there is no longer a belief in being able to predict a placebo effect by reference to the personality of the subject being "treated". In fact, it appears that the same individual may not always be a placebo reactor or resister, and that situation-specific or historical factors can influence the placebo response. This conceptual evolution parallels the evolution of ideas relating to psychosomatic illnesses. Beginning from the hypothesis of a psychosomatic personality, passing through the theory of specific conflicts, the notion of the psychosomatic response was arrived at, a response only arising in moments of depletion or of going beyond the more elaborated repertory of defensive behaviour-responses.

### *Usefulness of the Placebo*

To the extent that existence of the placebo effect is recognised, then *ipso facto* one also recognises that placebos are effective, something that is not without its consequences as regards ethical reflection.

The difficulty lies in the unforeseeable character of this effectiveness, the motivating forces of which are activated by agents that are difficult or impossible to control. Prediction seems reasonably possible for a group of subjects: taking into account the nature of the group, the placebo's physical properties (shape, colour, taste, method of administration, etc.), the quality of the relation between prescriber and subject, the orders given, and obviously the pathology, one could predict that around X% of subjects will present placebo effects. What does not seem capable of prediction is which subjects will be placebo reactors and what effects they will exhibit. In other words, for a given subject, and even though the response to the placebo would proceed essentially from his own system, we cannot predict what the administration of a placebo will bring about.

But does this unpredictability of effect constitute a radical difference between placebos and active substances? Would the latter, as opposed to the former, have an effect that is systematic, logical (in relation to its characteristics) and foreseeable? Empirically, that is not the case: most supposedly effective drugs tested in human therapy do not affect all subjects, and the observed effects are not identical in each subject.

The difference *vis-à-vis* placebos is thus rather of a quantitative nature: there are more subjects who present more effects with the active substance than with the placebo.

### Ethical Questions

### *Placebos and Active Substances: Advantages and Inconveniences*

If that is how things stand regarding the relative difference between the effectiveness of placebos and active substances, we have arrived at a first ethical reference point.

The placebo-substance distinction not being an absolute one, the fact of administering one or the other does not mean that the subject either is not being treated or is being treated effectively.

Developing this point further, one could even suppose that, pending proof to the contrary, the placebo would be preferable, because:

(1) it is not toxic
(2) its mode of action is more constructive since it mobilises the strategies of the individual rather than introducing foreign bodies into the organism.

This opinion might be met with the objections that:

(1) an inert substance cannot be deliberately administered to an individual in place of an active substance, even if the latter is of a dubious specificity and if the former has chances of working;
(2) an inert substance cannot be given to an individual if an effective treatment is believed to exist.

These two objections would hardly be debatable if it were definitively proved that a treatment is to be preferred to an absence of treatment and that the conviction of the prescriber is a sufficient argument in favour of the treatment that is the object of his conviction. The examples of drugs that are ineffective, indeed dangerous, prescribed for years on end do not allow for a complacent acceptance of such an argument. The conviction of a clinician can, certainly, be grounded on rational proofs or on intuitions generated by his practice, but it can also be grounded on the need to reassure himself, to protect his narcissism, or to expand his clientele, all of which sow the seeds of behaviours based on superstition or magic.

Regarding the *a priori* preference for an active substance, it should be nuanced by the recognition that the suppression of treatments that are often burdensome, complex and lasting a long time is, not exceptionally, followed by an improvement and not a worsening of the patient's state of health, as one might have feared.

It is worth noting that the conditions of this recognition have been criticised, indeed, even condemned for the very ethical reasons we are discussing: the prejudice in favour of active substances has commonly served as an argument for refusing the medicinal wash-out of sick people entering a clinical trial. It is a delicate matter to affirm, today, that wash-out was only practised after a critical analysis had demonstrated its ethical viability, or that such viability had become evident after wash-out had been carried out in accordance with the norms of pharmaceutical research and was followed by positive effects for the patient. Thus, by chance, an operation (wash-out) that is *a priori* debatable, as it goes against what is considered the good of the patient, has turned out to be beneficial to the patient and therefore preferable to treatment itself.

(1) If therefore we establish that:
(a) the prescription of an active substance is not necessarily effective;
(b) the prescription of an inert substance is not necessarily ineffective;

(2) If, moreover, the good of the subject overrides all other considerations;

(3) then the pursuit of the good — a good that is not defined abstractly but instead is concretely formulated as "the best possible" — cannot be reduced to the application of solutions that are not well-tried and/or given once and for all.

It follows from these propositions that the principle of using placebos is not, in and of itself, unacceptable, as its legitimacy is to be tested anew in each particular case.

This conclusion does not, however, put an end to the debate over the placebo's role in experimentation. Even if, as we have just tried to show, placebos do not violate the requirement of effectiveness, there remains an open question, concerning the relationship between the doctor's act of prescribing, the administration of the placebo, and the consent given by the patient.

### *Placebo, Benevolent Paternalism, and Benevolent Deception*

Rawlinson, retracing the evolution of ideas in medicine since Antiquity, has shown that until recently the duty of truth did not form part of a doctor's obligations.[8] On the contrary, due to his loyalty to the principle of benevolence, the doctor owed a duty to silence those truths that could be painful or harmful to the patient.

Thus, from Hippocrates and Galen to the modern era, the doctor-patient relationship was marked by benevolent paternalism. "Physicians", writes Rawlinson, "were urged to exploit their position of authority, the mystery of their esoteric knowledge, and the symbolic power of the language and implements of their art in obtaining the necessary compliance on the part of their patients".[9] To this day still, this compliance interferes with even the most "objective" treatments. In every culture, the rigorous submission to the demands of ritual (carrying out with exactitude the prescribed acts, the sacred gestures, etc.) has been treated as a primary condition of magic's efficacy. In our civilisation, the respect for the prescription, rationally justified by pharmaco-kinetics, is at one and the same time a sign of the sick person's allegiance to the healer and an act of faith in his powers.

However, and still according to Rawlinson, it was at the beginning of the nineteenth century that Percival (1803) discussed the duty to practice "benevolent deception". He was responding to a criticism by Gisborne who, appealing to St. Augustine, declared that one cannot use bad means even in order to obtain a good. Percival thought that the evaluation of good and bad cannot be performed in *abstracto*, that circumstances have to be taken into account. Thus, if deception is generally reprehensible, there are cases in medicine where it is justified:

(1) "When full disclosure of the facts of his or her conditions would be injurious to the patient, and

---

[8] M.C. Rawlinson, "Truth-telling and Paternalism in the Clinic: Philosophical Reflections on the Use of Placebos in Medical Practice" in L. White, B. Tursky & G.E. Schwartz, eds., *Placebo: Theory, research, and mechanism* (New York: Guilyard Press, 1985) 403.

[9] *Ibid.* at 405.

(2) when some ruse is necessary in order to ensure the success of the treatment".[10]

Percival also responds to a certain Dr. Johnson, who seems to be the father of the contemporary critique of the use of placebos.[11] Johnson in effect demanded that truth take precedence because falsehood respects neither the person nor his autonomy. It is worth noting that Johnson draws his argument from his own experience ("Of all lying I have the greatest abhorrence {of that practiced in medicine}, because I believe it has been frequently practiced against myself"),[12] which gives his remarks a human resonance even as it limits their scope. Indeed, if the argument expresses nothing but a personal opinion based on individual experience, it can be contradicted by another opinion based on different experiences.

Alert to Johnson's argument, Percival finds in it one more good reason to be well acquainted with his patient, so as to avoid affronting the patient by employing manoeuvres that could hurt his feelings. Percival adds that, "it is the moral substance binding doctor to patient that requires the paternalistic practice of benevolent deception",[13] since revealing the truth could harm the patient, which goes against the therapeutic goal that is the object of the contract of trust binding him to the doctor.

By virtue of this contract of trust, the patient recognises a specific professional skill in the doctor enabling him to take the soundest medical decisions possible on behalf of the patient. The patient's putting himself in the doctor's hands in this manner leads to the characterisation of the doctor, or at least his position, as paternalistic.

### Placebo and Informed Consent

To the contract of trust between doctor and patient, one can directly oppose the contract of mistrust that constitutes informed consent.

That is certainly not a common way to define informed consent, the essential justification for which is the respect of the subject and his autonomy. It is undeniable, however, that the development and generalisation of informed consent owe a great deal to American legal practices, which, as we know, have considerably broadened the possibilities of suing for medical malpractice. From this point of view, informed consent has the utilitarian function of protecting the doctor from the patient who would consider himself injured by not having been able to decide, on his own and with full knowledge of the facts, about the risks he was assuming in accepting the impugned treatment. This utilitarian function has little to do with ethics.

The mistrust that infiltrates informed consent goes in two directions: from the doctor to the patient and from the patient to the doctor. The sympathies of the doctor,

---

[10] *Ibid.* at 407.
[11] J. Boswell's *The Life of Samuel Johnson* is discussed in Rawlinson, *supra* note 4 at 407.
[12] *Ibid.*
[13] *Ibid.* at 408.

his assessment of the risks, his judgment, his decision-making strategies, his financial advantages, and his desire for prestige do not always, or entirely, coincide with the optimal conditions of care for his patient. The latter thus has reasons not to give himself over to the paternalism — no matter how benevolent — of the former; and no one may take the patient's place when what is at stake is a decision concerning his own health or the acceptance of measures that involve and commit him.

If informed consent plays a major role in critical reflection on the use of placebos, the question as to its validity takes on a more defined profile.

It is not a matter, here, of discussing the general conditions of informed consent (e.g., level of intelligence, state of mind, a subject's capacity to judge, quality and exactitude of information, etc.), but rather of certain of its pragmatic aspects.

*Consent and Unawareness*

First of all, what can be said about the possibility of giving informed consent when, by definition, the object (a placebo) of consent must be unknown to the consenting subject? The placebo is, as we discussed, the physical vector of a discourse (prescription) and of a belief, which issues from unawareness. If unawareness of the placebo is maintained, then what is there for the subject to consent to? The solution currently adopted in trials consists of explaining to the subject that he may receive a placebo, that his agreement must relate to this fact but that, in any case, if a placebo is given to him he won't know. Assuming that it has no major impact on the placebo effect and therefore on the usefulness of the research, this solution has the advantage of being based on correct information. It does not, however, put a definitive end to the debate — since the placebo is either considered to be a trick, and the fact of alerting the subject changes nothing, or rather it is considered as no more than inoffensive substance, in which case one could ask what more the subject has to consent to that he has not already accepted by participating in the experiment.

This question, regarding the validity and usefulness of the consent of a healthy volunteer, ceases to seem appropriate when the subject is a sick volunteer: for someone who is ill, receiving a placebo rather than a substance supposed to be active constitutes a real risk. The taking of that risk cannot be considered a mere experimental modality. It must be decided by the person who assumes the risk.

But what information will be provided him to guide his decision? The replacement by a placebo of a remedy that is certain to be effective being ruled out, the suggestion of experimental treatment cannot be made unless existing treatments produce mediocre or non-existent results, which reduces them to the status of (impure) placebos. Will the subject then be told that the known treatments are unsatisfactory, that a new treatment may be better but that it will only be given him if chance (aleatory allocation) decides in his favour? Can it be said, simply, that this is putting the subject in a position to exercise his right to self-determination? Or should the anguish, the revolt, that such a statement can trigger be taken into account? One can certainly say that these painful emotions are for the individual to assume, but they often generate psychosomatic reactions that are close, indeed, assimilable to, reactions to placebos (and "nocebos"). Is there, ethically speaking, a difference between a placebo reaction induced by the administration of a placebo and the placebo reaction arising from the suppression of belief that is constitutive of the placebo?

If, nonetheless, one recognises the primacy of the principle of self-determination over other values in human experimentation, then there arises the question as to the limits to this self-determination. For example, a prisoner who expects no more of life, a person condemned to execution, someone disillusioned with life, or a fanatical altruist — do they have the right to give their informed consent to experiments putting their life and health in danger?

It is easy to put such a question aside by turning to the ethical limits that the researcher may not cross. This return to the ethics of the experimenter is interesting and helpful; it underscores the dialectical and interactive nature of ethical debate. Furthermore, if the ethics of the experimenter constitutes such a solid guarantee, why give such a pre-eminence to informed consent?

*Consent and Competence*

In innumerable situations, the individual does not possess most of the facts of a problem confronting him. That is obviously the case when he is sick and must be cared for. He then turns to a specialist who, being supposed to know, is most capable of determining what should be done. But being supposed to know does not equal being certain. If the specialist does not offer certainty, then the patient has no peremptory reason to submit to his decision. Moreover, the subject is not fully capable of appreciating the weight of the arguments and the risks in a domain he is not well-versed in. He also cannot estimate the skill, the resources and preferences of the doctor in relation to possible treatments. These elements are among the risk factors, as they inevitably influence the treatment's application: a doctor will undoubtedly be more effective if he administers a treatment he favours and knows well than if he resorts to a technique that is less familiar, even if it is intrinsically superior. Under such conditions, how can informed consent be given?

*Consent, Illusion of Control, and Honest Information*

The doctor who seeks informed consent and the patient who gives it do not only exercise, jointly, the individual-patient's right to self-determination, they also sacrifice to the illusion (more grandiose in some cultures than others) of control exercised by the individual over his world and his fate. "Illusion" in the sense that consent is based more on ignorance than knowledge; an illusion harking back to benevolent paternalism now expressed at the level of "honest information" and no longer at that of the benevolent decision. Honesty of information in no way guarantees the justice of a decision; it only protects against a decision knowingly distorted by a deliberate skewing of information. Thus, honest information brings the doctor the peace of a clear conscience, whatever difficulties it may impose upon the patient forced to exercise his self-determination.

*Consent and Decision-Making Strategies*

These difficulties do not flow only from the complexity of the problem to be solved; they arise also from the fact that decision-making strategies are varied and unequally appropriate to the unique task presented by informed consent. Observation and research have brought to the fore the notions of the independence of the field, of locus of control, of coping strategies, of relational modalities, and of social interactions, all of which demonstrate that the determination of self is lived, represented, desired and decided in diverse ways, according to the encompassed and encompassing organisations of each individual. For example, in the obligation to give informed consent based on scientific information to be analysed, the individual whose adaptive strategy is avoidance is at a disadvantage in relation to another individual who resorts spontaneously to rationalisation and intellectualisation. The aptitudes required to give informed consent are not distributed equally among beings; some of them receive dual positive connotations, one psychological (it is a "skill" to have access to strategies that allow one to determine oneself by following criteria valorised by society), the other moral (it is seen as good, by a given culture at a given time, to be gifted with this skill). There thus arises discrimination among citizens, not all of whom are able to respond in the desired manner to the ideological obligation of self-determination, an obligation that takes the characteristic form of the double-bind: "be autonomous".

## Placebo, Healthy Volunteer, Sick Volunteer

It seems to go without saying that the individual involved in an experiment must have agreed to participate in it and must have done so knowing what he would be exposed to (informed consent), it being understood that whether sick or healthy he will benefit from the best methods known in the field (cf., the *Helsinki Declaration*). In this context, the situation of the healthy subject and that of the sick subject are different, especially as regards placebos.

*The Healthy Volunteer*

With respect to the healthy volunteer, the administration of an active molecule would seem to be the only condition that poses an ethical problem, because it amounts to making a subject, who (even hypothetically) has no need of it, absorb a substance capable of changing him. Things are not that simple, however, since we know that an inert substance (placebo) can give rise to effects (placebo effects) among a certain number of subjects (placebo reactors) and, even, that these effects can be for the worse (nocebo effects).

One of the elements to be considered in ethical evaluation of the use of placebos is related to the information given before and after an experiment. If one opts for "benevolent paternalism" the subject knows only (and one can imagine cases where he would not even know this much) that he is being used for research purposes and trusts the morality of the researcher to protect him from experimental risks. The placebo is a modality of the experiment, about which there is no more of an obligation to divulge information than about the nature of the active substance. Neither is there any need to explain to the subject the analysis of his reactions within the framework of the experiment.

Such an attitude provokes two principal objections:

(1) this benevolent paternalism opens the door to all kinds of abuses, and even if there is no abuse the subject may be exposed to undesired experimental conditions;

(2) the requirement of informed consent is not met.

This last criticism, besides the methodological difficulties (mainly, inducing effects by suggestion) it brings if followed, poses a psychological problem of no small importance, that is found in other experiments as well (Milgram's remains a prototypical reference in this respect): after the experiment, the subject learns from the mouth of the experimenter of certain aspects about himself that he was unaware of and that perhaps trouble him. One can ask, then, if the fact of accepting to participate in an experiment automatically implies the abandonment of the right not to know. Or, rather, in accordance with the moral valorisation of the αυτoζ (self), do we consider it good for the subject to know himself better, to have access to a more objective truth about himself revealed by science and not corresponding to the Socratic γνωθι σεαυτον (know thyself). In the face of the valorisation of extrinsic knowledge of self, and even though the subject has accepted to increase it, one can object that the subject is in the dark concerning whatever will be revealed to him and so cannot have truly consented to receive this information (this argument is used in the framework of professional secrecy, in order to justify the person in control of a secret being unable to allow the therapist to divulge the secret).

Moreover, healthy volunteers constitute a particular population (one that has always called for circumspection in the generalisation of research results), an observation that would exceed our focus if it had not been recently observed that, among them, placebo reactions seem to multiply over the course of time.[14] Thus, methodology requiring the use of placebos reveals a preference in selection that leads to a related ethical question: doesn't the participation of a healthy volunteer in a pharmacological trial result in economic pressures (free access to medical examinations, etc.) exerted on a fringe of socially unstable individuals? If that were the case, it would constitute a real exploitation of (even relative) destitution and it would be difficult to accept from an ethical point of view.

*The Sick Volunteer*

With the sick volunteer, the administration of a placebo is what poses a problem. On the one hand, there is an obvious concern about the lack of effectiveness of known treatments, though one can never say for certain that no cure can come of them. On the other hand, it is difficult to prescribe a substance one knows to be inert, even if

---

[14] M. Ansseau, Personal communication to the author (1995).

one recalls that placebos are embarrassing on account of their (unforeseeable) effectiveness and not their ineffectiveness. It remains the case that in circumstances where clearly effective medication exists, it is hardly defensible or practical (because superfluous) to administer a placebo. In general, a new molecule will be tested in comparison with an active drug.

Aside from the case of a tried and true drug, different situations present themselves, as a function of the curability of the illness, its stage of development, of the more or less aggressive nature of existing treatments, and of what is regarded as an improvement.

For example, the use of a placebo does not pose the same problems when dealing with: an individual suffering from an illness that goes away spontaneously (leaving the possibility of attempting to accelerate the process), one who suffers from a malady that is curable by a specific treatment, or yet another one who is incurable and for whom quality of life is more of a concern than a modification, without major consequences, of biological parameters. In each of these cases, the objective risk differs, and should give rise to adjusted precautions, but does not seem to bring up questions that have not yet been raised (consent, information, self-determination . . .).

Thus, the patient who has an illness that spontaneously goes away does not run serious risks in taking a placebo, and his consent to the experiment is automatic. The sick person who can be cured by a specific treatment exposes himself to an important risk by accepting the possible administration of a placebo. He cannot be asked to give his consent unless exceptional reasons justify this risk, and the experimenter cannot be content merely to provide honest information; he must make sure that the distress or confusion of the patient do not weigh too much in the decision to accept or refuse to take part in the experiment. Finally, things are less clear and depend more on the specific circumstances when one is dealing with the incurable: if the paramount value for such a patient is quality of life, one could say that any practice capable of furthering it should be looked into. If a placebo produces satisfactory or even superior results, should one, in deference to the principle of self-determination and of the demand for truth, reveal the experimental scheme and require an informed consent at the risk of destroying the beneficial effects of the placebo?

## FINAL COMMENTARY

### *What Is the Object of the Doctor-Patient Contract?*

Whether the experimenter tells the whole truth, obtains informed consent, or acts in a more paternalistic way, the same debate is repeated, and the same values are confronted. Without hoping to put an end to the debate, perhaps we may find it helpful nonetheless to return to the contract between doctor and patient. On the one hand, the contract can be conceived of as a list of terms and conditions enumerating acts and materials to be implemented and which the co-contractors agree on, just as an architect and his client would regarding the construction of a house. Seen in this way, the client, looking to enter into an exchange relationship, must know what the other party will give in return, in order to compare among offers, methods, results, and prices.

The object of the contract could, in contrast, be seen not as a set of terms and conditions but as consisting of a definition of the goal to be attained. The client tells the architect what his expectations are, his means, his resources and leaves the expert to do his work, even if that means modifying plans over the course of the project. Following this example, the sick person is interested less in the means employed by his doctor than in the intention of the latter to heal him in the most efficient way possible.

Depending on whether the object of the contract is defined one way or the other, telling the whole truth with a view to obtaining informed consent and administering a placebo are acts that will differ significantly in terms of consequences, values and necessity.

Ethical discussion cannot bring a definitive response to an interrogation. It testifies rather to a conscience alert to the risk of abuse to which is exposed anyone who holds power, as well as, more seriously, anyone who is submitted to it. He who exercises power cannot consider himself cleared of obligation upon having defined a course of conduct as "ethically correct", since he must still ask himself about the conditions that this conduct creates for the other, recognised in his existential reality and not considered as a being that is abstract and, consequently, fictitious.

CHAPTER 28

# RESOLVING THE INHERENT DISSONANCE BETWEEN THE DOCTOR'S ROLES AS HEALER AND RESEARCHER: A PROPOSAL*

DOMINIQUE SPRUMONT**

## INTRODUCTION

In a recent study of the doctor-patient relationship conducted in Switzerland's Canton of Ticino,[1] a group of 967 individuals were asked whether medicine is an exact science. Approximately 80% responded by "yes" or "fairly so". Paradoxically, when the same question was posed to practioners following a training program on public health, 92% replied that it is not.[2] Clearly, the latter opinion is more reflective of reality, for medicine is in large part characterized by uncertainty. Indeed, depending on the source, estimates of treatment devoid of any scientific foundation range from 20 to as high as 90%.[3]

Given this situation, one might expect doctors to proffer research as the most efficient and effective means of reducing medical uncertainty. Surprisingly, they do not. Not only do doctors hesitate to invite patients to participate in research projects, they, much more significantly, often fail to disclose the full extent of medical uncertainty to those they treat. This explains the public's high, and somewhat misguided, level of confidence in the efficacy of medicine.

---

* In memoriam of Benjamin Freedman.
** The author wishes to thank Colonel Michael Schmitt of the United States Air Force Academy Department of Law for his friendly and critical support in editing this paper.
[1] DOS — Sezione Sanitaria, *Relazione medico-paziente, studio pilota.* (Bellinzona: Sezione Sanitaria, 1995).
[2] Personal communication of the researcher Gianfranco Domenighetti, Sezione Sanitaria, CH - 6501 Bellinzona. A larger study on a group of more than 600 doctors tends to confirm this percentage.
[3] For a recent and authoritative discussion on this question, see E. Jonathan *et al.*, "The Inpatient General Medicine is Evidence Based" (1995) 346 Lancet 407.

What explains the discrepancy between perception and fact? In 1984 Taylor and his collegues published an enlightening study analyzing the reasons physicians tend to avoid including all eligible patients in relevant research.[4] Of particular import was their discovery that only 27% of participating physicians maximized inclusion. The authors cited the following as the most common reasons for this finding:

> (1) concern that the doctor-patient relationship would be affected by a randomized clinical trial (73%), (2) difficulty with informed consent (38%), (3) dislike of open discussions involving uncertainty (22%), (4) perceived conflict between the role of scientist and clinician (18%), (5) practical difficulties in following procedures (9%), and (6) feelings of personal responsibility if the treatment were found to be unequal (8%).

This research clearly points out the uneasiness of doctors when acting as investigators to the detriment of their role as healers. It is therefore of seminal importance to seek resolution of the dissonance between the doctor's two quintessential roles — healer and researcher. Employing randomized clinical trials (RCT) to illustrate how research and practice can be confused, this article will address that conflict and offer tentative suggestions for its solution.

### *The Randomized Clinical Trial (RCT)*

The RCT is regarded as the gold standard for the evaluation of therapeutic agents.[5] However, because there is a therapeutic component to the RCT, it has been queried whether the procedure is strictly research, or research and practice, or, as some have termed it, "therapeutic research".[6] In fact, as will be demonstrated, this question is based on an incorrect understanding of the terms "research" and "practice".

One of the congressionally mandated tasks of the National Commission for the Protection of Human Subjects of Biomedical and Behavioral Research (the Commission) was to establish "the boundaries between biomedical and behavioral research involving human subjects and the accepted and routine practice of medicine".[7] After extensive consideration of the topic, the Commission concluded that:

> For the most part, the term "practice" refers to interventions that are designed solely to enhance the well-being of an individual patient or client and that have a reasonable expectation of success. The purpose of medical and behavioral practice is to provide diagnosis, preventive treatment or therapy to particular individuals. By contrast, the term "research" designates an activity designed to test a hypothesis, permit conclusions to be drawn and thereby to develop or contribute to generalizable knowledge (expressed, for example, in theories, principles and statements of relationships). Research is usually described in a formal protocol that sets forth an objective and a set of procedures designed to reach that objective.

---

[4] K.M. Taylor, R.G. Margolese & C.L. Soskolne, "Physicians' Reasons for Not Entering Eligible Patients in a Randomized Clinical Trial for Breast Cancer" (1984) 310 New England Journal of Medicine 1363.

[5] R.J. Levine, "Uncertainty in Clinical Research" (1988) 16 Law, Medicine & Health Care 174. See also A.L. Blum, T.C. Chalmers, E. Deutsch *et al.* "The Lugano Statements on Controlled Clinical Trials" (1987) 15:1 Journal of International Medical Research 2.

[6] For an interesting discussion on that matter, see A.M. Capron & R.J. Levine, "A Concluding, and Possibly Final, Exchange about 'Therapy' and 'Research' " (1982) 4:1 IRB 10.

[7] *The National Research Act*, Pub. Law 93–348 (1974).

When a clinician departs in a significant way from standard or accepted practice, the innovation does not, in and of itself, constitute research. The fact that a procedure is "experimental", in the sense of new, untested or different, does not automatically place it in the category of research. Radically new procedures of this description should, however, be made the object of formal research at an early stage in order to determine whether they are safe and effective. Thus, it is the responsibility of medical practice committees, for example, to insist that a major innovation be incorporated into a formal research project.

Research and practice may be carried on together when research is designed to evaluate the safety and efficacy of therapy. This need not cause any confusion regarding whether or not the activity requires review; the general rule is that if there is any element of research in an activity, that activity should undergo review for the protection of human subjects.[8]

In view of these characterizations, there are two basic distinctions between practice and research. Firstly, the sole goal of practice is to enhance the health and/or the well-being of an individual patient. By contrast, the investigator's goals include those of the research itself. He or she is somehow a double agent whose two masters are research and practice. Even where one does not behave contrary to the interests of one's subjects, the investigator does not act exclusively in their interests. Secondly, the doctor-patient relationship is highly personal in the sense that all the activities of the practitioner are based exclusively on the specific needs and interests of the patient. By contrast, an investigator must strictly follow the procedures fixed in the research protocol.

### Unacceptable Terminology: Therapeutical Research

Research and therapy are fundamentally different. It has been successfully argued that it is a contradiction in terms to speak of "therapeutic research". Furthermore, such a term is ethically dangerous.[9] In particular, consider the fact that the investigator must respect the research protocol procedure. Even if the object of the research is to assess the efficacy or innocuousness of a treatment, the subject is not treated like a simple patient. As Levine has observed, the questions posed in research and practice are different: "What is the antihypertensive effect of administration of this thiazide diuretic in a specified dose range for six weeks to patients with moderately severe hypertension?" This is a typical issue an investigator must confront. A practitioner would instead ask: "What is the best way to control BP of this patient who not only has moderately severe hypertension but also has diabetes, congestive heart failure, and recently lost her job?".[10]

To achieve useful results, the investigator is required to administer the drug as planned, which makes it difficult to take into account the specific needs and interests

---

[8] National Commission for the Protection of Human Subjects of Biomedical and Behavioral Research, *The Belmont Report: Ethical Principles and Guidelines for the Protection of Human Subjects of Research* (Washington, D.C.: U.S Government Printing Office, 1978) at 2–3.

[9] For a review of the arguments to reject the notion of "therapeutic research" see K. Lebacqz, "Reflections on the Report and Recommendations of the National Commission: Research on the Fetus" (1976–1977) 22 Villanova Law Review 357; F. Rolleston & J.R. Miller, "Therapy or Research: A Need for Precision" (1981) 3:7 IRB. 1; R.J. Levine, *Ethics and Regulation of Clinical Research*, 2d ed. (Baltimore: Urban & Schwarzenberg, 1986) at 8–10; or D. Sprumont, *La protection des sujets de recherche* (Berne: Staempfli, 1993) at 33–37.

[10] R.J. Levine, "Informed Consent in Research and Practice, Similarities and Differences" (1993) 143 Archives of Internal Medicine 1231.

of his or her subjects. In fact, some research protocols mention severe side effects or adverse reactions as criteria to exclude a subject from an on-going research. It is therefore admitted in principle that the subjects involved may have to suffer mild adverse reactions, even though this might not be necessary if treated as patients. As Levine explains: "The individualized dosage adjustments and changes in therapeutic modalities are less likely to occur in the context of a clinical trial than they are in the practice of medicine. This deprivation of the experimentation ordinarily done to enhance the well-being of a patient is one of the burdens imposed on the patient-subject in a clinical trial".[11] This remark is based on the fact that the treatment is chosen randomly. From the beginning of the research, the subject is not considered as an individual patient, normally entitled to receive the best known therapy for his personal condition. Instead, he is administered a randomly chosen treatment.

### *The Notion of "Therapeutic Research" and RCT*

Even if it is widely accepted that the notion of "therapeutic research" is a contradiction in terms and dangerously misleading, an analysis of the theoretical basis of RCT demonstrates that this notion is still deeply rooted in the minds of doctors. To illustrate this point, the ethical arguments used to justify initiation of an RCT must be reviewed.

#### *The Null Hypothesis and the Clinical Equipoise*
It is generally accepted that a necessary ethical precondition for such a trial is that the investigator must be able to make an honest statement of equivalence between therapy A and therapy B, often called the null hypothesis.[12] One may also use the term "equipoise", first used by Charles Fried and which refers to the state of perfect therapeutic equivalence, between the treatments being compared, necessary to justify the conduct of an RCT.[13]

Among the first to defend the moral necessity of a null hypothesis were Shaw and Chalmers. Focusing on the principle that the physician-investigator's primary responsibility is to his patient, they argued that:

> If a physician knows, or has good reason to believe, that a new therapy (A) is better than another therapy (B), he cannot participate in a comparative trial of Therapy A versus Therapy B. Ethically, the clinician is obligated to give Therapy A to each new patient with a need for one of these therapies.
>
> If the physician (or his peers) has genuine doubt as to which therapy is better, he should give each patient an equal chance to receive one or the other therapy. The physician must fully recognize that the new therapy may be worse than the old. Each new patient must have a fair chance of receiving either the new and, hopefully, better therapy or the limited benefits of the old therapy.[14]

---

[11] See Levine (1986), *supra* note 9 at 10.

[12] *Ibid.* at 187–190.

[13] C. Fried, *Medical Experimentation: Personal Integrity and Social Policy* (New York: American Elsevier, 1974) at 51.

[14] L.W. Shaw & T.C. Chalmers, "Ethics in Cooperative Clinical Trials" (1970) 169 Annals New York Academy of Sciences 487.

Shaw and Chalmers recognized that a limit of their theory is to make it extremely difficult to pursue a RCT. Even if an honest null hypothesis may be posited at the beginning of the study, the accumulation of data during the research is likely to provide evidence favoring one therapy over another. There is accordingly an increasingly strong reason to abandon the research. The physician-investigator, being aware that one therapy is more favorable, is, thus, morally obliged to withdraw his or her patients from the RCT.

To prevent this from occurring two solutions were advocated. First, Shaw and Chalmers encouraged the introduction of new therapies on a random basis to the patients as early as possible. Yet, for several reasons, this procedure is highly impractical, the main barrier being the difficulty of assessing a null hypothesis without prior data on the effect of the new treatment on people.[15] Second, Shaw and Chalmers recommended keeping "the results confidential from the participating physician until a peer review group says that the study is over. When appropriate, an understanding of the need for confidence until a conclusion is reached, fortified by an explanation of the built-in safety mechanism, should be a part of the information imparted in obtaining informed consent".[16] The doctor-investigator, by accepting to remain unaware of the results of the research, is not confronted with the aforementioned dilemma, which, thus, allows the research to be completed as planned. The use of Data Monitoring Committees is indeed becoming more common, especially in large-scale studies.[17]

As Freedman has argued, these problems with equipoise arise from an overly theoretical understanding of that concept.[18] He then continued by developing an alternative notion, which he calls "clinical equipoise":

> To understand the alternative, preferable interpretation of equipoise, we need to recall the basic reason for conducting clinical trials: there is a current or imminent conflict in the clinical community over what treatment is preferred for patients in a defined population P. The standard treatment is A, but some evidence suggests that B will be superior (because of its effectiveness or its reduction of undesirable side effects, or some other reason). . . . Or there is a split in the clinical community, with some clinicians favoring A and others B. Each side recognizes that the opposing side has evidence to support its position), yet each still thinks that overall its own view is correct. There exists (or, in the case of a novel therapy, there may soon exist) an honest, professional disagreement among expert clinicians about the preferred treatment. A clinical trial is instituted with the aim of resolving this dispute.
>
> . . . We may state the formal conditions under which such a trial would be ethical as follows: at the start of the trial, there must be a state of clinical equipoise regarding the merits of the regimens to be tested, and the trial must be designed in such a way as to make it reasonable to expect that, if it is successfully concluded, clinical equipoise will be disturbed. In other words, the results of a successful clinical trial should be convincing enough to resolve the dispute among clinicians.

---

[15] For a detailed presentation of this problem, see Levine, *supra* note 10 at 187.

[16] See Shaw & Chalmers, *supra* note 14 at 493.

[17] L. Friedman & D. Demets, "The Data Monitoring Committee: How It Operates and Why" (1981) 3:4 IRB 6.

[18] B. Freedman, "Equipoise and the Ethics of Clinical Research" (1987) 317 New England Journal of Medicine 143.

> A state of clinical equipoise is consistent with a decided treatment preference on the part of the investigators. They must simply recognize that their less-favored treatment is prefered by colleagues whom they consider to be responsible and competent.[19]

I agree with Levine that this concept of "clinical equipoise" provides a clearer description of the state of knowledge about the comparative treatments at the beginning of a RCT.[20] It helps to ease communication between the doctor-investigator and his or her subjects by avoiding the contradictions of "theoretical equipoise". It also has the advantage of offering a strong moral argument for the participation in a RCT of a doctor who has a personal preference for one of the compared therapies. Yet, even though I recognize that clinical equipoise is a useful if not indispensable tool for resolving some ethical dilemmas posed by RCT, I shall now argue that the principles on which it is based are questionable as a justification to begin such a study.

*Confusion between Research and Practice*
As indicated, Shaw and Chalmers founded the concept of the null hypothesis on the principle that the doctor-investigator's primary responsibility is to his or her patients. This argument was further developed by Freedman whose concept of clinical equipoise is aimed at solving the "conflict between the requirement that a patient be offered the best treatment known (the principle underlying the requirement for equipoise) and the conduct of clinical trial".[21] It is in fact generally accepted that the necessity of an honest statement of the null hypothesis is based on two rationales: (1) the scientific obligation that the therapies tested are comparable and that there is some reasonable chance that the RCT produces useful data; and, (2) the professional obligation of the doctor to be advising his patient on what is the best treatment. In my opinion, these arguments illustrate a confusion between the actual role of investigator and that of practitioner, and exemplify the practical difficulties of abandoning the notion of "therapeutic research". They also underline the fact that the doctors-investigators reluctantly acknowledge the fact they do not always act in the best interests of their patients.[22]

As previously mentioned, the theory of clinical equipoise is intended to address the internal dissonance between the doctor's healing and research roles. In my view, there is an alternative interpretation of the need for clinical equipoise which better acknowledges the distinction between research and practice, while yielding more specific solutions to the conflict between the dual roles of the practitioner-investigator.

### *Ethical Justification for Initiating a RCT*

Formulation of an honest statement of equivalence between the therapies compared in a RCT is undoubtedly a moral obligation. It is important to understand that this moral obligation is based on the principle of beneficence rather than on the

---

[19] *Ibid.* at 143–144.
[20] See Levine (1988), *supra* note 10 at 176.
[21] See Freedman, *supra* note 18 at 142.
[22] See Taylor, Margolese & Soskolne, *supra* note 4 at 1363–1367.

incongruency between a doctor's responsibility to provide the best known treatment to his patients and the mandates of clinical research. If one admits that research is not performed in the interests of the subjects, but, instead, for scientific purposes, one is not directly considering the patient's right to receive the best known treatment when deciding whether to begin a particular RCT. By definition, the research is not designed in the interest of the patients. Yet, this fact does not mean that ultimately the patient's rights will not be taken into account. But, in order to do so, one has to make a distinction between the moral evaluation of the justification for beginning a RCT made by the doctor-investigator and the IRB and the actual decision to participate made by the doctor-investigator and his or her patients.

The Commission has formulated two general rules which are complementary expressions of the principle of beneficence: (1) do no harm and (2) maximize possible benefits and minimize possible harms.[23] The first consequence of this principle is to forbid any research which has no reasonable chance of success. It would be morally unacceptable to pursue a useless study which, by definition, puts an unnecessary burden on the subjects. Therefore, the first assessment which has to be made by the investigator is whether his research has some reasonable chance of success. In particular, are the tested therapies comparable? This is what I previously labeled the scientific obligation. Secondly the investigator must maximize possible benefits and minimize possible harms. This task is particularly difficult because of the special design of RCT.

When involved in a RCT, a patient is no longer treated in his best interests; the therapy he receives is not based exclusively on his individual needs but rather in conformity with the research protocol. As Schafer notes, a RCT implies "the need to sacrifice individualized treatment".[24] This loss of personal treatment is theoretically a factor of risk assumed by the subjects. To minimize this risk, it is therefore necessary that the compared therapies, first, are similar to the habitual treatment the subject would be entitled to receive if treated as a patient and, second, that they do not depart from each other (this second condition deriving directly from the first one). From this perspective, one can argue that the requirement of a null hypothesis is based on the principle of beneficence as applied in the field of research ethics, rather than on the obligation of the doctor to provide the best known treatment to his patient.

Interpreted through the principles of research ethics, the concept of clinical equipoise remains useful to justify a given research in abstracto. In particular, it gives the doctors moral authorization to participate, whatever their convictions about the compared therapies may be. As Freedman has stated: "(t)hey must simply recognize that their less-favored treatment is preferred by colleagues whom they consider to be responsible and competent". But this only means that it is morally acceptable to pursue that research; it does not automatically imply that one has to participate.

---

[23] See *The Belmont Report*, supra note 8 at 6.
[24] A. Schafer, "The Ethics of Randomized Trial" (1982) 307 New England Journal of Medicine 720.

## *Ethical Justification for a Doctor's Involvement in a RCT*

In order to decide whether to involve eligible patients in a study a doctor should make a further evaluation. Once the existence of clinical equipoise is admitted, the doctor-investigator must solve the following ethical dilemma: under which conditions can he include a patient in the RCT or, in other words, how can he square his role of investigator with his responsibilities toward his patient? It is indeed the question addressed by Freedman. In my opinion, this problem deals with the ethics of practice, not with the ethics of research. It is therefore appropriate to invoke at that point the right of the patient to receive the best known treatment for his or her individual condition. If a doctor estimates that the research is not important enough, it is then morally acceptable for him to refuse to get involved in the RCT.

In the event that a doctor does not want to participate in a RCT, he or she still has the obligation of informing patients about it. Yet, this obligation is limited to the case where the doctor has some good reason to believe that the interested patients will be included in the research if they wish to participate. For example, if a study is designed to involve twenty subjects and only ten are actually taking part in it, the research should be mentioned as an alternative. On the contrary, if there are enough subjects (in accordance with the research protocol), the existence of the study is no longer material because the patients have no chance of inclusion. In this case, it is questionable whether there is an interest in mentioning the study at all.

Weijer and Fuks have argued that "the clinical investigator has the duty to screen for, and exclude, potential research subjects who may be unduly vulnerable to the risks of a particular clinical trial. . . . To reinforce and make explicit this legal and moral duty, (they) propose that the investigator sign a statement, appended to each subject's consent, to attest that this duty has been responsibly discharged".[25] This position is perhaps too paternalistic in the sense that a well-informed patient — i.e., one who understands all the viable alternatives — may choose to participate in the RCT regardless of the risks. The duty of doctor-investigators to identify patients who would be at greater than average risk does not discharge their obligation to inform potential research subjects of the existence of a RCT as a possible option.

Several authors have expressed concerns about the information to be given to prospective subjects. Fearing that they would refuse to participate in a RCT if thoroughly informed about it, which would consequently preclude the research, these authors have come up with various theories. For example, Zelen has made a proposal of various prerandomization designs for RCTs,[26] Don Marquis suggested "that perhaps what is needed is an ethics that will justify the conscription of subjects for medical research"[27] and Schafer wondered if there should not be a "shift from a

---

[25] C. Weijer & A. Fuks, "The Duty to Exclude: Excluding People at Undue Risk from Research" (1994) 17 Clinical and Investigative Medicine 116.

[26] M. Zelen, "A New Design for Randomized Clinical Trials" (1979) 300 New England Journal of Medicine 1242. For a full presentation of the criticisms of these proposals, see L. Kopelman, "Randomized Clinical Trials, Consent and the Therapeutical Relationship" (1993) 31 Clinical Research 1.

[27] D. Marquis, "Leaving Therapy a Chance" (1993) 13:4 The Hasting Center Report 47.

patient-centered to a social-welfare-centered ethic".[28] These ideas are not only morally questionable, they are based on an improper assumption. In fact, the attitude of the people towards research has been proven to be rather positive.[29] Meier has rightfully declared that: "As a matter of normal social behavior, most of us would be quite willing to forego a modest expected gain in the general interest of learning something of value. However, we should want to be assured that what we agree to give up is indeed modest and not a truly large amount".[30]

In order to guarantee that the principles of beneficence and autonomy are both respected in the field of human research, this last remark leads us to make the following proposal. While the investigator elaborates his research protocol, he is primarily entitled to respect the principle of beneficence. He must ask whether it is morally acceptable to invite someone to participate in his study. The investigator must minimize the risks and maximize the hoped-for benefits. First, the design of the research must be scientifically sound there should be some reasonable evidence that this study is likely "to develop or to contribute to generalizable knowledge". Then, the risks have to be minimized without jeopardizing the validity of the research. We have just seen the specific conditions for randomized clinical trials, but the general rule applies to every kind of research. At that point, the investigator's sole preoccupation is the protection of the subjects' health and welfare, and the scientific interest of his research. If the investigator is not convinced of the favorable balance between the risks and the benefits, then he should abandon his project. Aware of the fact that he has a strong influence on the subjects' choice, he should be especially careful when making that assessment. It is only then that he may contact potential subjects.

At that moment, the investigator's responsibility is to respect the principle of autonomy, even to the detriment of his research. This means an absolute frankness about the experimental aspect of the procedure, the possible alternative, its risks, and the right of the subject to withdraw his consent at any time without penalty or loss of benefit to which the subject is otherwise entitled. Of course, the investigator still has to protect the subjects against any unnecessary harm. In doing so, one should no longer balance the risks of the research against the hoped-for benefits which are not directly related to the subjects' conditions. One's sole obligation is to protect the subjects, no matter what the consequences are for the research. In a sense, the scientific goal of the research becomes secondary when the study actually starts; the main concern being the health and welfare of the subjects.

## CONCLUSION

There is little doubt that it is taxing in practice to separate the doctors' role as healer from that of investigator, especially in a RCT. However, the difficulties lie mainly in

---

[28] See Schafer, *supra* note 24 at 724.
[29] On the willingness of the people to participate at a research, see for exemple Levine, *supra* note 9 at 192–193.
[30] P. Meier, "Terminating a Trial — The Ethical Problem" (1979) 25 Clin. Pharmacol. Ther. 637.

the reluctance of doctors to admit that they do not always act in their patients' best interests. Instead of accepting that fact, they have developed rules which take into account the right of the patients to receive the best known treatment for their condition. In doing so, a blurred vision of the research has been maintained; the question of the real nature of research has been avoided. Even if the term "therapeutical research" is rejected, this notion is unfortunately still used in the ethical approaches to RCTs.

In the relation between doctor and patient, there is theoretically only one interest: that of the patient. In the case of research, the doctor-investigator is a double agent. On one hand, he has to protect the research subjects, but, on the other hand, he is bound to respect the research protocol. These duties may seem contradictory when considered in light of the norms of medical practice, but, after all, research is not practice. It is fundamental that we more clearly acknowledge the difference between the doctor's two roles. In particular, it is indispensable to develop ethical rules which deal exclusively with the problems of clinical research and which are distinct from the norms of practice. I believe that in doing so, part of the ethical dilemma experienced by the doctors involved in a research would disappear and the relation between investigator and subject would be made clearer. Given a more accurate picture of what their involvement in research implies, potential subjects are more likely to participate positively. It may mean that some research may have to be abandoned, but, as Hans Jonas has said :

> Let us not forget that progress is an optional goal, not an unconditional commitment, and that its tempo in particular, compulsive as it may become, has nothing sacred about it. Let us always remember that a slower progress in the conquest of disease would not threaten society ... but that society would indeed be threatened by the erosion of those moral values whose loss, possibly caused by too ruthless a pursuit of scientific progress, would make its most dazzling triumphs not worth having.[31]

---

[31] H. Jonas, "Philosophical Reflections on Human Experimentation" in P.A Freund, ed., *Experimentation with Human Subjects* (New York: George Braziller, 1970) at 28.

CHAPTER 29

# ETHICAL ISSUES IN EPIDEMIOLOGICAL RESEARCH

JULIO ARBOLEDA-FLOREZ, DAVID N. WEISSTUB AND
HEATHER HOLLEY

## INTRODUCTION

Epidemiological research does not usually involve individuals in experiments, and may seem remote from the research issues pertaining to specific populations. However, depending on the definition used, "epidemiology" initiates research or may furnish the methodological tools required to conduct research among these populations. Epidemiological methods and results tend to impact on a large number of persons, many of whom are members of vulnerable groups, with findings having an effect on the subjects themselves or on the class to which they belong. At a much larger scale, epidemiological research (based on files, for example) could be conducted even without the knowledge of the subject or population involved.

Epidemiological findings that shed light on the causes and natural history of diseases and the extent of their distribution, are applied to set priorities for resource allocation on particular health and research issues, to evaluate treatment and diagnostic technologies, to initiate preventative measures, and to determine surveillance strategies. Epidemiological methodologies used to assess patterns of resource utilization usually serve to assist governments or private bureaucracies in making policy decisions. As such, epidemiological research has the potential for causing an entire population to become vulnerable, especially if decisions are made without proper disclosure about the research being conducted or about the quality of the methodologies used. Therefore, in exploring the impact of research with vulnerable populations, it is important to conduct an inquiry into epidemiology and the ethical issues involved.

557

## DEFINITIONS

Epidemiology is defined as "the study of the occurrence and distribution of diseases and other health-related conditions in populations".[1] The word "epidemiology" derives from the greek $\epsilon\pi\iota$ = upon, $\delta\epsilon\mu o\xi$ = people, and $\lambda o\gamma o\xi$ = reasoning. For centuries, the term was applied to the study of epidemics (outbreaks of contagious diseases in the population), but epidemics in the classical sense have been brought under control since the introduction of multiple public health strategies to improve the general level of nutrition and hygiene in the population. Primary prevention to forestall the development of infectious diseases, and secondary prevention through the use of powerful antibiotics and other drugs, have gone a long way to protect humans from the scourges of the past.[2] Consequently, epidemiology has moved from the battlefield of epidemics to a seemingly unending reconnaissance operation surveying risk factors for many medical conditions, especially those that are chronic and debilitating. In fact, the demographic shift of the population towards old age may have signaled to epidemiologists that the conditions of interest for study were no longer infections usually associated with children, but systemic illnesses most commonly found at older ages. With time, therefore, the word "epidemiology" has come to indicate not just the study of epidemics, but the study of an array of risk factors, socio-determinant elements of disease, and other health-related aspects of human groups. As Last has indicated: "[E]pidemiology includes surveillance, research, experiments, and applications in population screening and health care evaluation".[3]

Epidemiology, a basic science in medicine, can be divided into several components and along several dimensions. *Descriptive epidemiology* occupies itself with providing information on a particular condition at a population level, and is used to conduct studies when the epidemiology of a disease is not too well known. *Analytical epidemiology*, on the other hand, tries to establish associations on an individual level, and pertains to studies when leads about the etiology of a disease are already available.[4] Epidemiology can also be divided according to the subject matter under investigation. For example, *clinical epidemiology* applies specifically to sick people and their care, and is defined as the "[study of] groups of people to

---

[1] J.L. Kelsey, W.D. Thompson & A.S. Evans, *Methods in Observational Epidemiology* (Oxford: Oxford University Press, 1986) at 3.

[2] Apart from a "flesh eating disease" attributed to a most virulent form of streptococcus A, and which is usually confined to a few cases, modern epidemics are being caused usually by viruses such as AIDS, ebola, and hantovirus. These epidemics and the emergence of increasingly antibiotic-resistant strains of bacteria, however, are shaking the belief that infections have been conquered. See "Return to the Hot Zone" *Time* (22 May 1995) 44 at 44–45 for a journalistic description of the ebola outbreak in Central Africa, and CDC, *Morbidity and Mortality Weekly Report (MMWR)* (9 July 1993), for a description of the hantavirus outbreak in New Mexico.

[3] J.M. Last, "Epidemiology and Ethics" in Z. Bankowski, J.H. Bryant & J.M. Last, eds., *Ethics and Epidemiology: International Guidelines* (Proceedings of the XXVth CIOMS Conference, Geneva, 7–9 November 1990) (Geneva: CIOMS, 1991) 1 at 14. [hereinafter *Ethics and Epidemiology*]

[4] *Ibid.* at 4.

achieve the background evidence needed for clinical decisions in patient care".[5] This is contrasted with *classical epidemiology*, also referred to as population-based epidemiology, that was traditionally oriented towards public health and studies of health-related events among the general population. The *denominator* in clinical epidemiology contains the *clinical group*, while in classical epidemiology, the denominator contains the *general population*.[6] A particular branch of epidemiology, which for lack of a better term will hereinafter be called *administrative epidemiology*,[7] appears far removed from people as subjects, but uses information collected on people and contained in large data sets. Administrative epidemiology utilizes high level information technology, such as record linkage, to discern such things as patterns of utilization, costs, and variances among different groups or individuals.[8]

Epidemiological studies are sometimes labelled *observational studies*. These studies use analytical techniques aimed at testing hypotheses about suspected causes of disease. They are differentiated from *experimental epidemiology*, which sets up experiments on groups of individuals, ill or healthy. Experimental epidemiology, despite its promise to disentangle complex causal problems, cannot be so readily utilized. It requires large sample sizes, its methods cannot accommodate the large periods of time necessary for observation of events to happen, and, most importantly, we cannot ethically allow humans to be placed in situations of danger or harm. That is why, in epidemiology, most studies are observational.

The *raison d'être* of classical epidemiology is to discover the cause of disease. In this endeavor, epidemiology works from the cause to the effect or from the effect to the cause. Either way, it takes into account temporal relationships and associations; thus, observational studies are commonly classified according to their temporal directionality.[9] The three most common designs of epidemiological studies are cross-sectional, case-control, and cohort studies.

*Cross-sectional studies*, or surveys, identify populations and ask questions about health status, usually to measure the prevalence of a particular condition, or to measure levels of risk factors. This is regularly done on samples rather than on an

---

[5] A.R. Feinstein, *Clinical Epidemiology — The Architecture of Clnincal Research* (Toronto: W.B. Saunders, 1985).

[6] *Ethics and Epidemiology, supra* note 3 at 4.

[7] Some authors use the term *consequential epidemiology* to mean activities related to health research, such as research on health systems, policy development, etc. See "Outgoing SER President Calls for More 'Consequential Epidemiology' " (1994) 15:7 *The Epidemiology Monitor* 1.

[8] N.P. Roos *et al.*, "Using Administrative Data to Predict Important Health Outcomes: Entry to Hospital, Nursing Home, and Death" in K.L. White *et al.*, eds., *Health Services Research: An Anthology* (Washington, D.C.: Pan American Health Organization, 1992) 1022.

[9] Much controversy exists about classification schemes for these type of studies. Some authors have proposed discontinuation of the time element and advocate the use of a "study base" (source population of individuals enrolled in the study) as a better foundation for their classification. See O.S. Miettinen, "Striving to Deconfound the Fundamentals of Epidemiological Study Design" (1988) 41 *Journal of Clinical Epidemiology* 709; S. Greenland & H. Morgenstern, "Classification Schemes for Epidemiological Research Design" (1988) 41 *Journal of Clinical Epidemiology* 715; O.S. Miettinen, " 'Directionality' in Epidemiological Research" (1989) 42 *Journal of Clinical Epidemiology* 821; and P.H. Kass, "Converging Toward a 'Unified Field Theory' of Epidemiology" (1992) 3 *Epidemiology* 473.

entire population, and findings are then generalized from the sample to the population, with a degree of confidence in their validity. Cross-sectional studies are snapshots of the state of a population regarding a particular disease at a moment in history. They are useful to determine levels of pathology and for planning needed services for the population.

*Case-control studies* follow a paradigm that proceeds from effect to cause.[10] These studies attempt to identify patients who suffer from a particular disease or condition of interest. Their past history of exposure to a suspected risk factor is compared to the past history of exposure to the same factor among a group of persons who act as controls, and who resemble the cases in aspects such as age and sex, but who do not have the condition of interest. Case-control studies are retrospective, that is, they try to ascertain the proportion among the cases and the controls who have certain background characteristics of exposure to a particular substance.

*Cohort studies* are longitudinal, meaning that they extend forward over a period of time. These studies could be prospective, in which case the cohort is assembled in the present, and then followed into the future from that moment on. Cohort studies can also be carried out on historical cohorts, in which case the cohort is assembled in the present, but with the intent to begin the study at a time in the past when the members were suspected to have been exposed to a particular risk; the cohort is then followed up to the present, or into the future. In a cohort study, patients with different exposure levels to a particular risk factor are identified and observed over a period of time (commonly years), and the occurrence of disease is measured and compared in relation to the exposure levels.

Other *observational studies* include record linkage technology, studying medical records, occupational and environmental surveillance, and studying reactions to drugs in the field of pharmaco-epidemiology. Finally, contrary to most other epidemiological studies where the unit of analysis is the individual, *ecological studies* use an aggregate of individuals, such as a group or a whole community, as the unit of analysis.

Although findings in epidemiology are intended to apply to the group as a class, or to the population in general, the basic unit of analysis, except in ecological studies, remains the individual. For this reason, epidemiological research cannot be separated from biomedical research and the ethical issues arising from research on human subjects.

## ETHICAL FACTORS IN EPIDEMIOLOGICAL RESEARCH

It is also important to review ethical issues in the context of the patient/physician interaction, and to compare them to similar ones that arise in research undertaken by

---

[10] J.J. Schlesselman, *Case-Control Studies, Design, Conduct, Analysis* (New York: Oxford University Press, 1982) at 14.

physicians and epidemiologists on populations. To follow Gostin,[11] ethical issues arising from the patient/physician interaction will be referred to as *micro-ethics*, and those arising out of population research as *macroethics*. Similarities and disparities will then be identified and discussed.

"It is probably just a cold" was, reportedly, the diagnosis doctors gave their patients in 1347, when the Black Death epidemic started to spread, eventually killing one third of the population of Europe. This "lie" has major relevance to the contemporary discourse on bioethics. If this is what patients were indeed told about their afflictions, either physicians knew no better about the epidemic, or they were simply sparing their patients the bad news. In doing the latter, however, physicians were not breaching any ethical rules. In 1347, and up until over the latter part of this century, to hide a terrible diagnosis from the patient was the appropriate thing to do. Physicians were known to do it, relatives expected and even requested that they do it, and patients, as well, felt that that was the right thing to do. Within the paternalistic mindset of the time, which still prevails in many parts of the world, physician and patient colluded in the belief that further suffering could be prevented through concealing a bad prognosis. There was harmony, then, and no conflict in the therapeutic interaction. Today, however, patients and physicians do not necessarily have a shared view of the meaning and management of disease. Patients still have faith in their physicians, but such faith is now tempered by personal and active knowledge of their medical condition, a keen awareness of balancing rights in the therapeutic exchange, and suspicions about the physician's deeper motives. A shared nexus of belief in the sense of participating in a ritual of healing akin to a quasi-religious experience has long since departed from the now mechanical, highly techno-scientific interaction.[12]

### Developments in Biomedical Ethics

The Hippocratic Oath[13] had provided physicians with a model of how to approach the patient, and rules for handling the physician/patient relationship. The Hippocratic model was paternalistic along medico-ethical guidelines; the physician took control not only of the ailment, but also of the patient. This approach and the specific medico-ethical paradigm was also consequentialist in that it strove to enhance the benefits and to minimize the harm of a medical action.[14] Medical Hippocratic ethics served physicians until 1970, when Ramsey, in *The Patient as a Person*,[15] made the admonition Hippocrates gave physicians in his *Decorum*[16] (to

---

[11] L. Gostin, "Macroethical Principles for the Conduct of Research on Human Subjects: Population-Based Research and Ethics" in *Ethics and Epidemiology, supra* note 3 at 29.

[12] C.E. Rosenberg, *Explaining Epidemics and Other Studies in the History of Medicine* (Cambridge: Cambridge University Press, 1992).

[13] Hippocrates, "Precepts" in W.H.S.Jones, *Hippocrates* (Cambridge: Harvard University Press, 1962).

[14] See Chapter 1.

[15] P. Ramsey, *The Patient as a Person* (New Haven: Yale University Press, 1970).

[16] Hippocrates, "Decorum" as cited in J.Katz, "Disclosure and Consent in Psychiatric Practice: Mission Impossible?" in C.K. Holling, ed., *Law and Ethics in the Practice of Psychiatry* (New York: Brunner/Mazel, 1981) 91.

"perform [their duties] calmly and adroitly, concealing most things from the patient while you are attending to him") no longer tenable. Ramsey ushered in a new era of medical ethics which Bryant has described along four dimensions: *normative ethics, casuistry, institutionalization and metaethics.*[17]

*Microethics*

Since 1970, biomedical ethics has developed a normative frame, and a new vocabulary. Paternalism, for example, was felt to be detrimental to the well-being of the patient as a person. Normative changes in the understanding of bioethical issues have imposed a reformulation of the major medico-ethical questions from a solely patient-centered and physician-driven interaction to one of self-determination and rights, as viewed from the larger perspective of a social dimension. To the extent that bioethical issues are now seen as affecting social values, bioethics have entered into political agendas.

Developments at the level of casuistry have kept pace with those in normative ethics. For example, on the issue of informed consent, the rights of the patient over the paternalism of the physician is now a subject of common-law decisions and part and parcel of the work of any physician. Casuistry that altered the balance in the physician/patient relationship, in fact, predated Ramsey's book. Already in 1960, in the United States, *Nathanson* v. *Kline*[18] had given fair warning to physicians about what was ethical practice in regard to proposed treatments:

> Anglo-American law starts with the premise of thorough-going self determination. It follows that each man is considered to be master of his own body, and he may, if he be of sound mind, expressly prohibit the performance of life-saving surgery or other medical treatment. A doctor might well believe that an operation or form of treatment is desirable or necessary but the law does not permit to substitute his own judgment for that of the patient by any form of artifice or deception.

*Nathanson* v. *Kline* was further cemented in *Canterbury* v. *Spence*[19] in which a "standard set by law" was substituted for the centuries-old "reasonable doctor standard". The two leading Canadian cases of *Reibl* v. *Hughes*[20] and *Hopp* v. *Lepp*,[21] echoed the American cases and supported the patient's demands for information by enjoining physicians to disclose:

(1) the *nature* of the procedure;
(2) the *gravity* of the procedure;
(3) any "*material*" risks;
(4) answers to any specific *questions* asked by the patient.

---

[17] J.H. Bryant, "Trends in Biomedical Ethics — Forerunners of Ethical Questions for Epidemiology" in Ethics and Epidemiology, *supra* note 3 at 8.
[18] *Nathanson* v. *Kline*, 350 P.2d. 1093 (Kansas S.C. 1960).
[19] *Canterbury* v. *Spence*, 464 F.2d. 772, (1972).
[20] *Reibl* v. *Hughes*, [1980] 2 S.C.R. 880.
[21] *Hopp* v. *Lepp*, [1980] 2 S.C.R. 192.

The right to refuse treatment was also ruled on in the *Quinlan*[22] case in the United States, and in Canada in the *Nancy B*[23] case, which held that the patient had the right to request a discontinuation of the respirator which had kept her alive up to then. More recently, Canadian courts have also given patients the right to obtain the information in their medical records.[24] This legal codification of patients' rights creates a new paradigm for patient/physician interaction.

Finally, as Fulton states, there is a political nature to health care, where decision-making power on health matters has been dispersed from physicians to grass roots power and the power of other providers.[25] Health is a matter of national policy and is no longer an issue to be transacted only between the physician and the patient. In the United States, for example, the *President's Commission Report*[26] advised that people have an entitlement to an adequate minimum of care without undue burden; this is also a tenet of the *Canada Health Act*.[27]

Ethics has also become institutionalized. That is, ethical issues are no longer an individual matter between the patient and the physician, they are also the subject matter of institutional bioethics committees.[28] At the research level, local or national ethics review boards have the final authority to grant permission so that a research project may proceed. The role of bioethics has been to shape the way society views itself.

*From Microethics To Macroethics*
Gostin defines macroethics as "a set of principles designed to protect the human dignity, integrity, self-determination, confidentiality, rights and health of populations and the people comprising them",[29] which are therefore applicable to epidemiological studies.[30] How these principles apply could also be seen in Last's definition of epidemiology. He defines epidemiology as "the study of the distribution and determinants of health-related states or events in specified populations, and the application of this study to control of health problems".[31] He further qualifies these terms: *distribution* covers time sequences, places of occurrence, and persons affected; *determinants* are all physical, biological and behavioral factors that can influence health; *health-related events* include diseases, injuries, deaths, behavior such as smoking, and reactions to preventive, diagnostic or therapeutic regimes or procedures; *specified populations* are populations with identifiable characteristics

---

[22] *Re Quinlan*, 355 A.2d 625 (1976).
[23] *B (N) v. Hôtel-Dieu de Québec*, [1992] Q.J. 1 (C.A.).
[24] *McInerney v. MacDonald* (1992), 93 D.L.R. (4th) 415 at 430 (S.C.C).
[25] J. Fulton, *Canada's Health System: Bordering on the Possible* (New York: Faulkner & Gray, 1993) at 4.
[26] The White House Domestic Policy Council, *Final Reports: The President's Health Security Plan* (New York: Random House, 1993).
[27] *The Canada Health Act*, R.S.C. 1985, c. C–6.
[28] See G. Griener, "An Ethicist Reflects on the Role of Hospital Ethics Committees" (1995) 4:1 Health Law Institute 3; L. Sanchez-Sweatman "Hospital Ethics Committees: A Reflection of Medical Model Decision-Making" (1995) 4:1 Health Law Institute 5.
[29] See Gostin, *supra* note 11 at 30.
[30] See Kelsey, Thompson & Evans, *supra* note 1.
[31] See Last (1991), *supra* note 3 at 1.

and, in particular, defined numbers; and *application ... to control of health problems* makes explicit that epidemiology is a fundamental science essential to public health practice.

Microethical principles apply to individual issues of protection of the rights of patients or subjects, and to the clarification of duties of physicians and researchers. Parallels can be traced between microethical and macroethical principles. Micro-ethics helps to understand the ethical dilemmas encountered in epidemiological research. The protection of vulnerable subjects at this level echoes that owed to vulnerable groups in the population at large. In epidemiological research, these protections take place at the level of the individual subject and at the level of the community. With respect to the individual subject, for example, epidemiologists must be aware, and advise their subjects accordingly, that assurances of confidentiality may have statutory limitations, such as reporting suspected child abuse. In this respect, epidemiologists also have an obligation towards their lay interviewers to train them on ethical issues and situations that may arise, such as illegal activities, while conducting home interviews.[32] Feinleib[33] proposes that it is the responsibility of epidemiologists and statisticians "to ensure that the risks and burdens for the participants are minimized while the potential benefits for society are maximized".

Regarding the protections owed a group, epidemiological research on minorities defined by racial or sexual orientation, or by clearly identified disease categories such as patients suffering from AIDS, could reveal particular characteristics about those groups. Although individuals are not identified, findings could have a detrimental effect on a group as a whole, affecting members' reputations, diminishing their sense of self-worth and cultural identity, or providing justifications for discrimination. Thus, although at the level of the individual no harms may have been identified at the time the study was designed, it may still be that at the macroethical level the research could lead to major detrimental effects. The principle of non-maleficence would have, then, been violated. On the other hand, to exclude members of minorities may be to the detriment of their future health.[34]

---

[32] J.M. Last, "Respect for Privacy and Promises of Confidentiality" (1992) 13:7 *The Epidemiology Monitor* 4.

[33] M. Feinleib, "The Epidemiologist's Responsibilities to Study Participants" (1991) 44, Suppl. I, 738–798. In the Abstract for this paper, Feinleib further states:

> Safeguards for the subjects' welfare and privacy must be considered during the planning of the study, recruitment of participants, conducting the interviews or examinations, maintaining the records, and analyzing and disseminating the information. In order to achieve maximum participation and candor of study respondents, the importance and purposes of the study and the safeguards for protecting the respondents' privacy and minimizing risks must be clearly explained. All staff must be carefully trained as to their responsibilities for protecting the privacy and safety for the participants. Adherence to these principles will promote the collection and dissemination of quality data while minimizing the risks and burdens to study participants.

[34] See "NIH Issues Guidelines for Including Women and Minorities as Research Subjects" (1994) 15:4 *The Epidemiology Monitor* 1.

*Randomized Controlled Trials as Experimental Epidemiology*

Experimental epidemiology, as the term is utilized in present literature, usually means the use of randomized controlled trials (RCTs) to demonstrate the efficacy or effectiveness of a preventive, diagnostic or therapeutic regimen or procedure. Although experiments, or *tests*, are conducted in practically every field of inquiry, a *design experiment* involves purposeful manipulations of input variables in a system to observe and identify reasons for changes in the output variables.[35] Last defines RCTs as:

> ... an epidemiological experiment in which subjects in a population are randomly allocated into groups, usually called "study" and "control" groups, to receive or not to receive an experimental preventive or therapeutic procedure, manoeuvre, or intervention. The results are assessed by rigorous comparison of rates of disease, death, recovery, or other appropriate outcome in the study and control groups, respectively. Randomized controlled trials are generally regarded as the most scientifically rigorous method of hypothesis testing available in epidemiology.[36]

In RCTs, the study group is given the intervention, procedure, or drug, and the control group is given an existing conventional drug or procedure, and at times a placebo. The use of placebos is regularly encountered in RCTs and is one of the most controversial issues in these research designs. Placebos are justified on the basis that about one third of the population tend to react positively to any treatment from a psychological reaction to the treatment itself and to the atmosphere in which it is provided. Placebos are justified in trials where there is no available treatment, or where specific symptoms are not properly addressed by existing treatments. However, it would be unethical not to advise prospective subjects that, as a result of the randomization scheme, they may end up in the placebo arm of the study. It would be highly unethical to use placebos without prospective subjects' knowledge and voluntary consent.[37] Because experiments are sometimes hazardous, rigid exclusion and inclusion criteria have to be devised and adhered to, and subjects have to be advised about all potential risks. Informed consent is, therefore, a major prerequisite to participation in RCTs. A RCT, however, could be devised to include an entire community as opposed to just individuals. Examples include studies involving fluoridation of water supplies, heart-disease control programs, and massive vaccination trials. Ethical concerns in terms of community consent to participate have to be considered along with the benefits the community would draw from the study, and the relevance of the study to the needs of the community at the time. In this regard, the potential for ethical trespasses could not be overemphasized

---

[35] D.C. Montgomery, *Design and Analysis of Experiments* (New York: John Wiley & Sons, 1991) at 1.

[36] J.M. Last, *A Dictionary of Epidemiology* (New York: Oxford University Press, 1988) at 110. See, however, M.C. Klein *et al.*, "Physicians' Beliefs and Behavior During a Randomized Controlled Trial of Episiotomy: Consequences for Women in Their Care" (1995) 153 Canadian Medical Association Journal 769.

[37] For an example of the use of placebos where this rule was broken, see S. Goldzieher *et al.*, "A Placebo Controlled Double-Blind, Cross-Over Investigation of Side Effects Attributed to all Contraceptives" (1977) 22 Journal of Fertility and Sterility 609.

if the study is to be conducted among impoverished, non-literate groups, or otherwise marginalized populations.

*Ethical Conflicts Arising from Epidemiological Research*

The point was already made that epidemiology is essential not only to public health, but to medicine in general.[38] Epidemiological methods are applied to the screening of particular conditions among populations, both to calculate the frequency of the condition and the power of the test to detect the condition (in relation to its yield of false positive and false negative results). Similarly, and as indicated above, epidemiological methods are also used to evaluate the effectiveness of interventions, new technologies, or health programs in general.

Several ethical concerns arise from this context. The first pertains to the ethical duty of the epidemiologist to report findings accurately, unencumbered by managerial or political constraints. This, parenthetically, could develop into a major ethical dilemma for the researcher if a government, or a corporation, decided not to publish results that may be detrimental to their interests but that could warn the population about particular risks. Accurately reporting research findings should also remind the researcher that releasing research findings has an impact on the population, such as increasing, perhaps unnecessarily, anxieties about particular disease risks.[39]

As indicated earlier, a major interest in epidemiology is ascertaining the causes and risks of disease. However, determining the causes of conditions that are clearly multifactorial and complex can be extremely difficult. Armstrong, White and Saracci explain this:

- There may be no necessary cause for the disease under study, and any single component cause may make only a small contribution to aetiology,
- the interval from onset, and perhaps cessation, of exposure to appearance of disease is more often measured in years than days, weeks, or months,
- the agent of disease may leave no easily measurable indicator of past exposure, and
- the range of agents of interest has increased so much that it is no longer possible for one scientist to have an expert understanding of all those in which he or she may be interested.[40]

Thus, unless there are time imperatives in relation to an imminent risk, findings should not be released to the public until appropriate scientific reviews are

---

[38] See J.R. Paul, *Clinical Epidemiology* (Chicago: University of Chicago Press, 1966) at xii, in which he remarked that epidemiology should stand in the "same relationship to the practice of preventive medicine as do some of the more familiar basic medical sciences to curative medicine". See also D.L. Sackett, R.B. Haynes & P. Tugwell, *Clinical Epidemiology — A Basic Science for Clinical Medicine* (Toronto: Little, Brown, 1985).

[39] See "Gallup Releases Poll on Attitudes Towards Public Health Issues in America and the World" (1994) 15:11 *The Epidemiology Monitor* 1.

[40] B.K. Armstrong, E. White & R. Saracci, *Principles of Exposure Measurement in Epidemiology* (Oxford: Oxford University Press, 1994) at 1.

concluded to make sure that they are accurate. With regard to reporting, epidemiology also faces an ethical dilemma when it assesses the magnitude of risk that requires the notification of individual subjects. Legal entanglements may arise if the study determines whether the risks were postulated *a priori* or were determined as part of a finding *a posteriori*, or if the study establishes negligence by a third party.[41] Finally, researchers and publishers alike hold a responsibility not to publish research reports that do not provide proof of approval by recognized ethics committees,[42] whose importance is not clear (frivolous reasons), or whose results are not relevant.[43]

The second concern relates to the need for the epidemiologist to be aware of the potential ill effects of evaluations on those affected by the evaluation results. These include the fact that a government or a corporation may make decisions based on the findings of the research. Such decisions at a time of budgetary restraints may result in closing of programs, or discontinuation of particular diagnostic or treatment procedures. Thus, the result of an epidemiological evaluation, such as, for example, outcomes evaluations,[44] may affect a group of individuals because of criticisms about their work in a program. Individuals may face a potential loss of jobs if the program is closed, and the population at large may be deprived of a government service.

Finally, the third concern relates to the duties macroethics imposes not only on the epidemiologist, but also on public officers, health planners, and politicians. Epidemiological research helps these officers and government agencies to deploy health care resources more efficiently. Although comparisons of resource utilization (e.g. between regions or countries),[45] and the application of sensitive economics analysis techniques[46] to health expenditures are necessary and welcome tools in health economics, they could lead decision-makers to adopt policies contrary to public welfare. In fact, meta-analysis, decision analysis, and cost-effectiveness analysis are quantitative techniques developed to reduce uncertainty by providing a summary conclusion from often contradictory research findings on a particular medical or health system research problem. As Petitti indicates, meta-analysis is used to reduce uncertainty about the medical literature, decision analysis reduces uncertainty about the management of clinical problems, and cost-effectiveness

---

[41] S.H. Zahm, "On Epidemiology and the Obligation to Notify Study Subjects" (1992) 3:2 *The Epidemiology Monitor* 5.

[42] J. Arboleda-Florez, "Ethical Issues Regarding Research on Prisoners" (1991) 35:1 *International Journal of Offender Therapy and Comparative Criminology* 1.

[43] R. Ruback & C. Innes, "The Relevance and Irrelevance of Psychological Research — The Example of Prison Crowding (1988) 43 *American Psychologist* 683.

[44] For a good view of research on outcomes evaluation, see K.S. Warren & F. Mosteller, eds., *Doing More Good than Harm* (New York: Annals of the New York Academy of Sciences, 1993).

[45] H. Vayda, "A Comparison of Surgical Rates in Canada and in England and Wales" (1974) 289 *New England Journal of Medicine* 1224.

[46] M.C. Weinstein & W.B. Stason, "Foundations of Cost-Effectiveness Analysis for Health and Medical Practices" (1977) 296 *New England Journal of Medicine* 716.

analysis reduces uncertainty about how best to allocate resources.[47] Therefore, even methodological tools have to be devised to deal with the many issues raised by health research on populations. It would be indeed disastrous if policies were adopted based on studies whose methodology or scientific rigor are lacking, or whose results are not placed in the public domain for proper scrutiny. Unfortunately, this occurs frequently.

The duty of government officers, therefore, is to ensure that they do not take bureaucratic or political action on research findings unless the scientific quality of the research is unimpeachable.[48] Guidelines for "Good Epidemiological Practices" have been developed. The Epidemiology Task Group from the Chemical Manufacturers' Association (USA) proposed in 1990 to:

(1) promote sound epidemiological research by helping to ensure the quality of data collection and analysis;
(2) improve the acceptance for publication of studies using sound scientific methods;
(3) provide a framework for critiquing epidemiological studies;
(4) facilitate the continued development of improved epidemiological research methodology;
(5) improve the utility of epidemiological studies in the formulation of public policy; and
(6) provide a framework to ensure adherence to good research principles.[49]

The CIOMS, in its *Guidelines* for epidemiological studies, calls for independent ethical review of all epidemiological studies. The CIOMS reflects that the *Nuremberg Code* and the *Helsinki Declaration* are based on a model of clinical medicine and address the interests of the patient, or of the individual subject, whereas epidemiological research concerns groups of people.[50] A final point of concern about studies that use existing large data sets is that they are often undertaken without the knowledge of the individuals, the groups, or the communities. Those likely to be affected by this type of research should expect at least that the projects meet the highest and strictest scientific requirements.

Macroethical problems assume a different dimension when the research is carried out in Third World countries. Epidemiological investigations in these countries carry a large potential for ethical wrongdoing. For example, in 1982, Gopalan[51] reviewed

---

[47] D.B. Petitti, *Meta-Analysis Decision Analysis and Cost-Effectiveness Analysis* (Oxford: Oxford University Press, 1994).

[48] J.A. Baron *et al.*, "Internal Validation of Medicare Claims Data" (1994) 5 *Epidemiology* 541; J. Avorn, "Medicaid-Based Pharmacoepidemiology: Claims and Counterclaims" (1990) 1 *Epidemiology* 98.

[49] See "Guidelines for 'Good Epidemiology Research'" (1990) 11:4 *The Epidemiology Monitor* 1.

[50] Council for International Organizations of Medical Sciences, in collaboration with the World Health Organization, *International Guidelines for Ethical Review of Epidemiological Studies* (Geneva: CIOMS, 1991) [hereinafter *CIOMS Epidemiological Guidelines*].

[51] C. Gopalan, "Ethical Aspects in Community-Based Research with Particular Emphasis on Nutrition Research" in Z. Bankowski & N. Howard-Jones, eds. *Human Experimentation and Medical Ethics* (Proceedings of the XVth CIOMS Round Table Conference, Manila, 13–16 September 1981) (Geneva: CIOMS, 1982) 124.

ethical issues that arise out of epidemiologically-based nutrition research. He listed the following problems:

(1) matters of informed consent (who gives consent in villages where the concept of individual autonomy is taken over by the collective and invested on the elderly or on the leader of the village or tenement?);
(2) surveys not followed by action and hence raising unnecessary expectations;
(3) lack of research on alternative actions;
(4) validity of specific nutrition intervention programs, especially among children below the age of three; and
(5) problems with monitoring and evaluation of the program.

Furthermore, Gopalan points out several clearly objectionable practices, such as the deliberate maintenance of groups on subsistence diets so that they could act as control subjects, using food supplements rejected as unsuitable for human consumption in the developed world, using highly developed nutritional industrial products to the point that the results could not be compared in other localities or countries, and lack of involvement and confidence among the members of the community.

More recently, a major controversy has erupted regarding a study on the use of AZT (zidovudine) among pregnant HIV positive women in Thailand and the Ivory Coast (Côte d'Ivore) compared to a control group of pregnant uninfected women. This study has been supported by the United States through the Center for Disease Control (CDC) in Atlanta. Researchers will be using a "short-course" AZT régime during the last four weeks of pregnancy and during delivery as opposed to the "long-course" throughout the length of the pregnancy and during childbirth recommended by the National Institutes of Health for use in the United States. At issue is why the study is conducted abroad and not in the United States, whether any side effects could be detected, and whether the short-term course could lead to development of drug resistant strains of the HIV virus. The first concern, however, is the most pressing, "when is it ethical to do studies overseas that could not be done in the United States?" (or in any other developed country for that matter). CDC has explained that the cost of the long-course treatment is absolutely prohibitive in third world countries, whereas the abbreviated treatment, if proved effective, will cost only US$50. Countries where this major economical imperative exists should benefit from finding out whether the treatment is effective. As it is, AZT in any form, short or long course, is not even available in Thailand or the Ivory Coast or in many underdeveloped countries. CDC has responded to critics of the study by invoking Guideline #8 of the International Ethical Guidelines for Biomedical Research Involving Human Subjects prepared by the Council for International Organizations of Medical Sciences (CIOMS) which states "Diseases that rarely or never occur in economically developed countries or communities exact a heavy toll of illness, disability, or death in some communities that are socially and economically at risk of being exploited for research purposes. Research into the prevention and treatment of such diseases is needed and, in general, must be carried out in large part in the countries and communities at risk".[52]

As well as being sensitive to different cultural environments and customs, researchers should be aware of the danger of importing alien philosophical and ethical concepts that could be interpreted as ethical imperialism.[53] That is, they must refrain from imposing cultural norms or solutions from one society on another, based on the belief that these solutions represent ethical absolutes. However, neither should researchers fall into a state of ethical relativism whereby minimal ethical standards would be malleable according to place and time. From Nuremberg to Helsinki and after, the foundations of international human rights has been related to an irreducible set of international ethical standards common to all societies, unbending to local laws, politics or custom.[54]

Bryant and Khan[55] point out that the methods and findings of epidemiology are not sufficient to comprehend issues of justice in health care. A deeper understanding of the meaning of health and its value in our society is also required. Such issues of justice are beyond epidemiology. They include:

- political decisions on the setting of priorities;
- cultural values about the importance attributed to particular risk factors;
- social and economic judgments about what sub-population (demographic or clinical) should benefit from what level of care;
- community participants who may wish to make a decision at variance with the scientific advice of the epidemiologist, but consonant with local values, means and resources.

Finally, a word about conflicts between microethics and macroethics is necessary. In terms of the effects of research findings on special populations, for example, ethics still has to define what principles should hold sway when opposite rights clash as we move from microethical situations to macroethical social concerns. For example, how far would the public have to be exposed to an epidemic for the needs of the group to trigger an obligation to act, and act without infringing upon confidentiality concerns and civil rights of affected individuals or specific minorities? Raging controversies on such issues as AIDS, or a proposed but not funded American research project from Harvard University to study violence, call for a determination of these higher principles, and certainly for more dialogue between ethicists, researchers, and community leaders. It may be that, under certain conditions, the principles of microethics do not completely apply to macroethical

---

[52] Health Letter on the CDC (1997, July 28): "CDC Explains its Stand on Controversial Third World AZT Study, at 2–5.

[53] See A.M Rossignol & C.S. Campbell, "Are Epidemiological Research Standards Determined by Culture?" (1994) 15:3 *The Epidemiology Monitor* 5, for an interesting review of the ethical cultural issues raised by situations involving female circumcision, lead poisoning, and bride-burning. See also P.C. Gupta, "On Ethics, Epidemiology, and Culture" (1994) 15:7 *The Epidemiology Monitor* 3.

[54] See Chapter 1.

[55] J.H. Bryant & K.S. Khan, "Epidemiology and Ethics in the Face of Scarcity" in *Ethics and Epidemiology, supra* note 3 at 70.

situations, just as how informed consent, a construct of our Western belief structure, may not apply in some situations outside the Western world.[56]

To conclude this section, the foundational principles for population-based research, as summarized by Gostin, include: the overriding imperative to protect the health and well-being of a population; the right of populations to self-determination, including the right to refuse participation; protection of vulnerable populations and the need for special justification for research; protecting the privacy, integrity and self-esteem of populations; and the equitable distribution of benefits and burdens of research.[57]

### *Metaethics*

Up to now, the foregoing sections have dealt with what can be termed "descriptive ethics", that is, how ethical principles could be observed in operation from the surface, as they apply to medical interventions or research efforts. Yet, over the past thirty years, ethics has moved from the exclusive domain of the physician to include an interdisciplinary team of ethics experts. Bryant[58] has applied the term "metaethics" to the move of ethical reasoning from the physician to ethics experts, or to how medical-ethical issues have been put in the hands of the lay person. He argues that medical ethics, previously the sole domain of the physician, became an interdisciplinary concern. Under its new name of bioethics, medical ethics has kept its definition within the domain of the biomedical sciences, but the ethical framework and concepts are now provided by bioethical experts. This switch has spurred the development of bioethics institutes, the publication of specialized journals, and the formation of a new specialist, the bioethicist. Questions of ethics, originally framed in relation to the duties of physicians to their patients, are now reformulated in terms of rights and ethical choices. Physicians are no longer the experts on medico-ethical issues. Rather, such expertise has been vested in bioethicists, jurists, and other non-medical experts. Metaethics,[59] therefore, may have been the most important development in medical ethics during the past two decades.

However, the term "metaethics" could be extended to mean a deeper understanding of the impact of political attitudes, values, and ethical imperatives on research in general and epidemiological research in particular. Ethics consists of values in action, and values are the reflection of a cultural heritage which is reflected through political processes. Tancredi and Weisstub[60] write about value-laden concepts in epidemiological forensic psychiatric research, proposing that research-

---

[56] R.J. Levine, "Informed Consent: Some Challenges to the Universal Validity of the Western Model" in *Ethics and Epidemiology, supra* note 3 at 47.
[57] See Gostin, *supra* note 11 at 45.
[58] See Bryant, *supra* note 17
[59] *Ibid.*
[60] L.R. Tancredi & D.N. Weisstub, "Ideology and Power: Epidemiology and Interpretation in Law and Psychiatry" in L.Romanucci-Ross, D.E. Moerman & L.R. Tancredi, eds., *The Anthropology of Medicine, From Culture to Method* (New York: Bergin & Garvey, 1991) 301.

ers should identify their "ideology of error", or the values under-girding the conceptual aspects of their research. By extending the concept of social research to all research with an impact on social groups or populations, Tancredi and Weisstub make epidemiology a social science whose central function is to reconstitute social reality. Thus, the scientific accuracy of the research, important as it may be, takes secondary position to the social impact of the results inasmuch as these may upset a balance of power, clash with existing ideological positions, or create such dissonance as to overthrow existing social paradigms. Arboleda-Florez, in fact, considers these "ideologies of error" to act as "ethical confounders",[61] structuring the research effort in such a way that the values of the researcher impinge on the exposure of the variable to be measured and have an effect on the outcome of the research.[62]

In a historical review of epidemiology research on pesticides, for example, Levine, Hersh, and Hodder[63] call attention to potential "unconscious factors [as] important determinants of research directions" and point out that

(1) epidemiology research is seldom initiated by the researcher, but tends to respond to sponsoring agencies (that usually act under political pressures);
(2) unconscious factors may lead not only the researcher, but the social group as well, to seek reassurances, or assuage feelings of guilt, in relation to particular researchable subjects (such as making Agent Orange the "scape-goat" to deny guilt on the issue of Vietnam veterans); and
(3) greater feasibility of enrolling a particular group of persons as research subjects may dictate the type of reseach to be done.

As they call for an ethical inquiry on the determinants of research agendas, these authors conclude with a warning:

> ... [i]f ethical inquiry were to be conducted without sufficient understanding of such undercurrents, the unwary might be more likely to perpetuate primitive fears and emotions, rather than eliminate them from the scientific milieu.[64]

In the same vein, Hatch warns that "epidemiologists, like all scientists, need to be alert to the internal and external forces that influence our work — beginning with choices about what and what not to study".[65] Health is held as a high value in practically every society. Epidemiological research whose findings may be used to

---

[61] The word *confounder* is used in epidemiology to describe a situation in which the effects of two processes are mixed so that there is distortion of the apparent effect of exposure on risk brought about by the association with other factors that can influence the outcome. See Last (1988), *supra* note 36 at 28–29.

[62] J. Arboleda-Florez, "Methodological and Ethical Concerns Regarding Forensic Psychiatry Epidemiological Studies" (1993) 35:2 International Journal of Offender Therapy and Comparative Criminology 95.

[63] R.J. Levine, C.B. Hersh, & R.A. Hodder, "Historical Patterns of Pesticide Epidemiology Research from 1945 to 1988" (1990) 1 Epidemiology 181.

[64] *Ibid.* at 183.

[65] M. Hatch, "Pursuit of Improbable Hypotheses" (1990) 1 Epidemiology 97.

affect the health entitlements of any population is immediately seen as politically tainted.

### Arguments against Epidemiological Research

At the microethical level, concerns about the potential abuse of the rights of patients and/or of research subjects make the mainstay of the bioethical discourse. Yet, to the exception of ethical issues surrounding community RCTs, microethical issues do not often surface in epidemiological research. Ethical issues that may affect epidemiological research involving large population groups, or research based on data manipulation, refer more often to potential clashes with rules and tenets on privacy and confidentiality.[66] Epidemiological studies may also be affected by considerations of justice in relation to the burdens of research, or discrimination preventing specific populations from participating.

*Causality*

Any research project entails an element of uncertainty in the application of the methodology, the procedures, or the analysis. Often, judgment calls[67] have to be made as the research progresses. Kahneman and Tversky[68] have written about the fact that, when choices are not clear, people tend to make decisions on a limited number of automatic heuristic principles which, ultimately, work against making proper choices based on realistic probabilities and lead to errors in judgment. Tancredi points out that "the fact that one has expert knowledge of science and medicine does not mean that heuristic principles will not be employed in the face of uncertainty".[69] These types of error could affect the researcher, the research subject, or the policy-maker. A major criticism of epidemiological research, therefore, is the potential that socio-political pressures would increase the level of uncertainty and allow heuristic values to assume paramount importance in the choice of researchable issues. Similarly, as populations used in epidemiological observational studies cannot be kept under strict laboratory-like control, too many "judgment calls" could derail the proper conduct of the study and lead to erroneous conclusions.

Cause, by its very nature, has bedevilled science and philosophy and eluded clear definition throughout the centuries. Russell[70] suggested that the concept was of no use in science. More recently, Einhorn and Hogarth[71] have indicated that causal judgment involves inference and uncertainty and depends on the determination of a

---

[66] A.M. Capron (1991) "Protection of Research Subjects: Do Special Rules Apply in Epidemiology?" *Law, Medicine & Health Care* 19(3/4) 184–190.

[67] B.M. Staw, "Some Judgements on the Judgement Calls Approach" in J.E. McGrath, J. Martin & R.A. Kulka, eds., *Judgement Calls in Research* (Beverly Hills: SAGE Publications, 1982) 119.

[68] D. Kahneman & H. Tversky, "The Simulation Heuristic" in D. Kahneman, P. Slovic & A. Tversky, eds. *Judgement Under Uncertainty: Heuristics and Biases* (Cambridge: Cambridge University Press, 1982) 201.

[69] L. Tancredi, *Ethical Issues in Epidemiological Research* (New Brunswick, N.J.: Rutgers University Press, 1986).

[70] B. Russell, *Human Knowledge: Its Scope and Meaning* (New York: Simon & Schuster, 1948).

[71] H.J. Einhorn & R.M. Hogarth, "Judging Probable Cause" (1986) 99:1 *Psychological Bulletin* 3.

"causal field" of relevance and go on to say that the elucidation of "cues to causality" can conflict with probabilistic ideas, so that at the end, all these factors have to be combined to establish causality.

Causes are called *necessary* when, perforce, they precede an event, and *sufficient* when, inevitably, they trigger the event.[72] For example, a person may have come into contact with the TB bacillus; although such exposure is "necessary" to develop tuberculosis, by itself it is not a "sufficient" cause (host predisposition and environmental reasons also play a causal role). In the case of malnutrition research and its effects on the development of a particular country, it is not always easy to determine whether a malnourished population contributes to underdevelopment or *vice versa*. In part, difficulties in establishing causality in these cases may depend on limitations of statistical analytical techniques, since these techniques have usually been devised for experiments whereas most epidemiological studies are non-experimental in nature. As such, statistical techniques devised to analyse experiments provide "no basis for predicting or evaluating the uncertainty in the data that derives from hidden confounders".[73] Greenland has commented that while randomization is needed to justify causal inferences from conventional statistics, and random sampling to justify descriptive inferences, in most epidemiological studies, randomization and random sampling play little or no role in the assembly of study cohorts. He concludes that:

> . . . probabilistic interpretations of conventional statistics are rarely justified, and that such interpretations may encourage misinterpretation of nonrandomized studies. Possible remedies for this problem include de-emphasizing inferential statistics in favor of data descriptors, and adopting statistical techniques based on more realistic probability models than those in common use.[74]

Schlesselman notes that "observational and experimental studies in medicine and public health are designed to identify factors that cause or prevent the occurrence of disease".[75] In these types of studies the researcher tries to ascertain "cues to causality" such as temporal sequence, consistency, dose-response relationship, strength of any association, specificity of effect, and biological plausibility.[76] Most observational studies, however, can only advance an opinion on causality, not a claim of proof.[77]

If all of these difficulties are kept in mind, then, it would not be surprising that in large, community-based epidemiological studies, erroneous claims of identifications of dangerous risk factors could mislead the population, and cause damage to industrial interests or to the economies of whole countries. In bureaucratic

---

[72] K.J. Rothman, "Causes" (1976) 104 *American Journal of Epidemiology* 587.

[73] K.J. Rothman, "Statistics in Nonrandomized Studies" (1990) 1 *Epidemiology* 417.

[74] S. Greenland, "Randomization, Statistics, and Causal Inference" (1990) 1 *Epidemiology* 421.

[75] J.J. Schlesselman, *Case-Control Studies, Design, Conduct, Analysis* (New York: Oxford University Press, 1982) at 20.

[76] J.S. Mausner & S. Kramer, *Epidemiology: An Introductory Text* (Toronto: W.B. Saunders, 1985) at 185–186.

[77] W.G. Cochran, "The Planning of Observational Studies of Human Populations" (1965) 28 *Journal of the Royal Statistical Society* 234.

administrative epidemiological research, false claims arising from a study may lead a government to make decisions that will have major social impacts. Although some problems faced when establishing causality are more technical than ethical, vigilance in the quality and the motives of epidemiological research should be high on the agenda of ethics review committees.

*Abuse of Subjects*

As previously mentioned, abuse of the rights of subjects has a potential to happen more frequently at the microethical level, such as in RCTs, than in epidemiological research. The autonomy of individuals may be violated if informed consent to participate is not obtained; if it is obtained by misleading the subject about the benefits or risks of the study and about the use of placebos; or if subjects are not informed about their rights not to participate or to withdraw without incurring penalties in their regular medical care.

Discriminating against minorities by not including them in research projects, or by placing an unfair burden of research on them, is a matter of concern in epidemiology. In the past, prisoners have frequently carried the burden in relation to investigations of new drugs, and the reverse potential abuse to their autonomy when a paternalistic blanket prohibition to their involvement in research has been extended.[78] Abuse of subjects in relation to considerations of justice, however, has taken place in epidemiological investigations on other types of populations. The study by Goldzieher *et al.*[79] stands out as one of the most flagrant abuses of the right to informed consent and the fraudulent use of placebos. In this study, the researchers divided women who attended a clinic to prevent conception into two groups: women in the treatment group were placed on oral contraceptives and those in the control group were given placebos. However, women in the control group were not told that they were not receiving contraceptive medication, but rather, they were only advised to use vaginal cream. To add to the abuse, the real reason for the experiment was not disclosed, in that it was not so much to test the efficacy of the contraceptives, but to determine their side effects. Seven pregnancies resulted.

The major concern in relation to issues of justice pertains to (1) investigations in Third World countries where, as already reviewed, ethical standards and especially controls of research protocols, may not protect subjects of research, and (2) reverse discrimination against some minorities, notably women and black men. In this regard, it has been noted that many studies on new drugs are carried out on white male patients, for conditions that more frequently and severely affect black men, such as hypertension; black men, therefore, tend to be under-researched, or the findings may not be directly applicable to them. It was only in 1994 that NIH issued guidelines for the inclusion of women and members of minority groups in all NIH-supported biomedical and behavioral research involving human subjects, unless compelling reasons existed to the contrary. The new policies are based on the belief

---

[78] See e.g. K. Schroeder, "A Recommendation to the FDA Concerning Drug Research on Prisoners" (1983) 56 *Southern California Law Review* 969.

[79] See Goldzhieher, *supra* note 37.

that research should be of benefit to all persons at risk of the disease or condition regardless of gender or race. They seek to increase the information collected on diverse racial and ethnic groups, including sub-populations, and, in the case of randomized controlled trials, to examine differential effects of new drugs.[80]

*Privacy and Confidentiality*

Privacy issues may arise in relation to the impact that a particular research project could have on the way the social body at large perceives of a subgroup in the population, or an identified minority. Thus, research that may target minorities of any kind as carrying an inordinately high risk factor for a particular condition tends to encourage discrimination against the minorities so identified. In this respect, it could be assumed that an invasion of privacy owed to that minority has taken place. In AIDS research and violence research, investigators have to be extremely sensitive so as to prevent stigmatization of the group. Recently, for example, the link between mental illness and violence has become a contentious issue with researchers rapidly bringing evidence to demonstrate an association between the two.[81] But correlation or even association do not mean causality. A review of the literature applying rigorous causality rules has failed to establish conclusive scientific evidence that mental illness *per se* causes violence.[82]

Threats to privacy and confidentiality posed by epidemiological studies are also potentially damaging when they are conducted on large data sets. Concerns can develop at two levels. The first lies with immediate research subjects who fill out surveys or participate in interviews. Privacy can be violated if questions on forms or interviews are not carefully phrased, or when promises of confidentiality cannot be fulfilled because of obligatory reporting laws such as those regarding child abuse;[83] or when data on prisoners, for example, remains in the prison or is subject to subpoena.[84] The second area of concern lies with research involving large data banks, or biobanks, where investigators of every kind have a free hand in obtaining whatever data may be available because there are no data protection laws or regulations. The absence of laws prompted the European Community's Privacy Directive (first draft in 1991) to restrict the use of computerized information, such as medical, employment and insurance records.[85] Although originally interpreted as

[80] National Institutes of Health, *Guide to NIH Grants and Contracts* (Washington, D.C.: National Institutes of Health, 1994); S.M. Schwartz & M.E. Friedman, *A Guide to NIH Grant Programs* (Oxford: Oxford University Press, 1992) at 162.

[81] M. Swartz, J. Swanson & R. Borum, "Link Between Violence and Mental Illness May Be Real" *Psychiatric News* (7 July 1995) 2.

[82] J. Arboleda-Florez, H.L. Holley & A. Crisanti, "Mental Illness and Violence: Proof or Stereotype" (Paper presented to the Canadian Academy of Psychiatric Epidemiology, 1995) [unpublished].

[83] Some commentators indicate that to not inform study subjects of the possibility of reporting in cases of discovery of child abuse is unethical and not in keeping with requirements of informed consent. See comments by H. McGee & W.K. Holland in "Questions of Privacy and Confidentiality Applicable to Michigan Study" (1992) 13:9 *The Epidemiology Monitor* 4.

[84] J. Arboleda-Florez, *The Prevalence of Mental Illness in a Remanded Population and the Relationship between Mental Illness and Criminality* (Ph.D. Dissertation, University of Calgary, 1994).

[85] "Proposed European Data Protection Laws Threaten Exchange of Epidemiological Data" (1991) 12:8 *The Epidemiology Monitor* 1.

placing a major roadblock to the conduct of epidemiological studies by demanding informed consent of the individuals whose names were contained in the data set, the most recent proposal takes into account the need for epidemiological reseach and, therefore, the needs of medical researchers, especially epidemiologists. Consequently, the restrictions on the storage of data and the demands on informed consent have been lifted.[86] This being said, the threat of abuse of biobanks remains, especially because of record-linkage technologies. This may be potentially more so in the case of unregulated intramural research in government bureaucracies, in large insurance companies, or in medical corporations where proper, independent, peer-review may not take place. Such potential threats leave many persons wary of the intentions of governments or corporations when collecting data.

*Summary*

The most important reasons to oppose epidemiological research rest on concerns about threats to the autonomy of subjects and to their privacy and confidentiality. These criticisms are based on Kantian principles of respect for the person, groups, or communities, and generally flow from microethics postulates at the level of clinical practice. Criticisms leveled at the way epidemiological research utilizes human groups are based on considerations of justice in relation either to inequalities among the burdens of research or discriminatory practices against particular groups, such as women or minorities. In this respect, macroethical considerations apply to potentially discriminatory effects of results of epidemiological research that might identify subpopulations as posing risks to the population at large, or to abuse subjects when economic situations or socio-political structures in Third World countries permit violative research practices. Macroethics concerns also enter into play with respect to the protection of the privacy and confidentiality of large numbers of people affected by record-linkage research. In these cases, subjects, who may actually be the total population of a country or nation, may not even know that they are being researched.

### Arguments in Favor of Epidemiological Research

Ethical issues in epidemiological research involve conflicts in values between the rights of the person and the needs of the community. Epidemiology, in its entanglements with public health and health care systems, lies at the crossroads of ethical conflict in health. Does the state have an interest in the health of its citizens? Is health a right, or a commodity? If health is a right, what sort of entitlements does it provide and how do these entitlements relate to the allotment of scarce resources? Answers to these questions will inform the discourse of epidemiological inquiry.

*Scientific Rigor of Epidemiological Inquiry*

One major element in favour of epidemiological research is its scientific rigor. In determining causality, epidemiology studies patterns of disease occurrence and

---

[86] Council of Europe, *The Protection of Individuals with regards to the Processing of Personal Data and on the Free Movement of Such Data* (Brussels: Council of Europe, 1995).

health events among human populations and attempts to identify factors that cause or influence these patterns.[87] Given the multiple threats to the internal scientific validity[88] of epidemiological studies and the consequential invalidation of any claims to generalization (external validity),[89] epidemiological studies are known for their care in ruling our biases in the selection of samples, the information collected and the classification of subjects, and in the treatment of confounding variables.[90] Such rigor has made quantitative epidemiological methodologies the state of the art for statements of causality in human populations. In *Daubert* v. *Merrell Dow Pharmaceutical, Inc.,*[91] the U.S. Supreme Court rejected the Frye Rule[92] of "general acceptability" on scientific evidence and indicated that:

> ... scientific methodology today is based on generating hypotheses and testing them to see if they can be falsified; indeed, this methodology is what distinguishes science from other fields of human inquiry.... The criterion of the scientific status of a theory is its falsifiability, or refutability, or testability.

The court also stipulated that animal, pharmacological, or other types of scientific information, or research, was not sufficient to explain risk factors affecting large numbers of the population; epidemiological evidence was required to assess them properly. The court's digression into issues of scientific evidence and the testability of hypotheses mirrored Karl Popper's approach to scientific inquiry.[93] Popper based his approach on a hypothetical-deductive process whereby knowledge advances by testing hypotheses and rejecting those that fail. The application of Popper's approach to epidemiological research distinguishes it from the elaboration of global hypotheses, or the naturalistic qualitative collection of observations.[94] It is such a rigorous approach to hypothesis testing in epidemiology that influenced the court in *Daubert*.

## *Moral Rights to Health*

There are moral rights and obligations and there are legal rights and obligations. As individuals, we have moral obligations to ourselves and to our fellow human beings,

---

[87] D.E. Lilienfeld & P.D. Stolley, *Foundations of Epidemiology* (New York: Oxford University Press, 1994).

[88] D. Campbell & J.C. Stanley, *Experimental and Quasi-Experimental Designs for Research* (Chicago: Rand McNally, 1963).

[89] T.D. Cook & D.T. Campbell, *Quasi-Experimentation: Design and Analysis for Field Settings* (Chicago: Rand McNally, 1979).

[90] D.G. Kleinbaum, L.L. Kupper & H. Morgenstern, *Epidemiological Research: Principles and Quantitative Methods* (New York: Van Nostrand Reinhold, 1982).

[91] *Daubert* v. *Merrell Dow Pahrmaceuticals*, 113 S.Ct. 2786 (1993). The Daubert cases pertain to the putative reduction of limbs among some babies whose mothers had used bendectin during some time in their pregnancies.

[92] *Frye* v. *United States*, 293 F. 1013 (D.C. Cir 1923). The *Frey* Test simply states: "... expert opinion based on a scientific technique is inadmissible unless the technique is "generally accepted" as reliable in the general scientifc community ...".

[93] K.R. Popper, *The Logic of Scientific Discovery* (New York: Basic Books, 1959).

[94] For a full treatment of this issue, see C. Buck, "Popper's Philosophy for Epidemiologists" (1975) 4 *International Journal of Epidemiology* 159.

and we have moral rights that flow out of our entitlements as humans and from our social contract with the state; our legal obligations and legal rights flow from the laws that we give ourselves in our respective countries. Health is conceived by Dubos, not in the utopian way of the World Health Organization definition of "an ideal state of well-being achieved through the complete elimination of disease", but as a *modus vivendi* enabling imperfect men to achieve a rewarding and not too painful existence while they cope with an imperfect world".[95] In this sense, health is a requisite, a *sine qua non*: without health we cannot function, we cannot work, we cannot achieve, we cannot love, and cannot get rewards. Health, mental and physical, has an intrinsic value to humans.

In Canada, the National Forum on Health, established to advise the federal government on future directions for health and health care in the country, determined through extensive population-based research using focus-group method-ologies that Canadians hold as very important six health, and health system-related values: *equality* or *fairness* of access to health and an equal opportunity to achieve health; *compassion* and recognition of different risks and health status and concern for the protection of persons with particular health vulnerabilities; *dignity* and *respect* for the innate self-worth of individuals, their intelligence, and their ability to choose; *efficiency* in the system and *effectiveness* of interventions; personal responsibility in maintaining one's own health and in the appropriate use of the system; and *collective responsibility* in making sure that citizens, who have a stake in the integrity of the system, are properly informed and participate in decision-making about the system. Overall, Canadians desire high-quality health care, "the best that is reasonable".[96] It would not be inconceivable to think that similar values on health are held by populations everywhere.

If it is agreed that the state has an interest in keeping its people healthy and productive, then the state has a responsibility to provide some modicum of health, be it in the form of preventative health measures, or the basic of restorative, curative, measures. On the basis that the state takes on these responsibilities, then, the public would have to allow the state to further the ends of a healthy population by setting up mechanisms to monitor health and threats to health. The state could, then, mount preventative campaigns on conditions that pose a threat to the population, and devise ways to help those who have lost their health. In return for these services, the people would pay taxes to maintain the system, but more importantly, would agree to give up some rights, and allow some incursions into their privacy and confidentiality. A *quid pro quo* would allow the development of a balance between rights, expectations, and duties of the population, and the obligations of the state. In this situation, the state would need tools and methodologies to accomplish its part of the bargain. Tools and methodologies required to monitor population health, risk

---

[95] R. Dubos, *Determinants of Health and Disease in Man, Medicine and Environment* (New York: Frederick A. Praeger, 1968).
[96] National Forum on Health, *Health and Health Care Issues* (Ottawa: Minister of Public Works and Government Services Canada, 1997) at 19.

factors, and utilization of curative services are usually furnished by epidemiological research.[97]

### Legal Rights to Health

"Health" is defined in the *Constitution of the World Health Organization* as "a state of complete physical, mental and social well-being and not merely the absence of disease or infirmity".[98] Although this right is difficult, if not impossible, to define more precisely, it constitutes a fundamental human right that is protected by numerous international declarations and conventions.[99] The right to health, however, is often expressed as a goal rather than a guarantee, and exemplifies the "promotional approach" in international law.[100]

The *Universal Declaration of Human Rights* describes a person's right to:

> . . . a standard of living adequate for the health and well-being of himself and of his family, including food, clothing, housing and medical care and necessary social services, and the right to security in the event of unemployment, sickness, disability, widowhood, old age or other lack of livelihood in circumstances beyond his control.[101]

In the pursuit of this goal, states are expected not only to recognize the right to enjoy the "highest attainable standard of physical and mental health", but to take all necessary steps towards the realization of this right, including in particular:

- The prevention, treatment and control of epidemic, endemic, occupational and other diseases; and
- The creation of conditions which would assure to all medical service and medical attention in the event of sickness.[102]

The attainment of this goal is dependent upon a state's technological and economic advancement. Developing nations will be less able to allocate the resources necessary to achieve a standard of living (and health) comparable to that of more technologically advanced countries.[103] Although it has been asserted that all states, regardless of their economic status, must without exception utilize their resources to guarantee a minimum quality of health, which would include making

---

[97] J. Arboleda-Florez, "Editorial — Psychiatric Epidemiology in Canada" (1997) 42 Canadian Journal of Psychiatry 699–700.

[98] *Constitution of the World Health Organization*, 1948, 14 U.N.T.S. 185, preamble.

[99] V. Leay, "The Right to Health in International Human Rights Law" (1994) Health and Human Rights 25. See for example the *Convention on the Rights of the Child*, GA Res. 44/25, UN Doc. A/44/25 (1989). Furthermore, the right to health can be subsumed under the basic right to protection of the security of the person. See *Universal Declaration of Human Rights*, G.A. Res. 217 A(III), UN Doc. A/810 (1948), art. 3.

[100] M.J. Bernardi, "The Impact of Public International Law on Canadian Constitutional and Criminal Law and Child Health" in B.M. Knoppers, ed., *Canadian Child Health Law* (Toronto: Thompson Educational Publishing, 1992) 43 at 46, 53.

[101] See *Universal Declaration of Human Rights*, *supra* note 99, art. 25.

[102] *International Covenant on Economic, Social and Cultural Rights,* (1966) 1976 Can. T.S. No. 46, 993 U.N.T.S. 3, art. 12.

[103] See e.g. R.E. Robertson, "Measuring State Compliance with the Obligation to devote the 'Maximum Available Resources' to Realizing Economic, Social and Cultural Rights" (1994) 16 *Human Rights Quarterly* 693.

necessary appeals for humanitarian assistance and international cooperation,[104] it must be acknowledged that states are constrained by fiscal realities that may limit their ability to guarantee a right to health.

Secondly, although the ideal of a global right to health has been reaffirmed throughout the years, through various resolutions by the WHO and other international organizations,[105] the success of such efforts has been seriously questioned.[106] It is not clear whether the right to health truly exists in customary international law, and treaties such as the *International Covenant on Economic, Social and Cultural Rights* provide for programs to be implemented by member states rather than individual rights.[107] Therefore, it is with great difficulty that one could assert a "right to health" based solely on international law.

There is also the problem of enforcement. States have historically resisted the limitation of their sovereignty over what are perceived to be internal matters, including the rights of their citizens. Neither the International Court of Justice nor the United Nations Human Rights Commission have the power to enforce their rulings or recommendations; both are dependent upon the acquiescence of the state parties involved. This is not to imply that a universal ideal of health, or the global organizations that promote it, are without value. On the contrary, given that diseases do not respect political boundaries, public health is a matter of global concern and responsibility.[108] The WHO must continue in its supervisory role, perhaps even more actively than in the past, and also nourish the progressive evolution of "health" in all nations.[109] This includes, of course, encouraging individual states to entrench the right to health in legislation.

Finally, it must be asked whether it is even possible to develop a universal notion of a right to health:

> The challenge of drafting cogent international health norms reflects the general theoretical difficulty of developing human rights standards with universal applicability. Certain groups of relativists have argued that, even if a particular human right has universal application, the specific content of the norm depends on the circumstances in each nation, and that it cannot be legislated effectively at the international level.[110]

Proponents of a universal right to health must take into consideration the diverse cultural, social and economic conditions existing in different states. For example, one must question "whether or not it makes sense to put forward a human right in a society not yet able to provide that right".[111] However, one should not lose hope

---

[104] See Limburg Principles, reproduced in (1987) 9 *Human Rights Quarterly* 122.

[105] See *Health for all in the Year 2000*, GA Res. 34/58 (1979); See also World Health Organization, *From Alma-Ata to the Year 2000: Reflections at the Midpoint* (Switzerland: WHO, 1988).

[106] See e.g. A.L. Taylor, "Making the World Health Organization Work: A Legal Framework for Universal Access to Conditions for Health" (1992) 18 *American Journal of Law & Medicine* 301.

[107] See Bernardi, *supra* note 100 at 54–55.

[108] See Taylor, *supra* note 106 at 311.

[109] *Ibid.* at 302; M. Bélanger, "Réflexion sur la réalité du droit international de la santé" (1985) 18 *Revue québécoise de droit international* 54.

[110] See Taylor, *supra* note 106 at 332.

[111] A.D. Renteln, "The Unanswered Challenge of Relativism and the Consequences for Human Rights" (1985) 7 *Human Rights Quarterly* 514 at 517.

by yielding too readily to relativistic arguments; normative principles continue to serve a vital role in the evolution of legal documents that define human rights and entitlements.

It is not surprising, therefore, that the entrenchment of a right to health in the legislation of individual nations has occurred only to varying degrees. In the United States of America, health and health care do not enjoy constitutional status. This absence, however, has not precluded a widely recognized obligation to ensure that an adequate level of health care is provided to all.[112] The Constitution gives Congress the power to make laws that are necessary to protect public welfare. As costs of public health, including the development of regulatory standards, have increased, the federal government has instituted wide-reaching health care programs. Neither the Constitution nor any body of constitutional jurisprudence provides a simple explanation or justifications for the actions of the Federal Government in the area of health.[113]

Some countries, however, have enshrined a right to health in their laws. In Finland, Section 3 of the *Law on the Status and the Rights of Patients* gives Finnish nationals and permanent residents a "right to care":

> Every person who stays permanently in Finland is without discrimination entitled to health and medical care required by his state of health within the limits of those resources which are available to health care at the time in question . . .[114]

Kokkonen[115] indicates that this provision has drawn much attention at a time of diminishing resources to health care, but the law was enacted knowing that society "had the right to limit how much and what kind of care the society could provide for its citizens". The section has been invoked by individuals seeking to obtain treatment abroad; treatment and care that are not available in Finland, and by relatives or advocates of those incapable of asserting their rights, such as mental patients. While the law seems to protect the latter well, decision-makers have had much difficulty reconciling the excessive demands of the former.

In Mexico, the Constitution grants a "right to health protection" under Article 4:

> Every person has the right to health protection. The law shall define the ways and means to provide access to health services, and shall establish the participation of the Federation and of federal agencies concerning general health, in accordance with the provisions of Paragraph XVI or Article 73 of this Constitution

It appears that the Mexicans assert a difference between the right to the protection of health and the right to the enjoyment of health. This seems to be a more pragmatic

---

[112] A.M. Capron, "United States of America" in H.L. Fuenzalida-Puelma & S. Scholle Connor, eds., *The Right to Health in the Americas: A Comparative Constitutional Study* (Washington, D.C.: Pan American Health Organization, 1989) 498 [hereinafter *Right to Health in the Americas*].

[113] *Ibid.*

[114] P. Kokkonen, "Finns Legislate for Patients' Rights (Translation of the *Law on the Status and the Rights of Patients*)" (1992) 84 *Bulletin of the Medical Ethics* 8.

[115] *Ibid.*

and achievable position. The right to health protection is a universal right and comprehensive in its coverage.[116]

In Canada, Section 7 of the *Charter* gives Canadians the right to life, liberty and security of the person:

> Everyone has the right to life, liberty and security of the person and the right not to be deprived thereof except in accordance with the principles of fundamental justice.[117]

However, although an argument could be made that s. 7 provides a right to health,[118] in *Morgentaler, Smolin and Scott* v. *The Queen*, Dickson C.J.C. construed "security of the person" as a negative right (i.e. the right not to be interfered with):

> [T]he case-law leads me to believe that state interference with bodily integrity and serious state-imposed psychological stress, at least in the criminal law context, constitute a breach of security of the person.[119]

The European Court of Human Rights has held that there is no right under the European convention to demand a particular treatment,[120] and it is likely that Canadian courts would come to a similar conclusion.[121] In *Rodriguez*, for example, Lamer, C.J.C.[122] remarked in his dissenting judgement:

> [W]hile there may be no limitations on the treatments to which a patient may refuse or discontinue, there are always limits on the treatment which a patient may demand, and to which the patient will be legally permitted to consent.[123]

It appears, therefore, that in Canada, s. 7 of the *Charter*, or for that matter any other section of the Canadian Constitution, does not provide for a right to health for Canadians. Apart from the *Corrections and Conditional Release Act*,[124] no other Act makes this stipulation, including the *Canada Health Act*,[125] which does not define "health", but stipulates that

> It is hereby declared that the primary objective of Canadian health care policy is to protect, promote and restore the physical and mental well-being of residents of Canada and to facilitate reasonable access to health services without financial barriers.[126]

---

[116] J.F.R. Massieu, "Mexico" in *The Right to Health in the Americas*, *supra* note 112 at 372.

[117] *Canadian Charter of Rights and Freedoms*, Part I of the *Constitution Act, 1982*, being Schedule B to the *Canada Act, 1982*, c. 11, s. 7 [hereinafter *Charter*].

[118] I. Johnstone, "Section 7 of the *Charter* and Constitutionally Protected Welfare" (1993) 46:1 Toronto Faculty of Law Review 35.

[119] *R.* v. *Morgentaler*, [1988] 1 S.C.R. 30.

[120] *Winterwerp* v. *The Netherlands* (1979), 2 E.H.R.R. 387 at 407.

[121] J. Arboleda-Florez & M. Copithorne, *Mental Health Law and Practice* (Toronto: Carswell, 1994) at 3.12.

[122] *Rodriguez* v. *British Columbia (Attorney General)* (1993), 85 C.C.C. (3d) 15 (S.C.C.).

[123] *Ibid.* at 40.

[124] *Corrections and Conditional Release Act*, S.C. 1993, c. C–44.6, s. 86.

[125] *The Canada Health Act*, *supra* note 27, s. 3.

[126] *Ibid.* The legal enforceability of the *Canada Health Act* is probably limited to the federal government, and not likely to be effective against the provinces. However, it has been suggested that the *Act* can serve as a strong political tool towards ensuring the integrity of the Canadian health system. See S. Choudhry, "The Enforcement of the *Canada Health Act*" (1996) 41 *McGill Law Journal* 461.

As well, the Act contains five principles: non-profit public administration, comprehensiveness, universality, portability, and accessibility. The *Canada Health Act* does not define services that are "medically necessary". It may be, however, that the principle of "equality under the law" in section 15 of the *Charter* and the *Canada Health Act's* principle of accessibility give Canadians an equal right of access to health services. Access, however, does not mean specific treatment, nor the quality of such treatment. This would mean that to create "equality" in the present economic climate in which the country finds itself, a common lowest denominator in quality would have to be found.

A clarification of such a common denominator through definition of "medically necessary" interventions, or of "essential services", would require: (1) proper epidemiological research at the level of the population necessary to determine risk factors impacting on the health of the population, and the prevalence[127] and incidence[128] of particular medical conditions; and (2) proper health services research to determine which among those conditions presents a larger risk to the welfare of the population. It is necessary to identify those that are most prevalent and thus require more funds, and how much to allocate to the care of each of these conditions in the population while acknowledging competing interests for funds. This weighty agenda can be delivered only with high quality epidemiological research.

## ETHICAL GUIDELINES

### *The CIOMS Guidelines for Ethical Review of Epidemiological Studies*

The CIOMS' *International Guidelines for Ethical Review of Epidemiological Studies*[129] contains a total of 53 recommendations. Although the *CIOMS Guidelines* were controversial at the time of their promulgation, they are now considered to embody the present ethical standard for epidemiological research. These comprehensive recommendations are divided into two major sections. The first section in the *Guidelines* pertains to *Ethical Principles Applied to Epidemiology* developed in five subheadings: informed consent, maximizing benefit, minimizing harm, confidentiality, and conflict of interest. The second section relates to *Ethical Review Procedures.*

All epidemiological studies must be approved by ethical review committees that are constituted to review both scientific and ethical aspects, and that include epidemiologists, other health practitioners, and lay persons representing the community. Such a committee must strive to balance personal and social interests,

---

[127] C.H. Hennekens & J.E. Buring, *Epidemiology in Medicine* (Boston: Little, Brown, 1987) defines *prevalence* at 57 in the following terms: "[P]revalence quantifies the proportion of individuals in a population who have the disease at a specific instant and provides an estimate of the probability (risk) that an individual will be ill at a point in time".

[128] Hennekens & Buring, *Ibid.* at 57, defines *incidence* thus: "Incidence quantifies the number of new events or cases of disease that develop in the population of individuals at risk during a specified time interval".

[129] See *CIOMS Epidemiological Guidelines, supra* note 50.

assuring scientific soundness, safety and quality of research, and equity in subject selection, with particular care being taken with vulnerable groups. Control groups, multi-centre and externally sponsored studies require additional safeguards.

As a basic rule, the informed consent of subjects is required in epidemiological studies. In situations where obtaining such consent is impossible or impractical, strict safeguards must be maintained to protect confidentiality, and the study must be "aimed at protecting or advancing health". In certain cases, non-disclosure is essential and ethically permissible, but should not occur unless justified before an ethical review committee. When individual subject consent cannot be obtained, group or community representatives should be consulted. Even with such group consent, an individual refusal must be respected. Subjects must not be under undue influence to participate, including through excessive inducements which might vary according to the particular community.

As with experiments in general, risks should be minimized and benefits should be maximized.[130] Harm in epidemiological studies may be indirect, such as diverting scarce health personnel from their routine duties to serve the needs of a study. Ethical review must assess risks of "stigmatization, prejudice, loss of prestige or self-esteem, or economic loss as a result of taking part in the study", and be sensitive to different cultures and value-systems.

Protecting groups and individuals includes providing for confidentiality. In unlinked studies, there is little need for concern; however, linked studies raise different issues depending upon whether they are anonymous, non-nominal or nominal. Personal identifying information should not be used when a study can be conducted without it.

Findings should use language that "does not imply moral criticism of subjects' behaviour". Findings should be publicized by suitable and available means, and should not be misrepresented either by the researcher or the sponsors of the research. Where it is not possible to inform subjects of results that pertain to their health, they should be informed of this fact. However, where individual subjects cannot be advised to seek medical attention, pertinent health care advice should be made available to their communities.

Conflicts of interest should be disclosed, and investigators should avoid pressures from the various interest groups that may come into conflict throughout the course of a study. The highest standards of honesty and impartiality must be maintained, and where investigators become advocates, "they must be seen to rely on objective, scientific data".

### *The Ottawa Workshop on Computerized Record Linkage in Health Care*

In Canada, in 1986, a Workshop on Computerized Record Linkage in Health Care[131] held in Ottawa (May 21–23, 1986) approved several recommendations for the

---

[130] See Chapter 18.
[131] Workshop on Computerized Record Linkage in Health Research, *Proceedings of the Workshop on Computerized Record Linkage in Health Research* (Toronto: University of Toronto Press, 1986) at 21–23.

development of data sets and better methodologies for record-linkage. Section 2.3 deals with ethical guidelines in record-linkage research, which are as follows:

- Legislation and regulations regarding health data bases should contain a statement recognizing the legitimacy of using the data for research purposes, under appropriate safeguards.
- Health data banks should be designed to facilitate computerized record linkage.
- Agencies with custody of files containing personal identifying information should grant qualified researchers access to such files for computerized record linkage, under appropriate safeguards.
- Mechanisms should be developed to permit researchers to contact persons identified by record linkage, with safeguards that are considered appropriate by the custodians of the data provided.
- Agencies funding health research projects using computerized record linkage should ensure that appropriate safeguards of confidentiality for such projects have been met.
- Agencies under whose auspices record linkage projects are conducted (universities, governments, etc.) must ensure adequate physical and electronic security for the data.
- Every research project involving computerized record linkage should be reviewed by an appropriately constituted Human Experimentation Committee.
- Researchers must honor restrictions imposed by data custodians on the use of identified data.
- Researchers must accept responsibility for the consequences of any breach of security of identified data.
- Researchers must accept responsibility for the supervision and disciplining of their staff.
- Special studies should be conducted, when needed, to establish the safety of record linkage studies.
- Researchers should accept responsibility for public relations.

Both the CIOMS and the Ottawa Workshop *Guidelines* approve of epidemiological research and record-linkage methodologies as long as proper safeguards are put into place.

## CONCLUSION

The purpose here was to review the scientific and ethical issues related to epidemiological and health services research. The latter were analyzed from a microethical perspective in the belief that elements of autonomy, risk/benefits ratios, and considerations of justice apply equally to epidemiological research as they do to biomedical and behavioral research in general. The analysis was extended to review macroethical issues, those ethical problems that arise in epidemiological research because of its specific scientific approach to the study of populations. The point is made that record-linkage studies pose a risk to privacy and confidentiality and that, as they can be conducted without awareness of the public at large, they have the potential to convert the entire population into a vulnerable population, unless adequate safeguards are implemented. Finally, larger issues pertaining to a possible right to health tempered by a utilitarian need of the state to keep the population healthy and productive establish a need to balance citizens' rights and duties with the obligations of the state. To accomplish this balance, it is concluded that the state needs epidemiological data to make reasonable and reasoned decisions on health allocations.

In sum, epidemiological research and the use of record-linkage methodologies should be endorsed as long as proper safeguards are put into place in order to protect the autonomy of vulnerable persons and of the population at large at the time of epidemiological surveys or experiments. Access to biobanks should be properly regulated, and their data made accessible to *bona fide* researchers upon approval from constituted research ethics committees. Finally, epidemiological and health care research, including intramural research projects in Government, should undergo a process of scientific and ethical review and approval prior to initiation.

**Note:** Chapter 29 was previously published in an earlier form in <u>Acta Psychiatrica Belgica</u>, fascicule 3, 1997, p. 125-165.

CHAPTER 30

# ETHICAL GUIDELINES FOR EPIDEMIOLOGICAL RESEARCH

POVL RIIS

Epidemiology as a term has widened its scope gradually since it was coined as a discipline covering the studies of major contagious diseases. Today everyone will understand the word as covering all studies and all knowledge of disease-related factors in large population groups, the factors spanning from ecological ones, via lifestyle ones, to viruses and bacteria. In particular, studies that correlate societal factors to disease have increased immensely in number since World War II. Until rather recently such studies, most of which are by nature non-invasive, have been considered to belong to the innocent or sometimes forgotten part of the research ethical spectrum. But now this has changed.

Transgressing the integrity borders of citizens, even by non-invasive techniques resulting in distilled information, is today considered as much a case for public awareness and control as drug trials and experimental surgery.

Early epidemiological studies of societal factors were based on figures and words, referable to the test group as a whole, or subdivisions after stratification. Such collections of figures and words were usually named *registers*. Often the individual subjects were non-identifiable when the results were published. Applying such an anonymous technique created no, or few, research ethical problems, except in rare cases of so-called stigmatization (to be described later).

## THE NOVEL UNITS OF INFORMATION

Today person-referable information in biological materials supplements the classical figures and letters. In particular, very detailed information in *bound* form now constitutes a broader base for epidemiological research. Cells, tissue cultures, histological sections, DNA-fragments, samples of plasma or sera, etc. . . . have made

588

a redefinition of registers necessary and have changed them to *information banks* or *bio-banks*. Such bio-banks are very numerous in health science institutions (e.g. clinical departments, diagnostic departments, research institutes, national health service departments, etc. . . .). Nobody knows the exact number of such bio-banks, to say nothing of the total sum of information "units" contained therein. One has just to think of contemporary genetic research to imagine the rapidly increasing disengagement of bound information in already existing bio-banks. This development has stimulated the demand from the public and from politicians that such bio-banks have to be notified and controlled, and that the derived research projects, including epidemiological research, have to be ethically evaluated on a case-by-case basis.

## THE DILEMMA BETWEEN PROTECTION AND SCIENTIFIC USE

National data protection acts were introduced in many countries during the seventies and eighties. The computer age and the possible linkings between files of different kinds created a public demand for strong protection against non-consensual use of personal data. Even if research exploitation of data was not the primary target, the resulting restrictive legislation influenced epidemiological research and its potentials.

During the same period clinical research ethics experienced two counteracting forces: one asked for protection and respect, leading to safety measures and the necessity of informed consent; the other, less audible, asked for doctors to use only treatments of known value. With the publication of the *Second Helsinki Declaration* in 1975 biomedical research obtained a set of guidelines meeting the demand for protection of patients and healthy volunteers, besides respect in the form of an obligatory informed consent in the majority of cases. Epidemiology was by analogy included in the scope of research covered by the *Declaration*, but because clinical research was the main target, epidemiology became the *Declaration's* step child. Not until the eighties and nineties have data protection acts, epidemiological research and the corresponding control systems been co-ordinated sufficiently.

## INTERNATIONAL GUIDELINES AND NATIONAL LEGISLATION

Until World War II international and national control over biomedical research on man was patchy. In other words no international debate or consensus existed. Like other historical examples of developments favoring the protection of fundamental human rights, the initiative was also based on demonstrated severe transgressions, and not primarily derived from an attitude of steadily improving man's conditions by a continuous humanitarian evolution.

The severe crimes against mankind, including use of human research subjects against their will in lethal or invalidating experiments, reported during the postwar trials, emphasized the need for international guidelines, hopefully to be followed by national rules and control mechanisms.

The *Declaration on Human Rights* and the *Declaration of Geneva* directly led to the appearance of the *First Declaration of Helsinki* in 1964 and the second version from 1975.

Few countries have projected the *Second Helsinki Declaration* into national legislation, covering biomedical research including epidemiological studies. The Nordic countries represent the exceptions, with official control systems based on different structures, but resting on the same principles for protection.

Denmark enacted its law, covering biomedical research in 1992.[1] Here epidemiology is explicitly placed within the research spectrum covered by the law, whose paragraph 6 states in section 2: "The same (i.e. registration and permission) applies to research projects, in which biomedical research as defined in Section 1, constitutes a substantial part of the total project, *plus projects based on questionnaires and registers*, related to the areas mentioned".

Chapter 3, paragraph 8, section 1, point 2 states that "the patients or healthy volunteers, participating in the project, will be informed in writing and verbally about its content, foreseeable risks and advantages, and that their free and explicit consent will be obtained and given in writing, cf. Section 3". The comments to section 3 say: "Register research is not comprehended by the rule in Section 1, points 2 and 3, on informed consent. In special cases the Committee can demand that informed consent is obtained in projects based on data registers, in accordance with Section 1, point 2 or 3".

The essential point is that as a main rule, epidemiological research does not presuppose informed consent, *unless* the research ethics committee handling the protocol thinks it necessary. In evaluating such exceptions, the given regional committee will emphasize whether so-called *sensitive information* is comprised. Such sensitivity depends globally on the local culture and political tradition. In Denmark some of the sensitive areas are: ethnicity, sexual orientation, income, political party membership, criminal behavior, and religion.

Even if anonymous data are used, the committee system will consider possible stigmatization of small sensitive groups, recognizable as *groups* and not as individuals.

The Danish research ethics control system has issued a number of recommendations informing and guiding scientists on special procedures and precautions. These recommendations have no direct judicial power, but still will be taken into account in case of transgressions of the main principles of the *Declaration*.

Three recommendations, in particular, deal with epidemiological research. Recommendation No. 3 on multicenter studies states:

> A multicenter study is a research project which comprises the recruitment of patients from more than one scientific-ethical committee's region.
>
> By recruitment is understood that the contact with potential trial subjects takes place within the health and educational institutions of the region in question or through professionals in the health sector.

---

[1] *Act on a Scientific Ethical Committee System & the Handling of Biomedical Research Projects.* No. 503 of 24 June 1992.

A responsible co-ordinator (project leader) shall always be appointed among the investigators, if possible a person working within the region of one of the regional committees from which patients or healthy volunteers are recruited for the study.

Recommendation No. 9 has special relevance for epidemiological research. It says in its introduction to the research ethical control with such subjects:

Biomedical information banks that in the register legislative sense are considered data registers, comprise in a scientific-ethical context all collections of information on human beings in this area whether stored as words, numbers, DNA codes, antigens, antibodies, morphological, biochemical, physiological or biophysical characteristics as well as other information which may be related to a specific person.

All types of information banks are considered to be equal for a *scientific-ethical* evaluation, whether they are public or private or combinations thereof, are stored electronically, as images or as basic material for the liberation of stored information.

The information banks of the health sector serve several purposes. They may be established as collections of information on diagnostic results, or in other cases as solely scientific collections of information — primarily to serve diagnostic revision purposes or purely scientific purposes respectively.

However, it is characteristic that they have an inherent unpredictability concerning their possible later use. They may each serve purposes within the area of diagnostics and research, which were unforeseen at the time of their establishment, but they may also be used with other purposes than the original one as a consequence of the scientific development. Thus, e.g. an originally solely scientific information bank may serve a later diagnostic purpose which may benefit a given patient, or a solely diagnostic information bank may serve a later scientific purpose ranging from fundamental over clinical to epidemiological research.

Thus, definite and final deletion terms for such information banks or the linkage to detailed purposes established from the start are according to the views of the Central Scientific-Ethical Committee unacceptable in a health related and scientific connection. The register legislation requires that there are stipulated deletion terms, which may be prolonged, as well as the formulation of clauses for the use of the information bank.

Further under the notification procedure:

The scientific use of register/bank data in a specific research project shall always be notified to the scientific-ethical committee system according to the Act on a scientific-ethical committee system and the handling of biomedical research projects § 6, Section 1. Furthermore, all private research projects shall be notified to the Data Surveillance Authority pursuant to the Act on private registers, § 2, Section 3.

It has been agreed with the Data Surveillance Authority that research projects should *first* be submitted to the committee system. The research project is submitted to the Data Surveillance Authority when the scientific-ethical committee has granted its approval. A copy of the committee's approval shall be enclosed. The scientific-ethical committee's decision shall *always* be forwarded to the Data Surveillance Authority *in case of concomitant* submission to the Data Surveillance Authority and a scientific-ethical committee. The inspection stipulates its conditions for the project, and these constitute together with the requirements of the committee system the terms for the initiation and implementation of the project.

The linkage with a private research project to data from a public register including biomedical information banks presupposes that the investigator contacts the authority responsible for the register in question. The responsible authority will subsequently obtain a statement from the Data Surveillance Authority.

Many bio-banks within a national health system will serve a potential double purpose. On this the recommendation says:

> The use of information closely related to the original sampling situation, e.g. revision of diagnoses in the interest of the patients, method improvements with the same purpose, and other situations linked to the original physician-patient relationship, shall not be subject to a scientific-ethical evaluation. The regional scientific-ethical committee is consulted in case of doubt.

The third recommendation to be mentioned here is No. 12. It stresses the obligatory notification of all epidemiological research projects belonging to the health sciences:

> A register project shall *also be notified to the Data Surveillance Authority* according to the Act on private registers. The notification to the Data Surveillance Authority does not stay the project and documentation for the scientific-ethical approval shall be included. A research project comprising the *linkage* to other public registers requires evaluation by *both* the scientific-ethical committee system *and* by the Data Surveillance Authority and must *not be initiated* until this double approval has been granted.

If epidemiological research projects are multidisciplinary, i.e. comprise both bio-medicine and other research disciplines, for instance sociology, criminology, psychology etc., the existing law on biomedical research will be applied due to the inclusion of biomedical data. The reason is that control measures do not often exist in other research sectors (this is still the case in Denmark too). Should such parallel systems be established later, the ethical evaluation would naturally take place among the systems involved. Health sciences until now fill a vacuum and its scientists and research ethics committee members sometimes wonder why research ethical problems only seem to exist (or at least are recognized) within the health sciences.

## NATIONAL AND INSTITUTIONAL CONTROL COMMITTEES

The prevalent model for research ethics committees dealing with biomedical research is *the institutional committee*, related to university institutes, hospitals, ministries, firms, etc. They have varying membership structure, bylaws and rules of procedure. Consequently standards may vary, and especially multicenter trials' separate centers may meet different demands and degrees of acceptance.

A different model is the *regional* one, sometimes supplemented by a central committee for the whole country. By making the system double-stringed and nation-wide, much greater consistency is obtained, and especially large projects involving many or all regions benefit from the coordinated evaluation in such a system.

An example of a system of the last mentioned type is the Danish one: the Danish system of scientific-ethical committees serves to implement *The Second Declaration of Helsinki* (amended by the 35th World Medical Assembly, in 1983 and in 1989 and later), through a two-tier system of scientific-ethical committees (i.e. eight regional scientific-ethical committees and one central scientific-ethical committee).

The eight regional committees cover (1) the Municipalities of Copenhagen and Frederiksberg (two for this double-region), (2) the County of Copenhagen, (3) the Counties of Bornholm, Frederiksborg, Roskilde, Storstrøm and Western Sealand, (4) the Counties of Vejle and Funen, (5) the Counties of Southern Jutland, Ribe and Ringkøbing (the Southern and Western part of Jutland), (6) the County of Århus, and (7) the Counties of Northern Jutland and Viborg. Each Committee has a basic membership of 7 members and 7 substitutes, either group comprising 3 researchers and 4 lay members.

The researchers are appointed by the Danish Medical Research Council from among regionally proposed researchers. The lay members are appointed by the local Councils and the region in question. Both groups of members serve for four years, following the election term of the regional political bodies.

The regional committees cover all biomedical research within their region, comprising medicine, dentistry and pharmacy research, conducted in hospitals, research institutions, universities, industrial undertakings or within the primary health services for medicine, dentistry and pharmacy.

The Central Scientific-Ethical Committee of Denmark is composed according to a basic structure of two members from each regional committee, one researcher and one lay member, two lay members appointed by the Minister and further has a chairman and a vice-chairman appointed by the Danish Medical Research Council, this part also consisting of one researcher and one lay person.

Appeals from the regional committees lie with the Central Scientific-Ethical Committee, which also serves to establish contact between researchers and the public, the ministries, the government, the press, etc.

The activities of both levels of committees rest on a set of rules and standing orders.

The Ethical Council of Denmark, based on an act of 1987 by the Danish Parliament, serves the *principal* discussions on ethical matters related to research and the health professions. It has no direct influence on concrete research proposals represented by protocols sent to the scientific-ethical committee system. The Ethical Council corresponds to a parliamentary board of politicians, which in this way links the Council to the Danish Parliament.


## THE WORKING PROCEDURE IN THE COMMITTEES

Undoubtedly there are several ways of handling concrete research projects in a scientific-ethical committee. Consequently it cannot be described in a paradigmatic way, but only based on empirical observations.

The first phase is *an analysis of the methodology*, methodology defined as the art of planning, carrying through, evaluating and publishing a research project. Here of course the planning phase is *the* object for the analysis. Are the variables, the methods, the statistics, the researcher's ability and the institution's resources of equipment and scientific advice sufficient? If not the project will not advance to the next phase, because "a technically invalid project involving man is in itself unethical to proceed with". If on the other hand the project is sound, or can become so after minor corrections or additions, the handling moves on to the *analysis of its ethical components*. The corresponding questions are the following: Is the idea promising for the group involved? Is security and respect — in the shape of, for instance, informed consent — sufficient? Could a vulnerable group be stigmatized in an epidemiological project? Do conflicts of interest exist on the scientist's side?

After having located potential answers to these questions, they are weighted together and carried over to the *projection into the prevalent ethical norms*.

These norms obviously vary from country to country. But even within culturally homogenous nations pluralism grows, according to the greater freedom for citizens' private norms, attitudes, religious beliefs, etc. If, however, the lay representation within research ethics committees is large enough, substantial parts of the national pluralism will be represented. And if at the same time all members seek to obtain consensus, even in existential questions, then the projection mentioned above can take place onto a consistent pattern of ethical norms. If the weighted resultant falls outside these national norms, then the protocol is not accepted. On the other hand, if by changes of the protocol — or due to the qualities of the original protocol — the resultant falls within the norms, then the study is accepted.

Because research ethical analyses are so complex the ideal handling of the cases is based on *discussion with the aim of consensus*, not voting. Voting is by its nature an oversimplification of decision-making, especially when it compensates for discussion and debate. If disagreement emerges in research ethical evaluations, preventing a committee from reaching a consensus, then a serious attempt must be made to change the protocol in order to obtain consensus.

Examples of concrete cases of a clinical-epidemiological nature, dealt with by a research ethical committee, are given below:[2]

> The presence of a *dilatation of aorta* because of degenerative changes, a so-called aortic aneurysm, was planned to be assessed in a population study based on ultrasound in 65–75 year old men. It is a relatively common disease which may be cured through an operation. These patients will often die through the catastrophic event that takes place when the aneurysm ruptures if they are not identified in time. The Committee found that it was a fundamental ethical problem to label healthy persons as sick following a positive examination. For the participants who at the time of diagnosis were so sick that they would be operated upon and cured it would on the other hand mean approximately 10 extra years of life. The Committee concluded that the project had to be rejected in its present form but that it might be reformulated to make it acceptable for a scientific-ethical evaluation.
>
> A study of *ear infections caused by bathing in private swimming pools* was submitted by a regional committee for a fundamental discussion. An important question was the possible inclusion of foreign nationals with a temporary rented residence in Denmark. It should be possible to include the private swimming pools in question even with a majority of the participants being foreigners. The Act on a scientific-ethical committee system is valid for research projects performed in Denmark, while the nationality of the research participants is without legal importance.
>
> A regional committee submitted a project on *parvovirus B19 infections* during pregnancy for a fundamental discussion. The question was whether it could be approved from a scientific-ethical point of view to allow tissue samples from deceased embryos from the period 1980–1991 to be included in such a study when the relatives were not aware of the existence of the tissue samples. Parvovirus B19 may cause damage to the embryo and spontaneous abortion. The aim of the study was to clarify the full importance of this infection during pregnancy and would thereby prepare the way for possible future

---

[2] The Danish Central Scientific-Ethical Committee, *Annual Report 1993* (Copenhagen: Ministry of Science and Technology, 1994).

vaccinations. The committee agreed unanimously that anonymous tissue samples could be included in the study not least in the light of the possible preventive advantages.

An appeal concerned the project *3H-paroxetin binding and paroxetin treatment of a pathologic passion for gambling*. It was the purpose to study whether such treatment possibilities exist for persons with a severe passion for gambling. It was stressed during the discussions that the study was valuable because the legal system in the future might choose treatment as a sanction rather than penalty. The researchers had suggested to recruit the trial subjects directly at the gambling locations but this was unacceptable to the Committee. It was instead suggested to recruit participants through psychologists and psychiatrists and possibly through advertisements in the daily press, professional journals and the local press. The risk of freedom from responsibility for one's actions was balanced against the possible advantages that a small but heavily encumbered group may gain.

The project *The Psychiatry Legislation Study* was submitted for fundamental discussions. The study aimed at elucidating how the psychiatry legislation had functioned in 1991–1992 to create the basis for a planned revision. The study would be based on a nation-wide collection of data concerning the extent of use of constraint, and the number of complaints during the study period. The data collection should include an analysis of hospital records and possibly documents in complaint cases. A concomitant questionnaire study should be performed concerning the parties involved in the use of constraint excepting patients and relatives. While the committee system usually does not approve of access to information in hospital records without consent, it was found by the Committee that a request for explicit consent would invalidate the study and it was instead stressed that the information should become available in an anonymous form. The Committee approved the study and suggested that the group of researchers and the Central Scientific-Ethical Committee issued a joint press release.

*The efficacy and quality of prenatal care in rural districts in Tamil Nadu, Southern India*, was submitted for a guiding evaluation. The rationale for the study was the high mortality in pregnancy and birth especially in Southern Asia. The committee suggested changes in the Danish language patient information realizing that it would be translated into a language understandable to the rural population.

A project concerning *the frequency of depression in hospital admitted cancer patients* was submitted to the Central Scientific-Ethical Committee as a complaint. The study had the purpose of describing the pattern of depressions in 100 cancer patients by using a series of scales. It was stressed by several members during the discussion that the starting point for the study was not considered correct namely that cancer patients have a psychiatric disorder. On the contrary they have a life crisis that should be treated individually. Special stress was put on the aspect of attention during treatment including listening to the patients and learning to deal with cancer patients in a better way and thereby bring humanitarian elements forward in the relation. The Committee concluded that the study could not be approved on the present basis. A revised research protocol including ethical considerations could be forwarded to the Central Scientific-Ethical Committee for revaluation.

## TEMPORAL PERSPECTIVES IN CLINICAL AND EPIDEMIOLOGICAL RESEARCH

The now classical controlled clinical trial has a typical duration of a few weeks to a few months for individual patients, and the total trial usually does not last longer than 1–2 years. This means that the initial informed consent obtained from patients has a perspective that is easy to grasp, i.e. it does not easily expire psychologically during trial participation. Further, patients might benefit from the trial in an easily foreseeable future.

In epidemiological research the temporal perspective is quite different. Consent may be given years before follow-up examinations take place. Sometimes consent may be given by a pregnant mother and the large-scale epidemiological follow-up of the offspring may take place in young or middle-aged persons who for obvious reasons did not participate in the original consenting procedure. And turning to personal benefit from the results of a study, these are farfetched for subjects, whose data are involved in epidemiological research.

On the other hand these research subjects do not run any physical risks as trial patients sometimes do. Participants in epidemiological research projects might obtain knowledge of non-favourable relations between their genetic constitution or life-style and disease, but such information is only sometimes a mental burden.

These differences between clinical and epidemiological research necessitate a broad scope of analytic training for research ethics committee members. Consequently the conditions for initiating a project will vary accordingly, for instance, a demand for a renewed consent in case of long-range studies involving family members who did not consent originally.

## PROJECTS OUTSIDE THE SCIENTIST'S HOME COUNTRY

Biomedical research becomes more and more international, both in the form of international multicenter studies and, for instance, that of epidemiological research carried out by scientists in foreign countries. In such cases the World Health Organization advocates that research projects are handled both in the scientist's home country and in the country in which the research takes place. Especially when scientists from the First World work in the Second or Third World all measures must be taken to avoid even suspicions of exploitation of less privileged populations.

## INTERNATIONAL COOPERATION IN SETTING RESEARCH ETHICAL STANDARDS

Besides the *Helsinki Declaration* international initiatives to set research ethical standards have been rare until recently. But in the last few years both the Council of Europe and the European Union have initiated the necessary preparatory work with a convention and a directive. By nature they will of course both be European, but an influence on the global standards and control models will undoubtedly also be the result.

The Council of Europe's Convention on Bio-Ethics was adopted in 1997 and is now being ratified by the member states. It contains all fundamental principles for patients' and citizens' rights and for their participation in research.[3]

The conditions for research, for instance, presuppose that the project has been examined independently with regard to its ethical acceptability and scientific quality. Further, when research in human beings is the only way to obtain the necessary

---

[3] Council of Europe, Steering Committeee on Bioethics, *Convention for the Protection of Human Rights and Dignity of the Human Being with Regard to Biology and Medicine: Bioethics Convention* (Strasbourg: Council of Europe, Directorate of Legal Affairs, 1997).

information, that risks are kept to a minimum and are proportionate to the expected benefits and importance of the aims, while at the same time the researchers involved observe the highest standards of conduct.

If consent is not obtainable due to the research subject's condition, his or her inclusion in a research project is not acceptable, unless the project is expected to procure a significant benefit to his or her health. Yet exceptionally, even in the absence of a potential benefit a project is permitted, if the research may significantly improve the understanding of disease or disorder mechanisms. In such cases the following conditions must be met: (1) a participant's refusal shall always be respected, (2) the risk will have to be negligible, and (3) equally effective research cannot be carried out in individuals capable of consenting.

Very important for all biomedical research, including epidemiological research, is the introductory statement in the Article on research: "Scientific research in the field of biology and medicine shall be carried out freely" under the condition that the Convention and other legal provisions are obeyed.

The European Union's third proposal on "the protection of individuals with regard to the processing of personal data and on the free movement of such data" was published in Spring 1995.[4] Whereas research was not seriously considered in the first two proposals, the present third version takes much more into consideration the interests of medical scientists, and especially epidemiologists. The earlier restrictions of the storing of data have been lifted, and the demand for obligatory informed consent as well. Medical research has as a whole been inserted as an activity that represents an exemption to the strict rules of protection. The balance between protection of the citizens' privacy and freedom for research is now referred to the member states' national regulations. For those countries that have established control systems within the field of research ethics, this overall principle eases the work of epidemiology.

## RESEARCH ETHICAL ANALYSES AS INTEGRAL PART OF PUBLICATIONS

Until now ethical analyses have been demanded as a condition for publication in biomedical journals of projects involving humans. But the documentation has either been referred to the author's accompanying letter, or it has only been expressed in a rather lapidary way, as for instance: "The study was in accordance with the Second Helsinki Declaration and was approved by the local ethics committee". In the future this part of a biomedical publication involving humans deserves as much space within an article as many technical descriptions of methods applied. In other words, both the analysis, its major components, and the final projection to the framework of norms will have to be mentioned. The same is true for the research ethics committee: its official name and location should be mentioned, in order to make its

---

[4] Council of Europe, *The Protection of Individuals with Regards to the Processing of Personal Data and on the Free Movement of Such Data* (Brussels: Council of Europe, 1995).

co-responsibility more visible. Such a change of editorial policy will not only serve protection of research subjects, but also represent an important daily educational impact on readers and scientists.

## CONCLUSION

Epidemiological research is an integral part of biomedical research. Consequently, the fundamental principles underlying research ethical analyses and acceptance procedures in laboratory and clinical research in humans also apply to epidemiology. Even if consent procedures, possibilities of rendering data anonymous, and the duration of projects differ, epidemiological projects can easily be dealt with by research ethical committees covering biomedicine as such. Fortunately international conventions and directives seem to follow this line, in this way facilitating research on humans, both in its direct interventionist and in its epidemiological form.

CHAPTER 31

# EPIDEMIOLOGY AND THE OWNERSHIP OF HEALTH DATA: ETHICAL, LEGAL AND SOCIAL ASPECTS

CLAES-GÖRAN WESTRIN AND TORE NILSTUN

Epidemiology is a fundamental science that is essential to public health practice. As such, it should satisfy the basic requirements of research. It should reveal and convey information obtained by valid and reliable methods; serve the information needs of given audiences; be cost-effective; and be conducted with due regard for the welfare and rights of those involved in the study, as well as those affected by the results.[1]

It is sometimes impossible to satisfy all these requirements at the same time. This means that for some study designs ethical conflicts are unavoidable in epidemiological research. By "ethical conflicts" we mean situations in which one ought to fulfil incompatible requirements.[2] For instance, in longitudinal studies on large computerised case registers linking data from census and hospitals, individual informed consent would both impair scientific quality and increase costs. Thus rigorous claims as to individual informed consent would have hindered important public health research such as the studies showing the relation between tobacco smoking and cancer.[3] The same holds true for the recent evaluations of the effectiveness of mammography screening in the early detection of breast cancer implying a follow-up of 70 000 study subjects for several years.[4]

---

[1] J.M. Last, ed., *A Dictionary of Epidemiology*, 3rd ed. (New York and Oxford: Oxford University Press, 1995).

[2] C.W. Gowan, ed., *Moral Dilemmas* (New York and Oxford: Oxford University Press, 1987).

[3] R. Doll & A.B. Hill, "Smoking and Carcinoma of the Lung. Preliminary Report" (1950) 2 *British Medical Journal* 739; E.L. Wynder & E.A. Graham, "Tobacco Smoking as a Possible Etiologic Factor in Bronchogenic Carcinoma" (1950) 143 *Journal of the American Medical Association* 329.

[4] L. Nyström *et al.*, "Breast Cancer Screening with Mammography: Overview of Swedish Randomised Trials" (1993) 341 Lancet 973.

599

To avoid misunderstanding we would like to emphasise that much epidemiological research can and therefore should be conducted with due regard for obtaining individually informed consent. This is particularly true with regard to epidemiological studies where the researcher is in direct contact with the research subjects. But, as indicated above, this is not always possible.

With reference to such situations research guidelines give conflicting recommendations. For instance, the *International Guidelines for Ethical Review of Epidemiological Studies*[5] admits exceptions to the requirement of individually informed consent. "Access [to occupational records, medical records, etc.] may be ethical on such grounds as minimal risk of harm to the individual, public benefit, and investigators' protection of the confidentiality of the individuals whose data they study". But many research ethics committees ignore this code when assessing epidemiological studies. Instead they use the *Declaration of Helsinki* (1996) which requires that "the [competent] subjects' freely-given informed consent, preferably in writing", should be obtained.

The demand for individually informed consent brings forth the question of who actually owns the health data relevant to epidemiological research. This question has during later decades caused continuous controversies between the medical research community, the legal community and the political community.

In this paper we are going to discuss the ethical, legal and social questions associated with this controversy. Our main contention is that these three aspects should be clearly distinguished but, in practice, dealt with in a comprehensive way.

## A FRAME OF REFERENCE

In order to give structure to our discussion we will introduce a simple model inspired by Francoeur[6] and Hermerén.[7] The model has two dimensions. The first specifies different groups of persons, in this case those involved in or affected by epidemiological research. Each group should consist of persons who, in that particular situation, have similar interests. Groups of persons relevant to most discussions of the ownership of health data are (1) the study subjects, (2) the beneficiaries, (3) the community at large, and (4) the epidemiologists.

The second dimension specifies the relevant ethical principles. These should be based on some existing values, preferably the values of the different groups of persons involved in or affected by the study. Commonly accepted value premises in medical ethics are the principles of beneficence, autonomy and justice.[8]

---

[5] Council for International Organizations of Medical Sciences, *International Guidelines for Ethical Review of Epidemiological Studies* (Geneva: CIOMS, 1993).
[6] R.T. Francoeur, *Biomedical Ethics. A Guide to Decision Making* (New York: John Wiley & Sons, 1983).
[7] G. Hermerén, *Kunskapens pris* (The price of knowledge) (Stockholm: Forskningsrådsnämndens förlagstjänst, 1986).
[8] T.L. Beauchamp & J.F. Childress, *Principles of Biomedical Ethics* (New York and Oxford: Oxford University Press, 1994); CIOMS (1991), *supra* note 5, Council for International Organizations of Medical Sciences, in collaboration with the World Health Organization, *International Ethical Guidelines for Biomedical Research Involving Human Subjects* (Geneva: CIOMS, 1993).

(1) The principle of beneficence refers to the obligation to maximise benefits and minimise harms. The risks and burdens of research should be reasonable in the light of the benefits.

(2) The principle of autonomy requires respect for the right to self-determination. Those who are capable of deliberation about their personal goals should be treated with respect for their preferences.

(3) The principle of justice requires protection of vulnerable people. Those bearing the burden should receive an appropriate benefit, and those intended to benefit should bear a fair proportion of the burdens.

The principle of beneficence is inspired by the utilitarian tradition[9] and the principle of autonomy by the liberal tradition,[10] while the principle of justice has several sources.[11]

## OBLIGATIONS AND RIGHTS

The idea that certain acts are required, or ought to be done, is called an obligation, and the idea that one is entitled, or has a valid claim, to something is called a right. Such obligations and rights may be justified with reference to the three principles of beneficence, autonomy and justice.

The study subjects have an obligation, based on the principle of beneficence, to offer data about themselves to prevent or reduce illness among their neighbours. Since those who are ill often are vulnerable this obligation to offer data also has support from the principle of justice. However, the principle of autonomy requires respect for self-determination, implying that study subjects have a right to decide for themselves whether or not to allow the use of information about themselves.

Those who gain from epidemiological research, the beneficiaries, have rights according to the principles of beneficence and justice. The principle of autonomy is hardly relevant since the beneficiaries' right to self-determination is not affected, though new epidemiological knowledge may of course increase their options in the future.

The community, or, to be more precise, the representatives of an open and democratic society, are responsible for the welfare of the individual citizens. Based on the principles of beneficence and justice, they have an obligation to facilitate epidemiological research. It should be feasible to investigate such public health issues as deleterious environmental conditions, the cost-effectiveness of health care, and the possible discrimination of vulnerable groups. But there is also an obligation to consider the right to autonomy of the study subjects.

---

[9] J. Bentham, *An Introduction to the Principles of Morals and Legislation* (Edinburgh: Trait, 1843); J.S. Mill, Utilitarianism (London: Longmans, Green, 1871).

[10] R. Nozick, *Anarchy, State and Utopia* (New York: Basic Books, 1974).

[11] N. Recher, *Distributive Justice. A Construtive Critique of the Utilitarian Theory of Distribution.* (New York: Bobbs-Merill, 1966); J. Rawls, *A Theory of Justice* (Cambridge: Harvard University Press, 1971); R.C. Solomon & M.C. Murphy, eds., *What is Justice? Classic and Contemporary Readings* (Oxford: Oxford University Press, 1990).

**Table 1**

**Distribution of Rights and Obligations in Epidemiological Research Justified By the Three Principles of Beneficence, Autonomy and Justice**

| Persons Involved | Beneficence | Autonomy | Ethical principles Justice |
|---|---|---|---|
| Study subjects | Obligations | Rights | Obligations |
| Beneficiaries | Rights | Not relevant | Rights |
| Community at large | Obligations | Not relevant | Obligations |
| Epidemiologists | Obligations | Rights | Obligations |

Epidemiologists have similar obligations. They should seek knowledge relevant to the reduction of suffering and untimely deaths, with due regard for the welfare and rights of those involved in the study. Since most epidemiologists are eager to fulfill their social obligations, there is usually no conflict with their own right to autonomy.

Summarising the rights and obligations (Table 1) it becomes evident that there is one predominant ethical conflict in epidemiological research. The right to self-determination of the study subjects is not only in conflict with their own obligations (based on the principles of beneficence and justice) but also with most of the identified rights and obligations of other groups involved or affected.

How is this conflict to be solved? In our opinion the right to autonomy of study subjects does not always override the other ethical considerations. But there seems to be no way of proving this position. For instance, a libertarian will emphasise the right of individuals to self-determination based on the principle of autonomy. According to this position individuals own the health data related to themselves. If one does not accept the use of such data in epidemiological research, others may have reasons for saying that one is hardhearted or unfeeling, but not that one is blameworthy.

## AN EXAMPLE

However, such an argument in terms of abstract obligations and rights is hardly sufficient to convince a libertarian. We will therefore try another approach and use a study by Wall and Taube[12] to make our reasoning more concrete. They studied the working environment and its consequences for workers' health at the Rönnskär plant in Sweden.

Rönnskär is a copper smelter refining ore containing different metals. Production started in 1930 and the potential health risks soon became apparent. A first attempt

---

[12] S. Wall & A. Taube, *Fallet Rönnskär. En epidemiologisk studie av livslängd och dödsorsaksmönster bland smältverksarbetare* (The *Rönnskär* case. An Epidemiological Study of Survival and Mortality Patterns among Smelter Workers) (Stockholm: Rapport från Cancerkommittéen, 1983). S. Wall, "Survival and Mortality Pattern among Swedish Smelter Workers" (1980) 9 *International Journal of Epidemiology* 73; B. Haglund *et al.*, "Longitudinal Studies on Environmental Factors and Disease" (1991) 19 *Scandinavian Journal of Social Medicine* 81.

at a medical study at Rönnskär was conducted as early as the 1930s by Gunnar Inghe in cooperation with the trade union. The management of the company did not cooperate with the researchers and the results were never published since they were regarded as "unscientific" mainly due to high attrition rate.

Towards the end of the 1940s and at the beginning of the 1950s a few more studies were made, primarily on damage to the respiratory organs of Rönnskär workers, but on the whole no more research was conducted until the government in 1975 declared that the licence for operation would not be renewed unless studies of the effects of the working environment were carried out.

Thus, the study of Wall and Taube was planned in 1976 in collaboration with the union, management and industrial health authorities. The study included almost 4000 men who altogether had contributed about 50 000 years of work. Data on duration and type of work were collected from employment records. These data were supplemented with information from the population and cause-of-death registers. Nearly everybody was found in the public registers and almost 1000 had died before 1977.

Their study showed excess mortality in general among Rönnskär workers. The excess mortality was especially great in deaths from cancer and diseases of the circulatory organs. In particular, work in the roaster and arsenic departments was associated with an elevated risk of dying from lung cancer. As a result of this study considerable improvements in the working environment have been instituted and the high-risk areas of work have been closed.

## IDENTIFICATION OF COSTS AND BENEFITS

We will use our model (presented in Table 1) to make explicit and assess the ethical cost and ethical benefits to those involved in or affected by the Rönnskär study. In addition to the exposed workers, the future workers, the community at large, and the epidemiologists, two more groups have interests to protect. These are the company management and the trade unions.

The company management has an obligation to protect the safety and health of the workers. To fulfil this obligation knowledge is needed, which is one reason to cooperate with qualified epidemiologists. But the company management has to pay for improvements in working environment. Based on the principle of autonomy they also have a right to protect their economic interests.

The trade unions have an obligation to promote the many (and sometimes conflicting) interests of the workers: safe working environment, respect for individual self-determination and protection from unemployment.

In the Rönnskär study there are three possible alternatives. The first alternative is that no study is carried out, practically equivalent to what happened in the 1930s when the company management was opposed to research into health risks. The second alternative is a study of the kind made by Wall and Taube, wherein individual informed consent was not obtained from the studied workers. The third alternative, which is hypothetical, is a study carried out with individual informed consent as a prerequisite.

For the sake of argument we will assume that the first alternative, no study, is unethical. This means that we will only compare the second and the third alternative: a study without individually informed consent (as carried out by Wall and Taube) is compared to a similar (hypothetical) study with individually informed consent. In other words, we compare the ethical costs and ethical benefits between two different study-designs: one without and one with informed consent. The latter, i.e. the study-design with informed consent, is used as a baseline.

As to the principle of beneficence there are, with the possible exceptions of company management, only benefits. To do epidemiological research without individual informed consent (compared to do such research with individual informed consent) reduces costs and avoids problems with both attrition and selection bias. (Compare the first column in Table 2.) As to the principle of autonomy, there seem to be no benefits, only costs. The most important costs are the infringements on the exposed workers' right to informed consent. (Compare the second column in Table 2.) As to the principle of justice there are benefits to the future workers and possible costs to company management. (Compare the third column in Table 2.)

In this example the model demonstrates a typical dilemma of public-health research: valid scientific information may not be obtained without some infringements of personal autonomy. It also makes clear that the relevant ethical considerations may include the interests of concerned groups other than the studied subjects and also conflicts of interests other than between subjects and the community at large.

## BALANCING COSTS AND BENEFITS

So far our approach to the Rönnskär study has been descriptive and analytic, and our ambition has been to meet a minimal standard of intersubjectivity, i.e. competent

### Table 2

**Ethical Costs and Ethical Benefits of a Study of the Working Environment of Rönnskär Carried Out without Individually Informed Consent Compared to a Similar Study with Informed Consent (The Latter Being Used as a Base-Line)**

| Persons Involved | Beneficence | Ethical principles | |
| --- | --- | --- | --- |
| | | Autonomy | Justice |
| Exposed workers | Benefits | Costs | No costs |
| Future workers | Benefits | Not relevant | Benefits |
| Community at large | Benefits | Not relevant | No costs |
| Epidemiologists | Benefits | No costs | Not relevant |
| Company management | Costs? | Costs? | Costs? |
| Trade union | Benefits | Costs? | Not relevant |

persons, asking the same questions and using similar methods, should also reach similar conclusions.[13] The choice of value premises and the identification of ethical costs and benefits to those involved and affected are, we believe, intersubjective in this sense. But when cost and benefits are to be balanced difficulties arise. All costs are not equally serious and all benefits are not equally important. In such situations analogies may be used. This can be illustrated by comparing the ethics of data utilisation in epidemiology and in journalism.[14]

Much of the available knowledge about the harmful effects of working environment derives from epidemiology, often using case-registers and record-linkage. The purpose of such studies is to prevent harm to unidentified individuals. But the large number of individuals investigated often makes it practically impossible to obtain individually informed consent. To require such consent would increase costs, lead to a higher attrition rate and create problems with selection bias. But without such consent there is infringement on personal autonomy. This means that epidemiology is almost inconceivable without some ethical costs.

In the European countries there is at present a trend towards increased protection for individual autonomy at the expense of benefits to the whole population. Legal controls over data collection have adversely affected the prospects for epidemiology, especially in countries with previously favourable conditions for epidemiological research, such as Sweden.

By contrast, journalists have been allowed far greater freedoms. One reason is that the aims and tasks of journalism — with its emphasis on an open society — are not compatible with strict adherence to the principles of respect for individual autonomy and of doing no harm. Hence infringements on both the principle of beneficence and the principle of autonomy by journalism are unavoidable ethical costs in an open society, which requires freedom of information.

But is this difference between journalism and epidemiology justifiable? One can argue that it is not. In epidemiology, which also aims at benefits to the open society, there is no need to harm research subjects. But it is not possible to carry out case-register research and record-linkage without some infringement on individual autonomy. Compared with the ethical costs of journalism, however, the ethical costs of epidemiology are very modest. So if it is justified to give priority to the principle of beneficence over the the principle of autonomy with reference to journalism it should also be accepted with reference to epidemiology.

This analysis suggests that to demand individual informed consent as a necessary prerequisite in case register research is not reasonable. The ethical cost of slight infringements of personal autonomy caused by the utilisation of case registers without individual informed consent is low compared to the ethical cost of conducting such research with individual informed consent. Our conclusion is that from an ethical point of view, the study carried out by Wall and Taube is preferable to a hypothetical study with individual informed consent.

---

[13] G. Hermerén, *Värdering och objektivitet* (Valuation and objectivity) (Lund: Studentlitteratur, 1972).

[14] C.-G. Westrin & T. Nilstun, "The Ethics of Data Utilization: A Comparison between Epidemiology and Journalism" (1994) 308 *British Medical Journal* 522.

## SOCIAL AND LEGAL ASPECTS

However, to defend an ethical position is one thing, whereas to deal with the practical research conditions is something quite different. In this respect the situation was for a long time paradoxical in epidemiological research. On the one hand, there was triumphal scientific progress combined with an increasing awareness among clinicians and policy-makers of the decisive importance of epidemiology. On the other hand, epidemiologists encountered increasing legal and social difficulties in carrying out their research.

In Sweden, which is a country with a long tradition of population statistics and strong community participation, there was also a trend toward less participation in epidemiological field studies involving personal interviews. Thus the average non-response rate in Sweden in health surveys, mainly scientific epidemiological studies, was about 14% during the 60s, but during the 70s and 80s about 21 and 22%.[15]

A possible explanation is the increased demands for participation in decision-making among the citizens, and a tendency towards more individualistic behavior. Another explanation, most alarming for epidemiology, may be a general increased distrust of research as such. Indications of such distrust appear in two surveys undertaken by Statistics Sweden. A representative sample of Swedes were asked whether they would be willing to be enrolled in registers comprising personal identification, civil status, number of children and income. In 1976 registration in research registers was accepted by 45% but in 1984 only by 30% and 1986 only 14%.[16] As a matter of fact, at that time fewer approved of registration in research registers than in the registers of the credit information agencies.

However in a survey in 1995, carried out at the Department of Social Medicine in Uppsala, the proportion of individuals who accepted a transfer of individual data to research registers has increased to almost 30%, i.e. twice as many as nine years ago. According to our preliminary analysis the higher-educated in the big cities have swung towards a more positive attitude while the attitude of the lower-educated rural population seems to be even more negative.[17]

As to the legal aspects, conditions have also improved for the epidemiologists of Europe. An earlier proposal of the European Commission for legal provisions on data protection implied far-reaching demands on individual informed consent for almost every occasion of utilisation of computerised health data. These restriction were so severe that they were described by a leading British epidemiologist as "a whiff of legal pedantry, as deadly as the hazards that epidemiologists will no longer

---

[15] J. Ählfeldt, T. Burns & C.-G. Westrin, "Participation in Public Health Research and General Health Screening Programs in Sweden 1950 to 1990" (Uppsala: Department of Social Medicine, University of Uppsala, 1992) (Unpublished manuscript).

[16] J. Ählfeldt, "The Citizens and the Research. A Questionnaire Study Concerning the Swedish Population Aged 18-74, Spring 1995" in *Health Data Registers. Report by the Committee on Health Data* (Stockholm: SOU, 1995).

[17] *Data and Integrity: The Knowledge and Attitudes of the General Public* (Stockholm: Swedish Statistics, 1995) (In Swedish).

be able to investigate".[18] The proposal was also criticised because of the inconsistency in exempting journalists from the prohibition but not scientists. The criticism led to important changes with a more reasonable balance between the necessary protection of individuals, on the one hand, and the pursuit of health related research, on the other.

A number of specific derogations are stipulated. Thus, for reasons of important public interest national legislation may permit the processing of personal health data without requiring individual consent. However, this limitation in application should be exercised with due respect for prevailing ethical principles and rules such as the ones included in the *Helsinki Declaration* and in the *International Guidelines for Ethical Review of Epidemiological Studies* published by CIOMS. Informed consent should normally be required and if an investigator proposes not to seek such consent he or she has the obligation to explain to the Ethical Committee how the study would nevertheless be ethical. It would also be possible to keep data in a form allowing identification of the data subject as long as it would be needed for scientific use if national legislation provides for safeguards.

## CONCLUDING REMARKS

So far we have tried to show that infringements on the right to individual informed consent often are ethically justified in longitudinal studies on large computerised case registers. From a purely ethical point of view this means that the right to autonomy is overridden by considerations related to beneficence and justice. But social and legal considerations are also relevant to epidemiological research, and how should these different requirements be balanced when in conflict?

We believe that the proposal of the European Commission has established a useful frame of reference. It allows for consideration, in appropriate order, of legal, ethical and social demands on research projects. The system may be described as a funnel-shaped system of different filters.

Firstly, there is the legal filter. This provides the general aims and directions and also some, non-negotiable, demands. These demands, if too rigid and restrictive, may preclude the comprehensive ethical and social examination which should take into consideration the many different components involved.

Secondly, there will be the ethical filter. The basic tenet is that all derogations from individual informed consent require justification. The reasons advanced must be presented to, examined and approved by a research ethical committee.

Thirdly, there is yet another filter, namely, the different social, psychological and political considerations which a research community and the individual researchers must impose on themselves. The purpose is to prevent difficult conflicts with other groups in the community. It is worth considering that organ transplant surgeons, even with the law and ethical demands behind themselves, often refrain from

---

[18] E.G. Knox, "Confidential Medical Records and Epidemiological Research" (1992) 304 *British Medical Journal* 727.

making use of donated organs in cases of severe resistance from the relatives, to prevent harmful attacks from the media. Similarly, epidemiologists may sometimes have to protect the reputation of their profession. In the perspective of social psychology the opportunity must sometimes get the upper hand of what is legally permissible and ethically right.

# BIBLIOGRAPHY

## BACKGROUND AND HISTORICAL SOURCES

### Background

Bankowski, Z. & Levine, R.J., eds. *Ethics and Research on Human Subjects: International Guidelines* (Geneva: CIOMS, 1993).

Baylis, F. *et al. Contemporary Health Care Ethics in Canada* (Canada: HBJ-Holt, 1995).

Beauchamp, T.L. & Walters, L., eds. *Contemporary Issues in Bioethics*, 4th ed. (Belmont, CA: Wadsworth Publishing, 1978).

Beauchamp, T.L. & Childress, J.F. *Principles in Biomedical Ethics* (New York: Oxford University Press, 1994).

Bernard, J. *La bioéthique* (Paris: Flammarion, 1994).

Connor, S.S. & Fuenzalida, H.L., eds. *Bioethics: Issues and Perspectives (Scientific Publication No. 527)* (Washington: Pan American Health Organization, 1990).

De Deyn, P.P., ed. *The Ethics of Animal and Human Experimentation* (London: John Libbey, 1994).

Delfosse, M.-L. *L'expérimentation médicale sur l'être humain. Construire les normes, construire l'éthique* (Brussels: De Boeck University, 1993).

Durand, G. *La bioéthique. Nature, principes, enjeux.* (Montréal: Cerf, 1989).

Edwards, R.B. & Graber, G.C. *Bioethics* (New York: Harcourt Brace Jovanovich, 1988).

Erwin, E., Gendin, S. & Kleiman, L., eds. *Ethical Issues in Scientific Research: An Anthology* (New York: Garland, 1994).

Fried, C. *Medical Experimentation: Personal Integrity and Social Policy* (New York: American Elsevier Company, 1974).

Freund, P.A., ed. *Experimentation With Human Subjects* (New York: George Braziller, 1970).

Gillon, R., ed. *Principles of Health Care Ethics* (New York: Wiley, 1994).

Gray, B.H. *Human Subjects in Medical Experimentation* (New York: Wiley-Interscience, 1975).

Homan, R. *The Ethics of Social Research* (London: Longman, 1991).

Howard-Jones, N. & Bankowski, Z., eds. *Medical Experimentation and the Protection of*

*Human Rights* (Geneva: CIOMS & The Sandoz Institute for Health and Socio-Economic Studies, 1979).

Katz, J. *Experimentation With Human Beings* (New York: Russell Sage Foundation, 1972).

Kluge, E. *Biomedical Ethics in a Canadian Context* (Scarborough: Prentice-Hall Canada, 1992).

Levine, R.J. *Ethics and Regulation of Clinical Research*, 2d ed. (Baltimore: Urban and Schwarzenberg, 1986).

Mason, J.K. & McCall Smith, R.A. *Law and Medical Ethics* (London: Butterworths, 1991).

McNeill, P.M. *The Ethics and Politics of Human Experimentation* (Cambridge: Cambridge University Press, 1993).

Meyers, D.W. *The Human Body and the Law* (Edinburgh: Edinburgh University Press, 1990).

Munson, R. *Intervention and Reflection. Basic Issues in Medical Ethics*, 4th ed. (Belmont, CA: Wadsworth Publishing, 1992).

Ramsey, P. *The Patient as Person* (New Haven: Yale University Press, 1970).

Roy, D.J., Williams, J.R. & Dickens, B.M. *Bioethics in Canada* (Scarborough: Prentice-Hall Canada, 1994).

Sharpe, G. *The Law and Medicine in Canada*, 2d ed. (Toronto: Butterworths, 1987).

Silverman, W.A. *Human Experimentation: A Guided Step into the Unknown* (Oxford: Oxford University Press, 1985).

Spicker, S.F. *et al.*, eds. *The Use of Human Beings in Research* (Boston: Kluwer Academic Publishers, 1988).

Veatch, R.M. *The Patient as Partner: A Theory of Human Experimentation Ethics* (Bloomington: Indiana University Press, 1987).

Wettstein, R.M. *Research Ethics and Human Subject Issues* (Washington, DC: American Psychiatric Association, 1995).

### *Historical Sources and Perspectives*

Advisory Committee on Human Radiation Expirements. *Final Report* (Washington, DC: U.S. Government Printing Office, 1995).

Annas, G.J. "The Changing Landscape of Human Experimentation: Nuremberg, Helsinki and Beyond" (1992) 2:2 Health Matrix 119–140.

Annas, G.J. & Grodin, M.A., eds. *The Nazi Doctors and the Nuremberg Code: Human Rights in Human Experimentation* (Oxford: Oxford University Press, 1992).

Annas, G.J. "The Nuremberg Code in U.S. Courts: Ethics Versus Expediency" in G.J. Annas & M.A. Grodin, eds. *The Nazi Doctors and the Nuremberg Code: Human Rights in Human Experimentation* (New York: Oxford University Press, 1992) 201–222.

Bower, T. *The Paperclip Conspiracy: The Hunt for Nazi Scientists* (Boston: Little, Brown, 1987).

Beauchamp, T.L. "Looking Back and Judging Our Predecessors" (1996) 6:3 Kennedy Institute of Ethics Journal 251–270.

Beecher, H.K. "Ethics and Clinical Research" (1966) 274:24 New England Journal of Medicine 1354–1360.

Berger, R.L. "Nazi Science: Comments on the Validation of the Dachau Human Hypothermia Experiments" in A.L. Caplan, ed. *When Medicine Went Mad: Bioethics and the Holocaust* (Totowa, NJ: Humana Press, 1992) 109–133, 337–340.

Blomquist, C.D. "From the *Oath of Hippocrates* to the *Declaration of Hawaii*" (1977) 4:3–4 Ethics in Science & Medicine 139–149.

Brackman, A. *The Other Nuremberg* (London: Collins, 1989).

Brandt, A.M. & Freidenfelds, L. "Research Ethics After World War II: The Insular Culture of Biomedicine (1996) 6:3 Kennedy Institute of Ethics Journal 239–243.

Buchanan, A. "Judging the Past: The Case of the Human Radiation Experiments" (1996) 26:3 Hastings Center Report 25–30.

Buchanan, A. "The Controversy over Retrospective Moral Judgment" (1996) 6:3 Kennedy Institute of Ethics Journal 245–250.

Campbell, C.S. "It Never Dies: Assessing the Nazi Analogy in Bioethics" (1992) 13:1 Journal of Medical Humanities 21–29.

Caplan, A.L. "How Did Medicine Go So Wrong? Is Moral Inquiry into Nazi Medical Crimes Immoral?" in A.L. Caplan, ed. *When Medicine Went Mad: Bioethics and the Holocaust* (Totowa, NJ: Humana Press, 1992) 53–92, 334–335.

Caplan, A.L. "The Doctors' Trial and Analogies to the Holocaust in Contemporary Bioethical Debates" in G.J. Annas & M.A. Grodin, eds. *The Nazi Doctors and the Nuremberg Code: Human Rights in Human Experimentation* (New York: Oxford University Press, 1992) 258–275.

Caplan, A.L. "Twenty Years After. The Legacy of the Tuskegee Syphilis Study. When Evil Intrudes" (1992) 22:6 Hastings Center Report 29–32.

Caplan, A.L. *When Medicine Went Mad: Bioethics and the Holocaust* (Totowa, NJ: Humana Press, 1992).

Clouser, K.D. "Historical Relativism in Bioethics: Can We Judge the Standards and Conduct of Those Who Preceded Us? (1994) 5:1 APA Newsletter on Philosophy & Medicine 124–126.

Drobniewski, F. "Why Did Nazi Doctors Break Their 'Hippocratic' Oaths?" (1993) 86:9 Journal of the Royal Society of Medicine 541–543.

Edgar, H. "Twenty Years After. The Legacy of the Tuskegee Syphilis Study. Outside the Community" (1992) 22:6 Hastings Center Report 32–35.

Freedman, B. "Moral Analysis and the Use of Nazi Experimental Results" in A.L. Caplan, ed. *When Medicine Went Mad: Bioethics and the Holocaust* (Totowa, NJ: Humana Press, 1992) 141–154, 340–343.

Gamble, V.N. "A Legacy of Distrust: African Americans and Medical Research" (1993) 9:S6 American Journal of Preventive Medicine 35–38.

Geiger, J.H. "Medicine, Public Health, and the Lurid Ethics of the Cold War" (1994) 1:1 Medicine & Global Survival 33–36.

Gellhorn, A. "Medical Ethics in the Modern World" in N. Howard-Jones & Z. Bankowski, eds. *Medical Experimentation and the Protection of Human Rights* (Geneva: CIOMS & The Sandoz Institute for Health and Socio-Economic Studies, 1979) 6–12.

Glantz, L.H. "The Influence of the Nuremberg Code on U.S. Statutes and Regulations" in G.J. Annas & M.A. Grodin, eds. *The Nazi Doctors and the Nuremberg Code: Human Rights in Human Experimentation* (New York: Oxford University Press, 1992) 183–200.

Greene, V.W. "Can Scientists Use Information Derived From the Concentration Camps? Ancient Answers to New Questions" in A.L. Caplan, ed. *When Medicine Went Mad: Bioethics and the Holocaust* (Totowa, NJ: Humana Press, 1992) 155–170, 343.

Grodin, M.A. "Historical Origins of the Nuremberg Code" in G.J. Annas & M.A. Grodin, eds. *The Nazi Doctors and the Nuremberg Code: Human Rights in Human Experimentation* (New York: Oxford University Press, 1992) 121–144.

Hoedemen, P. *Hitler or Hippocrates: Medical Experiments and Euthanasia in the Third Reich*, trans. Ralph de Rijke (Sussex: The Book Guild, 1991).

Howard-Jones, N. "Human Experimentation in Historical and Ethical Perspectives" (1982) 16:20 Social Science & Medicine 1429–1448.

Institut National de la Santé de de la Recherche Médicale. *Histoire de la recherche biomédicale et droits de l'homme* (Paris: INSERM, 1990).

Institute of History of Medicine and Medical Research. *Theories and Philosophies of Medicine*, 2d ed. (New Delhi: Institute of History of Medicine and Medical Research, 1973).

Ivy, A.C. "The History and Ethics of the Use of Human Subjects in Medical Experiments" (1948) 108 Science 1–5.

Jones, J.H. *Bad Blood* (New York: Free Press, 1981).

Jones, J.H. "The Tuskegee Legacy: AIDS and the Black Community" (1992) 22:6 Hastings Center Report 38–40.

Jonsen, A.R. "The Birth of Bioethics" (1993) 23:S6 Hastings Center Report S1-S4.

Kater, M. *Doctors Under Hitler* (Chapel Hill, NC: University of North Carolina Press, 1989).

Katz, J. "Abuse of Human Beings for the Sake of Science" in A.L. Caplan, ed. *When Medicine Went Mad: Bioethics and the Holocaust* (Totowa, NJ: Humana Press, 1992) 233–270, 347–348.

Katz, J. ""Ethics and Clinical Research" Revisited: A Tribute to Henry K. Beecher" (1993) 23:5 Hastings Center Report 31–39.

Katz, J. "Reflections on Unethical Experiments and the Beginnings of Bioethics in the United States" (1994) 4:2 Kennedy Institute of Ethics Journal 85–92.

Katz, J. "The Consent Principle of the Nuremberg Code: Its Significance Then and Now" in G.J. Annas & M.A. Grodin, eds. *The Nazi Doctors and the Nuremberg Code: Human Rights in Human Experimentation* (New York: Oxford University Press, 1992) 227–239.

Katz, J. & Pozos, R.S. "The Dachau Hypothermia Study: An Ethical and Scientific Commentary" in A.L. Caplan, ed. *When Medicine Went Mad: Bioethics and the Holocaust* (Totowa, NJ: Humana Press, 1992) 135–139, 340.

Kor, E.M. "Nazi Experiments as Viewed by a Survivor of Mengele's Experiments" in A.L. Caplan, *When Medicine Went Mad: Bioethics and the Holocaust* (Totowa, NJ: Humana Press, 1992) 3–8.

Lederer, S. & Grodin, M.A. "Historical Overview: Pediatric Experimentation" in M.A. Grodin & L. Glantz, ed. *Children as Research Subjects: Science, Ethics, and Law* (New York: Oxford University Press, 1994) 3–28.

Lifton, R.J. *The Nazi Doctors: Medical Killing and the Psychology of Genocide* (New York: Basic Books, 1986).

Macklin, R. "Universality of the Nuremberg Code" in G.J. Annas & M.A. Grodin, eds. *The Nazi Doctors and the Nuremberg Code: Human Rights in Human Experimentation* (New York: Oxford University Press, 1992) 240–257.

McCally, M., Cassel, C. & Kimball, D.G. "U.S. Government-Sponsored Radiation Research on Humans 1945–1975" (1994) 1:1 Medicine and Global Survival 4–17.

McCarthy, C.R. "Historical Background of Clinical Trials Involving Women and Minorities" (1994) 69:9 Academic Medicine 695–698.

Mitscherlich, A. & Mielke, F. *Doctors of Infamy* (New York: Henry Schuman, 1949).

Moreno, J.D. & Lederer, S.E. "Revising the History of Cold War Research Ethics" (1996) 6:3 Kennedy Institute of Ethics Journal 223–237.

Mostow, P. ""Like Building on Top of Auschwitz": On the Symbolic Meaning of Using Data From the Nazi Experiments, and on Nonuse as a Form of Memorial" (1993–1994) 10:2 Journal of Law & Religion 403–431.

Muller-Hill, B. *Murderous Science: Elimination by Scientific Selection of Jews, Gypsies and Others, Germany 1933–1945* (Toronto: Oxford University Press, 1988).

O'Reilly, M. "Nazi Medicine: 'The Perversion of the Noblest Profession'". (1993) 148:5 Canadian Medical Association Journal 819–821.

Pappworth, M.H. "Human Guinea Pigs" (1990) 301:6766 British Medical Journal 1456–1460.

Perley, S. et al. "The Nuremberg Code: An International Overview" in G.J. Annas & M.A. Grodin, eds. *The Nazi Doctors and the Nuremberg Code: Human Rights in Human Experimentation* (New York: Oxford University Press, 1992) 149–173.

Petersson, B. "The Mentally Retarded as Research Subjects: A Research Ethics Study of the Vipeholm Investigations of 1945–1955" in M. Hallberg, ed. *Ideal and Reality: Applying Ethics in Theory and Practice* (Goteborg, Sweden: Centre for Research Ethics, 1993) 1–31.

Post, S.G. "Nazi Data and the Rights of Jews" (1988) 6:2 The Journal of Law and Religion 429–433.

Pozos, R.S. "Scientific Inquiry and Ethics: The Dachau Data" in A.L. Caplan, ed. *When Medicine Went Mad: Bioethics and the Holocaust* (Totowa, NJ: Humana Press, 1992) 95–108, 335–337.

Proctor, R.N. "Nazi Doctors, Racial Medicine, and Human Experimentation" in G.J. Annas & M.A. Grodin, eds. *The Nazi Doctors and the Nuremberg Code: Human Rights in Human Experimentation* (New York: Oxford University Press, 1992) 17–31.

Proctor, R.N. *Racial Hygiene: Medicine Under the Nazis* (Cambridge, Mass: Harvard University Press, 1988).

Pross, C. "Nazi Doctors, German Medicine, and Historical Truth" in G.J. Annas & M.A. Grodin, eds. *The Nazi Doctors and the Nuremberg Code: Human Rights in Human Experimentation* (New York: Oxford University Press, 1992) 32–52.

Reiser, S.J., Dyck, A.J. & Curran, W.J., eds. *Ethics in Medicine: Historical Perspectives and Contemporary Concerns* (Cambridge, Mass: MIT Press, 1977).

Rothman, D.J. "Ethics and Human Experimentation: Henry Beecher Revisited" (1987) 317:19 The New England Journal of Medicine 1195–1199.

Rothman, D.J. "Human Experimentation and the Origins of Bioethics in the United States" in G. Weisz, ed. *Social Science Perspectives on Medical Ethics* (Boston: Kluwer Academic Publishers, 1990) 185–200.

Rothman, D.J. "Radiation" 276:5 JAMA. 421–423.

Rothman, D.J. *Strangers at the Bedside: A History of How Law and Bioethics Transformed Medical Decision Making* (New York: Basic Books, 1991).

Rothman, D.J. & Rothman, S.M. *The Willowbrook Wars* (New York: Harper & Row, 1984).

Roy, B. "The Tuskegee Syphilis Experiment: Biotechnology and the Administrative State" (1995) 87:1 Journal of the National Medical Association 56–66.

Roy, B. "The Tuskegee Syphilis Experiment: Medical Ethics, Constitutionalism, and Property in the Body" (1994) 5:1 APA Newsletter on Philosophy & Medicine 11–15.

Toulmin, S. "Medical Ethics in its American Context: An Historical Survey" in D. Callahan & G.R. Dunston, eds. *Biomedical Ethics: An Anglo-American Dialogue* (New York: Annals of the New York Academy of Sciences, 1988) 7–15.

Vaux, K.L. & Schade, S.G. "The Search for Universality in the Ethics of Human Research: Andrew C. Ivy, Henry K. Beecher, and the Legacy of Nuremberg" in S.F. Spicker *et al.*, eds. *The Use of Human Beings in Research* (Boston: Kluwer Academic Publishers, 1988) 3–16.

Veatch, R.M., ed. *Cross Cultural Perspectives in Medical Ethics: Readings* (Boston: Jones & Bartlett, 1989).

Vigorito, S.S. "A Profile of Nazi Medicine: The Nazi Doctor -- His Methods and Goals" in A.L. Caplan, *When Medicine Went Mad: Bioethics and the Holocaust* (Totowa, NJ: Humana Press, 1992) 9–13.

Wikler, D. & Barondess, J. "Bioethics and Anti-Bioethics in Light of Nazi Medicine: What Must We Remember?" (1993) 3:1 Kennedy Institute of Ethics Journal 39–55.

Williams, P. & Wallace, D. *Unit 731: The Japanese Army's Secret of Secrets* (London: Hodder & Stughton, 1989).

Yeide, H. "Research Objectives and the Social Structuring of the Research Enterprise: An Historical and Ethical Perspective" in V.L. Melnick & N.N. Dubler, eds. *Alzheimer's Dementia: Dilemmas in Clinical Research* (Clifton, NJ: Humana Press, 1985) 79–97.

## PHILOSOPHICAL PERSPECTIVES

Ackerman, T.F. "Experimentalism in Bioethics Research" (1983) 8:2 Journal of Medicine & Philosophy 169–180.

Ackerman, T.F. "Medical Research, Society and Health Care Ethics" in R. Gillon, ed. *Principles of Health Care Ethics* (New York: Wiley, 1994) 873–884.

Batchelor, J.A. & Briggs, C.M. "Subject, Project or Self? Thoughts on Ethical Dilemmas for Social and Medical Researchers" (1994) 39:7 Social Science & Medicine 949–954.

Baudouin, J.L. "L'expérimentation sur les humains: un conflit de valeurs" (1981) 26:4 McGill Law Journal 809–846.

Bayley, C. "Our World Views (May Be) Incommensurable: Now What?" (1995) 20:3 Journal of Medicine & Philosophy 271–284.

Brody, B.A. "Autonomy and Paternalism – Some Value Problems; A Utilitarian Perspective; A Deontological Perspective" in B.A. Brody, *Ethics and its Applications* (New York: HBJ, 1983) 159–198.

Brody, B.A. *Ethics and its Applications* (New York: HBJ, 1983).

Brody, B.A. *Moral Theory and Moral Judgments in Medical Ethics* (Boston: Kluwer Academic Publishers, 1988).

Browne, A. "Morality and Medical Experimentation" in W.L. Robison & M.S. Pritchard, eds. *Medical Responsiblity: Paternalism, Informed Consent, and Euthanasia* (Clifton, NJ: Humana Press, 1979) 101–112.

Caplan, A.L. "Is There an Obligation to Participate in Biomedical Research?" in S.F. Spicker *et al.*, eds. *The Use of Human Beings in Research* (Boston: Kluwer Academic Publishers, 1988) 229–248.

Capron, A.M. "Human Experimentation" in R.M.Veatch, ed. *Medical Ethics* (Boston: Jones and Bartlett, 1989) 125–172.

Charo, R.A. "Principles and Pragmatism" (1996) 6:3 Kennedy Institute of Ethics Journal 319–322.

Childress, J.F. & Fletcher, J.C. "Individualism and Community: The Contested Terrain of Autonomy – Respect for Autonomy" (1994) 24:3 Hastings Center Report 34–35.

Clouser, K.D. & Gert, B. "A Critique of Principlism" (1990) 15:2 Journal of Medicine & Philosophy 219–236.

Cohen, C. "Moral Issues in Medical Experimentation on Humans" (1979) 2:5 Philosophic Exchange 37–51.

Darvall, L. "Autonomy and Protectionism: Striking a Balance in Human Subject Research Policy and Regulation" in K. Petersen, ed. *Law and Medicine. Law in Context.* (Victoria: Bundoora, 1994) 82–96.

Delkeskamp-Hayes, C. "Moral Appropriateness in Human Research (A Practical Suggestion and a Theoretical Interpretation)" in S.F. Spicker *et al.*, eds. *The Use of Human Beings in Research* (Boston: Kluwer Academic Publishers, 1988) 91–101.

Edsall, G. "A Positive Approach to the Problem of Human Experimentation" (1969) 98:2 Daedalus 463–479.

Elliott, C. "Doing Harm: Living Organ Donors, Clinical Research and *The Tenth Man*" (1995) 21:2 Journal of Medical Ethics 91–96.

Fagot-Largeault, A.M. "Epistemological Presuppositions Involved in the Programs of Human Research" in S.F. Spicker *et al.*, eds. *The Use of Human Beings in Research* (Boston: Kluwer Academic Publishers, 1988) 161–189.

Fethe, C. "Beyond Voluntary Consent: Hans Jonas on the Moral Requirements of Human Experimentation" (1993) 19:2 Journal of Medical Ethics 99–103.

Finsen, S. "Sinking the Research Lifeboat" (1988) 13:2 Journal of Medicine & Philosophy 197–212.

Furness, S.H. "Medical Ethics, Kant and Mortality" in R. Gillon, ed. *Principles of Health Care Ethics* (New York: Wiley, 1994) 159–171.

Glick, S.M. "Research in a Hierarchy of Values" (1992) 59:2 Mount Sinai Journal of Medicine 102–107.

Gray, B.H. & Osterweis, M. "Ethical Issues in a Social Context" in L.H. Aiken & D. Mechanic, eds. *Applications of Social Science to Clinical Medicine and Health Policy* (New Brunswick, NJ: Rutgers University Press, 1986) 543–564.

Gustafson, J.M. "Moral Discourse about Medicine: A Variety of Forms" (1990) 15:2 Journal of Medicine & Philosophy 125–142.

Holm, S. "Moral Reasoning in Biomedical Research Protocols" (1994) 22:2 Scandinavian Journal of Social Medicine 81–85.

Huibers, A.K. "Kant's Pain: Ethical Considerations on Medical Experiments in the General Practice" in P.P. De Deyn, ed. *The Ethics of Animal and Human Experimentation* (London: John Libbey, 1994) 259–268.

Jonas, H. *Philosophical Reflections on Experimenting With Human Subjects* (New York: George Braziller, 1970).

Lebacqz, K. "Justice and Human Research" in E.E. Shelp, ed. *Justice and Health Care* (Boston: D. Reidel, 1981) 179–191.

LeCourt, D. "Ethics, Politics, and Medical Obligations in Biomedical Research: The Philosophical Point of View" (1992) 8:5 AIDS Research & Human Retroviruses 853–858.

Lieberman, J.A. & Sloan, J. "The Moral Imperatives of Medical Research in Human Subjects" (1994) 5:1 APA Newsletter on Philosophy & Medicine 40-42.

Loewy, E.H. "Kant, Health Care and Justification" (1995) 16:2 Theoretical Medicine 215–222.

Lurie, N. & Shapiro, M. "Is the Pursuit of Scientific Truth Always the Greatest Good? Ethical Issues in Health Care Research" (1987) 35:6 Clinical Research 517-520.

Macklin, R. "Disagreement, Consensus, and Moral Integrity" (1996) 6:3 Kennedy Institute of Ethics Journal 289–311.

Macklin, R. & Sherwin, S. "Experimenting on Human Subjects: Philosophical Perspectives" (1975) 25:3 Case Western Law Review 434-471.

Mainetti, J.A. "Bioethics: A New Health Philosophy" in S.S. Connor & H.L. Fuenzalida, eds. *Bioethics: Issues and Perspectives (Scientific Publication No. 527)* (Washington: Pan American Health Organization, 1990) 208–212.

May, W.F. "Code and Covenant or Philanthropy and Contract?" in S.J. Reiser, A.J. Dyck & W.J. Curran, eds. *Ethics in Medicine: Historical Perspectives and Contemporary Concerns* (Cambridge, Mass.: MIT Press, 1977) 65–76.

McCullough, L.B. "Preventive Ethics, Professional Integrity, and Boundary Setting: The Clinical Management of Moral Uncertainty" (1995) 20:1 Journal of Medicine & Philosophy 1–11.

McCullough, L.B. & Wear, S. "Respect for Autonomy and Medical Paternalism Reconsidered" (1985) 6:3 Theoretical Medicine 295–308.

Meslin, E.M. *et al.* "Prinicpalism and the Ethical Appraisal of Clinical Trials" (1995) 9:5 Bioethics 399-418.

Minogue, B.P. *et al.* "Individual Autonomy and the Double-Blind Controlled Experiment: The Case of Desperate Volunteers" (1995) 20:1 Journal of Medicine & Philosophy 43-55.

Morison, R.S. "The Biological Limits on Autonomy" (1984) 14:5 Hastings Center Report 43-49.

Moros, D.A. "The Philosophy of Medicine: Clinical Science and its Ethics" (1987) 31:1 Perspectives in Biology & Medicine 134–150.

Mosher, D.L. & Bond, S.B. "Ethics -- Perceived or Reasoned From Principles?: A Rejoinder to Korn, Huelsman, and Reed" (1992) 2:3 Ethics & Behavior 203–214.

National Academy of Sciences, "Experiments and Research With Humans: Values in Conflict" (Washington, DC: National Academy of Sciences, 1975).

Nelkin, D. "Value Conflict and Social Controls Over Research" in H. Wechsler, R.W.

Lamont-Havers & G.F. Cahill, eds. *The Social Context of Medical Research* (Cambridge, Mass: Ballinger, 1981) 3–25.

Nicholas, J.M., ed. *Moral Priorities in Medical Research: The Second Hannah Conference* (Toronto: The Hannah Institute for the History of Medicine, 1984).

Nicholson, R.H. "Paternalism No Problem" (1992) 22:2 Hastings Center Report 4-5.

Ogletree, T.W. "Values, Obligations, and Virtues: Approaches to Bio-Medical Ethics" (1976) 4:1 Journal of Religious Ethics 105–130.

O'Neill, O. "Paternalism and Partial Autonomy" (1984) 10:4 Journal of Medical Ethics 173–178.

Ottenberg, P. "Dehumanization and Human Experimentation" in J.C. Schoolar & C.M. Gaitz, eds. *Research and the Psychiatric Patient* (New York: Brunner/Mazel,1975) 87–103.

Pardes, H. "Defending Humanistic Values" (1990) 147:9 American Journal of Psychiatry 1112–1119.

Parsons, T. "Research With Human Subjects and the æProfessional Complex'" (1969) 98:2 Daedalus 325–360.

Pellegrini, R.J. "Ethics and Identity: A Note on the Call to Conscience" (1972) 27:9 American Psychologist 896–897.

Pellegrino, E.D. "Beneficience, Scientific Autonomy and Self-Interest: Ethical Dilemmas in Clinical Research" (1992) 1:4 Cambridge Quarterly of Health Care Ethics 361–369.

Pfaltz, C.R. "The Impact of State and Society on Medical Research" (1990) 35 Progress in Drug Research 9–23.

Reiman, J. "Utilitarianism and the Informed Consent Requirement" in B.A. Brody, ed. *Moral Theory and Moral Judgments in Medical Ethics* (Boston: Kluwer Academic Publishers, 1988) 41-51.

Riis, P. "Clinical Freedom: Patients' and Physicians' Autonomy Versus the Demands of Society" in P. Allebeck & B. Jansson, eds. *Ethics in Medicine: Individual Integrity Versus Demands of Society* (New York: Raven Press, 1990) 103–113.

Ruddick, W. "Utilitarians among the Optimists" in B.A. Brody, ed. *Moral Theory and Moral Judgments in Medical Ethics* (Boston: Kluwer Academic Publishers, 1988) 33–39.

Schafer, A. "Experimentation With Human Subjects: A Critique of the Views of Hans Jonas" (1983) 9:2 Journal of Medical Ethics 76–79.

Schafer, A. "The Moral Autonomy and Ethical Pathology of the Randomized Clinical Trial" in P.P. De Deyn, ed. *The Ethics of Animal and Human Experimentation* (London: John Libbey, 1994) 269–276.

Sjoberg, G. & Vaughan, T.R. "A Moral Context for Social Research" (1983) 13:2 Hastings Center Report 44-46.

Sumner, L.W. "Utilitarian Goals and Kantian Constraints" in B.A. Brody, ed. *Moral Theory and Moral Judgments in Medical Ethics* (Boston: Kluwer Academic Publishers, 1988) 15–31.

Tabak, N. "Medical Experimentation on Humans: A Moral Point of View" (1992) 11:5–6 Medicine & Law 353–362.

Tannsjo, T. "The Morality of Clinical Research – A Case Study" (1994) 19:1 Journal of Medicine & Philosophy 7–21.

Thier, S.O. "Ethics, Physicians, and Public Policy" (1992) 40:4 Journal of the American Geriatrics Society 417-420.

Tranøy, K.E. "Ethical Problems of Scientific Research : An Action-Theoretic Approach" (1996) 79:2 The Monist 183–196.

Vermeersch, E. "Ethical-Philosophical Aspects of Human and Animal Experimentation" in P.P. De Deyn, ed. *The Ethics of Animal and Human Experimentation* (London: John Libbey, 1994) 3–12.

Weisstub, D.N. "Roles and Fictions in Clinical and Research Ethics" (1996) 4 Health Law Journal 259–282.

Whitbeck, C. "Ethics as Design. Doing Justice to Moral Problems" (1996) 26:3 Hastings Center Report 9–16.

Winick, B.J. "On Autonomy: Legal and Psychological Perpsectives" (1992) 37:6 Villanova Law Review 1705–1777.

## REGULATION OF THE RESEARCH ENDEAVOUR

### *General Ethical and Legal Requirements*

Ackerman, T.F. "Balancing Moral Principles in Federal Regulations on Human Research" (1992) 14:1 IRB 1–6.

American Psychological Association. "Ethical Principles of Psychologists and Code of Conduct" (1992) 47:12 American Psychologist 1597–1611.

Bankowski, Z. "International Ethical Considerations for Research on Human Subjects" in D. Cheney, ed. *Ethical Issues in Research* (Frederick, MD: University Publishing Group, 1993) 177–188.

Benson, P.R. & Roth, L.H. "Trends in the Social Control of Medical and Psychiatric Researchin D.N. Weisstub, ed. *Law and Mental Health: International Perspectives, Volume 4* (New York: Pergamon Press, 1988) 1.

Bulger, R.E. "Toward a Statement of the Principles Underlying Responsible Conduct in Biomedical Research" (1994) 69:2 Academic Medicine 102–107.

Byk, C. "The European Convention on Bioethics" (1993) 19:1 Journal of Medical Ethics 13–16.

Caplan, A.L. "Are Existing Safeguards Adequate?" (1994) 5:1 APA Newsletter on Philosophy & Medicine 36–38.

Capron, A.M. "Research Ethics and the Law" in K. Berg & K.E. Tranøy, eds. *Research Ethics*. (New York: Alan R. Liss, 1983) 13–23.

Carriero, A., Spinazzi, A. & Bonomo, L. "Ethics and Research" (1996) 6:2 European Radiology S11-S15.

Chen, M., Scarth, B.J. & Slack, C.J. "Good Clinical Research Practice in Canada" (1994) 151:9 Canadian Medical Association Journal 1255–1257.

Christakis, N.A. "Ethics are Local: Engaging Cross-Cultural Variation in the Ethics for Clinical Research" (1992) 35:9 Social Science & Medicine 1079–1091

Council for International Organizations of Medical Science, in collaboration With the World Health Organization, *International Ethical Guidelines for Biomedical Research Involving Human Subjects* (Geneva: CIOMS, 1993).

Curl, R.M. "Adherence to the Legal and Ethical Requirements of Human Research" (1982) 5:4 Behavior Therapist 123–128.

Curran, C.E. "Ethical Considerations in Human Experimentation" (1975) 13:4 Duquesne Law Review 819–840.

Curran, W.J. "Governmental Regulation of the Use of Human Subjects in Medical Research: The Approach of Two Federal Agencies" (1969) 98:2 Daedalus 542-594.

Deutsch, E. "Medical Experimentation: International Rules and Practice" (1989) 19:1 Victoria University of Wellington Law Review 1–10.

Dougherty, C.J. "Criteria for Morally Acceptable Research With Human Subjects" in D. Lamb, T. Davies & M. Roberts, eds. *Explorations in Medicine*, vol. 1 (Brookfield, VT: Gower, 1987) 1–21.

Dworkin, R.B. *Limits: The Role of the Law in Bioethical Decision Making* (Bloomington: Indiana University Press, 1996).

Elizalde, J. "Bioethics as a New Human Rights Emphasis in European Research Policy" (1992) 2:2 Kennedy Institute of Ethics Journal 159–170.

Freund, P.A. "Legal Frameworks for Human Experimentation" (1969) 98:2 Daedalus 314–324.

Friedman, P.J. "A Last Call for Self-Regulation of Biomedical Research" (1989) 64:9 Academic Medicine 502-504.

Garfield, S.L. "Ethical Issues in Research on Psychotherapy" (1987) 31:2 Counseling & Values 115–125.

Goldner, J.A. "An Overview of Legal Controls on Human Experimentation and the Regulatory Implications of Taking Professor Katz Seriously" 38:1 Saint Louis University Law Journal 63–134.

Griffin, A. & Sechzer, J.A. "Mandatory Versus Voluntary Regulation of Biomedical Research" (1983) 406 Annals of the New York Academy of Sciences 187–200.

Helmchen, H. "Ethical and Practical Problems in Therapeutic Research in Psychiatry" (1982) 23:6 Comprehensive Psychiatry 505-515.

Hodges, C. "Harmonization of European Controls over Research: Ethics Committees, Consent, Compensation and Indemnity" in A. Goldberg & I. Dodds-Smith, eds. *Pharmaceutical Medicine and the Law* (London: RCP, 1991).

Hoffmaster, B. "The Medical Research Council's New *Guidelines on Research Involving Human Subjects:* Too Much Law, Too Little Ethics" (1989) 10:1 Health Law in Canada 146–155.

Howell, T. & Sack, R.L. "The Ethics of Human Experimentation in Psychiatry: Toward a More Informed Consensus" (1981) 44:2 Psychiatry 113–132.

Hughes, J. "Legal and Ethical Implications of Clinical Research on Human Subjects" (1991) 22 The Cambrian Law Review 5–25.

Idanpaan-Heikkila, J.E. "WHO Guidelines for Good Clinical Practice (GCP) for Trials on Pharmaceutical Products: Responsibilities of the Investigator" (1994) 26:2 Annals of Medicine 89–94.

Kass, N.E. & Sugarman, J. "Are Research Subjects Adequately Protected? A Review and Discussion of Studies Conducted by the Advisory Committee on Human Radiation Experiments" (1996) 6:3 Kennedy Institute of Ethics Journal 271–282.

Katz, J. "Human Experimentation and Human Rights" (1993) 38:1 Saint Louis University Law Journal 7-54.

Kouri, R.P. "The Law Governing Human Experimentation in Québec". (1991) 22:1 Revue de Droit, Université de Sherbrooke 77–108.

Kozlowski, S. "Ethical Problems of Biomedical Research Involving Human Subjects" in D.J. Roy *et al.*, eds. *Medicine, Ethics, and Law: Canadian and Polish Perspectives* (Montreal: Center for Bioethics, Clinical Research Institute of Montreal, 1991) 353–360.

Ladd, J. "Ethical Issues in Human Experimentation" in Z. Bankowski & J.C. Bernardelli, eds. *Medical Ethics and Medical Education* (Geneva: CIOMS, 1981) 26-42.

Laufer, S. "The Regulation of Medical/Scientific Research Practices Involving Experimentation on Human Beings" (1990) 8 Law in Context 78–105.

Lawson, C. "Research Participation as a Contract" (1995) 5:3 Ethics & Behavior 205–215.

Lesneski, G.J. & Adler, B.S. "Ethical and Legal Dilemmas in Medical Research" (1992) 144 New Jersey Lawyer 37-42.

Levine, C. "AIDS and the Ethics of Human Subjects Research" in F.C. Reamer, ed. *AIDS and Ethics* (New York: Columbia University Press, 1991) 77–104.

Levine, C. "Building a New Consensus: Ethical Principles and Policies for Clinical Research on HIV/AIDS" (1991) 13:1–2 IRB 1–17.

Lieberman, J.A. "Ethical Dilemmas in Clinical Research With Human Subjects: An Investigator's Perspective" (1996) 32:1 Psychopharmacology Bulletin 19–25.

Lynch, D. & Graves, W. "Participants' Perceptions of Ethical Issues in Research With Humans" (1983) 52:1 Psychological Reports 231–238.

Maloney, D. M. *Protection of Human Research Subjects: A Practical Guide to Federal Laws and Regulations* (New York: Plenum Press, 1984).

Marke, J.J. "Human Experimentation and the Law" (1985) 194:4 New York Law Journal 41.

Medical Research Council of Canada. *Guidelines on Research Involving Human Subjects* (Ottawa: Supply & Services Canada, 1987).

Medical Research Council of Canada. *MRC Report No. 6: Ethics in Human Experimentation* (Ottawa: Supply & Services Canada, 1978).

Mehlman, M.J. "Fiduciary Contracting: Limitations on Bargaining Between Patients and Health Care Providers" (1990) 51:2 University of Pittsburgh Law Review 365-418.

Miller, J. & Crigger, B.-J. "Ethical Standards for Human Subject Research in Developing Countries"(1992) 14:3 IRB 7–8.

National Commission for the Protection of Human Subjects of Biomedical and Behavioural Research. *The Belmont Report: Ethical Principles for the Protection of Human Subjects of Research* (Washington DC: U.S. Government Printing Office, 1978).

National Council for Hospice and Specialist Palliative Care Services (Great Britain), "Guidelines on Research in Palliative Care (Official Statement)" (1995) 110 Institute of Medical Ethics Bulletin 19–20.

National Council on Bioethics in Human Research. *National Council on Bioethics in Human Research* (Ottawa: NCBHR, 1989).

National Health and Medical Research Council. *Statement on Human Experimentation and Supplementary Notes* (Canberra: NHMRC, 1992).

National Institutes of Health Intramural Research Program & Office of Human Subjects Research, *Guidelines for the Conduct of Research Involving Human Subjects at the National Institutes of Health* (Bethesda, MD: National Institutes of Health, 1995).

National Research Council of Canada. *Research Involving Human Subjects: Guidelines for Institutes* (Ottawa: NRC, 1995).

Norway. Ministry of Health and Social Affairs. Ethics Committee, *Man and Biotechonolgy: Report of the Ethics Committee [Summary]* (Oslo: The Ministry, 1991).

Nyapidi, T.J. "Legal Aspects of Medical Research and the Way Forward" (1995) 41:2 Central African Journal of Medicine 73–76.

Prentice, E.D. "Bill of Rights for Research Subjects" (1993) 15:2 IRB 7–9.

President's Commission for the Study of Ethical Problems in Medicine and Biomedical and Behavioural Research. *Protecting Human Subjects: The Adequacy of Federal Rules and Their Implementation* (Washington, DC: U.S. Government Printing Office, 1981).

President's Commission for the Study of Ethical Problems in Medicine and Biomedical and Behavioural Research. *Summing Up: Final Report on Studies of the Ethical and Legal Problems in Medicine and Biomedical and Behavioural Research* (Washington, DC: U.S. Government Printing Office, 1983).

President's Commission for the Study of Ethical Problems in Medicine and Biomedical and Behavioural Research. *Implementing Human Research Legislation: Second Biennial Report on the Adequacy and Uniformity of Federal Rules and Policies, and Their Implementation, for the Protection of Human Subjects* (Washington, DC: U.S. Government Printing Office, 1983).

Riis, P. "Medical Ethics in the European Community" (1993) 19:1 Journal of Medical Ethics 7–12.

Rolleston, F.S. "Revision of the Medical Research Council of Canada Guidelines" in D.J. Roy *et al*, eds. *Medicine, Ethics, and Law: Canadian and Polish Perspectives* (Montreal: Center for Bioethics, Clinical Research Institute of Montreal, 1991) 343–351.

Royal College of Physicians of London. *Research Involving Healthy Volunteers* (London: RCP, 1986)

Royal College of Physicians of London. *Research Involving Patients* (London: RCP, 1990).

Sieber, J.E. "Ethical Considerations in Planning and Conducting Research on Human Subjects" (1993) 68:S9 Academic Medicine S9-S13.

Smith, H.L. "Ethical Considerations in Research Involving Human Subjects" (1979) 6:3 Ethics in Science & Medicine 167–175.

South African Medical Research Council, *Guidelines on Ethics for Medical Research* (Tygerberg, South Africa: The Council, 1993).

Taylor, J.L. "Ethical and Legal Aspects of Non-Therapeutic Clinical Investigations" (1975) 43:2 Medico-Legal Journal 53–68.

Teff, H. "The Law and Ethics and Medical Experimentation" (1987) 3:6 Professional Negligence 182–186.

Tri-Council Working Group, *Code of Conduct for Research Involving Human Subjects* (Ottawa: Tri-Council Working Group, 1997).

Vennell, M.A. "Medical Research and Treatment: Ethical Standards in the International Context" (1995) 2:1 Medical Law International 1–21.

Verdun-Jones, S.N. & Weisstub, D.N. "The Regulation of Biomedical Research Experimentation in Canada: Developing an Effective Apparatus for the Implementation of Ethical Principles in a Scientific Milieu" (1996–97) 28 Ottawa Law Review 297–341.

Voges, M.A. "Human Experimentation: Medical, Ethical and Legal Aspects" (1980) 73:4 Law Library Journal 986–996.

Walters, L. "Some Ethical Issues in Research Involving Human Subjects" (1977) 20:2 Perspectives in Biology & Medicine 193–211.

World Medical Association, *Declaration of Helsinki*. Adopted at the 18th World Medical Assembly in Helsinki in June 1964. Amended at the 29th World Medical Assembly in Tokyo in October 1975; the 35th World Medical Assembly in Venice in October 1983; and the 41st World Medical Assembly in Hong Kong in September 1989.

Williams, R.N. "Statutory Regulations and Ethical Conduct" in N. Howard-Jones & Z. Bankowski, eds. *Medical Experimentation and the Protection of Human Rights* (Geneva: CIOMS & The Sandoz Institute for Health and Socio-Economic Studies, 1979) 21–30.

### *Legal Reform*

Alberta Law Reform Institute. *Advance Directives and Substitute Decision-Making in Personal Health Care*. (Edmonton: Alberta Law Reform Institute, 1993).

Law Reform Commission of Canada. *Toward a Canadian Advisory Board on Biomedical Ethics* (Ottawa: Law Reform Commission of Canada, 1990).

Law Reform Commission of Canada. *Working Paper No. 26: Medical Treatment and the Criminal Law* (Ottawa: Law Reform Commission of Canada, 1980).

Law Reform Commission of Canada. *Working Paper No. 61: Biomedical Experimentation Involving Human Subjects* (Ottawa: Law Reform Commission, 1989).

Law Reform Commission of Victoria. *Medicine, Science and Law: Informed Consent* (Melbourne: Globe Press, 1987).

Newfoundland Law Reform Commission. *Discussion Paper on Advance Health Care Directives and Attorneys for Health Care* (St. John's: Newfoundland Law Reform Comission, 1988).

Queensland Law Reform Commission. *Assisted and Substituted Decisions: Decision-Making By and For People With a Decision-Making Disability* (Brisbane: Q.L.R.C., 1996).

The Law Commission. *Mentally Incapacitated Adults and Decision-Making: A New Jurisdiction (Consultation Paper No. 128)* (London: HMSO, 1992).

The Law Commission. *Mentally Incapacitated Adults and Decision-Making: An Overview (Consultation Paper No. 119)* (London: HMSO, 1991).

The Law Commission. *Mentally Incapacitated Adults and Decision-Making: Medical Treatment and Research (Consultation Paper No. 129)* (London: HMSO, 1993).

The Law Commission. *Mentally-Incapacitated and Other Vulnerable Adults: Public Law Protection (Consultation Paper No. 130)* (London: HMSO, 1993).

The Law Commission. *Mental Incapacity* (London: HMSO, 1995).

## SPECIFIC LEGAL ISSUES

### *Decision-Making*

Alexander, D. "Decision Making for Research Involving Persons With Severe Mental Retardation: Guidance From the National Commission for the Protection of Human Subjects of Biomedical and Behavioral Research" in P.R. Dokecki & R.M. Zaner, eds. *Ethics of Dealing With Persons With Severe Handicaps: Toward a Research Agenda* (Baltimore, MD: Paul H. Brookes, 1986) 39–52.

Battin, M.P. "Non-Patient Decision-Making in Medicine: The Eclipse of Altruism" (1985) 10:1 Journal of Medicine & Philosophy 19–44.

Brown, B. "Proxy Consent for Research on the Incompetent Elderly" in J.E. Thornton & E.R. Winkler, eds. *Ethics and Aging: The Right to Live, the Right to Die* (Vancouver: University of British Columbia Press, 1988) 183–193.

Capron, A.M. "The Authority of Others to Decide about Biomedical Interventions With Incompetents" in W. Gaylin & R. Macklin, eds. *Who Speaks for the Child: The Problems of Proxy Consent* (New York: Plenum Press, 1982) 115–152.

DeRenzo, E. "Surrogate Decision Making for Severely Cognitively Impaired Research Subjects: The Continuing Debate" (1994) 3:4 Cambridge Quarterly of Health Care Ethics 539–548.

Dickens, B.M. "Substitute Consent to Participation of Persons With Alzheimer's Disease in Medical Research: Legal Issues" in J.M. Berg, H. Karlinsky & F.H. Lowy, eds. *Alzheimer's Disease Research: Ethical and Legal Issues.* (Toronto: Carswell, 1991) 60–75.

Ellis, J.W. "Decisions By and For People With Mental Retardation: Balancing Considerations of Autonomy and Protection" (1992) 37:6 Villanova Law Review 1779–1809.

Engelhardt, H.T. "Proxy Consent" in *Ethics and Public Policy: An Introduction to Ethics* (Englewood Cliffs, NJ: Prentice-Hall, 1983) 365–369.

Franzi, C., Orgren, R.A. & Rozance, C. "Informed Consent by Proxy: A Dilemma in Long Term Care Research (1994) 15:2 Clinical Gerontologist 23–35.

Gutheil, T.G. & Appelbaum, P.S. "Substituted Judgment: Best Interests in Disguise" (1983) 13:3 Hastings Center Report 8–11.

Harmon, L. "Falling off the Vine: Legal Fictions in the Doctrine of Substituted Judgment" (1990) 100:1 Yale Law Journal 1–71.

Kapp, M.B. "Proxy Decision-Making in Alzheimer Disease Research: Durable Powers of Attorney, Guardianship, and Other Alternatives" (1994) 8:S4 Alzheimer Disease & Associated Disorders 28–37.

Kline, S.A. "Substitute Decision-Making and the Substitute Decision-Maker" (1992) 13:1 Health Law in Canada 125–128.

May, W.E. "Proxy Consent to Human Experimentation" (1976) 43:2 Linacre Quarterly 73–84.

Miller, B.L. "Autonomy and Proxy Consent" in V.L. Melnick & N.N. Dubler, eds. *Alzheimer's Dementia: Dilemmas in Clinical Research.* (Clifton, NJ: Humana Press 1985) 239–263.

Somerville, M.A. "Labels Versus Contents: Variance Between Philosophy, Psychiatry and Law in Concepts Governing Decision-Making" (1994) 39:1 McGill Law Journal 179–199.

Tibbles, L. "Derived Consent, Proxy Consent: Legal Issues" in V.L. Melnick & N.N. Dubler, eds. *Alzheimer's Dementia: Dilemmas in Clinical Research.* (Clifton, NJ: Humana Press 1985) 265–294.

Tomlinson, T. *et al.* "An Empirical Study of Proxy Consent for Elderly Persons" (1990) 30:1 Gerontologist 54–64.

Veatch, R.M. "Persons With Severe Mental Retardation and the Limits of Guardian Decision Making" in P.R. Dodecki & R.M. Zaner, eds. *Ethics of Dealing With Persons With Severe Handicaps* (Baltimore: Paul H. Brookes, 1986) 239–256.

Wierenga, E. "Proxy Consent and Counterfactual Wishes" (1983) 8:4 Journal of Medicine & Philosophy 405–416.

### *Advance Directives*

Advance Directives Seminar Group, Centre for Bioethics, University of Toronto. "Advance Directives: Are They An Advance?" (1992) 146:2 Canadian Medical Association Journal 127–134.

Danis, M. "Following Advance Directives" (1994) 24:S6 Hastings Center Report S21-S23.

Danis, M. et al. "A Prospective Study of Advance Directives for Life-Sustaining Care" (1991) 324:13 New England Journal of Medicine 882–888.

Downie, J. "Where There is a Will, There May Be a Better Way: Legislating Advance Directives" (1992) 12:3 Health Law in Canada 73–80.

Emanuel, L.L. "What Makes A Directive Valid" (1994) 24:S6 Hastings Center Report S27-S29.

Emanuel, L.L. "Advance Care Planning as a Process: Structuring the Discussions in Practice" (1995) 43:4 Journal of the American Geriatrics Society 440–446.

High, D.M. "Families' Roles in Advance Directives" (1994) 24:S6 Hastings Center Report S16-S18.

Hughes, D.L. & Singer, P.A. "Family Physicians' Attitudes Toward Advance Directives" (1992) 146:11 Canadian Medical Association Journal 1937–1944.

Macklin, A. "Bound to Freedom: The Ulysses Contract and the Psychiatric Will" (1987) 45:1 University of Toronto Faculty Law Review 37–68.

Moorhouse, A. & Weisstub, D.N. "Advance Directives for Research: Ethical Problems and Responses (1996) 19:2 International Journal of Law & Psychiatry 107–41.

Rasooly, I. et al. "Hospital Policies on Life-Sustaining Treatments and Advance Directives in Canada" (1994) 150:8 Canadian Medical Association Journal 1265–1270.

Sachs, G.A. "Advance Consent for Dementia Research" (1994) 8:S4 Alzheimer Disease & Associated Disorders 19–27.

Silberfeld, M., Madigan, K.V. & Dickens, B.M. "Liability Concerns about the Implementation of Advance Directives" (1994) 15:9 Canadian Medical Association Journal 285–289.

Silberfeld, M., Nash, C. & Singer, P.A. "Capacity to Complete An Advance Directive" (1993) 41:10 Journal of the American Geriatrics Society 1141–1143.

Singer, P.A. "Disease-Specific Advance Directives" (1994) 344:8922 Lancet 594–596.

Teno, J.M., Nelson, H.L. & Lynn, J. "Advance Care Planning. Priorities for Ethical and Empirical Research" (1994) 24:S6 Hastings Center Report S32-S36.

Winick, B.J. "Advance Directive Instruments for Those with Mental Illness" (1996) 51 University of Miami Law Review 57–95.

Zinberg, J.M. "Decisions for the Dying: An Empirical Study of Physicians' Responses to Advance Directives" (1989) 13:2 Vermont Law Review 445–492.

### *Responsibility, Liability and Compensation*

Ackerman, T.F. "Compensation of Human Volunteers" in Y. Champey, R.J. Levine & P.S. Lietman, eds. *Development of New Medicines: Ethical Questions* (New York: Royal Society of Medicine, 1989) 51–69.

Barton, J.M., Macmillan, M.S. & Sawyer, L. "The Compensation of Patients Injured in Clinical Trials" (1995) 21:3 Journal of Medical Ethics 166–169.

Childress, J.F. "Compensating Injured Research Subjects: I. The Moral Argument" (1976) 6:6 Hastings Center Report 21–27.

Department of Health, Education & Welfare, *Task Force on the Compensation of Injured Research Subjects* (Washington, DC: DHEW, 1977).

Diamond, A.L. & Laurence, D.R. "Compensation and Drug Trials" (1983) 287:6393 British Medical Journal 675–677.

Freedman, B. & Glass, K.C. "*Weiss v. Solomon*: A Case Study in Institutional Responsibility for Clinical Research" (1990) 18:4 Law, Medicine, & Health Care 395–403.

Glass, J.C. "No-Fault Compensation for Human Subjects Injured in Biomedical Research: A Public Policy Conflict" (1985) 48:3 Pharos of Alpha Omega Alpha Honor Medical Society 2–7.

Harvey, I. & Chadwick, R. "Compensation for Harm: The Implications For Medical Research" (1992) 34:12 Social Science & Medicine 1399–1404.

Holder, A.R. "Medical Insurance Payments and Patients Involved in Research" (1994) 16:1–2 IRB 19–22.

Ladimer, I. "Protection of Human Subjects: Remedies for Injury" in S.F. Spicker *et al.*, eds. *The Use of Human Beings in Research* (Boston: Kluwer Academic Publishers, 1988) 261–271.

McCarthy, C.R. "When OPRR Comes Calling: Enforcing Federal Research Regulations" (1995) 5:1 Kennedy Institute of Ethics Journal 51–55.

Picard, E.I. *Legal Liability of Doctors and Hospitals in Canada*, 2d ed. (Toronto: Carswell, 1984).

Sava, H., Matlow, P.T. & Sole, M.J. "Legal Liability of Physicians in Medical Research" (1994) 17:2 Clinical & Investigative Medicine 148–184.

Scheuer, R. "Research in the Hospital Setting on Human Subjects: Protecting the Patient and the Institution" (1993) 60:5 Mount Sinai Journal of Medicine 391–398.

Weiler, P.C. *et al.* "Proposal for Medical Liability Reform" (1992) 267:17 J.A.M.A. 2355–2358.

### *The Therapeutic - Non-Therapeutic Distinction*

Annas, G.J. "Experimentation and Research" (1994) 5:1 APA Newsletter on Philosophy & Medicine 9–11.

Annas, G.J. "Questing for Grails: Duplicity, Betrayal and Self-Deception in Postmodern Medical Research" (1996) 12:2 Journal of Contemporary Health Law & Policy 297–324.

Bamberg, M. & Budwig, N. "Therapeutic Misconceptions: When the Voices of Caring and Research are Misconstrued as the Voice of Curing" (1992) 2:3 Ethics & Behavior 165–184.

Beauchamp, T.L. "The Boundary Between Therapeutic and Non-Therapeutic Research. B. The Intersection of Research and Practice" in A. Goldworth *et al.*, eds. *Ethics and Perinatology* (New York: Oxford University Press, 1995) 231–244.

Capron, A.M. & Levine, R.J. "A Concluding, and Possibly Final, Exchange about "Therapy" and "Research" (1982) 4:1 IRB 10.

Cowan, D.H. "Innovative Therapy Versus Experimentation" (1985) 21:4 Tort & Insurance Law Journal 619–633.

De Groot, J.M. & Kennedy, S.H. "Integrating Clinical and Research Psychiatry" (1995) 20:2 Journal of Psychiatry & Neuroscience 150–154.

Dickens, B.M. "What is a Medical Experiment?" (1975) 113:7 Canadian Medical Association Journal 635–639.

Freedman, B., Fuks, A. & Weijer, C. "Demarcating Research and Treatment: A Systematic Approach for the Analysis of the Ethics of Clinical Research" (1992) 40:4 Clinical Research 653–660.

Gaze, B. & Dawson, K. "Distinguishing Medical Practice and Research: The Special Case of IVF" (1989) 3:4 Bioethics 301–319.

Goldiamond, I. "Protection of Human Subjects and Patients: A Social Contingency Analysis of Distinctions Between Research and Practice, and its Implications" (1976) 4:1 Behaviorism 1–41.

King, N.M.P. "Experimental Treatment: Oxymoron or Aspiration?" (1995) 25:4 Hastings Center Report 6–15.

Lantos, J. "Ethical Issues. How Can we Distinguish Clinical Research From Innovative Therapy?" (1994) 16:1 American Journal of Pediatric Hematology-Oncology 72–75.

Marquis, D. *et al.* "When Research is Best Therapy" (1988) 18:2 Hastings Center Report 24–26.

McCarthy, C.R. "Regulatory Aspects of the Distinction Between Research and Medical Practice" (1984) 6:3 IRB 7–8.

Moore, F.D. "Therapeutic Innovation: Ethical Boundaries in the Initial Clinical Trials of New Drugs and Surgical Procedures" (1969) 98:2 Daedalus 502–522.

Norton, M.L. "When does an Experimental/Innovative Procedure become an Accepted Procedure?" (1975) 38:4 Pharos of Alpha Omega Alpha Honor Medical Society 161–165.

Reiser, S.J. "Human Experimentation and the Convergence of Medical Research and Patient Care" in B. Barber, ed. *Medical Ethics and Social Change* (Philadelphia: American Academy of Political and Social Science, 1978) 8–18.

Rolleston, F. & Miller, J.R. "Therapy or Research: A Need for Precision" (1981) 3:7 IRB 1.

Schuchardt, E.J. "Walking a Thin Line: Distinguishing Between Research and Medical Practice During Operation Desert Storm" (1992) 26:1 Columbia Journal of Law & Social Problems 77–115.

Somerville, M.A. "Does the Aim of Human Medical Experimentation Affect its Legal or Ethical Validity?" (1979) 3:2 Legal Medical Quarterly 83–92.

Somerville, M.A. "Therapeutic and Non-Therapeutic Medical Procedures - What are the Distinctions?" (1981) 2:4 Health Law in Canada 85–90.

Tyson, J. "The Boundary Between Therapeutic and Non-Therapeutic Research. Dubious Distinctions Between Research and Clinical Practice Using Experimental Therapies: Have Patients Been Well Served?" in A. Goldworth *et al.*, eds. *Ethics and Perinatology* (New York: Oxford University Press, 1995) 214–230.

Ward, C.M. "Surgical Research, Experimentation and Innovation" (1994) 47:2 British Journal of Plastic Surgery 90–94.

## THE APPROVAL PROCESS

### *The Role and Function of Research Ethics Committees*

Adams, J. "Institutional Review Boards: The Public Safety Net" (1995) 12:2 Journal of Pediatric Oncology Nursing 93–95.

Ahmed, A.H. & Nicholson, K.G. "Delays and Diversity in the Practice of Local Research Ethics Committees" (1996) 22:5 Journal of Medical Ethics 263–266.

Alberti, K.G.M.M. "Local Research Ethics Committees" (1995) 311:7006 British Medical Journal 639–640.

Alderson, P. "A National Research Ethics Committee?" (1995) 107 Institute of Medical Ethics Bulletin 13–16.

Aldrige, H.D. & Walport, M.J. "Ethical Issues in Clinical Research: The Role of the Research Ethics Committee" (1995) 76:S2 British Journal of Urology 23–28.

Ayer, S. "Submitting a Research Proposal for Ethical Approval" (1994) 9:12 Professional Nursing 805–806.

Beauchamp, T.L. "Ethical Issues in Funding and Monitoring University Research" (1992) 11:1 Business & Professional Ethics Journal 5–16.

Bendell, C. *Standard Operating Procedures for Local Research Ethics Committees: Comments and Examples* (London: McKenna, 1994).

Benster, R. & Pollock, A.M. "Guidelines for Local Research Ethics Committees: Distinguishing Between Patient and Population Research in the Multicentre Research Project" (1993) 107:1 Public Health 3–7.

Bergkamp, L. "American IRBs and Dutch Research Ethics Committees: How Do They Compare?" (1988) 10:5 IRB 1–6.

Bergkamp, L. "Research Ethics Committees and the Regulation of Medical Experimentation With Human Beings in the Netherlands" (1988) 7:1 Medicine & Law 65–72.

Brudermuller, G. & Seelmann, K. "How to Discuss Ethics in Institutional Review Boards?" (1992) 14:9 Methods & Findings in Experimental & Clinical Pharmacology 737–742.

Cassidy, V.R. & Oddi, L.F. "Nurses on Hospital IRBs: A Critical Voice in Protecting Human Subjects" (1993) 6:1 Nursingconnections 31–38.

Cattorini, P. & Mordacci, R. "Ethics Committees in Italy" (1992) 4:3 HEC (Hospital Ethics Committee) Forum 219–226.

Ceci, S.J., Peters, D. & Plotkin, J. "Human Subjects Review, Personal Values, and the Regulation of Social Science Research" (1985) 40:9 American Psychologist 994–1002.

Chalmers, D. *et al. Report of the Review of the Role and Functioning of Institutional Ethics Committees: A Report to the Minister of Health and Family Services* (Canberra: Department of Health and Family Services, 1996).

Cohen, J. "The Costs of IRB Review" in R.A. Greenwald, M.K. Ryan & J.E. Mulvihill, eds. *Human Subjects Research: A Handbook for Institutional Review Boards* (New York: Plenum Press, 1982) 39–47.

Colombo, J. "Cost, Utility, and Judgments of Institutional Review Boards" (1995) 6:5 Psychological Science 318–319.

Cookson, J.B. "Auditing a Research Ethics Committee" (1992) 26:2 Journal of the Royal College of Physicians of London 181–183.

Corwan, D.H. "Scientific Design, Ethics, and Monitoring: IRB Review of Randomized Clinical Trials" (1980) 2:9 IRB 1–4.

Cowan, D.H. "Human Experimentation: The Review Process in Practice" (1975) 25:3 Case Western Law Review 533–564.

Creutzfeld, W. "Functions and Duties of an Ethics Committee at a University Faculty of Medicine: 13 Years of Personal Experience" (1994) 69:3 Forensic Science International 307–314.

Curran, W.J. "Evolution of Formal Mechanisms for Ethical Review of Clinical Research" in N. Howard-Jones & Z. Bankowski, eds. *Medical Experimentation and the Protection of Human Rights* (Geneva: CIOMS & The Sandoz Institute for Health and Socio-Economic Studies, 1979) 11–20.

Danis, M. "Should Ethics Committees Study Themselves?" (1994) 5:2 Journal of Clinical Ethics 159–162.

De Deyn, P.P. & Verhaegen, H. "Review Procedures and Review Committees for Research Protocols Involving Humans" in P.P. De Deyn, ed. *The Ethics of Animal and Human Experimentation* (London: John Libbey, 1994) 193–205.

Department of Health (U.K.). *Local Research Ethics Committees* (London: HMSO, 1991).

De Wachter, M.A.M. "Ethics Committees: Regulatory, Scientific or Ethical Institutions?" in Y. Champey, R.J. Levine & P.S. Lietman, eds. *Development of New Medicines: Ethical Questions* (New York: Royal Society of Medicine Services, 1989) 71–79.

Diamond, A.L. & Laurence, D.R. "Better Research Ethics Committees: Comments on Guidance From the Department of Health and the European Community" (1993) 27:2 Journal of the Royal College of Physicians of London 141–146.

Dillman, R.J.M. "Alzheimer's Disease: Necessary Elements in the Review of Research Protocols" in P.P. De Deyn, ed. *The Ethics of Animal and Human Experimentation* (London: John Libbey, 1994) 207–226.

Dockerty, J. & Elwood, M. "The Difficulties and Costs of Ethical Assessment of a National Research Project" (1992) 105:939 New Zealand Medical Journal 311–312.

Dodds, S., Albury, R. & Thomson, C. *Ethical Research and Ethics Committee Review of Social and Behavioural Research Proposals: Report to the Department of Human Services and Health* (Canberra: Australian Government Publishing Service, 1994).

Doyal, L. "Towards a Standard Application Form for LRECs" (1994) 101 Institute of Medical Ethics Bulletin 15–28.

Drury, M. "Ethical Appraisal for Multi-Centre Medical Research" (1990) 16:3 Journal of Medical Ethics 148–149.

Dunne, J.F. "Ethical Review Procedures for Research Involving Human Subjects" in Z. Bankowski & J.C. Bernardelli, eds. *Medical Ethics and Medical Education* (Geneva: CIOMS, 1981) 77–103.

Eaton, W.O. "Reliability in Ethics Reviews: Some Initial Empirical Findings" (1983) 24:1 Canadian Psychology 14–18.

Edgar, H. & Rothman, D.J. "The Institutional Review Board and Beyond: Future Challenges to the Ethics of Human Experimentation" (1995) 73:4 Milbank Quarterly 489–506.

Eichler, H.G. "Hazards of Misguided Ethics Committees" (1995) 346:8983 Lancet 1115–1116.

Evans, D. & Evans, M. *A Decent Proposal: Ethical Review of Clinical Research* (Chichester: John Wiley & Sons, 1996).

Evans, M.E. "The Legal Background of the Institutional Review Board" in R.A. Greenwald, M.K. Ryan & J.E. Mulvihill, eds. *Human Subjects Research: A Handbook for Institutional Review Boards* (New York: Plenum Press, 1982) 19–27.

Farnsworth, D.L. "Review Committees and Screening Groups: Attempts to Accommodate the Interests" in R.L. Bogomolny, ed. *Human Experimentation* (Dallas: Southern Methodist University Press, 1976) 122–131.

Fischer, F.W. & Breuer, H. "Influences of Ethical Guidance Committees on Medical Research – A Critical Reappraisal" in N. Howard-Jones & Z. Bankowski, eds. *Medical Experimentation and the Protection of Human Rights* (Geneva: CIOMS & The Sandoz Institute for Health and Socio-Economic Studies, 1979) 65–71.

Fletcher, J.C. & Hoffmann, D.E. "Ethics Committees: Time to Experiment With Standards" (1994) 120:4 Annals of Internal Medicine 335–338.

Fletcher, J.C. & Boverman, M. "Evolution of the Role of an Applied Bioethicist in a Research Hospital" in K. Berg & K.E. Tranøy, eds. *Research Ethics* (New York: Alan R. Liss, 1983) 131–158.

Foster, C.G. *Manual for Research Ethics Committees*, 2d ed. (London: Centre for Medical Law and Ethics, King's College, 1993).

Foster, C.G. "The Development and Future of Research Ethics Committees in Britain" in A. Grubb, ed. *Choices and Decisions in Health Care* (New York: Wiley, 1993) 161–181.

Foster, C.G. "Why Do Research Ethics Committees Disagree With Each Other?" (1995) 29:4 Journal of the Royal College of Physicians of London 315–318.

Foster, C.G., Fulford, K.W.M. & Parker, C. "Diversity in the Practice of District Ethics Committees" (1989) 299:6713 British Medical Journal 1437–1439.

Foster, C.G., Marshall, T. & Moodie, P. "The Annual Reports of Local Research Ethics Committees" (1995) 21:4 Journal of Medical Ethics 214–219.

Freedman, B. "Multicenter Trials and Subject Eligibility: Should Local IRBs Play a Role?" (1994) 16:1–2 IRB 1–6.

Giertz, G. "Scope of Review Procedures of Ethical Review Committees" in N. Howard-Jones & Z. Bankowski, eds. *Medical Experimentation and the Protection of Human Rights* (Geneva: CIOMS & The Sandoz Institute for Health and Socio-Economic Studies, 1979) 90–103.

Glantz, L.H. "Contrasting Institutional Review Boards With Institutional Ethics Committees" in R.E. Cranford & E.A. Doudera, eds. *Institutional Ethics Committees and Health Care Decision Making* (Ann Arbor, MI: Health Administration Press, 1984) 129–137.

Graf, H.-P. & Cole, D. "Ethics-Committee Authorization in Germany" (1995) 21:4 Journal of Medical Ethics 229–233.

Gray, B.H. "An Assessment of Institutional Review Committees in Human Experimentation" (1975) 13:4 Medical Care 318–328.

Gray, B.H. "The Functions of Human Subjects Review Committees" (1977) 134:8 American Journal of Psychiatry 907–910.

Greenwald, R.A. "General Principles of IRB Review" in R.A. Greenwald, M.K. Ryan & J.E. Mulvihill, eds. *Human Subjects Research: A Handbook for Institutional Review Boards* (New York: Plenum Press, 1982) 51–62.

Gutteridge, F. *et al.* "The Structure and Functioning of Ethical Review Committees" (1982) 16:20 Social Science & Medicine 1791–1800.

Hall, D. "The Research Imperative and Bureaucratic Control: The Case of Clinical Research" (1991) 32:3 Social Science & Medicine 333–342.

Hammerschmidt, D.E. & Keane, M.A. "Institutional Review Board (IRB) Review Lacks Impact on the Readability of Consent Forms for Research" (1992) 304:6 American Journal of the Medical Sciences 348–351.

Harries, U.J. *et al.* "Local Research Ethics Committees: Widely Differing Responses to a National Survey Protocol" (1994) 28:2 Journal of the Royal College of Physicians of London 150–154.

Hayes, G.J., Hayes, S.C. & Dykstra, T. "A Survey of University Institutional Review Boards: Characteristics, Policies, and Procedures" (1995) 17:3 IRB 1–6.

Horner, J.S. "Criteria for Decision Making in Local Research (Ethics) Committees" (1993) 107:6 Public Health 403–411.

Horwitz, R. "Ethical Review of Multi-Centre Research" (1994) 102 Institute of Medical Ethics Bulletin 13–15.

Hotopf, M. *et al.* "Are Ethical Committees Reliable?" (1995) 88:1 Journal of the Royal Society of Medicine 31–33.

Idanpaan-Heikkila, J. "WHO Views on Responsibilities of Research Ethics Boards and Good Clinical Practice (GCP)" (1993) 4:2 NCBHR Communiqué 7–10.

Isambert, F.-A. "Ethics Committees in France" (1989) 14:4 Journal of Medicine & Philosophy 445–456.

Jamrozik, K. "Ethics Committees: Is the Tail Wagging the Dog?" (1992) 157:9 Medical Journal of Australia 636–637.

Jarvis, R. "Ethical Review of Medical Research and Health Policy" in D. Greaves & H. Upton, eds. *Philosophical Problems in Health Care* (Brookfield, VT: Avebury, 1996) 175–190.

Jean, A., Paré, S. & Parizeau, M.-H. *Les Comités d'Éthique au Québec: Guide des Ressources en Centres Hospitaliers* (Québec: Gouvernment du Québec, 1991).

Kohn, A. & Putterman, C. "Problems and Conflicts in Peer Review" (1993) 5:3 International Journal of Impotence Research 133–137.

Koren, G. "The Process of Ethics Review in Pediatric Research: The Toronto Model" in G.Koren, ed. *Textbook of Ethics in Pediatric Research* (Malabar, FL: Krieger, 1993) 197–220.

Lackey, D.P. "Which Subjects Should an IRB Protect? Two Moral Models" (1982) 4:7 IRB 5–6.

Levine, R.J. "AIDS Research and Ethical Review Boards" in H. Fuenzalida-Puelma, A.M. Linares Parada & D.S. LaVertu, eds. *Ethics and Law in the Study of AIDS* (Washington, DC: Pan American Health Organization, 1992) 170–177.

Levine, R.J. "The Value and Limitations of Ethical Review Committees for Clinical

Research" in Z. Bankowski & J.C. Bernardelli, eds. *Medical Ethics and Medical Education* (Geneva: CIOMS, 1981) 43–63.

Lind, S.E. "The Institutional Review Board: An Evolving Ethics Committee" (1992) 3:4 Journal of Clinical Ethics 278–282.

Lock, S. "Monitoring Research Ethical Committees" (1990) 300:6717 British Medical Journal 61–62.

Macara, A.W. "Ethical Review of Multi-Centred Trials" (1990) 16:3 Journal of Medical Ethics 150–152.

Magnusson, R.S. "Confidentiality and Consent in Medical Research: Some Recurrent, Unresolved Legal Issues Faced by IECs" (1995) 17 The Sydney Law Review 549–572.

Mander, T. "The Legal Standing of Local Research Ethics Committees" (1996) 2 Medical Law International 149–168.

McCarthy, C.R. "Experience With Boards and Commissions Concerned With Research Ethics in the United States" in K. Berg & K.E. Tranøy, eds. *Research Ethics* (New York: Alan R. Liss, 1983) 111–122.

McLean, S.A. "Research Ethics Committees: Principles and Proposals" (1995) 53:5 Health Bulletin 243–248.

McNeill, P.M. "Research Ethics Review in Australia, Europe and North America" (1989) 11:3 IRB 4–7.

McNeill, P.M., Berglund, C.A. & Webster, I.W. "Do Australian Researchers Accept Committee Review and Conduct Ethical Research?" (1992) 35:3 Social Science & Medicine 317–322.

McNeill, P.M., Berglund, C.A. & Webster, I.W. "How Much Influence Do Various Members Have Within Research Ethics Committees?" (1994) 3:4 Cambridge Quarterly of Healthcare Ethics 522–532.

McNeill, P.M., Berglund, C.A. & Webster, I.W. "Reviewing the Reviewers: A Survey of Institutional Ethics Committees in Australia" (1990) 152:6 Medical Journal of Australia 289–296.

Meade, T.W. "The Trouble With Ethics Committees" (1994) 28:2 Journal of the Royal College of Physicians of London 102–104.

Meslin, E.M. "Ethical Issues in the Substantive and Procedural Aspects of Research Ethics Review" (1993) 13:3 Health Law in Canada 179–191.

Meslin, E.M. *et al.* "Judging the Ethical Merit of Clinical Trials: What Criteria Do Research Ethics Board Members Use?" (1994) 16:4 IRB 6–10.

Micetich, K.C. "Reflections of an IRB Chair" (1994) 3:4 Cambridge Quarterly of Healthcare Ethics 506–509.

Miller, C. "Protection of Human Subjects of Research in Canada (Ethics Committees)" (1995) 4:1 Health Law Review 8–16.

Miller, J.N. "Ethics Review in Canada: Highlights From a National Workshop: Part 1" (1989) 22:7 Annals of the Royal College of Physicians and Surgeons of Canada 515–523.

Miller, J.N. "Ethics Review in Canada: Highlights From a National Workshop: Part 2" (1990) 23:1 Annals of the Royal College of Physicians and Surgeons of Canada 29–33.

Moher, D. "Trial Registration: A New Standard for REBs" (1993) 4:2 NCBHR Communiqué 6–7.

Moline, M.L. & Aisen, M.W. "Perspectives of Protocol Reviewers" (1994) 5:1 APA Newsletter on Philosophy & Medicine 59–60.

Montgomery, J. "Improving Review of Multi-Centre Trials" (1994) 95 Institute of Medical Ethics Bulletin 19–22.

Moodie, P. "The Role of Local Research Ethics Committees: Maintaining the Pressure for Improvement" (1992) 304:6835 British Medical Journal 1129–1130.

Moodie, P. & Marshall, T. "Guidelines for Local Research Ethics Committees" (1992) 304:6837 British Medical Journal 1293–1295.

Morehouse, R.L. "Dilemnas of the Clinical Researcher: A View From the Inside" (1994) 15:2 Health Law in Canada 52–54.

National Council on Bioethics in Human Research, "Ethics of Clinical Trials for Research Ethics Boards: Proceedings of a National Workshop" (1991) 2:2 NCBHR Communiqué 4–23.

National Council on Bioethics in Human Research. "Protecting and Promoting the Human Research Subject: A Review of the Function of Research Ethics Boards in Canadian Faculties of Medicine" (1995) 6:1 NCBHR Communiqué 3–32.

National Health and Medical Research Council, *Guidelines for the Monitoring of Research by Institutional Ethics Committees* (Canberra: NHMRC, 1992).

Neuberger, J. *Ethics and Health Care: The Role of Research Ethics Committees in the United Kingdom* (London: King's Fund Institute, 1992).

Neuhauser, D. "More Tales From Institutional Review Boards" (1994) 4:1 Health Matrix 153–158.

New Zealand Department of Health, *Standard for Ethics Committees Established to Review Research and Ethical Aspects of Health Care* (Wellington: New Zealand Department of Health, 1991).

Nicholson, R. "Ways Forward for Research Ethics Review" (1993) 94 Institute of Medical Ethics Bulletin 13–16.

Office for Protection From Research Risks, National Institutes of Health. *Protecting Human Subjects: Institutional Review Board Guidebook* (Washington, DC: U.S. Government Printing Office, 1993).

Olson, C.M. & Jobe, K.A. "Reporting Approval by Research Ethics Committees and Subjects' Consent in Human Resuscitation Research" (1996) 31:3 Resuscitation 255–263.

Ortega, R. & Dal-Re, R. "Clinical Trials Committees: How Long is the Protocol Review and Approval Process in Spain? A Prospective Study" (1995) 17:4 IRB 6–9.

Palca, J. "Institutional Review Boards: A Net too Thin" (1996) 26:3 Hastings Center Report 4.

Pettit, P. "Instituting a Research Ethic" (1993) 14:4 Controlled Clinical Trials 261–265.

Penn, Z.J. & Steer, P.J. "Local Research Ethics Committees: Hindrance or Help?" (1995) 102:1 British Journal of Obstetrics & Gynaecology 1–2.

Porter, J.P. "What are the Ideal Characteristics of Unaffiliated/Nonscientist IRB Members?" (1986) 8:3 IRB 1–6.

Redshaw, M.E., Harris, A. & Baum, J.D. "Research Ethics Committee Audit: Differences Between Committees" (1996) 22:2 Journal of Medical Ethics 78–82.

Reiser, S.J. & Knudson, P. "Protecting Research Subjects After Consent: The Case for the æResearch Intermediary'"(1993) 15:2 IRB 10–11.

Riis, P. "Composition, Authority and Influence of Ethical Review Committees" in N. Howard-Jones & Z. Bankowski, eds. *Medical Experimentation and the Protection of Human Rights*. (Geneva: CIOMS & The Sandoz Institute for Health and Socio-Economic Studies, 1979) 85–89.

Riis, P. "Experiences With Committees and Councils for Research Ethics in Scandinavia" in K. Berg & K.E. Tranøy, eds. *Research Ethics*. (New York: Alan R. Liss, 1983) 123–129.

Riis, P. "What a National Research Ethics Committee Does?" (1992) 84 Institute of Medical Ethics Bulletin 13–18.

Rosner, F. "Hospital Medical Ethics Committees: A Review of Their Development" (1985) 253:18 JAMA. 2693–2697.

Rosnow, R.L. et al. "The Institutional Review Board as a Mirror of Scientific and Ethical Standards" (1993) 48:7 American Psychologist 821–826.

Royal College of Physicians of London. *Guidelines on the Practice of Ethics Committess in Medical Research Involving Human Subjects*, 3d. ed. (London: RCP, 1996).

Ryan, M.K. "General Organization of the IRB" in R.A. Greenwald, M.K. Ryan & J.E.

Mulvihill, eds. *Human Subjects Research: A Handbook for Institutional Review Boards* (New York: Plenum Press, 1982) 29–38.

Ryan, M.K. "IRB Procedures" in R.A. Greenwald, M.K. Ryan & J.E. Mulvihill, eds. *Human Subjects Research: A Handbook for Institutional Review Boards* (New York: Plenum Press, 1982) 63–77.

Saito, T. "Ethics Committees in Japanese Medical Schools" (1992) 4:4 HEC (Hospital Ethics Committee) Forum 281–287.

Scott, G. & Goode, K. "Reporting the Outcome of Ethics Review" (1995) 110 Institute of Medical Ethics Bulletin 16–18.

Sieber, J. *Planning Ethically Responsible Research: A Guide for Students and Internal Review Boards* (Newbury Park, NJ: Sage Publications, 1992).

Sugarman, J. "Should Hospital Ethics Committees Do Research" (1994) 5:2 Journal of Clinical Ethics 121–125.

Thevos, J.-M. "Research and Hospital Ethics Committees in Switzerland" (1992) 4:1 HEC (Hospital Ethics Committee) Forum 41–47.

Tomenson, J.A. & Paddle, G.M. "Better Quality Studies Through Review of Protocols" (1991) 33:12 Journal of Occupational Medicine 1240–1243.

Veatch, R.M. "Lay Participation in Medical Policy Making. Part I. Human Experimentation Committees: Professional or Representative" (1975) 5:5 Hastings Center Report 31–40.

Waldron, H.A. & Cookson, R.F. "Avoiding the Pitfalls of Sponsored Multicentre Research in General Practice" (1993) 307:6915 British Medical Journal 1331–1334.

Watling, M.I.L. & Dewhurst, J.K. "Current Experience of Central Versus Local Ethics Approval in Multicentre Studies" (1993) 27:4 Journal of the Royal College of Physicians of London 399–402.

Webb, W.B. "Human Subjects Review Boards: A Modest Proposal" (1987) 42:5 American Psychologist 516–517.

Weijer, C. *et al.* "Monitoring Clinical Research: An Obligation Unfulfilled" (1995) 152:12 Canadian Medical Association Journal 1973–1980.

Williams, J. "Ethical Review of Clinical Research" (1993) 98 Institute of Medical Ethics Bulletin 28–30.

Williams, P.C. "Success in Spite of Failure: Why IRBs Falter in Reviewing Risks and Benefits" (1984) 6:3 IRB 1–4.

Wollman, S. & Ryan, M.K. "Continuing Review of Research" in R.A. Greenwald, M.K. Ryan & J.E. Mulvihill, eds. *Human Subjects Research: A Handbook for Institutional Review Boards* (New York: Plenum Press, 1982) 125–136.

Yesley, M.S. "Federal Commissions and Local IRBs" (1983) 13:5 Hastings Center Report 11–12.

### *Validity of Research Design*

Freedman, B. "Equipoise and the Ethics of Clinical Research" (1987) 317:3 New England Journal of Medicine 141–145.

Freedman, B. "Scientific Value and Validity as Ethical Requirements for Research: A Proposed Explication" 9:6 IRB 7–10.

McGrath, P.J. "Research Design and Research Ethics" in K.J.S. Anand & P.J. McGrath, eds. Pain in Neonates: Pain Research and Clinical Management, vol. 5 (Amsterdam: Elsevier Science Publishers, 1993) 287–305.

Moraczewski, A.S. "Scientific Methodology and Ethical Imperatives" in J.C. Schoolar & C.M. Gaitz, eds. Research and the Psychiatric Patient (New York: Brunner/Mazel, 1975) 142–149.

Naquet, R. "Ethical and Moral Considerations in the Design of Experiments" (1993) 57:1 Neuroscience 183–189.

Pendergast, M.K. "Assuring the Integrity, Accuracy, and Quality of Scientific Information" (1993) 2:1–2 Quality Assurance 57–61.

Rutstein, D.D. "The Ethical Design of Human Experiments" (1969) 98:2 Daedalus 523–541.

Veatch, R.M. "Justice and Research Design: The Case for a Semi-Randomized Clinical Trial" (1979) 300 New England Journal of Medicine 1242–1245.


### *Risk and Risk Assessment*

Castronovo, F.P. "An Attempt to Standardize the Radiodiagnostic Risk Statement in an Institutional Review Board Consent Form" (1993) 28:6 Investigative Radiology 533–538.

Forrow, L., Wartman, S.A. & Brock, D.W. "Science, Ethics, and the Making of Clinical Decisions: Implications for Risk Factor Intervention" (1988) 259:21 JAMA. 3161–3167.

Harvey, M. & Levine, R.J. "The Risk of Research Procedures: Metholodologic Problems and Proposed Standards" (1983) 31:2 Clinical Research 126–139.

Kopelman, L. "Estimating Risk in Human Research" (1981) 29:1 Clinical Research 1–8.

Marsh, F.H. "Planning for the Unknown in Research: Ethical Dilemmas Confronting the Clinician-Investigator" (1994) 7:3 Journal of the American Board of Family Practice 263–266.

Martin, D.K. "The Incommensurability of Research Risks and Benefits: Practical Help for Research Ethics Committees" (1995) 17:2 IRB 8–10.

Meslin, E.M. "Philosophical Considerations about Risk and Risk Assessment in Medical Research" in G. Koren, ed. *Textbook of Ethics in Pediatric Research* (Malabar, FL: Krieger, 1993) 37–55.

Meslin, E.M. "Protecting Human Subjects From Harm Through Improved Risk Judgments" (1990) 12:1 IRB 7–10.

Oberst, M.T. "Research Ethics. Part 2: The Concept of Risk in Clinical Studies" (1979) 2:6 Cancer Nursing 481–482.

Sherertz, R.J. & Streed, S.A. "Medical Devices: Significant Risk vs. Nonsignificant Risk" (1994) 272:12 JAMA. 955–956.

Spicker, S.F. "Research Risks, Randomization, and Risks to Research: Reflections on the Prudential Use of "Pilot" Trials" in S.F. Spicker *et al.*, eds. *The Use of Human Beings in Research* (Boston: Kluwer Academic Publishers, 1988) 143–160.

Thompson, R.A. "Vulnerability in Research: A Developmental Perspective on Risk" (1990) 61:1 Child Development 1–16.

Weijer, C. & Fuks, A. "The Duty to Exclude: Excluding People at Undue Risk From Research" (1994) 17:2 Clinical and Investigative Medicine 115–122.

Wiatrowski, W.A., Giles, E.R. & Cooke, E.P. "Development of a System to Evaluate and Communicate Radiation Risk" (1996) 70:1 Health Physics 111–117.


### *Informed Consent*

Agich, G.J. "Human Experimentation and Clinical Consent" in J.F. Monagle & D.C. Thomasma, eds. *Medical Ethics: A Guide for Health Professionals* (Rockville, MD: Aspen Publishers, 1988) 127–139.

Alfidi, R.J. "Informed Consent: A Study of Patient Reaction" (1971) 216:8 JAMA. 1325–1329.

Annas, G.J., Glantz, L.H. & Katz, B.F. *Informed Consent to Human Experimentation: The Subject's Dilemma* (Cambridge, Mass: Ballinger, 1977).

Appelbaum, P.S. "Informed Consent" in D.N. Weisstub, ed. *Law and Mental Health: International Perspectives, Volume I* (New York: Pergamon Press, 1984) 45.

Appelbaum, P.S., Lidz, C.W. & Meisel, A.J. *Informed Consent: Legal Theory and Clinical Practice* (New York: Oxford University Press, 1987).

Appelbaum, P.S. & Roth, L.H. "The Structure of Informed Consent in Psychiatric Research" (1983) 1:4 Behavioural Sciences & the Law 9–20.

Applebaum, P.S., Roth, L.H. & Lidz, C.W. "The Therapeutic Misconception: Informed Consent in Psychiatric Research" (1982) 5:3–4 International Journal of Law & Psychiatry 319–330.

Arboleda-Florez, J. *"Reibl v. Hughes*: The Consent Issue" (1987) 32:1 Canadian Journal of Psychiatry 66–70.

Barber, B. *Informed Consent in Medical Therapy and Research* (New Brunswick, NJ: Rutgers University Press, 1980).

Benson, P.R., Roth, L.H. & Winslade, W.J. "Informed Consent in Psychiatric Research: Preliminary Findings From an Ongoing Investigation" (1985) 20:12 Social Science & Medicine 1331–1341.

Benson, P.R. *et al.* "Information Disclosure, Subject Understanding, and Informed Consent in Psychiatric Research" (1988) 12 Law & Human Behavior 455–475.

Berscheid, E. *et al.* "Anticipating Informed Consent: An Empirical Approach" (1973) 28:10 American Psychologist 913–925.

Biros, M.H. *et al.* "Informed Consent in Emergency Research: Consensus Statement From the Coalition Conference of Acute Resuscitation and Critical Care Researchers" (1995) 273:16 JAMA. 1283–1287.

Bonnie, R.J. & Hoffman, P.B. "Regulation of Human Experimentation: A Reappraisal of Informed Consent" in R.L. Bogomolny, ed. *Human Experimentation* (Dallas: Southern Methodist University Press, 1976) 52–79.

Boverman, M. "Mental Health Aspects of the Informed Consent Process" in K. Berg & K.E. Tranẏ, eds. *Research Ethics.* (New York: Alan R. Liss, 1983) 229–241.

Brady, J.V. "A Consent Form Does Not Informed Consent Make" (1979) 1:7 IRB 6–7.

Brazier, M. "Patient Autonomy and Consent to Treatment: The Role of Law?" (1987) 7:2 Legal Studies 169–193.

Bryne, P. "Medical Research and the Human Subject: Problems of Consent and Control in the UK Experience" in D. Callahan & G.R. Dunstan, eds. *Biomedical Ethics: An Anglo-American Dialogue* (New York: Annals of the New York Academy of Sciences, 1988) 144–153.

Burgess, M.M. "Informed Consent in Medical Experimentation" (1987) 20:1 Annals of the Royal College of Physicians and Surgeons of Canada 29–31.

Capron, A. "Informed Consent in Catastrophic Disease Research in Treatment" (1974) 123:2 University of Pennsylvania Law Review 340–438.

Cocking, D. & Oakley, J. "Medical Experimentation, Informed Consent and Using People" (1994) 8:4 Bioethics 293–311.

Delgado, R. & Leskovac, H. "Informed Consent in Human Experimentation: Bridging the Gap Between Ethical Thought and Current Practice" (1986) 34:1 UCLA Law Review 67–130.

Dickens, B.M. "Criteria of Adequately Informed Consent" in N. Howard-Jones & Z. Bankowski, eds. *Medical Experimentation and the Protection of Human Rights* (Geneva: CIOMS & The Sandoz Institute for Health and Socio-Economic Studies, 1979) 200–21.

Engelhardt, H.T. "Free and Informed Consent, Refusal of Treatment, and the Health Care Team: The Many Faces of Freedom" in H.T. Engelhardt, *The Foundations of Bioethics*, 2d ed. (New York: Oxford University Press, 1996) 288–374.

Faden, R.R. & Beauchamp, T.L. *A History and Theory of Informed Consent* (Oxford: Oxford University Press, 1986).

Faden, R.R. & Beauchamp, T.L. "Decision-Making and Informed Consent: A Study of the Impact of Disclosed Information" (1980) 7 Social Indicators Research 313–336.

Faden, R.R. *et al.* "Disclosure Standards and Informed Consent" (1981) 6:2 Journal of Health Politics, Policy and Law 255–284.

Faulder, C. *Whose Body Is It? The Troubling Issue of Informed Consent* (London: Virago Press, 1985).

Feinberg, J. "Informed Consent in Medicine" in J. Feinberg, *The Moral Limits of the Criminal Law, Volume Three: Harm to Self* (New York: Oxford University Press, 1986) 305–315, 406–407.

Fellner, C.H. & Marshall, J.R. "Kidney Donors – The Myth of Informed Consent" (1970) 176:9 American Journal of Psychiatry 1245–1251.

Finkelstein, D., Smith, M.K. & Faden, R.R. "Informed Consent and Medical Ethics" (1993) 111:3 Archives of Ophthalmology 324–326.

Fletcher, J.C. "The Evolution of the Ethics of Informed Consent" in K. Berg & K.E. Tranøy, eds. *Research Ethics* (New York: Alan R. Liss, 1983) 187–228.

Fost, N. "Consent as a Barrier to Research" (1979) 300:22 New England Journal of Medicine 1272–1273.

Freedman, B. "A Moral Theory of Informed Consent" (1975) 5:4 Hastings Center Report 32–39.

Freedman, B. "The Validity of Ignorant Consent to Medical Research" (1982) 4:2 IRB 1–5.

Frielander, W.J. "The Evolution of Informed Consent in American Medicine" (1995) 38:3 Perspectives in Biology and Medicine 498–510.

Gillon, R. "Medical Treatment, Medical Research and Informed Consent" (1989) 15:1 Journal of Medical Ethics 3–5.

Gochnauer, M. & Fleming, D. "Tort Law – Informed Consent – New Directions for Medical Disclosure – Hopp v. Lepp and Reibl v. Hughes" (1981) 15:2 University of British Columbia Law Review 475–497.

Goldworth, A. "Standards of Disclosure in Informed Consent" in A. Goldworth *et al.*, eds. *Ethics and Perinatology* (New York: Oxford University Press, 1995) 263–278.

Gorman, H.M. & Dane, F.C. "Balancing Methodological Rigour and Ethical Treatment: The Necessity of Voluntary, Informed Consent" in P.P. De Deyn, ed. *The Ethics of Animal and Human Experimentation* (London: John Libbey, 1994) 35–41.

Greenwald, R.A. "Informed Consent" in R.A. Greenwald, M.K. Ryan & J.E. Mulvihill, eds. *Human Subjects Research: A Handbook for Institutional Review Boards* (New York: Plenum Press, 1982) 79–90.

Grudner, T.M. "How to Make Consent Forms More Readable" (1981) 3:7 IRB 9–10.

Haber, J.G. "Patients, Agents, and Informed Consent" (1985–86) 1 Journal of Law & Health 43–59.

Hall, R.C.W. & Gardner, E.R. "Informed Consent: Elements for Crisis" in R.C.W. Hall, ed. *Psychiatry in Crisis* (New York: Spectrum Publications, 1982) 55–63.

Hawkins, I.A. & Bullock, S.L. "Informed Consent and Religious Values: A Neglected Area of Diversity" (1995) 32:2 Psychotherapy 293–300.

Hooker, E.Z. "Can Your Research Subjects Read Your Study's Informed Consent Form" (1995) 12:2 Scientific Nursing 57–58.

Hooker, E.Z. "Informed Consent" (1992) 9:3 Scientific Nursing 86–91.

Hopper, K.D., TenHave, T.R. & Hartzel, J. "Informed Consent for Clinical and Research Imaging Procedures: How Much Do Patients Understand?" (1995) 164:2 American Journal of Roentgenology 493–496.

Isambert, F.A. "Le consentement: le point de vue d'une éthique rationnelle" (1986) 44 Médecine et Hygiène 2021–2023.

Karlawish, J.H. & Hall, J.B. "The Controversy over Emergency Research. A Review of the Issues and Suggestions for a Resolution" (1996) 153:2 American Journal of Respiratory and Critical Care Medicine 499–506.

Katz, J. "Informed Consent – A Fairy Tale? Law's Vision" (1977) 39:2 University of Pittsburgh Law Review 137–174.

Katz, J. "Informed Consent – Must it Remain a Fairy Tale?" (1994) 10:1 Journal of Contemporary Health Law & Policy 69–91.

Katzner, L.I. "The Ethics of Human Experimentation: The Information Condition" in W.L. Robinson & M.S. Pritchard, eds. *Medical Responsibility: Paternalism, Informed Consent and Euthanasia* (Clifton, NJ: Humana Press, 1979) 43–56.

Kennedy, B.J. & Lillehaugen, A. "Patient Recall of Informed Consent" (1979) 7:2 Medical & Pediatric Oncology 173–178.

Kimsma, G.K. "Dissent on Informed Consent. Tension Between Medical, Legal and Communication Science Related Conceptions on Informed Consent: An Argument to Reformulate the Role and Responsibilities of Physicians" in P.P. De Deyn, ed. *The Ethics of Animal and Human Experimentation* (London: John Libbey, 1994) 249–257.

King, N.M.P. & Henderson, G. "Treatments of Last Resort: Informed Consent and the Diffusion of New Technology" (1991) 42:3 Mercer Law Review 1007–1050.

Kleinig, J. "The Ethics of Consent" (1982) 8 Canadian Journal of Philosophy 91–117.

Koch, H.G., Reiter-Theil, F. & Helmchen, H., eds. *Informed Consent in Psychiatry* (Baden-Baden: Nomos, 1996).

Kouri, K.P. "L'influence de la Cour Suprême sur l'obligation de renseigner en droit médical québécois" (1984) 44:5 Revue du Barreau 851–868.

Lantos, J. "Informed Consent. The Whole Truth for Patients?" (1993) 72:S9 Cancer 2811–2815.

Lebacqz, K. & Levine, R.J. "Respect for Persons and Informed Consent to Participate in Research" (1977) 25:3 Clinical Research 101–107.

LeBlang, T.R. "Informed Consent and Disclosure in the Physician-Patient Relationship: Expanding Obligations for Physicians in the United States" (1995) 14:5–6 Medicine & Law 429–444.

LeBlang, T.R. "Informed Consent – Duty and Causation: A Survey of Current Developments" (1983) 18:2 The Forum: American Bar Association Tort and Insurance Practice Section 280–289.

Levine, R.J. "Guidelines for Negotiating Informed Consent With Prospective Human Subjects of Experimentations" (1974) 22:2 Clinical Research 42–46.

Levine, R.J. "Informed Consent in Research and Practice: Similarities and Differences" (1983) 143:6 Archives of Internal Medicine 1229–1231.

Levine, R.J. "Informed Consent: Some Challenges to the Universal Validity of the Western Model" (1991) 19:3–4 Law, Medicine & Health Care 207–213.

Lidz, C.W. *Informed Consent: A Study of Decisionmaking in Psychiatry* (New York: Guilford Press, 1984).

Lowe, C.U. & Alexander, D.F. "Informed Consent and the Rights of Research Subjects" in H. Wechsler, R.W. Lamont-Havers & G.F. Cahill, eds. *The Social Context of Medical Research* (Cambridge, Mass: Ballinger, 1981) 97–126.

Lurie, K.G. & Benditt, D. "Regulated to Death: The Matter of Informed Consent for Human Experimentation in Emergency Resuscitation Research" (1995) 18:7 Pacing & Clinical Electrophysiology 1443–1447.

McCarthy, C.R. "To Be or Not to Be: Waiving Informed Consent in Emergency Research" (1995) 5:2 Kennedy Institute of Ethics Journal 155–162.

McDonald, R.M. "Do Human Rights Interfere With Progress: Problems of Consent in Medical Research" in L.M. Hodges, ed. *Social Responsibility: Journalism, Law, Medicine*, vol. 4 (Lexington, VA: Washington and Lee University, 1978) 72–80.

Meisel, A. "Assuring Adequate Consent: Special Considerations in Patients of Uncertain Competence" in V.L. Melnick & N.N. Dubler, eds. *Alzheimer's Dementia: Dilemmas in Clinical Research*. (Clifton, NJ: Humana Press 1985) 205–225.

Meisel, A. & Roth, L.H. "Toward an Informed Discussion of Informed Consent: A Review and Critique of the Empirical Studies" (1983) 25 Arizona Law Review 265.

Miller, B.L. "Informed Consent to Research on Emergency Care (1991) 20:11 Annals of Emergency Medicine 1267–1269.

Montague, J. "Balancing Caution & Courage. Physicians and Regulators Weigh Informed Consent Issues in Clinical Research" (1994) 68:18 Hospital & Health Networks 50, 52–54.

Morgenbesser, S. "Experimentation and Consent: A Note" in S.F. Spicker & H.T. Engelhardt, eds. *Philosophical Medical Ethics: Its Nature and Significance. Proceedings* (Boston: D. Reidel, 1977) 97–110.

Murphy, J.G. "Therapy and the Problem of Autonomous Consent" (1979) 2:4 International Journal of Law & Psychiatry 415–430.

Murphy, J.G. "Total Institutions and the Possibility of Consent to Organic Therapies" (1975) 5:1 Human Rights 25–45.

O'Donnell, T.J. "Ethical Concepts of Consent" in F.J. Ayd, ed. *Medical, Moral and Legal Issues in Mental Health Care* (Baltimore: Williams & Wilkins, 1974) 1–6.

Owens, J.F. "Informed Consent in the Clinical Research Setting: Experimentation on Human Subjects" (1987) 33 Medical Trial Technique Quarterly 335–350.

Peckham, M.J. *et al.* "Informed Consent: Ethical, Legal, and Medical Implications for Doctors and Patients who Participate in Randomised Clinical Trials" (1983) 286:6371 British Medical Journal 1117–1121.

Perrin, K.M. "Informed Consent: A Challenge for Patient Advocacy" in G.B. White, ed. *Ethical Dilemmas in Contemporary Nursing Practice* (Washington, DC: American Nurses Publishing, 1992) 105–116.

Phillipson, S.J. *et al.* "Informed Consent for Research: A Study to Evaluate Readability and Processability to Effect Change" (1995) 43:5 Journal of Investigative Medicine 459–467.

Picard, E.I. "Consent to Medical Treatment in Canada" (1981) 19:1 Osgoode Hall Law Journal 140–151.

Robertson, G.B. "Informed Consent Ten Years Later: The Impact of *Reibl v. Hughes*" (1991) 70:3 Canadian Bar Review 423–447.

Robertson, G.B. "Overcoming the Causation Hurdle in Informed Consent Cases: The Principle in *McGhee v. National Coal Board*" (1984) 22:1 University of Western Ontario Law Review 75–93.

Romano, J. "Reflections on Informed Consent" (1974) 30:1 Archives of General Psychiatry 129–135.

Roth, L.H. & Appelbaum, P.S. "Obtaining Informed Consent for Research With Psychiatric Patients: The Controversy Continues" (1983) 6:4 Psychiatric Clinics of North America 551–565.

Roth, L.H. *et al.* "Informed Consent in Psychiatric Research" (1987) 39:2–3 Rutgers Law Review 425–441.

Roy, D.J. "Informed Consent: Obligations and Limits" in Y. Champey, R.J. Levine & P.S. Lietman, eds. *Development of New Medicines: Ethical Questions* (New York: Royal Society of Medicine Services, 1989) 23–49.

Rozovsky, L.E. & Rozovsky, F.A. *The Canadian Law of Consent to Treatment* (Toronto: Butterworths, 1990).

Samuels, A. "Informed Consent: The Law" (1992) 32:1 Medicine, Science & the Law 35–42.

Saunders, C.M., Baum, M. & Houghton, J. "Consent, Research and the Doctor-Patient Relationship" in R. Gillon, ed. *Principles of Health Care Ethics* (New York: Wiley, 1994) 457–469.

Schafer, A. "Achieving Informed Consent in Clinical Trials" in B.A. Stoll, ed. *Ethical Dilemmas in Cancer Care* (Basinstoke, England: Macmillan, 1989) 29–38.

Schuck, P.H. "Rethinking Informed Consent" (1994) 103:4 Yale Law Journal 899–959.

Sheybani, S., Bovet, J. & Seywert, F. "Informed Consent in the Framework of Biomedical Research: A Survey" (1995) 115:5 Revue Medical de la Suisse Romande 411–420.

Shimm, D.S. & Spece, R.G. Jr. "Conflict of Interest and Informed Consent in Industry-Sponsored Clinical Trials" (1991) 12:4 Journal of Legal Medicine 477–513.

Silva, M.C. "Informed Consent in Human Experimentation: The Scientist's Responsibility — The Subject's Right" (1980) 16:12 Trial 37–41.

Silva, M.C. & Sorrell, J.M. "Enhancing Comprehension of Information for Informed Consent: A Review of Empirical Research" (1988) 10:1 IRB 1–5.

Silverman, H.J. "Ethical Considerations of Ensuring an Informed and Autonomous Consent in Research Involving Critically Ill Patients" (1996) 154:3 American Journal of Respiratory & Critical Care Medicine 582–586.

Silverman, W.A. "Informed Consent in Customary Practice and in Clinical Trials" in A. Goldworth *et al.*, eds. *Ethics and Perinatology* (New York: Oxford University Press, 1995) 245–262.

Sloan, J. & Resnick, G.D. "The Consent Form Revisited" (1993) 153:10 Archives of Internal Medicine 1170–1173.

Smith, M.L. "Power, Advocacy, and Informed Consent Forms" (1994) 5:1 APA Newsletter on Philosophy & Medicine 25–27.

Spivey, W.H. *et al.* "Informed Consent for Biomedical Research in Acute Care Medicine" (1991) 20:11 Annals of Emergency Medicine 1251–1265.

Stuart, R.B. "Protection of the Right to Informed Consent to Participate in Research" (1978) 9:1 Behavior Therapy 73–82.

Sutherland, H.J., Lockwood, G.A. & Till, J.E. "Are We Getting Informed Consent From Patients With Cancer?" (1990) 83:7 Journal of the Royal Society of Medicine 439–443.

Szczygiel, A. "Beyond Informed Consent" (1994) 21:1 Ohio Northern University Law Review 171–262.

Taub, H.A. "Comprehension of Informed Consent for Research: Issues and Directions for Future Study" (1986) 8:6 IRB 7–10.

Teff, H. "Consent to Medical Procedures: Paternalism, Self-Determination or Therapeutic Alliance?" (1985) 101 Law Quarterly Review 432–453.

Tomamichel, M. *et al.* "Informed Consent for Phase I Studies: Evaluation of Quantity and Quality of Information Provided to Patients" (1995) 6:4 Annals of Oncology 363–369.

Veatch, R.M. "Abandoning Informed Consent" (1995) 25:2 Hastings Center Report 5–12.

Weiss, A.H. "Asking about Asking: Informed Consent in Organ Donation Research" 18:1 IRB 6–10.

Wilkinson, J. "Ethical Problems of Human Experimentation and Consent" in J. Wilkinson, *Christian Ethics in Health Care* (Edinburgh: Handsel Press, 1988) 339–374.

Woody, K.J. "Legal and Ethical Concepts Involved in Informed Consent to Human Research" (1981) 18:1 California Western Law Review 50–79.

## VULNERABLE POPULATIONS

Abrams, N. "Medical Experimentation: The Consent of Prisoners and Children" in S.F. Spicker & H.T. Engelhardt, eds. *Philosophical Medical Ethics: Its Nature and Significance. Proceedings.* (Boston: D. Reidel, 1977) 111–124.

Arboleda-Florez, J. & Weisstub, D.N. "Problaciones vulnerables: protección de su condición como sujetos de investigación" (1996) 27 Quirón 54–67.

Bowker, W.F. "Minors and Mental Incompetents: Consent to Experimentation, Gifts of Tissue and Sterilization" (1981) 26:4 McGill Law Journal 951–977.

Brazier, M. & Lobjoit, M., eds. *Protecting the Vulnerable: Autonomy and Consent in Health Care* (London: Routledge, 1991).

Cusveller, B.S. & Jochemsen, H. "The New Dutch 'Medical Experimentation Bill' and Incompetent Patients" (1993) 9:2 Ethics & Medicine 18–20.

Dworkin, G. "Law and Medical Experimentation: Of Embryos, Children and Others With Limited Legal Capacity" (1987) 13:3–4 Monash University Law Review 189–208.

Fletcher, J.C., Dommel, F.W. Jr. & Cowell, D.D. "Consent to Research With Impaired Human Subjects" (1985) 7:6 IRB 1–6.

Hill, J.L. "Exploitation" (1994) 79:3 Cornell Law Review 631–699.

Iserson, K.V. & Lindsey, D. "Research on Critically Ill and Injured Patients: Rules, Reality, and Ethics" (1995) 13:4 Journal of Emergency Medicine 563–567.

Lasagna, L. "Special Subjects in Human Experimentation" in P.A. Freund, ed. *Experimentation With Human Subjects* (New York: George Braziller, 1970) 262–275.

Neki, J.S. "Selection and Recruitment of Institutionalized Subjects" in N. Howard-Jones & Z. Bankowski, eds. *Medical Experimentation and the Protection of Human Rights* (Geneva: CIOMS & The Sandoz Institute for Health and Socio-Economic Studies, 1979) 154–159.

Reid, G., Dunn, J.M. & McClements, J. "People With Disabilities as Subjects in Research" (1993) 10:4 Adapted Physical Activity Quarterly 346–358.

Verdun-Jones, S.N. & Weisstub, D.N. "Consent to Human Experimentation in Quebec: The Application of the Civil Law Principle of Personal Inviolability to Protect Special Populations" (1995) 18:2 International Journal of Law & Psychiatry 163–182.

Wecht, C.H. "Medical, Legal, and Moral Considerations in Human Experiments Involving Minors and Incompetent Adults" (1976) 4:2 Journal of Legal Medicine 27–30.

Weisstub, D.N., Arboleda-Florez, J. & Tomossy, G.F. "Establishing the Boundaries of Ethically Permissible Research with Special Populations" (1996) 17 Health Law in Canada 45–63.

### *Children*

Abramovitch, R. *et al.* "Children's Capacity to Consent to Participation in Psychological Research: Empirical Findings" in G. Koren, ed. *Textbook of Ethics in Pediatric Research* (Malabar, FL: Krieger, 1993) 11–21.

Ackerman, T.F. "Fooling Ourselves With Child Autonomy and Assent in Nontherapeutic Clinical Research" (1979) 27:5 Clinical Research 345–348.

Ackerman, T.F. "Moral Duties of Investigators Toward Sick Children" (1981) 3:6 IRB 1–5.

Ackerman, T.F "Moral Duties of Parents and Nontherapeutic Clinical Research Procedures Involving Children" (1980) 2:2 Bioethics Quarterly 94–111.

Ackerman, T.F. "Nontherapeutic Research Procedures Involving Children With Cancer" (1994) 11:3 Journal of Pediatric Oncology Nursing 134–136.

Ackerman, T.F. "Protectionism and the New Research Imperative in Pediatric AIDS" (1990) 12:5 IRB 1–5.

Akers, J.A. & Bell, S.K. "Should Children be Used as Research Subjects?" (1994) 29:3 Nursing Forum 28–33.

Armstrong, M.P. "Ethics of Research in Children: A Parent's Perspective" in G. Koren, ed. *Textbook of Ethics in Pediatric Research* (Malabar, FL: Krieger, 1993) 271–280.

Arnold, L.E. *et al.* "Ethical Issues in Biological Psychiatric Research With Children and Adolescents" (1995) 34:7 Journal of the American Academy of Child & Adolescent Psychiatry 929–939.

Australian College of Paediatrics, "Report on the Ethics of Research in Children" (1981) 17:3 Australian Paediatrics Journal 162.

Bala, N., Hornick, J.P. & Vogl, R. *Canadian Child Welfare Law: Children, Families and the State* (Toronto: Thompson Educational Publishing, 1991).

Bartholome, W.G. "Central Themes in the Debate over Involvement of Infants and Children in Biomedical Research: A Critical Examination" in J. Van Eys, ed. *Research on Children:*

*Medical Imperatives, Ethical Quandaries, and Legal Constraints* (Baltimore: University Park Press, 1978) 69–76.

Bartholome, W.G. "Parents, Children, and the Moral Benefits of Research" (1976) 6:6 Hastings Center Report 44–45.

Baylis, F. & Downie, J. "An Ethical and Criminal Law Framework for Research Involving Children in Canada" (1993) 1 Health Law Journal 39–64.

Berglund, C.A. "Children in Medical Research: Australian Ethical Standards" (1995) 21:2 Child: Care, Health and Development 149–159.

Bernard, C. & Knoppers, B.M. "Legal Aspects of Research Involving Children in Canada" in B.M. Knoppers, ed. *Canadian Child Health Law: Health Rights and Risks of Children* (Toronto: Thompson Educational Publishing, 1991) 259–330.

Berryman, J. "Discussing the Ethics of Research on Children" in J. Van Eys, ed. *Research on Children: Medical Imperatives, Ethical Quandaries, and Legal Constraints* (Baltimore: University Park Press, 1978) 85–101.

Bersoff, D.N. "Children as Research Subjects: Problems of Competency and Consent" in J.S. Henning, ed. *The Rights of Children: Legal and Psychological Perspectives* (Springfield, IL: Charles C. Thomas, 1982) 186–214.

Biggar, W.D. "Ethics of Research With the Disabled Child" in G. Koren, ed. *Textbook of Ethics in Pediatric Research* (Malabar, FL: Krieger, 1993) 117–123.

British Paediatric Association. "Guidelines to Aid Ethical Committees Considering Research Involving Children"(1980) 55:1 Archives of Disease in Childhood 75–77.

Brock, D.W. "Ethical Issues in Exposing Children to Risks in Research" in M.A. Grodin & L.H. Glantz, eds. *Children as Research Subjects: Science, Ethics, and Law* (New York: Oxford University Press, 1994) 81–101.

Broome, M.E. & Stieglitz, K.A. "The Consent Process and Children" (1992) 15:2 Research in Nursing & Health 147–152.

Committee on Bioethics, American Academy of Pediatrics. "Informed Consent, Parental Permission and Assent in Pediatric Practice" (1995) 95:2 Pediatrics 314–317.

Choudry, S. "Review of Legal Instruments and Codes on Medical Experimentation With Children" (1994) 3:4 Cambridge Quarterly of Healthcare Ethics 560–572.

Clara, R. "Research Involving Children: Scientific and Ethical Aspects" in P.P. De Deyn, ed. *The Ethics of Animal and Human Experimentation* (London: John Libbey, 1994) 303–320.

Cooke, R.E. "Vulnerable Children" in M.A. Grodin & L.H. Glantz, eds. *Children as Research Subjects: Science, Ethics, and Law* (New York: Oxford University Press, 1994) 193–214.

Coulter, D.J., Murray, T.H. & Cerreto, M.C. "Practical Ethics in Pediatrics" (1988) 18:3 Current Problems in Pediatrics 143–195.

Dickens, B.M. "Medicine and the Law – Withholding Pediatric Medical Care (1984) 62:1 Canadian Bar Review 196–210.

Dorn, L.D., Susman, E.L.J. & Fletcher, J.C. "Informed Consent in Children and Adolescents: Age, Maturation and Psychological State" (1995) 16:3 Journal of Adolescent Health 185–190.

Doucet, H. "L'experimentation sur les enfants" in G. Durand & C. Perrotin, eds. *Contribution à la réflexion bioéthique,* 3 (Quebec: Presses de l'Université Laval, 1991) 119–132.

Edwards, J.E. & Coghlan, A.L. "The Role of the Nurse in Pediatric Research" in G. Koren, ed. *Textbook of Ethics in Pediatric Research* (Malabar, FL: Krieger, 1993) 257–270.

Eiser, C. & Patterson, D. " 'Slugs and Snails and Puppy-Dog Tails': Children's Ideas about the Inside of Their Bodies" (1983) 9:3 Child: Care, Health, & Development 233–240.

Evans, M. "Conflicts of Interest in Research on Children" (1994) 3:4 Cambridge Quarterly of Healthcare Ethics 549–559.

Everson-Bates, S. "Research Involving Children: Ethical Concerns and Dilemmas" (1988) 2:5 Journal of Pediatric Health Care 234–239.

Ferguson, L.R. "The Competence and Freedom of Children to Make Choices Regarding Participation in Research: A Statement" (1978) 34:2 Journal of Social Issues 114–121.

Fisher, C.B. "Integrating Science and Ethics in Research With High-Risk Children and Youth" (1993) 7:4 Social Policy Report-Society for Research in Child Development 1–27.

Fisher, C.B. "Reporting and Referring Research Participants: Ethical Challenges for Investigators Studying Children and Youth" (1994) 4:2 Ethics & Behavior 87–95.

Fletcher, J.C., van Eys, J. & Dorn, L.D. "Ethical Considerations in Pediatric Oncology" in P.A. Pizzo & D.G. Poplack, eds. *Principles and Practice of Pediatric Oncology*, 2d ed. (Philadelphia: J.B. Lippincott, 1993) 1179–1191.

Forman, E.N. & Ladd, R.E. *Ethical Dilemmas in Pediatrics: A Case Study Approach* (New York: Springer-Verlag, 1991).

Frankel, M.S. "Social, Legal, and Political Responses to Ethical Issues in the Use of Children as Experimental Subjects" (1978) 34:2 Journal of Social Issues 101–113.

Freedman, B., Fuks, A. & Weijer, C. "*In Loco Parentis*: Minimal Risk as an Ethical Threshold for Research upon Children" (1993) 23:2 Hastings Center Report 13–19.

Freeman, W.L. "Research With Radiation and Healthy Children: Greater than Minimal Risk" (1994) 16:5 IRB 1–5.

Fried, C. "Children as Subjects for Medical Experimentation" in J. van Eys, ed. *Research on Children: Medical Imperatives, Ethical Quandaries, and Legal Constraints* (Baltimore: University Park Press, 1978) 107–115.

Friman, P.C. "Take Away Their Hammer: Logical and Ethical Problems in Range and Cotton's "Reports of Assent and Permission in Research With Children: Illustrations and Suggestions" (1995) 5:4 Ethics & Behavior 349–353.

Gaylin, W. "Competence No Longer All or None" in W. Gaylin & R. Macklin, eds. *Who Speaks for the Child: The Problems of Proxy Consent* (New York: Plenum Press, 1982) 27–56.

Gaylin, W. & Macklin, R., eds. *Who Speaks for the Child* (New York: Plenum Press, 1982).

Glantz, L.H. "The Law of Human Experimentation With Children" in M.A. Grodin & L.H. Glantz, eds. *Children as Research Subjects: Science, Ethics, and Law* (New York: Oxford University Press, 1994) 103–130.

Goldberg, S. "Some Costs and Benefits of Psychological Research in Pediatric Settings" in G. Koren, ed. *Textbook of Ethics in Pediatric Research* (Malabar, FL: Krieger, 1993) 63–73.

Gray, J.N. "Pediatric AIDS Research: Legal, Ethical, and Policy Influences" in J.M. Seibert & R.A. Olson, eds. *Children, Adolescents, and AIDS* (Lincoln, NB: University of Nebraska Press, 1989) 179–227.

Grisso, T. "Minors' Assent to Behavioural Research Without Parental Consent" in B. Stanley & J. Seiber, *Social Research on Children and Adolescents: Ethical Issues* (Newbury Park, NJ: Sage Publications, 1992) 109.

Grisso, T. & Vierling, L. "Minor's Consent to Treatment" (1982) 9:3 Professional Psychology 412–427.

Grodin, M.A. & Alpert, J.J. "Children as Participants in Medical Research" (1988) 35:6 Pediatric Clinics of North America 1389–1401.

Grodin, M.A. & Alpert, J.J. "Informed Consent and Pediatric Care" in G. Melton, G. Koocher & M.J. Saks, eds. *Children's Competence to Consent* (New York: Plenum Press, 1983) 93.

Grodin, M.A. & Glantz, L., eds. *Children as Research Subjects: Science, Ethics, and Law* (New York: Oxford University Press, 1994).

Group for the Advancement of Psychiatry, Committee on Child Psychiatry. *How Old is Enough? The Ages of Rights and Responsibilites: Report 126* (New York: Brunner/Mazel, 1989).

Hall, D. "Reviewing Research Involving Children: The Practice of British Research Ethics Committees" (1988) 10:2 IRB 1–5.

Harth, S.C. & Thong, Y.H. "Parental Perceptions and Attitudes About Informed Consent in Clinical Research Involving Children" (1995) 40:11 Social Science & Medicine 1573–1577.

Harth, S.C., Johnstone, R.R. & Thong, Y.H. "The Psychological Profile of Parents Who Volunteer Their Children for Clinical Research: A Controlled Study" (1992) 18:2 Journal of Medical Ethics 86–93.

Haslam, R.H.A. "Research in Children – Issues of Risk and Harm" in G. Koren, ed. *Textbook of Ethics in Pediatric Research* (Malabar, FL: Krieger, 1993) 25–36.

Holder, A. "Can Teenagers Participate in Research Without Parental Consent?" (1981) 3:2 IRB 5–8.

Holder, A. "Constraints on Experimentation: Protecting Children to Death" (1988) 6:1 Yale Law & Policy Review 137–156.

Holder, A. "Disclosure and Consent Problems in Pediatrics" (1988) 16:3–4 Law, Medicine & Health Care 219–228.

Holder, A. "Mental Illness and Parental Rights" (1971) 216:13 JAMA. 575–576.

Holder, A.R. "The Minor as Research Subject or Transplant Donor" in A.R. Holder, *Legal Issues in Pediatrics and Adolescent Medicine*, 2d ed. (New Haven, CT: Yale University Press, 1985) 146–178.

Ingelfinger, F.J. "Ethics of Experimentation on Children" (1973) 288:15 New England Journal of Medicine 791–792.

Jaffe, D. & Matsui, D. "Ethics of Pediatric Research in the Emergency Department" in G. Koren, ed. *Textbook of Ethics in Pediatric Research* (Malabar, FL: Krieger, 1993) 147–153.

Janofsky, J. & Starfield, B. "Assessment of Risk in Research on Children" (1981) 98:5 Journal of Pediatrics 842–846.

Kapp, M.B. "Children's Assent for Participation in Pediatric Research Protocols: Assessing National Practice" (1983) 22:4 Clinical Pediatrics 275–278.

Kaufmann, R. "Drug Trials in Children: Ethical, Legal, and Practical Issues" (1994) 34:4 Journal of Clinical Pharmacology 296–299.

Kaufmann, R. "Scientific Issues in Biomedical Research With Children" in M.A. Grodin & L. Glantz, eds. *Children as Research Subjects: Science, Ethics and Law* (New York: Oxford University Press, 1994) 29–45.

Keens, T.G. *et al.* "Does Radiation Research in Healthy Children Pose Greater than Minimal Risk?" (1994) 16:5 IRB 5–10.

Keith-Spiegel, P. "Children and Consent to Participate in Research" in G.B. Melton, G.P Koocher & M.J. Saks, eds. *Children's Competence to Consent* (New York: Plenum Press, 1983) 179.

Kister, M.C. & Patterson, C.J. "Children's Conception of the Cause of Illness: Understanding of Contagion and Use of Immanent Justice" (1980) 51:3 Child Development 839–846.

Kopelman, L. & Moskop, J., eds. *Children and Health Care: Moral and Social Issues* (New York: Kluwer Academic Publishers, 1989)

Kopelman, L.M. "When is the Risk Minimal Enough for Children to be Research Subjects?" in L.M. Kopelman and J.C. Moskop, eds. *Children and Health Care: Moral and Social Issues* (New York: Kluwer Academic Publishers, 1989) 89–99.

Koren, G. "A Practical Approach to Risk-Benefit Estimation in Pediatric Research" in G. Koren, ed. *Textbook of Ethics in Pediatric Research* (Malabar, FL: Krieger, 1993) 57–62.

Koren, G. "Ethical Boundaries of Medical Research in Infants and Children in the 80s: Analysis of Rejected Protocols and a New Solution for Drug Studies" (1990) 15:3–4 Developmental Pharmacology & Therapeutics 130–141.

Koren, G. "Ethics of Drug Research in Pregnancy, Infancy and Childhood" in G. Koren, ed. *Textbook of Ethics in Pediatric Research* (Malabar, FL: Krieger, 1993) 171–181.

Koren, G. "Informed Consent in Pediatric Research" in G. Koren, ed. *Textbook of Ethics in Pediatric Research* (Malabar, FL: Krieger, 1993) 3–10.

Koren, G., ed. *Textbook of Ethics in Pediatric Research* (Malabar, FL: Krieger Publishing, 1993).

Koren, G. et al. "Maturity of Children to Consent to Medical Research: The Babysitter Test" (1993) 19:3 Journal of Medical Ethics 142–147.

Langer, D.H. "Medical Research Involving Children: Some Legal and Ethical Issues" (1984) 36:1 Baylor Law Review 1–39.

Lantos, J.D. "ECMO, Innovative Therapy, and the Ethos of Critical Care Medicine" in G. Koren, ed. *Textbook of Ethics in Pediatric Research* (Malabar, FL: Krieger, 1993) 99–106.

Laor, N. "Toward Liberal Guidelines for Clinical Research With Children" (1987) 6:2 Medicine & Law 127–137.

Lee, L. "Ethical Issues Related to Research Involving Children" (1991) 8:1 Journal of Pediatric Oncology Nursing 24–29.

Leikin, S. "Minors' Assent, Consent, or Dissent to Medical Research" (1993) 15:2 IRB 1–7.

Levine, C. "Children in HIV/AIDS Clinical Trials: Still Vulnerable After All These Years" (1991) 19:3–4 Law, Medicine & Health Care 231–237.

Levine, R.J. "Adolescents as Research Subjects Without Permission of Their Parents or Guardians: Ethical Considerations" (1995) 17:5 Journal of Adolescent Health 287–297.

Levine, R.J. "Children as Research Subjects" in L.M. Kopelman & J.C. Moskop, eds. *Children and Health Care: Moral and Social Issues* (Boston: Kluwer Academic Publishers, 1989) 73–87.

Lewis, C.E., Lewis, M.A. & Ifekwunigue, M. "Informed Consent by Children and Participation in an Influenza Vaccine Trial" (1978) 68:11 American Journal of Public Health 1079–1082.

Lewis, M. "Organ Transplants, Research, and Children" in D.H. Schetky & E.P. Benedek, eds. *Emerging Issues in Child Psychiatry and the Law* (New York: Brunner/Mazel, 1985) 44–56.

Lewis, M. "Comments on Some Ethical, Legal, and Clinical Issues Affecting Consent in Treatment, Organ Transplants, and Research in Children" (1982) Annual Progress in Child Psychiatry & Child Development 651–666.

Lewis, M. "Nonbeneficial, Risk-Involved, Psychiatric Research in Children: A Conflict of Interests" in J.C. Schoolar & C.M. Gaitz, eds. *Research and the Psychiatric Patient* (New York: Brunner/Mazel, 1975) 153–164.

Lynch, A. "Ethics of Research Involving Children" (1992) 25:6 Annals of the Royal College of Physicians & Surgeons of Canada 371–372.

Lynch, A. "Research Involving Adolescents: Are They Ethically Competent to Consent/ Refuse on Their Own?" in G. Koren, ed. *Textbook of Ethics in Pediatric Research* (Malabar, FL: Krieger, 1993) 125–136.

McCormick, R. "Experimentation in Children: Sharing in Sociality" (1976) 6:6 Hastings Center Report 41–46.

McLean, S. "Medical Experimentation With Children" (1992) 6:1 International Journal of Law and the Family 173–191.

Melton, G.B. "Ethical and Legal Issues in Research and Intervention" (1989) 10:3S Journal of Adolescent Health Care 36S–44S.

Melton, G.B., Koocher, G.P. & Saks, M.J. *Children's Competence to Consent* (New York: Plenum Press, 1983).

Millstein, S.G., Adler, N.E. & Irwin, C.E. "Conceptions of Illness in Young Adolescents" (1981) 68:6 Pediatrics 834–839.

Munir, K. & Earls, F. "Ethical Principles Governing Research in Child and Adolescent

Psychiatry" (1992) 31:3 Journal of the American Academy of Child & Adolescent Psychiatry 408–414.

Nassiry, L. "Health Science Research Ethics and the Law: Research Involving Children" (1993) 4:2 NCBHR Communiqué 11–13.

National Commission for the Protection of Human Subjects of Biomedical and Behavioral Research. *Research Involving Children: Report and Recommendations* (Washington, DC: U.S. Government Printing Office, 1977).

National Council on Bioethics in Human Research, "Ethics of Research Involving Children: Proceedings of a National Workshop" (1993) 4:1 NCBHR Communiqué 12–41.

National Council on Bioethics in Human Research, Consent Panel Task Force. *Reflections on Research Involving Children* (Ottawa: NCBHR, 1993).

Nicholson, R.H. *Medical Research With Children: Ethics, Law, and Practice* (Oxford: Oxford University Press, 1986).

O'Sullivan, G. "Studies Involving Children" in R.A. Greenwald, M.K. Ryan & J.E. Mulvihill, eds. *Human Subjects Research: A Handbook for Institutional Review Boards* (New York: Plenum Press, 1982) 139–150.

Outwater, K.M. "Ethics of Research in Pediatric Critical Care" in G. Koren, ed. *Textbook of Ethics in Pediatric Research* (Malabar, FL: Krieger, 1993) 107–115.

Parkin, P. "The Relationship Between Ethics and Science in Pediatric Research" in G. Koren, ed. *Textbook of Ethics in Pediatric Research* (Malabar, FL: Krieger, 1993) 221–234.

Pearn, J. "A Classification of Clinical Paediatric Research With Analysis of Related Ethical Themes" (1987) 13:1 Journal of Medical Ethics 26–30.

Pearn, J. "The Child and Clinical Research" (1984) 2:8401 Lancet 510–512.

Pence, G.E. "Children's Dissent to Research – A Minor Matter?" (1980) 2:10 IRB 1–4.

Perlman, M. & Koren, G. "Social Control of Pediatric Research" in G. Koren, ed. *Textbook of Ethics in Pediatric Research* (Malabar, FL: Krieger, 1993) 281–292.

Perrin, E.C. & Gerrity, P.S. "There's a Demon in Your Belly: Children's Understanding of Illness" (1981) 67:6 Pediatrics 841–849.

Pinkus, R.L. & Haines, S.J. "The Rights of Children Involved in Research" in M.D. Hiller, ed. *Medical Ethics and the Law: Implications for Public Policy.* (Cambridge, Mass: Ballinger, 1981) 421–440.

Rae, W.A., Worchel, F.F. & Brunnquell, D. *Ethical and Legal Issues in Pediatric Psychology* (New York: Guilford Press, 1995).

Ramsey, P. "Ethical Dimensions of Experimental Research on Children" in J. Van Eys, ed. *Research on Children: Medical Imperatives, Ethical Quandaries, and Legal Constraints* (Baltimore: University Park Press, 1978) 57–68.

Ramsey, P. "The Enforcement of Morals: Nontherapeutic Research on Children — A Reply to Richard McCormick" (1976) 6:4 Hastings Center Report 21–30.

Range, L.M. & Cotton, C.R. "Assent and Permission Rejoinder" (1995) 5:4 Ethics & Behavior 345–347.

Range, L.M. & Cotton, C.R. "Reports of Assent and Permission in Research With Children: Illustrations and Suggestions" (1995) 5:1 Ethics & Behavior 49–66.

Redmon, R.B. "How Children Can Be Respected As 'Ends' Yet Still Be Used As Subjects in Non-Therapeutic Research" (1986) 12:2 Journal of Medical Ethics 77–82.

Richie, E. "Biomedical Research: For the Patient or On the Patient?" in J. Van Eys, ed. The *Truly Cured Child: The New Challenge in Pediatric Cancer Care* (Baltimore: University Park Press, 1977) 25–37.

Roberts, M.C. & Buckloh, L.M. "Five Points and a Lament about Range and Cotton's "Reports of Assent and Permission in Research With Children: Illustrations and Suggestions" (1995) 5:4 Ethics & Behavior 333–344.

Robinson, G. "Ethics Committees and Research in Children" (1987) 294:6582 British Medical Journal 1243–1244.

Rogers, A.S., D'Angelo, L. & Futterman, D. "Guidelines for Adolescent Participation in Research: Current Realities and Possible Solutions" (1994) 16:4 IRB 1–6.

Rowell, M. & Zlotkin, S. "The Ethical Boundaries of Drug Research in Pediatrics" (1997) 44 The Pediatric Cloinics of North America 27–40.

Sann, L. "L'éthique de l'éxperimentation. Reflexions critiques d'un pédiatre" in G. Durand & C. Perrotin, eds. *Contribution à la réflexion bioéthique* (Montreal: Dialogue France-Quebec, 1991) 133–141.

Schoeman, F. "Children's Competence and Children's Rights" (1982) 4:6 IRB 1–6.

Scott, E.S., Reppucci, N.D. & Woolard, J.L. "Evaluating Adolescent Decision Making in Legal Contexts" (1995) 19:3 Law & Human Behavior 221–244.

Shirkey, H.C. "Therapeutic Orphans" (1968) 72:1 Journal of Pediatrics 119–120.

Simeon, J.G. & Wiggins, D.M. "The Placebo Problem in Child and Adolescent Psychiatry" (1993) 56 Acta Paedopsychiatrica 119–122

Smith, B.M. "Conflicting Values Affecting Behavioral Research With Children" (1967) 22:5 American Psychologist 377–382.

Smith, W.C., Eisenberg, L. & Halpern, C.R. "Individual Risks vs. Societal Benefits: What Consent is Needed? The Child" in National Academy of Sciences, *Experiments and Research With Humans: Values in Conflict* (Washington, DC: National Academy of Sciences, 1975) 90–117.

Susman, E., Dorn, L. & Fletcher, J. "Participation in Biomedical Research: The Consent Process As Viewed by Children, Adolescents, Young Adults and Physicians" (1992) 121:4 Journal of Pediatrics 547–552.

Thomasma, D.C. & Mauer, A.M. "Ethical Complications of Clinical Therapeutic Research on Children" (1982) 16:8 Social Science & Medicine 913–919.

Thompson, R.A. "Child Development and Research Ethics: A Changing Calculus of Concerns" (1990) 9:1–2 Business and Professional Ethics Journal 193–206.

Thurber, F.W., Deatrick, J.A. & Grey, M. "Children's Participation in Research: Their Right to Consent" (1992) 7:3 Journal of Pediatric Nursing 165–170.

Vitiello, B. & Jensen, S. "Medication Development and Testing in Children and Adolescents: Current Problems, Future Directions" (1997) 54 Archives of General Psychiatry 871–876.

Wecht, C.H. "Medical, Legal and Moral Considerations in Human Experiments Involving Minors and Incompetent Adults" in C.H. Wecht (Chair), *Medical Ethics and Legal Liability* (New York: Practising Law Institute, 1976) 7–26.

Weithorn, L. "Children's Capacities to Decide About Participation in Research" (1983) 5:2 IRB 1–5.

Weithorn, L. "Developmental Factors and Competence to Make Informed Treatment Decisions" (1982) 5:1–2 Child & Youth Services 85–100.

Weithorn, L. & Campbell, S. "The Competency of Children and Adolescents to Make Informed Treatment Decisions" (1982) 53:6 Child Development 1589–1598.

Weithorn, L. & Scherer, D.G. "Children's Involvement in Research Participation Decisions: Psychological Considerations" in M.A. Grodin & L.H. Glantz, eds. *Children as Research Subjects: Science, Ethics, and Law* (New York: Oxford University Press, 1994) 133–179.

Wellman, C. "Consent to Medical Research on Children" in T.L. Beauchamp & T.P. Pinkard, eds. *Ethics and Public Policy: An Introduction to Ethics* (Englewood Cliffs, NJ: Prentice-Hall, 1983) 369–386.

Williams, P.C. "Ethical Principles in Federal Regulations: The Case of Children and Research Risks" (1996) 21:2 Journal of Medicine & Philosophy 169–186.

Wilson, L.S. "Regulation of Research Involving Children: Origins, Costs, and Benefits" (1982) 1 Advances in Law & Child Development 153–179.

Working Party on Research on Children. *The Ethics of Research on Children* (London: MRC, 1991).

Yaffe, S.J. "Problems of Drug Testing in Children in the United States" (1983) 3:3–4 Pediatric Pharmacology 339–348.

### *The Elderly*

Alt-White, A.C. "Obtaining "Informed" Consent From the Elderly" (1995) 17:6 Western Journal of Nursing Research 700–705.

American Psychiatric Association Task Force on Alzheimer's Disease. "The Alzheimer's Disease Imperative — The Challenge for Psychiatry" (1988) 145:12 American Journal of Psychiatry 1550–1551.

Ancill, R.J. "Medication Studies in Alzheimer's Disease: Methodological and Consent Concerns" in J.M. Berg, H. Karlinsky & F.H. Lowy, eds. *Alzheimer's Disease Research: Ethical and Legal Issues* (Toronto: Carswell, 1991) 296–304.

Annas, G.J. & Glantz, L.H. "Rules for Research in Nursing Homes" (1986) 315:18 New England Journal of Medicine 1157–1158.

Berg, J.M. *et al.* "Ethical and Legal Guidelines for Alzheimer's Disease Research: Progress, Problems and Current Recommendations" in J.M. Berg, H. Karlinsky & F.H. Lowy, eds. *Alzheimer's Disease Research: Ethical and Legal Issues* (Toronto: Carswell, 1991) 333–354.

Berg, J.M., Karlinsky, H. & Lowy, F.H., eds. *Alzheimer's Disease Research: Ethical and Legal Issues* (Toronto: Carswell, 1991).

Berghmans, R.L.P. & Ter Meulen, R.H.J. "Ethical Issues in Research With Dementia Patients" (1995) 10:8 International Journal of Geriatric Psychiatry 647–651.

Bernstein, J.E. & Nelson, F.K. "Medical Experimentation in the Elderly" (1975) 23:7 Journal of the American Geriatrics Society 327–329.

Bowsher, J. et al. "Methodological Considerations in the Study of Frail Elderly People" (1993) 18:6 Journal of Advanced Nursing 873–874.

Brod, M.S. & Feinbloom, R.I. "Feasibility and Efficacy of Verbal Consents" (1990) 12:3 Research on Aging 364–372.

Campion, E.W. "Ethical Issues in the Care of the Patient Involved in Alzheimer's Disease Research" in V.L. Melnick & N.N. Dubler, eds. Alzheimer's Dementia: Dilemmas in Clinical Research (Clifton, NJ: Humana Press, 1985) 71–78.

Cassel, C.K. "Ethical Issues in the Conduct of Research in Long Term Care" (1988) 28S Gerontologist 90–96.

Cassel, C.K. "Informed Consent for Research in Geriatrics: History and Concepts" (1987) 35:6 Journal of the American Geriatrics Society 542–544.

Cassel, C.K. "Research in Geriatrics" (1985) 10:2 Generations 45–48.

Cassel, C.K. "Research in Nursing Homes: Ethical Issues" (1985) 33:11 Journal of the American Geriatrics Society 795–799.

Cassel, C.K. "Research on Senile Dementia of the Alzheimer's Type: Ethical Issues Involving Informed Consent" in V.L. Melnick & N.N. Dubler, eds. *Alzheimer's Dementia: Dilemmas in Clinical Research* (Clifton, NJ: Humana Press, 1985) 99–108.

Cohen, E.S. "Realism, Law and Aging" (1990) 18:3 Law, Medicine & Health Care 183–192.

Congress of the United States, Office of Technology Assessment. *Losing a Million Minds: Confronting the Tragedy of Alzheimer's Disease and other Dementias* (Washington, DC: U.S. Government Printing Office, 1987).

Council of Scientific Affairs. "Elder Abuse and Neglect" (1987) 257:7 Journal of the American Medical Association 966–971.

Cutler, N.R. & Sramek, J.J. "Scientific and Ethical Concerns in Clinical Trials in Alzheimer's Patients: The Bridging Study" (1995) 48:6 European Journal of Clinical Pharmacology 421–428.

Dubler, N.N. "Legal Issues in Research on Institutionalized Demented Patients" in V.L.

Melnick & N.N. Dubler, eds. *Alzheimer's Dementia: Dilemmas in Clinical Research.* (Clifton, NJ: Humana Press, 1985) 149–173.

Dubler, N.N. "Legal Judgments and Informed Consent in Geriatric Research" (1987) 35:6 Journal of the American Geriatrics Society 545–549.

Dyer, A.R. "Assessment of Competence to Give Informed Consent" in V.L. Melnick & N.N. Dubler, eds. *Alzheimer's Dementia: Dilemmas in Clinical Research* (Clifton, NJ: Humana Press, 1985) 227–237.

Eisch, J.S. *et al.* "Issues in Implementing Clinical Research in Nursing Home Settings" (1991) 22:3 Journal of the New York State Nurses Association 18–22.

Geiselmann, B & Helmchen, H. "Demented Subjects' Competence to Consent to Participate in Field Studies: the Berlin Aging Study" (1994) 13:1–2 Medicine & Law 177–184.

Glass, K.C. "Informed Decision Making and Vulnerable Persons: Meeting the Needs of the Competent Elderly Patient or Research Subject" (1993) 18:1 Queens Law Journal 191–238.

Glass, K.C. & Somerville, M.A. "Informed Consent to Medical Research on Persons With Alzheimer's Disease: Ethical and Legal Parameters" in J.M. Berg, H. Karlinsky & F.H. Lowy, eds. *Alzheimer's Disease Research: Ethical and Legal Issues.* (Toronto: Carswell, 1991) 30–59.

Hayley, D.C. *et al.* "Ethical and Legal Issues in Nursing Home Care" (1996) 156:3 Archives of Internal Medicine 249–256.

Helmchen, H. "The Problem of Informed Consent in Dementia Research" (1990) 9:6 Medicine & Law 1206–1213.

High, D.M. "Advancing Research With Alzheimer Disease Subjects: Investigators' Perceptions and Ethical Issues" (1993) 7:3 Alzheimer Disease & Associated Disorders 165–178.

High, D.M. "Research With Alzheimer's Disease Subjects: Informed Consent and Proxy Decision Making" (1992) 40:9 Journal of the American Geriatrics Society 950–957.

High, D.M. & Doole, M.M. "Ethical and Legal Issues in Conducting Research Involving Elderly Subjects" (1995) 13:3 Behavioral Sciences & the Law 319–335.

High, D.M., Post, S.G. & Whitehouse, P.J. "Ethical and Legal Issues in Alzheimer Disease Research. Introduction" (1994) 8:S4 Alzheimer Disease & Associated Disorders 1–4.

High, D.M. *et al.* "Guidelines for Addressing Ethical and Legal Issues in Alzheimer Disease Research: A Position Paper" (1994) 8:S4 Alzheimer Disease & Associated Disorders 66–74.

Hoffman, P.B. & Libow, L.S. "The Need for Alternatives to Informed Consent by Older Patients: Psychological and Physical Aspects of the Institutionalized Elderly" in V.L. Melnick & N.N. Dubler, eds. *Alzheimer's Dementia: Dilemmas in Clinical Research.* (Clifton, NJ: Humana Press 1985) 141–148.

Jameton, A. "An Alternative Approach to Informed Consent in Research With Vulnerable Patients" in V.L. Melnick & N.N. Dubler, eds. *Alzheimer's Dementia: Dilemmas in Clinical Research* (Clifton, NJ: Humana Press, 1985) 109–122.

Jarvik, L.F. *et al.* "Clinical Drug Trials in Alzheimer Disease. What are Some of the Issues?" (1990) 4:4 Alzheimer Disease & Associated Disorders 193–202.

Kapp, M.B. "Introduction: Law and Aging" (1990) 18:3 Law, Medicine & Health Care 181–182.

Kapp, M.B. "Elder Mistreatment: Legal Internventions and Policy Uncertainties" (1995) 13:3 Behavioral Sciences and the Law 319–335.

Kapp, M.B. & Bigot, A. *Geriatrics and the Law: Patient Rights and Professional Responsibilities* (New York: Springer, 1985).

Karlinsky, H. & Lennox, A. "Assessment of Competency of Persons With Alzheimer's Disease to Provide Consent for Research" in J.M. Berg, H. Karlinsky & F.H. Lowy, eds. *Alzheimer's Disease Research: Ethical and Legal Issues.* (Toronto: Carswell, 1991) 76–90.

Kendell, R.E. "Ethics of Research With Dementia Sufferers: Comment" (1989) 4:4 International Journal of Geriatric Psychiatry 239–241.

Keyserlingk, E.W. *et al.* "Proposed Guidelines for the Participation of Person With Dementia as Research Subjects" (1995) 38:2 Perspectives in Biology & Medicine 319–361.

Kitwood, T. "Exploring the Ethics of Dementia Research: A Response to Berghmans and Ter Meulen: A Psychosocial Perspective" (1995) 10:8 International Journal of Geriatric Psychiatry 655–657.

Kristjanson, L.J., Hanson, E.J. & Balneaves, L. "Research in Palliative Care Populations: Ethical Issues" (1994) 10:3 Journal of Palliative Care 10–15.

Lane, L.W., Cassel, C.K. & Bennet, W. "Ethical Aspects of Research Involving Elderly Subjects: Are We Doing More Than We Say?" (1990) 1:4 Journal of Clinical Ethics 278–285.

Lonergan, E.T. & Krevans, J.R. "A National Agenda for Research on Aging" (1991) 324:25 New England Journal of Medicine 1825–1828.

Lynch, A. "Research Use of Individuals With Alzheimer's Disease: The Ethical Challenges" in J.M. Berg, H. Karlinsky & F.H. Lowy, eds. *Alzheimer's Disease Research: Ethical and Legal Issues* (Toronto: Carswell, 1991) 3–19.

Lynch, A. & Ratzan, R.M. " 'Being Old Makes You Different': The Ethics of Research With Elderly Subjects" (1980) 10:5 Hastings Center Report 32–42.

Marshall, T.D. & Dickens, B.M. "Basic Canadian Legal Principles Concerning Research Involving Human Subjects, With Implications for Alzheimer's Disease" in J.M. Berg, H. Karlinsky & F.H. Lowy, eds. *Alzheimer's Disease Research: Ethical and Legal Issues* (Toronto: Carswell, 1991) 20–29.

Marson, D.C. *et al.* "Determining the Competency of Alzheimer Patients to Consent to Treatment and Research" (1994) 8:S4 Alzheimer Disease & Associated Disorders 5–18.

McIntosh, H.D. "Ethical Considerations in the Conduct of Cardiovascular Research in the Elderly: An American Perspective" in A. Martin & J.A. Camm, eds. *Geriatric Cardiology: Principles and Practice* (New York: Wiley, 1994 ) 771–780.

Melnick, V.L. et al. "Clinical Research in Senile Dementia of the Alzheimer Type: Suggested Guidelines Addressing the Ethical and Legal Issues" (1984) 32:7 Journal of the American Geriatrics Society 531–536.

Miller, S.T., Applegate, W.B. & Perry, C. "Clinical Trials in Elderly Persons" (1985) 33:2 Journal of the American Geriatrics Society 91–92.

Moody, H.R. "Issues of Equity in the Selection of Subjects for Experimental Research on Senile Dementia of the Alzheimer's Type" in V.L. Melnick & N.N. Dubler, eds. *Alzheimer's Dementia: Dilemmas in Clinical Research* (Clifton, NJ: Humana Press, 1985) 175–189.

Moorhouse, A. "Ethical and Legal Issues Associated With Alzheimer's Disease Research and Patient Care: To Do Good Without Doing Harm" in S.N. Verdun-Jones & M. Layton, eds. *Mental Health Law and Practice Through the Life Cycle: Proceedings From the XVIIth International Congress on Law and Mental Health* (Burnaby, BC: Simon Fraser University, 1994) 43.

Nilstun, T. "Theory and Methods for Research on Ethical Issues in Dementia Care" (1992) 6:3 Scandinavian Journal of Caring Sciences 173–177.

Ozanne, E. "Informed Consent and the Elderly: Professional Defence or Consumer Right?" in Law Reform Commission of Victoria, *Medicine, Science and the Law: Informed Consent* (Melbourne: Globe Press, 1987) 50–67.

Paschall, N.C. "Advocacy for Persons With Senile Dementia" in V.L. Melnick & N.N. Dubler, eds. *Alzheimer's Dementia: Dilemmas in Clinical Research* (Clifton, NJ: Humana Press, 1985) 59–69.

Perry, C.B. & Miller, S.T. "Ethical Consideration of Clinical Research" (1986) 34:1 Journal of the American Geriatrics Society 49–51.

Podnieks, E. *et al. National Survey on Abuse of the Elderly in Canada* (Toronto: Ryerson Polytechnical Institute, 1990).

Post, S.G. "Alzheimer's Disease: Ethics and the Progression of Dementia" (1994) 10:2 Clinics in Geriatric Medicine 379–384.

Post, S.G., Ripich, D.N. & Whitehouse, P.J. "Discourse Ethics: Research, Dementia and Communication" (1994) 8:S4 Alzheimer Disease & Associated Disorders 58–65.

Proctor, A.W. "Ethical Issues in Research With Dementia Patients: A Neuroscience Perspective: A Response to Berghmans and Ter Meulen" (1995) 10:8 International Journal of Geriatric Psychiatry 653–654.

Ratzan, R.M. "Technical Aspects of Obtaining Informed Consent From Persons With Senile Dementia of the Alzheimer's Type" in V.L. Melnick & N.N. Dubler, eds. *Alzheimer's Dementia: Dilemmas in Clinical Research* (Clifton, NJ: Humana Press, 1985) 123–139.

Reisberg, B. "Clinical Symptoms Accompanying Progressive Cognitive Decline and Alzheimer's Disease: Relationship to "Denial" and Ability to give Informed Consent" in V.L. Melnick & N.N. Dubler, eds. *Alzheimer's Dementia: Dilemmas in Clinical Research* (Clifton, NJ: Humana Press, 1985) 19–39.

Reseau, L.S. "Obtaining Informed Consent in Alzheimer's Research" (1995) 27:1 Journal of Neuroscience Nursing 57–60.

Sachs, G.A. & Cassel, C.K. "Biomedical Research Involving Older Human Subjects" (1990) 18:3 Law, Medicine & Health Care 234–243.

Sachs, G.A. & Cassel, C.K. "Ethical Aspects of Dementia" (1989) 7:4 Neurologic Clinics 845–858.

Sachs, G.A., Rhymes, J. & Cassel, C.K. "Biomedical and Behavioral Research in Nursing Homes: Guidelines for Ethical Investigations" (1993) 41:7 Journal of the American Geriatrics Society 771–777.

Schwartz, R.L. "Informed Consent to Participation in Medical Research Employing Elderly Human Subjects" (1985) 1 Journal of Comtemporary Health Law & Policy 115–131.

Silver, H.M. "Alzheimer's Disease: Ethical and Legal Decisions" (1987) 6:6 Medicine & Law 537–551.

Stanley, B., ed. *Geriatric Psychiatry: Clinical, Ethical and Legal Issues* (Washington, DC: American Psychological Association Press, 1985).

Stanley, B. "Senile Dementia and Informed Consent" (1982) 1:4 Behavioral Sciences & the Law 57–72.

Stanley, B., Stanley, M. & Pomara, N. "Informed Consent and Geriatric Patients" in B. Stanley, ed. *Geriatric Psychiatry: Clinical, Ethical and Legal Issues* (Washington, DC: American Psychological Association Press, 1985) 17–36.

Stanley, B. *et al.* "The Elderly Patient and Informed Consent: Empirical Findings" (1984) 252:10 J.A.M.A. 1302–1306.

Swift, C.G. "Ethical Aspects of Clinical Research in the Elderly" (1988) 40 British Journal of Hospital Medicine 370–373.

Taub, H.A. & Baker, M.T. "A Reevaluation of Informed Consent in the Elderly: A Method of Improving Comprehension through Direct Testing" (1984) 32:1 Clincial Research 17–21.

Taub, H.A., Kline, G.E. & Baker, M.T. "The Elderly and Informed Consent: Effects of Vocabulary Level and Corrected Feedback" (1981) 7:2 Experimental Aging Research 137–146.

Tymchuk, A.J. *et al.* "Medical Decision Making Among Elderly People in Long-Term Care" (1988) 28S Gerontologist 59–63.

Tymchuk, A.J., Ouslander, J.G. & Rader, N. "Informing the Elderly: A Comparison of Four Methods" (1986) 34:11 Journal of the American Geriatrics Society 818–822.

Warren, J.H. *et al.* "Informed Consent by Proxy. An Issue in Research With Elderly Patients" (1986) 315:18 New England Journal of Medicine 1124–1128.

Wicclair, M.R. *Ethics and the Elderly* (New York: Oxford University Press, 1993).

Wichman, A. & Sandler, A.L. "Research Involving Subjects With Dementia and Other Cognitive Impairments: Experience at the NIH, and Some Unresolved Ethical Considerations" (1995) 45:9 Neurology 1777–1778.

Zimmer, A.W. *et al.* "Conducting Clinical Research in Geriatric Populations" (1985) 103:2 Annals of Internal Medicine 276–283.

### *The Mentally Disordered*

Appelbaum, P.S. & Roth, L.H. "Competency to Consent to Research: A Psychiatric Overview" (1982) 39:8 Archives of General Psychiatry 951–958.

Arboleda-Florez, J.A. & Weisstub, D.N. "Ethical Research with the Mentally Disordered" (1997) 42 Canadian Journal of Psychiatry 485–491.

Barchas, J.D. & Barchas, I.D. "The Imperative for Research on Severe Mental Disorders" (1994) 5:1 APA Newsletter on Philosophy & Medicine 18–19.

Cosyns, P. "Psychiatric Patients as Research Subjects" in P.P. De Deyn, ed. *The Ethics of Animal and Human Experimentation* (London: John Libbey, 1994) 295–301.

Curran, W.J. "Ethical and Legal Considerations in High Risk Studies of Schizophrenia" (1974) 10 Schizophrenia Bulletin 74–92.

Davis, A.J. & Mahon, K.A. "Research With the Mentally Retarded and Mentally Ill: Rights and Duties Versus Compelling State Interest" (1984) 9:1 Journal of Advanced Nursing 15–21.

Delano, S.J. & Zucker, J.L. "Protecting Mental Health Research Subjects Without Prohibiting Progress" (1994) 45:6 Hospital & Community Psychiatry 601–603.

DeRenzo, E.G. "The Ethics of Involving Psychiatrically Impaired Persons in Research" (1994) 16:6 IRB 7–9, 11.

Dworkin, R.J. *Researching Persons With Mental Illness* (Newbury Park, NY: Sage Publications, 1992).

Eichelman, B., Wikler, D. & Hartwig, A.C. "Ethics and Psychiatric Research: Problems and Justification" (1984) 141:3 American Journal of Psychiatry 400–405.

Fulford, K.W.M. & Howse, K. "Ethics of Research With Psychiatric Patients: Principles, Problems and the Primary Responsibilities of Researchers" (1993) 19:2 Journal of Medical Ethics 85–91.

Gallant, D.M. "Clinical and Methodological Considerations in Psychotropic Drug Evaluation With Schizophrenic Patients" (1970) 6:4 Psychopharmacology Bulletin 4–24.

Ghaemi, S.N. & Hundert, E.M. "The Ethics of Research in Mental Illness" (1994) 5:1 APA Newsletter on Philosophy & Medicine 47–49.

Hollister, L.E. "The Use of Psychiatric Patients as Experimental Subjects" in F.J. Ayd, ed. *Medical, Moral and Legal Issues in Mental Health Care* (Baltimore: Williams & Wilkins, 1974) 28–36.

Irwin, M. *et al.* "Psychotic Patients' Understanding of Informed Consent" (1985) 142:11 American Journal of Psychiatry 1351–1354.

Kane, J.M. "Obstacles to Clinical Research and New Drug Development in Schizophrenia" (1991) 17:2 Schizophrenia Bulletin 353–356.

Kane, J.M., Robbins, L.L. & Stanley, B. "Psychiatric Research" in R.A. Greenwald, M.K. Ryan & J.E. Mulvihill, eds. *Human Subjects Research: A Handbook for Institutional Review Boards* (New York: Plenum Press, 1982)193–205.

Milliken, A.D. "The Need for Research and Ethical Safeguards in Special Populations" (1993) 38:10 Canadian Journal of Psychiatry 681–685.

Monahan, J. *et al.* "Ethical and Legal Duties in Conducting Research on Violence: Lessons From the MacArthur Risk Assessment Study" (1993) 8:4 Violence & Victims 387–396.

Rose, L.E. "Ethical Considerations of Patient Involvement in Clinical Psychiatric Research" (1986) 34:2 Canada's Mental Health 8–11.

Stanton, A.L. & New, M.J. "Ethical Responsibilities to Depressed Research Participants" (1988) 19:3 Professional Psychology - Research & Practice 279–285.

Tomossy, G.F. & Weisstub, D.N. "The Reform of Adult Guardianship Laws: The Case of Non-Therapeutic Experimentation" (1997) 20 International Journal of Law and Psychiatry 118–139.

Wing, J. "Ethics and Psychiatric Research" in S. Bloch & P. Chodoff, eds. *Psychiatric Ethics* (New York: Oxford University Press, 1981) 277–294.

Winick, B.J. "Psychotropic Medication and the Criminal Trial Process: The Constitutional and Therapeutic Implications of *Riggins v. Nevada*" (1993) 10:3 New York Law School Journal of Human Rights 637–709.

Winick, B.J. "The Right to Refuse Mental Health Treatment: A Therapeutic Jurisprudence Analysis" (1994) 17:1 International Journal of Law & Psychiatry 99–117.

### *The Developmentally Disabled*

Alexander, D. "Decision Making for Research Involving Persons With Severe Retardation: Guidance From the National Commission for the Protection of Human Subjects of Biomedical and Behavioral Research" in P. Dokecki & M. Zaner, eds. *Ethics of Dealing With Persons With Severe Handicaps: Toward a Research Agenda* (Baltimore: Brooks, 1986) 39–52.

American College of Physicians. "Cognitively Impaired Subjects" (1989) 111:10 Annals of Internal Medicine 843–848.

Appelbaum, P.S. & Grisso, T. "Assessing Patients' Capacities to Consent to Treatment" (1988) 319:25 New England Journal of Medicine 1635–1638.

Baudouin, J.-L. "Biomedical Experimentation on the Mentally Handicapped: Ethical and Legal Dilemmas" (1990) 9 Medicine & Law 465.

Brakel, S.J., Perry, J. & Weiner, B.A. *The Mentally Disabled and the Law*, 3rd ed. (American Law Foundation, 1985).

Berg, J.W. "Legal and Ethical Complexities of Consent With Cognitively Impaired Research Subjects: Proposed Guidelines" (1996) 24:1 Journal of Law, Medicine & Ethics 18–35.

Brock, D.W. "Good Decisionmaking for Incompetent Patients" (1994) 24:S6 Hastings Center Report S8-S11.

Crissey, M.S. "Vignettes in Mental Retardation" (1983) 18:2 Education and Training of the Mentally Retarded 117–119.

Dresser, R. "Mentally Disabled Research Subjects. The Enduring Policy Issues" (1996) 276:1 J.A.M.A. 67–72.

Dresser, R. "Missing Persons: Legal Perceptions of Incompetent Patients" (1994) 46:2 Rutgers Law Review 609–719.

Haywood, H.C. "The Ethics of Doing Researchàand of Not Doing It" (1977) 81:4 American Journal of Mental Deficiency 311–317.

Jenkinson, J.C. "Who Shall Decide? The Relevance of Theory and Research to Decision-Making by People With an Intellectual Disability" (1993) 8:4 Disability, Handicap & Society 361–375.

Jonas, C. & Soutoul, J.H. "Biomedical Research on Incapacitated People According to French Law" (1993) 12:6–8 Medicine & Law 567–572.

Kopelman, L. & Moskop, J.C. *Ethics and Mental Retardation* (Boston: D. Reidel Publishing, 1984).

Levenbook, B.B. "Examining Legal Restrictions on the Retarded" in L.M. Kopelman & J.C. Moskop, eds. *Ethics and Mental Retardation* (Boston: D. Reidel, 1984) 195–234.

Morris, R.J. & Hoschouer, R.L. "Current Issues in Applied Research With Mentally Retarded Persons" (1980) 1:1–2 Applied Research in Mental Retardation 85–93.

National Commission for the Protection of Human Subjects of Biomedical and Behavioural

Research. *Research Involving Those Institutionalized as Mentally Infirm: Report and Recommendations* (Bethesda, MD: The Commission, 1977).

Quarrington, B. "Approaches to the Assessment of Mental Competency" (1994) 15:2 Health Law in Canada 35–37.

Robertson, G.B. *Mental Disability and the Law in Canada*, 2d. ed. (Toronto: Carswell, 1994).

Schaefer, G.F. "Limited Guardianship: Additional Protection for Mentally Disabled Research Subjects Used in Biomedical and Behavioral Research" (1981) 16:4 Forum 796–824.

Siegel, P.S. & Ellis, N.R. "Note on the Recruitment of Subjects for Mental Retardation Research" (1985) 89:4 American Journal of Mental Deficiency 431–433.

Silberfeld, M. "Legal Standards and the Threshold of Competence" (1993) 14:4 Advocates Quarterly 482–487.

Silberfeld, M. "The Mentally Incompetent Patient: A Perspective From the Competency Clinic" (1990) 11:2 Health Law in Canada 33–37.

Silva, M.C. "Assessing Competency for Informed Consent With Mentally Retarded Minors" (1984) 10:4 Pediatric Nursing 261–265, 306.

Thomasma, D.C. "Obtaining Consent for Research in the Neurobiologically Impaired" (1994) 5:1 APA Newsletter on Philosophy & Medicine 54–55.

Turnbull, H.R. "Consent Procedures: A Conceptual Approach to Protection Without Overprotection" in P. Mittler, ed. *Research to Practice in Mental Retardation. Volume 1: Care and Intervention* (Baltimore: University Park Press, 1977) 65–70.

Tymchuk, A.J., Andron, L. & Rahbar, B. "Effective Decision-Making/Problem-Solving Training With Mothers who have Mental Retardation" (1988) 92:6 American Journal of Mental Retardation 510–516.

Weisstub, D.N. & Arboleda-Florez, J. "Ethical Research with the Developmentally Disabled" (1997) 42 Canadian Journal of Psychiatry 492–496.

Wikler, D. "Reflections on Research on Mentally Disabled Human Subjects" (1987) 23:3 Psychopharmacology Bulletin 372–374.

Working Party on Research on the Mentally Incapacitated. *The Ethical Conduct of Research on the Mentally Incapacitated* (London: Medical Research Council, 1991).

## *Prisoners*

Arbodela-Florez, J. "Ethical Issues Regarding Research on Prisoners" (1991) 35:1 International Journal of Offender Therapy & Comparative Criminology 1–5.

Ayd, F.J. "Drug Studies in Prisoner Volunteers" in T.L. Beauchamp & L. Walters, eds. *Contemporary Issues in Bioethics* (Belmont, CA: Wadsworth Publishing Company, 1978) 492–496.

Bach-y-Rita, G. "The Prisoner as an Experimental Subject" (1974) 229:1 Journal of the American Medical Association 45–46.

Benfell, C. "Abusing the Bodies and Minds of Prisoners: How They are Turned into Guinea Pigs Without Their Knowledge" (1977) 4:2 Barrister 22–25.

Brakel, S.J. "Considering Behavioral and Biomedical Research on Detainees in the Mental Health Unit of an Urban Mega-Jail" (1996) 22 New England Journal on Criminal & Civil Confinement 1–27.

Burt, R.A. "Reflections on the Detroit Psychosurgery Case. Why We Should Keep Prisoners From the Doctors" (1975) 5:1 Hastings Center Report 25–34.

Capron, A. "Medical Research in Prisons: Should a Moratorium Be Called?" in T.L. Beauchamp & L. Walters, ed. *Contemporary Issues in Bioethics* (Belmont, CA: Wadsworth Publishing, 1978) 497.

Erez, E. "Randomized Experiments in Correctional Context: Legal, Ethical, and Practical Concerns" (1986) 14:5 Journal of Criminal Justice 389–400.

Fowles, A.J. *Prisoner's Rights in England and the United States* (Brookfield: Gower Publishing Company, 1989).

Hammett, T.M. & Dubler, N.N. "Clinical and Epidemiologic Research on HIV Infection and AIDS Among Correctional Inmates: Regulations, Ethics, and Procedures" (1990) 14:5 Evaluation Review 482–501.

Hawkins, R. & Alpert, J.P. *American Prison Systems - Punishment and Justice* (Englewood Cliffs, NJ: Prentice Hall, 1989).

Johnson, D.G. et al. "Prisoners and Consent to Experimentation" in L.T. Sargent, ed. *Consent: Concept, Capacity, Conditions, and Constraints* (Wiesbaden, Germany: Steiner, 1979) 167–200.

Klein, M. "Problems Arising From Biological Experimentation in Prisons" in G.E.W. Wolstenholme & M. O'Connor, eds. *Medical Care of Prisoners and Detainees* (New York: Elsevier, 1973) 65–78.

Livingstone, S. & Owen, T. *Prison Law: Text and Materials* (Oxford: Clarendon Press, 1993).

McCarthy, C.M. "Experimentation on Prisoners" (1989) 15:1 New England Journal on Criminal & Civil Confinement 55–88.

National Commission for the Protection of Human Subjects of Biomedical and Behavioural Research. *Research Involving Prisoners* (Bethseda, MD: The Commission, 1976).

Moore, J. "Informed Consent With Prisoners in Research: Issues, Dilemmas and Regulations" (1995) 9:2 APA Newsletter on Philosophy & Medicine 14–15, 18.

Potler, C., Sharp, V.L. & Remick, S. "Prisoners' Access to HIV Experimental Trials: Legal, Ethical and Practical Considerations" (1994) 7:10 Journal of Acquired Immune Deficiency Syndrome 1086–1094.

Richardson, G. *Law, Process and Custody: Prisoners and Patients* (London: Weidenfeld & Nicholson, 1993).

Sabin, A.B., Bronstein, A.J. & Hubbard, W.N. "Individual Risks vs. Societal Benefits: How are the Risks Distributed? The Military/The Prisoner" in National Academy of Sciences, *Experiments and Research With Humans: Values in Conflict* (Washington, DC: National Academy of Sciences, 1975) 127–149.

Schroeder, K. "A Recommendation to the F.D.A. Concerning Drug Research on Prisoners" (1983) 56:4 Southern California Law Review 969–1000.

Schrogie, J.J. *et al.* "Evaluation of the Prison Inmate as a Subject in Drug Assessment" (1977) 21:1 Clinical Pharmacology & Therapeutics 1–8.

Weisstub, D.N. "La recherche médicale en milieu carcéral" (1997) 6 International Journal of Bioethics 87–93.

Winkleman, L. *et al.* "Drug Testing in Prisons" in M.D. Basson, R.E. Lipson & D.L. Ganos, eds. *Troubling Problems in Medical Ethics: The Third Volume in a Series on Ethics, Humanism, and Medicine* (New York: Alan R. Liss, 1981) 49–93.

## EPIDEMIOLOGY

### *General Ethical Prinicples*

Andrews, J.S. "Does Good Peer Review Assure Good Epidemiology?" (1991) 44:1S Journal of Clinical Epidemiology 131S–134S.

Arboleda-Florez, J. "Methodological and Ethical Concerns Regarding Forensic Psychiatry Epidemiological Studies" (1993) 37:2 International Journal of Offender Therapy & Comparative Criminology 95–98.

Bankowski, Z. "Epidemiology, Ethics and "Health for All" (1991) 19:3–4 Law, Medicine & Health Care 162–163.

Bayer, R., Lumey, L.H. & Wan, L. "The American, British, and Dutch Responses to Unlinked

Anonymous HIV Seroprevalence Studies: An International Comparison" (1991) 19:3–4 Law, Medicine & Health Care 222–230.

Beauchamp, T.L. "Ethical Theory and Epidemiology" (1991) 44:1S Journal of Clinical Epidemiology 5S–8S.

Beauchamp, T.L. & Coughlin, S.S., eds. *Ethics in Epidemiology* (Oxford: Oxford University Press, 1995).

Beauchamp, T.L. et al. "Ethical Guidelines for Epidemiologists" (1991) 44:1S Journal of Clinical Epidemiology 151S–169S.

Bond, G.G. "Ethical Issues Relating to the Conduct and Interpretation of Epidemiologic Research in Private Industry" (1991) 44:1S Journal of Clinical Epidemiology 29S–34S.

Brett, A.S. & Grodin, M.A. "Ethical Aspects of Human Experimentation in Health Services Research" (1991) 265:14 JAMA. 1854–1857.

Bryant, J.H. "Trends in Biomedical Ethics - Forerunners of Ethical Questions in Epidemiology" in Z. Bankowski, J.H. Bryant & J.M. Last, eds. *Ethics and Epidemiology: International Guidelines* (Geneva: CIOMS, 1991) 8.

Bryant, J.H. & Khan, K.S. "Epidemiology and Ethics in the Face of Scarcity" in Z.Bankowski, J.H. Bryant & J.M. Last, eds. *Ethics and Epidemiology: International Guidelines* (Geneva: CIOMS, 1991) 70.

Capron, A.M. "Protection of Research Subjects: Do Special Rules Apply in Epidemiology?" (1991) 44:1S Journal of Clinical Epidemiology 81S–89S.

Christakis, N.A. & Panner, M.J. "Existing International Ethical Guidelines for Human Subjects Research: Some Open Questions" (1991) 19:3–4 Law, Medicine & Health Care 214–221.

Coburn, D. "Individual and Community Rights in Anonymous Unlinked Seroprevalence Research: A Response to Dr. Emson" (1992) 12:4 Health Law in Canada 97–100.

Coughlin, S.S., ed. *Ethics in Epidemiology and Clinical Research: Annotated Readings* (Newton, Mass: Epidemiology Resources, 1995).

Coughlin, S.S. & Beauchamp, T.L. "Ethics, Scientific Validity, and the Design of Epidemiologic Studies" (1992) 3:4 Epidemiology 343–347.

Council for International Organizations of Medical Science. International Guidelines for the *Ethical Review of Epidemiological Studies* (Geneva: CIOMS, 1991).

Dickens, B.M. "Issues in Preparing Ethical Gudelines for Epidemiological Studies" (1991) 19:3–4 Law, Medicine & Health Care 175–183.

Dickens, B.M., Gostin, L. & Levine, R.J. "Research on Human Populations: National and International Ethical Guidelines" (1991) 19:3–4 Law, Medicine & Health Care 157–161.

Dicker, K. "Physician Consent and Researchers' Access to Patients" (1990) 1:2 Epidemiology 160–163.

Emson, H.E. "The Ethics and Legality of HIV Prevalence Studies: A Contrary View" (1992) 12:4 Health Law in Canada 95–96, 114.

Feinlib, M. "The Epidemiologist's Responsibilities to Study Participants" (1991) 44:1S Journal of Clinical Epidemiology 73S–79S.

Feinstein, A.R. "Scientific Paradigms and Ethical Problems in Epidemiologic Research" (1991) 44:1S Journal of Clinical Epidemiology 119S–123S.

Gordis, L. "Ethical and Professional Issues in the Changing Practice of Epidemology" (1991) 44:1S Journal of Clinical Epidemiology 9S–13S.

Gostin, L. "Ethical Principles for the Conduct of Human Subject Research: Population-Based Research and Ethics" (1991) 19:3–4 Law, Medicine & Health Care 191–201.

Gupta, P.C. "On Ethics, Epidemiology, and Culture" (1994) 15:7 Epidemiology Monitor 3.

Hermeren, G. "Ethical Problems in Register Based Medical Research" (1988) 9:2 Theoretical Medicine 105–116.

Khan, K.S. "Epidemiology and Ethics: The People's Perspective" (1991) 19:3–4 Law, Medicine & Health Care 202–206

Kleinbaum, D.G., Kupper, L.L. & Morganstern, H. *Epidemiologic Research, Principles and Quantitative Methods* (New York: Van Nostrand Reinhold, 1982).

Khan, K.S. "Epidemiology and Ethics: The Perspective of the Third World" (1994) 15:2 Journal of Public Health Policy 218–225.

Last, J.M. "Epidemiology and Ethics" (1991) 19:3–4 Law, Medicine & Health Care 166–174.

Last, J.M. "Obligations and Responsibilities of Epidemiologists to Research Subjects" (1991) 44:1S Journal of Clinical Epidemiology 95S–101S.

Lilienfeld, D.E. & Stolley, P.D. *Foundations of Epidemiology* (New York: Oxford University Press, 1994).

Mausner, J.S. & Kramer, S. *Epidemiology. An Introductory Text* (Toronto: W.B. Saunders, 1985).

Nakajima, H. "The Responsibilities of Epidemiologists" (1991) 19:3–4 Law, Medicine & Health Care 164–165.

Roberts, F.D. *et al*. "Perceived Risks of Participation in an Epidemiological Study" (1993) 15:1 IRB 8–10.

Sharpe, G. "The Ethics and Legality of HIV Seroprevalence Studies" (1991) 11:4 Health Law in Canada 102–118.

Soskolne, C.L. "Epidemiological Research, Interest Groups, and the Review Process" (1985) 6:2 Journal of Public Health Policy 173–184.

Susser, M., Stein, Z. & Kline, J. "Ethics in Epidemiology" in B. Barber, ed. *Medical Ethics and Social Change* (Philadelphia: American Academy of Political and Social Science, 1978) 128–141.

Tancredi, L.R. *Ethical Issues in Epidemiologic Research* (New Brunswick, NJ: Rutgers University Press, 1986).

Tancredi, L.R "The New Technology of Psychiatry: Ethics, Epidemiology, and Technology Assessment" in L.R. Tancredi, ed. *Ethical Issues in Epidemiologic Research* (New Brunswick, NJ: Rutgers University Press, 1986) 1–36.

Tancredi, L.R. & Weisstub, D.N. "Ideology and Power: Epidemiology and Interpretation in Law and Psychiatry" in L. Romanucci-Ross, D.E. Moorman & L.R. Tancredi, eds. *The Anthropology of Medicine, From Culture to Method* (New York: Bergin & Garvey, 1991) 301.

Vandenbroucke, J.P. "How Trustworthy is Epidemiologic Research?" (1990) 1:1 Epidemiology 83–84.

Westrin, C.-G. "Epidemiology and Moral Philosophy" (1992) 18:4 Journal of Medical Ethics 193–196.

### *Confidentiality and Privacy*

Bentley-Cooper, J.E. "Protecting Human Research From an Invasion of Privacy: The Unintended Results of the Commonwealth Privacy Act 1988" (1991) 15:3 Australian Journal of Public Health 228–234.

Bravender-Coyle, P. "The Law Relating to Confidentiality of Data Acquired by Researchers in the Biomedical and Social Sciences" (1986) 8:3 University of Tasmania Law Review 333–360.

Buchan, H. & Paul, C. "When Important Social Interests Compete With Privacy" (1992) 105:947 New Zealand Medical Journal 492–493.

Clayton, E.W. "Why the Use of Anonymous Samples for Research Matters" (1995) 23:4 Journal of Law, Medicine & Ethics 375–377.

Cooper, J.E. "Balancing the Scales of Public Interest: Medical Research and Privacy" (1991) 155:8 Medical Journal of Australia 556–560.

Council of Europe, *The Protection of Individuals With Regards to the Processing of Personal Data and on the Free Movement of Such Data* (Brussels: Council of Europe, 1995).

Delamothe, T. "Whose Data are They Anyway? Raw Data From Research on Patients Should be Available, Anonymised, to Whoever Wants Them" (1996) 312:7041 British Medical Journal 1241–1242.

Fienberg, S.E. "Sharing Statistical Data in the Biomedical and Health Sciences: Ethical, Institutional, Legal, and Professional Dimensions" (1994) 15:1 Annual Review of Public Health 1–18.

Gordis, L., Gold, E. & Selger, R. "Privacy Protection in Epidemiological and Medical Research: A Challenge and a Responsibility" (1977) 105:3 American Journal of Epidemiology 163–169.

Helgason, T. "Epidemiological Research Needs Access to Data" (1992) 20:3 Scandinavian Journal of Social Medicine 129–133.

Kelsey, J.L. "Privacy and Confidentiality in Epidemiological Research Involving Patients" (1981) 3:2 IRB 1–4.

Knox, E.G. "Confidential Medical Records and Epidemiological Research" (1992) 304:6829 British Medical Journal 727–728.

Lane, M.J. "Privacy Protection: Implications for Public Health Researchers" (1991) 155:11–12 Medical Journal of Australia 831–833.

Last, J.M. "Individual Privacy and Health Information — An Ethics Dilemma?" (1986) 77:3 Canadian Journal of Public Health 168–170.

Last, J.M. "Respect for Privacy and Promises of Confidentiality" (1992) 13:7 The Epidemiology Monitor 4.

McCarthy, C.R. & Porter, J.P. "Confidentiality: The Protection of Personal Data in Epidemiological and Clinical Research Trials" (1991) 19:3–4 Law, Medicine & Health Care 238–241.

Melton, G.B. & Gray, J.N. "Ethical Dilemmas in AIDS Research: Individual Privacy and Public Health" (1988) 43:1 American Psychologist 60–64.

Schulte, P.A. & Sweeney, M.H. "Ethical Considerations, Confidentiality Issues, Rights of Human Subjects, and Uses of Monitoring Data in Research and Regulation" (1995) 103:S3 Environmental Health Perspectives 69–74.

Sibthorpe, B., Kliewer, E. & Smith, L. "Record Linkage in Australian Epidemiological Research: Health Benefits, Privacy Safeguards and Future Potential" (1995) 19:3 Australian Journal of Public Health 250–256.

Smallwood, R.A. "Privacy Laws and Their Effect on Medical Research" (1993) 47:5 Journal of Epidemiology and Community Health 342–344.

Van der Leer, O.F. "The Use of Personal Data for Medical Research: How to Deal With New European Privacy Standards" (1994) 35S International Journal of Bio-Medical Computing 87–95.

Wald, N. *et al.* "Use of Personal Medical Records for Research Purposes" (1994) 309:6926 British Medical Journal 1422–1424.

Wallace, R.J. "Privacy and the Use of Data in Epidemiology" in T.L. Beauchamp, ed. *Ethical Issues in Social Science Research* (Baltimore: Johns Hopkins University Press, 1982) 274–291.

Zahm, S.H. "On Epidemiology and the Obligation to Notify Study Subjects" (1992) 3:2 The Epidemiology Monitor 5.

## RANDOMIZED CLINICAL TRIALS

Arpaillange, P., Dion, S. & Mathe, G. "Proposal for Ethical Standards in Therapeutic Trials" (1985) 291:6499 British Medical Journal 887–889.

Baum, M. "New Approach for Recruitment into Randomised Controlled Trials" (1993) 341:8848 Lancet 812–813.

Botros, S. "Equipoise, Consent and the Ethics of Randomised Clinical Trials" in P. Byrne, ed. *Ethics and Law in Health Care and Research* (New York: Wiley, 1990) 9–24.

Brody, B.A. "Conflicts of Interest and the Validity of Clinical Trials" in R.G. Spece *et al.*, eds. *Conflicts of Interest in Clinical Practice and Research* (New York: Oxford University Press, 1996) 407–417.

Buck, B.A. "Ethical Issues of Randomized Clinical Trials" (1990) 61:3 Radiologic Technology 202–205.

Corbett, F., Oldham, J. & Lilford, R. "Offering Patients Entry in Clinical Trials: Preliminary Study of the Views of Prospective Participants" (1996) 22:4 Journal of Medical Ethics 227–231.

Cotton, D.J. *et al.* "Guidelines for the Design and Conduct of AIDS Clinical Trials" (1993) 16:6 Clinical Infectious Diseases 816–822.

Coughlin, S.S. *et al.* "Ethical Issues in the Design of a Randomized Trial for the Prevention of HIV Infection" (1992) 7:7 Human Research Report 2–3.

Deutsch, E. "Controlled Clinical Trials in Drug Research: Starting by Permission v. Notification" (1985) 4:6 Medicine & Law 493–497.

Dodds-Smith, I. "Clinical Research" in C. Dyer, ed. *Doctors, Patients and the Law* (Oxford: Blackwell Scientific, 1992) 140–166.

Durant, J.R. "Overview. Current Status of Clinical Trials" (1990) 65:10S Cancer 2371–2375.

Elander, G. & Hermeren, G. "Placebo Effect and Randomized Clinical Trials" (1995) 16:2 Theoretical Medicine 171–182.

Elbourne, D. "Subjects' Views about Participation in a Randomized Controlled Trial" (1987) 5 Journal of Reproductive & Infant Psychology 3–8.

Freedman, B. "A Response to a Purported Ethical Difficulty With Randomized Clinical Trials Involving Cancer Patients" (1992) 3:3 Journal of Clinical Ethics 231–234.

Gifford, F. "Community-Equipoise and the Ethics of Randomized Clinical Trials" (1995) 9:2 Bioethics 127–148.

Gifford, F. "The Conflict Between Randomized Clinical Trials and the Therapeutic Obligation" (1986) 11:4 Journal of Medicine & Philosophy 347–366.

Gordon, R.S. & Fletcher, J.C. "Can Strict Randomization be Ethically Acceptable?" (1983) 31:1 Clinical Research 23–25.

Hellman, S. & Hellman, D.S. "Of Mice but Not Men: Problems of the Randomized Clinical Trial" (1991) 324:22 New England Journal of Medicine 1585–1589.

Helmchen, H. "Ethical Problems and Design of Controlled Clinical Trials" (1990) 8 Psychopharmacology Series 82–87.

Herman, J. "The Demise of the Randomized Controlled Trial" (1995) 48:7 Journal of Clinical Epidemiology 985–988.

Herxheimer, A. "Clinical Trials: Two Neglected Ethical Issues" (1993) 19:4 Journal of Medical Ethics 211, 218.

Johnson, N., Lilford, R.J. & Brazier, W. "At What Level of Collective Equipoise Does a Clinical Trial Become Ethical?" (1991) 17:1 Journal of Medical Ethics 30–34.

Kardinal, C.G. "Ethical Issues in Cancer Clinical Trials" (1994) 146:8 Journal of the Louisiana State Medical Society 359–361.

Kodish, E., Lantos, J.D. & Siegler, M. "Ethical Considerations in Randomized Controlled Clinical Trials" (1990) 65:10S Cancer 2400–2404.

Kopelman, L. "Consent and Randomized Clinical Trials: Are there Moral or Design Problems?" (1986) 11:4 Journal of Medicine & Philosophy 317–345.

Lantos, J.D. "Ethics, Randomization, and Technology Assessment" (1994) 74:9S Cancer 2653–2656.

Lawrence, W. Jr. & Bear, H.D. "Is there Really an Ethical Conflict in Clinical Trials?" (1995) 75:10 Cancer 2407–2409.

Levine, R.J. "Clinical Trials and Physicians as Double Agents" (1992) 65:2 Yale Journal of Biology & Medicine 65–74.

Levine, R.J. "Controlled Clinical Trials: Ethical Problems" in Y. Champey, R.J. Levine & P.S. Lietman, eds. *Development of New Medicines: Ethical Questions* (New York: Royal Society of Medicine Services, 1989) 7–22.

Levine, R.J. "Randomized Clinical Trials: Ethical Considerations" in E.E. Bittar & N. Bittar, eds. *Bioethics* (Greenwich, CT: JAI Press, 1994) 37–63.

Levine, R.J. "Referral of Patients With Cancer for Participation in Randomized Clinical Trials: Ethical Considerations" (1986) 36:2 Ca: A Cancer Journal for Clinicians 95–99.

Levine, R.J. "Uncertainty in Clinical Research" (1988) 16:3–4 Law, Medicine & Health Care 174–182.

Lilford, R.J. & Jackson, J. "Equipoise and the Ethics of Randomization" (1995) 88:10 Journal of the Royal Society of Medicine 552–559.

Lumley, J. & Bastian, H. "Competing or Complementary? Ethical Considerations and the Quality of Randomized Trials" (1996) 12:2 International Journal of Technology Assessment in Health Care 247–263.

Mackillop, W.J. & Johnston, P.A. "Ethical Problems in Clinical Research: The Need for Empirical Studies of the Clinical Trials Process" (1986) 39:3 Journal of Chronic Diseases 177–188.

Macklin, R. & Friedland, G. "AIDS Research: The Ethics of Clinical Trials" (1986) 14:5–6 Law, Medicine & Health Care 273–280.

Marquis, D. "An Argument That All Prerandomized Clinical Trials are Unethical" (1986) 11 Journal of Medicine and Philosophy 367–383.

Meslin, E.M. "Judging the Ethical Merit of Clinical Trials: What Criteria to Research Ethics Board Members Use?" (1994) 16:4 IRB 6–10.

Miller, B. "Experimentation on Human Subjects: The Ethics of Random Clinical Trials" in D. Van De Veer & T. Regan, eds. *Health Care Ethics: An Introduction* (Philadelphia: Temple University Press, 1987) 127–159.

Palter, S.F. "Ethics of Clinical Trials" (1996) 14:2 Seminars in Reproductive Endocrinology 85–92.

Passamani, E. "Clinical Trials — Are They Ethical?" (1991) 324:22 New England Journal of Medicine 1589–1592.

Rawlings, G. "Ethics and Regulation in Randomised Controlled Clinical Trials of Therapy" in A. Grubb, ed. *Challenges in Medical Care* (New York: Wiley, 1992) 29–58.

Roginsky, M. "Clinical Trials of New Drugs: Special Problems" in R.A. Greenwald, M.K. Ryan & J.E. Mulvihill, eds. *Human Subjects Research: A Handbook for Institutional Review Boards* (New York: Plenum Press, 1982) 181–192.

Rosner, F. "The Ethics of Randomized Clinical Trials" (1987) 82:2 American Journal of Medicine 283–290.

Schafer, A. "The Ethics of the Randomized Clinical Trial" (1982) 307:2 New England Journal of Medicine 719–724.

Schaffner, K.F. "Ethical Problems in Clinical Trials" (1986) 11:4 Journal of Medicine & Philosophy 297–315.

Segelov, E., Tattersall, M.H.N. & Coates, A.S. "Redressing the Balance: The Ethics of <not> Entering an Eligible Patient on a Randomised Clinical Trial" (1992) 3:2 Annals of Oncology 103–105.

Senn, S. "A Personal View of Some Controversies in Allocating Treatment to Patients in Clinical Trials" (1995) 14:24 Statistics in Medicine 2661–2674.

Shimm, D.S. & Spece, R.G. "Ethical Issues and Clinical Trials" (1993) 46:4 Drugs 579–584.

Sniderman, A.D. "The Governance of Clinical Trials" (1996) 347:9012 Lancet 1387–1388.

Sutherland, H.J., Meslin, E.M. & Till, J.E. "What's Missing From Current Clinical Trial